McGRAW-HILL YEARBOOK OF
Science &
Technology

2000

McGRAW-HILL YEARBOOK OF
Science &
Technology

2000

Comprehensive coverage of recent events and research as compiled by the staff of the McGraw-Hill Encyclopedia of Science & Technology

McGraw-Hill
New York   San Francisco   Washington, D.C.   Auckland   Bogotá   Caracas   Lisbon   London   Madrid
Mexico City   Milan   Montreal   New Delhi   San Juan   Singapore   Sydney   Tokyo   Toronto

Library of Congress Cataloging in Publication data

McGraw-Hill yearbook of science and technology.
1962-   . New York, McGraw-Hill Book Co.

  v. illus. 26 cm.
  Vols. for 1962-   compiled by the staff of the
McGraw-Hill encyclopedia of science and technology.
    1. Science—Yearbooks.   2. Technology—
Yearbooks.   1. McGraw-Hill encyclopedia of
science and technology.
Q1.M13   505.8   62-12028

ISBN 0-07-052771-7
ISSN 0076-2016

McGraw-Hill

A Division of The McGraw·Hill Companies

McGRAW-HILL YEARBOOK OF SCIENCE & TECHNOLOGY
Copyright © 1999 by The McGraw-Hill Companies, Inc.

1 2 3 4 5 6 7 8 9 0   DOW/DOW   9 0 4 3 2 1 0 9

This book was printed on acid-free paper.

*It was set in Garamond Book and Neue Helvetica Black Condensed by
TechBooks, Fairfax, Virginia. The art was prepared by TechBooks.
The book was printed and bound by R. R. Donnelley & Sons Company,
The Lakeside Press.*

Editing, Design, & Production Staff

Roger Kasunic, Director of Editing, Design, and Production

Joe Faulk, Editing Manager

Ron Lane, Art Director

Thomas G. Kowalczyk, Production Manager

Consulting Editors

Dr. Milton B. Adesnik. *Department of Cell Biology, New York University School of Medicine, New York.* CELL BIOLOGY.

Prof. Eugene A. Avallone. *Consulting Engineer; Professor Emeritus of Mechanical Engineering, City College of the City University of New York.* MECHANICAL AND POWER ENGINEERING.

A. E. Bailey. *Formerly, Superintendent of Electrical Science, National Physical Laboratory, London, England.* ELECTRICITY AND ELECTROMAGNETISM.

Prof. William P. Banks. *Chairman, Department of Psychology, Pomona College, Claremont, California.* PHYSIOLOGICAL AND EXPERIMENTAL PSYCHOLOGY.

Prof. Gregory C. Beroza. *Department of Geophysics, Stanford University, California.* GEOPHYSICS.

Dr. Eugene W. Bierly. *American Geophysical Union, Washington, D.C.* METEOROLOGY AND CLIMATOLOGY.

Prof. Carrol Bingham. *Department of Physics, University of Tennessee, Knoxville.* NUCLEAR AND ELEMENTARY PARTICLE PHYSICS.

Dr. Chaim Braun. *Bechtel Corporation, Gaithersburg, Maryland.* NUCLEAR ENGINEERING.

Robert D. Briskman. *President, CD Radio, Inc., Washington, D.C.* TELECOMMUNICATIONS.

Prof. Wai-Fah Chen. *Head, Structural Engineering, Purdue University, West Lafayette, Indiana.* CIVIL ENGINEERING.

Dr. John F. Clark. *Director, Graduate Studies, and Professor, Space Systems, Spaceport Graduate Center, Florida Institute of Technology, Satellite Beach.* SPACE TECHNOLOGY.

Prof. David L. Cowan. *Chairman, Department of Physics and Astronomy, University of Missouri, Columbia.* CLASSICAL MECHANICS AND HEAT.

Dr. Michael R. Descour. *Optical Sciences Center, University of Arizona, Tucson.* ELECTROMAGNETIC RADIATION AND OPTICS.

Tom Destree. *Graphic Arts Technical Foundation, Sewickley, Pennsylvania.* GRAPHIC ARTS.

Dr. Derek Enlander. *Internal Medicine, New York, New York.* MEDICINE AND PATHOLOGY.

Prof. Turgay Ertekin. *Chairman, Department of Petroleum and Natural Gas Engineering, Pennsylvania State University, University Park.* PETROLEUM ENGINEERING.

Peter A. Gale. *Chief Naval Architect, John J. McMullen Associates, Inc., Arlington, Virginia.* NAVAL ARCHITECTURE AND MARINE ENGINEERING.

Dr. John Gordon. *School of Forestry and the Environment, Yale University, New Haven, Connecticut.* FORESTRY.

Dr. Richard L. Greenspan. *The Charles Stark Draper Laboratory, Cambridge, Massachusetts.* NAVIGATION.

Contributors

A list of contributors, their affiliations, and the titles of the articles they wrote appears in the back of this volume.

Preface

The 2000 *McGraw-Hill Yearbook of Science & Technology* provides the reader with a wide overview of the most significant recent developments in science, technology, and engineering, as selected by our distinguished board of consulting editors. At the same time, it satisfies the reader's need to stay informed about important trends in research and development that will fundamentally influence future understanding and practical applications of knowledge in fields ranging from astronomy to zoology. Readers of the *McGraw-Hill Encyclopedia of Science & Technology* will find the *Yearbook* to be a valuable companion publication, enhancing the timeliness and depth of the *Encyclopedia*.

Each contribution to the *Yearbook* is a concise yet authoritative article authored by one or more specialists in the field. We are pleased that noted researchers have been supporting the *Yearbook* since its first edition in 1962 by taking time to share their knowledge with our readers. The topics are selected by our consluting editors in conjunction with our editorial staff based on present significance and potential applications. McGraw-Hill strives to make each article as readily understandable as possible for the nonspecialist reader through careful editing and the extensive use of graphics, much of which is prepared specially for the *Yearbook*.

Librarians, students, teachers, the scientific community, and the general public continue to find in the *McGraw-Hill Yearbook of Science & Technology* the information they want and need in order to follow the rapid pace of advances in science and technology and to understand the developments in these fields that will shape the world of the twenty-first century.

Mark D. Licker
PUBLISHER

McGRAW-HILL YEARBOOK OF
Science &
Technology

2000

A–Z

Agribiotechnology

In applying biotechnology to agriculture, the principal areas involve manipulation of crop plants and manipulation of microbial or invertebrate associates of crop plants and weeds. Technological objectives may be achieved by conventional selection and breeding or by genetic engineering of plants, microbes, or invertebrates, or via manipulation of the environment. Such technologies can be combined in various agronomic systems with traditional practices such as chemical pesticides or fertilizers. Fungi play an expanding role in agribiotechnology as biological control agents, as mycorrhizal associates of plant roots, as agents for the degradation of lignocellulose and other forms of bioremediation (the application of microorganisms to biodegrade pollutants), or as food crops which utilize agricultural waste as a substrate. In food and beverage processing, fungi supply an array of enzymes plus natural flavor and color additives.

Fungi. Fungi have been used to control insects, weeds, nematodes (roundworms), and other fungi. Commercial products are available for control of whiteflies and thrips with the fungus *Verticillium lecanii* or *Beauveria bassiana*; whiteflies with *Paecilomyces fumosoroseus*; cockroaches and termites with *Metarhizium anisopliae*; and Colorado potato beetle, corn borer, and grasshoppers with *B. bassiana*. These fungi are in the class Hyphomycetes, many of which are comparatively easy to mass-produce and whose spores often survive conditions of storage and application. Other fungi pathogenic for insects, such as those in the order Entomophthorales, sometimes achieve notable reductions of insect pests in natural epidemics. These fungi are not readily produced in artificial culture, or their spores are less adaptable to industrial production. However, fungi in the Entomophthorales may slowly spread disease in host populations, or the crop environment may be manipulated to promote the natural infection. Both methods have proven feasible for the control of insect pests of alfalfa by *Zoophthora* species.

Fungi are also employed as mycoherbicides against weeds. Mycoherbicides registered in North America contain one of several fungi: *Colletotrichum gloeosporioides* f.sp. *aeschynomene* (used against northern jointvetch), *Puccinia canaliculata* (used against yellow nutsedge), *Phytophthora palmivora* (used to control strangler vine), and *C. gloeosporioides* f.sp. *malvae* (employed against round-leaved mallow). Classical instances of biological control of weeds by fungi include *Puccinia chondrilla* against skeletonweed in Australia and *Entyloma ageratinae* against mistflower in Hawaii. These fungi are pathogens specific to the target weed and are nonpathogenic to crop plants.

Progress has also been made in exploiting the ability of fungi to control other fungi. The mechanism for control of insects and weeds is direct pathogenesis, and control of fungi may be achieved by analogous parasitism of the target fungus. However, mechanisms such as the competition for nutrients, the production of antifungal metabolites, and the stimulation of host defenses are also important. The registered agents include *Peniophora gigantea* used against the wood-rotter *Heterobasidium annosum* (in the United Kingdom), *Pythium oligandrum* used against other soil-inhabiting *Pythium* species (in the former Soviet Union), *Trichoderma viride* and *T. harzianum* used against wood-rotting fungi (Europe and the United States, respectively), and *Gliocladium virens* used against fungi causing diseases of seedlings (United States). In Europe, some chestnut blight has been successfully controlled with strains of *Cryphonectria parasitica* which are much less virulent compared to the pathogenic strains. These hypovirulent strains are able to invade the cankers (localized necrosis of bark, stem, branches, or twigs) produced by pathogenic strains, combine with these strains, and permit gradual healing.

Although mycopesticides are now commercially available for the control of certain insects, weeds, and pathogenic fungi, control of nematodes by fungi remains largely experimental. There is a large volume of literature and experimental work on the interactions between nematophagous (nematode-devouring) fungi and nematodes. Fungi in the genera *Arthrobotrys*, *Dactylaria*, and *Dactylella* (class Hyphomycetes) capture nematodes by adhesive knobs, by nets, or by nooselike rings. Fungi in *Nematoctonus* (a basidiomycete) also capture nematodes by adhesive knobs. Fungi in the zygomycete genus *Stylopage* capture nematodes with adhesive hyphae, then penetrate and digest them. Members of *Catenaria* (class Chytridiomycetes) or *Myzocytium* (class Oomycetes) grow and reproduce inside the host. Experimental work continues, especially with agents such as the hyphomycete *Verticillium chlamydosporum*, which can be grown in culture and which produces large numbers of relatively durable spores.

Mycorrhizal symbiosis. Mycorrhizal fungi function as active, growth-promoting partners of crop plants by establishing a symbiotic relationship with plant roots. The fungus utilizes plant sugars that are produced via photosynthesis. In return, the fungus acts as an ancillary system for the plant, transferring minerals and water to the plant roots. A mycorrhizal fungus is classified by the formal taxonomic category to which the fungus belongs. The fungus-root complex (called a mycorrhiza) is variously classified as ectomycorrhizal, endomycorrhizal, and other designations, depending on structure.

Considerable experimental work has been done on the production of inoculum for application to the roots of crop plants with the intention of producing mycorrhizal plants with enhanced growth potential, but there are presently few commercial products. An exception is the inoculum produced from a mixture of fungal mycelium, vermiculite, peat moss, and other ingredients and applied to the roots of pine seedlings in the southern United States. The fungi used are a puffball, *Pitholithus tinctorius*, and two mushrooms, *Hebeloma crustuliniforma* and *Laccaria laccata*. These fungi produce an ectomycorrhizal association. Although the objective of these inoculations is to produce improved crop plants, commercial production of seedlings colonized by mycorrhizal fungi can have a different objective such as the production of edible fungal fruiting bodies. Truffles (*Tuber* species in the class Ascomycetes) are now produced by this method with plantings of oak and hazelnut. Endomycorrhizae result from the union of the plant root and obligately mycorrhizal fungi in the class Zygomycetes. These fungi can be raised only in conjunction with growing plants, but inoculation of high-value greenhouse crops is sometimes practical.

Bioremediation. Crop residues and forest wastes are high in lignin, a component of plant cell walls. Such wastes are typically unutilized or underutilized. Certain fungi, primarily many wood-rotting fungi in the class Basidiomycetes, are termed white rot fungi because they degrade the lignin prior to utilization of the polysaccharides in the cell wall. *Phanerochaetae chrysosporium* and related basidiomycetes have been utilized for experimental degradation of lignin wastes, as well as for experimental pulping processes in the paper industry. The metabolic capabilities that enable these and other fungi to degrade lignin also allow the degradation of various agricultural pesticides, polychlorinated biphenyls (PCBs), and similar chemicals. Large-scale exploitation of these fungi for bioremediation is pending. Current commercial ventures center on the exploitation of lignocellulose wastes for the production of edible mushrooms, particularly shiitake (*Lentinula edodes*) and oyster mushrooms (*Pleurotus* species). These mushrooms, particularly the latter, can use or recycle a large array of lignocellulose wastes such as straw, wood chips, and sawdust.

Experimental approaches. Mycopesticides must employ fungi which are resistant to the conditions of manufacture, storage, and application. They must contain protectants and adjuvants to improve the survival and performance of these living fungi. Many plant pathogenic fungi produce toxins, some host-specific, that are essential to disease in the host. There has been interest in using fungi for the manufacture of host-specific toxins which could be employed in a manner analogous to chemical herbicides, thereby omitting the need to design formulations that protect living active agents. An example of a host-specific toxin is maculosin from *Alternaria* sp. and specific for spotted knapweed. Another approach is the utilization of toxin-deactivating enzymes produced by genes found in the pathogen itself. This approach has been demonstrated by introducing into the tobacco plant the detoxifying enzyme gene from the bacterium causing wildfire disease to make the plant resistant to the disease. The pursuit of similar strategies is contemplated using fungal genes.

Experimental genetic manipulation of fungal biocontrol agents has been achieved through various means. Protoplast fusion has produced strains of *Trichoderma* that act against fungal pathogens and are simultaneously tolerant of fungicides. Deoxyribonucleic acid (DNA) capable of bestowing hypovirulence in strains of the chestnut blight fungus has been introduced into new strains of the fungus, making them potential agents of biological control. It may be possible to introduce genes for the production or enhancement of antibiotic activity, or to delete genes for pathogenicity. In the field of food and beverage processing, fungi which are generally recognized as safe (GRAS) by the U.S. Food and Drug Administration are increasingly being used as "factories" for the production of recombinant proteins. Fungi which produce mycotoxins cannot attain GRAS status. Genetic engineering can potentially be used to disable genes critical for mycotoxin production, thereby greatly expanding the number of GRAS fungal strains. The increasing sophistication of molecular techniques and the natural versatility of the fungi will rapidly enlarge the role of fungi in agribiotechnology.

For background information *see* ECOLOGY; FUNGI; INSECTICIDE; MYCORRHIZAE; PESTICIDE; PLANT PATHOLOGY in the McGraw-Hill Encyclopedia of Science & Technology.   Frank M. Dugan

Bibliography. A. Altman (ed.), *Agricultural Biotechnology*, Marcel Dekker, 1998; I. Chet (ed.), *Biotechnology in Plant Disease Control*, John Wiley, 1993; F. R. Hall and J. J. Menn (eds.), *Biopesticides: Use and Delivery*, Humana Press, 1999; R. K. Upadhyay and K. G. Mukerji (eds.), *Toxins in Plant Disease Development and Evolving Biotechnology*, Science Publishers, 1997; R. G. Van Driesche and T. S. Bellow, Jr., *Biological Control*, Chapman & Hall, 1996.

Air-traffic control

The United States and international civil aviation communities have intensive activities under way to modernize the air-traffic control service. A principal objective is to safely provide aircraft operators greater flexibility in selecting their routes, altitudes, and departure and arrival times. The objective has been given the name free flight, and will be achieved by introducing new technologies and associated air-traffic control procedures and airspace designs. Increased flexibility to fly user-preferred trajectories will provide significant benefits for airspace users, including fuel and time savings, improved productivity for aircraft and flight crews, fewer flight delays, and more efficient use of airspace and airport capacity resources.

Within the United States, the Federal Aviation Administration (FAA) is the principal provider of air-traffic control service, and RTCA, Inc., provides a forum wherein members of the aviation community collaborate on future directions for air-traffic control system development. Efforts to modernize the air-traffic control system in Europe are coordinated by the European Organization for the Safety of Air Navigation (EUROCONTROL) in Haren, Belgium.

Decision support tools. Much of the effort to modernize air-traffic control services is focused on the development and implementation of effective computer-based decision support tools for controllers. At the tactical level, the tools are intended to help controllers to safely separate aircraft and to meter and sequence aircraft entering congested airspace in the vicinity of major airports. At a more strategic level, tools are being implemented to help controllers manage national traffic flows in collaboration with aircraft operators, notably the airlines. These so-called automation tools exploit advanced information processing and data communications technologies.

Conflict probes. These are computer-based aids to help en route controllers and controller teams plan aircraft movements in the high-altitude cruise phase of flight. A flight crew will sometimes request an amendment to its currently approved flight plan which specifies the route of flight and altitude. The request may be to shorten the route by flying directly to the destination airport or to change altitude for a smoother ride when there is turbulence. Prior to approving such a request, the controller must be sure that the modified flight plan will not result in conflicts with other aircraft. A conflict probe estimates the future four-dimensional trajectory (in three spatial dimensions and time) of each aircraft based on its performance characteristics (such as cruising speed), current position, and flight plan, including any proposed amendments. It then projects the route of all aircraft forward along their expected trajectories over a time interval of 15–20 min to determine whether or not aircraft separations will become less than prescribed minima. Any such predicted conflict is displayed to the controller for resolution. For example, the User Request Evaluation Tool (URET) is a conflict probe under development by the FAA.

Arrival traffic managers. These are computer-based automation tools designed to aid controllers in sequencing, merging, and spacing aircraft for landing. When an aircraft approaches a major metropolitan airport, the en route controller team responsible for the aircraft during the cruise phase of flight hands off the flight to a terminal area controller (or controller team) who directs the aircraft to its landing runway. Major United States airports commonly are fed by four separate en route controller teams, and there may be as many as nine terminal area controller teams handling arrivals. The purpose of an arrival traffic manager is to establish an efficient, coherent plan that assigns aircraft to runways and determines the landing sequence for each runway. The tool displays the plan to all of the controller teams involved for coordinated execution. It is recognized that controllers will not always direct aircraft in accordance with the plan. For example, a flight crew may request a different landing runway and the controllers may oblige. A principal challenge in tool design is the real-time modification of the plan when these exceptions arise. The Center-TRACON (Terminal Radar Approach Control) Automation System includes two arrival traffic managers: Traffic Management Advisor and Final Approach Spacing Tool. Both were developed by the National Aeronautics and Space Administration (NASA) in collaboration with the FAA.

Traffic flow management. This is the strategic process of allocating national and regional traffic flows to capacity-constrained resources in the national airspace. When the weather is good, there is little need to limit traffic flows. The national airspace generally is able to accommodate the traffic demand imposed by the airlines, the military, and general aviation (corporate and private aircraft operators). However, when inclement weather sets in, for example, at major hub airports such as Chicago, airport acceptance rates are reduced, and FAA traffic managers must reduce traffic flows accordingly. These reductions generally are implemented by holding inbound

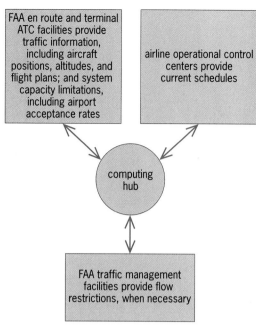

Collaborative Decision Making distributed information system.

aircraft on the ground; they may impact air travel nationwide. The art of traffic flow management is to implement flow control strategies that prevent unsafe levels of traffic congestion at system bottlenecks while least disrupting air travel. The new approach to traffic flow management is called Collaborative Decision Making. It is based on a distributed information system that supports both FAA traffic managers, who are responsible for flow control, and airline dispatchers, who are responsible for the management of their fleets. The information system (see **illus.**) assures that these operational specialists have an accurate, shared understanding of the national traffic situation, and it provides computer-based decision support tools that can predict system capacity, traffic demand, and the effects of candidate flow control strategies. The Collaborative Decision Making computing hub is located in Cambridge, Massachusetts, and the FAA traffic managers are located in a central flow control facility in northern Virginia and in regional traffic management units distributed throughout the 50 states. The dispatchers are located in airline operational control centers throughout the country. The FAA and airline specialists access the information system from their respective locations and through it collaborate to develop and implement effective flow control strategies.

Human factors design. A common challenge in the application of decision support tools to air-traffic control is to assure an adequate human factors design. Human factors can be defined as a multidisciplinary effort to generate and compile information about human capabilities and limitations and to apply that information during system development, implementation, and operation to assure that the system safely and effectively performs its intended function. Its realm of application includes hardware and software design (including the human-system interfaces), operational procedures, the facility work environment, and training. Poor human factors design has resulted in the demise of a number of otherwise superior decision support tools intended for use in air-traffic control.

The decision support tools described above will be implemented by the FAA as elements of Free Flight Phase 1. EUROCONTROL's PHARE project is developing similar capabilities.

Utilities. Air-to-ground communications, navigation, surveillance, and aviation weather reporting capabilities are the essential utilities of the air-traffic control service. Substantial efforts are under way to enhance these system elements.

Air-ground communications. The term "data link" refers to communicating data between ground stations and aircraft. The aircraft may be on the ground (for example, at an airport gate) or airborne. Data link initiatives are focused on developing modulation techniques and communications protocols that use the available radio frequency spectrum efficiently. It is expected that the data link soon will be a mainstay of the air-traffic control system, providing current weather observations and forecasts to aircraft along with other airspace status information such as air-traffic control equipment outages and current and projected traffic delays. Traffic advisories will be transmitted to aircraft in flight to aid pilots in seeing and avoiding conflicts with other aircraft. Controller clearances and instructions will be uplinked, and aircraft will downlink their positions and flight-path intentions to aid ground automation equipment in predicting conflicts among flight paths and in planning traffic flows. In time, ground air-traffic control computers and flight management computer systems in aircraft will use the data link to negotiate conflict-free flight paths that best correspond to user-preferred trajectories.

Navigation. The constellation of Global Positioning System (GPS) satellites provides a highly accurate position determination capability worldwide to users who are equipped to receive these signals. A number of ground-based augmentations of GPS are being developed and implemented in the United States and elsewhere to assure that the resulting navigation capability will meet the stringent accuracy, availability, reliability, and integrity requirements of civil aviation. It is clear that GPS will be a cornerstone of the future air-traffic control system. The use of GPS is not mandated today. However, it is a strong candidate for certification as a stand-alone system. *See* GLOBAL POSITIONING SYSTEM (GPS).

Surveillance. There is considerable interest in developing an automatic dependent surveillance broadcast (ADS-B) capability, wherein an aircraft's position will be determined from an on-board GPS receiver or other navigation sensor and the position reports will be broadcast from the aircraft periodically. The position reports will be augmented with additional relevant information about the aircraft, including the radio call sign (for example, Northern Airlines Flight

34) and the planned route of flight. This technique will provide accurate aircraft position information to simple receiving equipment both on the ground and in other aircraft. It is expected that the current heavy reliance on expensive ground-based air-traffic control radar will lessen over time, and inexpensive avionics receivers and displays will provide flight crews an accurate understanding of the positions, radio call signs, and flight intentions of the aircraft in their vicinity. The initial large-scale applications of ADS-B are likely to be in areas of the world where extensive ground-based air-traffic control equipment infrastructures are not yet in place and would not be cost-effective to install.

Aviation weather capabilities. The improvement of aviation weather information is an essential element of air-traffic control system modernization. Adverse weather is a leading cause of aviation accidents and a principal contributor to system delays, reroutings, and other inefficiencies. Aircraft operators and air-traffic controllers require accurate current information and forecasts concerning ceiling and visibility, winds aloft and at airports, precipitation (especially snow and ice accumulations), storm growth and decay, turbulence, icing conditions aloft, windshear, and microbursts. The FAA and the National Weather Service are making substantial research investments in this area.

For background information *see* AERONAUTICAL-METEOROLOGY; AIR-TRAFFIC CONTROL; DECISION SUPPORT SYSTEM; HUMAN-FACTORS ENGINEERING; SATELLITE NAVIGATION SYSTEMS in the McGraw-Hill Encyclopedia of Science & Technology.   Clyde A. Miller

Bibliography. Federal Aviation Administration, *National Airspace System Architecture*, updated annually; M. Kayton and W. R. Fried, *Avionics Navigation Systems*, Wiley-Interscience, 1997; RTCA, Inc., *Government/Industry Operational Concept for the Evolution of Free Flight*, 1997, C. D. Wickens et al., *Flight to the Future: Human Factors in Air Traffic Control*, vols. I and II, National Academy Press, 1997.

Aircraft laminar flow control

For flying, an upward lift force must be generated to overcome the weight of the aircraft, and the engines must generate a forward thrust force to overcome the drag force produced by the air flowing over the aircraft. These lift and drag forces are closely coupled, and an improvement in one can lead to benefits in the other. A principal contributor to the drag force is boundary-layer skin-friction drag, which results from friction forces generated as air passes over the aircraft surfaces.

The flow of air over the aircraft skin may be classified as laminar near the leading edges, then transitional, then turbulent over the aft sections (**Fig. 1**). Laminar flow consists of smooth streamlines and little unsteadiness, hence low viscous stresses. Turbulent flow, further downstream on the skin, contains continuous, superimposed random fluctuations

which greatly increase friction. Transitional (intermittent) flow is the bridge region between the two. Laminar flow has the lowest skin friction and therefore is desirable. Unfortunately, on typical aircraft at cruise speeds and natural conditions, the laminar region is confined to a small region near the leading edge and most of the flow is turbulent. Thus, the goal of laminar flow control is to reduce the drag force of an aircraft by maintaining laminar flow over large regions of the aircraft by means of a mechanical system.

Boundary-layer theory. Although laminar flow is desired, it is sensitive to a number of parameters existing in most practical applications. For example, if the surface finish is not smooth, the flow can undergo a rapid transition to turbulent flow, yielding higher, undesirable skin-friction penalties. The laminar boundary-layer flow on aircraft with engine nacelles and low-sweep (less than 25°) wings and tails will undergo transition due to traveling-wave instabilities excited in the boundary layer. These waves, or instabilities, are analogous to strong winds blowing over water. Surface waves will form which increase in amplitude (size) as they move downstream. Similarly, boundary-layer waves in the laminar region become unstable, grow in size, and cause transition to turbulence downstream. One way to reduce these waves, or even cause them to decay, is to design regions of favorable gradient, where the flow over the surface is accelerating. For aircraft with high-sweep (greater than 25°) wings, the laminar-to-turbulent-flow transition occurs very near the leading edge of the wing. This transition is caused by vortices (instabilities) which are generated by the three-dimensional behavior of the velocity profiles resulting from the high wing sweep. Hence, some means of active control system must be used to prevent these naturally occurring instabilities from growing in amplitude and causing the transition to turbulent flow.

Natural laminar flow. Laminar flow may be naturally maintained in some applications without the use of an active mechanical control system. Low-sweep wings and tails, engine nacelles, and the nose of a fuselage are areas that can benefit from the low skin-friction drag associated with laminar flow. Present-day manufacturing technology can provide

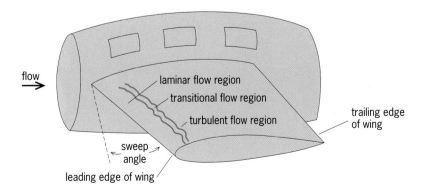

Fig. 1. Flow regions on a typical aircraft lifting surface.

surface tolerances that favor laminar flow. For these applications, the laminar-to-turbulent transition is dominated by traveling-wave instability. To suppress the growth of these traveling waves (and the onset of turbulent flow), the shape of the surface should be such that a favorable (accelerating flow) pressure gradient will delay the occurrence of transition and yield a considerable region of laminar flow. For example, on a business jet, proper use of favorable gradients can maintain laminar flow on about 50% of the wing surface, reducing drag by 12%. Smaller aircraft and gliders may approach 100% laminar flow on their wings. Further drag reductions (up to 22%) can be achieved by proper design of horizontal and vertical tail surfaces, the fuselage, and engine nacelles.

Active control system. For applications where natural laminar flow designs are not viable, obtaining laminar flow through an active control system may be an alternative to stabilize the flow and hence prevent the higher drag associated with turbulent flow. More than 50 years of research have demonstrated that suction can be an effective basis for a laminar flow control system. The suction system requires that air be extracted through the skin of the wing or nacelle, pass into underlying ducts or flutes and on to the suction pump, and then be discharged in some region of the aircraft where an adverse aerodynamic effect will not result. As such, suction over the entire configuration or even a small portion of the leading-edge region (combined with favorable pressure distributions) has led to considerable regions of laminar flow. While the suction pump and ducts are conventional hardware, the extraction of air through the skin has been the subject of considerable research. Porosity, slots, as well as perforations have been used as the interface between the external flow and the internal ducts. Porosity is achieved by using a fabriclike material to cover holes in the aircraft skin leading to the ducts; slots are achieved using a saw to cut through the aircraft skin; and perforations are made using holes drilled through the skin with electron beams or lasers (**Fig. 2**).

By stabilizing traveling-wave instabilities, the con-

Fig. 3. F-16XL Ship 2 with supersonic laminar flow control. The perforated regions are the wing portions emphasized in black.

trol has successfully led to as much as 100% laminar flow on nacelles and low-sweep wings in windtunnel and flight experiments. For aircraft with high-sweep wings (normally associated with transonic or supersonic cruise flight conditions), vortex instabilities cause the flow to become turbulent very near the leading edge of the wings. For this situation, a control system must be used to enable a region of laminar flow. The use of suction systems has led to significant laminar flow on high-sweep wings, with more than 50% laminar flow achieved in windtunnel and flight experiments. The first significant extent of laminar flow obtained in supersonic flight using a suction system was demonstrated with an F-16XL test aircraft (**Fig. 3**).

Benefits versus penalties. For laminar flow control to be considered economically feasible, the benefits of reduced drag due to laminar flow must overcome the penalties imposed by additional system weight, manufacturing costs, and potential operational costs. The drag reduction achieved with regions of laminar flow ultimately leads to an increase in aerodynamic performance and a reduction in the fuel burn for any given flight. Because the laminar flow control system is designed for cruise flight, an aircraft with a large cruise range mission, such as a large transport on an intercontinental flight, will incur fuel savings offsetting other costs. A typical reduction in fuel burn for a turbulent long-range aircraft may be of the order of 20%. This would either enhance the range of the aircraft or permit the manufacturer to reduce its size (since less fuel is needed for a given mission) and achieve additional cursory benefits. This reduction in aircraft size also leads to smaller requirements in engine thrust and size (10% less thrust required plus 6% weight savings) and reduced emissions and noise. Systems studies have shown that the overall takeoff gross weight (TOGW) of the aircraft may be reduced by more than 10%. For a typical turbulent transport configuration with a TOGW of 750,000 lb (340,000 kg), a comparable laminar flow control aircraft would be 75,000 lb (34,000 kg) lighter, which is equivalent to the total weight of the payload (passengers and freight).

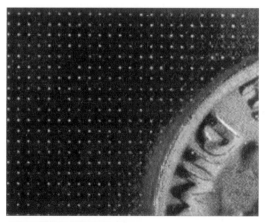

Fig. 2. Perforated surface for laminar flow control compared with U.S. dime.

For background information *see* AERODYNAMIC-FORCE; AERODYNAMICS; AERONAUTICS; AIRCRAFT-DESIGN; AIRFOIL; BOUNDARY-LAYER FLOW; LAMINAR-FLOW; TURBULENT FLOW in the McGraw-Hill Encyclopedia of Science & Technology.   Ronald D. Joslin

Bibliography. R. D. Joslin, Aircraft laminar flow control, *Annu. Rev. Fluid Mech.*, 30:1–29, 1998; R. D. Joslin, *Overview of Laminar Flow Control*, NASA TP-208705, October 1998; H. Schlichting, *Boundary Layer Theory*, 7th ed., McGraw-Hill, 1979.

Airplane trailing vortices

The production of lift results in the generation of vortices trailing an aircraft. The airplane wing applies pressure to the band of air through which it sweeps. The natural tendency of the fluid to escape sideways creates an overturning motion on each side of the aircraft, and therefore the wake consists primarily of a pair of large, powerful counterrotating vortices (**Fig. 1**). Until these vortices have sufficiently weakened or broken up, they present a hazard and must be avoided by following aircraft. This sets limits on airport operations and requires detailed regulations for the spacing and relative flight paths of aircraft. Air-traffic controllers frequently warn pilots of wake turbulence. There is great interest in locating, predicting, and, if possible, controlling the trailing vortices in order to maintain safety and potentially increase the capacity of the airports without adding runways.

Description. The circular motions of air around each vortex have their highest magnitudes near the vortex centers and give rise to downward motion between the vortices (**Fig. 2**). The curvature of the trajectories lowers the pressure near the vortex centers, thus lowering the temperature and sometimes causing visible condensation. The velocity component along the wake, in the flight direction (not shown in Fig. 2), may be in either direction depending on the region, may change sign as the wake ages, and is not as well understood. There is recent interest in wakes which contain two or more vortex pairs, created by the tips of the high-lift flaps and horizontal tail (Fig. 1), in addition to the wing tips. (Only one such pair is shown in Fig. 2.) It is not easy to predict how long the pairs will remain distinct and "tumble" together.

A following airplane that cuts through the wake from the side experiences a series of bumps as it encounters an upward gust, then a downward gust, and then another gust upward. In extreme cases, such gusts can cause structural failures. In contrast, a following airplane that is roughly aligned with the wake experiences a rolling motion as one wing is in a downdraft and the other in an updraft. Such a roll input can cause a loss of control and be hazardous at low altitudes. Essentially, the direct encounter with the intact wake left by an airplane of the same or larger size poses a serious threat. The solution is either to avoid the wakes (flight-path control) or to encounter only wakes old enough to be much weaker (spacing control).

The principal characteristics of the wake are the strength of the vortices and the distance between them. The strength is well measured by the circulation Γ, which is the line integral of the velocity vector $\vec{U}$ along a contour that surrounds the vortex (Fig. 2), as expressed in Eq. (1). To a good approximation,

$$\Gamma \equiv \oint \vec{U} \cdot \vec{dl} \qquad (1)$$

Eq. (2) is satisfied, where b_0 is the distance between

$$W = \rho V \Gamma b_0 \qquad (2)$$

vortices, ρ is the density of the air, V is the velocity of the airplane, and W is the airplane's weight. This formula shows that the vortices are strongest from an airplane that is at high altitude (lower ρ), heavy, slow, and has a short span b. [In general, the distance between vortices is related to the span by Eq. (3).] The kinetic energy per unit length is

$$b_0 \cong \frac{\pi b}{4} \qquad (3)$$

approximately $0.4\rho\Gamma^2$. This energy corresponds to the induced drag of the wing. Gliders have large b and b_0 to reduce Γ and the energy loss. Wing design makes a compromise between this favorable trend and the higher structural weight of a long-span wing. Winglets slightly increase b_0 for a given b, thus lowering the induced drag.

A typical airliner has a span of 50 m (165 ft), mass of 250,000 kg (550,000 lb), and approach speed of 75 m/s (250 ft/s or 145 knots), and the air density near sea level is 1.3 kg/m^3 (0.08 lb/ft^3). The vortex

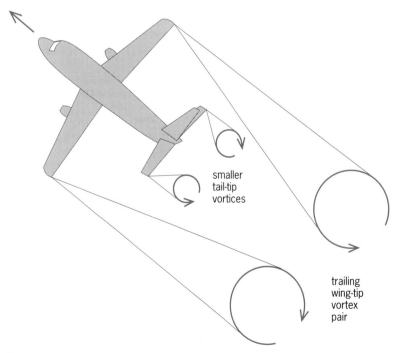

smaller tail-tip vortices

trailing wing-tip vortex pair

Fig. 1. Airplane generating trailing vortices.

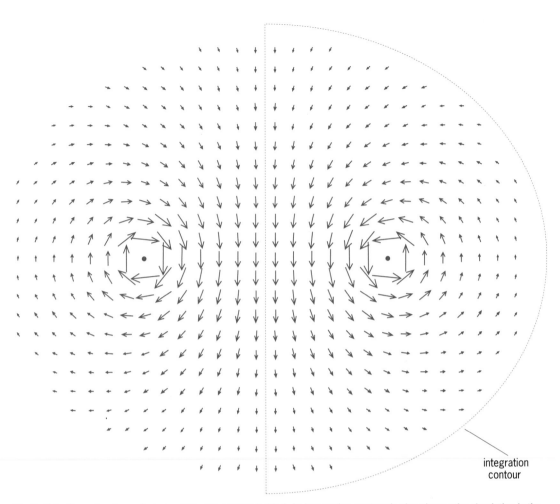

integration
contour

Fig. 2. Velocity vectors in an airplane wake, viewed from behind. An integration contour is also shown; the circulation is the line integral of the velocity vector along this contour.

characteristics are then approximately $\Gamma = 630 \text{ m}^2/\text{s}$ (6800 ft²/s) and $b_0 = 40$ m (130 ft). The downward velocity on the centerline is 10 m/s (33 ft/s or 20 knots). The roll rate induced on a 25-m-span (82-ft) airplane in the very worst location is nearly 80°/s.

Motion and evolution. In the simplest model of a vortex pair, the vortices drift straight down at the velocity $\Gamma/(2\pi b_0)$, or 2.5 m/s (8 ft/s or 5 knots) in the above example. They are carried by the atmosphere and therefore obey its winds and vertical currents. In addition, atmospheric stratification (upward decrease of air density) can have a noticeable effect, particularly with larger airplanes. Some theories predict that stable stratification slows and arrests the descent, but other theories predict a reverse effect: a faster descent and a reduction of the pair spacing. When the vortex system approaches the ground, it has a mild rebound and mingles with vortices which the vortex system itself caused to rise from the ground. All these phenomena are present in two dimensions, but the vortex system also has three-dimensional instabilities which, in their pure form, turn the parallel vortices into wavy lines and then a train of rings. These rings can be seen on condensation trails, usually at high altitudes on clear days. This instability is seeded by atmospheric turbulence

and then amplified by the dynamics of the vortices, which gives it a random character. Once rings have formed, the breakdown into small harmless eddies is more rapid. On windy days the atmosphere can also contain turbulence on small scales, intense enough to gradually mix the vortices with each other and the surrounding fluid. In practice, a useful time frame in which to dispose of the vortices is of the order of 1 min, 2 min at the most. If it were not for instabilities and turbulence, pure (molecular) viscous effects would take years to dissipate the vortices.

It is now clear that trailing vortices are subjected to numerous perturbations, most of which change by the hour if not by the minute, and have a random character. For instance, flight tests have proven that wakes can in fact rise during part of their lifetime, even when they are out of reach of the ground effect. The speculation about this anomaly invokes stratification, or simply atmospheric currents. Flight tests also reveal apparently unpredictable events in which the region marked with smoke at the center of the vortex suddenly shrinks, expelling fluid in both directions which eventually accumulates in disks. No theory exists for this phenomenon. There is still no consensus regarding an issue as basic as whether the circulation Γ remains constant or experiences a predictable decay, even in a quiet and uniform atmosphere.

Prediction. The present state of understanding is related to the vital question of whether the motion and the decay or collapse of the vortex systems can be predicted to a useful level of accuracy, even with improved theories and numerical methods. A fairly well accepted measure of the intensity of atmospheric turbulence is known as energy dissipation rate. Measurements of the time to destruction of wakes at various values of turbulence intensity, in flight tests, laboratory tests, and simulation, as well as theoretical work, all show a correlation, namely, that more intense turbulence (higher energy dissipation rate) causes earlier destruction. However, there is a large scatter in the results, even with a modest number of samples (much smaller than would be needed to establish a rule in safety matters). Either the turbulence level (energy dissipation rate) is insufficient to describe the physics, and the way to improve predictability is to introduce additional parameters (for instance the stratification), or else predictions of vortex destruction are simply too unreliable to be of use.

As a result of the low predictability of the vortices and the lack of consensus regarding their theory, practices and regulations are based almost entirely on experience. A much improved prediction capability would make it possible to determine those wind or weather conditions in which the current regulations are so conservative that they could be safely relaxed to reduce delays and increase airport capacity. The National Aeronautics and Space Administration (NASA) is developing such a dynamic spacing system. There is also an intense need to predict the spacing behind very large commercial transports, with weights in excess of 500,000 kg (1,100,000 lb). This information is needed, for airport planning and productivity estimates, before the programs to develop these aircraft are even launched.

Detection and control. Wake vortices would be much less disruptive if accurate knowledge were available of their position, extending several miles back, and even a rough measure of their remaining strength. At least in clear weather, both can be provided by a ground-based lidar system, an optical analog of radar, in which laser pulses are directed at the object of investigation, and the reflected pulses are picked up by sensitive optical detectors. The lidar responds to the motion of particles, including water droplets, in the vortices. However, an all-weather lidar with a range of the order of 1 km (0.6 mi) is not available. Obtaining access to build lidar systems at enough locations under the approach corridor to survey incoming aircraft would also pose problems, especially for an approach over water. Airborne detectors of the velocity field or of the exhaust chemicals have also been proposed, but so far no devices have demonstrated the required capability.

The control of trailing vortices originates in the wing system. Passive systems alter the circulation, distance, kinetic energy, and velocity profiles of the vortices, while active systems seed the instabilities

through cyclic motion of the control surfaces. Again, there is controversy regarding vortex control systems, and the extensive flight-test campaigns needed to demonstrate them will be delicate and costly. Analytical and scale model studies of such systems have been undertaken. *See* MICRO-ELECTRO-MECHANICAL SYSTEMS (MEMS).

For background information *see* AERODYNAMIC FORCE; CALCULUS OF VECTORS; LIDAR; TURBULENT FLOW; VORTEX; WAKE FLOW in the McGraw-Hill Encyclopedia of Science & Technology.   Philippe R. Spalart

Bibliography. J. D. Anderson, *Fundamentals of Aerodynamics*, 2d ed., 1991; P. R. Spalart, Airplane trailing vortices, *Annu. Rev. Fluid Mech.*, 30:107–138, 1998.

Alzheimer's disease

Alzheimer's disease is a degenerative brain disease of unknown cause that leads to the form of dementia most common in old age. It was first described in 1907 by Alois Alzheimer, a Viennese physician. It is characterized by a rapidly increasing loss of memory, disorientation with respect to time and place, and perplexed behavior.

For decades Alzheimer's disease was thought to strike during middle age and to lead to memory loss termed presenile dementia. Senile dementia was defined as affecting old people and was thought to be a fairly common part of aging. It is now known that most senile dementia is actually Alzheimer's disease and that this disorder is extremely common. It affects about 10% of people by age 65 and 30–40% of those over age 85.

The brain. Alzheimer described the neuropathological appearance of a patient's brain after autopsy. In certain areas involved in cognition and memory, there was a loss of neurons and numerous neuritic plaques composed of a dense core of amyloid material, surrounded by a halo of degenerating nerve cell processes (the long thin arms of a neuron that carry signals from one cell to another), neurofibrillary tangles that were stongly staining deposits inside neurons, and a "rampant gliosis" (see **illus.**). Gliosis is evidenced by the presence of numerous activated glia, the brain cells that carry out support functions such as guidance of neuronal movement and process pathfinding during brain development and the clearing away of debris after brain damage. Clearly the neuronal loss was partly responsible for the cognitive and memory deficits. The pathology alone, however, could not reveal which of the other features of the disease were causes and which were effects of the neuronal cell death. A clear understanding of the complete Alzheimer pathogenic pathway has not been possible until recently.

Aβ peptide. The breakthrough in Alzheimer's disease research came when the major component of the amyloid deposits that form the core of the neuritic plaques (and that also deposit in some blood vessel walls) was identified as a 28–amino acid (later

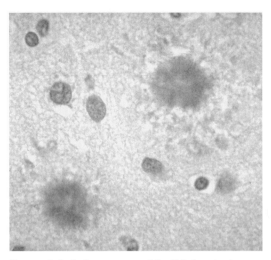

Neuropathological appearance of the Alzheimer brain, showing neuritic plaques.

found to be 40–42) peptide. This discovery marked the beginning of a highly successful biochemical and genetic attack on the disease by researchers. The gene corresponding to the amyloid peptide was cloned and found to encode a large protein from which the amyloid peptide (termed Aβ for Alzheimer beta to designate its largely beta pleated sheet structure) was derived by cleavage with proteolytic enzymes.

The finding of additional proteins such as antichymotrypsin and apolipoprotein E in the amyloid deposits together with Aβ and of the protein tau in the neurofibrillary tangles suggests that they may contribute to the pathological process. The other possibility was that any or all of these proteins might be peripheral or even irrelevant and that their investigation would not lead to an understanding of Alzheimer's disease but only of the response of the brain to the disease. For instance, the amyloid deposits could be debris from dying neurons; and the inflammation, most clearly shown by the fact that the inflammatory molecules antichymotrypsin and interleukin-1 (a signaling molecule) were greatly overproduced in affected areas of the Alzheimer brain, could merely reflect a secondary response of the brain to damage. This uncertainty was primarily solved by genetics.

Role of genetics. Almost 20% of Alzheimer's disease, especially involving an early age of onset (before age 60), is inherited in a dominant mode; this inheritance is seen when half of the children of an Alzheimer's patient also develop the disease. At least another 30% of the disease is influenced by inherited variations in single genes (polymorphisms). The rest is termed sporadic (in contrast to familial) Alzheimer's disease, but it could still be influenced by complex genetic factors not yet identified.

The advantage of studying a genetic disease is that the product of any gene in which a mutation causes, or increases the risk of, the disease must participate in the disease mechanism. That is, the encoded protein must be part of the pathogenic pathway. For example, the reason it is known that the APP protein and probably the Aβ peptide are directly involved in Alzheimer's disease is that a few mutations were found in the APP gene that cause inherited Alzheimer's disease. These mutations appear either to increase the amount of Aβ peptide produced or to alter its structure, for example its length (tending to make more of the 42-amino acid form) or its sequence, so that it more readily polymerizes into the long fibers that make up the amyloid deposits. Other studies demonstrated that the Aβ peptide is toxic to neurons, especially in its polymerized filamentous form found in amyloid deposits. Thus, any mutation that increased the production of amyloid filaments would increase the extent of neuronal cell death in the brain.

Two other genes, termed presenilin 1 and 2, have been shown to carry mutations that also cause inherited Alzheimer's disease. Like the APP mutations, the presenilin mutations are dominant (only one bad copy is necessary for the disease to develop with 100% certainty).

Finally, simple inherited changes in the genes encoding antichymotrypsin and apolipoprotein E have been shown to increase the probability that a person carrying the alteration (polymorphism) will develop Alzheimer's disease. Many more of these risk-factor genes are probably waiting to be discovered (see **table**).

The mutations in APP appear to be easily explained: they directly affect the amount or amyloidogenic potential of the major component of Alzheimer amyloid, the Aβ peptide. Similarly, the fact that all Down syndrome individuals develop Alzheimer's disease by age 30–40 is usually explained by the fact that the APP gene is on chromosome 21; they have 50% more of this protein and hence at least 50% more Aβ peptide. This conclusion was reinforced when mice were generated that carried the mutant APP gene and developed Alzheimer amyloid plaques.

Inflammation. Antichymotrypsin and apolipoprotein E appear to be part of an inflammatory cascade in the Alzheimer brain that is necessary for the efficient polymerization of the Aβ peptide into neurotoxic amyloid filaments. For example, it has been found that the inflammatory protein, antichymotrypsin, is

Genetics of Alzheimers disease	
Causes and risk factors	Location
Trisomy 21 (Down syndrome)	Chromosome 21
Genes with dominant, causal mutations	
Amyloid precursor protein (APP)	Chromosome 21
Presenilin 1	Chromosome 14
Presenilin 2	Chromosome 1
Genes with risk-enhancing polymorphisms	
Apolipoprotein E (apoE4 in particular)	Chromosome 19
Antichymotrypsin (ACT-A in particular)	Chromosome 14

part of the amyloid deposits, and binds directly to the Aβ peptide, and is overexpressed in astrocyte glial cells in those areas of the Alzheimer brain in which amyloid formed; this overexpression is in response to the inflammatory signaling molecule interleukin-1 (IL-1). Findings further suggest that antichymotrypsin promotes or stabilizes the formation of the amyloid filaments. IL-1 overproduction in microglial cells in Alzheimer's patient's brain provides strong confirmatory evidence for the presence of inflammation. It has been established that IL-1 also increases the expression of APP in human astrocytes through an effect on the translation of the APP messenger ribonucleic acid (mRNA). Thus, inflammation is now implicated in the upregulation of the central gene in Alzheimer's disease—APP itself. The discovery that apolipoprotein E is also part of the amyloid deposits led researchers to propose a helper function for this protein in the disease, and to coin the term pathological chaperone to describe such proteins as antichymotrypsin and apolipoprotein E. The E4 allele of apolipoprotein E has been shown to be a strong risk factor for late-onset inherited Alzheimers's disease; a similar effect is seen for the A allele of antichymotrypsin. These genetic findings clearly indicated that the two proteins must be involved in the pathogenic pathway.

The direct demonstration of the amyloid-promoting power of antichymotrypsin and apolipoprotein, especially the E4 form, was first carried out in vitro. The results showed that Aβ by itself would polymerize into neurotoxic filaments in the test tube, but the reaction was accelerated more than 10-20-fold by adding a very small amount of antichymotrypsin or apolipoprotein E. The form of apolipoprotein E identified as harmful by genetics, apoE4, was the most effective amyloid promoter or pathological chaperone.

An in vivo demonstration that apoE is an amyloid promoter was provided by a series of elegant mouse experiments. Specifically, a set of mouse strains were developed that expressed transgenic human APP and lacked one or both of their apoE genes (because they had been knocked out). The animals showed a variable amount and speed of amyloid deposition that was completely dependent on the number of copies of the apoE gene. If there was no apoE, mature, filamentous amyloid did not form even up to 2 years of age, compared to massive amyloid formation by 7 months in the presence of the normal two copies of the apoE gene. In short, human Aβ by itself is incapable of forming amyloid in the mouse without the promoting effect of apoE.

Therapeutic intervention. The pathogenic pathway in Alzheimer's disease begins with small amounts of Aβ peptide production, is accelerated by an inflammatory cascade, and ends with antichymotrypsin- or apolipoprotein E–promoted amyloid filament formation and neurotoxicity. The importance of inflammation is strongly supported by the finding that general anti-inflammatory drugs such as indomethecin slow the development of Alzheimer's disease. In addition, there are potential points of more specific therapeutic intervention. For example, it may be possible to develop molecules which inhibit the interaction between antichymotrypsin and Aβ or between apolipoprotein E and Aβ and thus prevent the accelerated formation of amyloid filaments. Researchers have found that blocking the IL-1 receptor on astrocytes in vitro prevents their expression of the amyloid promoter antichymotrypsin. If such blockade can be accomplished in the brain, it should eliminate the accelerating effect of the inflammatory cascade, therefore preventing or drastically slowing the development of Alzheimer's disease.

For background information *see* AGING; ALZHEIMER'S DISEASE; AMYLOIDOSIS; BEHAVIOR GENETICS; DOWN SYNDROME; HUMAN GENETICS; INFLAMMATION; NEURON; SENILE DEMENTIA in the McGraw-Hill Encyclopedia of Science & Technology.

Huntington Potter

Bibliography. J. Li et al., Alzheimer presenilins in the nuclear membrane, interphase kinetochores, and centrosomes suggest a role in chromosome segregation, *Cell*, 90:917–927, 1997; J. Ma et al., Amyloid-associated proteins a₁-antichymotrypsin and apolipoprotein E promote assembly of Alzheimer β-protein into filaments, *Nature*, 372:92–94, 1994; D. L. Price et al., Alzheimer's disease: Genetic studies and transgenic models, *Annu. Rev. Genet.*, 32:461–493, 1998; D. J. Selkoe, The cell biology of beta-amyloid precursor protein and presenilin in Alzheimer's disease, *Trends Cell Biol.*, 8:447–453, 1998.

Angiogenesis

Normal development of all organs and maintenance of their physiological functions are critically dependent on vascularization. The quantity and quality of the vascularization are tightly controlled to meet the specific anatomical and physiological requirements of each organ. In fine-tuning the vascularization, blood vessels and their corresponding organ exchange chemical signals to control morphogenesis and homeostasis. Failure or malfunctioning of this "crosstalk" between the blood vessels and the organ can lead to disease conditions. Therefore, it is important to understand the molecular nature of this crosstalk, which underlies the development of future therapies.

Blood vessel formation. Significant advances have been made in understanding the molecular nature of blood vessel formation under normal as well as disease conditions. This scientific progress may be mostly attributed to the discovery of a number of unique molecules that control blood vessel formation, or angiogenesis. This growing list of molecules demonstrates that the process of blood vessel formation is far more complex than ever imagined.

Vascular-cell-type specification. Mainly three types of cells compose blood vessels: endothelial cells, smooth muscle cells, and pericytes (cells of the connective tissue of capillaries or other small vessels).

Endothelial cells occupy the innermost part of the vessels and have direct contact with circulating blood. Smooth muscle cells form one or more continuous layers around endothelial cells in medium to large-size blood vessels. Pericytes form a discontinuous layer around endothelial cells of small vessels called microvessels. In recent years, it has become clear that each of these cell types consists of even more heterogeneous populations of cells. For example, it has been shown that endothelial cells in veins and arteries are distinct and this differentiation is genetically predetermined. Furthermore, endothelial cells of various vessels localized in different organs are functionally and phenotypically distinct.

Process of vasculogenesis. The circulatory system is one of the first organ systems to develop in the fetus. Initial vessels are formed by endothelial cells. These cells assemble and form tubelike structures in a process called tubulogenesis. These fragments of vessel tubes are eventually connected and form a seamless network of vessels. This initial vessel network is called the primary capillary plexus, and the formation and differentiation of the vascular system is called vasculogenesis. Vascular endothelial growth factor (VEGF) is a soluble factor that acts through specific cell surface receptors on endothelial cells to critically regulate vasculogenesis. Vasculogenesis precedes the blood circulation, and the process is independent of the circulation.

Process of angiogenesis. As the primary capillary plexus forms, more vascular branches are produced to extend the initial vessel network in a process called angiogenesis. Therefore, angiogenesis is a process of generating the secondary vessel network from the preexisting network, which is formed by vasculogenesis. This process begins when endothelial cells are stimulated by regulatory molecules. The

formation of new vessels requires the migration and proliferation of cells.

Maturation. As new vessels are formed by angiogenesis, smooth muscle cells and pericytes are recruited to the vasculature, primarily consisting of endothelial cells, for more mature vessels. Platelet-derived growth factor-β (PDGF-β) has been shown to interact with specific cell surface receptors on smooth muscle precursor cells and mediate the recruitment of these cells to the endothelial cells of the blood vessels. Furthermore, specialized extracellular matrices are established between endothelial cells and nonendothelial cells, that is, smooth muscle cells and pericytes. Once the smooth muscle precursor cells are recruited to the vessels, endothelial cells secrete transforming growth factor-β (TGF-β) that acts on the specific receptor on these precursor cells and induces their differentiation to mature smooth muscle cells. Peri-endothelial cells, which reside next to and support the endothelial cells (smooth muscle cells, mesenchymal cells, and pericytes), secrete angiopoietin-1. Angiopoietin-1 acts on its specific cell surface receptors expressed by endothelial cells and induces and maintains the integrity of the blood vessels. It has been suggested that angiopoietin-1 acts in cooperation with vascular endothelial growth factor and other angiogenic factors to induce the maturation of the blood vessels. These associations assist in the homeostasis stabilization, and overall maintenance of blood vessels.

Remodeling. One emerging model is that the degree of vascularization in each organ is determined by a net balance of formation and regression of the vessels. This means that when the formation outweighs the regression, more vessels are formed, and when the regression outweighs the formation, vessel density decreases. It has been shown that in many organs the combination of formation and regression is continuously occurring during normal development and it is the net balance that determines the degree of vascularization and the pattern of vascular network. This process, called remodeling, is an essential step toward the final determination of the vascular pattern and the vessel density in the body.

A participant in this process is the ligand angiopoietin-2. It is secreted by the peri-endothelial cells and inhibits the action of angiopoietin-1. It has been proposed that this blockage induces the regression of the blood vessels. However, in the presence of endothelial growth factors such as VEGF, the angiopoietin-2-mediated blockage allows the remodeling of the blood vessel network.

Regulatory molecules. A number of molecules that critically regulate various aspects of blood vessel formation have been discovered in the past several years. They can be categorized into four groups: extracellular factors, cell surface receptors, intracellular signaling molecules, and transcription factors.

Extracellular factors are synthesized and secreted by cells located near endothelial cells, such as smooth muscle cells, or by endothelial cells themselves to

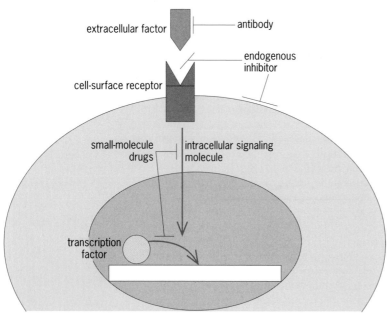

Targeting to block angiogenesis.

control the proliferation, differentiation, and other behaviors of endothelial cells. Growth factors, cell survival factors, cytokines, and extracellular matrix molecules belong to this group.

Cell surface receptors are the proteins that are expressed on the surface of endothelial cells and that sense extracellular factors; they transduce the chemical signals in endothelial cells. This receptor-mediated sensing mechanism of the extracellular environment is one of the most essential mechanisms underlying endothelial cell proliferation, survival, and other behaviors.

Intracellular signaling molecules convey the chemical message received by the cell surface receptors. These molecules regulate endothelial cell proliferation, survival, and other behaviors.

Transcription factors are a group of molecules that control gene expression in the nucleus of a cell. They may control the expression of other genes whose functions regulate the endothelial cell proliferation, migration, and other behaviors.

Coordinated regulation of all of these molecules is required for the normal formation, maturation, and remodeling of blood vessels.

Tumor angiogenesis. Just like a normal organ, tumors require nourishment from blood vessels for growth and survival. Tumors secrete a number of factors that stimulate the formation of blood vessels (tumor angiogenesis). Therefore, one therapeutic strategy to inhibit tumors is to block the formation of blood vessels that specifically feed them. Understanding how normal vessels form provides a basis upon which to devise different types of therapies to accomplish this goal (see **illus.**). As described above, several groups of molecules critically regulate the formation of blood vessels. Antiangiogenic therapies currently being tested in clinical trials attempt to block these molecules. These therapies include (1) antibodies against the extracellular stimuli for angiogenesis, (2) a number of small synthetic molecules to block the intracellular signaling or the activities of transcription factors critical for tumor angiogenesis, and (3) a number of specific endogenous inhibitory molecules for tumor angiogenesis. The effectiveness of these inhibitory factors, in the form of recombinant proteins or gene therapy, such as virus-mediated gene delivery, remains inconclusive.

Therapeutic angiogenesis. Inhibiting angiogenesis brings therapeutic benefits to the treatment of many types of cancer. However, in other human disease conditions, such as ischemic heart, limb, and brain conditions, promoting angiogenesis, or revascularization, is expected to bring therapeutic benefits. This is a far more challenging task since all new vessels are not necessarily the same. Therapeutically induced vessels need to be just like normal vessels of the same region in terms of stability, maturity, and physiological function. Achieving this goal requires assembly of everything needed to form the normal vessels. While inhibition of vessel formation may require blocking an activity of only one or two

components, promoting angiogenesis may require many more targets to be correctly assembled.

For background information *see* CARDIOVASCULAR SYSTEM: CIRCULATION; ONCOLOGY in the McGraw-Hill Encyclopedia of Science & Technology.

Thomas N. Sato

Bibliography. J. Folkman, Angiogenesis in cancer, vascular, rheumatoid and other disease, *Nature Med.*, 1:27–31, 1995; J. Folkman, Antiangiogenic gene therapy, *Proc. Nat. Acad. Sci. USA*, 95:9064–9066, 1998; W. Risau, Mechanisms of angiogenesis, *Nature*, 386: 671–674, 1997; T. N. Sato, A new approach to fighting cancer?, *Proc. Nat. Acad. Sci. USA*, 95:5843–5844, 1998.

Animal waste management

Animal waste is generated in beef cattle feedlots, swine operation sites, dairy barns, poultry houses, and other sites of livestock operations. This renewable resource is a source of macronutrients (such as nitrogen, phosphorus, and potassium) and micronutrients (including zinc, copper, iron, and magnesium) that are essential for growing plants. Therefore, manure can be an excellent substitute for synthetic fertilizers. Manure application can also increase the soil organic matter, an important factor influencing physical characteristics and chemical and biological activities in the soil. Manure was used for centuries all over the world for improving soil fertility and enhancing crop productivity. However, with the advent of synthetic fertilizers after World War II, manure was considered more of a liability than a great soil nutrient resource.

When animals are grazing on pastures and rangelands, manure is dispersed across a large area. Little management is needed because the material is not concentrated and decomposes on the soil. However, when animals are restricted to a small feeding area (**Fig. 1**), the quantity of manure requiring proper management increases significantly. Farmers are reluctant to use manure because of factors such as hauling and spreading costs, potential introduction of weed seeds, insufficient availability of manure where needed, uncertainty about availability of manure nutrients to plants, and variable application uniformity. In addition, environmental concerns regarding manure application must be addressed before manure is widely utilized in an area.

Waste characteristics. Several factors that influence mineral composition of manure are animal species and size, housing and rearing management, diets, manure storage, and climate. For example, nonruminant animals (such as swine and poultry) cannot absorb most of the phosphorus in the feed because they lack the enzyme phytase to break down the organic materials and release the phosphorus. Inorganic phosphorus is added to the diet of the nonruminants to compensate. Preliminary results from studies on addition of phytase enzyme to nonruminant

Fig. 1. Cattle on a feedlot in Nebraska.

diets have indicated significant reduction of phosphorus excretion in manure, which in turn reduces water pollution by excess phosphorus when this manure is applied to the land. Decreasing the levels of nitrogen and other nutrients in the diet can also reduce excretion of these with little effect on animal performance.

Storage and handling of manure has a great effect on its composition. A significant amount of nitrogen is lost, primarily by ammonia (NH_3) volatilization, from cattle feedlots and swine lagoons holding manure. This ammonia loss not only reduces the value of manure as a nitrogen fertilizer but also pollutes the air.

Environmental effects. Manure can be a source of water, air, and land pollution because of its potential for excess phosphorus, nitrate, salts, undesirable microorganisms, pathogens, and greenhouse gases. Manure application in excess of crop needs can cause a significant buildup of phosphorus, nitrogen, metals (arsenic, cobalt, copper, iron, magnesium, selenium, and zinc), and salt in soils. The elevated phosphorus and nitrogen levels in the soil are an environmental concern when these nutrients are carried by runoff to streams and lakes and cause eutrophication, which is promotion of algae growth and depletion of dissolved oxygen in water. This oxygen is essential for aquatic animals. Pathogens (such as bacteria, viruses, and parasites) in runoff from fields receiving manure can be another source of water pollution. Pathogens and odorous materials can also be carried by wind from the feeding operations to neighboring areas. Excess manure application can contaminate the ground water with nitrate, a water-soluble ion that moves into the soil and can reach the ground water even within a few days following application. The critical level of 10 mg L^{-1} (10 ppm) has been set for drinking water by the U.S. Environmental Protection Agency (EPA).

Alternative uses of waste. Manure can be used for algae and fish production in lagoons, reclamation of sandy and mined soils, composting, recovery of energy by anaerobic methane gas production, and refeeding. Use of manure in aquaculture has been practiced for centuries in the Far East. About one-third of the manure nitrogen can be used in animal refeeding depending on the type of manure and animal consuming the manure. In refeeding manure, presence of pathogens, heavy metals, and drugs used in the source animal needs to be monitored. Other uses of manure are pyrolysis, hydrogasification, and oil conversion process. Pyrolysis is treatment of animal manure by thermochemical processes in a closed system at temperatures of 204–800°C (400–1472°F) to produce oil. In the hydrogasification process, organic material in the presence of hydrogen under high pressure and temperature is partially converted to methane gas. A process similar to liquefaction of coal can be used to convert manure to an oillike product. However, very little manure is used for any of the foregoing purposes.

Composting is the aerobic treatment of manure in the thermophilic temperature range (40–65°C or 104–149°F). The composted material is finely textured with an earthy smell, is devoid of pathogens and fly larvae, and has a low moisture content. It can be bagged and sold for use in gardens, potting, and nurseries or can be used as fertilizer on cropland.

The heat generated during composting is also a product that can be harvested. Disadvantages of composting organic residues include loss of nitrogen and other nutrients, time necessary for processing, costs for handling equipment and labor, land needed for composting, odors generated, marketing of the composted materials, diversion of manure or residue from cropland, and reduced availability of nitrogen to plants (**Fig. 2**).

Programs. In February 1998, President Clinton released the Clean Water Action Plan, which provides a blueprint for restoring and protecting water quality across the United States. As part of this effort, the plan calls for the Department of Agriculture (USDA) and the EPA to develop a unified national strategy to minimize the water quality and public health impacts of animal feeding operations. All animal feeding operations are expected to develop and implement technically sound and economically feasible Comprehensive Nutrient Management Plans (CNMPs). The actions contained in such a plan include feed management, manure handling and storage, land application of manure, land management (tillage, crop residue, and grazing management), and record keeping. The program is voluntary for more than 85% of animal feeding operations, but operations with more than 1000 animal units (one animal unit is equal to a 454-kg or 1000-lb beef cow or equivalent of other kinds of animals) would still require National Pollutant Discharge Elimination System (NPDES) permits and a CNMP. The permits require that an animal feeding operation construct a holding area for all the generated wastewater plus runoff from a storm that has a probability of happening every 25 years for a period of 24 hours. The operations with more than 1000 animal units would also need to have a CNMP. Operations that have more than 300 animal units are required to have an NPDES permit if they are discharging pollutants directly into waters that originate outside and pass over, across, or through the facility and come in contact directly with the confined animals.

There are numerous sources of technical and financial assistance, such as the USDA, EPA, Soil and Water Conservation Districts, state agencies, and the private sector to assist animal feeding operators in developing and implementing CNMPs. Cost-share and loan programs can help defray the costs of approved or needed structures (for example, waste storage facilities for small operations) or to implement other practices such as installing conservation grass or tree buffers to protect water quality. An increasing number of states have financial assistance programs that supplement or enhance federal assistance.

Research needs. Many questions and problems remain in determining the proper application rate of manure to the land. Suitable methodology is not available for making rapid and economically acceptable field determinations of the nutrient content of manure, a necessary step in calculating rates of application. Dependable and practical equipment to accurately spread manure on soils at the desired rates is lacking. Considerable basic research on the soil microbiology associated with manure decomposition is needed to accurately predict availability and release rates of nutrients in manure. Effects of manure

Fig. 2. Composting windrows of beef cattle feedlot and dairy manure.

on air (odor, ammonia, greenhouse gases) and water quality need further study. Acceptable upper limits for soil phosphorus resulting from repeated manure application must be defined. Effects of diet manipulation on nutrients excreted in manure need to be studied. Complementary to the research, a technology transfer program is required to get the information into the hands of the users.

For background information *see* AGRICULTURAL BUILDINGS; EUTROPHICATION; FERTILIZER in the McGraw-Hill Encyclopedia of Science & Technology.

Bahman Eghball

Bibliography. B. Eghball and J. F. Power, Beef cattle feedlot manure management, *J. Soil Water Conserv.*, 49:113–122, 1994; National Research Council, *Nutrient Requirement of Swine*, U.S. Government Printing Office, 1988; USDA-EPA, *Unified National Strategy for Animal Feeding Operation*, U.S. Government Printing Office, 1998.

Antibiotic resistance

The emergence and spread of antibiotic-resistant bacteria has generated increasing interest in recent years. In some cases, resistance involves the emergence of bacteria which were previously thought to be of low pathogenicity but whose intrinsic resistance to specific antimicrobials provides a selective advantage for survival in an antibiotic-rich environment. Such species are, in general, of only modest concern, and exist in very specific settings. Of greater significance are those pathogens that are intrinsically susceptible to a given antibiotic but have acquired resistance in association with exposure to it. These species present particular clinical problems because they appear to retain most of the virulence factors that made them important pathogens in the first place, but may no longer respond to therapy with antimicrobial agents. Examples of species that have acquired resistance determinants abound, and include some prominent human pathogens such as *Escherichia coli*, *Pseudomonas aeruginosa*, *Salmonella* spp., *Bacteroides fragilis*, *Staphylococcus aureus*, *Enterococcus faecium*, and *Streptococcus pneumoniae*.

Emergence. The emergence of antimicrobial resistance in individual species results from mutations in genes encoding intrinsic cellular processes (these may be structural or regulatory), the acquisition of foreign deoxyribonucleic acid (DNA), or some combination of these two processes. The mutations are not caused by exposure to antibiotics, and appear to occur at a specific rate under normal circumstances. Antibiotic exposure provides the environment in which antibiotic-resistant mutants will survive in numbers that exceed their sensitive brethren. Common examples of the selection of such mutants include the emergence of fluoroquinolone-resistant *E. coli* [caused by mutations of bacterial topoisomerase genes] or rifampin-resistant strains of *S. aureus* [caused by point mutations of genes encoding ribonucleic acid (RNA) polymerase]. Resistance may

also emerge as a result of normally silent genes being turned on by mutations in promoter sequences. It has been shown that 1–3% of *B. fragilis* strains possess a chromosomal β-lactamase gene that effectively hydrolyzes the broad-spectrum β-lactam antibiotic imipenem. Expression of this enzyme results from the transposition of an insertion sequence (IS) element upstream of the gene, creating a new and much more active promoter sequence that increases expression of the β-lactamase.

Perhaps the most common and efficient mechanism of resistance emergence is the acquisition of foreign DNA. By importing resistance genes, bacteria can in many instances express immediate, high-level resistance with no obvious cost in terms of the other normal cellular functions. Circumstances which favor the sharing of antibiotic-resistance genes include the presence of multiple organisms in settings where antimicrobial selection pressure is strong (the human and animal gut are prime examples), and the existence of a prodigious and widespread exclusive group of mobile elements that facilitate the transfer of DNA within and between species. These mobile elements include plasmids, transposons, and bacteriophages, many of which can function independently or together to foster the dissemination of antimicrobial resistance determinants.

Most antibiotics represent natural products or derivatives of natural products that are manufactured by the microorganisms themselves. The manufacture of an antibiotic without the associated resistance determinant would be a suicidal undertaking. Since pure cultures of single microorganisms rarely if ever exist in nature, it is likely that resistance determinants for most, if not all, of the naturally occuring antibiotics already exist. The likelihood that these resistance determinants will emerge in human pathogens is, therefore, a function of the efficiency with which the bacteria can scavenge the resistance determinants from the species that normally encode them.

Multiresistance. Enterococci (spherical bacteria in short chains) are important causes of infection in the hospital (bloodstream, urinary-tract, and wound infections) and the community (endocarditis). The clinically important enterococci are *E. faecalis* (responsible for 85–90% of clinical enterococcal infections) and *E. faecium* (responsible for most of the remainder). Enterococci are intrinsically resistant to a wide variety of commonly used antimicrobial agents. As a result, ampicillin and vancomycin have served as the cornerstones of effective therapy of enterococcal infections. The emergence of vancomycin resistance (conferred primarily by one of two determinants, *vanA* or *vanB*) in the United States has occurred almost exclusively in *E. faecium* strains that express high levels of resistance to ampicillin, making infections caused by these strains difficult, if not impossible, to treat effectively with antibiotics. Emergence of these multiresistant strains represents an impressive convergence of environmental pressures. These multiresistant strains are excellent examples of the resourcefulness of the bacteria and the consequences of indiscriminate antibiotic use.

Bacterial cell-wall synthesis is accomplished primarily through the activity of transglycosylases and transpeptidases known as penicillin-binding proteins (PBPs). *Enterococcus faecium* expresses ampicillin resistance by synthesizing the low-affinity (for β-lactam antibiotics) PBP5, thought to be intrinsic to this species. Most *E. faecium* strains express low levels of ampicillin resistance because PBP5 is present in relatively small amounts. Increased levels of ampicillin resistance can be achieved by overproduction of PBP5, or by mutations within the PBP5 gene that further decrease ampicillin-binding affinity. Recent years have seen an increase in the incidence of high-level resistance in *E. faecium* in United States hospitals (even in vancomycin-susceptible strains). Clinical studies have tied this emergence to the use of extended-spectrum cephalosporins, to which the ampicillin-resistant enterococci are particularly resistant.

Despite the fact that vancomycin has been in clinical use since 1958, no vancomycin-resistant enterococci were reported before 1986. The sudden appearance and rapid dissemination of these strains is perhaps best explained by the use of oral vancomycin (which is not absorbed and achieves high concentrations in the gastrointestinal tract) to treat antibiotic-induced pseudomembranous enterocolitis. Pseudomembranous colitis is due to overgrowth of toxin-producing *Clostridium difficile* in the colon following treatment with antibiotics such as clindamycin or ampicillin. Orally administered and nonabsorbed vancomycin (topical therapy, essentially) appeared to be the ideal treatment against *C. difficile*.

The early 1980s saw the introduction of a variety of extended-spectrum cephalosporins, as well as the use of local vancomycin therapy in the gastrointestinal tract. The stage was perfectly set for multiresistance to emerge in enterococci. The gastrointestinal tracts of hospitalized patients were already colonized with ampicillin-resistant *E. faecium*. Now, most of the other gram-positive flora were being wiped out by the administration of oral vancomycin (enterococci are not killed, only inhibited, by vancomycin). The microbiologic void that was created in the gastrointestinal tract of hospitalized patients was soon filled by species that could thrive in the presence of vancomycin. Among these species were various streptomycetes. It now appears that these *Streptomyces* species were the source of the vancomycin-resistance genes prevalent in enterococci (perhaps not surprising, since the streptomycetes are the microbiologic manufacturers of vancomycin and other glycopeptides). Genetic exchange events eventually occurred that allowed the ampicillin-resistant *E. faecium* to acquire these resistance genes, and since that time these genes have spread widely among *E. faecium*.

The vancomycin-resistance genes do not appear to be native to the enterococci because the guanine and cytosine ratios of these resistance genes differ significantly from typical enterococcal genes. Acquisition of the genes involved mobile DNA elements. Two transposons encode the *vanA* (Tn*1546*) and *vanB* (Tn*5382*) determinants. In one instance, Tn*5382* is integrated immediately downstream of the *pbp5* gene within the enterococcal chromosome, and the two resistance determinants transfer together during mating events. This linkage relationship suggests that selection for vancomycin-resistant enterococci by antibiotics other than vancomycin could occur. In fact, extended-spectrum cephalosporins are among the most common antibiotics associated with the emergence of these troublesome multiresistant pathogens, an association explainable by the cephalosporin resistance conferred by the low-affinity PBP5.

For background information *see* ANTIBIOTICS; BACTERIAL GENETICS; DRUG RESISTANCE in the McGraw-Hill Encyclopedia of Science & Technology.

Louis B. Rice

Bibliography. L. L. Carias et al., Genetic linkage and co-transfer of a novel, *vanB*-encoding transposon (Tn*5382*) and a low-affinity penicillin-binding protein 5 gene in a clinical vancomycin-resistant *Enterococcus faecium* isolate, *J. Bacteriol.*, 180: 4426–4434, 1998; V. A. Chirurgi et al., Nosocomial acquisition of β-lactamase-negative, ampicillin-resistant enterococcus, *Arch. Int. Med.*, 152:1457–1461, 1992; J. W. Chow et al., *Enterobacter* bacteremia: Clinical features and emergence of antibiotic resistance during therapy, *Ann. Int. Med.*, 115:585–590, 1991; C. Jacobs, J.-M. Frere, and S. Normark, Cytosolic intermediates for cell wall biosynthesis and degradation control inducible β-lactam resistance in gram-negative bacteria, *Cell*, 88:823–832, 1997; L. B. Rice, The theoretical origin of vancomycin-resistant enterococci, *Clin. Microbiol. Newslett.*, 17:189–192, 1995.

Antidepressant

St. John's wort (*Hypericum perforatum* L., Hypericaceae; see **illus.**) has been used for millennia for its many medicinal properties, including wound healing and treatment of kidney and lung ailments, insomnia, and depression. St. John's wort has become popular in Germany, where it is approved for use in the treatment of mild to moderate

St. John's wort plants.

depression, and outsells all other antidepressants combined. Interest in its use as an antidepressant in the United States has also increased in recent years as evidenced in publications. This rise in popularity began with a 1996 meta-analysis which contained an evaluation of 23 randomized clinical trials of St. John's wort, mostly from Germany, in 1757 outpatients with mild to moderately severe depression. Improvement was seen in depressive symptoms in all the trials. St. John's wort was significantly better than placebo and not significantly different from standard antidepressants, while its side effects were much less.

Regulation of botanical medicines. Unlike the United States, most other industrialized countries have regulatory procedures for licensing and using botanical medicines. Such medicines have official status in many countries such as Germany, where they are registered for specific medical uses and are reimbursed by the health care system. Herbal products such as St. John's wort are subjected to experimental clinical trials like those performed for their synthetic counterparts. The diagnostic criteria used in most other countries were developed by the World Health Organization and are contained in the International Classification of Disease (Version 9 or 10), while the Diagnostic and Statistical Manual of Mental Disorders IV is used in the United States. In addition, the German studies are performed by a relatively large number of primary care physicians, each providing a relatively small number of patients from within his or her practice. Recruiting patients through advertising, as in the United States, is not popular in Europe since these patients may not be representative of those treated in general practice. Finally, the comparative drugs used are those commonly used in Germany rather than the ones American physicians use.

St. John's wort. The mechanism of action of St. John's wort is still unclear. Early studies suggesting monoamine oxidase inhibition are no longer considered pertinent since this phenomenon appears to be limited only to in vitro assays. Other proposed mechanisms involve other effects on monoamines. Inhibition of the uptake of serotonin, norepinephrine, and dopamine, with IC_{50} (the concentration necessary to inhibit 50% of the receptors) of 2.4, 4.5, and 0.9 micrograms/ml, respectively, has been reported. Attaining blood concentrations high enough to see these effects, however, seems unlikely. The most potent effect thus far reported is for the GABA (gamma-aminobutyric acid) A and B receptors, with binding affinities of approximately 75 and 6 nanograms/ml, respectively.

While *Hypericum* extracts, or simply hypericum, are currently standardized by their content of hypericin (usually 0.3%), clinical data suggest that another major component of the extract, called hyperforin, may be more important for therapeutic activity. A significant advantage in efficacy was found with a hyperforin-rich (5%) formulation versus a hyperforin-poor (0.5%) formulation. Animal data also suggest

that many of the pharmacologic properties associated with the whole extract can be replicated by hyperforin alone. The hyperforin component has been shown in pharmacokinetic studies to be well absorbed. After oral administration to rats of 300 mg/kg of the hyperforin-rich formulation, maximum plasma levels of approximately 370 ng/ml ($\sim$690 nM) were reached after 3 h, as quantified by a high-pressure liquid chromatography and ultraviolet (HPLC/UV) detection method. The estimated half-life was 6 h. In human volunteers, plasma levels of hyperforin were followed for up to 24 h after administration of 300 mg of the extract. Maximum plasma levels of approximately 150 ng/ml were reached in 3.5 h. The half-life was 9 h. Hyperforin pharmacokinetics were linear up to administration levels of 600 mg of the extract. Increasing the doses to 900 or 1200 mg of extract produced a plateau effect. The estimated steady-state plasma concentrations of hyperforin after three doses of 300 mg of the extract per day (that is, after a normal therapeutic dose regimen) was approximately 100 ng/ml.

Adverse effects of St. John's wort. The side effects reported for St. John's wort are generally mild. Gastrointestinal symptoms, allergic reaction, and fatigue have been reported in less than 1% of patients. Photosensitization may occur, especially in fair-skinned people. Photosensitization was demonstrated in a placebo-controlled clinical trial involving hypericin and exposure to metered doses of ultraviolet irradiation. Each of 13 volunteers received placebo or 900, 1800, or 3600 mg of a standardized extract (LI 160) containing 0, 2.81, 5.62, and 11.25 mg of total hypericin (hypericin plus pseudohypericin). Maximum total hypericin plasma concentrations were observed about 4 h after dosage and were 0, 28, 61, and 159 ng/ml, respectively. Before and 4 h after drug administration, the subjects were exposed to increasing doses of simulated solar irradiation (SSI, with combined UV-A and UV-B light) on small areas of the back, and another group was exposed to selective UV-A light irradiation. Minimal erythema (reddening of the skin) dose was determined 5, 20, and 68 h after irradiation. Comparison of SSI sensitivity before and after the hypericum dose showed no difference. Sensitivity to selective UV-A light was increased only after the highest dose of hypericum. After multiple doses of hypericum (50 volunteers received 600 mg of extract three times per day), UV light sensitivity before administration of the drug as compared with day 15 after beginning treatment showed a slightly increased SSI sensitivity. This increase in cutaneous light sensitivity could be compensated for by reducing irradiation time by 21%. Doses used in this study were higher than typical doses of currently used commercial preparations. In spite of these high doses, frequency of side effects was no different from placebo.

Formal animal toxicology studies have been performed with LI 160. The single dose producing no effect was reported to be greater than 5000 mg/kg. In a 26-week study in rats and dogs that received oral doses of 900 and 2700 mg/kg, only nonspecific symptoms of toxicity were seen. These included reduced body weight, some slight pathological changes in liver and kidneys from the large metabolic load, and some histopathologic changes in the adrenals. There were no effects on fertility or reproduction. A toxicity study using dietary administration of St. John's wort to rats (5% of the daily diet for 119 days) found no effect on various hepatic drug-metabolizing enzymes. No mutagenic potential was evident in standard assays. Two-year carcinogenicity studies have recently been completed but have not yet been reported. The toxicity of the isolated hypericin appears to be low in mice as well.

St. John's wort is a remarkably safe antidepressant with an apparently unique mode of action. While it has demonstrated efficacy in mild and moderate depression when compared with placebo or tricyclic antidepressants, further research is needed. Its effects should be compared with the current therapeutic standards, the serotonin reuptake inhibitors. Studies in severely depressed patients are lacking, as are investigations on the utility of St. John's wort as a therapeutic adjunct to standard antidepressants.

For background information *see* MONOAMINE OXIDASE; NORADRENERGIC SYSTEM; PSYCHOPHARMACOLOGY; SEROTONIN in the McGraw-Hill Encyclopedia of Science & Technology.   Jerry M. Cott

Bibliography. A. Biber et al., Oral bioavailability of hyperforin from hypericum extracts in rats and human volunteers, *Pharmacopsychiatry*, 31(suppl. 1): 36–43, 1998; H. Bloomfield, M. Nordfors, and P. McWilliams, *Hypericum (St. John's Wort) & Depression*, Prelude Press, 1997; J. Brockmöller et al., Hypericin and pseudohypericin: Pharmacokinetics and effects on photosensitivity in humans, *Pharmacopsychiatry*, 30(suppl. 2):94–101, 1997; H. Cass, *St. John's Wort: Nature's Blues Buster*, Avery Publishing, 1998; S. S. Chatterjee et al., Hyperforin as a possible antidepressant component of hypericum extracts, *Life Sci.*, 63(6):499–510, 1998; J. M. Cott, In vitro receptor binding and enzyme inhibition by *Hypericum perforatum* extract, *Pharmacopsychiatry*, 30(suppl. II):108–112, 1997; G. Laakman et al., St. John's wort in mild to moderate depression: The relevance of hyperforin for the clinical efficacy, *Pharmacopsychiatry*, 31(suppl):54–59, 1998; K. Linde et al., St. John's wort for depression: An overview and meta-analysis of randomized clinical trials, *Brit. Med. J.*, 313:253–258, 1996; W. E. Müller et al., Effects of *Hypericum* extract (LI 160) in biochemical models of antidepressant activity, *Pharmacopsychiatry*, 30(suppl. II):102–107, 1997.

Antisense drugs

Antisense drugs, involving short strands of genetic material modified to enhance drug properties, represent a new class of materials referred to as oligonucleotides. These materials target (bind to) ribonucleic acid (RNA), which occupies a central position in the process of synthesizing proteins (gene ex-

pression). The binding of the two strands of nucleic acid to form a drug or inhibitor-receptor complex is achieved by the application of Watson-Crick base-pair hydrogen bonding rules (**Fig. 1**).

Antisense therapeutics is relatively new. The drugs are based on the genetics of viral, bacterial, and human genomes. A strict definition of antisense drug discovery describes the inhibition of gene expression, by targeting a predetermined sequence, in RNA. Drugs employed in antisense therapeutics are the first to be derived from the emerging genetic revolution.

Antisense molecules prevent the synthesis of abnormal protein by specifically binding to the message, or RNA, that is responsible for its formation. Antisense drugs are gene-based or informational drugs in that they contain information in their sequence about which nucleic acid target they can bind. Because of the many binding sites available between two nucleic acids, antisense drugs have an extraordinary level of specificity of action. The first antisense oligonucleotide drug, Vitravene™, is used for treatment of cytomegalovirus retinitis in acquired immunodeficiency syndrome (AIDS) patients. A number of oligonucleotides with indications for antiviral, anticancer, and anti-inflammatory chemotherapy are in late phase II and III clinical studies.

Bases of drug action. In general terms, to derive a useful pharmacological effect the drug must bind with a receptor on the pathogen or a cellular component, forming a drug-receptor complex, which elicits the desired action. Assuming that the drug and the receptor are able to specifically interact (complementary binding), formation of a useful complex depends on the pharmacokinetics of the drug, that is, its absorption, distribution, metabolism, and excretion within the body. What the body does to the drug impacts on whether an appropriate concentration of the drug is obtained at the site of the receptor and at an appropriate time. Once at the site of action, the drug must specifically bind to the target receptor, forming the drug-receptor complex. This process is referred to as the pharmacodynamics of the drug, or what the drug does to the body. It is imperative that the drug binds to only the desired receptor. Binding to bystander receptors may elicit undesirable pharmacological actions, or toxicity. The drug must also bind with sufficient affinity to the receptor such that a sufficiently long-lived complex is available to elicit the desired pharmacological action. Effective binding of two molecules is primarily described by nonconvalent hydrogen bond interactions, which have very specific distance and directional requirements. This aspect is particularly important with nucleic acid interactions. Thus, a pharmacologically useful drug requires appropriate pharmacokinetics, pharmacodynamics, and an acceptable level of specificity, relating to toxicity. Economic factors include how long it takes to discover and develop the drug and issues of large-scale production. The first generation of antisense drugs, phosphorothioates, is currently undergoing a number of clinical trials and has

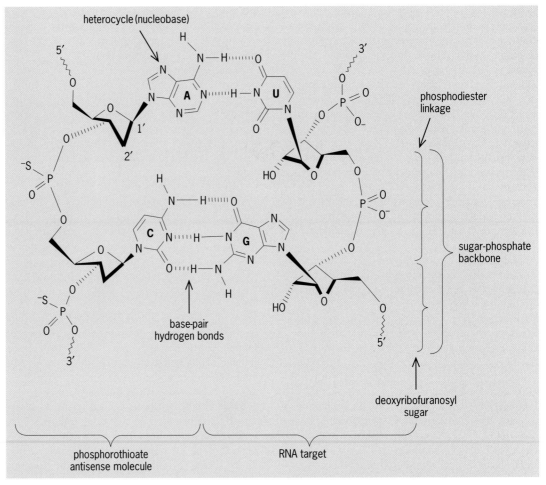

Fig. 1. Phosphorothioate antisense molecule targeting an RNA receptor. According to Watson-Crick base-pair hydrogen bonding rules, nucleobase A hydrogen-bonds to nucleobase U, and nucleobase C hydrogen-bonds to nucleobase G.

demonstrated acceptable pharmacokinetics and toxicity levels.

Significance of antisense therapeutics. The current processes to research, discover, and develop new or improved drugs are long and expensive endeavors, which primarily focus on the interference of proteins functioning in an abnormal manner. Abnormal protein occurrence is typically caused by a specific disease state. More important than the cost of drug research and development is the problem of toxicity. Antisense drugs are designed from the knowledge of the sequence of the RNA producing the abnormal protein. The Watson-Crick base-pair hydrogen bonding rules allow the rapid selection of the sequence structure of a potential inhibitor of gene expression at the message level (Fig. 1). Using these rules, any molecular message that expresses an abnormal protein can be targeted.

Antisense materials are oligomeric, providing many specific binding sites to their RNA targets. Typically, a 20-mer oligonucleotide (containing 20 consecutive nucleobases or heterocycles) has approximately 50 specific hydrogen bonds to its RNA target, compared to only several binding sites that a small molecule (nonoligomeric) may possess for its target protein. Antisense materials, being short strands of nucleic acids, can be targeted to any RNA by simply changing their base sequence. Thus, they have essentially the same chemical structure as their target (receptor), but a different ordering of the base recognition components, and are greatly reduced in size compared to the RNA receptor.

Antisense materials currently provide the only example of truly gene-based, informational drugs. An important characteristic of such materials is that their pharmacokinetic properties and toxicity profiles are very similar (generic). For example, a 20-mer oligonucleotide may be designed to have a sequence that allows the specific targeting of a message essential for human immunodeficiency virus (HIV) replication, but by simply changing only the sequence of the oligonucleotide, an entirely new message can be specifically targeted, such as a human oncogene or an intracellular adhesion molecule. The pharmacokinetics and the toxicity profile of these pharmacologically different drugs are similar. The chemical synthesis of these drugs is also similar. Thus, in comparison to traditional drug discovery processes, antisense drugs are rationally and rapidly designed for any disease having abnormal protein, and they have unprecedented specificity of action, resulting in relatively low toxicity levels. Furthermore, because of the similar structures of pharmacologically different oligonucleotides, discovery and manufacturing costs of different drugs are greatly reduced compared to traditional drug discovery.

Basic structural biology concepts. A single strand of nucleic acid has a polymeric sugar-phosphate backbone—ribofuranosyl (in RNA) or deoxyribofuranosyl (in deoxyribonucleic acid; DNA) residues—connected at their 3′ and 5′ hydroxyls by phosphoric acid to form phosphodiester linkages (Fig. 1). Each sugar residue has one of five types of heterocycles (nucleobases) attached at the C-1′ position of either the deoxyribofuanosyl or ribofuranosyl sugar. These heterocycles are the purines guanine (G) and adenine (A) and the pyrimidines cytosine (C) and thymine (T), with uracil (U) replacing thymine in RNA.

Complementary DNA is a chemical complex of two strands of deoxyribonucleic acids that are bound together by Watson-Crick base-pair hydrogen bonding (binding motif). This "essences of life" bonding of nucleobases specifies that the guanine in one DNA strand specifically binds to the cytosine in the other strand, and the adenine in one DNA strand specifically binds to the thymine in the other strand. A sequence (or a specific ordering of the nucleobases) of an RNA or a DNA strand will bind specifically (hydrogen-bond) to another sequence only if the nucleobases match up (base-pair according to Watson-Crick rules). On applying the binding rules, one DNA strand is said to be complementary to another DNA strand via Watson-Crick base-pairing. The complementary DNA strands twist into a helical motif with one strand having its bases ordered in the 3′ to 5′ direction and the other strand having its bases ordered in the 5′ to 3′ direction (antiparallel orientation; **Fig. 2**). The double-stranded DNA serves as the templates either for complementary DNA molecules, to provide a process of self-replication, or for complementary RNA molecules (from the sense strand), to provide a process of transcription. The complementary DNA strand (the antisense strand) rarely transcribes an RNA molecule. In turn, the transcribed RNA molecules serve as the templates that allow the ordering of the amino acids within the polypeptide chain of proteins during the process of translation. In translation, the nucleotide sequence information (language) of nucleic acids is translated into the amino acid sequence (language) of protein.

The complete process has been referred to as the central dogma: DNA makes RNA makes protein. The common term for this process is gene expression. Thus, the stages of gene expression are transcription (the conversion of DNA to RNA) and translation (the conversion of RNA to proteins). DNA also self-replicates to maintain an organism's genome. Furthermore, regulation of gene expression in living organisms involves extensive recognition and binding to specific nucleic acid sequences by nucleic acid binding proteins.

Disease states (genetic, oncogenic, or infective) are typically a result of abnormal protein, the end-stage product of gene expression. Historically, drug discovery focused on the interference with the functions of abnormal protein rather than the prevention of the formation of abnormal protein. This is the basis of the familiar phrase "treatment of the symptoms of the disease rather the cause." The use of gene-based, informational chemicals, such as antisense oligonucleotides, attempts to prevent the formation (gene expression) of abnormal protein by targeting DNA, RNA, or regulatory proteins that are required for transcription and translation of the abnormal protein. A set of binding rules that allows rapid and precise selection of a molecule to synthesize which would specifically inhibit gene expression at the DNA or RNA level and offer exquisite specificity for its receptor, is unique in the history of drug research. The employment of gene-based, informational drugs to complement genomic target validation in modern drug discovery is revolutionary.

In the process of transcription, complementary RNA (message RNA) is derived from the sense DNA strand by Watson-Crick base-pair recognition and binding. The antisense approach targets the initial sense complementary RNA strand (primary transcript) as well as many downstream sites that are available as RNA is metabolized in the process of protein production (Fig. 2).

Enabling technologies. The thought that oligonucleotides could be made into drugs stems from several enabling technologies. In the late 1950s and during the 1960s, the Watson-Crick rules provided simple, precise knowledge of how nucleic acids bind specifically to each other. During the 1970s and 1980s, traditional biology evolved into understanding biology at the molecular level, molecular biology, as well as structural biology. This provided the knowledge of how genetic information is stored, transcribed, and translated into essential proteins—molecular biology of gene expression. Next, the discovery and development of methods to rapidly determine the ordering of the nucleobases guanine, adenine, cytosine, and thymine—that is, the sequence of genetic materials—via automated machines was developed. The knowledge of the precise base sequence of the genomes of pathogens and humans provided the required knowledge to now target messages of abnormal protein. Commercial, automated oligonucleotide synthesizers, reagents, and purification protocols were developed during the late 1980s, setting the stage for antisense oligonucleotide drug discovery. At this point in the genetics revolution, the genetics of life processes were well known. This gave researchers the knowledge to rapidly determine the sequences of genes expressing abnormal protein, target these messages (Watson-Crick rules), and readily synthesize many potential oligonucleotide antisense inhibitors. During the 1990s, oligonucleotide medicinal chemistry and pharmacology, which form the basis of structure-activity-relationship studies, have been highly developed in the process of converting oligonucleotides into drugs. These enabling technologies, along with the entrepreneurial nature of

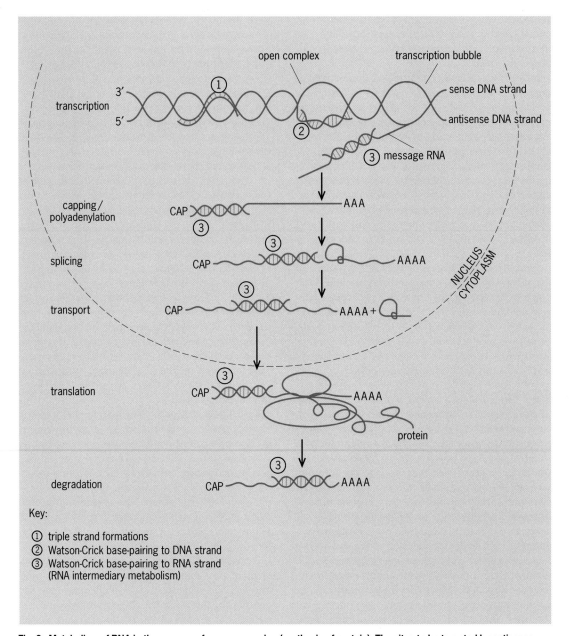

Fig. 2. Metabolism of RNA in the process of gene expression (synthesis of protein). The sites to be targeted by antisense molecules are shown.

several biotechnology companies, have spurred great interest in making drugs out of oligonucleotides.

Chemical issues and perspective. A phosphorothioate oligonucleotide, P=S (phosphor and sulfur double bond), is now an available drug (Fig. 1). However, to continually impove this novel and exciting drug class and to overcome certain limitations, structural changes are required. In the 1990s a diverse range of modifications, at all possible modification sites of an oligonucleotide, were reported. This application of traditional medicinal chemistry (structure-activity-relationship studies) to drug discovery in antisense oligonucleotides and in oligonucleotides in general has answered many important biophysical and biochemical questions. A large body of data sup-

ports the notion that enhancing biochemical and biophysical properties correlates with enhanced antisense biological activity in animals. For example, changing the structure of phosphorothioate oligonucleotides provides an opportunity to alter their pharmacokinetic profile. In structural changes, which remove sulfur (as thiophosphate) or change lipophilicity (for example, by 2′-O modifications), more favorable toxicity profiles have resulted. In addition, antisense oligonucleotides that are orally available or penetrate the blood-brain barrier represent the most current, important deficiency of antisense oligonucleotides. Recent reports of antisense oligonucleotides doubly modified at the 3′ and 5′ ends with 2′-O-methyl or 2′-O-methoxyethyl and P=S phosphorothioates, to provide a high level of nuclease resistance, have provided encouraging results that these pharmacokinetic deficiencies will soon be solved by appropriate chemical modifications. The types of modified oligonucleotides currently being pursued (going beyond P=S oligonucleotides) possess a combination of modifications. This trend will certainly continue as small molecules are attached (conjugated) to active oligonucleotides to allow modulation of pharmacokinetics proper to provide completely optimized oligonucleotide drugs.

For background information *see* BIOTECHNOLOGY; DEOXYRIBONUCLEIC ACID (DNA); DEVELOPMENTAL GENETICS; GENETIC ENGINEERING; GENETIC MAPPING; MOLECULAR BIOLOGY; NUCLEIC ACID; OLIGONUCLEOTIDE; PHARMACOLOGY; PROTEIN; RIBONUCLEIC ACID (RNA) in the McGraw-Hill Encyclopedia of Science & Technology.   P. Dan Cook

Bibliography. P. D. Cook, Antisense medicinal chemistry, in S. Crooke (ed.), *Handbook of Experimental Pharmacology*, 131:51–101, 1998; H. M. Weintraub, Antisense RNA and DNA, *Sci. Amer.*, 262: 40–46, 1990.

Astrometry

The *Hipparcos* space astronomy mission was dedicated to the very accurate measurement of star positions and star motions in space. The satellite was designed, built, and operated under the responsibility of the European Space Agency. It was launched in August 1989 and operated until mid-1993. The two resulting star catalogs were completed and published in 1997: the *Hipparcos Catalogue* is a high-accuracy compilation of nearly 120,000 star positions, distances, and motions; the *Tycho Catalogue* is a lower-accuracy catalog of slightly more than 1 million stars. Together they represent astronomers' best understanding of the spatial distribution, space motion, and corresponding physical properties of stars in the solar neighborhood. The experiment was carried out by a collaboration of about 200 scientists.

Star position measurements. Catalogs of star positions have been compiled for more than 2000 years. Hipparchus, for whom the mission is named, prepared a naked-eye catalog of about 1000 stars around the first century B.C. Over the past few hundred years, improved instruments and observational techniques have led to star catalogs of continuously improving accuracy and size. This astronomical discipline, referred to as astrometry, has led to a greatly improved picture and comprehension of the Earth's place in the Milky Way Galaxy. Directly and indirectly, the accurate measurement of star positions has provided profound insights into the content and structure of the universe as a whole.

Efforts in the seventeenth century were directed at compiling catalogs to assist sea navigation and the determination of longitude, but measurements were also aimed at furthering scientific understanding of the Earth's motion in space. During the eighteenth and nineteenth centuries, positional improvements led to the detection of small changes in the relative positions of stars, attributed to the motions of the stars themselves through space. The realization that stars lie at finite, albeit very large, distances from the solar system led to intensified efforts at measuring the effect of parallax, which is the apparent displacement of nearby stars compared to more distant stars as the Earth orbits around the Sun (**Fig. 1**). Because of the enormous distances of the stars, the angular displacements are very small, and even the nearest stars have parallaxes of only about 1 second of arc (1/3600 degree). For more distant stars the effect is much smaller. Consequently, to measure stellar distances and motions within the Milky Way Galaxy, angular accuracies very much better than 1 second of arc are required.

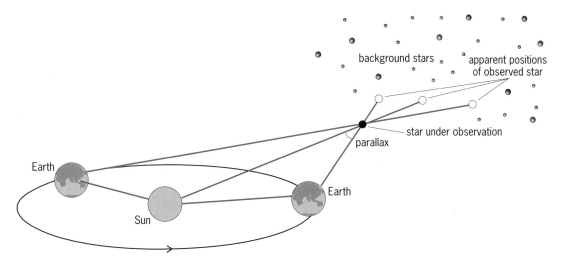

Fig. 1. Concept of parallax measurements. The Earth's orbital motion around the Sun allows the rigorous determination of stellar distances. The direction of the line of sight from the Earth to a particular star is slightly different at different times of year because of this orbital motion. The nearer a star is to the Sun, the greater is the change in its apparent position relative to more distant background stars.

Fig. 2. *Hipparcos* satellite being prepared for testing at the European Space Agency's Research and Technology Center, Noordwijk, The Netherlands, in 1988. (*European Space Agency*)

Space-based measurements. Accurate star position measurement from ground-based telescopes is complicated by the Earth's atmosphere, and by gravitational and thermal effects, which are difficult to correct for. The idea for a space experiment to overcome these difficulties dates back to the 1960s. Recognition of the fundamental scientific importance of

obtaining greatly improved distances and space motions of the stars led to the development of *Hipparcos*, the first satellite devoted to star position measurements. The scientific success of that mission has led to plans for future space missions aiming at significantly higher angular accuracies, of a few millionths of a second of arc.

Operational principle of Hipparcos. The satellite comprised a small but highly accurate optical telescope, with a primary mirror diameter of about 30 cm (12 in.), and with two distinct viewing directions (**Fig. 2**). The entire sky was scanned in a continuous and regular manner, and the observations were used to build up a network of one-dimensional angular measurements between the stars. The on-ground processing of the observations, taken at different orientations and at many different epochs throughout the operational lifetime of the mission, led to the reconstruction of the positions, distances, and space motions of each star. Following the scientific analysis phase, the results of the project were published in June 1997. The *Hipparcos Catalogue* contains the positions distances, movements, and many other properties of nearly 120,000 stars, brighter than a magnitude of about 12, with typical positional errors of about 0.001 second of arc (**Fig. 3**). The *Tycho Catalogue* contains more than 1 million stars with a median astrometric accuracy of about 25 milliarcseconds.

Applications of Hipparcos data. The resulting stellar distances and motions impact on all areas of astronomy and astrophysics, but particularly on the understanding of stellar structure and evolution, and on the structure and dynamics of the Milky Way Galaxy. More than 20,000 stars out to distances of 200–300 light-years now have their distances known to better than 10%, and another 30,000 are known to better than 20%.

Stellar properties. Stars are very complex constituents of the Milky Way Galaxy and therefore of the universe as a whole. They differ from each other in many important respects, including mass, chemical composition, size, rotation rate, and age. Understanding their composition and structure, how and where they were born, how they are evolving, and how they will eventually die means determining their individual properties, and comparing the stars with detailed models of their initial conditions and subsequent evolution, involving a very comprehensive understanding of many areas of physics. Knowing the luminosity of a star, for example, requires knowledge of the star's distance, in order to convert its apparent magnitude into an absolute luminosity. Comparing theoretical models with measurements of the brightnesses, temperatures, masses, and ages of the stars provides direct information on whether the star formation process and subsequent evolution are correctly understood. Ultimately, this comparison leads to an understanding of the internal structure and other hidden properties of the stars, since only when the correct physical ingredients and processes are included will the models agree with the ever-improving

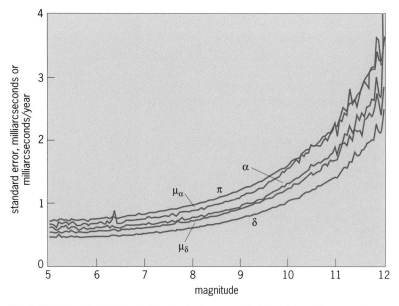

Fig. 3. *Hipparcos Catalogue* median standard errors of the five astrometric parameters as a function of magnitude. The unit of the standard error is the milliarcsecond for the positional components (α and δ) and parallax (π), and milliarcsecond/year for the proper-motion components (μ_α and μ_δ).

observations. The precise stellar luminosities that accurate distance measurements yield also provide a direct check on models of internal stellar structure. Such models are fundamental to the emerging discipline of asteroseismology, in which stars other than the Sun are scrutinized for observational signatures of solarlike oscillations.

Galactic evolution. The 10^{11} stars constituting the Milky Way Galaxy are linked in their collective positional and kinematical distribution through their gravitational environment, and via the history of star formation throughout the Galaxy. The initial properties of this distribution were subsequently modified, often substantially, by small- and large-scale dynamical processes. Thus present-day star positions and motions encode the details of the initial formation and subsequent evolution of the Galaxy.

Solar neighborhood and dark matter. The *Hipparcos* mission results have led to important progress in understanding the spatial distribution of nearby stars. This knowledge is used to define the distribution of stellar luminosities, the distribution of stellar masses, and the kinematical behavior of stars in the solar neighborhood. Additionally, the results of observing the distances and motions of stars perpendicular to the plane of the Milky Way Galaxy have also been used to probe the distribution of dark matter in the universe, which is inferred to exist from cosmological and other arguments. This method is possible because stars act as kinematic tracers, probing the gravitational signature of material irrespective of its physical manifestation. The results support a growing consensus concerning the main structural parameters of the galactic disk populations of stars, and the possible distribution of dark matter in the Milky Way Galaxy. They demonstrate, for example, that any dark matter in the Galaxy must have a distribution following that of the Galaxy's halo, rather than being in the form of a disk.

Hyades observations. The Hyades is the nearest moderately rich cluster of stars; its distance has been the subject of intense interest and investigation over a very long period, due to its central role in defining the entire distance scale used in astronomy. More generally, clusters are important because they represent a gravitationally bound collection of typically a few hundred stars, born at the same time, and with a given chemical composition. The development of such a coeval system, both dynamically and with respect to the individual stellar properties such as temperature and luminosity, therefore provides an ideal laboratory for testing models of stellar evolution. Open star clusters are also subject to evaporation and eventual disruption over time scales of tens or hundreds of millions of years due to gravitational encounters between stars within the cluster, or between the cluster and the galactic disk or giant molecular clouds. *Hipparcos* has provided a reliable direct distance measurement to about 200 stars in the Hyades cluster, allowing its internal structure, internal motion, mass segregation, and disintegration processes to be observed.

Distance scales. The problem of the determination of distances to objects throughout the Milky Way Galaxy and beyond is one of the central themes in present-day astronomy. On the largest scales, distances to other galaxies affect knowledge of the Hubble constant. While most astronomers do not question the fact that the universe is expanding, the actual expansion rate is still vigorously debated. The precise value, as given by the Hubble constant, is of more than quantitative interest, since theories seeking to explain the formation and subsequent evolution of the early universe now predict properties of the present-day universe which are becoming ever more stringently constrained by observations. A class of stars referred to as Cepheid variables has played a very crucial role in these discussions, due to the regular pulsations of these stars with a period considered to depend primarily only on their luminosity. The *Hipparcos* results on these and other objects are leading to more secure distance scales within and beyond the Milky Way Galaxy.

Globular cluster observations. *Hipparcos* results are facilitating efforts to determine stellar ages and, in particular, to date the oldest stellar populations in the Milky Way Galaxy, the globular clusters. These ages therefore provide a lower limit on the age of the universe. Pre-*Hipparcos* estimates had put some globular cluster ages as high as 18×10^9 years, older than current cosmological age estimates of the universe itself. Revised ages, derived from the new distance measurements, of perhaps $12–14 \times 10^9$ years, may have alleviated this paradox. *See* GLOBULAR CLUSTERS.

For background information *see* ASTROMETRY; CEPHEIDS; COSMOLOGY; HYADES; MILKY WAY GALAXY; STAR; STAR CLUSTERS; STELLAR EVOLUTION in the McGraw-Hill Encyclopedia of Science & Technology.

Michael A. C. Perryman

Bibliography. European Space Agency, *The Hipparcos and Tycho Catalogues*, ESA SP-1200, 17 vols., 1997; M. A. C. Perryman, The *Hipparcos* astrometry mission, *Phys. Today*, 51(6):38–43, June 1998; C. Turon et al., From Hipparchus to *Hipparcos*, *Sky Telesc.*, 94(1):28–34, July 1997.

Atmospheric chemistry

Ozone, O_3, an allotropic form of oxygen containing three atoms, is an important chemical constituent of the Earth's atmosphere (**Fig. 1**). Ozone serves different functions in various regions of the atmosphere. Its fractional concentration ranges from parts per billion (ppb) to parts per million (ppm), and reaches its maximum abundance between roughly 20 and 25 km (12 and 15 mi) in altitude. This region of the stratosphere, called the ozone layer, acts as an ultraviolet (UV) radiation shield for the Earth. In the troposphere, ozone concentration is lower; yet, ozone plays many critical roles. The photolysis of ozone is the main source of the highly reactive hydroxyl (OH•) radical which acts as a cleansing agent in the troposphere. This radical allows for

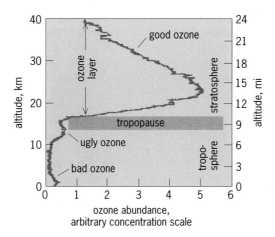

Fig. 1. Rough distribution of atmospheric ozone and the variation of temperature as a function of altitude. Most of the ozone is in the stratosphere. If all the ozone in the atmosphere was brought down to sea level, it would occupy a layer only a few millimeters thick.

most emissions in the atmosphere to be removed. However, direct contact with large concentrations of ozone is toxic and is major cause of urban air pollution (smog). When present in high concentrations, ozone negatively affects the health of animals and plants, and also damages structures. Further, ozone can absorb infrared (IR) radiation at wavelengths not greatly affected by carbon dioxide (CO_2) and act as a greenhouse gas. Thus, increases in ozone in the troposphere, especially in the upper troposphere, can lead to a greater greenhouse effect. Ozone is needed in the stratosphere to protect biological organisms from the harmful effects of ultraviolet radiation and in the upper troposphere to cleanse the atmosphere. However, depletion of ozone in the stratosphere and increases in the troposphere will have deleterious effects.

In the stratosphere, dissociation of molecular oxygen (O_2) produces O atoms [reaction (1)], which combine with O_2 to make ozone [reaction (2)]. The

$$O_2 \xrightarrow{h\nu} O + O \qquad (1)$$

$$O + O_2 \longrightarrow O_3 \qquad (2)$$

harsh ultraviolet radiation that is needed to break the O_2 molecule apart is available in the stratosphere. Chemical and photochemical reactions also destroy ozone. For example, it has been known for many decades that O atoms react with ozone and convert it to O_2 [reaction (3)]. Reactions (1)–(3) constitute

$$O + O_3 \longrightarrow 2O_2 \qquad (3)$$

the Chapman mechanism for the presence of ozone in the stratosphere. Most of the ozone is made in the tropics, the region with the most intense sunlight. The poleward and downward transport combined with the destruction of ozone in other regions of the stratosphere leads to the observed distribution of ozone as a function of latitude. It became clear during the middle of the twentieth century that the

Chapman mechanism overpredicted the abundance of ozone. It was later discovered that homogeneous gas-phase catalysis involving reactive species such as OH•, nitric oxide (NO), chlorine (Cl), and bromine (Br) can destroy significant amounts of ozone. This key finding satisfactorily explained the abundance of ozone in the stratosphere. This discovery also led to the realization that, if the abundances of these catalysts were to increase above their natural levels because of human activity, stratospheric ozone could decline. Such a concern was first raised in connection with the environmental impact of supersonic aircraft, which emit nitric oxide in the stratosphere, and later with the impact of chlorofluorocarbons (CFCs). CFCs are very stable in the lower atmosphere and therefore eventually reach the stratosphere, where they decompose to release chlorine. Ozone has decreased dramatically in the polar regions and also visibly at midlatitudes. Ozone depletions were larger soon after large volcanic eruptions that injected sulfur compounds into the stratosphere and enhanced stratospheric sulfuric acid aerosol abundance. Based on a combination of laboratory experiments, measurements of key stratospheric species (from ground, aircraft, balloons, and satellites), and modeling studies, it is now clear that the depletion in the stratospheric ozone over the last few decades is due to the increases in the CFCs and bromine compounds. It is also clear that these compounds have increased in concentration due mostly to human industrial activities.

There are some important differences between the ozone changes in the polar regions and those at midlatitudes. The ozone depletion in the Antarctic region, popularly called the ozone hole, is seasonal, very large, easily visible over the natural fluctuations, and has been observed to occur during late winter and spring every year since the late 1970s. The extent of the ozone hole increased tremendously during the 1970s and early 1980s and is now nearly constant. During the ozone hole period, ozone is essentially completely depleted between roughly 18 and 25 km (about 11 and 15 mi) in altitude. Seasonal depletions have been seen in the Arctic region which are similar to what has been observed in Antarctica, but less drastic and shorter-lived. The ozone depletion over midlatitudes since the late 1970s is much smaller and is discernible via statistical analyses. Though the extent varies with season, midlatitude depletions have been observed for all seasons.

Most of the chlorine released from CFCs and other sources are always in chemical forms that cannot destroy ozone. Presence of condensed matter converts the inactive chlorine chemicals to active forms that can destroy ozone. Of course, many other reactions convert the active form back to inactive forms. The competition between activation and deactivation determines the fraction of chlorine-containing compounds that can be in the ozone-destroying active form. In Antarctica, it is cold enough in winter and early spring for ice (such as water ice and mixtures of nitric acid and water ice) to form even though water vapor is scarce. (In general, clouds do not form in the stratosphere.) The condensed matter could also be cold sulfuric acid aerosol. Ozone destruction by chlorine also requires sunlight. Thus, ozone can be chemically removed following reactions on condensed matter in the presence of sunlight, that is, during spring. During winter the sunlight is limited, and during fall and summer the temperatures are not cold enough to "activate" inactive chlorine compounds. Over Antarctica, a vortex is formed during winter and persists well into spring. Therefore, stratospheric air over Antarctica is reasonably well contained and prevents replenishment of ozone-rich air from the tropics during winter and spring. The dynamical phenomena of vortex formation and containment have been always present. Around the middle of the 1970s the level of chlorine increased to levels sufficient to make the Antarctic ozone hole.

Over the Arctic, the vortex does not always persist throughout the winter and spring, and is not as cold. The temperature of the vortex and its longevity are interrelated. Also, the Arctic vortex is usually smaller and wobbles around the pole more than its Antarctic counterpart. The strength and persistence of the vortex is highly variable. It has been such that not a great deal of ozone was destroyed until the late 1980s. During this time, the level of chlorine was not increasing very rapidly, but the vortex became more conducive to ozone destruction. Therefore, larger and more persistent but highly variable ozone losses were observed over the Arctic during the last decade.

The eruption of Mount Pinatubo in 1991 greatly enhanced the level of sulfuric acid droplets in the stratosphere and showed that the reactions that take place on ice (and lead to ozone removal) can also take place in sulfuric acid. The trend in ozone levels observed over midlatitudes can also be attributed to chlorine and bromine increases due to human activity.

Based on scientific evidence and the desire of the world community to take action, the Montreal Protocol and its many amendments were enacted. These actions have already resulted in decreases in the shorter-lived chlorine-containing species, such as methyl chloroform (**Fig. 2**), and have led to a leveling-off of the longer-lived gases, such as CFCs, in the atmosphere. Based on the known atmospheric lifetimes of these compounds, it is expected that the chlorine levels will decrease to pre–ozone hole values around the middle of the twenty-first century. If there are no other alterations, the ozone hole can be expected to go away during the latter part of that century. This recovery will not be directly measurable for quite some time, however.

It was long believed that the ozone in the troposphere came only from the stratosphere. The harsh ultraviolet radiation necessary to break the O_2 bond and make ozone is not available in the troposphere. High levels of ozone in the Los Angeles smog was a clear indicator that ozone could be produced even in the troposphere. Since the 1970s, it has become clear that ozone is made in the troposphere by chemical reactions involving hydrocarbons (or carbon monox-

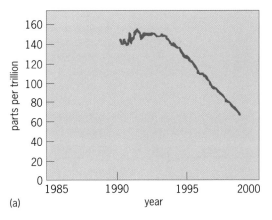

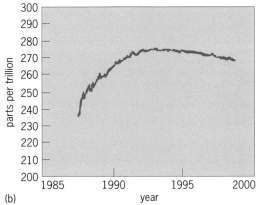

Fig. 2. Plots of (*a*) methyl chloroform, a short-lived gas, and (*b*) chlorofluorocarbon-11, a long-lived gas, as a function of year. Both emissions were curtailed following the Montreal Protocol and its amendments. Methyl chloroform is already decreasing in the atmosphere. Such data show that the Montreal Protocol is reducing human-produced chlorine-containing gases in the atmosphere. (*Data from National Oceanic and Atmospheric Administration, Climate Monitoring and Diagnostics Laboratory*)

ide), nitrogen oxides, and sunlight. Ozone can be produced via the oxidation of methane, the simplest hydrocarbon (**Fig. 3**). All the information needed to quantitatively predict the amount of ozone in a given region is not available and has led to difficulties in deciding what emissions should be controlled to mitigate urban air quality. For example, if enough hydrocarbons are emitted by vegetation (as in the southeastern United States), controlling hydrocarbon emissions from cars will not be as effective as controlling nitrogen oxides. Also, the amount of ozone formed due to an emission is not linearly dependent on the amount of that emission. Therefore, careful thought has to be given to control strategies. In addition, other factors such as particulate matter could be important consideration for emission controls.

What was believed to be urban air pollution has spread to adjacent areas and become regional air pollution. It is clear that as industrial emissions increase, there is a continuity in going from urban to regional to global scale. Therefore, understanding regional air quality and ozone production will be important for global-scale effects. As the size of cities increases in the future, the distinction between urban, regional, and global ozone production will be increasingly

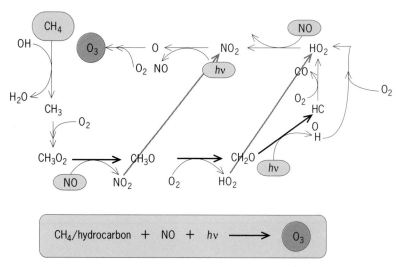

Fig. 3. Chemical pathway by which a combination of a hydrocarbon, nitrogen oxides, and sunlight can generate ozone in the troposphere. In this region, molecular O_2 cannot be split apart by light, unlike in the stratosphere. Yet, because of the chemical reactions, ozone can be chemically generated. The scheme is shown for methane, the simplest atmospheric hydrocarbon. Similar mechanisms work for other hydrocarbons. Depending on the conditions, control of hydrocarbons or nitrogen oxides may be needed to reduce ozone production.

difficult. Therefore, mitigation and adaptation are important strategies that need to be developed by policy makers with full knowledge of the environmental effects of ozone-producing emissions.

For background information *see* AEROSOL; AIR POLLUTION; ATMOSPHERIC CHEMISTRY; ATMOSPHERIC OZONE; HALOGENATED HYROCARBON; OXYGEN; OZONE; SMOG; STRATOSPHERE; TROPOSPHERE; ULTRAVIOLET RADIATION (BIOLOGY) in the McGraw-Hill Encyclopedia of Science & Technology.

A. R. Ravishankara

Bibliography. R. E. Benediek, *Ozone Diplomacy: New Directions in Safeguarding the Planet*, Harvard University Press, 1998; R. L. Johnson, *Investigating the Ozone Hole*, Lerner Publications, 1993; *Montreal Protocol on Substances That Deplete the Ozone Layer: 1994 Scientific Assessment of Ozone Depletion*, United Nations Publications, 1995; A. Nilsson, *Ultraviolet Reflections: Life Under a Thinning Ozone Layer*, John Wiley, 1996.

Autonomous navigation

Autonomous navigation means that a vehicle is able to plan its path and execute its plan without human intervention. In some cases remote navigation aids are used in the planning process, while at other times the only information available to compute a path is based on input from sensors aboard the vehicle itself. An autonomous robot is one which not only can maintain its own stability as it moves but also can plan its movements. Autonomous robots use navigation aids when possible but can also rely on visual, auditory, and olfactory cues. Once basic position information is gathered in the form of triangulated signals or environmental perception, machine intelligence must be applied to translate some basic motivation (reason for leaving the present position) into a route

and motion plan. This plan may have to accommodate the estimated or communicated intentions of other autonomous robots in order to prevent collisions, while considering the dynamics of the robot's own movement envelope.

The following examples illustrate missions that are being developed for autonomous vehicles. A Mars probe lands in the caldera of the solar system's largest volcano, Olympus Mons. Scientists wish to have the probe map the caldera by deploying a small aerial robot that will fly a pattern across the base of the caldera and take photographs. How will the aerial robot know where it is? The magnetic field of Mars is very weak, there is no ground-based or orbital standard emitting signals by which to triangulate a position, and the terrain is unfamiliar and precludes the use of landmarks.

Another example is a micro air vehicle (MAV) operating inside a building. Though Global Positioning System (GPS) signals are available just outside the building, they are effectively blocked by the materials used to construct the walls and ceiling. Moreover, the vehicle may be too small to carry an antenna that could receive any radio navigation aids. Without prior knowledge of the building interior, how could such a tiny reconnaissance vehicle find its target?

Principles. Autonomous vehicles provide superior solutions for many mission requirements, but the challenge of making the vehicle navigationally robust in all situations is formidable. A robust navigator must possess six attributes: (1) a mission goal (the motivation to move), (2) the ability to perceive its environment (for obstacle avoidance), (3) understanding of its present location, (4) the ability to plan a path to achieve its goal, (5) self-actuating mobility, and (6) the ability to replan as it moves (to compensate for unexpected situations).

Before an autonomous vehicle can intelligently plan a path to its goal, it either must have a stored map of its world or must create one as it moves based on what it perceives. In the case of the micro air vehicle, a map of the interior of the building in which it will fly would be of significant value, but even were such a map stored onboard, furniture and other unbriefed threats to the vehicle could block its path. In most cases, a map is not available and the micro air vehicle must sense the path to its target from other cues.

Knowledge of "up." In most systems, particularly those used in flying robots, it is critical to know where "up" is. Knowing the orientation of the vertical gravity vector allows an aerial robot to remain in flight parallel to the surface of the Earth, or to return to that orientation after completing a maneuver. Knowing where "up" is can also affect the calibration of various onboard sensors.

Electronic pendulums can be constructed from accelerometer arrays, but they are not reliable indicators of the vertical direction on a moving platform because they cannot distinguish the acceleration produced by gravity from the acceleration produced by vehicle motion, as in a banked curve. Unlike the pendulum, a properly placed vertical gyroscope is not affected by centrifugal "force." For this reason vertical gyroscopes, which are gimbaled to allow the spin axis to freely rotate about its spin center, are often used to remember where "up" is. A vertical gyroscope can indicate offsets in yaw, roll, and pitch relative to its calibrated starting position if placed at the center of rotation (often the center of gravity) of an autonomous aerial robot. This starting position is usually the vertical gravity vector as derived from a pendulum sensor when the vehicle is at rest.

Vertical gyroscopes tend to drift over a period as short as tens of minutes. Therefore, designers prefer to combine vertical gyroscopes with another sensor, such as the GPS, that is less accurate over short intervals but exhibits little or no long-term drift.

Route and motion planning. If a global map is available, an autonomous vehicle can plan its entire route from its present position to its goal. Some route planners search for the optimum path based on rules which attempt to minimize transit time, fuel consumption, threat exposure, or other factors. Thousands of routes are examined based on a selection of available way points, with the one best conforming to the mission rules being chosen. As the selected path is traversed, unbriefed threats which would cause the autonomous vehicle to violate the mission rules may be encountered and the route must be recomputed from the vehicle's current position. If no solution is possible under the rules, then certain rules must be relaxed. For example, a higher degree of threat exposure may be acceptable, but other rules may be inviolate such as those concerning mission endurance. A route requiring the vehicle to exceed its remaining fuel allotment is obviously an unacceptable alternative.

If no global map is available, the optimum route cannot be predicted, and a combination of dead reckoning and seek/avoid behaviors must be used. Dead reckoning uses time-in-motion at a certain speed along a given heading to extrapolate a new position based on a known starting point. Odometry is a form of dead reckoning often used in factory robots to count the revolutions of a drive wheel of known circumference in order to determine distance traveled independent of time. Visual odometry is also possible from aerial robots in which the passage of objects on the ground is noted. By knowing the altitude of the aerial robot and the field of view of its vision sensor, a measure of distance traveled can be deduced.

Dead reckoning is plagued by cumulative errors which arise from inaccuracies in the measurement of time, speed, and heading. These inaccuracies may be due to the inherent resolution of the sensors used, or to drift caused by unpredictable changes in the environment. Dead-reckoning errors grow as the mission progresses unless there is some standard to periodically recalibrate the absolute position of the vehicle. Dead-reckoning sensors include devices such as accelerometers (to measure acceleration), rate gyroscopes (to measure rate of change of velocity), and magnetometers (to measure heading). By integrating acceleration the navigation system can determine velocity, and by integrating velocity it can determine position.

In contrast, seek/avoid systems are as accurate as the resolution of the sensors used to seek the goal. Accuracy improves as the seeking sensor is brought nearer to its goal because the error signals provided by the sensor are greater for smaller vehicle heading deviations when near the target than when far from it. Larger error signals are less susceptible to noise, and the heading can be maintained more accurately.

The avoidance signal serves as a warning to override the seeking behavior when a threat to the vehicle is encountered. After successfully diverting the vehicle from the desired seeker path by changing heading or altitude to avoid a detected obstacle by means of a preprogrammed (reflexive) or calculated (cognitive) maneuver, the avoidance sensors no longer detect the obstacle. Control is returned to the seeking sensors, whereupon the robot continues toward its goal on a new path.

An example of the use of sensors is a mission in which a tiny autonomous air vehicle is launched through an air vent from the outside to search for hostages being held somewhere in an abandoned building. No recent map of the interior is available, though intelligence reports indicate that the building has a group of central rooms accessible by hallways off a main corridor. A reasonable sensor suite would include ranging devices to avoid obstacles in front of, and to the sides of, the vehicle. In addition, a downward-looking ranging device would provide altimetry information. Active radio frequency, optical, or acoustic transceivers similar to radar or sonar would be suitable and would serve only to keep the vehicle out of harm's way during its ingress.

Another kind of sensor would be used to provide motivation. This might be an electronic nose, which detects small quantities of molecular species (pheromones, ammonia, or other chemicals given off by the skin) that indicate the presence of human beings. A pair of these molecular sensors placed on opposite sides of the "head" of the micro air vehicle could then indicate that concentration of the target molecules is greater to the right, left, or, if equal, straight ahead. Thus, a motivation to move in a particular direction is provided.

Flight competitions. Work is under way to develop fully autonomous micro air vehicles that use seek/avoid navigation strategies to conduct indoor operations. However, the smallest, most intelligent fully autonomous robots are currently those in the International Aerial Robotics Competition, all less than 3 m (10 ft) in any dimension. Since 1990, competing collegiate teams have devised autonomously navigating aerial robots capable of environmental perception, intelligent motion over an arena, and manipulation of objects on the ground while in flight. Some of these automatons replan their routes based on information gleaned from their environment. In 1997 a fully autonomous self-navigating entry completed a mission which required it to locate a toxic waste dump, map randomly oriented partially buried drums, read the labels on the drums from the air to determine the contents of each, and retrieve a sample from a particular drum.

Potential applications. Autonomous robots excel at applications which are too dull, dirty, or dangerous for human beings, or for missions which extend beyond a human life-span such as space travel. These mobile robots will autonomously navigate the Earth's surface, oceans, and skies without human intervention, albeit with assistance from navigation aids such as the GPS, which will take on greater importance as mobile robots proliferate throughout society. Future applications will include automated delivery services, continuous high-altitude weather measurement, maintenance of homes and offices by service robots, and autonomous farming machinery, personnel transports, and machines of war.

For background information *see* DEAD RECKONING; DRONE; GUIDANCE SYSTEM; GYROSCOPE; NAVIGATION; ROBOTICS; SATELLITE NAVIGATION SYSTEMS in the McGraw-Hill Encyclopedia of Science & Technology.   Robert C. Michelson

Bibliography. A. Halme and K. Koskinen, Intelligent autonomous vehicles, *International Federation of Automatic Control Conference*, Espoo, Finland, June 1995; A. L. Meyrowitz, D. R. Blidberg, and R. C. Michelson, Autonomous vehicles, *Proceedings of the IEEE*, 84(8):1147–1164, 1996; R. C. Michelson, Update on flapping wing micro air vehicle research—Ongoing work to develop a flapping wing, crawling entomopter, *13th Bristol International RPV/UAV Systems Conference Proceedings*, Bristol, England, March–April 1998.

Biocavity laser

Lasers are an integral part of today's health care delivery system. They are used in the laboratory analysis of human blood samples; serve as surgical tools that kill, burn, or cut tissue; and are used in oph-

Fig. 1. Compact biocavity laser device consists of a photo-pumped microlaser system and an associated laptop computer with an internal spectrum analyzer. A syringe for sample injection is in in front of the microlaser.

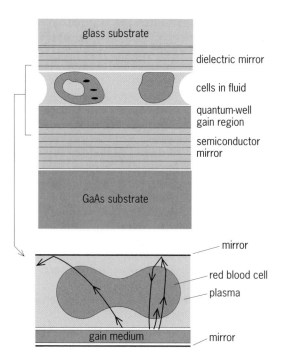

Fig. 2. Cross section of the biocavity laser. The red blood cell acts as a waveguide to resonate light between the two mirrors.

thalmology to weld torn retinas. Semiconductor microtechnology has reduced the diameter of a laser to a fraction of a human hair to create micro-electromechanical systems (MEMS). MEMS devices are expected to advance health care delivery.

Components and operation. One MEMS device is the biological microcavity laser, or biocavity laser for short (**Fig. 1**). It is the result of research conducted at the Sandia National Laboratories, Albuquerque, New Mexico.

The biocavity laser employs a novel method to analyze various substances such as blood, cells, or tissue. It uses a semiconductor wafer in which the sample acts as an internal component and produces lasing. Although it is simple in design, the profound number of applications that can be derived from this instrument is considerable.

The biocavity laser is based on vertical-cavity surface-emitting laser (VCSEL) technology. VCSELs play a role in the rapidly expanding fields of communications, information systems, and biomedical technology. The VCSEL utilizes the element silicon. Silicon is readily available and ideally suited for the manufacture of semiconductor chips. The biocavity device uses two mirrors that enclose a cavity and a gain medium (**Fig. 2**). One mirror surface is the top of an AlGaAs/GaAs (aluminum gallium arsenide/gallium arsenide) surface-emitting semiconductor wafer comprising a laser gain region atop a multilayer reflector. Hence, this component amplifies light signals. The other surface is a dielectric mirror atop a glass substrate that forms a laser resonator. Between these mirrors is gap where a sample is placed. The sample may be whole blood, cells from culture, tissue from a biopsy, or a scraping of body tissue.

When placed into the biocavity laser, most samples display nearly ideal optical elements as light traverses the specimen near 850 nanometers. At this wavelength, light scatter and absorption are minimal, allowing the cell to be transparent. As the chamber of the biocavity laser is excited by an external source, light resonates between the mirrors, and ultimately the sample acts as a lens focusing light (similar to a magnifying glass). This concentration of light assists a light-generating process to such a degree that light is liberated through the top mirror. The discharged light can then be analyzed. Thus, the sample becomes an intrinsic component of the laser and renders information about itself by acting as an optical waveguide; that is, it guides the light waves to a coherent state. The sample acts like a laser by producing coherent light.

Sample waveguiding is distinctive and assigns a signature to each specimen. This unique signature is due to differences in the dielectric constants between the cell components and the surrounding fluids. Each specimen has differing protein or nucleic acid concentrations in the various cellular compartments, such as the nucleus, cytoplasm, or membranes. When the specimen is placed in the biocavity laser, the variations create information that can be captured and analyzed.

Resonated light escapes from the sample chamber, and it is detected using a high-resolution spectrometer that captures the emitted lasing frequencies. These frequencies are analyzed as a light spectrum displaying well-defined peaks. The spectral peaks have both an intensity and a spacing distribution that provide individual spectral signatures for each specimen tested. Computer analysis of the spectral signature identifies the sample.

Clinical advantages and applications. Semiconductors operate efficiently with high throughput and can process large numbers of samples rapidly. A great amount of information can be gained using a fraction (1/10 to 1/1000) of the routine amount of sample needed for traditional testing. This is accomplished by rastering (repeated scanning) of cells or by flowing of cells through etched channels in the semiconductor wafer. Holding great promise is this cell-flow approach (**Fig. 3**). Researchers have been able to precisely fashion grooves in the silicon material as small as 15 micrometers (approximately twice the diameter of a red blood cell). Technical advances with flow technology can route cells in single file through the channels. In short, this semiconductor wafer acts as a flow cytometer.

Flow cytometry is used in research and medical testing. It is used in counting the number of CD4+ cells (a type of lymphocyte) in patients with HIV, which serves as an indicator of the severity of disease because the count decreases as AIDS worsens. Presently, blood samples are obtained by standard venipuncture, then stained with special reagents to visualize the cells. This entire process is labor-inten-

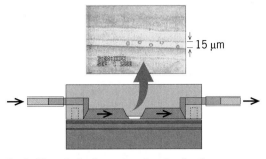

15 μm

Fig. 3. Biocavity for flow cytometry, showing the cross section of the semiconductor wafer where a 15-μm groove has been etched. The red blood cells are shown flowing in single file.

sive and time-consuming. It is anticipated that devices such as the biocavity laser will be more rapid, more efficient, and less costly. Furthermore, the biocavity laser uses smaller sample volumes, and this translates into less invasive sampling from a patient; that is, a finger puncture can provide information similar to that in venipuncture.

The biolaser can be used in cellular analysis to find rare events, such as a malignant cell, dying cell, or contaminating cell in a large volume of sample. Applications may include examining tissue cells from a PAP smear to find precancerous cells or sampling a blood specimen to test for leukemia or other cancer. The biocavity laser can advance stem-cell transplantation, a therapy used in treating blood cancers and other serious disease. This technique relies on the isolation of stem cells, which have the potential to give rise to a large number of mature blood cells. The stem cells constitute a very small percentage of circulating blood cells, hence large volumes of blood are processed to collect small numbers of them. The biocavity laser can assist in finding these rare cells more rapidly and efficiently.

The point-of-care concept in medicine, convenient to the patient, is unfortunately weakening. A visit to a physician's office may result in the patient's having to make trips to other facilities in order to provide a blood or urine sample. Samples that are obtained in the doctor's office are sent to a large reference laboratory for analysis. Turnaround time for the delivery of results from the blood tests can be days. An instrument that can provide near-instantaneous results on a specimen at the point of care is of great assistance. MEMS devices can be manufactured relatively inexpensively in compact sizes, allowing testing to be accomplished in a physician's office and promptly providing important and often anxiety-relieving information to the patient.

With the advantages of miniaturization, cost containment, decreased time to test results, improved point-of-care delivery, and increased diagnostic yield, biocavity lasers can also be optimally applied in third world countries or in areas devastated by war or other disasters.

For background information see LASER; LIGHT in the McGraw-Hill Encyclopedia of Science & Technology.   Mark F. Gourley; Paul L. Gourley

Bibliography. P. L. Gourley, Microstructured semiconductor lasers for high speed information processing, *Nature*, 371:571–577, 1994; P. L. Gourley, Semiconductor microlasers: A new approach to cell-structure analysis, *Nature Med.*, 2:942–944, 1996; P. L. Gourley et al., A semiconductor microlaser for intracavity flow cytometry, *Proceedings of the SPIE Conference*, San Jose, January, 1999; P. L. Gourley et al., Ultrasensitive detection of red blood cell lysing in a microfabricated semiconductor laser cavity, in *Progress in Biomedical Optics: Proceedings of the Conference on Micro- and Nanofrabricated Structures and Devices for Biomedical and Environmental Applications*, San Jose, SPIE Conf. 3258, January 26–27, 1998; H. M. Shapiro, in *Practical Flow Cytometry*, 2d ed., Liss, 1988.

Biodiversity

Life on Earth is experiencing its sixth major extinction event. This event differs from previous ones because it is caused by human activity. There is increasing scientific and public concern over the massive loss in biodiversity (all aspects of biological diversity, including species richness, ecosystem complexity, and genetic variability). Convincing evidence since the 1960s shows that ecosystem functioning is dependent on attributes of individual species. Only recently have ecologists examined how species diversity affects ecosystem processes. The importance of ecosystem complexity and genetic variability remains largely unexplored.

Ecosystem functioning. An ecosystem consists of living and nonliving parts interacting to form a stable system that perpetuates indefinitely. The functioning of an ecosystem is often described by the flow of energy via food chains and webs, and the cycling of elements biogeochemically. Ecosystems provide a number of other functions and services, including habitat diversity, inhibition of soil erosion, and purification of air and water. A recent study estimated that 17 services that ecosystems provide to humans are equal in value to the entire global gross national product. Because Earth's biodiveristy is rapidly declining, understanding how species affect ecosystem processes is important.

Scientific methods. To scientifically address a question, there are different methodological approaches, including theoretical, which seeks to explain phenomena based on first principles; observational, which seeks to explain phenomena by studying intact natural systems; and experimental, which seeks to explain phenomena by isolating cause-and-effect mechanisms. Each approach has advantages and disadvantages. In addressing ecological questions, there is often a lack of first principles. Experimental studies are valuable because they are used to explore fundamental factors controlling processes. However, many ecosystems are difficult or impossible to manipulate experimentally, and ecologists are often limited to observational studies.

Functional roles. Some species can play similar functional roles with respect to aspects of the ecosystem. Therefore, the loss of a single species, even when it plays an important role, can be compensated for by another species with similar attributes. In these cases, the loss of a species will not have a measurable effect on ecosystem processes. However, species loss can be cumulative, and a gradual decline in diversity may ultimately lead to rapid ecosystem restructuring. In some instances the function of one species cannot be compensated for by another species. These species are described as keystone because their loss results in significant changes in ecosystem structure and function (see **illus.**).

Structure and function can respond very differently. An ecosystem can undergo large changes in structure, but in some cases, rates of energy flux and element cycling remain unaltered. Another issue is to determine the trophic level that controls ecosystem function. If a predator has a strong influence on ecosystem processes, control is called top-down because it comes from the top of the food chain. If control stems from primary producers, it is called bottom-up. Ecosystems are complex and often exhibit features of both top-down and bottom-up control.

Tropical deforestation. Slash-and-burn agriculture has been practiced by human populations in the tropics for thousands of years. Surprisingly, low-diversity agroecosystems can be nearly as productive as high-diversity forests. In tropical forests with nutrient-deficient soils, however, initial productive years are soon replaced with a chronic decline. There are many possible explanations for this decline, including competition by weeds, increased soil erosion, consumption by insects, and a decline in soil pH. Change in nutrient cycling dynamics, primarily phosphorus availability, is probably the most important factor. In historical times, land was allowed to recover its productivity during a fallow period while forest trees species recolonized the site. Since the industrial revolution, an increase in land-use intensity has resulted in shorter fallow periods. When a forest ecosystem is kept in agricultural production for long periods of time, land degradation occurs. Many years are required to recover the forest when the land is abandoned. A tropical forest often contains hundreds of plant species, whereas a pasture contains just a few. Tropical deforestation and land degradation are conspicuous examples demonstrating that a loss of species diversity can have dramatic effects on ecosystem processes.

Sea otters. By the early 1900s, commercial exploitation had eliminated sea otters (*Enhydra lutris*) from most of their geographic range, extending from the western Aleutian Islands to the southern California coast. Before otters were extirpated, they controlled the numbers of their preferred prey, herbivorous sea urchins (*Stongylocentrotus* spp.). Urchins feed primarily on kelp, so predation pressure on them allowed kelp beds to flourish. When otters were removed, urchin populations expanded, drastically reducing the geographic extent of kelp

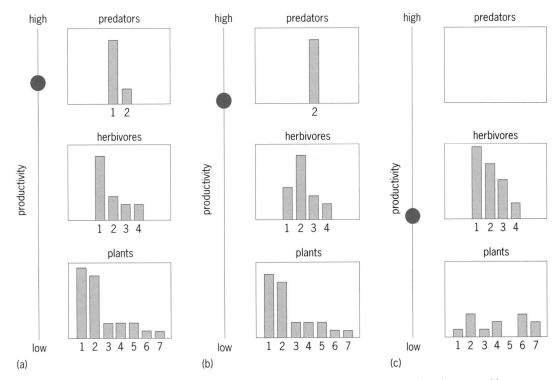

Changes in species composition can modify ecosystem structure and function. (*a*) **Example of species composition.** (*b*) **Removal of a predator species has little effect because predator species 2 partially compensates by increasing in numbers.** (*c*) **When both predators are removed, the ecosystem changes considerably. In this scenario, a single predator does not act as keystone species, but the predator functional group acts as a keystone guild (group of species).**

forests. In recent years, otters have recolonized much of their historic range, and kelp forests are returning. Kelp forests have high productivity and high species diversity, whereas urchin barrens have low productivity and low species diversity. The otters demonstrate that the attributes of a single keystone species can control the structure and function of an entire ecosystem.

Desert rodents. Observational studies are valuable for investigating species' roles in ecosystem functioning. However, to develop a detailed understanding of underlying processes, controlled experiments are required, where just one or a few factors are manipulated, and ecosystem response is quantified. A number of experimental studies have investigated community structural changes in response to changes in species diversity. Scientists are only beginning to study how changes in species diversity affect ecosystem processes.

Scientists in the United States' Chihuahuan Desert, for example, found that the elimination of small rodents (primarily kangaroo rats) from experimental plots decreased the abundance of desert shrubs and increased perennial grasses. This study demonstrates top-down control by keystone herbivores and suggests that the removal of small rodents can transform a desert into a grassland. *See* RESTORATION ECOLOGY.

Coastal rocky ecosystems. In a classic study in the Pacific Northwest, R. T. Paine examined the effect of the experimental removal of a predatory seastar (*Pisaster ochraceus*) on community composition.

When the seastar was present, 16 species of seaweed and invertebrates were found. When *Pisaster* was removed, the ecosystem was ultimately populated by only one species (a mussel). *Pisaster* is not always a keystone species. In some areas a predatory whelk partially compensates for the loss of *Pisaster*. Although ecosystem processes were not directly measured in Paine's studies, the elimination of seaweed species probably reduced rates of local primary productivity.

Plant community structure. Model plant communities have recently been established at the Ecotron in Silwood Park in southern England, and at the Cedar Creek Reserve in Minnesota. In these experiments, effects of changes in species diversity on ecosystem properties such as primary productivity, nitrogen mineralization, and litter decomposition are quantified. In general, these studies have not provided convincing evidence that ecosystem processes are crucially dependent on high species diversity. Instead, they suggest that the biological characteristics of dominant plant species control productivity and nutrient cycling. When dominant species are removed, ecosystem processes often experience little change because other species can compensate. These studies demonstrate that variables describing ecosystem function work in complex ways with respect to species diversity. Overall, it appears that species composition, rather than species diversity per se, is more consistently important in determining ecosystem response to perturbations.

Conclusions. Only a few studies have experimentally manipulated species diversity and monitored key ecosystem variables such as productivity and nutrient cycling. Given the paucity of concrete evidence, results must be taken as preliminary. Studies have been limited to ecosystems that are easily manipulated, and to time spans that are short compared to the life-span of many organisms and the time scale of evolutionary processes. Many ecosystem components are not amenable to experimental manipulation, and the response time of individuals can be very slow.

Even if a loss of biodiversity has minimal impact on ecosystem processes, there are still esthetic and ethical reasons for the preservation of biodiversity that go beyond productivity and nutrient cycling. The loss of a species is irreversible because the evolutionary events that lead to the origin of a species will never be repeated. The loss of a species may not cause changes in ecosystem function but can still dramatically change ecosystem structure. A fertilized tropical pasture may be just as productive as the forest it replaces, but the value of a forest goes beyond ecosystem processes.

For background information *see* ECOLOGY; ECOSYSTEM; FOREST ECOSYSTEM; FRESH-WATER ECOSYSTEM; GRASSLAND ECOSYSTEM; SYSTEMS ECOLOGY in the McGraw-Hill Encyclopedia of Science & Technology.   Jeff Chambers

Bibliography. F. S. I. Chapin et al., Ecosystem consequences of changing biodiversity, *BioScience*, 48:45–52, 1998; H. A. Mooney et al. (eds.), *Functional Roles of Biodiversity: A Global Perspective*, John Wiley, 1996; G. A. Polis and K. O. Winemiller (eds.), *Food Webs: Integration of Patterns and Dynamics*, Chapman and Hall, 1996; E.-D. Schulze and H. A. Mooney (eds.), *Biodiveristy and Ecosystem Function*, Springer-Verlag, 1993.

Bioethics

Bioethics is a broad term that encompasses the environmental, ethical, legal, and social implications of agriculture, life sciences, and medicine. The rapidity of progress in genetics and genome mapping, spurred by the Human Genome Project, is reforming the whole process of science and providing new tools for use in technology, but also raising many questions for society. Bioethical issues arising from recent applications of science include genetic engineering of microorganisms, plants, and animals; in vitro fertilization; cloning; xenotransplantation; gene therapy; and genetic testing.

Origins. The word "bioethics" originated in Van R. Potter's book *Bioethics, Bridge to the Future*, but the concept comes from human heritage over thousands of years. Abortion, euthanasia, communication between doctors and patients, and conservation of resources and biodiversity have been discussed for millennia in almost all cultures, philosophies, and religions. However, the rapid advances in life sciences and the global environmental crisis since the 1960s have led to the development of the academic field of bioethics. These scientific advances included life-sustaining respirators in intensive care, organ transplantation, assisted reproductive technologies, and genetic engineering. Also, in recent decades more people have felt they should be involved in the formation of public policy, in line with the United Nations Declaration of Human Rights.

A set of four principles or ideals has been developed as a common ground for bioethical decision-making. In all things people do, the ideal is to avoid doing harm (nonmaleficence) and to try to do good (beneficence), while respecting autonomy and preserving justice. Macer argues in *Bioethics Is Love of Life* that these principles arise from the definition that bioethics is love of life. These four principles of bioethics involve loving self (autonomy), others (justice), life (nonmaleficence), and good (beneficence). Love is not only a universally recognized goal of ethical action but also the foundation of normative principles of ethics.

While bioethics is the concept of love of life, balancing the benefits and risks of choices and decisions, it encompasses numerous other fields such as technology assessment, medical ethics, and environmental ethics. In addition, there are at least three ways to view bioethics: (1) Descriptive bioethics is the way that people view life, affecting their moral interactions and responsibilities with living organisms. (2) Prescriptive bioethics is telling others what is ethically good or bad, or what principles are most important in making decisions. It says that someone has rights, and others have a duty to respect those rights. (3) Interactive bioethics is discussion and debate between people, groups within society, and communities.

Biosafety. The principle of beneficence is that technology should progress if it will not cause harm or injustice. Biotechnology is the use of living organisms, or parts of them, to provide goods or services. The principle of "do no harm" has led to both biosafety and socioeconomic concerns about drug use and genetic engineering. Several hundred medicinal products have been approved for clinical use since 1982, when human growth hormone—still controversial—was implemented. A series of safety levels was established in the mid-1970s to minimize risks of harm to the environment and organisms (including human beings) from research using genetic engineering. In the 1990s genetically modified animals and plants have been used as research tools and for toxicology testing.

By the mid-1980s, some genetically modified organisms were released into the environment in small-scale field trials to assess the environmental impact. By the mid-1990s, certain genetically modified organisms were grown openly as agricultural crops, particularly in China and North America, and by 1997 such organisms were being commercially grown on all inhabited continents. Genetically modified cotton, soybean, and maize plants have already become standard in North America. The advantages of the crops include disease resistance, insect resistance, altered or novel products, different varieties, and tolerance of adverse soil, water, or environmental conditions. Reductions in the use of chemical pesticides are also associated with some varieties. However, there are concerns whether the introduced genes might transfer to other species and cause harm to them or adversely alter the ecosystem. This outcome has been called genetic pollution, and scientific experiments generally have found the probabilities of adverse events to be low, but varying between species. Continual monitoring is called for, such as food safety tests to monitor allergies and side effects.

Animal rights. Concurrent with the emergence of bioethics has been the animal rights movement. Protests have led to reduced use of animals for toxicity testing and research. Understanding of evolutionary links and the common gene pool has also increased the arguments to regard animals as ethical beings in their own right, a biocentric view of life. Alternative tests have been devised which may not involve animals or may reduce the number of animals used and the level of suffering. Technology has allowed development of cell lines which can be used instead; or genetically modified animals, especially transgenic mice, that have increased susceptibility to develop cancer, so less animals are needed. Cloning genetically identical animals to be used for control and test animals can also reduce the numbers needed.

Nevertheless, concerns remain about crossing species boundaries in genetic manipulation, and regarding animals as commodities, tools, and bioreactors for making products. Xenotransplants are expected to provide a new source of organs, such as humanized hearts grown in pigs. However, the safety of humans exposed to hidden viruses and diseases in animals is being questioned. This is especially a concern with primates that are closely related to humans, since chimpanzees are considered to be the origin of the human immunodeficiency virus (HIV).

Bioethicists have stated that all creatures, whatever their genes or origin, and especially sentient ones, are fellow creatures and should be included in the reverence for life. There have even been moves to grant great apes (chimpanzees, gorillas, and orangutans) the same equal and inalienable rights as humans.

Biotechnology. Genes are a fundamental resource of biotechnology, and biodiversity is important for maintenance of the gene pool. The Convention on Biological Diversity gave intellectual property rights to discoveries of new drugs in native species to the country of origin, with the major exception of any biological materials collected before 1992, that is, what is already in botanic gardens, aquariums, or zoos. Some would argue that no person has a right to ownership of world biological heritage, which should be open to all. This debate with the Trade-Related Intellectual Property Provisions (TRIPPS) and the World Trade Organization that called for universal recognition of patents should intensify in the twenty-first century. Over half of the fully sequenced genomes have been products of industry research, and while most data are shared openly soon after discovery, selected genes and features are subject to patent protection. Article 4 of the Universal Declaration on the Human Genome and Human Rights, accepted by all countries of UNESCO in 1997, reads, "The human genome in its natural state shall not give rise to financial gains." The questions are whether complementary DNA is natural in the spirit of the Declaration, whether a single nucleotide makes something an invention, and what is a just reward for industrial investment in biotechnology.

Another emerging issue is the use of "terminator" techniques to stop a plant seed from being viable in the second generation, so that the farmer cannot grow it. This controls the supply for seed providers (usually commercial). The approach has been criticized as killing the principle of life itself, but companies argue it is no different from hybrid seeds. Hybrid seeds come from two unlike parents and, as a result seed from the first-generation cross hybrid will not produce the same yield or quality as the parent plants. The approach may have strong ethical support only in protecting pharmaceutical-producing plants that are special products of research and could be dangerous if growth was uncontrolled. However, there are biosafety questions if the terminator feature is passed to other plants of the same or different species. From the viewpoint of justice, there is no general scientific reason why genetically modified organisms need to be made unfertile, unlike hybrids, so several countries (including India) have banned import of such seeds. Termination would also conflict with the ideal that food is a basic human right.

Human reproductive technologies. The options for assisting human reproduction include use of a sperm donor, egg donor, or surrogate mother. Some sperm banks allow the selection of sperm based upon information about the donor's race, hair color, educational level, height, weight, and character. The donor usually remains anonymous to the parents, but in Sweden and New Zealand children can trace their genetic fathers through records. Generally in other countries tracing is difficult, and in some countries maintenance of reliable records is not required. The debate remains over the right of a child to know its genetic parents versus privacy of donors. In the case of known donors, in Belgium for example, stored sperm can be utilized even after the death of a husband, whereas in the United Kingdom it cannot.

Embryo freezing has become commonplace, meaning that twins can be born at different times, which could lead to a new type of developmental study. The question of whether to dispose of the excess embryos after the parents have stopped trying to use in vitro fertilization, to perform destructive research on them until up to 14 days of in vitro growth, or to donate them to prospective parents seeking implants is usually solved by having the genetic parents specify their choice at the time of deposition. Following a divorce, either partner may still want to utilize their frozen embryo. This issue has been discussed in law courts with various conclusions.

Surrogacy remains controversial. Some commercial services are offered on the Internet in the United States. International approaches are needed; for example, Germany and Japan forbid surrogacy so their citizen's may contract for the services of surrogate mothers in the United States. In some countries a surrogate mother can change her mind and refuse to hand over the child to the contracting parents. Further considerations include whether access to these services for reproduction should be government-funded, and whether any potential parent can contract for such services.

Human cloning has been banned by Article 11 of the Universal Declaration on the Human Genome and Human Rights: "Practices which are contrary to human dignity, such as reproductive cloning of human beings, shall not be permitted...." There is debate, however, over what is human dignity and whether the law can really restrict individual liberty for a family that wants to clone only a single child. Greater pressure to change policy is expected when mammalian cloning in agriculture becomes safe and efficient.

Genetic testing. Genetic tests can be offered to people to confirm medical diagnoses or to find presymptomatically the risk of developing certain diseases. DNA chips are allowing multiple screening for risks to chemicals, of side effects from pharmaceuticals, and of onset of inherited diseases. It is generally considered that children should not be tested until they can understand the results, and Article 5c of the Universal Declaration on the Human Genome and Human Rights reads, "The right of each individual to decide whether or not to be informed of the results of genetic examination and the resulting consequences should be respected." There have been laws to protect privacy and to forbid genetic discrimination in many countries.

Prenatal genetic testing allows the fetus to be evaluated, which can reassure parents who are at high risk to pass on inherited disorders or chromosomal abnormalities, but it is also linked to abortion choice. In some countries, abortion is legally permitted at later stages for reasons of fetal disease. With in vitro fertilization, preimplantation diagnosis allows 2-day-old embryos to be tested and then the healthy ones selectively reimplanted. This removes some concern about fetal rights, but still creates division among disease support groups concerned about discrimination. As techniques develop for testing fetal cells in maternal blood, tests are likely to be conducted throughout the population and can be done earlier and more cheaply.

Gene therapy. Not yet a successful clinical technique, gene therapy has become a primary topic for bioethics. In all countries that have used somatic-cell gene therapy, there are separate ethical review committees to monitor it because of the perceptions about the powerful nature of genes. It is debated whether enhancement gene therapy should be allowed once it is technically safe and effective. Cosmetic surgery is allowed as a type of enhancement without medical reason and is generally not considered unethical.

Germ-line gene therapy is outlawed in the Council of Europe Convention on Biomedicine and Human Rights, signed by over 20 countries in 1996. Other countries have regulations against it, but some are debating when it might be justified to alter the genes that children inherit. The result might be human genetic engineering for no medical reason. The issue is international, as people born with altered genes can move to foreign countries and disseminate their germ cells. As bioethicists realize, gene therapy is a science and technology that can change the human population.

For background information *see* AGRICULTURAL SCIENCE (PLANT); BIOTECHNOLOGY; DEOXYRIBONUCLEIC ACID (DNA); GENETIC ENGINEERING; GENETIC MAPPING; GENETICS; HUMAN GENETICS; MOLECULAR BIOLOGY; PATENT; REPRODUCTIVE TECHNOLOGY in the McGraw-Hill Encyclopedia of Science & Technology.   Darryl Macer

Bibliography. T. L. Beauchamp and J. F. Childress, *Principles of Biomedical Ethics*, 4th ed., Oxford University Press, 1994; D. R. J. Macer, *Bioethics Is Love of Life*, Eubios Ethics Institute, Christchurch, 1998; V. R. Potter, *Bioethics, Bridge to the Future*, Prentice Hall, 1971.

Biomarkers

The term biomarker is a contraction of biological marker. Biomarkers are complex organic compounds composed of carbon, hydrogen, and other elements which are found in oil, bitumen, rocks, and sediments, and show little or no change in structure from their parent organic molecules in living organisms. Other terms that have been used synonymously with biomarker are molecular fossil and chemical fossil.

Biomarkers are used to provide paleo-historical information about petroleum, sedimentary rocks, and recent sediments. In combination with other geochemical and geological parameters, they are used to correlate oil with source rocks; determine the type of organic matter input, depositional environment, and age of the source rock; and evaluate the extent of thermal maturation. They are among the more resistant molecules toward microbial degradation and are used to measure the extent of biodegradation of oil found either in a petroleum reservoir or on the Earth's surface. Environment of deposition and climatic record can also be assessed for recent sediments.

Chemical structures. With few exceptions, biomarkers are derived from lipids in organisms that are initially formed by biochemically coupled iso-prene units. They retain the structural features of isoprenoids. Examples of biomarkers include phy-tane (C_{20}), *bis*-phytane (C_{40}), bacteriohopane (C_{35}), cholestane (C_{27}), and phyllocladane (C_{20}) [**Fig. 1**]. Certain biomarkers retain heteroatoms as exempli-fied by deoxophylloerythroetioporphyrin (DPEP). Biomarkers can also contain aromatic rings such as triaromatic dinosteroid. Usually all chemical func-tionality is removed from the natural product lipids produced by organisms during diagenesis, in which the lipid molecules are transformed into stable hydro-carbons. Thus, the molecule chlorophyll *a* is trans-formed into the biomarkers DPEP and phytane, cholesterol into cholestane, bacteriohopanetetrol into bacteriohopane, biphytol diethers into *bis*-phytane, and phyllocladene into phyllocladane. In the early stages of diagenesis the free natural prod-ucts and intermediate structures that have been par-tially converted to the hydrocarbons are present. Thus, analysis of recent or ancient thermally unal-tered sediments reveals a host of biomarkers that re-tain chemical functionality. For example, cholesterol, an intermediate cholestanol, and cholestane may be found in the analysis of a sediment at the stage of early diagenesis.

Further structural alteration can occur in biomark-ers due to the effects of catalysis in the sediments and by virtue of thermal alteration during catagen-esis. An example is the conversion of cholestene to a rearranged cholestene (diacholestene) and eventu-ally diacholestane by the catalytic action of acidic clay minerals. Thermal alteration converts biomark-ers into stable stereoisomers. Thus, in petroleum, which is always liberated from the source rock by heating during catagenesis, numerous stereoisomers of cholestane can invariably be found.

Analysis. Computerized gas chromatography–mass spectrometry (GC-MS) is the principal method used to measure biomarkers. The GC-MS approach is successful because (1) the GC separates biomark-ers in elution time according to their volutility, and (2) the MS detects and provisionally identifies com-pounds through mass spectral fragmentation pat-terns characteristic of their structures. Recent ad-vances in biomarker technology rely strongly on the added selectivity of GC-MS-MS or a relatedmethod called metastable reaction monitoring(MRM-GC-MS), which allows detection of additional key biomarkers and more accurate quantification.

The basic interpretive display obtained in bio-marker analysis by GC-MS is a mass chromatogram (**Fig. 2**) which provides a "fingerprint" of structurally related biomarkers in a single graph. Commonly, a display of mass-to-charge (*m/z*) fragments 191 and 217 is used to provide a fingerprint of biomarkers structurally related to hopanes and steranes, respec-tively.

Applications. Biomarkers are used in a wide array of applications, such as (1) obtaining information from petroleum systems and basins, (2) evaluating relationships between oil and rocks to ascertain oil

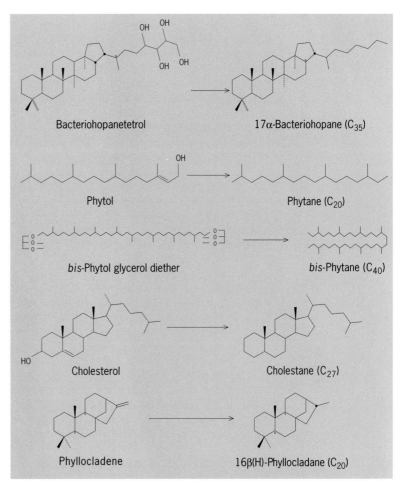

Fig. 1. Precursors as natural products in organisms and their diagenetic products as biomarkers in petroleum and rock extracts. Arrows may represent diagenetic processes of more than one step. Bacteriohopanetetrol, common in bacteria, undergoes a stereoisomerization to 17α-bacteriohopane, found in petroleum and thermally altered rocks. Phytol is attached as an ester in the side chain of chlorophyll characteristic of photosynthetic organisms. *bis*-Phytol glycerol diethers are found in Archea. Cholesterol is common in eukaryotes (plants and animals). Phyllocladene is found in terrestrial plants.

reserves, (3) determining geologic age by analyzing oil, (4) measuring thermal maturity of oil and bitu-men, (5) monitoring biodegradation of oil reserves, and (6) observing the bioremediation of environmen-tal processes.

Petroleum systems and basin analysis. The regional distri-bution and quality of petroleum in a basin is largely dependent on variations in source-rock distribution and quality. Certain biomarkers in oil and bitumen can provide this information. For example, the pres-ence of certain C_{30} steranes (24-*n*-propylcholestanes) proves that at least part of the oil was generated from a marine source rock because these C_{30} steranes ap-pear to be derived from only certain types of marine algae. Oleanane is a marker of input from higher plants (angiosperms) found only in petroleum that is Cretaceous or younger in age. Homohopane dis-tributions can be used to describe the level of oxic-ity in marine source-rock depositional environments. Some biomarkers provide very specific information. For example, botryococcane is a lacustrine marker produced from only one organism, *Botryococcus*

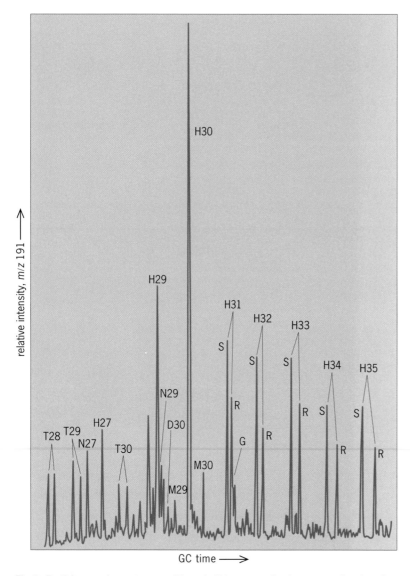

Fig. 2. Partial mass chromatogram of the *m/z* 191 response for the saturate fraction of an oil from the North Caspian Basin, Kazakhstan, shows peaks indicating biomarkers of microbial and algal origin. Numbers refer to carbon atoms in the chemical structures. T's are tricyclic terpanes (cheilanthanes) possibly related to tasmanites, a single-celled alga. G is gammacerane, attributed to bacterivorous ciliates. N's are neohopanes, hopanes with rearranged structures. D is diahopane, a rearranged hopane. M's are moretanes. H's are 17β-hopanes, diastereomers of hopane which is 17β, with various side chains accounting for the range in carbon numbers 27–35. Hopane homologs $\geq$ H31 show stereochemical antipodes at C-22, indicated by an *R* or *S*.

excellent source-dependent correlation parameters. When supported by additional analyses, such as stable carbon isotopes, ternary diagrams of C_{27}, C_{28}, and C_{29} sterane composition can be used to group related oils and source rocks. Similar ternary diagrams can be prepared for other compound classes, such as diasteranes or monoaromatic steroids, thus providing multiple lines of evidence to support correlation.

Distributions of bacterial triterpanes (hopanes, gammacerane) also can be used to characterize and correlate oils and source rocks. Due to their origination from bacterial sources, hopanoids are ubiquitous in oils and rocks, but the distributions of their numerous homologs, isomers, and stereoisomers are determined mostly by factors in the depositional environment, with additional changes due to thermal maturation. The oxicity and acidity of the basin waters during source-rock deposition determine to a large extent the hopanoid fingerprint that will be seen in the mature source rock and oil. The fingerprint will be overprinted by the effects of thermal maturation and possibly biodegradation. Thus, the hopanoid fingerprint is indicative of the organic facies of the source rock and can be used for correlation, provided the effects of thermal maturation and biodegradation are taken into account.

Geologic age indicators. Application of biomarkers to determine the geologic age of the source of an noncorrelated oil is a goal of ongoing fundamental research. Recent advances have shown that the biomarker oleanane correlates with the evolution and radiation of flowering plants, and its occurrence in oils can be used to indicate Cretaceous and younger or Tertiary sourced oils, depending on the relative amounts present. That is, oils with detectable amounts of oleanane have received input from sources that are Cretaceous or younger, while those oils with greater than 20% oleanane relative to hopane are likely Tertiary-sourced.

Progress has been made in developing parameters based on the occurrence of markers related to dinoflagellates, prymnesiophytes, diatoms, and sponges. These markers show potential in distinguishing various segments of geologic time, depending on the evolution and radiation of (1) dinoflagellates, Triassic and younger; (2) prymnesiophytes, Triassic and younger; (3) diatoms, Upper Jurassic and younger; and (4) sponges, Vendian and Cambrian. *See* GEOLOGICAL TIME SCALE.

Measurements of thermal maturity. Regional assessment of petroleum thermal maturity is useful in ranking the desirability of exploration plays. By measuring the progress of various maturation-dependent biomarker reactions, detailed information on the relative thermal maturity of oil and source-rock bitumen can be obtained. Isomerization is one example of the type of thermally dependent biomarker reactions that can be monitored by GC-MS.

Saturated carbon atoms are linked to their substituents by means of four, single covalent bonds which radiate outward toward the corners of a tetrahedron. If all four substituents differ, the carbon atom at the center is "asymmetric," and two mirror

braunii. This type of detailed information can be used to evaluate subtle changes in the character of organic matter or the oxidation potential in a single source rock, thus allowing the geologist to predict areas where oil of the best quality is likely to be found. Further, the character of the source rock can be evaluated from biomarkers in oil that may have migrated far from its source.

Oil-oil and oil–source-rock correlation. Proof of genetic relationships between widely spaced oil accumulations and source rocks allows the geologist to better understand migration routes and predict where additional reserves may be located. Steranes and several other compound classes show no major changes in C_{27}, C_{28}, and C_{29} distribution with thermal maturation in the oil generative window, and thus represent

image structures or stereoisomers of the compound can be formed by transposing any two of the substituents. These stereoisomers show the same molecular formulas but differ in the same manner as right- and left-handed gloves. Because of their complexity, biomarkers typically contain numerous asymmetric carbon atoms.

The two possible configurations of an asymmetric carbon atom in a side chain of a given biomarker are called R or S, depending on a chemical convention which describes the orientation of the substituents in space. If the asymmetric carbon atom is in a ring, the two configurations are called α and β. For example, a sterane can show an R or S configuration at C-20 and an α or β configuration at C-14.

The configurations at asymmetric carbon atoms imposed by enzymes in living organisms are not necessarily stable at the higher temperatures in buried sediments. Although the mechanism is unclear, configurational isomerization in biomarkers occurs at asymmetric carbons where one of the four substituents is a hydrogen. Isomerization can result in the conversion of a left- to a right-handed asymmetric center or vice versa, depending on the kinetics of the reaction. GC-MS measurements of the relative abundance of left- versus right-handed carbons in various biomarkers can be used for multiple assessments of the thermal maturity of any sample, thus improving confidence in interpretations. *See* CHIRAL DRUGS.

Biodegradation. A major factor in petroleum quality is biodegradation, affecting the world's heavy oil reserves. Understanding regional variations in the severity of biodegradation allows the geologist to rank prospects based on the predicted quality of the petroleum. Because of the differential resistance of biomarkers to biodegradation, comparisons of the relative amounts of biomarker classes can be used to rank oil as to the extent of biodegradation. Microbe-resistant biomarkers also allow correlation between biodegraded oil samples that would be difficult or impossible to relate by other methods. The established order of resistance to biodegradation for petroleum normal alkanes and biomarkers in the reservoir is n-alkanes < isoprenoids < steranes < hopanes < diasteranes < aromatic steroids. *See* OIL RESERVES.

Environmental processes. Bioremediation is the natural process by which petroleum and other materials deposited anthropogenically are removed from the environment or altered in such a way as to be rendered innocuous. Because the polycyclic biomarkers are recalcitrant compared to most other constituents of petroleum, biomarker analysis can be used to monitor bioremediation and recognize the final stages of microbial activity on oil residues. The biomarker, hopane, that occurs in Prudhoe Bay oil was used to monitor bioremediation of the major oil spill that occurred during the grounding of the *Exxon Valdez* supertanker in 1989. Hopane was found to be unaltered in the oil residues at any level of bioremediation, and increases in hopane concentration signaled the removal of other components in the oil. However, other work has shown that hopane is not the most recalcitrant biomarker.

Quaternary and Recent sediments. Biomarkers are used to characterize the environment of Recent sediments by characterizing the flora in and around lakes and ocean margins. Biomarkers can be observed to change in the sedimentary layers recording climatic variations, long-term trends, or cycles of various durations. Thus, the finding of abundant biomarkers for land plants, such as oleanenes and certain diterpenes, in a lake sediment would signal a lush nearby vegetation during the time of sediment deposition. In contrast, absence of land plant biomarkers and the abundance of compounds associated with halophyllic bacteria (such as β-carotane) or bacterivorus ciliates (such as gammacerane) would signal an arid climate with evaporitic conditions and hypersalinity of the lake waters. *See* TURBIDITE.

Direct evidence for paleoclimate has recently been developed by the use of alkenones which comprise a series of C_{37}–C_{39} di-, tri-, and tetraunsaturated methyl and ethyl ketones. These compounds are abundant biomarkers derived from the widely occurring abundant phototrophic marine haptophyte *Emiliania huxleyi*. Extensive studies in laboratory and field trials have resulted in a series of equations by which a value can be derived for near-surface water temperature during the time of sediment deposition, where such alkenones (unaltered) can be analyzed. The method seems to work best over a moderate temperature range of 4-25°C (39-77°F).

For background information *see* GAS CHROMATOGRAPHY; GEOLOGICAL TIME SCALE; LIPID; MARINE SEDIMENTS; MASS SPECTROMETRY; OIL ANALYSIS; ORGANIC GEOCHEMISTRY; PETROLEUM; PETROLEUM GEOLOGY; SEDIMENTARY ROCKS; SEDIMENTOLOGY in the McGraw-Hill Encyclopedia of Science & Technology.

J. Michael Moldowan

Bibliography. M. H. Engel and S. A. Macko, *Organic Geochemistry: Principles and Applications*, 1993; K. E. Peters and J. M. Moldowan, *The Biomarker Guide: Interpreting Molecular Fossils in Petroleum and Recent Sediments*, 1993.

Biosensor

Biosensors combine the recognition power of biological systems with advanced transducers and microelectronics to yield analytical instruments with unique measurement capabilities. The new generation of electrochemical blood-glucose monitors are widely used by diabetic people in their homes. While the measurement of glucose currently dominates commercial activity in the biosensor field, simple inexpensive instruments can be constructed to measure thousands of other analytes. These instruments have applications in medicine, the pharmaceutical industry, the environment, the food and process industries, as well as in security and defense.

Biological recognition. Biosensors comprise two principal components (see **illus.**): a biological or biologically derived (isolated from or produced by a living system, or synthesized as an analog of a biological molecule) recognition element capable

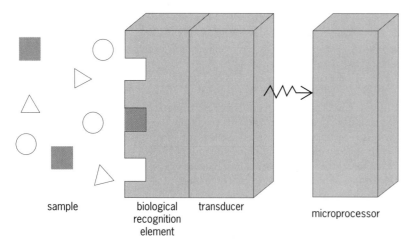

sample   biological   transducer
                recognition
                element   microprocessor

Diagram of a biosensor.

of interacting specifically with the target analyte; and a transducer capable of converting the molecular recognition event into a signal which can be processed. This output is usually, therefore, a digital electronic signal which is proportional to the concentration of the analyte.

Biological systems offer unrivaled sensitivity and specificity to particular chemicals or groups of chemicals. Enzymes and antibodies are well-established analytical reagents providing the basis for many biochemical analyses. The principle of bioassay has been widely exploited, especially in the discovery of pharmaceuticals and in the evaluation of toxicity. Enzymes can have highly specific catalytic properties capable of distinguishing between stereoisomers of the same compound or can offer a broad-based recognition element either acting on or being inhibited by a general class of compound. Antibodies can be formed against an almost limitless range of targets to yield binding partners with a high affinity for a specific analyte. Microorganisms are complete metabolic pathways capable of performing complex transformations or reflecting the perturbations that occur in the presence of noxious chemicals. All these biological elements have found widespread application in biosensors.

Transducers. Four principal classes of physicochemical transducers—electrochemical, optical, thermometric, and piezoelectric—have traditionally been used in biosensors to convert the biological recognition of an analyte into a signal that can be processed. A few groups of workers have reported use of a fifth type of transducer to detect magnetism. The most well-characterized transducers are electrochemical. The electrochemical transducer, combined with enzymes such as glucose oxidase, gave rise to the first biosensor, the enzyme electrode, described by Leyland C. Clark in 1962. Early work used the Clark oxygen electrode or a simple pH probe combined with oxidases to measure reactions, catalyzed by enzymes immobilized on the electrode tip, by virtue of the oxygen consumed or the acid produced; the former configuration constituted the first amperometric enzyme electrode and the latter,

the first potentiometric enzyme electrode or biosensor. This work expanded to measure another product of the glucose oxidase reaction, hydrogen peroxide, at a simple platinum electrode and formed the basis for the first commercial biosensor in 1975. The next landmark in the evolution of the amperometric biosensor was the introduction of immobilized mediators such as ferrocene and its derivatives to replace oxygen as the electron acceptor. Today, electrochemical biosensors dominate the biosensor market, with sensors for glucose accounting for more than 90% of sales. Now, all electrochemical biosensors used by diabetics in their homes utilize mediators, while instruments designed for continuous operation, for example, in hospital laboratories, exploit hydrogen peroxide electrochemistry. Very significant success has also been achieved by applying whole-cell amperometric biosensors to environmental monitoring as an alternative to traditional methods for measuring biochemical oxygen demand (BOD). Research has continued into electrochemical biosensors, with hundreds of different analytes being detected using amperometric, potentiometric, conductimetric, or impedimetric transducers coupled with enzymes, microorganisms, tissue slices, organelles, antibodies, antibody fragments, or deoxyribonucleic acid (DNA).

Optical transducers have been used as an alternative to electrodes, giving rise to enzyme optrodes. Chemical sensors, for example, for pH, oxygen, and carbon dioxide, have been created by immobilizing various indicators at the tip of an optical fiber. The addition of an appropriate enzyme results in an optical biosensor. The most important breakthrough in the development of optical biosensors, however, has been the combination of optical transducers with affinity elements such as antibodies and single-stranded DNA. In the early 1990s, commercial instruments capable of real-time label-free analysis of antibody-antigen interactions became available. The first and most successful device was based on the phenomenon of surface plasmon resonance. When light undergoes total internal reflection within a prism, a certain amount of energy propagates a small distance beyond the optical interface into the surrounding medium, forming an evanescent field. If a thin layer of metal is coated onto the surface, this energy can be absorbed by the free electrons at the metal surface and give rise to surface plasmon resonance. The angle of incidence at which this energy absorption occurs is critically dependent on the refractive index of the external medium. If a binding partner, such as an antibody, is immobilized on the surface of the prism, the angle at which surface plasmon resonance occurs can be used to monitor rate and extent of the biological interaction in real time. Surface plasmon resonance biosensors and related devices have found widespread applications in research, especially in the pharmaceutical industry. A second class of evanescent-field immunosensor involves the use of fluorescent labels, which may be displaced or bound at an optical surface by virtue of their conjugation with either a second antibody or an analyte analog. Fluorescent immunosensors show promise as robust devices capable of operating on crude samples, which can generate nonspecific binding effects with direct, nonlabeled systems such as surface plasmon resonance.

The remaining classes of transducer have been reported for use in biosensors but have not, to date, been widely applied for routine analysis. Piezoelectric transducers exploit the oscillations of a quartz crystal or the propagation of a wave through a quartz substrate to measure mass and viscoelastic changes resulting from biological interactions. Research on thermometric devices has concentrated on the measurement of temperature changes in enzyme reactors, but recent research on microfabricated devices has brought them closer to mainstream biosensor technology, where the sensing element is either integrated within or intimately associated with the transducer or transducing microsystem.

Applications. Biosensors have been applied to a wide variety of analytical problems in medicine, the environment, food process industries, security, and defense. The output from a biosensor is, in principle, continuous, but devices may be configured to yield single measurements to meet specific market requirements. Continuous monitoring using biosensors has been a long-term objective for clinical application, and limited success has been achieved in the case of glucose measurement. The most significant achievement so far, however, has been the production of one-shot biosensors capable of making a single determination either of a single analyte or of a menu of analytes such as glucose, lactate, urea, and creatinine. Hundreds of millions of such hygienic disposable enzyme electrodes are used each year. So-called stat machines have also been successful. These are used in intensive-care wards to give rapid information about the state of health of critically ill patients. Originally, these machines measured blood electrolytes and blood gases, but the menu has been expanded to include biosensors for the metabolites listed above. A wide variety of prototype devices have been created for potential applications in medicine, including the DNA chip for the detection of genetic disorders, immunosensors to identify infectious diseases or disease markers, and ion-channel sensors which mimic the action of cell receptors.

The practical application of biosensors in environmental monitoring has proceeded at a slower pace. Arguably, the greatest appeal of biosensors is their ability to reflect effects of a group of pollutants on living systems, although there are also examples where highly specific and sensitive measurements can be made of a single chemical using simple, inexpensive equipment. BOD sensors incorporating intact microorganisms have been available for years and offer a swift (20 min) alternative to the standard 5-day BOD test. They offer plant managers immediate and relevant information, although the standard test still has to be performed to meet legislative requirements in most parts of the world. Pesticide monitors based on either microorganisms or antibodies have been built. Increasingly, genetic engineering is being applied to generate diversity in environmental biosensors to produce devices capable of measuring, for example, heavy metals and a wider range of organic pollutants. Enzyme electrodes for the measurement of phenols are at an advanced stage of development and have been applied in aqueous, organic, and gaseous environments.

Laboratory-based biosensors have been used in food analysis to a limited extent for many years. Analytes such as sugars, amino acids, and alcohols can be conveniently measured directly in complex matrices. A great diversity of possible measurements appear in the literature, but a combination of low profit margins, variability of the sample, and conservatism of this industry have conspired to restrict the use of biosensors for food analysis. Similarly, instruments are available for on-line measurements of processes and fermentations. Commercially available instruments aseptically extract samples from the process line and introduce them into an automated biosensor-based analyzer. In situ probes have been applied to optimize commercial processes but are not used in routine operation.

The use of biosensors in national defense has a long history and shares some common technical advantages with environmental applications. The ability to detect poisons has an obvious relevance to the military, and there has been an intense interest in the detection of biological warfare agents. Chemical weapon detection is attractive because it relates directly to the action of nerve agents on the human body by monitoring, for example, the enzymatic activity of acetylcholinesterase. Enzyme electrodes and enzyme reactors coupled with electrochemical detectors have been available for years, but their requirement for consumables and their somewhat restricted analyte range are leading to their replacement by more physical techniques. In contrast, the challenge of detecting biological species in the field is leading to the design and trial of new biosensors.

Challenges. The success of biosensors has resulted from the ability to combine biological recognition with modern microelectronics or optoelectronics to yield inexpensive analytical devices that are easy to use and can be mass-produced. Continued advance in miniaturization and mass-production technology is essential to underpin the trend toward larger menus of analytes packaged in more convenient multichannel formats. The greatest strength of biosensor technology, the exquisite sensitivity and specificity of biological molecules, is also its greatest weakness. The complexity and size of proteins makes them susceptible to degradation, especially when isolated from their natural environment. Considerable effort is being expended to stabilize proteins incorporated into bioelectronic devices such as biosensors, and the existence of commercial devices attests to the success of these attempts. However, many molecules remain insufficiently stable or otherwise suboptimal for practical application. One promising avenue of research is the use of semisynthetic or synthetic analogs of biological molecules. Using advanced computational techniques in conjunction with combinatorial chemistry and molecular biology, new receptor elements are being created for biosensors. The hope is that stable, small molecules can be created that retain the molecular recognition characteristics of their larger counterparts but that are much more compatible with the fabrication of high-density microelectronic sensor arrays.

For background information *see* BIOELECTRONICS; TRANSDUCER in the McGraw-Hill Encyclopedia of Science & Technology.   Anthony P. F. Turner

Bibliography. U. Bilitewski and A. P. F. Turner, *Biosensors for Environmental Monitoring*, Harwood, Amsterdam, 1999; F. R. Hall and J. J. Menn (eds.), *Biopesticides: Use and Delivery*, Humana Press, 1999; E. Kress-Rogers, *Handbook of Biosensors and Electronic Noses: Medicine, Food and the Environment*, CRC Press, 1996; A. P. F. Turner, *Advances in Biosensors*, vols. 1–4 (including suppl. 1), JAI Press, London, 1991–1999; A. P. F. Turner, I. Karube, and G. S. Wilson, *Biosensors: Fundamentals and Applications*, Oxford University Press, 1989.

Biotechnology (yeast)

Yeasts are microorganisms which have had a close relationship with humankind for millennia. Some yeasts are beneficial to human life in the provision of foods, beverages, and medicines, while others play detrimental roles as spoilage organisms and agents of human disease. In general, however, the relationship between yeasts and human beings is harmonious, and yeasts have frequently been referred to as the most domesticated group of microbes. Yeasts are ubiquitous in the natural world, being found in many aquatic and terrestrial environments as well as in association with plants, animals, and insects. The preferred niches for yeasts in nature are on the surface of fruits and tree exudates and in dead and decaying vegetation, where they thrive on sugar-rich material. Around 100 yeast genera and 800 species have been isolated and characterized, but this may represent only a fraction of yeast biodiversity.

Yeasts are single-celled members of the fungal kingdom, and the majority of yeast species encountered in biotechnology are ascomycetous fungi which reproduce vegetatively by budding and sexually by forming spores within an ascus. In budding, daughter cells, or buds, emanate from surface outgrowths of mother cells and eventually enlarge to split off at cell division. An example is found in *Saccharomyces cerevisiae*, or baker's yeast, which is one of the most exploited and best-studied microorganisms. Such yeasts readily convert sugars into alcohol and carbon dioxide in a metabolic process known as fermentation.

Human welfare. Yeasts are of major economic, social, and health significance. They have been exploited, albeit unwittingly, since ancient times for the provision of food (leavened bread) and alcoholic beverages (beer, mead, wine, cider, sake, and distilled spirits). Beer brewing, thought to have originated in Egypt around 6000 B.C., may represent the first biotechnology. Yeast biomass itself is highly nutritious and a rich source of vitamins, proteins, essential amino acids, and minerals for the human diet. In nutritional terms, some foods directly benefit from the presence of yeasts and yeast extracts. Yeasts were used in early medicine to combat bacterial infections, and in more recent times they have been administered as biotherapeutic agents, or probiotics, to stabilize intestinal microflora in humans and stimulate rumen metabolism in livestock. *Saccharomyces boulardii* has been patented for use as a human probiotic agent. *Saccharomyces cerevisiae* has been advocated even in the treatment of acne and premenstrual stress. In relation to modern health-care biotechnology, genetically manipulated yeasts have been recruited by commercial organizations for research and development of novel pharmaceuticals for the prevention and treatment of human disease. The wide utility of yeasts in both old and new biotechnologies makes them the world's premier industrial microorganisms.

However, yeasts also exhibit detrimental aspects (**Table 1**). Some yeasts spoil certain foods and deleteriously affect the nutritive value. Common spoilage yeasts (such as *Zygosaccharomyces* spp.) are osmotolerant and can grow well and ferment in high-sugar-containing foods such as concentrated fruit juices. Some species are quite resistant to food preservatives

TABLE 1. Some detrimental aspects of yeasts in human life		
Area	Species	Examples
Food spoilage	*Cryptococcus laurentii*	Frozen poultry
	Zygosaccahromyces spp.	Fruit juices, wine
	Kluyveromyces, Rhodotorula, and *Candida* spp.	Milk, cheese, yogurt
	Torulopsis, Pichia, Candida, Hansenula, and "wild" *Saccharomyces* spp.	Beer, cider, and so on
Yeast infections	*Candida albicans*	Candidiasis
	Cryptococcus neoformans	Cryptococcosis
	Blastomyces dermatidis	Blastomycosis
	Histoplasma capsulatum	Histoplasmosis
	Sporotrichum schenckii	Sporotrichosis

such as weak acids (including benzoate and sorbate).

Some yeasts impact more directly on human well-being. For example, excessive consumption of yeast-derived ethanol leads to liver damage and socially disruptive behavior. However, moderate alcohol intake may actually benefit human health, for example, helping to prevent cardiovascular disease. Some individuals, particularly the immunocompromised, may exhibit allergies to dietary yeasts, resulting in intestinal and other disorders. A few yeasts are opportunistically pathogenic toward humans (Table 1). Yeast infections (mycoses) caused by *Candida albicans* are the most common in humans and are collectively referred to as candidiasis (candidosis in the United Kingdom). *Candida albicans* is a dimorphic organism and can exist as a budding yeast or can develop, via germ tubes, into hyphal (filamentous) forms. The hyphal forms are often associated with the invasive virulent properties of *C. albicans*. Three types of candidiasis are recognized: superficial (vaginal thrush), locally invasive (intestinal ulcerations), and deep-seated (disseminated infective lesions of many organs, including heart, lungs, kidneys, and brain). These deep-seated infections, which can be fatal, are common in immunocompromised individuals such as AIDS patients. Yeast infections are frequently nosocomial (hospital-derived) and can arise following surgery. Treatment of yeast infections is often difficult since few antifungal drugs are able to effectively discriminate between the eukaryotic pathogen (yeast cells) and the eukaryotic host (human cells).

Industrial exploitation. In traditional food and beverage fermentation processes, the common factor is the ability of yeast cells, principally *S. cerevisiae*, to ferment organic substrates (sugars) under anaerobic conditions to ethanol and carbon dioxide. In the absence of oxygen, fermentative yeasts need to replenish the coenzyme nicotinamide adenine dinuclueotide (NAD), which is needed for continued sugar breakdown during glycolysis, and to maintain the redox balance in the cells. They do so by reoxidation of the reduced coenzyme, NADH, in terminal step reactions derived from pyruvate, as shown below. The final reductive step is catalyzed

$$\text{Glucose} \xrightarrow{\text{glycolysis}} 2 \text{ Pyruvate} \longrightarrow$$
$$2CO_2$$

$$2 \text{ Acetaldehyde} \longrightarrow 2 \text{ Ethanol}$$
$$2NADH + 2H^+ \qquad 2NAD^+$$

by alcohol dehydrogenase, which is the enzyme ultimately responsible for the production of all alcoholic beverages. Beer, whiskey, and sake manufacture require fermentable sugars to be released from plant starch sources, such as barley and rice, following the action of amylase enzymes which degrade starch to maltose, glucose, and oligosaccharides. In the brewing process, small, branched starch molecules called dextrins often remain in the final product because they are unfermentable by the yeasts employed (ale and lager strains of *S. cerevisiae*). Low-carbohydrate beers (so-called lite beers) can, however, be produced by the addition of dextrinase enzymes or possibly by using yeasts which have been genetically manipulated to utilize all residual sugars present in the fermentation medium (the wort).

Winemaking is a much simpler process than the cereal-based fermented beverage technology. Here, grape sugars (glucose and fructose) are released by physically crushing the fruit before fermentation proceeds. Yeasts responsible for wine production may be either those which are part of the natural microflora of the grape (for example, *Kloeckera apiculata* and *Candida stellata*) and the winery environment (*S. cerevisiae*), as in traditional winemaking, or specially selected starter cultures of *S. cerevisiae*, as in modern, large-scale winemaking. In each case, the fermenting yeasts (predominantly *S. cerevisiae*) release major metabolites (ethanol and carbon dioxide) and minor metablites (including glycerol, succinic acid, acetic acid, esters, aldehydes, ketones, and higher alcohols) into the wine. Therefore, yeast metabolism of grape sugars plays an important role in determining the final organoleptic properties (taste, aroma, body) of wine.

The yeast fermentation industry is enormous, with an estimated annual production of over 30 billion liters providing millions of dollars profit. About two-thirds of this alcohol is for nonpotable industrial use, principally as a renewable energy source, or biofuel. This bioethanol can be produced from the fermentation of cheap and readily available carbon substrates such as sugarcane molasses, hydrolyzed cereal starch, or lignocellulosic hydrolysates. Bioethanol can be used in internal combustion engines either in pure, anhydrous form or in mixtures with gasoline (petroleum), as in gasohol. Compared with hydrocarbon-based liquid fuels, bioethanol is cleanburning, liberating less noxious gases into the environment.

The fermentation process also liberates carbon dioxide, which is the yeast metabolite essential for leavening of cereal doughs in the breadmaking process. Yeast is invaluable in texturizing bread and in improving its nutritive value, for example, by enhancing the vitamin content. The industrial production of baker's yeast and yeast biomass for yeast extracts for use as savory food additives relies on respiratory, rather than fermentative, metabolism. Thus, baker's yeast is propagated under fully aerobic conditions, and the rate of sugar (molasses) feeding into the bioreactors is carefully controlled to prevent the cells from converting sugar to ethanol. In this way, respiration predominates, energy generation in the form of adenosine triphosphate (ATP) is enhanced, and production of yeast biomass is maximized. Controlled sugar feeding is called fed-batch cultivation, and it is also used in the production of yeasts as a source of single-cell protein (SCP).

In recent years, genetically modified yeasts and their products have gained approval for human food use. *Saccharomyces cerevisiae* and *Kluyveromyces lactis*, which have GRAS (generally regarded as safe) status, have been exploited in this area. For example, in 1991 and 1994 the United Kingdom approved genetically modified baker's and brewer's strains of *S. cerevisae*. The modified baking strain contains promoter genes from another *S. cerevisiae* strain upstream of the maltose-utilizing genes, resulting in yeast with improved dough fermentation attributes. The brewing strain contains genes (from another yeast) which enable it to more completely ferment the available sugars in malt wort. Food enzymes from modified strains of *K. lactis* include recombinant chymosin, which is the protease used to coagulate milk casein in the manufacture of cheese.

Ethanol and carbon dioxide produced by *S. cerevisiae* represent two of the most lucrative biotechnological commodities worldwide. Moreover, other yeast products and processes are finding increasing applications in industry, and several of these center on the exploitation of non-*Saccharomyces* (or nonconventional) yeasts. For example, *Pichia pastoris* and *Hansenula polymorpha* can grow to very high cell densities on methanol as the sole carbon and energy source (they are methylotrophic), and they have great potential as cell factories for the production of pharmaceutical proteins. **Table 2** summarizes the applications and future potential of yeasts in industrial biotechnology.

Emerging yeast technologies are especially evident in the environmental and health-care sectors of biotechnology. Many of the novel uses of yeasts employ recombinant deoxyribonucleic acid (DNA) technology (genetic engineering), because yeasts possess several distinct advantages over bacteria for the expression of foreign genes and the secretion of their encoded proteins. Being eukaryotic, yeast cells carry out gene transcription and translation in a manner similar to mammalian cells. Therefore yeasts have become the preferred hosts for the production of many human therapeutic proteins for clinical use. The first of these was a vaccine against hepatitis B in 1986, and since then many more yeast-derived products have been developed. For example, insulin (for the treatment of diabetes), leukine GM-CSL (for autologous bone marrow transplants), and tumor necrosis factor, all produced by recombinant yeasts, have gained clearance from regulatory bodies (such as the Food and Drug Administration) for clinical use. Many more therapeutic products from yeast are in clinical trials or are at the research and development stage.

Advances. Yeasts are easily cultivated unicellular eukaryotes which are amenable to biochemical, cytological, genetic, and molecular biological analyses. Therefore, they are important experimental model organisms in biological and biomedical research. For over a hundred years, yeast cells have been studied to enhance the fundamental understanding of genetics, biochemistry, and cell biology. Indeed, the pioneering work of the Buchner brothers in the late 1800s on sugar metabolism in cell extracts of yeast led to the foundation of biochemistry as a new scientific

TABLE 2. Applications of yeasts in biotechnology		
Class of product	Type of product	Examples
Biomass-directed	Whole yeast cells	Baker's yeast, starter cultures for beer, wine, soy sauce, miso
		Animal feed, single-cell protein, probiotics, biological control agents, heavy-metal biosorption
		Reductive biocatalysis
	Yeast cell extracts	Nutritional supplements and savory food additives
		Microbiological culture media
	Yeast cell walls	Nonnutritive food bulking agents
		Immuno-stimulants
		Cell wall polysaccharides (phosphomannans)
	Yeast enzymes	Food processing (invertase, lactase, pectinase, lipase)
		Clinical diagnoses (phenylalanine ammonia lyase)
	Yeast pigments	Astaxanthin from *Phaffia rhodozyma* as a feed colorant for farmed salmonids
	Yeast nucleic acids, coenzymes, and so on	Ribonucleic acid (RNA) and its derivatives, FAD, NAD, ATP for pharmaceutical use
	Heterologous proteins	Hormones (insulin), viral vaccines (hepatitis B), antibodies (IgE receptor), growth factors (tumor necrosis factor), interferons (leukocyte IF-alpha), blood proteins (human serum albumin), enzymes (gastric lipase, chymosin)
Metabolite-directed	Ethanol	Potable alcoholic beverages (beer, wine, cider, sake, mead, whiskey, cognac, rum, vodka, gin)
		Bioethanol (fuel, industrial solvent)
	Carbon dioxide	Dough leavening
		Soft drink carbonation
	Glycerol	Nitroglycerine explosives (historical)
	Higher alcohols	Isopropanol for perfumes, cosmetics
	Organic acids	Citric, itaconic, malic, and fumaric acids (food acidulants, chemical and pharmaceutical uses)
	Fatty acids	Stearic acid and long-chain dicarboxylic acids
	Amino acids	Lysine, tryptophan, phenylalanine, glutamic acid, and methionine for pharmaceutical use

TABLE 3. Contributions of yeasts to biomedical research

Research field	Examples
Oncology	Cell cycle control: molecular mechanisms which regulate checkpoints at onset of DNA synthesis and mitosis are elucidated in yeasts. Oncogenes: regulation of human oncogenes (for example, *Ras*) is effectively studied in yeast cells. Tumor suppressor genes: *p53* mutations can be studied in a yeast functional assay. Telomeres: the function of telomere-binding proteins can be studied in yeast cells. Apoptosis: *S. cerevisiae* can be used as a model system for studying programmed cell death.
Pharmacology	Multidrug resistance, mode of drug action, drug metabolism, drug-drug interactions, pharmacokinetics and drug-screening assays can all be studied and performed using yeast cells.
Toxicology	Mutagens, genotoxic agents, mycotoxins, xenoestrogens, and toxic chemical pollutants can all be assayed using yeast cells and novel yeast biosensors.
Virology	Yeast can contribute to human virology research by providing a virus model (killer plasmids) and a model for studying viral protein action and interaction. Genetically modified yeasts are also used to produce viral surface antigens for antiviral vaccines.
Genetics	Yeast artificial chromosomes (YACs) are invaluable in cloning human DNA and in mapping the human genome.

discipline. The budding yeast, *Saccharomyces cerevisiae*, and the fission yeast, *Schizosaccharomyces pombe*, are now the best-understood unicellular eukaryotes, and both organisms have made major contributions to molecular biology in recent years (**Table 3**). For example, investigations into cell cycle control in *S. cerevisiae* and *Sch. pombe* have revealed regulatory mechanisms over the onset of mitosis and DNA synthesis which are significant for understanding the proliferation of cancer cells. Molecular genetic studies of yeast have also provided insight into aspects of human hereditary diseases. This is because human genes, such as those implicated in cystic fibrosis, can be functionally expressed and studied in yeast cells. Such research has been greatly enhanced following the complete sequencing in 1996 of the *S. cerevisiae* genome. This international collaborative effort represented a milestone in biotechnology because it was the first time the entire genetic blueprint of a eukaryotic organism had been unraveled. The goal for scientists now is to provide a complete understanding of how a simple eukaryotic organism works by assigning function to all yeast genes, 40% of which (around 2000 genes) have, as yet, unknown roles within the cell. Eventually, knowledge of the structure and function of yeast genomes, and their relationship with the human genome, will likely provide novel practical approaches for the diagnosis, prevention, and treatment of human diseases.

For background information *see* FERMENTATION; FOOD MANUFACTURING; GENETIC ENGINEERING; MALT BEVERAGE; YEAST; YEAST, INDUSTRIAL in the McGraw-Hill Encyclopedia of Science & Technology.

Graeme M. Walker

Bibliography. C. P. Kurtzman and J. W. Fell (eds.), *The Yeasts: A Taxonomic Study*, Elsevier Science, 1998; J. F. T. Spencer and D. M. Spencer (eds.), *Yeasts in Natural and Artificial Habitats,* Springer-Verlag, 1997; G. M. Walker, *Yeast Physiology and Biotechnology*, John Wiley, 1998; K. Wolf (ed.), *Nonconventional Yeasts in Biotechnology: A Handbook*, Springer-Verlag, 1996.

Blue lasers

Light-emitting diodes (LEDs) are ideal solid-state light sources. At present, incandescent and fluorescent lamps are used as light sources for many applications. However, these conventional glass-vacuum-type light sources have poor reliability and durability and low luminous efficiency. In the past, electronic circuits used glass vacuum tubes in spite of their poor reliability and durability. Now, all electronic circuits use highly reliable solid-state semiconductor components. Thus, only the light sources are still based on traditional technology rather than solid-state semiconductors. Recently, the development of indium gallium nitride (InGaN)–based compound semiconductors has opened the way to all-solid-state semiconductor light sources. The brightness and durability of solid-state light-emitting diodes make them ideal for use in displays and light sources, while semiconductor laser diodes have been used in everything from optical communications systems to compact-disk players. These applications have been limited by the lack of materials that can efficiently emit blue light.

Full-color displays, for example, require at least three colors, usually red, green, and blue, to produce the whole range of visible colors. Such a combination is also needed to make a white-light-emitting device that would be more durable with less power consumption than conventional incandescent bulbs or fluorescent lamps. For laser diodes, the shorter wavelength means that the light can be focused more sharply, so the storage capacity of optical disks can be increased. Digital versatile disks (DVDs), which entered production in 1996, rely on red aluminum indium gallium phosphide (AlInGaP) semiconductor lasers. They have a data capacity of about 4.7 gigabytes compared to 0.65 gigabyte for compact disks (CDs), which use infrared aluminum gallium arsenide (AlGaAs) laser diodes. By moving to bluish-purple wavelengths through the use of InGaN-based compound semiconductors, the capacity can be

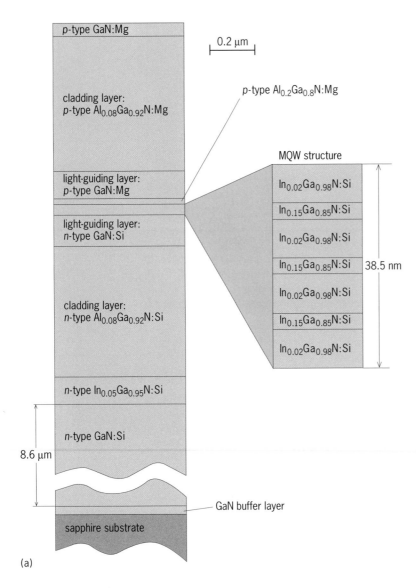

0.2 μm

p-type GaN:Mg

cladding layer:
p-type Al$_{0.08}$Ga$_{0.92}$N:Mg

light-guiding layer:
p-type GaN:Mg

light-guiding layer:
n-type GaN:Si

cladding layer:
n-type Al$_{0.08}$Ga$_{0.92}$N:Si

n-type In$_{0.05}$Ga$_{0.95}$N:Si

n-type GaN:Si

8.6 μm

GaN buffer layer

sapphire substrate

(a)

p-type Al$_{0.2}$Ga$_{0.8}$N:Mg

MQW structure

| In$_{0.02}$Ga$_{0.98}$N:Si |
| In$_{0.15}$Ga$_{0.85}$N:Si |
| In$_{0.02}$Ga$_{0.98}$N:Si |
| In$_{0.15}$Ga$_{0.85}$N:Si |
| In$_{0.02}$Ga$_{0.98}$N:Si |
| In$_{0.15}$Ga$_{0.85}$N:Si |
| In$_{0.02}$Ga$_{0.98}$N:Si |

38.5 nm

n-type GaN:Si

sapphire substrate

2 μm

(b)

Fig. 1. InGaN-MQW/GaN/AlGaN separate-confinement-heterostructure laser diode grown directly on a sapphire substrate. (a) Diagram of layer structure. (b) Cross-sectional transmission electron micrograph, showing threading dislocations.

increased to more than 15 gigabytes. Thus, these semiconductors are indispensable for next-generation digital versatile disks. The bluish-purple InGaN-based laser diodes may also improve the performance of laser printers and undersea optical communications.

Principle of LEDs and laser diodes. A light-emitting diode essentially consists of an active layer of semiconducting material sandwiched between n-type and p-type semiconductor cladding layers. When a voltage is applied to the junction, electrons from the n-type material move into the conduction band of the active layer, while holes from the p-type semiconductor are injected into the valence band. Light emission takes place in the active layer when electrons at the bottom of the conduction band spontaneously recombine with holes in the top of the valence band.

If many more electrons and holes are injected into the active layer, the energy difference between the quasi-Fermi level of electrons and the energy level related to the recombination with holes becomes larger, and there is an increased number of injected electrons. This phenomenon is called population inversion. For laser diodes, if the thickness of the active layer is too small to confine the light within it, the active layer is sandwiched between additional layers which are called light-guiding layers (**Fig. 1a**). The light generated by the spontaneous recombination between electrons and holes in the active layer is confined within the light-guiding layers due to the difference of the refractive index between the light-guiding layers and adjacent layers of cladding. This structure is called a separate-confinement heterostructure because the light is generated in the active layer and guided in the separate light-guiding layers. The light is guided parallel to the light-guiding layers, which form a structure called the laser cavity. The ends of the light-guiding layer are mirrors, which reflect the light back and forth. The reflected light causes other electrons to recombine with holes, stimulating the release of further light and amplifying the light whenever a condition of population inversion exists. At a certain current (the threshold current), strong stimulated emission is observed at the edges of the light-guiding layers of the laser diode.

InGaN-based emitting devices. Gallium nitride (GaN) and other III–V nitrides have a direct band gap that is suitable for light-emitting devices. The band gap energy of aluminum gallium indium nitride (AlGaInN) varies between 6.2 and 2.0 eV, depending on its composition at room temperature. However, there is no lattice-matched substrate for the growth of gallium nitride. Sapphire is commonly used despite a large lattice mismatch of 13.5%. Such large differences in the lattice constant create misfit dislocations at the interface between the gallium nitride and the substrate. In spite of the large number of dislocations, major developments in wideband-gap III–V nitride compound semiconductors have recently led to the commercial production of high-efficiency ultraviolet/blue/green/amber light-emitting diodes and to the demonstration of room-temperature bluish-purple laser light emission from InGaN/GaN/AlGaN-based heterostructures under pulsed and continuous-wave operation. All of these light-emitting devices have an active layer made of InGaN rather than gallium nitride because it is difficult to fabricate a highly efficient light-emitting device using a gallium nitride active layer, and the addition of indium to the layer is vital for achieving strong light emission.

The performance of many conventional optoelectronic devices has been limited by the control of both point defects and structural defects in these materials. However, it appears that InGaN-based emitting devices are less sensitive to dislocations than conventional III–V compound semiconductor-based devices. Indeed, the number of dislocations is as high as 10^{10}–10^{12} in these emitting devices, and yet these dislocations seem not to reduce the efficiency of light emission and have not prevented the development of practical devices. The localization of carriers at a certain potential minimum induced by the fluctuation of indium composition in the InGaN active layer seems to be a key factor in the high efficiency of the InGaN-based light-emitting diodes and laser diodes.

Laser diodes. The first laser emission from diodes based on III–V nitrides was reported in 1995. These devices were based on multiple quantum wells (MQWs) of InGaN and produced bluish-purple light efficiently at room temperature. Although early versions required high voltages and currents and could only generate laser pulses, continuous-wave lasers have now been developed. By 1997, the lifetime of continuous-wave blue lasers had improved to 300 hours, and by December 1998 research had further extended the lifetime to 10,000 h.

The basic laser diode was an InGaN-MQW/GaN/AlGaN separate-confinement heterostructure grown directly on the sapphire substrate (Fig. 1a). The lifetime of these laser diodes was only 30–300 h because of their high threshold current density, which was due to the large number of threading dislocations present on the device (Fig. 1b). Such dislocations form at the interface between the first layer of gallium nitride to be deposited and the sapphire substrate and can thread their way into the InGaN active layer. As discussed above, they do not seem to degrade the device properties severely. However, in order to attain a lifetime of more than 10,000 h, the dislocation density had to be decreased. An epitaxially laterally overgrown gallium nitride (ELOG) substrate was used to achieve this reduction.

Also, thick cladding layers with high aluminum content and freedom from cracks are required for optical confinement. It was difficult to grow thick AlGaN cladding layers for this purpose due to the formation of cracks during growth in the layers. These cracks are caused by the stress introduced in the AlGaN cladding layers by lattice mismatch and by the difference in thermal expansion coefficients between the AlGaN cladding layer and gallium nitride

layers. The mechanism of cracking is the relaxation of the strain energy due to the lattice mismatch between AlGaN and gallium nitride during growth. However, in the case of a thin AlGaN layer within a critical thickness, the elastic strain is not relieved by the formation of cracks and dislocations, and thus the crystal quality of the AlGaN cladding layer improves. Therefore, GaN/AlGaN modulation-doped strained-layer superlattices (MD-SLSs) within the range of critical thickness were used to grow the required cladding layers.

To fabricate the ELOG substrate, selective growth of gallium nitride was performed on a 2-μm-thick gallium nitride layer grown on a sapphire substrate. A 2-μm-thick silicon dioxide (SiO_2) mask was patterned to form 4-μm-wide stripe windows with a periodicity of 12 μm. After a 15-μm-thick gallium nitride layer was grown on the silicon dioxide mask pattern, the coalescence of the selectively grown gallium nitride made it possible to obtain a flat gallium nitride surface over the entire substrate. This gallium nitride is called the ELOG substrate.

Figure 2 is a cross-sectional transmission electron microscope image of the ELOG substrate. Threading dislocations, originating from the gallium nitride/ sapphire interface, propagate to the regrown gallium nitride layer within the window regions of the mask. The threading dislocations extend above only the window areas, and none are observable in the overgrown layer. Plan-view transmission electron micrograph observation of the surface confirmed that the number of dislocations on the silicon dioxide mask area was almost zero, and that on the window area was approximately 10^7 cm^{-2}.

After the 15-μm-thick ELOG substrate was obtained, the same laser structure as shown in Fig. 1, except for the cladding layers, was grown on top of

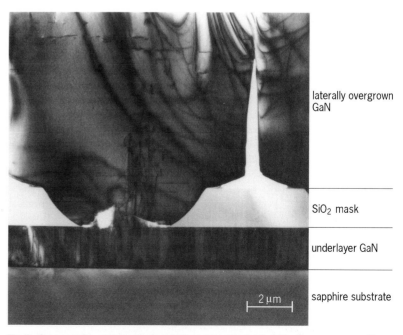

laterally overgrown GaN

SiO$_2$ mask

underlayer GaN

sapphire substrate

2 μm

Fig. 2. Cross-sectional transmission electron micrograph of a laterally overgrown gallium nitride (GaN) layer on a silicon dioxide (SiO$_2$) mask and window area.

it. Ridge-geometry laser diodes with a cavity length of 450 μm and a ridge width of 4 μm were formed on the gallium nitride layers above both the silicon dioxide region without dislocations and the window region with a high dislocation density. When the laser diode was formed on the gallium nitride layer above the silicon dioxide mask region without any dislocations, the threshold current was 53 mA, which corresponds to a threshold current density of 3 kA cm^{-2}. Laser diodes formed on the window region with the high dislocation density had a much higher threshold current density of 6 to 9 kA cm^{-2}. The higher threshold current density is probably caused by the large number of threading dislocations in the window region. The lifetime of the laser diodes formed on the silicon dioxide mask was longer than 10,000 h. The laser diodes formed on the window region showed the lifetimes of 1000–2000 h due to the high threshold current density. At a current of 50 mA, longitudinal modes with a mode separation of 0.04 nm, which is determined by the length of the laser cavity, were observed. At a current of 60 mA, a single-mode emission was observed at an emission wavelength of 396.6 nm.

Prospects. The advances in light-emitting diodes and laser diodes based on InGaN have progressed at a remarkable rate. InGaN-based ultraviolet/blue/green/amber light-emitting diodes are already being used in full-color displays, traffic lights, and lighting. The rapid progress in the development of bluish-purple laser diodes, from pulsed operation to continuous-wave operation with a lifetime of 10,000 h, suggests that bluish-purple lasers will soon enter commercial production. There are numerous potential applications, including optical data storage and laser printers.

For background information *see* ARTIFICIALLY LAYERED STRUCTURES; COMPACT DISK; CRYSTAL DEFECTS; CRYSTAL GROWTH; ELECTRON-HOLE RECOMBINATION; LASER; LIGHT-EMITTING DIODE (LED); QUANTIZED ELECTRONIC STRUCTURE (QUEST); SEMICONDUCTOR HETEROSTRUCTURES in the McGraw-Hill Encyclopedia of Science & Technology.   Shuji Nakamura

Bibliography. S. Nakamura, The roles of structural imperfections in InGaN-based blue light-emitting diodes and laser diodes, *Science*, 281:956–961, 1998; S. Nakamura et al., InGaN/GaN/AlGaN-based laser diodes with modulation-doped strained-layer superlattices grown on an epitaxially laterally overgrown GaN substrate, *Appl. Phys. Lett.*, 72:211–213, 1998; S. Nakamura and G. Fasol, *The Blue Laser Diode*, 1997; A. Usui et al., Thick GaN epitaxial growth with low dislocation density by hydride vapor phase epitaxy, *Jap. J. Appl. Phys.*, 36:899–901, 1997.

Candida (industry)

The genus *Candida* encompasses an extremely diverse group of asexual yeast species, some of which are of great industrial and medical relevance.

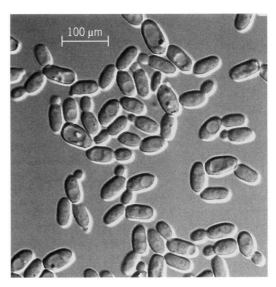

Candida utilis.

Candida utilis (see **illus.**) is used industrially in the production of single-cell protein for food and fodder. Dried *C. utilis* cells have been approved by the U.S. Food and Drug Administration for use as a food additive along with *Saccharomyces cerevisiae* and *Kluyveromyces fragilis*. Recently, a transformation system that allows the genetic engineering of *C. utilis* has become available for the production of recombinant proteins of commercial value.

Industrial utilization. In addition to its use in single-cell protein production, *C. utilis* has been used industrially in the treatment of wastewater and in the production of fine chemicals such as glutathione, ribonucleic acids, amino acids, and several enzymes. *Candida utilis* utilizes both pentoses and hexoses (5- and 6-carbon sugars) as carbon sources and can assimilate a broad spectrum of compounds for use as a nitrogen source. This ability to assimilate pentoses makes it the yeast of choice in the utilization of wood hydrolysate substrates and sulfite waste liquor. Sulfite waste liquor is produced in the sulfite pulping process of paper production. It contains approximately 2.5% fermentable sugars, 80% of which are hexoses and 20% pentoses, in addition to a variety of organic acids. *Candida utilis* can assimilate both sugars, and many of the organic acids present in sulfite waste liquors. *Saccharomyces cerevisiae* and other Crabtree-positive yeasts carry out alcoholic fermentation under aerobic culture conditions unless under growth-limiting conditions of low sugar supply rate. In contrast, *C. utilis* is a Crabtree-negative yeast and does not produce ethanol under strict aerobic conditions. It is thus not affected by the negative effects on biomass yield and growth rate induced by ethanol production. Therefore, *C. utilis* has been chosen for the large-scale production of single-cell protein under continuous culture conditions using cheap biomass-derived sugars such as sulfite waste liquor, a waste product of the paper industry, and molasses, a by-product of sugar production. *Candida* *utilis* cultivated on sulfite waste liquor and wood-derived sugars was used as a food and fodder supplement in Germany during World War II. *Candida utilis* has also been used to reduce the biological oxygen demand (BOD) of the distillery effluent from sugarcane molasses production, a major contributor to environmental pollution in some tropical countries. Production of single-cell protein using *C. utilis* continues today for food and fodder supplements. In the United States, most *C. utilis* produced is used in fabricated foods. *Candida utilis* is also used as a source of the 5′-nucleotides, inosine monophosphate and guanosine monophosphate, which are commercially important flavor enhancers for food and can be produced by enzymatic hydrolysis of the ribonucleic acids prepared from the cells.

Recombinant technology. Yeasts are an attractive system for the production of recombinant proteins. The most popular expression host at the present time is the baker's yeast *S. cerevisiae*. In general, the production yield of heterologous proteins in *S. cerevisiae* has been found to be low, and the high-cell-density culture of cells is laborious due to the Crabtree-positive effects. A growing number of non-conventional or non-*Saccharomyces* yeasts are becoming available as alternative systems for high-yield expression of heterologous proteins. *Candida utilis* has many advantageous characteristics as a host for heterologous protein production. It has been used for the past 70 years for the production of single-cell protein and has obtained the generally regarded as safe (GRAS) status. It has the ability to utilize a wide range of sugars as carbon sources and has no vitamin requirements for growth. Large-scale production of cells is possible with high-cell-density cultivation due to the yeast's Crabtree-negative character. In spite of its industrial importance and its attractive characteristics, knowledge of the genetics of *C. utilis* is still limited, and the possible use of the yeast as a host for protein production was not explored due to the lack of a transformation system. The diploid nature and lack of a sexual life cycle in *C. utilis* makes it difficult to obtain a suitable auxotrophic mutant that could be used as a host for transformation.

However, the first reproducible transformation system for the *C. utilis* wild-type strain has been developed using a gene conferring cycloheximide resistance as a selectable marker. Cycloheximide is an antibiotic that inhibits protein synthesis in eukaryotic cells by acting on the L41 protein, a component of the ribosome. The gene encoding this cycloheximide-sensitive ribosomal protein was cloned from *C. utilis*. Substitution of a proline (the fifty-sixth amino acid residue of the protein) with glutamine rendered the protein insensitive to cycloheximide. Thus, it became possible to use the modified gene as a selectable marker for transformation. Transformants were obtained by electroporation, a method that physically introduces the deoxyribonucleic acid (DNA) into the host cell and has been successfully applied to many yeast species. The ribosomal DNA (rDNA) was used as an integration target for plasmid DNA in order to maximize transformation efficiency. Since the rDNA units exist as multiple repeated copies in yeast genomes, integration of vector DNA at the rDNA locus has resulted in a higher transformation efficiency compared with that at a single-copy locus. Recently, a transformation system based on the complementation of an auxotrophic mutation was developed. An auxotrophic *C. utilis* mutant strain that requires uracil for growth was obtained by nitrosoguanidine treatment of the wild-type cell. The mutant has been used as a host for transformation. The *URA3* gene isolated from *C. utilis* was used as a selectable marker gene for plasmid integration.

Protein production. The transformation system has led to investigation of the potential use of *C. utilis* as a host for protein production. Vectors for the production of heterologous genes in *C. utilis* have been constructed. The essential components of the expression vector are the promoter and terminator sequences for transcription of the heterologous gene, a marker gene for selection of the transformants, and a DNA fragment used as a target for integration of the vector sequence into the host chromosome. The *C. utilis* gene encoding glyceraldehyde-3-phosphate decarboxylase, an enzyme in the glycolysis pathway and one of the most abundant proteins in yeast cells, was cloned, and its transcriptional regulatory sequences were used for gene expression. This transcriptional promoter sequence turned out to function extremely effectively in *C. utilis*. Use of the cycloheximide-resistance gene as a marker permits multiple copy integration of the vector sequences in the transformants. To acquire cycloheximide resistance, it is necessary for the yeast host to have several copies of the marker gene since the host also possesses endogenous genes encoding the cycloheximide-sensitive L41 protein. A defective cycloheximide-resistance gene constructed by deleting part of the promoter region of the gene allowed higher copy number integration of the vector sequences. The vectors have a structure that permits integration of the expression cassette and the marker gene, without the accompanying bacterial sequences which are not necessary for gene expression and may make the integrated vector sequences unstable in yeast. The integrated vectors were stably maintained on the chromosome for more than 50 generations of nonselective growth.

The combination of a powerful promoter sequence for heterologous gene expression with the defective marker gene that allows multiple copy integration of the expression cassette (more than 20 copies) resulted in outstanding production yields of industrially important proteins in *C. utilis*. One example is monellin, a sweet protein obtained from the African plant *Dioscorephyllum cumminisii*. Natural monellin consists of two polypeptide chains and is 3000 times sweeter than sucrose on a weight basis. A single-chain variant of monellin that is stable at boiling temperature was expressed at high levels in *C. utilis* using the above vector system. The single-chain monellin produced in the cell was as sweet as the natural monellin and accounted for more than 50% of the soluble protein in *C. utilis*. Another example is the thermostable α-amylase from *Sulfolobus solfataricus*, a commercially important enzyme because of its ability to hydrolyze glycosyltrehalose, and it can be used for the production of trehalose (a nonreducing disaccharide of two glucose residues bound by an α-1,1 linkage) from soluble starch. Expression of this gene using the vector system described also resulted in an extremely high level of production of the biologically active enzyme, accounting for more than 50% of the soluble protein in *C. utilis*. The high concentration of recombinant proteins and their stability at high temperature made purification from the host proteins easy and significantly reduced the production costs.

High-cell-density cultivation of the recombinant *C. utilis* strain has been attempted. The strain producing α-amylase was grown to a density of over 90 grams of dry cells per liter in a synthetic medium, yielding 12.3 g/L α-amylase which accounts for up to 27% of total cell protein. This production level is the highest obtained in heterologous protein production using a yeast expression system. Further optimization of the fermentation conditions might result in even higher production levels of the recombinant protein, since cell density cultivation of up to 120 g/L dry cells has been attained using a *C. utilis* strain. Thus, the efficient production of recombinant proteins and the ease of high density culture make *C. utilis* an excellent host for the commercial production of important enzymes.

For background information *see* BIOTECHNOLOGY; GENETIC ENGINEERING; YEAST; YEAST, INDUSTRIAL in the McGraw-Hill Encyclopedia of Science & Technology.   Keiji Kondo

Bibliography. S. C. Jong, J. M. Birmingham, and K. Kondo, *Candida utilis*. From genetics to biotechnology, *SIM News*, 48:60–65, 1998; R. D. Klein and M. A. Favreau, The *Candida* species: Biochemistry, molecular biology, and industrial applications, in Y. H. Hui and G. G. Khachatourians (eds.), *Food Biotechnology: Microorganisms*, pp. 297–371, VCH Publishers, 1995; K. Kondo et al., High-level expression of a sweet protein, monellin, in the food yeast *Candida utilis, Nat. Biotech.*, 15:453–457, 1997; K. Kondo et al., A transformation system for the yeast *Candida utilis*: Use of a modified endogenous ribosomal protein gene as a drug-resistant marker and ribosomal DNA as an integration target for vector DNA, *J. Bacteriol.*, 177:7171–7177, 1995.

Carbon dioxide

Environmental controls on chlorofluorocarbons and other solvents have led to industrial development of new cleaning technologies, including carbon dioxide (CO_2) cleaning methods. CO_2 offers new methods for removing both organic and particulate contamination. The four main cleaning methods are (1) hard CO_2 pellets impacting a surface for gross material removal; (2) liquid CO_2 acting as a solvent for removing hydrocarbons and oils; (3) supercritical CO_2 cleaning with superior solvent and penetration powers; and (4) CO_2 snow, a hybrid of soft dry ice and gas flow for cleaning both particle and organic contamination in a gentle and nondestructive way. CO_2 cleaning is environmentally friendly. No new CO_2 is generated for this process; instead, the CO_2 originates from an industrial or food fermentation process.

Properties. At standard temperature and pressure (STP) conditions ($20°C$, 1 atm, or $68°F$, 14.7 psi), CO_2 is a colorless, odorless, nonflammable gas about 1.5 times denser than air. Liquid CO_2, present at higher pressures, is a nonpolar liquid with excellent solvent properties for organics (see **illus.**). The solid phase, dry ice, is unstable at room temperature. It transforms (sublimates) directly into the gas phase. Above $31.05°C$ ($87.8°F$) and 71.85 atm (1055.87 psi), CO_2 enters the supercritical phase. At this point, CO_2 has the combined properties of both liquid and gas phases: the excellent liquid solvent properties, and the penetration power of a gas, such as low viscosity and zero surface energy.

Pellets. Cleaning with CO_2 pellets has been compared to sandblasting without sand. Macroscopic dry-ice pellets propelled by compressed air strike and clean a surface. There is no abrasion or residue. Pellets clean by fracturing off the surface layers. Pellet cleaning involves thermomechanical shock along with high surface shear forces that can break off overlayers having different thermal properties. Pellets can also push greases and grime off a surface. After impact, what is left of the pellet falls and sublimates while the contaminant is airborne or on the floor.

In manufacturing, CO_2 pellets have many applications, including (1) foundry operations for coreboxes, molds, and tooling; (2) rubber/tire industry for tire molds, gaskets and O-rings; (3) food industry for ovens, conveyor belts, molds, and mixers; (4) paint removal (rate dependent); (5) electrical machinery, wiring, printing equipment, tubing scaling, tooling, and machinery (with minimal disassembly); (6) nuclear decontamination; and (7) grease and oil removal.

Pellets, as in all CO_2 cleaning methods and other physical cleaning processes, cannot replace chemical and electrochemical cleaning methods in which surface dissolution is the primary goal. Further, pellet cleaning is generally a "line of sight" process, as is CO_2 snow cleaning. Damage to soft (plastic, copper) or brittle (glass, thin ceramics) or thin substrates may be a concern. Glass can be cleaned, but testing is required, especially of untempered glass. Loosely adherent oxides can be removed from rusted surfaces while adherent oxides stay. Many other material factors and properties can influence the cleaning. Surface cooling occurs, but if the object is heated, the temperature decrease is not a serious problem unless thermal cycling damage can occur (for example, hot thin molds).

Stationary or portable units are available. Systems consist of a dry-ice pellet source, means of acceleration, and a nozzle to direct the pellets. Pellet sources vary: some machines use delivered pellets that are then fed into the machine, or dry-ice blocks that are shaved or machined into pellets, or pellets made within the machine. Each method has its advantages. Next, the pellets must be accelerated to high velocities. The most common methods use dry compressed air at pressures ranging from approximately 2 atm up to and over 20 atm (30 to 300 psi). Another method accelerates the pellets by centrifugal action. Nozzle technologies vary between manufacturers.

Liquid CO$_2$. Conceptually, liquid CO$_2$ cleaning is a solvent cleaning process. It has excellent solvent properties for many oils and greases, along with low viscosity and low surface tension, which are useful for cleaning within crevices and blind holes. Liquid CO$_2$ cleaning has the potential for high-volume batch cleaning of many parts at once. Recent improvements have made liquid CO$_2$ a possible replacement for perchloroethylene in fabric dry cleaning.

The major focus is on removing oils and grime from metals, plastics, and fabrics. In recent years, the emphasis in liquid CO$_2$ cleaning has been to add solvents, surfactants, and enzymes and to develop the process as a replacement for perchloroethylene in dry cleaning operations. The major limitation of liquid CO$_2$ cleaning compared with the other CO$_2$ processes is the requirement for large equipment and the relatively poor particle removal. The addition of ultrasonics, agitation, spray nozzles, and other means has improved particle removal. In general, the process is meant for grime removal, not overlayer removal.

The process requires a cleaning chamber designed for high-pressure operation, CO$_2$ input and exit systems, and a means to separate out the contaminant before CO$_2$ is recovered or vented. The unit has associated compressors and pumps for high-pressure operation. There are CO$_2$ feed tanks and a capture-separator tank for removing grime from the CO$_2$, which can then be reused.

One process uses a centrifuge for agitation and spinning and gets enhanced particle removal. In this dynamic process, a combination of chemistry, centrifugal energy, and multidirectional shear forces separates the contamination from the part. Another method aimed toward fabric dry cleaning injects liquid CO$_2$ at about 50–57 atm (700–800 psi) to create fabric rotation and agitation. Another uses direct agitation. There is ongoing work in using additives in liquid CO$_2$ with the aim to develop detergents soluble in liquid CO$_2$ and then exploit the combination of enzymes, surfactants, and other additives in the liquid CO$_2$ for improved cleaning.

CO$_2$ snow. The CO$_2$ snow cleaning process is simple and straightforward; it can remove particles of all sizes (to below 0.1 micrometer) and organic contamination from surfaces. CO$_2$ snow cleaning is nondestructive, nonabrasive, and residue-free. A stream of small, less dense dry ice and high-velocity gas strikes the surface and removes particles via mo-

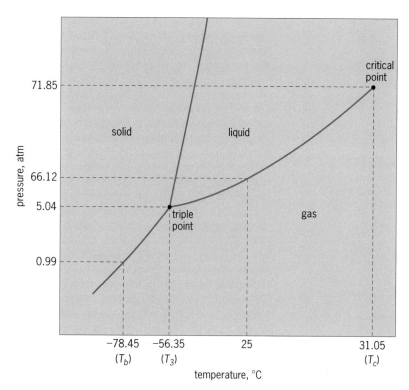

Carbon dioxide phase diagram (axes not to scale). The supercritical phase is above the critical point.

mentum transfer, and hydrocarbons via a transient liquid phase (solvency) or freeze fracture process. The high-velocity gas blows the contaminants away.

CO$_2$ snow cleaning is a softer version of pellets. It is formed by an expansion of either gas or liquid CO$_2$ through an orifice. The most efficient snow formation occurs with a constant enthalpy process. The exit stream from the nozzle should be a focused high-velocity stream directed at the surface.

CO$_2$ snow cleaning is versatile, ranging from simple laboratory applications to production contamination control problems. The main applications are usually within high-technology manufacturing and within research and development, where critical cleaning is of paramount importance. Applications include (1) removing contamination from metals, ceramics, polymers, and glasses; (2) cleaning silicon, epitaxial silicon, indium phosphide, diamond, sapphire, and gallium arsenide wafers; (3) cleaning optics, such as coated lenses, lasers, infrared spectroscopy and ultraviolet optics, fiber optics, gyroscopes, and optical assemblies; (4) preparing samples for surface analysis (Auger, x-ray photoelectron spectroscopy, secondary ion mass spectroscopy) and atomic force microscopy; (5) general cleaning in the laboratory or cleanroom, or for manufacturing; (6) cleaning process tools, disk drive parts, and medical equipment; (7) cleaning electrical contacts and fuel injectors; and (8) cleaning vacuum systems, components, electron and ion optics, and mass spectrometers.

Snow cleaning, as with any cleaning process, requires attention to process parameters, such as moisture condensation, CO_2 purity, and static charge. Addressing these issues makes it possible to scale a benchtop system to systems with automation and process control environments. Major concerns are: (1) Recontamination from the CO_2 cleaning equipment or process can occur. Proper testing, design, and procedures can eliminate the problem. (2) Moisture condensation from the cold CO_2 snow can interfere with many applications, but there are many ways to avoid it. (3) Static charge builds up. Grounding and static neutralizer equipment assist in keeping static charge to a minimum. (4) Damage can occur, although CO_2 snow cleaning is considered nondestructive. Atomic force microscopy studies have shown damage to soft polymers and coatings. Damage is rarely seen with an optical microscope. As with pellets, the parts must be supported and sturdy enough to withstand cleaning forces.

The first device for CO_2 snow cleaning had liquid CO_2 expanded to form soft, low-velocity snowflakes. Later, multiple-orifice nozzles, tubes, and Venturi nozzles were introduced. These improvements led to high-velocity streams and effective cleaning. CO_2 snow cleaning systems are straightforward. They consist of a CO_2 source, a nozzle, and an on/off device. A typical system comes with flexible tubing, an on/off gun or valve, and a nozzle. Available on/off controls include solenoid, pneumatic, or manual valves, and hand-operated guns.

The nozzle design is an important factor in performing CO_2 snow cleaning. The most efficient nozzles are a variation of a Venturi orifice. Other nozzle designs are available, including small-diameter-orifice tubes and leak valves. With these latter nozzles, sudden expasions can occur that violate the constant enthalpy conditions, and the nozzles may not work effectively with smaller orifices or with a gaseous CO_2 source. Different cleaning abilities can result from different designs.

Supercritical CO_2. Supercritical CO_2 cleaning is an extension of liquid CO_2 cleaning at higher pressures and temperatures above the critical point of $31.05°C$ ($87.8°F$) and 71.85 atm (1055.87 psi). Usually, the conditions involve several hundred degrees Celsius and over a hundred atmospheres for enhanced cleaning rates and abilities. With the combined improved solvency and penetrating powers, supercritical CO_2 can clean a wider range of contamination from more complex geometries than liquid CO_2, even down to atomically clean surfaces.

Supercritical CO_2 cleaning is a batch process in which the parts are sealed in a bomb filled with hot high-pressure CO_2. Units can be expensive and tend to be smaller than the liquid CO_2 units. Usually, liquid CO_2 is withdrawn from a cylinder, compressed above its critical pressure by a pump, and then heated above its critical temperature. The cleaning chamber may include an impeller to promote mixing. Supercritical CO_2 containing dissolved contaminants is then bled off to a separator vessel, where it is decompressed and returned to a gaseous state. The contaminants are collected out of the bottom of the separator, while the gaseous CO_2 is vented or sent through a liquefier for storage and reuse.

Supercritical CO_2 applications are similar to liquid CO_2, but they can involve removal on the atomic scale and are used on a wide variety of organics, silicones, and other compounds. Cleaning by supercritical CO_2 is considered a critical cleaning application in comparison to liquid CO_2, and the parts cleaned are usually higher tech. The main limitations of supercritical CO_2 are its inability to remove particles, dust, or inorganic compounds, and generally the small size of the units (although one has been made with a capacity of 60 liters).

Safety. Safety precautions for all processes include eye protection. For pellets, breathing masks can protect the worker from flying debris and airborne items, and ear plugs screen out excessive noise. The pellet and snow streams are cold ($-78.3°C$ or $-109°F$), can cause burns, and must be handled properly. Static charges can exist, so the units and workpieces should be grounded together. Pellet workers should use rubber boots and gloves. CO_2 buildup is a problem, and adequate ventilation is required. It is suggested that CO_2 and oxygen (O_2) meters be placed nearby, with alarms set. Adequate ventilation should be provided, and air quality should be monitored for CO_2, O_2, and other forms of contamination. The liquid CO_2 and supercritical CO_2 systems involve larger volumes of high-pressure CO_2 and require their own safety measures.

For background information *see* CARBON DIOXIDE; DRY ICE; LIQUID; SOLVENT; SUPERCRITICAL FLUID; TRIPLE POINT in the McGraw-Hill Encyclopedia of Science & Technology.   Robert Sherman

Bibliography. *Proceedings of Precision Cleaning*, Witter Publishing Corp., Flemington, NJ, pp. 215–257, 1997.

Cell senescence

Almost all cancer cells have overcome the normal cellular signals that prevent continued division, and thus are immortal. Normal cells from young individuals can divide many times, but they have not accumulated all the other changes needed to make a cell malignant. In most instances, cells become senescent (aged) before they become malignant. Cellular aging and cancer appear to be linked, and the mechanisms regulating aging may act as a cancer brake. A major challenge is to find out how to make cancer cells mortal to inhibit their growth, and how to increase the lifespan of healthy cells.

Cellular aging. Cellular aging reflects the limited capacity of normal cells to divide beyond a finite number of times. There is an irreversible growth-arrest state that depends on the number of replications that a cell has undergone, not on time itself. Cells that have undergone aging in culture (replicative senescence) can be maintained in a viable, non-dividing state for long periods of time. There are various indications that cellular aging may have biological significance. These include the correlation of cellular lifespan with the age of the donor and with average life expectancy of the species, as well as the reduced lifespan of cells from patients afflicted with premature aging syndromes (such as Hutchinson-Gilford progeria).

Cellular immortalization. At one end of the spectrum are diseases which result in premature aging and shortened lifespan. The opposite end of the spectrum is characterized by cells that do not age: malignant cancer cells. Since many commonly used tumor-derived cell lines are immortal, clearly it is possible to overcome cellular aging. In order to evolve into a fully malignant tumor, a normal human cell must progress through certain molecular steps: oncogene activation of signaling pathways, overcoming negative growth signals such as contact inhibition, and acquiring the ability to invade and metastasize. The ability of human cells to overcome the normal limitations to continuous growth is also essential for the proliferation that is acquired by most advanced cancers.

Normal human cells do not spontaneously immortalize when grown in a laboratory culture. Rarely, introduction of viral oncogenes into human cells can give rise to immortalized derivatives by neutralizing the effects of negatively acting cellular factors such as those in the p53 and p16/pRb cell cycle path-break ways. The introduction of viral oncogenes that block the action of negatively acting effectors can result in a bypass of senescence, but it does not lead directly to immortalization. This bypass of senescence has been referred to as extended lifespan, ultimately leading to another proliferative decline called crisis. The main difference is that at senescence cells are truly nondividing while at crisis there is a period of balance between cell growth and cell death. In general, this crisis state results in the loss of the cell culture. Rarely, there is a sporadic immortalization event. In experiments, a viral oncogene with an inducible promoter (for example, the viral onco-gene could be turned on or off at will) was introduced into normal human cells. In the occasional cell that escaped crisis and immortalized, the ability to proliferate depended on the continuous expression of the viral oncogene, with its senescence-bypass function. These immortalized human cells could divide for years in cell culture. Upon the removal of the viral oncogene, a rapid growth-arrest and reexpression of many of the characteristics of senescent cells was observed. Included were cell shape changes, increased senescence-associated or stress-activated β-galactosidase (an enzyme that breaks down lactose into its component sugars), and changes in the expression of key proteins such as p21 and p16. These experiments and the evidence that senescence was correlated with the number of cell divisions rather than with chronological time led to a two-stage model of cellular aging. The model is based on the idea that there are two independent mechanisms (mortality stage 1, normal senescence, and mortality stage 2, crisis), both of which must be overcome for normal human cells to immortalize.

Telomeres. The limited proliferative capacity of normal cells is believed to be controlled by a generational clock that resides in the nucleus of each cell. At the ends of each chromosome is an area known as the telomere which comprises repeated deoxyribonucleic acid (DNA) sequences. (For example, the hexameric sequence TTAGGG is repeated several thousand times in the DNA at the end of each human chromosome.) Due to the cell's inability to complete the replication at the ends of linear DNAs (end replication problem), some telomeric DNA is lost with each cell division (**Fig. 1**). Telomeres not only may protect chromosome ends from degradation and recombination but also may promote correct mitotic separation of sister chromatids. In addition, telomeres serve as a platform for telomere-binding proteins and provide a buffer of expendable noncoding DNA to accommodate the end replication

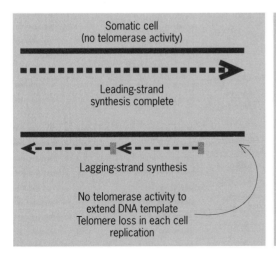
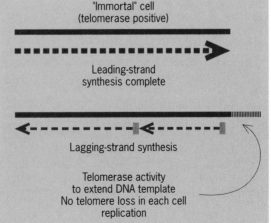

Fig. 1. Telomerase compensates for incomplete lagging-strand synthesis. Normal (somatic) human cells that lack telomerase activity progressively lose telomere sequences. Cancer cells and malignant tumors appear to maintain telomere length through telomerase activity.

problem. When a critically shortened telomere length is reached, entry into senescence is initiated, perhaps by inducing a DNA damage checkpoint pathway. There is mounting evidence that the sequential shortening of telomeric DNA may be an important molecular timing mechanism.

Telomerase. At birth, telomeres in human cells consist of about 15,000 base pairs of repeated TTAGGG DNA sequences. They become shorter with each cell division, owing to the end replication problem. Every time a cell divides, it loses 25–200 DNA base pairs from the telomere ends. Once this pruning has occurred about 100 times, a cell ages and does not continue dividing. Reproductive tissues have a mechanism to compensate for the progressive telomere erosion; otherwise the species would end. Nature's solution for germline cells is telomere terminal transferase (telomerase), a ribonucleoprotein reverse transcriptase enzyme [an enzyme that makes a DNA copy from a single-stranded ribonucleic acid (RNA) template molecule]. Telomerase uses its RNA component (with an internal template complementary to the telomeric TTAGGG repeats) to bind to telomeres and its catalytic protein component to synthesize (via reverse transcription) telomeric DNA directly onto the ends of the chromosomes. Telomerase is present in most fetal tissues, normal adult male reproductive cells, inflammatory cells, proliferative cells of renewal tissues, and most tumor cells. After adding six TTAGGG bases, the enzyme is thought to pause while it repositions (translocates) the template RNA for the repeated synthesis of the next six base pairs. This extension of the DNA template in turn permits additional replication, compensating for the end replication problem (**Fig. 2**).

Aging. The evidence that telomere shortening leads to cellular aging has only recently been demonstrated. Introduction of the catalytic component of telomerase into cells that do not express the enzyme (telomerase silent) results in telomerase detection and extension of the cellular lifespan at least fourfold and possibly much more. Human cells with introduced telomerase maintain telomere stability, retain normal p53/pRb cell cycle checkpoints, and preserve a normal complement of chromosomes. Initial concerns that the introduction of telomerase into normal cells may actually increase the risk of cancer have not been proven. In addition, reproductive tissues maintain high levels of telomerase throughout life, and there is no increased incidence of cancers. Since cancer requires the accumulation of many alterations that occur over a lifetime, introducing only telomerase in normal cells should not be expected to cause them to become malignant.

Biomedical applications. It has been shown that human cells are mortal and grow older because their telomeres shorten each time the cells divide. Manipulating telomere length in cells or tissues to change the rate of cellular aging should allow the removal of a person's cells, manipulation and rejuvenation of the cells, and their return without using up their lifespan. Some areas of cell engineering that may develop in the future include improving bone marrow transplants, having an unlimited supply of skin cells for grafts for burn patients, and making products for cosmetic applications (for example, for aging skin). Telomerase extension of muscle satellite-cell lifespan to facilitate therapy may be feasible in patients with muscular dystrophy. Finally, improving general immunity for old patients or those with blood disorders such as AIDS, or perhaps treating macular degeneration (a leading cause of age-related blindness) are other areas to be explored. In the past, many of these medical conditions were not treatable due to the limited proliferation potential of the patient's cells.

Treating cancer. Telomerase activity is detected in almost all advanced tumors. A therapy that inhibits telomerase activity might interfere with the growth of many types of cancer cells. Studies on pediatric neuroblastoma demonstrate that some tumors, without telomerase activity, do not continue growing and eventually die. These data support the idea that telomerase or another mechanism to provide telomere stability is required to sustain tumor cell growth and that telomerase inhibitors may have efficacy in treating cancer. Following conventional treatments (surgery, radiotherapy, chemotherapy), antitelomerase agents would be given to limit the proliferative

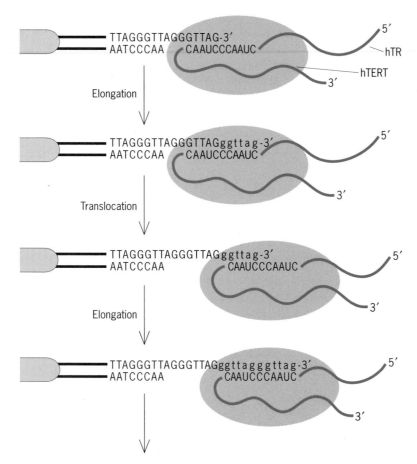

Fig. 2. Telomerase is composed of at least two subunits, hTR (human telomerase ribonucleic acid) and hTERT (human telomerase reverse transcriptase). Critical in determining the enzymatic activity of human telomerase, hTERT probably plays a role in the progression of human cancers by contributing to the maintenance of telomere length (elongation).

capacity of the rare surviving tumor cells in the hope of preventing cancer recurrence. One consideration in this proposed treatment regimen is the prolonged time potentially required for a telomerase inhibitor to be effective, since the mode of action of telomerase inhibitors may require telomeric shortening before inhibition of cell proliferation. Methods may have to be devised to increase the rate of telomere shortening when telomerase inhibitors are used therapeutically. Since preventing cancer recurrence after conventional treatments is the goal, it is believed that antitelomerase agents may be effective since there would be a large number of cell divisions required for the regrowth of rare resistant cancer cells and telomerase inhibitors may limit their proliferation before a palpable tumor reappears. Telomerase inhibitors could also be used in early-stage cancer to prevent overgrowth of metastatic cells.

There are potential risks in the use of such therapy, for example, the effects of inhibitors on telomerase-expressing stem cells. It is likely that this approach would be less toxic than conventional chemotherapy which affects all proliferating cells, including stem cells.

Summary. Aging and cancer research appear to be completely separate fields of scientific study with the only connection being that the incidence of cancer generally increases with age. With the recent knowledge that shortening of telomeres causes cellular aging and that in cancer the maintenance of telomere stability is critical, the pathways connecting cellular aging and cancer are certain to be intensely investigated. That manipulating telomere length could change the rate of cellular senescence and affect the degenerative diseases of aging, or that inhibition of telomerase could provide a novel approach to cancer therapeutics, are exciting possibilities. Overall, research suggests that telomerase is a potential target for the treatment of cancer and of age-related disease. Aging is a complex process, and it is unlikely that any single mechanism will explain all the molecular and physiological changes that occur.

For background information *see* CANCER (MEDICINE); (CELL BIOLOGY); CELL CYCLE; CELL DIVISION; CHROMOSOME in the McGraw-Hill Encyclopedia of Science & Technology.   Jerry W. Shay

Bibliography. A. G. Bodnar et al., Extension of life-span by introduction of telomerase in normal human cells, *Science*, 279:349–352, 1998; C. W. Greider and E. H. Blackburn, Telomeres, telomerase and cancer, *Sci. Amer.*, 274:92–97, 1996; J. W. Shay, Telomerase in cancer: Diagnostic, prognostic and therapeutic implications, *Cancer J. Sci. Amer.*, 4:26–34, 1998.

Chemical information sources

Chemistry has been called the central science because it overlaps a number of other scientific disciplines, bridging physics and the biosciences. Published findings about modern chemistry date from the eighteenth century and have increased tremendously in volume since World War II. In 1946, Chemical Abstracts Service (CAS) included in its flagship publication, *Chemical Abstracts*, references to 49,578 documents; in 1997, the number was 716,564. Chemists pride themselves on their skill in using the literature of chemistry despite its enormous size and complexity. They have made extraordinary contributions to the organization of chemical information, which has had a major impact on the information retrieval practices in other scientific disciplines.

Commercial databases and vendors. Much of the control of the literature of chemistry after the mid-1960s was due to partnerships between the abstracting and indexing (A&I) services and the on-line service vendors which leased their databases. The partnerships formed by Lockheed Information Systems (later Dialog) and System Development Corporation (later Orbit) with various private, not-for-profit, and governmental abstracting and indexing services led to the rapid adoption of on-line searching in the early 1970s.

Over 200 databases are available on the STN International search system, in which CAS is a major stakeholder. STN offers information on a broad range of topics, including chemistry, engineering, life sciences, biotechnology, regulatory compliance, patents, and business. STN International is the only vendor that provides the abstract data from *Chemical Abstracts* in addition to full structure searching capabilities (in the Chemical Abstracts and Registry databases, respectively). Utilizing the unique Registry Number assigned to compounds and mixtures indexed by CAS, it is possible to search in more than 60 databases on the STN system. For example, the CAS Registry Number for isatin is 91-56-5. Chemical Registry databases enable the unique identification of a chemical substance and help to avoid errors in indexing chemical compounds in related databases. Searches can be performed by utilizing parameters such as chemical name, molecular formula, and structure. The Registry database identifier for a chemical (usually a number) is used to index documents in other databases that deal with chemical substances. The main CAS databases are the Chemical Abstracts (CA) and Registry files. At the end of 1998, they included about 14 million document records and 18 million substance records, respectively. CAS also produces a database that covers chemical reactions (CASREACT), a database that combines the catalogs of chemical suppliers (CHEMCATS), and a database that indexes regulated chemicals (CHEMLIST).

The CA file has images of all issues back to the beginnings in 1907. However, the database is fully searchable only from 1967, with limited searching by author and subject to 1907 and by CAS Registry Number in the CAOLD (Chemical Abstracts Old) file between 1957 and 1966. The Registry file contains the records of all substances indexed by CAS from 1965 onward, as well as most older compounds. The

CA database is available in a number of formats, including a CD-ROM from 1987 to the present. STN also has a Web interface via STN*Easy*, and direct searching is possible through STN International and other vendors of database services. With such a large and complex database, CAS saw the need to develop a software tool that would allow chemists who are infrequent or novice searchers of the database to effectively utilize the information source. The solution was SciFinder, an easy-to-use tool that now has an academic counterpart, SciFinder Scholar. The new release of SciFinder is integrated with STN's ChemPort Connection, which in turn links to participating publishers' Web sites for the full-text electronic journals. Thus, a search can lead from a reference in the CA database to the text of the original article without the need to visit a library. Such links to electronic journals are becoming more common as abstracting and indexing databases and journal vendors (aggregators) forge new partnerships with primary journal publishers.

Other major database vendors maintain files for subjects as diverse as competitive intelligence and chemical reaction searching. The Institute for Scientific Information (ISI) has a Web version of Science Citation Index in its Web of Science product. Reaction searches can be done on the ISI Chemistry Server.

A special offering on the Questel-Orbit system is their Generic DARC and Markush DARC chemical databases. Markush structures are imprecisely defined chemical structures found in the patent literature. They cover a large number of related compounds for which patent protection is sought. Each compound that could be constructed from the list is covered by the claims in a patent. Generic DARC works with subsets of the CA databases that have exactly defined structures. Questel-Orbit is the only vendor besides STN to offer full structure searching of the entire CAS Registry system data. Markush DARC searches the Markush formulae databases MPHARM and IPAT (pharmaceutical and patent databases from the French Patent Office, INPI) and Derwent Information Ltd.'s comprehensive patent databases, WPIM and WPAT.

Though impressive, the databases mentioned above cover only the last three decades or so. The appearance of the Beilstein CrossFire system in the mid-1990s reawakened interest in the early literature of chemistry. Beilstein covers the literature of modern chemistry back to its beginnings in the eighteenth century. Users can search for facts, perform structure searches, and construct reaction searches in a database of over 7 million organic and approximately 1 million inorganic and organometallic compounds.

The Cambridge Structural Database (CSD) from the Cambridge Crystallographic Data Centre is the largest searchable database of experimentally determined crystal structures in the world. The CSD contains crystal structure information for over 190,000 organic and organometallic compounds analyzed using x-ray or neutron diffraction techniques. Network access is available to European users, but others must license and mount the files locally.

Producing specialized files on chemically related subjects as diverse as corrosion, biotechnology, cancer research, and materials science, Cambridge Scientific Abstracts makes available its own database as well as such databases as AGRICOLA (agriculture), MEDLINE (medicine), and TOXLINE (toxicology).

Ovid Technologies, Inc., provides access to bibliographic and full-text databases for academic, biomedical, and scientific research. In 1994 the company acquired BRS (Bibliographic Retrieval Services) Online and now has more than 90 databases, including MEDLINE and a growing collection of full-text electronic journals. Ovid links the references from certain of its databases directly to the original articles in the journals.

The Chemical Information System (CIS) contains over 30 databases covering a variety of subjects related to chemistry and the environment. Such topics as site assessment, hazardous materials, material safety data sheets (MSDSs), chemical and physical properties, biodegradation and bioremediation, toxicology and carcinogenicity, regulations, pharmaceuticals, and spectroscopy can be found on the CIS. The system allows both structure and nomenclature searching.

Technical Database Services, Inc. (TDS), is a provider of technical scientific information in the areas of chemistry, biology, environmental science, and medicine. Included is the American Institute of Chemical Engineers' DIPPR (Design Institute for Physical Property Data) Pure Component Data Compilation. DIPPR covers 29 fixed-value properties and 13 temperature-dependent properties for about 1600 industrial chemicals.

One of the earliest database vendors was the National Library of Medicine (NLM), whose MEDLINE database covers the literature from 1965 to the present. In 1998, versions of MEDLINE and other databases produced by NLM became available free on the Internet. There are two avenues to the Internet files: PubMed and Internet Grateful Med. PubMed has linkages to publishers' Web sites for approximately 250 journals and links to molecular biology databases of DNA/protein sequences and three-dimensional structure data. Internet Grateful Med connects to 15 NLM databases. Among those is ChemID, a chemical registry database with over 339,000 compounds of biomedical and regulatory interest.

Free Internet sources. The most reliable databases are generally those which charge a fee for searching. However, many databases can be searched free on the Internet. A good way to find such resources is to consult CHEMINFO and its SIRCh (Selected Internet Resources for Chemistry) guide.

For specific chemicals, CambridgeSoft's ChemFinder can locate hundreds of Internet sites by searching a chemical name, CAS Registry Number, molecular formula, or molecular weight. In addition, with CambridgeSoft's ChemDraw plug-in software, structure searching is possible on ChemFinder.

Among the sites indexed by ChemFinder is the NIST (National Institute of Standards and Technology) Chemistry WebBook. It contains thermochemical data, reaction thermochemistry data, mass spectra, ultraviolet/visible spectra, electronic and vibrational spectra, and constants of diatomic molecules, among other data.

CHEMCYCLOPEDIA gives several avenues to commercial sources of chemicals. A search can be made by chemical name or supplier name to find trade names, packaging, special shipping requirements, potential applications, and CAS Registry Numbers. Chemicals are divided into categories such as surfactants or specialty gases, thus making it easy for users to locate specific chemicals.

Chemical Patents Plus from CAS has the full text for all classes of patents issued by the U.S. Patent and Trademark Office from 1975 to the present, as well as partial coverage from 1971 to 1974. From January 1, 1995, the patent page images are available. Searching of the database is free, as is the display of patent titles and abstracts.

Handbooks, encyclopedias, and data compilations. CD-ROM versions of various handbooks and larger compilations of data and facts have become available, for example, the *CRC Handbook of Chemistry and Physics* and the *Merck Index*. Encyclopedias, such as the *Kirk-Othmer Encyclopedia of Chemical Technology*, will migrate to the Internet. What is missing from the mix of CD-ROM and Internet data sources is a comprehensive, authoritative scientific database of the scope of Landolt-Börnstein. Although impressive in their coverage, Web sources, such as WebElements and the NIST Chemistry WebBook, are of much narrower scope.

It will be a long time before computer databases and the Internet eliminate most needs to consult a traditional chemistry library. Nevertheless, the sources already available provide enough useful information that many chemists are turning to them first. As networks become faster and computers more robust, the traditional chemistry library will likely serve primarily an archival function. The Internet has fostered a new communication process that simply did not exist a decade or so ago. News groups and listserves, such as CHMINF-L (Chemical Information Sources Discussion List), provide almost instantaneous answers to questions that formerly would have taken days or weeks of research. Even more exciting developments are appearing, as the marriage of databases with molecular modeling and visualization techniques (for example, using MDL's Chime) becomes more widely applied in Internet chemical information sources.

For background information *see* DATABASE MANAGEMENT SYSTEMS; LITERATURE OF SCIENCE AND TECHNOLOGY; MICROCOMPUTER; PATENT in the McGraw-Hill Encyclopedia of Science & Technology.

Gary D. Wiggins

Bibliography. J. E. Ash, W. A. Warr, and P. Willett, *Chemical Structure Systems: Computational Techniques for Representation, Searching, and Process of Structural Information*, Ellis Horwood, 1991; R. E. Maizell, *How To Find Chemical Information: A Guide for Practicing Chemists, Educators, and Students*, 3d ed., John Wiley, 1998; D. D. Ridley, *Online Searching: A Scientist's Perspective; A Guide for the Chemical and Life Sciences*, John Wiley, 1996; G. Wiggins, *Chemical Information Sources*, McGraw-Hill, 1991.

Chemical nomenclature

Chemical nomenclature expresses the structure of compounds. Various nomenclature systems are used. Hence, a compound may have several acceptable names. Which name is preferred depends on the context. Each nomenclature type has several dialects due to different languages or other national variations. Even though the language of science communication is increasingly English, there are still differences such as aluminium versus aluminum and sulfur versus sulphur for the names of elements, icosane versus eicosane for the C_{20} hydrocarbon chain, and but-2-ene versus 2-butene for the locant.

Nomenclature systems. One role of the International Union of Pure and Applied Chemistry (IUPAC) is to document the different nomenclature systems and make recommendations on how they should be applied. The four nomenclature commissions have published their work in a series of color-key books: blue for organic nomenclature, red for inorganic, purple for macromolecular, and white for biochemical. The biochemical volume is prepared jointly with the International Union of Biochemistry and Molecular Biology (IUBMB). In addition, there are a green book for chemical terminology and other recommendations on physical chemistry, a silver book for clinical chemistry, an orange book for analytical chemistry, and a gold book as a compendium of terms from all branches of chemistry. However, like any language, chemical nomenclature does not remain static. New types of compounds (for example, fullerenes) and more complex molecules require new nomenclature and extensions of existing systems for naming. These new recommendations are published in the IUPAC journal *Pure and Applied Chemistry*.

Trivial names. Many compounds were recognized as pure substances well before their structure was determined. As a result, trivial names, containing little or no structural information, were given. Some of these names are still used, and indeed many form the basic building blocks for the construction of systematic names. Examples include the names of the elements, the first few alkanes (methane, ethane, propane, and butane), ring systems (for example, benzene, quinoline, and furan), and a few widely used, familiar trivial names (for example, chloroform and acetic acid). Natural products commonly are known by their trivial name (for example, morphine, limonene, and rotenone). However, only specialists can recognize the structure from the trivial name. The new transuranic elements, 104–109, presented a major problem in naming. Deciding who first identified each element and agreeing internationally on a suitable name took many years and was finally settled in 1997.

Organic compounds. The earliest international effort to standardize organic chemical nomenclature was the Geneva Conference of 1892. It considered aliphatic compounds in some detail and briefly considered aromatic compounds. Since then many changes have occurred, but some of the old names are still used. For example, diethyl ether is a radico-functional name based on the functional group ether with the attached ethyl groups.

The preferred method for most organic compounds is substitutive nomenclature. This requires identifying a parent molecule—usually a hydrocarbon or heterocyclic ring system—and then substituting it by replacing one or more hydrogen atoms by a group. Thus, with diethyl ether the name is ethoxyethane with ethane as the parent. Most functional groups can be expressed in both a suffix and a prefix form. With several different functional groups, there is a selection procedure to decide which group is preferred as the suffix. Then all other groups are expressed as prefixes. For example, with 3-hydroxy-cyclohexanone (structure **2**) the parent is cyclohexane (**1**), the preferred functional group is the ketone

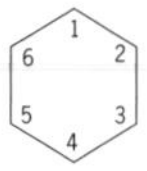

(**1**) 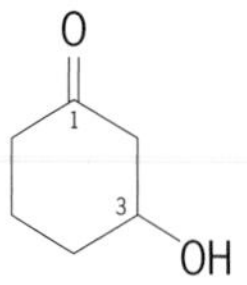   (**2**)

(suffix -one), and the hydroxyl group is indicated by the prefix hydroxy-. The 3- is needed to specify which isomer is present. The cyclohexane ring is numbered from 1 to 6 starting at the ketone carbon.

Ring systems. Ring systems are named in different ways. Monocyclic hydrocarbons are named as either cycloalkanes, for example, cyclohexane (**1**), or annulenes, for example, [10]annulene (**3**). Het-

(**3**)

erocyclic rings of up to 10 members use Hantzsch-Widman nomenclature, for example, 1,3-oxazole (**4**),

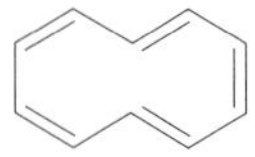

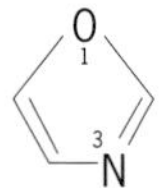

(**4**)

and larger rings use replacement nomenclature with the corresponding hydrocarbon, for example, 1,3-dioxacyclododecane (**5**). With polycylic systems,

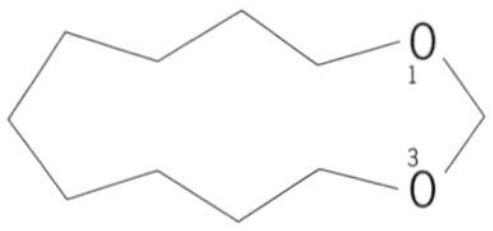

(**5**)

naming depends on the way the rings are linked together. Where two rings have two atoms and one bond in common, fusion nomenclature is used. This system is based on a limited number of mono- and polycyclic ring stems used as components in the construction of more complex systems, for example, furo[3,2-*b*]quinoline(**6**). Some atoms may be treated

(**6**)

as a bridge spanning a fused ring system, for example, 1,4-ethanoanthracene (**7**). If fusion nomen-

(**7**)

clature is not possible, von Baeyer nomenclature is used, for example, bicyclo[3.2.1]octane (**8**). Spiro

(**8**)

compounds have two rings with only one atom and no bonds in common. A form of von Baeyer nomenclature is used here if each ring is a monocyclic component, for example, spiro[4.5]decane (**9**);

(**9**)

otherwise a separate spiro-fusion procedure is used, for example, spiro[piperidene-4,9'-xanthene] (**10**).

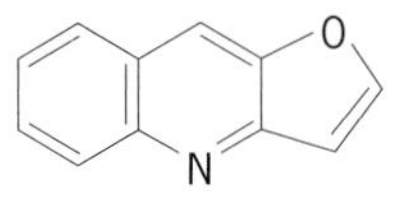

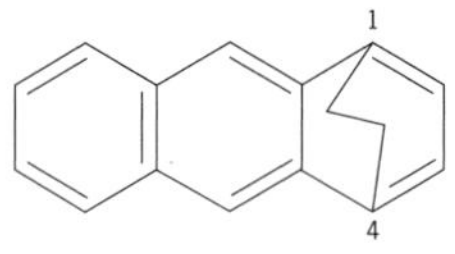

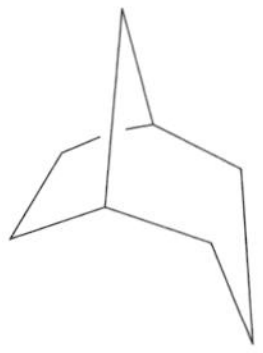

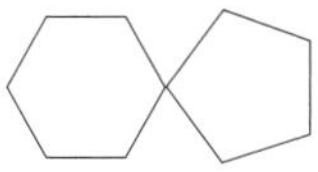

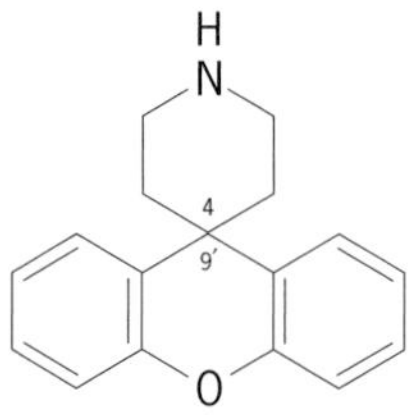

(**10**)

Finally, identical rings linked by a bond are named as a ring assembly, for example, 2,2'-bipyridine (**11**).

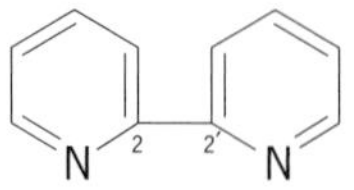

(**11**)

Three-dimensional nomenclature. The compounds mentioned so far have been named only in a two-dimensional sense. To indicate the third dimension, it is necessary to include stereochemistry in the name. That is largely conveyed by the Cahn-Ingold-Prelog (CIP) system, whereby groups are ranked by a set of priority rules and this ranking is used to decide if a double bond is *E* or *Z* (highest priority on opposite sides or the same side, respectively) or, for a chirality center, whether *R* or *S* (clockwise or counterclockwise priority order, respectively). Examples are (*E*)-but-2-ene (**12**) and (*S*)-2-hydroxy-propanoic acid (L-lactic acid; **13**).

(**12**)   (**13**)

Natural products. Molecules from natural products are frequently very complex. A systematic name would be cumbersome so selected parents are used to provide the basic structure, often with the stereo-chemistry implied. These parents are then modified by normal substitutive techniques, for example, cholesta-5,7-diene-3β-ol (**14**), with parent choles-

(**14**)

tane. Many classes of natural products have their own specialized systems of nomenclature, such as amino acids and carbohydrates. These two classes of compounds are the only ones where the old system of D and L is still used to identify configuration.

Inorganic compounds. With inorganic compounds, there are several different nomenclature systems. Many simple compounds are named by a binary-type system which specifies the stoichiometry (for example, sodium chloride). The electropositive portion is usually the element name, and the electronegative portion is normally the anion name. More complex examples are named similarly but may involve more than two components and may require information on how many atoms (or groups) are present, for example, aluminium potassium bis(sulfate), $AlK(SO_4)_2$. The nature of the interaction between the components is not implied by the name.

Coordination compounds. Coordination compounds require specification of the central atom and the ligand or ligands. Each ligand may be a single atom or a groups of atoms with one or more of the atoms coordinating the central atom. If necessary, the name needs to specify the coordinating atom, but otherwise it is assumed to be the electronegative atom, for example, amminedichloro(pyridine)

platinum(II). In this name the central atom, platinum with oxidation state II, is coordinated to an ammonia molecule (ligand name ammine), two chlorine atoms, and a pyridine molecule. It is assumed that the nitrogen atom of pyridine coordinates the platinum. The pyridine is put in parentheses to make it clear that the two chlorine atoms are ligands and not substituents of the pyridine. Once again, a complete name would require the specification of stereochemistry. Although the traditional *cis* and *trans* might be used here, a more general system using the CIP priority order of the ligands is preferred.

Organometallic compounds. Organometallic compounds have a direct carbon-to-metal bond. Although they can often be named by coordinative nomenclature, there is a separate organometallic system based on substitution of the central atom or atoms treated as a parent in a way similar to organic molecules. For example, hexaethyldiplumbane has as a parent diplumbane (H_3PbPbH_3) with a direct Pb—Pb bond.

Macromolecular compounds. Macromolecular nomenclature is concerned with polymers and usually is based on the repeating unit or units. Traditionally, names are based on the starting monomer (for example, polyethylene or polystyrene). The more systematic approach is based on the structure of the polymer, not what was used to make it. The basis is the constitutional repeating unit which is cited in parentheses. Thus, polystyrene is named poly-(1-phenylethylene), where ethylene is —CH_2CH_2—. Names of irregular polymers derived from a mixture of monomers are based on the two constitutional repeating units but do not indicate their ratio or how the units are linked together. For example, a copolymer between styrene and vinylchloride is called poly(1-chloroethylene/1-phenylethylene).

For background information *see* CHEMICAL SYMBOLS AND FORMULAS; COORDINATION COMPLEXES; HETEROCYCLIC COMPOUNDS; ORGANIC NOMENCLATURE in the McGraw-Hill Encyclopedia of Science & Technology.   Gerard P. Moss

Bibliography. G. J. Leigh (ed.), *Nomenclature of Inorganic Chemistry: Recommendations 1990*, Blackwell Scientific Publications, 1990; G. J. Leigh, H. A. Favre, and W. V. Metanomski, *Principles of Chemical Nomenclature: A Guide to IUPAC Recommendations*, Blackwell Science, 1998; C. Lièbecq (ed.), *Biochemical Nomenclature and Related Documents*, 2d ed., Portland Press, 1992; W. V. Metanomski (ed.), *Compendium of Macromolecular Nomenclature 1991*, Blackwell Scientific Publications, 1991; R. Panico, W. H. Powell, and J.-C. Richer (eds.), *A Guide to IUPAC Nomenclature of Organic Compounds: Recommendations 1993*, Blackwell Scientific Publications, 1993.

Chemical reactions

Until the late 1990s, measurements of chemical reaction rates were largely confined to ambient (room) temperature and above. Interest in combustion stimulated studies at high temperature in devices such as shock tubes. In the mid-1980s, experimentalists began to study the kinetics of gas-phase reactions at temperatures down to about 200 K ($-73°$C; $-99°$F) because of an increased interest in the chemistry of the Earth's atmosphere. Only in the last few years have such studies been carried out to even lower temperatures. This work at very low temperatures has been stimulated by three factors: the wish to explore kinetics under conditions previously unachievable, the belief that the results of such studies will provide new and demanding theoretical challenges, and the desire to understand the chemistry which leads to molecular synthesis in the dense interstellar clouds located in distant regions of space. These huge aggregates of molecules have temperatures between 10 and 100 K (-263 and $-173°$C; -441 and $-279°$F), and it is clear that molecular formation and destruction continually occur within them.

Experimental methods. The first kinetic studies in the gas phase at temperatures below 77 K ($-196°$C; $-321°$F) were carried out on ion-molecule reactions. Such systems offer an element of control—the manipulation of the ions in electromagnetic fields—not available when both reagents are electrically neutral. Furthermore, it was not a real surprise to learn that many ion-molecule reactions remain rapid as temperatures are lowered toward absolute zero. The dynamics of such collisions are governed by the long-range attractive forces that act between a charged ion and a polarizable (or dipolar) molecule. The lower the collision energy, the higher the cross section for the mutual capture that leads to the close collisions in which reaction can occur.

However, the method developed by Bertrand Rowe and his coworkers in France to study ion-molecule reactions did not rely on the charged nature of ions. This CRESU (*cinétique de réaction en ecoulement supersonique uniforme*) method has been adapted to study the rates of reactions and energy transfer occurring in collisions between neutral species at temperatures as low as 7 K ($-266°$C; $-447°$F).

The method relies on the fact that expansion of a gas (or gas mixture) through a properly defined Laval nozzle produces a uniform, relatively dense flow of gas moving at supersonic speed with a well-defined temperature. To measure reaction rates in this medium, small concentrations of two gases are included in the inert carrier (helium, argon, or nitrogen). One gas serves as a precursor for the free radicals whose reaction is to be studied. A pulsed laser is used to dissociate these species and produce radicals, such as cyano (CN) from nitrosyl cyanide (NCNO), hydroxyl (OH) from hydrogen peroxide (H_2O_2), or methylidene (CH) from bromoform ($CHBr_3$). The second added species is the co-reagent whose reaction with the radicals is to be studied. The rate of reaction is observed by following the disappearance of the radicals. Generally a second pulsed laser is used which causes the radical to fluoresce. The intensity of this fluorescence is proportional to the concentration of free radicals. The decrease in signal as the delay between pairs of laser pulses is increased provides a measure of the rate of reaction under the conditions of a particular experiment.

Rate constants versus temperature. Two CRESU apparatuses exist: one in Rennes, France, and one in Birmingham, England. Together they have been used to measure rate constants for about 30 reactions between neutral species at temperatures, in some cases, as low as 13 K ($-260°$C; $-436°$F). These reactions include those of the free radicals CN, OH, CH, and ethynyl (C_2H), and of the atoms carbon (C), aluminum (Al), and silicon (Si). The co-reagents include other radicals (such as dioxygen, O_2), unsaturated molecules (such as acetylene or ethyne, C_2H_2), and even saturated molecules (such as ethane, C_2H_6).

The trend for all these reactions is for the rate constants, $k(T)$, to increase as the temperature, T, is lowered below room temperature. In a significant fraction of the cases studied, the rate reaches a value at low temperature which can simplistically be described as "reaction at every collision." However, the details of the variation of rate constants with temperature differ markedly from reaction to reaction (see **illus.**).

Implications of experimental results. The finding that a large number of reactions occur very rapidly at very low temperatures was unexpected. The flow in the CRESU apparatus is so fast that only fast rates of reaction can be observed. Naturally, the reactions chosen for study have been those where it is reasonable to suppose, from studies at higher temperature, that rates may remain rapid as the temperature is lowered. This will not be the case if there is a significant energy barrier along the path for reaction, which is the fundamental reason why the rate constants for many reactions rise steeply with increasing temperature and are said to possess activation energies.

Conversely, the growing number of reactions which remain fast, indeed accelerate, at low temperatures must have no barrier along the reaction path. Rather it seems that the rate is controlled by attractive forces between reagents and their ability to capture the reagents and bring them into close contact where they react—just as for ion-molecule reactions. Such a situation is the norm for ion-molecule reactions and for reactions between pairs of free radicals. It is more surprising that this description extends to reactions between radicals and molecules (that is, species with fully paired electrons).

Within the framework of transition-state theory, it seems that the temperature dependence of rate constants must depend on a subtle interplay of factors. As the temperature is raised, the distribution of both the collision energy and the internal energies of the reagents becomes broader and shifted to higher energies. This means that more channels leading from reagents to the transition-state region come into play and the transition states move to closer separations in a manner which depends in detail on the underlying forces between the reagents. It is believed that, in this way, the observed temperature

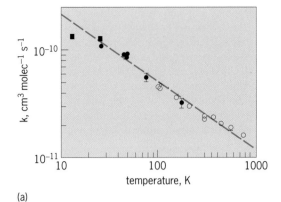

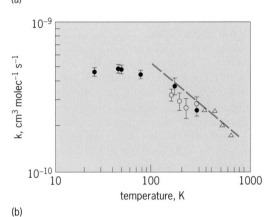

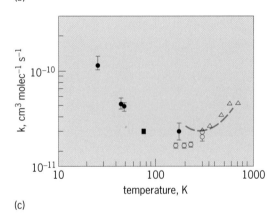

(a)

(b)

(c)

Some ways in which the rate constants for reactions which are rapid at low temperature can vary with temperature. Both the rate constants and the temperature are plotted on logarithmic scales. (a) Rate constants for the reaction of CN radicals with O_2. (b) Rate constants for $CN + C_2H_2$. (c) Rate constants for $CN + C_2H_6$.

dependences reflect differences in the intermolecular potentials for different reactions in the crucial regions where new bonds start to form and old bonds break.

The second major interest in kinetic data at very low temperatures is their relevance to the remarkable chemistry that clearly occurs in dense interstellar clouds. Until recently, it was felt that this chemistry predominantly involved electronically charged species. However, the results reviewed briefly in this article have demonstrated that many neutral-neutral reactions remain intrinsically rapid at the temperatures which characterize these huge astronomical

objects, and these reactions must now be included in models of their chemistry.

An example can illustrate this point. Based on the observation that reaction (1) between CN radicals

$$CN + C_2H_2 \longrightarrow HC \equiv CCN + H \qquad (1)$$

and C_2H_2 is rapid and the fact that CN and C_2H are isoelectronic, the suggestion was made that reactions of both CN and C_2H radicals with alkynes and cyanoalkynes might be fast, so that combinations of reactions such as (2)-(4) might create the long-chain

$$C_2H + HC_{2n}H \longrightarrow HC_{2n+2}H + H \qquad (2)$$
$$CN + HC_{2n}H \longrightarrow HC_{2n}CN + H \qquad (3)$$
$$C_2H + HC_{2n}CN \longrightarrow HC_{2n+2}CN + H \qquad (4)$$

cyanopolyynes that are the largest, and in many ways most remarkable, molecules definitively identified in interstellar clouds. The recent observation that C_2H radicals do indeed react rapidly with C_2H_2 appears to confirm this hypothesis.

For background information *see* CHEMICAL DYNAMICS; CHEMICAL THERMODYNAMICS; COLLISION (PHYSICS); FREE RADICAL; GAS DYNAMICS; LASER; REACTIVE INTERMEDIATES; ULTRAFAST MOLECULAR PROCESSES in the McGraw-Hill Encyclopedia of Science & Technology.   Ian W. M. Smith

Bibliography. I. R. Sims and I. W. M. Smith, Gas-phase reactions and energy transfer at very low temperatures, *Annu. Rev. Phys. Chem.*, 46:109-137, 1995; I. W. M. Smith, Molecules in space: The chemical laboratory at the end of the universe, *Chem. Rev.*, 7(3):2-6, 1998; I. W. M. Smith, Neutral-neutral reactions without barriers: Comparisons with ion-molecule systems and their possible role in the chemistry of interstellar clouds, *Int. J. Mass Spectrom. Ionic Proc.*, 149/150:231-249, 1995; I. W. M. Smith, B. R. Rowe, and I. R. Sims, Gas-phase reactions at low temperatures: Towards absolute zero, *Chem. Eur. J.*, 3:1925-1928, 1997; I. W. M. Smith, B. R. Rowe, and I. R. Sims, Kinetics at ultra-low temperatures: Non-Arrhenius behaviour and applications to the chemistry of interstellar clouds, in J. Wolfrum et al. (eds.), *Gas-Phase Chemical Reaction Systems: Experiments and Models 100 Years after Max Bodenstein*, Springer Ser. Chem. Phys., pp. 190-200, 1996.

Chiral drugs

Chiral compounds are those for which the mirror image of the molecular structure cannot be superposed on the original. If such superposition is possible, the compound is achiral. For example, the drug amphetamine is chiral, but aspirin is achiral (**Fig. 1**). The carbon atom in amphetamine with four different ligands is called a chiral center (stereogenic center). Stereoisomers related as object and mirror image are termed enantiomers. Most chiral drugs contain one or more chiral centers; less commonly, chirality results from an axis or plane of chirality. The actual

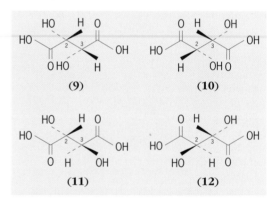

Key: R = CH₃   R′ = C₆H₅

Fig. 1. The vertical lines between structures (1) and (2) and between (4) and (5) represent plane mirrors at right angles to the plane of the paper. (S)-amphetamine (1) has mirror image (2), (R)-amphetamine. Structure (3) is obtained by rotation of (2) through 180°; it is not superposable on (1), hence (1) and (2) are enantiomers. Aspirin (4) has mirror image (5). Structure (6) is obtained by rotation of (5) through 180°; it is superposable on (4), hence aspirin is achiral.

molecular arrangements (configurations) at a chiral center are specified by the *RS* system, which defines a right-handed (*R*, rectus) or left-handed (*S*, sinister) ordering of groups arranged in an arbitrary priority sequence (**Fig. 2**).

Racemates versus enantiomers. Chemical synthesis of a chiral drug (in the absence of chiral reagents or chiral influences) leads to a mixture of equal parts of the two enantiomers—a racemic mixture. Chiral drugs have attracted considerable attention since the two enantiomers may have different physiological actions. Drugs once used as racemic mixtures have been reinvestigated as single enantiomers (racemic switch), and for new drugs stereospecific synthesis of one enantiomer is usually considered. In 1992 the U.S. Food and Drug Administration (FDA) issued a policy statement almost entirely concerned with the development of new chiral drugs; it was concluded that development of racemates might sometimes be

appropriate but only with the use of proper precautions. Chemists have developed many ingenious methods either to separate racemic mixtures or to carry out stereospecific synthesis of one enantiomer using chiral reagents or catalysts. Also important are analytical methods to detect the presence of one specific enantiomer during the preparation of the other. For chiral analysis, a chromatographic technique such as high-pressure or high-performance liquid chromatography (HPLC) is particularly useful; in many cases, direct resolution of racemates is possible with chiral stationary phases. Of the top 100 drugs worldwide, 50 are sold as single enantiomers, with a 1997 value of $90 billion; many are natural products usually derived by fermentation as a single form.

In vivo differentiation. Components of living cells are mainly chiral and usually recognize differences between the enantiomers of a chiral compound or drug. The asparagine enantiomers were found to have different tastes—the *R* form was sweet, while the *S* enantiomer was tasteless. Enantiomers may also differ in odor; (*R*)-carvone has a spearmint odor, and (*S*)-carvone an odor of caraway. An early finding of enantiomeric differences for a chiral drug was that (*S*)-hyoscyamine was more active than the *R* form when acting on motor nerve endings.

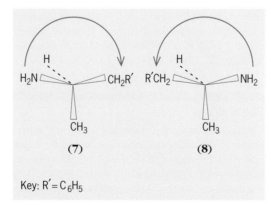

Key: R′ = C₆H₅

Fig. 2. *RS* system, exemplified by (*R*)-amphetamine (7) and (*S*)-amphetamine (8). The structure is viewed so that the group of lowest priority (that is, H) is farthest from the viewer. The priority sequence is NH₂ > CH₂—C₆H₅ > CH₃ > H. In (7) the ordering NH₂ → CH₂—C₆H₅ → CH₃ is clockwise, hence the configuration is *R*, and in (8) it is counterclockwise, hence the configuration is *S*.

Fig. 3. Stereochemistry of tartaric acid. Structures (9) and (10) are in a nonsuperposable, mirror image arrangement and are enantiomers; they are 2*R*,3*R* and 2*S*,3*S*, respectively. Structures (11) and (12) are superposable mirror images and hence either overall is an achiral, *meso* structure. Structures (9) and (11), and (10) and (11), are diastereoisomers. Structure (11) is 2*R*,3*S*, and (12) is 2*S*,3*R*. To check superposition, transpose structures (10) and (12) around the x axis.

The differentiation between two drug enantiomers by a receptor molecule is an example of the broad phenomenon of chiral recognition. In simple terms, this complex process requires diastereoisomeric intermediates or products to be present in a reaction or binding situation. Diastereoisomers are stereoisomers that are not in an object/mirror image relationship. Thus, tartaric acid, with two chiral centers, exists in three possible forms as well as a racemic mixture (**Fig. 3**). The 2*R*,3*S meso* structure has a superposable mirror image and overall is achiral; it has a diastereoisomeric relationship with each of the two enantiomeric forms. While enantiomers have identical properties in ordinary processes (not involving chiral reagents or influences), diastereoisomers have different properties; each tartaric acid enantiomer has a melting point of 171–174°C (340–345°F), but for the *meso* form the melting point is 146–148°C (295–298°F).

In a simple case of chiral recognition, diastereoisomer formation is used to separate (resolve) a racemic mixture. For racemic amphetamine, a basic compound containing the NH_2 group, salt formation proceeds with the 2R,3R enantiomer of tartaric acid. For example, if A is amphetamine and T is tartaric acid, various salt formations occur with the two enantiomers, such that the salts I [reaction (1)] and II [reaction (2)] are diastereoisomers with different

$$(R)\text{-A} + (2R,3R)\text{-T} \rightarrow (R)\text{-A}^+(2R,3R)\text{-T}^- \quad \text{[Salt I]} \quad (1)$$

$$(S)\text{-A} + (2R,3R)\text{-T} \rightarrow (S)\text{-A}^+(2R,3R)\text{-T}^- \quad \text{[Salt II]} \quad (2)$$

solubilities. Hence, they can be separated by crystallization, and each separate diastereoisomer on acid treatment yields a pure amphetamine enantiomer; the separated S enantiomer is known as dextroamphetamine.

Similarly, the two products formed by interaction of a cell receptor molecule and each enantiomer of a drug would be in a diastereoisomeric relationship and different physiological activities would be predicted. In many cases, however, it is likely that only one enantiomer will bind to a receptor such as a protein. In simplified terms, this possibility is explained by postulating three binding sites on the receptor for specific groups of the drug molecule leading to a three-point attachment (**Fig. 4**).

Physiological actions. The in vivo behavior of a drug enantiomer is influenced by many different factors. Not only may there be differences in the actual physiological effects produced by enantiomers (pharmacodynamics), but there may be differences in transport rates to receptor sites, in adsorption rates at receptors, and in metabolism to other products (pharmacokinetics). Moreover, animal experiments may not extrapolate to the human condition, and there are also possible sex differences. Individuals may have different abilities to metabolize a drug as a genetic consequence; for some drugs, the population will show two phenotypes—extensive metabolizers and poor metabolizers. A further consideration is that under certain conditions some enantiomers undergo racemization (inversion of configuration). There is much variation in the stability of drug enantiomers both in vitro and in vivo. For example, when hyoscyamine is isolated from *Atropa belladonna*, a partial racemization occurs. For general clinical use, this racemization is completed by alkali treatment, and the racemic mixture is sold as atropine.

In very general terms there are three possible categories of physiological effects obtained with chiral drugs: type A—both enantiomers have similar pharmacological actions; type B—one enantiomer may be active in a particular process, the other not; type

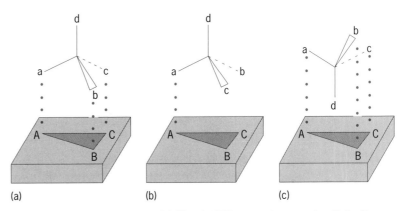

Fig. 4. Three-point attachment model. Triangle ABC represents a receptor site for the chiral drug abcd; a binds to A, b to B, and c to C. The priority sequence is assumed to be a > b > c > d. The site is on the exterior surface of a protein, and abcd cannot bind from below (from the interior). (*a*) Three-point attachment for the R enantiomer of abcd. (*b*) A one-point attachment for the S enantiomer; a two-point attachment (not shown) is also possible, but not a three-point attachment. (*c*) An alternative binding of the S enantiomer. However, this is not possible, since the presence of d makes the distances A-a, B-b, and C-c too large.

C—the two enantiomers have different pharmacological actions.

Type A. The nonsteroidal anti-inflammatory drug ibuprofen contains a single chiral center (**Fig. 5**) and has usually been marketed as a racemic mixture under many trade names. The S enantiomer, however, is far more active than the R form and lacks certain side effects. In the early 1990s, enthusiastic efforts were made to market (S)-ibuprofen in the United States (that is, to carry out a racemic switch). However, the FDA has refused approval for this pharmaceutical, although it is sold as a prescription drug in Austria and Switzerland. Ibuprofen is an example of a drug that undergoes inversion of configuration (racemization) during metabolism.

Type B. For propranolol, one enantiomer is effective as a β-adrenoreceptor antagonist while the other

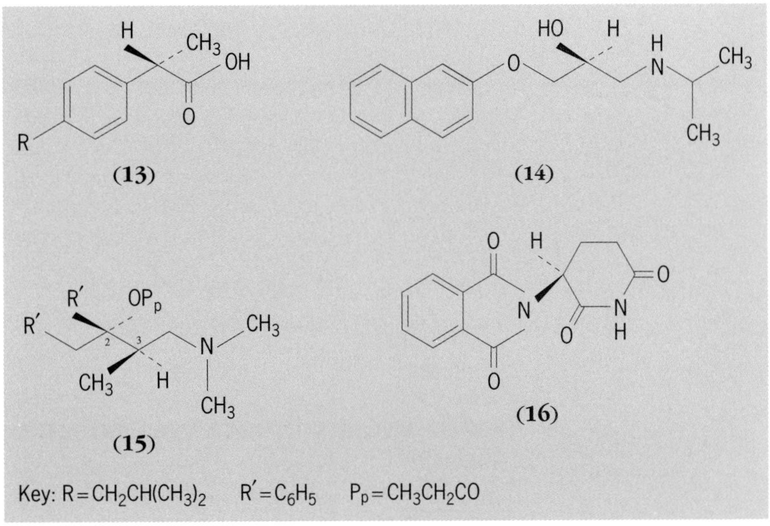

![Chemical structures of chiral drugs (13), (14), (15), and (16)]

Fig. 5. Chiral drugs: (*S*)-ibuprofen (13); (*S*)-propranolol (14); 2S,3R enantiomer of propoxyphene (15) [One trade name is Darvon. For the 2R,3S enantiomer, one trade name is Novrad. Note that Novrad is Darvon written backwards— that is, the two names are palindromes; they are not, however, exact mirror images in two dimensions.]; (*S*)-thalidomide (16).

Key: R = $CH_2CH(CH_3)_2$ R′ = C_6H_5 P_p = CH_3CH_2CO

is almost inactive; the activity ratio $S{:}R = 100{:}1$ (Fig. 5). Propranolol is normally marketed as the hydrochloride of the racemate. This drug is a good example of the complexities that may be encountered since it has other physiological actions in which the enantiomers are essentially equipotent, for example, as a local anesthetic, in decreasing plasma levels of triiodothyronine, and in membrane stabilization.

Type C. If enantiomers have different pharmacological properties, two practical situations may arise: if one form has a desirable activity, the activity of the other may be either desirable or undesirable. Propoxyphene exemplifies the first alternative. It is a material with two chiral centers (Fig. 5), and so there are four stereoisomers. One enantiomeric pair shows different but useful pharmacological activities for the two forms. The $2S,3R$ enantiomer has analgesic properties, while the enantiomeric $2R,3S$ form is an antitussive. The other diastereoisomers, the enantiomeric pair $2R,3R$ and $2S,3S$, are inactive.

Thalidomide, a drug responsible for some 10,000 cases of phocomelia (a birth defect where limbs are missing) when administered as the racemic mixture to pregnant women, is an example of the second alternative. There is some evidence that the sedative action rests in (R)-thalidomide and that (S)-thalidomide (Fig. 5) is teratogenic (in mice and rats). Contributing to the problem is the fact that thalidomide readily undergoes racemization under physiological conditions; in rabbits, plasma thalidomide obtained 2 h after administration of an enantiomer was completely racemic. It has been asserted categorically that it is practically impossible to show stereospecific actions of thalidomide enantiomers in vivo. Racemic thalidomide has been approved for use in a limited number of conditions with very precise precautions to prevent its use by pregnant women.

For background information *see* AMPHETAMINE; ASYMMETRIC SYNTHESIS; ATROPINE; ENANTIOMER; MOLECULAR ISOMERISM; RACEMIZATION; STEREOCHEMISTRY in the McGraw-Hill Encyclopedia of Science & Technology.   Ronald Bentley

Bibliography. H. Y. Aboul-Enein and I. W. Wainer (eds.), *The Impact Of Stereochemistry on Drug Development and Use*, John Wiley, 1997; A. S. Casy, *The Steric Factor in Medicinal Chemistry: Disymmetric Probes of Pharmaceutical Receptors*, Plenum Press, 1993; S. S. Stinson, Special reports on chiral drugs, *Chem. Eng. News*, 75(42):38–70, 1997, and 76(38):83–104, 1998; I. W. Wainer (ed.), *Drug Stereochemistry: Analytical Methods and Pharmacology*, M. Dekker, 2d ed., 1993.

Cholesterol

Cholesterol is a steroid alcohol that is essentially insoluble in aqueous solutions. In mammals it is normally solubilized by its association with other lipids, such as phospholipids or bile acids; thus, most cholesterol is found in cell membranes, plasma lipoproteins, and bile. Cholesterol can be esterified with a fatty acid to form cholesteryl esters. The latter form discrete lipid droplets in cells, especially in cells of steroidogenic tissues, and in the lipid core of low-density lipoproteins in the blood.

Functions. Cholesterol has many functions in mammals. It is an essential component of cell membranes, including specialized areas in the plasma membrane called caveolae, and the myelin sheath surrounding nerves. It is the precursor of steroid hormones (such as estrogen, progesterone, testosterone, and aldosterone) and bile acids, which are involved in the absorption of fats and certain vitamins from the intestine. Normal embryonic signaling controlled by the protein Hedgehog is dependent upon the covalent attachment of cholesterol. Studies have demonstrated that certain oxidized derivatives of cholesterol exhibit a variety of biological activities. The oxysterols, for example, can act either as signaling molecules or as hormones that activate nuclear receptors. They thus can modulate transcription of specific genes or can induce apoptosis. It is evident that cholesterol plays a central role in many important cellular events.

Metabolic defects causing reduced cholesterol synthesis result in various neurological disorders, such as the Smith-Lemli-Opitz syndrome. Excess blood cholesterol is associated with atherosclerosis, the leading cause of death in the Western world. Excess cholesterol in the bile can lead to gallstones.

Diets rich in cholesterol and saturated fatty acids result in increased levels of blood cholesterol, largely contained in low-density lipoproteins (LDL). Many studies have demonstrated a direct correlation between blood LDL cholesterol levels and atherosclerosis and heart attacks. The blood levels of high-density lipoproteins (HDL), which also contain cholesterol, albeit at lower levels, are inversely correlated with atherosclerosis and heart attacks.

Thus, the control of cellular and blood cholesterol levels is critical for many processes. This article emphasizes the transcriptional mechanisms that regulate the expression of genes that control cholesterol synthesis and endocytosis of LDL via the LDL receptor. Recent results demonstrated that the same mechanisms that control the transcription of genes also regulate fatty acid and phospholipid biosynthesis.

Sources. Cells obtain cholesterol in two ways: endogenous synthesis and endocytosis of LDL. Cells in the mammalian body, with few exceptions, have the capacity to synthesize the 27-carbon cholesterol molecule from 2-carbon acetyl coenzyme A units, a process that requires more than 30 different enzymes (**Fig. 1**). Acetyl coenzyme A is also a precursor of saturated and unsaturated fatty acids, which can be utilized in the synthesis of phospholipids, triglycerides, and cholesteryl esters. Fatty acids and phospholipids can be converted into a variety of signaling molecules.

Cellular cholesterol also can be obtained from blood lipoproteins. These lipoproteins bind to various cell surface receptors prior to either uptake into the cell (endocytosis) or selective delivery of their cholesteryl ester lipid moieties. A number of distinct receptors that bind lipoproteins have been cloned. These include the LDL receptor, the LDL receptor–related protein, the very low density lipoprotein receptor, CD36, and various scavenger receptors. Unlike the other lipoprotein receptors, the LDL receptor is expressed on the surface of virtually all mammalian cells, and its expression is regulated by the cholesterol content of the cell. Since the endocytosis of each LDL particle delivers approximately 1500 molecules of cholesteryl ester and significant amounts of phospholipids to the cell, it is important that this endocytic pathway be regulated.

In order to control cellular cholesterol levels within limits that are not deleterious, cells have developed a unique mechanism that controls both cholesterol synthesis and LDL-receptor-mediated endocytosis by means of end-product repression of genes. Thus, the accumulation of sterols (cholesterol and oxysterols) in cells results in decreased expression of cholesterogenic enzymes (such as HMG-CoA synthase, HMG-CoA reductase, farnesyl diphosphate synthase, and squalene synthase) and repression of the gene encoding the LDL receptor required for the endocytosis of LDL (Fig. 1 and **Fig. 2a**). In contrast, when cells are deprived of cholesterol, the expression of these same genes is increased in order to increase both the synthesis of cholesterol and the endocytosis of cholesterol-rich LDL and thus maintain cellular cholesterol homeostasis (Fig. 2b).

The sterol-regulated transcription of the genes encoding the LDL receptor and the enzymes involved in cholesterol synthesis involves an unusual mechanism that requires the binding of a transcription factor,

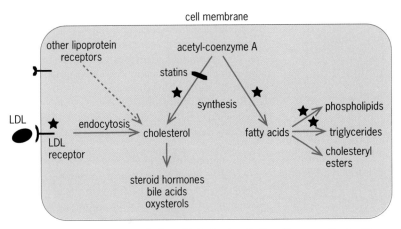

Fig. 1. Cellular cholesterol metabolism. Cholesterol and fatty acids are synthesized from acetyl coenzyme A. HMG-CoA reductase and cholesterol synthesis are inhibited by statins. Cholesterol-rich LDL is endocytosed by the LDL receptor. "Other lipoprotein receptors" include a number of scavenger receptors, the very low density lipoprotein receptor, CD36, and the LDL receptor–related protein. Pathways that are regulated in response to changing levels of nuclear SREBP are indicated by a star.

called the sterol regulatory element binding protein (SREBP), to the promoter of each sterol-regulated gene (Fig. 2b). The sequence bound by SREBP is termed a sterol regulatory element (SRE). Three proteins, SREBP-1a, SREBP-1c, and SREBP-2, have been identified, and all of them can bind to sterol regulatory elements. However, each SREBP is synthesized as a large precursor molecule of approximately 125 kilodaltons that is tethered to the endoplasmic reticulum and the nuclear envelope membranes via two transmembrane domains (Fig. 2a). The membrane-bound precursors have three distinct domains:

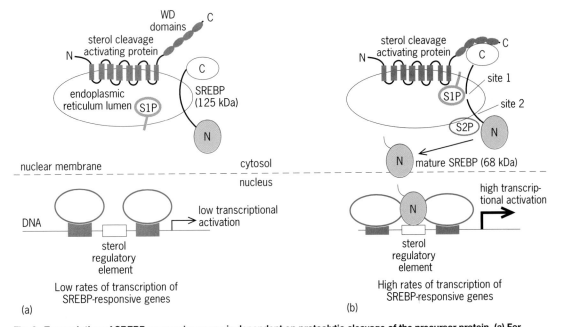

(a)   (b)

Fig. 2. Transcription of SREBP-responsive genes is dependent on proteolytic cleavage of the precursor protein. (a) For sterol-loaded cells, the precursor SREBP (125 kDa) is shown bound to the endoplasmic reticulum. The site-1 protease (S1P) and the sterol cleavage activating domain, containing specific WD domains, are localized in the same membranes. N and C refer to the NH₂—, and COOH—terminus of the indicated protein. (b) In sterol-deprived cells, the WD domains of the sterol cleavage activating domain interact with SREBP, and the site-1 and site-2 proteases (S2P) are activated. Transcriptional activation of SREBP-target genes requires that mature SREBP bind to the sterol regulatory element and that other transcription factors [NF-Y or Sp1 (open ellipsoids)] bind to the DNA at specific sites (solid rectangles) adjacent to the SRE. Transcriptional activation requires a large number of other proteins, including RNA polymerase II, that are not shown.

an NH$_2$-terminal domain that consists of a basic helix-loop-helix leucine zipper, a central domain that contains two transmembrane segments that are linked via a 31-amino-acid chain localized in the lumen of the endoplasmic reticulum, and a regulatory COOH-terminal domain. In cholesterol- or sterol-loaded cells, the SREBPs are localized in membranes outside the nucleus, and thus cannot function as deoxyribonucleic acid (DNA) binding transcription factors. Consequently, transcription of SREBP-responsive genes, such as the LDL receptor and HMG-CoA reductase gene, is low. This situation occurs in the livers of mammals that consume a diet rich in cholesterol and saturated fatty acids. Normally, approximately 75% of the LDL receptors in mammals are expressed on the surface of liver cells, where cholesterol, derived from the LDL, can be converted to bile acids and excreted into the intestine. Decreased transcription of the hepatic LDL receptor gene, as a consequence of a high fat diet, results in decreased expression of the LDL receptor on the surface of liver cells, decreased clearance of LDL cholesterol, and its accumulation in the blood.

A decrease in cellular cholesterol levels results in the proteolytic cleavage of SREBPs with the release of a soluble, mature 68-kDa NH$_2$-terminal fragment (Fig. 2*b*). The proteolysis of SREBPs requires at least three other proteins: sterol cleavage activating protein (SCAP), the site-1 protease, and the site-2 protease. Low cellular cholesterol levels result in an interaction between SCAP and SREBP and the activation of the site-1 protease. The site-1 protease cleaves the SREBPs within the endoplasmic reticulum luminal loop, leaving the two halves of SREBP still bound to the membrane. However, cleavage of SREBPs at site-1 results in activation of the site-2 protease, which cleaves the NH$_2$-terminal fragment within its transmembrane domain. This second proteolytic event releases the mature 68-kDa fragment from the membrane. The mature SREBP enters the nucleus, binds to sterol regulatory elements, and activates transcription. The result is increased expression of the LDL receptor and the enzymes involved in cholesterol synthesis. The increased expression of the LDL receptor results in increased clearance of the cholesterol-rich LDL from the blood into the cell. This, coupled with increased synthesis of cholesterol within the cell, results in a normalization of cellular cholesterol levels.

With this knowledge, it is easy to understand how the statin class of drugs reduce the level of blood LDL cholesterol. Statins are currently taken by millions of people. The drug is cleared from the bloodstream into the liver, where it partially inhibits the activity of HMG-CoA reductase, the rate-limiting enzyme of cholesterol synthesis (Fig. 1). The result is a decrease in cholesterol synthesis and cellular cholesterol level. As described above, such a decrease in cell cholesterol levels results in activation of the site-1 protease and an increase in the cleavage of SREBPs, in the nuclear levels of mature SREBP, and in the expression of a number of SREBP-responsive genes that include

the LDL receptor. Thus, statins result in elevated expression of the LDL receptor on the surface of liver cells, an increase in the rate of clearance of LDL from the blood, and a decrease in blood LDL cholesterol levels. This decrease in the blood LDL cholesterol level can decrease the risk of heart attacks.

Recently, mature SREBPs have been shown to bind to sterol regulatory elements in the promoters of genes that encode enzymes involved in fatty acid, phospholipid, and triglyceride synthesis (Fig. 1). Thus, mature SREBPs regulate the expression of genes involved not only in cholesterol synthesis and LDL endocytosis but also in fatty acid, phospholipid, and triglyceride synthesis. It is clear that other mechanisms not involving SREBP are also important in the regulated expression of genes involved in the biosynthesis of cholesterol, fatty acids, phospholipids, and triglycerides.

The identification and elucidation of the mechanism by which sterols regulate the transcription of genes involved in lipid homeostasis comes 45 years after dietary cholesterol was first shown to inhibit cholesterol synthesis. Future studies are likely to identify the natural physiological regulator of the site-1 protease and exactly how this activation process occurs. Such insights may lead to the development of new hypolipidemic drugs.

For background information *see* ARTERIOSCLEROSIS; CELL MEMBRANES; CHOLESTEROL; METABOLIC DISORDERS; STEROL in the McGraw-Hill Encyclopedia of Science & Technology.   Peter A. Edwards

Bibliography. P. A. Edwards and J. Ericsson, Sterols and isoprenoids: Signaling molecules derived from the cholesterol biosynthetic pathway, *Annu. Rev. Biochem.*, 68:157–185, 1999; J. L. Goldstein and M. S. Brown, The SREBP pathway: Regulation of cholesterol metabolism by proteolysis of a membrane-bound transcription factor, *Cell*, 89:331–340, 1998; M. Kreiger, The best of cholesterols, the "worst" of cholesterols: A tale of two receptors, *Proc. Nat. Acad. Sci. USA*, 95:4077–4080, 1998; J. Sakai et al., Molecular identification of the sterol-regulated luminal protease that cleaves SREBPs and controls lipid composition of animal cells, *Mol. Cell*, 2:505–514, 1998.

Circadian rhythm

Daily rhythms of activity and rest in animals are usually well coordinated with environmental cycles of light and dark. However, research in the past proved that environmental cues are insufficient to determine this behavior. Work in rodents has identified a specific brain region, the suprachiasmatic nucleus (SCN), that possesses circadian (approximately 24-h) clocks. Loss of these tissues produces arrhythmic behavior even in the presence of a day-night cycle. Cycles of sleep and wakefulness are restored by transplanting the suprachiasmatic nucleus from a healthy brain; environmental cycles are not needed for the production of the rhythmic behavior. When individual cells of the suprachiasmatic nucleus are propagated outside the animal, they produce approximately 24-h cycles of the electrical and metabolic activity, indicating a molecular mechanism that can generate a cellular circadian rhythm.

Clock genes. The molecules establishing these cellular and ultimately behavioral rhythms have been identified by genetic studies of the fruitfly, *Drosophila*. These insects produce easily monitored locomotor activity rhythms that resemble vertebrate sleep-wake cycles, and mutations have been detected that either eliminate or change the period of these rhythms. The search for mutants has identified five genes that collaborate to produce circadian rhythmicity. To find these five genes, many tens of thousands of individual flies were tested for evidence of altered behavior. Comparable genetic screening would be an enormous and expensive undertaking in a more complex organism such as a rodent but could provide clues about the underlying mechanism regulating human circadian behavior. Fortunately, alternative methods allowed rodent and human clock genes to be isolated based on their similarity to the five clock genes of *Drosophila*.

Two *Drosophila* genes, *period* (*per*) and *timeless* (*tim*), can be thought of as the mainspring of the fly clock. These genes are activated and repressed with a circadian rhythm. Each day, activation occurs around 10 a.m. and the genes are repressed together at about 8 p.m. Consequently, ribonucleic acid (RNA) produced by the *per* and *tim* genes rises and falls, and PER and TIM proteins encoded by the RNA also cycle with a circadian rhythm.

Mutations that block the fly's ability to make the TIM protein stop the cycling of both *per* and *tim* RNA. Mutations that halt PER protein synthesis also stop both RNA cycles and rhythmic production of TIM proteins. This genetic cross talk indicates a close working relationship of the *per* and *tim* genes. These mutations also affect the final destination of the PER and TIM proteins within the cell. Usually PER and TIM proteins accumulate in cell nuclei, and this accumulation is associated with times of *per* and *tim* gene inactivation. In mutants that cannot produce PER proteins, TIM proteins are no longer transported to nuclei and *per* and *tim* gene activity remains high. TIM protein mutants likewise block transport of PER proteins to the nucleus and also fail to properly inactivate the *per* and *tim* genes.

Protein interactions. These interdependent functions of *per* and *tim* were found to reflect physical associations that take place between the PER and TIM proteins. The two proteins have an affinity for each other and will dimerize to form a PER-TIM molecular complex. Interaction of the two proteins is required for transport to the nucleus. Before this pairing occurs, each protein is sequestered in the cytoplasm by the action of a specialized protein segment referred to as cytoplasmic localization domain (CLD). This inhibitory element prevents nuclear entry, and it also forms a motif for recognizing and binding to a partner protein. Thus, the CLD carried by PER is

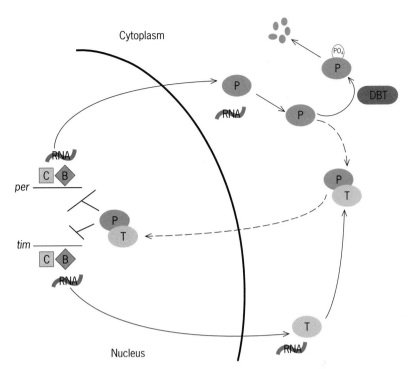

Molecular model of the *Drosophila* clock. In the absence of PER (P) and TIM (T) proteins, CLOCK (C) and BMAL (B) promote transcription of the *per* and *tim* genes, leading to gradual accumulation of *per* and *tim* RNA. The RNAs are translated to form monomeric PER and TIM proteins that are retained in the cytoplasm. If PER and TIM bind to each other, the resulting dimers are transported to the nucleus. Nuclear PER and TIM dimers inhibit *per* and *tim* gene activity by interfering with the function of CLOCK and BMAL. The rate of PER and TIM dimerization is delayed by DBT, a kinase which appears to stimulate phosphorylation and degradation of PER when it is not bound to TIM. Delayed PER and TIM dimerization promotes molecular cycling of *per* and *tim* transcription.

bound by a segment of TIM when the proteins associate. When bound in a PER-TIM complex, the two CLDs no longer promote cytoplasmic accumulation; amino acid sequences elsewhere on the linked PER and TIM proteins signal transport to the nucleus.

The strong regulation of nuclear transfer of the proteins is significant because, once in the nucleus, they affect *per* and *tim* gene expression. PER and TIM do this by interfering with the activities of two transcription factors, CLOCK and BMAL (also known as CYCLE). In the absence of PER and TIM, the latter proteins bind to an E-Box, a short regulatory sequence on *per* and *tim* deoxyribonucleic acid (DNA) [see **illus.**]. This DNA binding stimulates RNA synthesis. However, when present in the nucleus, PER-TIM dimers appear to physically bind CLOCK and BMAL and suppress further *per* and *tim* transcription.

Molecular response to light. These molecular interactions are strongly influenced by cycles of light and dark. Light promotes rapid degradation of the TIM protein, and the mechanism for this destruction involves, at least in part, an unusual photoreceptor called cryptochrome. *Drosophila* maintained in darkness and subsequently exposed to daylight lose their TIM proteins within a few minutes. TIM accumulation continues to be suppressed until the flies are returned to darkness. This response allows light and dark cycles to impose rhythmic patterns of *per* and *tim* gene activity that exactly match the period of the environmental oscillation. Gene activity will be elevated during the day because of the absence of PER-TIM suppressors, but gene expression will be inhibited at night when both TIM and PER can accumulate.

The reponse of TIM to light also demonstrates how a fly's rhythms would be reset by travel across time zones. In a 24-h environmental cycle (12 h of light and 12 h of darkness), highest levels of *per* and *tim* RNA will occur just after sunset when TIM starts to accumulate and to bind to PER. If sunset is delayed, as it would be for a westward bound traveler, TIM will not accumulate on schedule, and the molecular mechanism will freeze, starting again (with TIM accumulation) only when sunset finally occurs in the destination. The westward travel delays the molecular sequence and in so doing realigns the mechanism to the phase of the light and dark cycle at the destination.

Eastward travel has a different effect. For travel beginning late at night, at the time of departure high levels of nuclear PER and TIM proteins would be suppressing *per* and *tim* gene activity. As the traveler moves toward the destination, sunrise occurs prematurely. This curtailed night and early exposure to light will eliminate TIM and its suppressive effects ahead of schedule, allowing the *per* and *tim* genes to resume their activity with a new, advanced phase.

Absence of environmental cues. A mutation that altered the structure of the PER protein, so that it bound poorly to TIM, was found to produce a significant behavioral change. The binding became worse with increasing temperature and gave rise to a progressively slower clock in constant darkness. The results suggested that the rate of physical association of the PER and TIM proteins is a central factor in establishing the pace of the clock in the absence of environmental cues.

Other experiments indicated that TIM not only was required to transport PER proteins to nuclei but also had an additional protective function, shielding the PER protein from an activity that would destroy it if it were not bound to TIM. This suggested involvement of a mechanism that delays the accumulation of PER proteins until high protective levels of TIM have been achieved. Such delays would promote cycles of *per* and *tim* gene expression in the absence of an environmental rhythm of PER and TIM nuclear transport. The gene responsible for this regulation, and the fifth clock gene to be isolated in *Drosophila*, is called *double-time* (*dbt*).

Loss of the *dbt* gene causes clock cells to accumulate high levels of PER proteins, most of which are not bound to TIM proteins. In normal *Drosophila*, PER proteins tend to become heavily phosphorylated by the addition of phosphate groups to certain amino acids in the protein. In *dbt* mutants this too changed as PER proteins showed little or no phosphorylation. Transcription of the *per* and *tim* genes was also affected by elimination of *dbt*. In constant darkness, the activity of both genes was strongly suppressed at all times. This indicated (1) that PER proteins were overproduced in the mutants not be-

cause more *per* RNA was made but because PER proteins had somehow become resistant to degradation, (2) that the stabilized PER proteins could still associate with TIM, travel to nuclei, and downregulate gene expression, and (3) that instead of cycling, the clock in *dbt* mutants settled at an equilibrium: high levels of stable PER proteins constantly suppressed gene activity, but such high PER levels were nevertheless maintained and balanced by extremely low levels of *per* and *tim* transcription.

Physical characterization of the *dbt* gene showed how it may promote molecular cycling. The gene encodes a protein kinase, an enzyme that adds phosphates to target proteins. The kinase (DBT), like TIM, will physically bind to PER, but in contrast to TIM, DBT is made at all times of day. Early in the molecular cycle, when PER and TIM are first made, this PER-DBT partnership is probably favored and stimulates PER phosphorylation. This is a signal for protein degradation that is shared by many cell processes. As high levels of TIM protein accumulate later in the cycle, PER-TIM dimers are increasingly favored. PER-TIM dimers are resistant to the destabilizing effects of the kinase, allowing PER and TIM nuclear transport and the regulation of gene activity. Thus, DBT delays the time of PER-TIM binding by requiring the cell to first accumulate a high level of TIM.

The kinase also appears to influence the stability of PER after it has been transferred to nuclei. After a few hours, PER-TIM dimers dissociate and PER is once again phosphorylated in response to DBT, and then degraded. Although the events triggering production of free PER proteins from PER-TIM dimers are unknown, by restricting the duration of nuclear activity, DBT again facilitates cycling of gene expression.

Homologs of clock genes in mice and humans. Relatives of all five *Drosophila* clock genes have been discovered in mammals. Three *per* genes have been cloned from mice, and these appear to have human relatives. Each mouse gene produces a cycling RNA, albeit with a different phase. Robust cycling is seen in the suprachiasmatic nucleus, pointing to an evolutionarily conserved role in circadian behavior. One *tim* gene has been recognized in mice and humans. Little or no *tim* RNA cycling is seen in the mouse suprachiasmatic nucleus, but a rhythm has been reported in the retina, which also houses a circadian clock and expresses oscillating *per* RNA. Mammalian TIM and PER proteins will physically interact and will suppress activity of mammalian *per* genes in cultured cells. *Clock* was first identified in the mouse, and along with *BMAL*, is active in the suprachiasmatic nucleus. The kinase encoded by *double-time* is highly homologous to the human casein kinase Iε.

Behavioral response. The progression of this molecular clock is used to regulate activity of clock controlled genes (CCGs). Experimental work in mice and *Drosophila* has indicated that a first line of regulation is achieved by the clock proteins themselves: some responding genes are activated by the binding of CLOCK and BMAL proteins, so the presence of nuclear PER and TIM proteins directly times the activity of these clock controlled genes. In mice, vasopressin gene expression is controlled in this fashion. This cycling peptide hormone influences salt and water balance, electrical activity of neurons of the suprachiasmatic nucleus, and the daily accumulation of certain stress-related hormones such as corticosterone.

For background information *see* BIOLOGICAL CLOCKS; ENDOCRINE SYSTEM (INVERTEBRATE); GENE; MIGRATORY BEHAVIOR; REPRODUCTIVE BEHAVIOR in the Mcgraw-Hill Encyclopedia of Science & Technology.   Michael W. Young

Bibliography. R. Allada et al., A mutant *Drosophila* homolog of mammalian *Clock* disrupts circadian rhythms and transcription of *period* and *timeless*, *Cell*, 93:791-804, 1998; T. K. Darlington et al., Closing the circadian loop: CLOCK-induced transcription of its own inhibitors *per* and *tim*, *Science*, 280:1599-1603, 1998; J. C. Dunlap, Molecular bases for circadian clocks, *Cell*, 96:271-290, 1999; X. Jin et al., A molecular mechanism regulating rhythmic output from the suprachiasmatic circadian clock, *Cell*, 96: 57-68, 1999; J. Price et al., *double-time* is a novel *Drosophila* clock gene that regulates PERIOD protein accumulation, *Cell*, 94:83-95, 1998; S. M. Reppert, A clockwork explosion!, *Neuron*, 21:1-4, 1998; M. W. Young, The molecular control of circadian behavioral rhythms and their entrainment in *Drosophila*, *Annu. Rev. Biochem.*, 67:135-152, 1998.

Climate

Temperature records, maintained since about 1860, show an overall rise in the Earth's average temperature of about 0.75°C (1.35°F). Recent predictions suggest that this average temperature will rise another 1-4°C (1.8-7.2°F) over the next 100 years. This rate of change is ten times faster than any other climate change in the last 100,000 years, during which time the Earth has been repeatedly covered and then uncovered by glaciers. Even though the Earth has certainly been colder and hotter in the past, the abruptness of the climate change of the late 1990s is unprecedented.

Physical background. There is more consensus among scientists about global climate change than is reflected by most popular press coverage. Admittedly, it is very difficult to prove that recently observed temperature changes are due to human activities and nothing else, for the Earth's temperature fluctuates considerably even without any human input. Thus, the problem in detecting global climate change is to distinguish the human-induced component from the background signal. Nonetheless, there is broad agreement over the increase in temperature that has occurred since 1860 (most of this change occurring since 1960) and its cause—the amounts of greenhouse gases in the atmosphere that collectively serve to trap infrared radiation within the Earth's atmosphere, preventing atmospheric cooling. *See* GLOBAL WARMING.

Impact on individuals. Animals and plants commonly are able to tolerate a broader range of temperatures and other physical parameters than ordinarily encountered in the wild. For example, pigs can be raised in Texas, where the temperature is much higher than that encountered by ancestral wild boars in temperate Europe and North America. Many laboratory studies show that organisms can adapt to a certain extent. This adaptability has both a physiological and genetic component. Even though organisms can tolerate a change in average temperature, they may not reproduce very well in a warmer environment, and therefore their populations will be impacted. For this reason, climate alteration is unlikely to kill organisms outright; rather it is probable that organisms will respond to a warming climate by shifting the centers of their distribution. Shifts in distribution are common and have been extensively documented.

Most discussions of climate change focus on increasing carbon dioxide concentration in the atmosphere and the resulting increase in temperature. Another aspect of climate transition is the increasing level of ultraviolet radiation reaching the Earth. The increase in ultraviolet radiation is the result of depletion of atmospheric ozone. Ozone depletion is triggered by elevated levels of chlorofluorocarbons that are derived from aerosols and refrigerator coolants. The depletion of ozone is most pronounced over Antarctica, where an ozone hole has been identified. This increased ultraviolet radiation has been implicated in developmental abnormalities in toad embryos and increased incidence of skin cancers in South American sheep.

Impact on populations. Change in the Earth's mean temperature will likely have dramatic impacts on populations; indeed, there is strong evidence that changes have already occurred. Sooty shearwaters (*Puffinus griseus*), which nest on islands off New Zealand and Chile, are the most abundant seabird of the North American west coast in the northern summer. These birds feed on rockfishes (*Sebastes* sp.), squids, and planktonic invertebrates such as euphausiids in the highly productive California Current. The number of sooty shearwaters was estimated at 5 to 10 million during the mid-1970s. The population declined by approximately 90% between 1987 and 1994. This decline is closely correlated with an increase in ocean temperature. It is likely that a relatively small shift in sea temperature resulted in a major change in oceanic circulation. The changes in circulation have yielded smaller quantities of dissolved nutrients in the surface layers of the California Current, and consequently there has been less food for the zooplankton upon which the shearwaters and other pelagic predators feed. This example illustrates how a relatively small variation in a climatic parameter can have a profound, nonlinear consequence for the biota. Thus, the changing climate of the Earth is likely to have profound, perhaps unpredictable, consequences beyond physiologically stressing organisms.

The long-term change described above is distinct

from short-term episodes of ocean warming associated with El Niño–Southern Oscillation. Periods of unusually warm surface waters associated with the El Niño phenomenon of the west coast of South America can have devastating impacts upon marine animals, but these episodes are of short duration (less than 1 year). The decline of productivity within the California Current is related to a long-term warming trend. The long-term warming and El Niño may be linked, for El Niño episodes appear to be occurring with increasing frequency since the 1970s.

Computer models of climate change predict that the interiors of continents are likely to experience increased frequency of droughts. In central Canada, reproductive success of waterfowl is closely linked to the number and extent of prairie pothole ponds during the nesting season in summer. If the models are correct, North American waterfowl populations are likely to decline by 50%.

Populations can respond to a changing climate either by dispersing or by adapting physiologically or genetically to the new temperature regime. There is disagreement about whether populations can evolve in as short a time span (less than 100 years) as that predicted for ongoing climate changes. There is some evidence that organisms can adapt rapidly; for example, house sparrows (*Passer domesticus*) introduced into North America from Europe in the late 1800s have evolved morphologically over a time span of tens of years, or close to the scale at which atmospheric warming is anticipated to take place. In addition, there are a number of laboratory studies suggesting that at least some organisms can evolve to tolerate a new climate. Of course, it is the familiar larger vertebrates, which have long life-spans and long generation times, that are least likely to evolve in a short time. For populations confined to reserves, or patches of habitat that have become isolated due to human-induced destruction outside the reserve, dispersal may not be a realistic option. Thus, specialized species confined to isolated reserves are at the highest risk from climate change.

Other changes in population attributes linked to climatic warming include earlier arrival and later departure of migrant birds and butterflies, range expansion, and earlier dates of reproduction for birds and amphibians.

Impact on communities. To assess the impact on communities, interactions between species must be measured. Quantifying interactions such as competition, predation, and parasitism lies at the heart of ecology. Yet ecologists have found it extremely difficult to quantify these interactions—that is, to document the impact, in terms of population growth, of one species upon another. Therefore, it is especially challenging to anticipate how these interactions may change as the Earth's climate warms. There are, nevertheless, some patterns of past changes that suggest how ecological communities may change in the future.

The fossil record of pollen shows how distributions of tree species changed as the Earth's climate varied from glacial to interglacial periods. In general,

the pattern is not one of concurrent shifts of entire communities of species. Instead, each species of tree seems to have shifted its distribution independent of the shifts of other species. This independence of shifts resulted in unique assemblages of species appearing in different places at different times.

In the California Current, it seems likely that the reduction in the number of seabirds and other upper-level predators is a consequence of the decline of zooplankton stocks. However, it is not possible to say whether the nature of the bird-zooplankton interaction itself has changed as a consequence of climate change. As sooty shearwaters have declined in abundance, their congener the pink-footed shearwater (*P. creatopus*) has increased. This could be an instance of a climatic shift favoring one species over another. It is not clear whether an interaction (competition) among these species has been changed or whether each species has independently responded to the changing climate.

Invasions. Invasions of species into previously unoccupied environments have had devastating impacts upon communities, especially those on islands. It seems that invasive species often have an advantage in the new environment in that they are less susceptible to the local parasites and predators. Examples include the zebra mussel (*Dreissena polymorpha*) that now clogs water pipes in the Great Lakes and the brown tree snake (*Boiga irregularis*) that has caused the extinction or near-extinction of the avifauna of Guam and many other islands in the Indo-Pacific region. Global climate change will likely increase the incidence of natural invasions, because dispersal will be favored as environmental conditions deteriorate in some areas and improve in others.

For background information *see* CLIMATE HISTORY; CLIMATIC PREDICTION; CLIMATOLOGY; ENVIRONMENT; PACIFIC OCEAN in the McGraw-Hill Encyclopedia of Science & Technology.   Richard R. Veit

Bibliography. D. M. Gates, *Climate Change and Its Biological Consequences*, Sinauer, 1993; P. Kareiva, J. G. Kingsolver, and R. B. Huey (eds.), *Biotic Interactions and Global Change*, Sinauer, 1993; J. A. McGowan, D. R. Cayan, and L. M. Dorman, Climate-ocean variability and ecosystem response in the northeast Pacific, *Science*, 281:210–217, 1998; R. R. Veit et al., Apex marine predator declines ninety percent in association with changing oceanic climate, *Global Change Biol.*, 3:23–28, 1997.

Climate change

Analysis of the geothermal gradient near the surface of the Earth is extending knowledge of land surface temperatures hundreds of years into the past before the advent of the instrumental meteorological record. Vertical temperature versus depth profiles of the upper crust (down to 600 m; 2000 ft) contain a recoverable record of surface temperature change over the past few centuries due to particular thermophysical properties of the continental crust, low thermal diffusivity, and a steady-state surface heat flux from the interior. This new method of climate change research provides a direct record of past temperatures, unlike proxy data such as tree rings, paleobotany, stable isotopes, and lake salinity, which rely on analogy or transfer functions. Continued data collection and analysis on all continental areas has the potential to yield a global temperature record that may enable scientists to understand different forcing mechanisms in global climate change.

Geothermal gradient studies. Geophysicists studying the Earth's thermal regime beneath the continents routinely record the geothermal gradient, that is, the rate of change of temperature versus depth underground (**Fig. 1**), using temperature probes lowered into boreholes up to 600 m (2000 ft) deep. Their interest is in the documentation of heat flow from the Earth's interior and application of the results in a variety of studies such as Earth history, global tectonics, petroleum source rock maturation, and geothermal energy.

A common problem in these studies is that spatial and temporal variability in the ground surface temperature causes curvature or "noise" in the near-surface geothermal gradient. Spatial variability may be caused by a number of factors such as different microclimates resulting from topography, proximity to water bodies, and variability in ground cover. These spatial variability conditions and their effects are well known and are avoided or accounted for in data collection and analysis.

The temporal noise signal arises from transient changes in surface temperature which occur regularly on daily and annual scales and irregularly in response to synoptic weather patterns, year-to-year climate variability, and long-term climate change. These transient signals diffuse downward in the subsurface, and their amplitudes decrease exponentially with depth such that the disturbance to the geothermal gradient penetrates only to a vertical depth proportional to the period of variation and to the thermal diffusivity of the ground. The thermal diffusivity of rocks and soils is of the order of 10^{-11} m^2 s^{-1} so that annual signals remain detectable only within the upper 20–30 m (66–100 ft), signals from

the past decade are detectable within the upper 70–80 m (230–260 ft), and century-scale signals are detectable over lengths of hundreds of meters. Since heat flow research has largely focused on steady-state heat flux from the interior, geothermal gradient measurements routinely have been made in the deepest sections of boreholes to avoid the climate noise in the data. Ironically, heat flow researchers are now seeking this noise signal.

In 1986, A. Lachenbruch and V. Marshall used temperature-depth (T-z) profiles from the upper few hundred meters of petroleum exploration boreholes on the north slope of Alaska to show that the surface of the permafrost has warmed by 2–4°C (3.6–7.2°F) during the past century. The surface temperature increase recovered from the T-z profiles was linked to the concept of global warming, and launched a new field of research on climate change.

Geothermal gradient theory. The geothermal gradient (Γ) is determined by the ratio of heat flow (Q) to thermal conductivity (κ) and may be expressed by rearranging Fourier's law of heat conduction as $\Gamma = Q/\kappa$. On a planetary surface, the controlling quantities in Fourier's law are thermal conductivity and heat flow from the interior. Thermal conductivity is a physical property of rocks and generally falls within the range of 1.0–3.5 W m^{-1} K^{-1}, depending on rock composition and fabric. Heat flow from the interior is partly acquired during formation of the planet and partly generated by long-lived radioactive isotopes of potassium, thorium, and uranium that are present in trace amounts. Due to the nature of these sources, surface heat flow is constant on time scales of hundreds of millions of years, and the background geothermal gradient is linear. However, this applies only in stable continental interiors. Therefore, in the absence of heat transfer by groundwater flow, curvature in the near-surface T-z profile indicates a recent change in the bounding surface temperature. Cooling trends cause the surface intercept to swing toward cooler temperatures, and warming trends cause the surface intercept to swing toward warmer temperatures. Due to the exponential diffusion of the signal with depth and the thermal diffusivity values for crustal rocks, surface temperature changes occurring 100 years ago affect the T-z profile at about 100 m (330 ft) depth, and changes occurring 500 years ago affect the T-z profile at about 500 m (1650 ft) depth.

A useful property of the exponential diffusion of the signal with depth is that short-period signals, for example, interannual variability, are filtered out and long-term changes become readily identifiable. The filtering effect (**Fig. 2**) uses an air-temperature time series for the period 1900–1998 as a forcing function to generate time series of temperatures at 10, 20, and 30 m (33, 66, and 100 ft) depths. Although the increase in air temperature is obvious, linear regression of the data yields a low correlation coefficient: $r = 0.51$. By comparison, the correlation coefficients for the subsurface temperature time series at 10, 20, and 30 m are 0.84, 0.94, and 0.98, respectively. Thus, diffusion in the subsurface smooths the noise signal

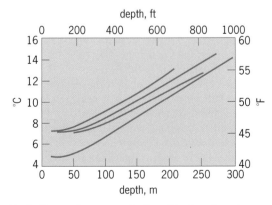

Fig. 1. Temperature versus depth profiles from four places in North America. The near-surface portion of each profile shows curvature toward warmer surface temperatures. The deeper portions of the profiles are relatively linear and provide a measure of the geothermal gradient at the borehole sites.

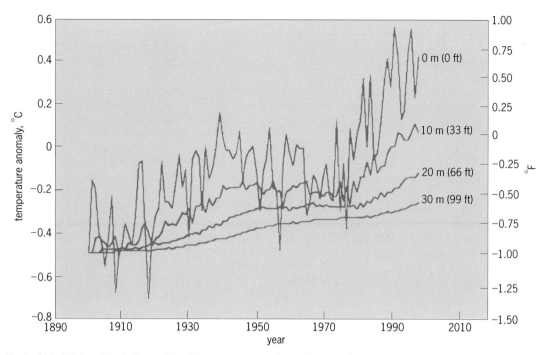

Fig. 2. Global History Climate Network land temperature anomaly used as a forcing signal to generate time series of subsurface temperatures at 10, 20, and 30 m depths. The reduction in signal-to-noise ratio increases with depth.

inherent in the interannual variability of surface temperatures and yields a long-term trend with a high degree of confidence.

Ground surface temperature history. Methods for determining the magnitude and timing of surface temperature change from the *T*-*z* profiles have evolved significantly since 1986. Lachenbruch and Marshall used the error function solution to the heat flow equation to develop forward models for different warming scenarios, for example, a step function, a linear increase, and warming rates that increase or decrease exponentially in time.

Although possible, it is unlikely that surface temperature changes obey such exact mathematical models. Thus, forward models cannot yield a precise record of ground surface warming, but they can provide an estimate of the magnitude of change and the probable time period for the change.

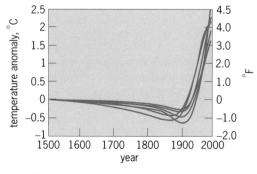

Fig. 3. Ground surface temperature history curves generated by inversion of six temperature depth profiles from sites in North Dakota. The cool period during the late 1800s commonly appears in North American borehole data and is a remnant of the Little Ice Age.

A more precise determination can be obtained by solving the inverse problem, that is, to use the *T*-*z* profiles to extract the ground surface temperature history (**Fig. 3**). The preferred approach is to remove the background heat flow signal from the data and treat the temperature anomaly that remains. In this approach, it is commonly assumed that heat transfer is solely by conduction in a one-dimensional system that may be heterogeneous. Two inverse methods have been widely used to infer the ground surface temperature history from *T*-*z* profiles: a linear formulation solved by the method of singular value decomposition, and a nonlinear bayesian formulation based on the method of least squares called functional space inversion. In the absence of noise, both methods produce essentially equal results. However, the functional space inversion method has been widely adopted by researchers attempting to assemble a global record of ground surface temperature histories.

Results of borehole climatology. Since the work of Lachenbruch and Marshall in 1986, researchers have applied the analysis of borehole temperature profiles to assess climate change. Many of the early reports addressed individual borehole sites or small regions and focused on showing that the record existed. These reports were largely confined to the northern portions of North America and Europe. A variety of methods for analysis were used, and the reported temperature changes varied in magnitude from slight cooling at some sites to 1–2°C (1.8–3.6°F) warming at other sites. Subsequently, data sets representing larger regions have been assembled, for example eastern, central, and western Canada and central North America, and efforts have been made to adopt a

Temperature change in the past century in midcontinent North America						
	Historical climatology network			Ground surface temperature history		
Latitude	°F	°C	Standard deviation	°F	°C	Standard deviation
41–42.9°	1.06	0.59	0.55	0.88	0.49	0.31
43–44.9°	1.30	0.72	0.62	1.86	1.03	0.66
45–46.9°	2.05	1.14	0.54	3.60	2.00	—
47–49.6°	2.41	1.34	0.40	4.39	2.44	0.52

uniform method of analysis.

There are two ongoing efforts to synthesize the data on a global scale. The project Climate Change Inferred from the Analyses of the Underground Temperature Field involves approximately 150 scientists collaborating on a range of issues, including paleoclimate reconstruction based on the interpretation of the inverted temperature logs, theoretical modeling, application of various techniques to the measured temperature profiles, comparison of the geothermal reconstruction and long-term meteorological series, proxy data, and soil-air temperature coupling.

The project Global Database of Borehole Temperatures and Climate is being used to compile global T-z profiles and thermophysical property data with the goal of producing a global assessment of surface temperature change during the past 500 years. The T-z data are restricted to 20–600 m (65–1970 ft) depths. Data above 20 m are omitted because they include annual variability, and data below 600 m are omitted because they contain no information about the past 500 years. As of late 1998, a database of 358 borehole analyses had been assembled in eastern North America, central Europe, southern Africa, and Australia. The results show that the averageBOTTOM surface temperature of the continental regions has increased by 0.5°C (0.9°F) this century and that the Earth's mean surface temperature has increased by about 1.0°C (1.8°F) over the past five centuries.

In a regional study designed to compare the ground surface temperature history with predictions of global circulation models of warming, the functional space inversion scheme was used on a group of specially drilled heat flow boreholes in a 1000-km-wide (620-mi) transect in midcontinent North America. The results show a century-long warming trend that increases systematically with latitude as predicted by global circulation models (see **table**).

These results, however, suggest a complicated relationship between the surface air temperature record and the reconstructed ground surface temperature history. Although this history is a valid climate indicator in itself, use of the method would be greatly expanded if the relationship between the ground surface temperature and surface air temperature records could be more clearly delineated. Most climate reconstructions using borehole temperature profiles have assumed a 1:1 relationship between the magnitude of the climate change inferred from the ground surface temperature history and that recorded in the surface air temperature history. If accurate, climate reconstruction based upon the ground surface temperature history would have enormous potential for filling in the global pattern of historic climate change in regions where the surface air temperature history is either spatially sparse or temporally limited. Further research on this aspect of borehole paleoclimatology is needed.

For background information *see* BAYESIAN STATISTICS; CLIMATE HISTORY; CONDUCTION (HEAT); EARTH, HEAT FLOW IN; EARTH CRUST; ESTIMATION THEORY; GEOLOGIC THERMOMETRY; HEAT RADIATION; HEAT TRANSFER; PALEOCLIMATOLOGY; RADIOACTIVE MINERALS in the McGraw-Hill Encyclopedia of Science & Technology.   Will Gosnold

Bibliography. H. Beltrami, A. M. Jessop, and J.-C. Mareschal, Ground temperature histories in eastern and central Canada from geothermal measurements: Evidence of climate change, *Palaeogeog. Palaeoclimatol. Palaeoecol. (Global Planet. Change Sec.)*, 98:167-183, 1992; D. Deming, Climatic warming in North America: Analysis of borehole temperatures, *Science*, 268:1576-1577, 1995; W. D. Gosnold, P. E. Todhunter, and W. Schmidt, The borehole temperature record of climate warming in the mid-continent of North America, *Global Planet. Change*, 15:33-45, 1997; J. T. Houghton et al., *Climate Change 1994: Radiative Forcing of Climate Change and an Evaluation of the IPCC IS92 Emission Scenarios*, Cambridge University Press, 1994; A. H. Lachenbruch and K. Marshall, Changing climate: Geothermal evidence from permafrost in the Alaskan Arctic, *Science*, 234:689-696, 1986; T. Lewis (ed.), Climatic change inferred from underground temperatures, *Palaeogeog. Palaeoclimatol. Paleoecol. (Global Planet. Change Sec.)*, 98:78-282, 1992; J. C. Mareschal and H. Beltrami, Evidence for recent warming from perturbed geothermal gradients: Examples from eastern Canada, *Clim. Dynam.*, 6:135-143, 1992; H. N. Pollack and D. S. Chapman, Underground records of changing climate, *Sci. Amer.*, 268:44-50, 1993; P.-Y. Shen, H. N. Polack, and S. Huang, Inference of ground surface temperature history from borehole temperature data: A comparison of two inverse methods, *Global Planet. Change*, 14:49-57, 1996.

Combinatorial chemistry

The main objective of combinatorial chemistry is synthesis of arrays of chemical or biological compounds called libraries. These libraries are screened

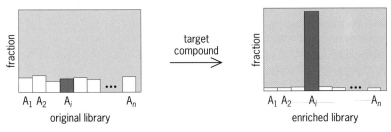

Fig. 1. Dynamic equilibrium for mixture components ($A_1, \ldots A_n$), following Le Chatelier's principle. T indicates the target.

to identify useful components, such as drug candidates. Synthesis and screening are often treated as separate tasks because they require different conditions, instrumentation, and scientific expertise. Synthesis involves the development of new chemical reactions to produce the compounds, while screening aims to identify the biological effect of these compounds, such as strong binding to proteins and other biomolecular targets.

Dynamic combinatorial chemistry. Dynamic combinatorial chemistry is in its early developmental stage. It integrates library synthesis and screening in one process, potentially accelerating the discovery of useful compounds.

In the dynamic approach the libraries are not created as arrays of individual compounds, but are generated as mixtures of components, similar to natural pools of antibodies. One important requirement is that the mixture components (**Fig. 1**) exist in dynamic equilibrium with each other. According to basic laws of thermodynamics (Le Chatelier's principle), if one of the components (A_i) is removed from the equilibrated mixture, the system will respond by producing more of the removed component to maintain the equilibrium balance in the mixture.

The dynamic mixture, as any other combinatorial library, is so designed that some of the components have potentially high affinity to a biomolecular target. These high-affinity (effective) components can form strong complexes with the target. If the target is added to the equilibrated mixture, when the effective components form complexes with the target they are removed from the equilibrium. This forces the system to make more of these components at the expense of other ones that bind to the target with less strength. As a result of such an equilibrium shift, the combinatorial library reorganizes to increase the amount of strong binders and decrease the amount of the weaker ones (Fig. 1). This reorganization leads to enrichment of the library with the effective components and simplifies their identification.

Although the general idea of the equilibrium shift is not new, its combinatorial application was demonstrated only recently by several groups. As an example, researchers described a dynamic library that produced inhibitors for the enzyme carbonic anhydrase. The components of this library were formed by a reversible reaction between two series of compounds: amines and aldehydes (**Fig. 2a**). The products of this reaction, Schiff bases, existed in the dynamic equilibrium with the starting amines and aldehydes (building blocks) and with each other. Functional groups in the building blocks (R^x and R^y) could recognize binding pockets of the target enzyme, and the best combination of those groups could make a particularly strong binder. To render the library components stable to hydrolysis, they were converted to similar amines (Fig. 2a) with retention of their basic target-binding properties. The researchers showed that the addition of carbonic anhydrase to the library led to the preferred formation of a sulfonamide compound (Fig. 2b), as compared to other library components. Structures similar to the sulfonamide compound are known to inhibit carbonic anhydrase via strong binding to its active site.

This example shows that the dynamic combinatorial approach requires a reversible reaction that can keep multiple components in equilibrium. A number of such reactions have been used, leading to various types of dynamic libraries. These reactions include peptide bond formation and cleavage to form libraries of peptides that bind to an antibody, photoisomerization of alkenes that changes their shape to fit a target amino acid, transesterification resulting in dynamic libraries of macrocycles that can selectively recognize metal ions, thiol-disulfide exchange that organizes a library of artificial peptide receptors, and scrambling of ligands in zinc complexes

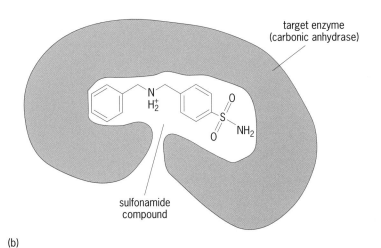

(a)

(b)

Fig. 2. Eqilibrium shift utilized to produce a dynamic library. (a) Reversible reaction between amines and aldehydes produces a Schiff base. (b) Addition of the enzyme carbonic anhydrase leads to the formation of the preferred sulfonamide compound.

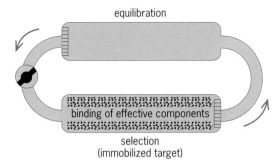

equilibration

binding of effective components

selection
(immobilized target)

Fig. 3. Separation of equilibration and binding sites in a dynamic combinatorial experiment.

to find an effective deoxyribonucleic acid (DNA) binder.

All the above examples, dealing with relatively small libraries, have been designed to prove the concept of the dynamic approach. Remarkably, however, all the systems described so far use different base reactions, target compounds, reaction media, and instrumentation. This fact indicates that numerous variations of the dynamic approach can be expected to develop in the future.

Problems. The dynamic combinatorial approach has some inherent problems. One is that it combines two intrinsically different processes: equilibrium between the library components and molecular recognition (selective binding) of the target, which generally require different conditions. One way to overcome this problem is to physically separate the equilibration and binding sites. Such separation has been accomplished by a stepwise approach to the organization of the dynamic libraries (chemical evolution). The target compound is immobilized by attachment to solid beads placed in a continuous-flow column, similar to a chromatography column (selection site). The immobilization does not allow the target to leave the column, and retains the target's properties. When the combinatorial mixture passes through the selection site, the effective components remain, bound to the target. Next, the depleted mixture enters another column (equilibration site), where special conditions, such as heating, catalysis, and irradiation, restore the equilibrium in the combinatorial mixture (**Fig. 3**). During this restoration, part of the mixture converts back into the effective components. The combinatorial mixture then returns to the selection site. After a number of such cycles, the original mixture evolves into a high-affinity subset accumulated on the target-containing beads. Later, this subset can be washed off the target and analyzed to identify the effective components.

Dynamic combinatorial chemistry is an emerging method that can potentially grow into a powerful tool for the discovery of new and useful compounds in molecular libraries. This approach may find its niche, for example, in the drug discovery process. The success of this method will depend upon the development of new reversible reactions that can be used to generate the dynamic libraries, and analytical methods that help identify the structure of components in complex mixtures.

For background information *see* CHEMICAL EQUILIBRIUM; LE CHATELIER'S PRINCIPLE in the McGraw-Hill Encyclopedia of Science & Technology.

Alexey V. Eliseev

Bibliography. P. A. Brady and J. K. M. Sanders, Thermodynamically-controlled cyclisation and interconversion of oligocholates: Metal ion templated living macrolactonisation, *J. Chem. Soc. Perkin Trans.*, 1:3237–3253, 1997; A. V. Eliseev and M. I. Nelen, Use of molecular recognition to drive chemical evolution, 1. Controlling the composition of an equilibrating mixture of simple arginine receptors. *J. Amer. Chem. Soc.*, 119:1147–1148, 1997; H. Hioki and W. C. Still, Chemical evolution: A model system that selects and amplifies a receptor for the tripeptide (D)Pro(L)Val(D)Val, *J. Org. Chem.*, 63:904–905, 1998; I. Huc and J. M. Lehn, Virtual combinatorial libraries: Dynamic generation of molecular and supramolecular diversity by self-assembly, *Proc. Nat. Acad. Sci. USA*, 94:8272, 1997; B. Klekota, M. H. Hammond, and B. L. Miller, Generation of novel DNA-binding compounds by selection and amplification from self-assembled combinatorial libraries, *Tetrahed. Lett.*, 38:8639–8642, 1997; P. G. Swann et al., Nonspecific protease-catalyzed hydrolysis synthesis of a mixture of peptides: Product diversity and ligand amplification by a molecular trap, *Biopolymers* 40:617–625, 1996.

Communications satellite

For many years satellite communications mostly utilized satellites in geostationary orbits, in which the satellite remains directly overhead at a fixed location on the Equator at an altitude of 35,786 km (22,236 mi). More recently, alternative satellite orbits have been seen to have distinct advantages and have found increasing use. The first two sections of this article discuss global communications networks based on constellations of satellites in low Earth orbit. The Iridium system makes it possible to make or receive calls from anywhere in the world using a handheld mobile telephone, while the SkyBridge system will provide worldwide broadband communications services. The third section discusses the advantages of satellite orbits that are geosynchronous but not geostationary in providing services to geographical regions at high latitudes.

Iridium System

The Iridium global communications system is based on a network of 66 low-orbiting satellites that are linked to land-based telephone networks. It provides comprehensive wireless communications by integrating landline, cellular, and other terrestrial wireless systems with satellites.

As an alternative to geostationary satellites, the Iridium system is one of a wide range of global services being developed by an increasing number of privately financed, low-Earth-orbit satellite networks. Until the 1990s, satellites were considered too expensive and too closely linked to national security to be owned and operated by anyone other than governments or international government cooperatives. But sophisticated, lower-cost systems have made new telecommunication ventures possible.

Principles of operation. In cellular telephone technology, a user moves through cells of coverage provided by land-based antennas. The Iridium system, however, relies on satellite-based antennas traveling overhead. In effect, the satellites function as cellular towers in the sky. A critical distinction is that a network or constellation of low-Earth-orbiting satellites is required to provide the same coverage that one geostationary satellite offers at a much higher altitude. However, the lower altitude of the constellation allows for the use of smaller, handheld phones and negligible transmission delays.

The Iridium system is distinguished by intersatellite links, known as cross-links, which connect each satellite by radio transmission to four others in the constellation of 66 spacecraft. Each satellite is in communication with the leading and trailing satellite in its plane (fore and aft) and a satellite in each adjacent co-rotating plane (right and left). The primary reason for cross-links is to minimize the number of ground stations required to support the communications network. With each satellite connected to its neighbors in the constellation, a call originating in one part of the world can be relayed from satellite to satellite to its destination in another region (**Fig. 1**).

Cross-links have a long history of use in government and military systems. The National Aeronautics and Space Administration's (NASA) Tracking and Data Relay Satellite System and the U.S. Army's Milstar system employ satellite-to-satellite links. Cross-links also make it possible for the Iridium system to provide full global coverage.

Fig. 1. Iridium global communications system.

Fig. 2. Iridium satellite.

The Iridium system relies on a "switchboard in the sky" to route calls between its satellites and to ground stations on Earth. The heart of an Iridium satellite is the communications payload, which is the key enabler for the Iridium system, and includes the equivalent of a sophisticated digital telephone switch. The on-board switches, in concert with Iridium satellite cross-links, enable a caller to be quickly connected to a called party anywhere in the world without the call being routed through terrestrial, public-switched telephone networks.

Structure. The Iridium system, in addition to having the largest currently orbiting satellite network, includes an extensive ground control network. Together, they permit communication with even the most remote locations on Earth, including the vast ocean surface.

The satellite network consists of six planes of 11 operational satellites in near-polar orbits. (Each plane also includes one backup satellite for a total of 72 satellites.) The satellites (**Fig. 2**), each weighing approximately 689 kilograms (1500 lb), complete an orbit every 100 min.

The ground control network serves as the central management component of the Iridium system. Two ground control facilities (primary and backup) monitor the satellite network. Four tracking stations send command signals to, and receive telemetry from, the satellites, serving as the conduit for commands from the primary ground control facility. A terrestrial operations support network is also used for communicating among these components.

The ground control network includes a dozen or so ground station gateways located in different regions of the world, connecting the Iridium satellite network to public-switched telephone networks. Four gateway antennas on each Iridium satellite link the constellation and the ground station gateways.

Capabilities. A handheld Iridium telephone works in two ways. First, as a cellular phone, it uses the available local cellular network. Second, if a cellular network is not available, calls are routed directly from the handheld phone to the Iridium satellite network. The result is that any type of telephone transmission—voice, data, fax, or paging—will be able to reach its destination anywhere in the world.

Unlike geostationary satellites, which have an altitude of 35,786 km (22,236 mi), the Iridium low-Earth-orbit satellites are deployed some 700 km (435 mi) above the Earth. The lower altitude, in combination with recent advances in semiconductors, microprocessors, and other technologies, reduces the transmission delay typically associated with geostationary satellite calls, and makes it possible to use a handheld telephone rather than the bulky equipment required by geostationary satellite systems.

The antenna of an Iridium phone communicates directly with an Iridium satellite's main mission antenna. Consisting of three phased-array antennas, the main mission antenna of each satellite produces 48 beams of coverage, which encompass a circular area with a diameter of approximately 4450 km (2765 mi). The beams of coverage of all 66 satellites in the Iridium network combine to form a continuous pattern on the Earth's surface.

The Iridium system also provides a solution for the gaps in cellular coverage and incompatible technology standards around the world. Cellular service now covers only 10% of the Earth's surface. Moreover, the three major digital wireless standards—code division multiple access (CDMA), time division multiple access (TDMA), and Global System for Mobile Communications (GSM)—and the advanced analog wireless standards (AMPS) are largely incompatible with each other. As a complement to cellular phone service, the Iridium system allows its customers to "roam" across any type of cellular network.

Satellite assembly. To accommodate a 66-satellite network, innovative mass-production techniques were developed. Unlike geostationary satellites, which traditionally have been assembled vertically and in one location, an Iridium satellite is assembled on a specially designed dolly that is rolled through 15 separate stations. Compared with the previous industry standard of 12–18 months, this assembly-line approach makes it possible to produce an Iridium satellite in less than 1 month.

Applications. As the first global wireless telephone network, Iridium is facilitating global connectivity. In a world where less than one-fifth of the land surface is wired for telecommunications services, the Iridium system will contribute, directly and indirectly, to major economic and social benefits.

The Iridium system improves the efficiency of international business travel by allowing travelers to contact their home office, no matter where they are located, even aboard a jet or ship. It also addresses the needs of a variety of industries, including multinational companies in sectors such as oil and gas production and exploration, mining, cargo shipping, power generation, commercial fishing, construction, and the media.

The Iridium system is expected to be useful for disaster relief, especially in countries with overburdened or incomplete cellular coverage. Unaffected by weather and damage to local telephone systems

and power lines, the Iridium system will be indispensable for use in damage assessment and coordination of resources. Governments, international relief organizations, and development institutions can use the Iridium system to coordinate their operations more closely, allocate resources more accurately, and deliver assistance more effectively.   Robert W. Kinzie

SkyBridge System

The SkyBridge system is a satellite-based global communications network that will provide business and residential users access to a variety of broadband communications services. Scheduled to commence service in 2001, the SkyBridge system will be optimized for highly interactive services, such as high-speed Internet access and video conferencing. The system is a commercial venture backed by an international group of technology companies. *See* MULTISERVICE BROADBAND NETWORK TECHNOLOGY.

A fundamental challenge to providing broadband services is the expense of the "last-mile" broadband link to the home or business. Typically, individual users obtain access to networks and servers, such as the Internet, by means of narrowband infrastructure, such as copper wire, which imposes severe limitations on the flow of data.

Satellite systems are capable of overcoming the local access challenge. Indeed, compared with providing such access by means of deploying fiber-optic infrastructure to homes and businesses throughout a large service region, the cost advantages of satellite systems are enormous. An appropriately designed satellite system can efficiently provide local access to users all over the world. Once the satellite system is deployed, each new user can be added to the network for a relatively small incremental cost corresponding to the cost of a user terminal, which includes a small antenna and related communications equipment.

The SkyBridge system will efficiently provide global access by means of a constellation of 80 satellites in low Earth orbit (LEO; as in **Fig. 3**). Each satellite in the constellation will transmit data, voice, and video between user terminals in the home or business and SkyBridge "gateway" Earth stations connected to terrestrial networks and servers (**Fig. 4**).

Communications links. The maximum data rates to and from the user are asymmetric, in conformance with the traffic patterns associated with access to services such as the Internet. The SkyBridge System will be able to transmit data to users at rates up to at least 100 megabits per second, and from users at rates up to at least 10 megabits per second, depending on the type of user terminal employed. The overall global capacity for the SkyBridge System will exceed 200 gigabits per second.

Signal transmission delay will be very low (about 20 milliseconds) because of the relatively short propagation distances between the Earth's surface and the LEO satellites. This is a key consideration for the highly interactive services that SkyBridge will

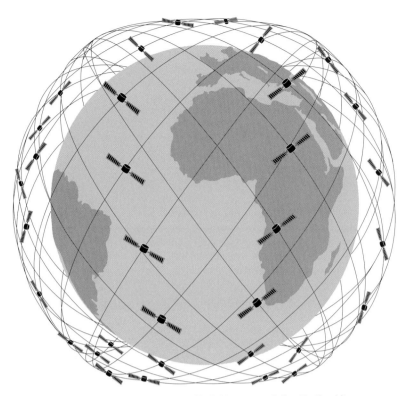

Fig. 3. Constellation of 80 satellites of the SkyBridge system in low Earth orbit.

Centralized management centers will monitor and control the satellites, and plan the most efficient use of the satellite resources to serve the customers.

Low Earth orbits. In contrast to geostationary satellites, placed in a high Earth orbit over the Equator that permits the satellite to remain at a fixed position with respect to the Earth, LEO satellites move with respect to the Earth. Accordingly, multiple LEO satellites are required to provide continuous service to a given service region on the Earth's surface. Traffic is then handed off from satellite to satellite as the satellites fly over the service region.

Constellation. The 80 SkyBridge satellites will be positioned in 20 orbital planes, with four equally spaced satellites per plane. The orbital planes will be inclined at 53° with respect to the Equator, and will be equally spaced around the Equator. All of the satellites will operate in circular orbits at an altitude of 1469 km (913 mi). Each satellite will circle the Earth every 115 min.

This constellation will permit continuous real-time service to users located within at least the ±68° latitudes, encompassing most of the world's population. At any location within this region, at least one satellite, and generally more, will be available at all times for communication with a given gateway Earth station or user terminal.

Space stations. The SkyBridge satellites are of a "bent-pipe" design, and will perform no signal processing, other than amplification and frequency translation. Each satellite employs multiple spotbeam antennas to simultaneously provide coverage to multiple gateways. Each SkyBridge satellite will be identical and therefore fully interchangeable with any other.

Terrestrial segment. The terrestrial segment of the SkyBridge system includes the gateway Earth stations and the user The gateways will be operated by local service providers and connected to local servers and terrestrial broadband and narrowband networks. The gateway operators or other local

be providing. Communications involving satellites in the high Earth orbits, including geostationary orbits, are (by comparison to LEO systems) significantly delayed, a sorce of annoyance in voice communications, for example.

The SkyBridge system uses automatic power control to minimize satellite and Earth station power and maintain quality of service. For example, power control permits compensation for fading attenuation due to rain.

Space segment. The SkyBridge space segment will comprise a constellation of 80 LEO satellites.

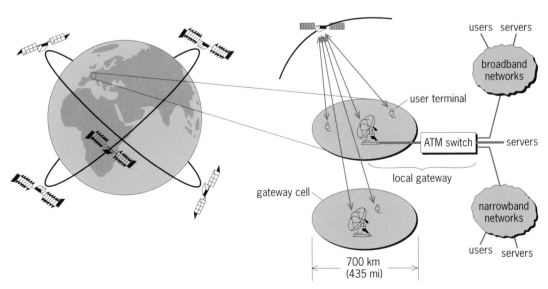

Fig. 4. Architecture of the SkyBridge system.

service providers will provide the user terminals and services over the SkyBridge system on a retail basis to consumers.

Gateway Earth stations. Each gateway Earth station will be located within a 700-km-diameter (435-mi) cell and will serve all of the user terminals located within the cell. At any given time, the gateway Earth station and user terminals in each cell will be served by at least one satellite. Approximately 200 gateway Earth stations will be required to serve the entire planned service region of the SkyBridge system.

Each gateway Earth station will be interconnected to local servers and to terrestrial broadband and narrowband networks by means of an asynchronous transfer mode (ATM) switch. This switch will provide high-speed data packet switching between the terrestrial network elements and components of the SkyBridge system.

User terminals. Each user terminal will include an outdoor and indoor component. The outdoor unit will feature a small antenna and related electronics for establishing the satellite link with the gateway Earth station. Linked to the outdoor unit will be an indoor device for delivering and collecting the user traffic from personal computers and television set-top boxes.

Two different types of user terminals will be used with the SkyBridge system. Residential user terminals, designed for the home, will offer low cost and easy installation and operation. Professional user terminals will offer higher capacity for office and multiunit residential buildings, including major corporate sites.

Frequency allocations. The SkyBridge system will use frequencies in the Ku-band (approximately 10–18 GHz) for all communications links. As the Ku-band has been used for satellite communications for many years, equipment for operating in this band is fully tested and less expensive than equipment for the higher-frequency bands to be used by other proposed broadband satellite systems. Furthermore, as compared to higher-frequency bands, the Ku-band is less susceptible to signal attenuation caused by rain and other atmospheric conditions.

Frequency sharing. The SkyBridge system employs a novel technique that permits it to operate without causing interference to geostationary satellite systems with which it will share spectrum. The SkyBridge technique exploits the directionality of the transmitters and receivers of geostationary systems. By avoiding transmissions in-line with the transmissions of geostationary systems, the SkyBridge system is able to employ the frequencies used by geostationary systems without causing interference to, or imposing operating constraints on, those systems. Thus, the SkyBridge system can operate without an exclusive allocation of spectrum, thereby promoting the efficient use of the scarce spectrum resource.

To implement this technique, each SkyBridge satellite will cease transmissions to a given gateway cell, and the gateway Earth station within that cell will cease transmission to the SkyBridge satellite, when the satellite enters the cell's "nonoperating zone." The nonoperating zone for each cell includes all SkyBridge satellite positions to or from which transmissions could potentially cause interference to a geostationary satellite system, and can be visualized as a region spanning $\pm 10°$ on either side of the geostationary orbital arc as seen from any point on Earth's surface within the cell. At the time of a shutdown of a beam to a given cell, the communications traffic of the cell will be seamlessly handed over to another satellite in the constellation that is not in the cell's nonoperating zone. Thus, the shutdown technique will not cause service interruptions. Use of this frequency-sharing technique ensures that the SkyBridge system will comply with applicable regulations, including those promulgated by the International Telecommunication Union and the U.S. Federal Communications Commission.

Regulation and deployment. The SkyBridge system requires regulatory approval from both international and national regulatory authorities. In November 1997, the World Radiocommunication Conference of the International Telecommunication Union provided international approval for nongeostationary satellite systems to use those portions of the Ku-band that will be used by the SkyBridge system.

The components for the SkyBridge system are being developed and manufactured during 1998–2000. Satellite launches and initial service offerings are scheduled to begin in 2001.

Phillip L. Spector; Diane C. Gaylor

Inclined, Elliptical, Geosynchronous Orbits

In the design of a communications satellite orbit, numerous alternatives and tradeoffs must be considered. Factors that depend on orbit geometry include the area of coverage, the dwell time, the angle of elevation, and the number of satellites simultaneously visible. Nongeometrical factors, such as orbital perturbations and launch vehicle constraints, must also be considered.

Using the concepts summarized in this section, an alternative type of geosynchronous communications satellite orbit will be described that provides better communication services to geographical regions and countries not near the Equator than the commonly used geostationary orbit.

Orbital elements. According to Kepler's first law, the orbit of a satellite is an ellipse with the center of the Earth at one focus (**Fig. 5**). The point of closest approach is the perigee, and the point farthest away is the apogee. The line formed by the intersection of the orbital plane with the equatorial plane is called the line of nodes, which connects the ascending and descending nodes and passes through the center of the Earth.

The orbit is described by six numbers, called the orbital elements. The semimajor axis a and eccentricity e determine the size and shape of the

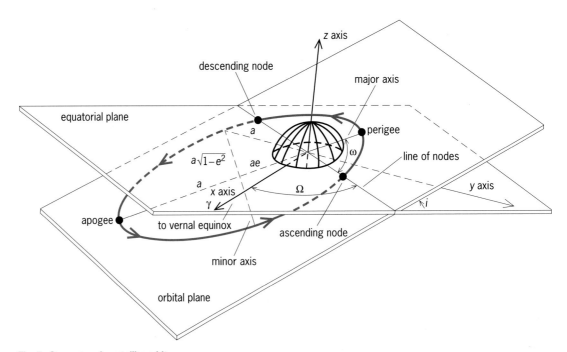

Fig. 5. Geometry of a satellite orbit.

orbit. The inclination i and right ascension of the ascending node (RAAN) Ω orient the orbital plane in inertial space. The argument of perigee ω orients the orbit within its own plane. The mean anomaly M_0 at a specified epoch establishes the time table of the orbit.

The semimajor axis a is one-half of the major axis. It is related to the period of revolution T by Kepler's third law, given by Eq. (1),

$$T^2 = \left(\frac{4\pi^2}{\mu}\right)a^3 \qquad (1)$$

where μ is the gravitational constant of the Earth (398,600.5 km^3/s^2).

The eccentricity e is the ratio of the distance of the focus from the center of the ellipse and the semimajor axis. For circular orbits, $e = 0$ and a satellite moves at uniform speed. For elliptical orbits, $0 < e < 1$ and the satellite moves at a variable speed governed by Kepler's second law, which states that the line joining the center of the Earth and the satellite sweeps out equal areas in equal times. The speed is slowest at apogee and fastest at perigee. The eccentricity is chosen to optimize the dwell time over a given region.

The inclination i is the angle between the orbital plane and the equatorial plane, measured at the ascending node where the satellite crosses the equatorial plane from south to north.

The right ascension of the ascending node Ω is the angle (right ascension) between the vernal equinox on the celestial sphere and the ascending node, measured in the equatorial plane.

The argument of the perigee ω is the angle between the ascending node and the perigee, measured in the plane of the orbit.

The mean anomaly M_0 at an epoch t_0 is related to the time of perigee τ. It is given by Eq. (2),

$$M_0 = n(t_0 - \tau) \qquad (2)$$

where $n = 2\pi/T$ is the mean motion.

A geosynchronous orbit is any orbit with a period of revolution equal to the period of rotation of the Earth (86,164 s). Thus it has a semimajor axis of 42,164 km (26,200 mi) but can have arbitrary eccentricity and inclination. The geostationary orbit is a special case with zero eccentricity (circular orbit) and zero inclination (orbit in the equatorial plane).

Ground trace. An important property of a communications satellite orbit is its ground trace. The ground trace is the path followed by the subsatellite point over the surface of the Earth. It results from the combined effects of the satellite motion in its orbit and the rotation of the Earth. If all the satellites in a constellation follow the same ground trace, each satellite must occupy a different orbital plane. For N equally spaced satellites, the difference in the mean anomalies of successive satellites is $\Delta M_0 = -360°/N$. If these satellites are in geosynchronous orbits, the separation of their orbital planes is $\Delta\Omega = 360°/N$.

Orbital perturbations. A keplerian orbit is an idealization that results from the inverse square law of gravitational force on a satellite due to an isolated, spherically symmetric central body. In practice, however, there are a variety of orbital perturbations that must be corrected occasionally by stationkeeping maneuvers.

In the geostationary orbit, the principal perturbations are third-body gravitational attractions due to the Sun and Moon. These perturbations produce a change in orbit inclination and account for about

95% of the stationkeeping fuel budget needed to keep the satellite in its desired orbit. (Stationkeeping maneuvers involve velocity adjustments that total about 50 m/s or 160 ft/s per year.) There are also small perturbations from the longitudinal acceleration produced by the Earth's triaxial shape and from solar radiation pressure.

Nongeostationary orbits are influenced by the Earth's oblateness, which has three principal effects. (1) The ascending node drifts to the west for inclinations less than 90° and to the east for inclinations greater than 90°. The ground trace can be maintained by a small compensating adjustment to the orbital period. (2) There is a rotation of the major axis. However, the major axis is stable for inclinations of 63.4° and 116.6°. At these angles, the perturbative forces that cause rotation cancel out, so that the apogee remains over its intended position. (3) The effect of oblateness is a small change in the orbital period.

Tundra orbit. The geostationary orbit is attractive because it offers fixed Earth-satellite geometry. The Earth station antenna looks in a constant direction and does not have to track the satellite. Typical elevation angles from the contiguous United States (CONUS) are between 30° and 55°, while those from major population centers in Europe are smaller.

A low elevation angle does not present a serious problem for point-to-point systems using large antennas deployed at Earth stations situated with little blockage from buildings and topography. However, higher elevations are desirable for small antennas located in urban areas and for mobile, low-gain antennas.

High angles of elevation to a satellite can be achieved using inclined orbits. To maximize the dwell time over a given region, an elliptical orbit may also be selected. If the region is confined to a particular country or area of the globe, a geosynchronous orbit is appropriate.

An interesting inclined, elliptical geosynchronous orbit is the 24-h *Tundra* orbit, which has an eccentricity of 0.2684 and, like the 12-h *Molniya* orbit with eccentricity 0.722, is inclined at 63.4°. For service coverage of the Northern Hemisphere, the argument of perigee is 270°, which places the satellite perigee in the Southern Hemisphere and the apogee in the Northern Hemisphere.

The satellite ground trace of the *Tundra* orbit (**Fig. 6**) resembles a figure-8 with small top loop and large bottom loop. With the eccentricity of 0.2684, the time intervals between perigee, ascending node, figure-8 node, apogee, figure-8 node, descending node, and perigee are each 4 h. The satellite spends a total of 16 h over the Northern Hemisphere.

To obtain continuous coverage over the contiguous United States, a constellation of satellites is required. Thus for three equally spaced geosynchronous satellites all following the same ground trace, the initial mean anomalies are offset by −120° and the plane separations are 120°.

Although the inclination of 63.4° is chosen to eliminate the effect of Earth oblateness that causes the major axis to rotate, there are nevertheless significant gravitational perturbations from the Sun and Moon that produce changes in all of the orbital elements. The total stationkeeping requirement is comparable to the requirement for the geostationary orbit. Since the perigee is in the Southern Hemisphere,

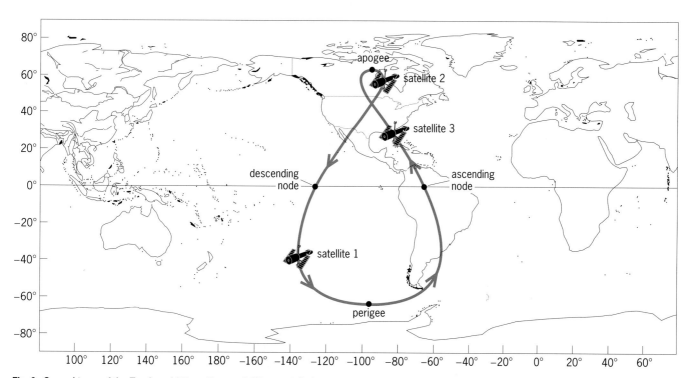

Fig. 6. Ground trace of the *Tundra* orbit for a three-satellite constellation.

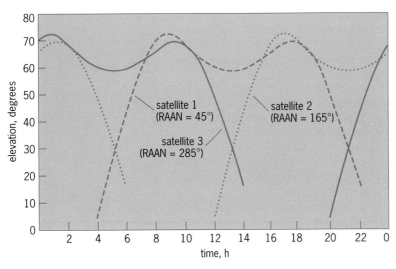

Fig. 7. Elevation versus time for a three-satellite constellation in *Tundra* orbit as seen from New York City.

the launch vehicle must follow a trajectory with an azimuth toward the south. This requirement places some constraints on the launch site and type of launch vehicle that may be selected.

The stationkeeping budget is minimized if the right ascensions of the ascending nodes are chosen to be initially in the vicinity of 45°, 165°, and 285°. Then for an epoch of January 1, 2000, 0 h Universal Time (UT), the mean anomalies that place the ground trace over the contiguous United States (Fig. 6) are about 50°, 290°, and 170°, respectively.

CD Radio will employ a satellite constellation of this type. This satellite system will provide digital audio radio service in the S-band (2.3 GHz) to mobile receivers in automobiles throughout the United States. The high angles of elevation achieved by this constellation ensure a high-quality signal and mitigate losses due to blockage from buildings and terrain, multipath effects, and attenuation by foliage. From New York City, for example, continuous single satellite coverage is obtained with a minimum elevation angle of about 60°, and dual satellite coverage is obtained with a minimum elevation angle of about 30° (**Fig. 7**).

For background information *see* CELESTIAL MECHANICS; COMMUNICATIONS SATELLITE; MOBILE RADIO; ORBITAL MOTION; PERTURBATION (ASTRONOMY); RADIO SPECTRUM ALLOCATIONS; SATELLITE (SPACECRAFT) in the McGraw-Hill Encyclopedia of Science & Technology.   Robert A. Nelson

Bibliography. G. D. Gordon and W. L. Morgan, *Principles of Communications Satellites*, John Wiley, 1993; R. A. Nelson, Iridium: From concept to reality, *Via Satellite*, 13(9):62–70, September 1998; J. N. Pelton, Telecommunications for the 21st century, *Sci. Amer.*, 278(4):80–85, April 1998; W. L. Pritchard, H.G. Suyderhoud, and R.A. Nelson, *Satellite Communication Systems Engineering*, 2d ed., Prentice Hall, 1993.

Computer-aided circuit design

Most electronic products employ very large scale integration (VLSI) circuits, often called microelectronic circuits. Such circuits may have over 1 million transistors. The large scale of integration is made possible by advances in semiconductor technologies that allow manufacture of transistors and wires with dimensions much smaller than 1 micrometer (micron). Design with such technologies is referred to as deep submicron design. Raising the scale of circuit integration is advantageous, because increasingly complex functions can be realized on a single integrated circuit, with a corresponding increase in reliability and decrease in overall cost.

Designing integrated circuits is extremely challenging because of the large amount of detailed technical data associated with transistors and their interconnections. Moreover, the ready availability of transistors allows many alternative implementations. Optimizing an integrated circuit means choosing an implementation that maximizes (or minimizes) an objective function of interest, such as performance, power consumption, area, and ease of testing. Due to competition in the integrated circuit marketplace, circuits must be optimal in some respect to be profitable. Thus, the design task is necessarily coupled with an optimization task. Because of this complexity, integrated circuits are designed using computer-aided design (CAD) tools.

Microelectronic circuits are manufactured using a photolithographic process, by which patterns corresponding to transistors and connection wires are transferred onto a silicon wafer, which is then customized by local physical or chemical reactions. The layout of an integrated circuit is a detailed draft of the geometries of the transistors and wires on several layers. Thus, the final objective of a computer-aided design system is to produce the circuit layout in a form suitable to be used for manufacturing.

There are four stages in the creation of an integrated circuit: design, fabrication, testing, and packaging (**Fig. 1**). Design is the refinement of a functional specification into the geometries needed for fabrication. Fabrication means manufacturing the dies, that is, the unpackaged circuits. Testing means checking for manufacturing defects that may lead to erroneous operation. Working dies are placed into ceramic or plastic packages. Computer-aided design tools support design and testing, while computer-aided manufacturing tools support the remaining stages.

Most electronic circuits are digital; that is, the basic information is quantized into two levels, corresponding to true and false. By contrast, information in analog circuits is represented in a continuous range. Since digital microelectronic circuits constitute the majority of designs, most computer-aided design tools are conceived to support digital circuit design.

Digital circuit design. Design can be split into three major tasks: conceptualization and modeling, synthesis and optimization, and validation. Conceptualization consists of casting an idea into a model that captures the function the circuit will perform. Synthesis consists of refining the model from an abstract one to a detailed one with all the features required for fabrication. Validation consists of verifying the consistency of the models used during design, as well as some properties of the original model.

Modeling. Circuit modeling plays a major role in microelectronic design, because it represents the vehicle used to convey information. Modeling must be rigorous as well as general, transparent to the designer, and machine-readable. Most modeling is now done using hardware description languages (HDLs). Graphic models are sometimes used, such as flow and schematic diagrams. The very large scale nature of the problem forces the modeling style, both textual and graphical, to support hierarchy and abstractions, allowing a designer to concentrate on a particular portion of the model at any given time.

Synthesis. The overall goal of circuit synthesis is to generate a detailed model of a circuit, such as a geometric layout, that can be used to fabricate the chip. This objective is achieved by means of a stepwise refinement process, during which the original abstract model provided by the designer is iteratively detailed. As synthesis proceeds in refining the model, more information is needed regarding the technology and the desired design implementation style. A functional model of a circuit may be fairly independent from the implementation details, while a geometric layout model must incorporate all technology-dependent specifics, such as wire widths.

Circuit synthesis is usually layered in three stages, related to the modeling abstraction level involved: high-level synthesis, logic-level synthesis, and physical design (**Fig. 2**). High-level synthesis, also called architectural-level synthesis, consists of generating the macroscopic structure of a circuit. This activity corresponds to determining an assignment of the circuit functions to operators, called resources, as well as their interconnection and the timing of their execution. Logic-level synthesis is the task of generating the gate-level structure of a circuit. It entails the manipulation of logic specifications to create logic models as interconnections of logic primitives. Logic design is often eased by the use of libraries of logic gates. The task of transforming a logic model into an interconnection of instances of library cells, that is, the back end of logic synthesis, is often referred to as library binding or technology mapping. Physical design consists of creating the geometric layout of the circuit. It entails the specification of all geometric patterns defining the physical layout of the chip, as well as their position.

Circuit optimization is often combined with synthesis. It entails the selection of some particular choices in a given model, with the goal of raising one or more figures of merit of the design. The role of optimization is to enhance the overall quality of the circuit. Most designers equate quality with circuit performance. Performance involves the time re-

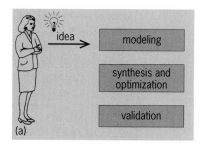
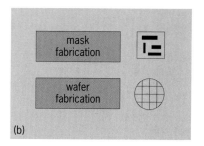
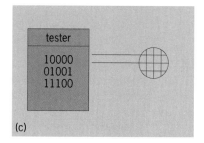
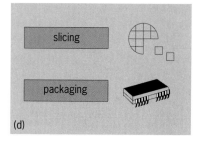

Fig. 1. Four phases in creating a microelectronic chip. (*a*) Design. (*b*) Fabrication. (*c*) Testing. (*d*) Packaging. (*After G. De Micheli, Synthesis and Optimization of Digital Circuits, McGraw-Hill, 1994*)

quired to process some information, as well as the amount of information that can be processed in a given time period. Circuit performance is essential to competitive products in many application domains. Power consumption is an important quality measure for microelectronic circuits to be used for portable applications, where the autonomous operation time depends on the battery capacity and discharge rate. Another measure of circuit quality is the overall circuit area. An objective of circuit design is to minimize the area, for three main reasons: (1) smaller circuits allow more dies to be placed on each wafer, and therefore enable lower manufacturing costs to be achieved; (2) the manufacturing yield decreases with an increase in chip area; and (3) large chips are more expensive to package. The circuit quality also involves the circuit testability, that is, the ease of testing the chip after manufacturing. Testable chips are clearly desirable because earlier detection

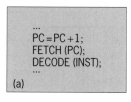
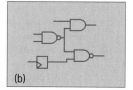
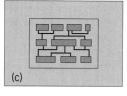

Fig. 2. Three abstraction levels of a circuit representation. (*a*) Architectural level. (*b*) Logic level. (*c*) Geometrical level. (*After G. De Micheli, Synthesis and Optimization of Digital Circuits, McGraw-Hill, 1994*)

of malfunctions in electronic systems enables overall costs to be lowered.

Validation. Circuit validation consists of acquiring a reasonable certainty level that a circuit will function correctly, under the assumption that no manufacturing fault is present. Circuit validation is motivated by the need to remove all possible design errors before proceeding to expensive chip manufacturing. It can be performed by simulation and by verification methods. Circuit simulation consists in analyzing the circuit variables over an interval of time for one specific set (or more sets) of input patterns. Simulation can be applied at different levels, corresponding to the model under consideration. Simulated output patterns must then conform to the expected ones. Even though simulation is the most commonly used way of validating circuits, it is often ineffective for large circuits except for detecting major design flaws. Indeed, the number of relevant input pattern sequences needed to analyze a circuit grows with the circuit size, and designers must be satisfied with monitoring a small subset. Verification methods, often called formal verification methods, consist of comparing two circuit models and detecting their consistency. Another facet of circuit verification is checking some properties of a circuit model, such as whether there are deadlock conditions.

Testing. Microelectronic circuits are tested after manufacturing to screen fabrication errors. Thus, testing is the verification of correct manufacturing. Since impurities are unavoidable in the air of manufacturing plants, some defects in random positions may be present on some circuits.

There are several testing techniques, and their relation to design is important. In particular, circuits may be designed with a self-testing feature. After manufacturing, the circuit can be set in testing mode, where it will generate, apply, and analyze testing patterns to check proper operation.

Design for testability means addressing the ease of testing during the design stage. Circuit testability affects the circuit quality. A circuit that is not fully testable is less valuable than another one that is fully testable, because faulty units may be released on the market. Enhancing testability can be done by appropriate logic and architectural synthesis techniques.

Analog circuit design. Despite the fact that most circuits are digital, analog circuits and analog components of mixed-signal circuits are necessary to perform specific functions, such as radio transmission. Analog functions are captured with appropriate languages and graphic formalisms. Analog synthesis tools address mainly the physical assembly of layouts from parametrized components. Analog circuit validation is usually achieved by circuit simulation. Specialized tools for analog design address design problems for radio-frequency and microwave components.

Hardware/software codesign. Microelectronic circuits may contain complex building blocks, such as processors, controllers, and memories. Processors and controllers may be designed to execute specific programs stored in memory, which are referred to as embedded software. For integrated circuits comprising programmable components executing software programs, hardware/software codesign techniques are needed to exploit the synergism between hardware and software to achieve the design objectives. Computer-aided codesign tools support a variety of tasks, including performance analysis of hardware/software systems, support for deciding on the hardware/software implementation of system functions, and support for developing software compilers that are retargetable to various processor architectures.

For background information *see* COMPUTER-AIDED DESIGN; COMPUTER-AIDED MANUFACTURING; INTEGRATED CIRCUITS; OPTIMIZATION in the McGraw-Hill Encyclopedia of Science & Technology.

Giovanni De Micheli

Bibliography. G. De Micheli, *Synthesis and Optimization of Digital Circuits*, McGraw-Hill, 1994; G. De Micheli and M. Sami, *Hardware/Software Co-Design*, Kluwer, 1996; H. Fujiwara, *Logic Testing and Design for Testability*, MIT Press, 1985; M. Sarrafzadeh and C. K. Wong, *An Introduction to VLSI Physical Design*, McGraw-Hill, 1996.

Computer-aided engineering

The evolution of computer-aided engineering has paralleled that of computer technology. Engineering was one of the first disciplines to take advantage of the computer's ability to work with great precision, reliability, and ever-increasing speed, first with numbers, then with nonnumerical symbols, and then with graphic and visual images. The term "computer-aided engineering" understates the significance of the field. The computer has become such an indispensable part of modern engineering that the term "computer-based engineering" might be more appropriate. This article illustrates the impact of computer technology on the process of designing complex systems by focusing on the design of the most complex mobile system ever developed, the large modern warship. These vessels have lengths up to 1000 ft (300 m) and displacements up to 100,000 tons (90,000 metric tons), and are capable of sustaining over 6000 people for extended periods while surviving severe weather and seas and enemy attack.

Ship design process. The design process starts when the need for a new ship system is established as the result of short falls in the operational capability in the fleet, leading to the formulation of the system requirements (the requirement phase). These requirements must be realistic, that is, based on feasible and affordable systems alternatives. Concept exploration follows and leads to the selection of the preferred concept. During system design, all subsystems are developed followed by detail design, where all parts are defined to permit manufacture. Each phase involves iterations of design definition, analysis, and review, followed by the decision either to revise the design or to proceed to the next, lower level of detail in the process of design and analysis.

Computer-aided engineering technologies have profoundly changed this process, affecting all phases. These changes could initially be characterized as evolutionary. Then high-performance computing and high-performance visualization made virtual prototyping possible. In parallel, advances in communication and networking led to virtual co-location of widely dispersed contributors to the design process and created the virtual enterprise. Together, these advances are revolutionizing the way that future ships will be designed.

Analysis tools. The evolutionary part of this transformation involves replacing manual methods with computer-aided methods. This is accomplished by first automating individual steps in the process and then gradually increasing the scope of automation by integrating related steps. Specifically, the computer's ability to solve first simple algebraic and then complex differential equations, and to perform matrix inversions, soon led to the ability to solve complex engineering problems. Large-scale finite element analysis became possible, and the discipline of computational fluid dynamics was born. Analysis tools specifically for the ship engineering and design process include programs to estimate weights; to help analyze stability, survivability, and reliability/maintainability/availability; to predict a ship's radar cross section; or to perform multispectral signature analysis. *See* COMPUTATIONAL FLUID DYNAMICS; SHIP DESIGN.

Computer-aided design. First functioning primarily as a tool for analysis, the computer was eventually "programmed" to control device movements. The control of a pen resulted in computer-aided drafting. Controlling the head of a flame-cutting (or laser-cutting or -welding) machine led to numerical control and computer-aided manufacturing. When event sequences could be digitally defined, processes could be modeled and simulated. This process simulation was then used to refine and optimize the process modeled. Furthermore, combining analysis tools and coded parametric design algorithms led to design synthesis models. As the computer's ability to manage the three-dimensional geometry of a ship's evolving definition grew, computer-aided design was born. The U.S. Navy's design program, ASSET, makes it possible to develop dozens of alternative early-stage design concepts in weeks. Using the hull form design system (HFDS), it is possible to perform in a matter of days the complete detailed definition (and fairing) of a ship's hull and to predict the ship motion behavior.

Product models. Ever-increasing data storage capabilities now make it possible to deal efficiently with the digital definition of a product as large and complex as a large aircraft carrier or a nuclear submarine. Such a database, commonly referred to as a product model, contains not only the complete three-dimensional definition of the ship and all of its parts but also the latest estimates of its aggregate attributes such as weight and volume, energy supply and con-

sumption, and logistics support requirements, that is, all data relevant to the design, manufacture, testing, and operation of the product. The huge volume of data mandates efficient data management, including ease of extraction and update. Associativity of data is increasingly provided. As one data element is changed in the course of the design evolution, other, dependent data can be updated automatically. Such databases are sometimes called smart product models. Nevertheless, for more complex changes, the designer may want to continue providing active, decision-making input to the feedback loop of the design process, rather than rely on artificial (that is, programmed) intelligence.

The objectives of underlying product models are (1) to define data only once, (2) to store it where it can be found and used by all members of the design team, and (3) to define it in a format that is directly usable by all design and analysis programs. Creation of an integrated product data environment for ship design remains a great challenge since many individual computer-aided design/computer-aided engineering software vendors have developed their own product- or company-specific environment.

Soft prototypes. These computer-aided engineering capabilities provide the means for creating a digital mock-up, the software equivalent of a complete three-dimensional model of the ship being designed. This model, coupled with advanced visualization techniques, allows validation-by-inspection to be performed. A digital mock-up can be viewed, much like a physical mock-up. Interferences can be detected, walk-throughs can be performed, and adequacy of access can be ascertained. Such a product definition, since it exists entirely in software, has been referred to as a soft prototype (not to be confused with a virtual prototype, discussed below).

Shortening of design process. The impact of the computer-aided engineering capabilities described so far permit dramatic reductions in the duration of the individual design phases and, thereby, the entire design process. More rapid evaluation and accelerated design convergence and optimization can be achieved. Actual reductions, however, have been less dramatic for two reasons. First, the advanced capabilities have been applied to greatly enlarge the trade space; that is, many more alternative design solutions are now developed and evaluated. Second, in the past, design data relating to the ship's hull form existed in parametric form derived from empirical data well into the design process. Today, even early analyses are generally based on the actual detailed definition of the hull, the so-called offsets. As a result, the depth of the analysis has greatly increased. Finally, the system development and acquisition process for warships is highly regulated by defense procurement policies, with milestones for high-level review. Still, goals for up to 50% cycle time reduction have been established.

Virtual prototyping. This technology is about to revolutionize the design process, less by shortening the process duration than by dramatically affecting the nature and the quality of design. Virtual prototyping grew out of the idea of not only designing a system entirely in the computer but also testing it there, that is, operating it in order to verify that performance requirements were met. Inherent in this capability is the opportunity for operators and users to participate in the design process to an unprecedented degree. Operator experience with the ship system design—albeit gained by "operation" in virtual, that is, computer-based, synthetic environments—can be factored into the design before proceeding to build it. Such a virtual prototype can even be "operated" without being designed in detail. Concurrently, builders can simulate construction processes and feed the experience gained back into the product definition and design process. In other words, concurrent engineering becomes a reality as virtual prototypes, operating in a virtual environment, become the focal point for all aspects of a product's life cycle (see **illus.**).

The development and application of virtual environments precedes virtual prototyping. This technology has been used successfully in training applications, which also require very high fidelity. Simulation-based training is indispensable to the training of airplane pilots; it is also used extensively in training ship operators. Virtual environments, however, are suitable for vitual prototyping (as for simula-

tion-based training) only if four conditions are met: (1) virtual worlds, and systems in them, must be modeled correctly concerning dimensions and spatial relationships; (2) they must behave in accordance with the laws of physics; (3) they should appear realistic to the human eye; and (4) they must respond in real time. The first three conditions can be met with advanced computer-aided design/computer-aided engineering tools. To meet the fourth condition, dramatic advances in interaction and display technology were required, generally referred to as high-performance visualization. The basic building blocks in high-performance visualization are polygons, surface segments (primarily traingles and rectangles), that are linked together and colored, shaded, or texture-mapped with additional images. These simplifications are necessary to enable even today's powerful computers to perform the task of recalculating the changing perspective views of a complex virtual scene at a frequency of about 30 times per second in order to achieve real-time responsiveness while creating the illusion of smooth motion.

Virtual prototyping, therefore, is the culmination of computer-aided engineering. Virtual prototyping and the virtual co-location afforded by Web-based technologies, including distributed interactive simulation, have advanced computer-aided engineering to a point that all stakeholders in a systems life cycle

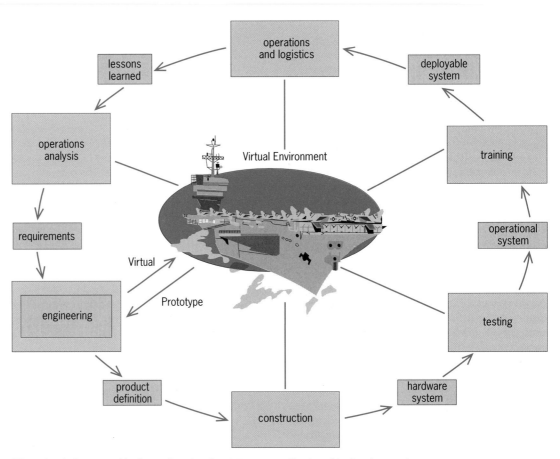

Life cycle of a large warship, focused on virtual prototypes operating in a virtual environment.

can actively participate in its development. These technologies strengthen the collective problem-solving abilities of human beings. People can act as teams, learning and improving, whether they are design teams, manufacturing teams, training teams, education teams, or war-fighting teams. Within the U.S. Department of Defense, an initiative called Simulation-based Acquisition, adopted in 1997, is being pursued to coordinate related efforts. An initial goal of cutting delivery time for new systems by 25% was stretched to 50% with the additional goal of reducing total ownership cost. A worldwide network offering real-time communication and sharing of virtual environments is under development, supporting not only engineering during all design phases but all related activities involved in a product's life cycle.

For background information *see* COMPUTATIONAL FLUID DYNAMICS; COMPUTER-AIDED DESIGN AND MANUFACTURING; COMPUTER-AIDED ENGINEERING; COMPUTER NUMERICAL CONTROL; FINITE ELEMENT METHOD; PROTOTYPE; SIMULATION in the McGraw-Hill Encyclopedia of Science & Technology.   Otto P. Jons

Bibliography. E. P. Andert, Jr., and D. Morgan, Collaborative, "virtual prototyping and test," *Nav. Eng. J.*, 110(6):17–24, November 1998; G. W. Jones et al., Using virtual environments in the design of ships, *Nav. Eng. J.*, 106(3):91–106, May 1994; B. Tibbits and R. G. Keane, Jr., Making design everybody's job: The warship design process, *Nav. Eng. J.*, 107(3):283–301, May 1995.

Computer-to-plate printing

The term computer-to-plate (CTP) is used within the printing industry to refer to devices, technologies, and work flow related to imaging aluminum or polyester printing plates without the use of film. In the 1990s, computer-to-plate matured rapidly from a concept of questionable validity to a well-established, practical manufacturing method. Although the current number of users represents only a small fraction of the total number of printing firms worldwide, it is expected that computer-to-plate will become the predominant method of imaging printing plates in the next decade.

Offset lithography. Offset lithography is a popular method for the mass reproduction of documents using ink on paper. The modern lithographic process depends on an image master called a printing plate, which is typically made of aluminum coated with a light-sensitive emulsion. The process known as imaging converts this emulsion into the hydrophilic and oleophilic surfaces that determine the printed and nonprinted areas of a page. Throughout most of the twentieth century, these printing plates were imaged photomechanically, meaning that a mechanical device (typically a sheet of film) was used to control which portions of the plate were exposed to light. Depending on the type of plate in use (positive-working or negative-working), the ultraviolet content of this light either caused the exposed surface of

the printing plate to become attracted to ink (oleophilic; also called the image area) or attracted to water, which in turn repels ink (hydrophilic; also called the background or nonimage area).

Although these basic tenets of offset lithography are still in place, most printshops are actively seeking ways to improve efficiency and streamline the production process of creating plates—especially if these goals can be accomplished while also improving reproduction quality. Computer-to-plate promises such benefits to the graphic arts industry. As pressure increases to produce printed work on a shorter deadline and for smaller budgets, printers are looking to computer-to-plate as a method by which the amount of work performed by manual processes (stripping, photomechanical platemaking, and proofing) can be drastically reduced.

Direct-to-plate. Direct-to-plate is often used as a synonym for computer-to-plate; however, direct-to-plate technologies do not have to be digital. Projection platemaking systems for newspapers have been around for decades and are still in widespread use, and photodirect methods of imaging electrostatic plates are very popular among small-format "quick printers." Both of these methods allow plates to be exposed directly from the original, without the use of an intermediate piece of film. It would be more accurate to think of computer-to-plate as an entirely digital subset of the larger number of direct-to-plate options.

Direct imaging. Direct-imaging presses are increasingly popular in today's graphic arts marketplace. The term "direct-imaging" (DI) refers to printing presses such as the Heidelberg QuickMaster DI or the Omni-Adast DI; these lithographic presses have been outfitted with digital platemaking equipment, allowing the plates to be made on the press itself. Plates are imaged via laser while the press is at rest; when the press begins printing, the plates are already mounted on the plate cylinders in register.

Platesetter. The term "platesetter" refers to any device capable of imaging a printing plate without the use of film. From the advent of the first known platesetter (the Lasergraph, 1974), the mechanism of choice for carrying out the imaging process has been the laser beam. The goal of actually imaging commercial-quality lithographic plates with a laser beam became a reality in the late 1980s with the debut of a polyester plate material that featured a light-sensitive coating. This material, introduced by 3M (now Imation Corp.), was designed to be exposed with an imagesetter (a laser-equipped device which is used to expose high-contrast film). In 1990 the DuPont-Howson Silverlith plate was introduced, featuring a similar light-sensitive silver emulsion, but coated over an aluminum substrate. The development of an aluminum plate that could be imaged via a laser beam sparked the creation of the modern computer-to-plate marketplace.

Although the number of choices presented by hardware vendors seems endless, platesetters can be divided into six categories, based on their physical geometry: flatbed, internal drum, external drum, imagesetters (for polyester plates), inkjet imaging, and all others. Vendors consider many pros and cons to distinguish each platesetter from its competition; there are also many successful users of each variety of computer-to-plate device. In the end, the differences between the devices seem less important than the reliability of the chosen device and the comfort level provided in regards to service and support. Since most adopters of computer-to-plate work flow begin with only a single platesetter, reliability is of paramount importance.

Independent of the choice of physical configuration, most platesetters have one or more laser light sources chosen from one of the following categories: laser diode, light-emitting diode (LED), YAG (yttrium-argon gas laser), argon-ion, thermal laser diode, and thermal YAG.

Flatbed. Flatbed platesetters were among the first commercially available products. This configuration holds the plate against a flat tablelike surface, which may travel in one direction or remain motionless, depending on the manufacturer. In the case of the moving table, a mirror is suspended above the table into which a single laser beam of light is projected. As the mirror swings back and forth to reflect the light from one side of the plate to the other, the table moves the plate forward until the entire plate is imaged. This configuration results in a platesetter with a large footprint (the amount of floor area occupied by the machine). In addition, the choice of light sources is limited due to the optical restrictions imposed to keep the narrow laser beam in focus from center to edge.

Internal drum. Internal drum platesetters use a curved drum into which a plate is placed. The plate is held tightly against the drum via vacuum channels; both the plate and the drum remain stationary during exposure. Similar to the flatbed concept, a rotating mirror projects a single beam of laser light around the inside of the drum. This mirror travels from one end of the drum to the other, until the entire plate is imaged. The curved surface of the drum keeps the mirror-to-plate surface distance constant, while the movement of the mirror in both directions results in a reduced footprint.

External drum. External drum platesetters wrap the flexible aluminum plate around the exterior of a large drum. This drum revolves around its axis while a number of light sources travel in a straight line across the length of the drum. The presence of multiple light sources (typically 8 to 48) allows the entire drum to be imaged in a brief period of time. Special precautions must be taken to ensure that the plate does not fly off the drum during the imaging process.

Imagesetters. Imagesetters are often used to produce polyester plates, especially in the smaller sizes convenient for duplicator presses. Imagesetter varieties include both internal drum (similar to the platesetters previously mentioned) and capstan. These devices do not require any modification to image polyester plate material, since polyester plates are sold in rolls of the same width as imagesetter film and are sensitive to the imagesetter's helium-neon laser. However, a few manufacturers offer specialized output devices that are designed just for the purpose of creating polyester printing plates, including the AB Dick/ITEK DPM2000 and Printware's PlateStream.

Inkjet. Inkjet imaging of plates for computer-to-plate remains a seldom-seen technology. IRIS is introducing a platesetter based on its popular inkjet proofing devices. The IRIS platesetter utilizes an uncoated, anodized sheet of aluminum onto which is sprayed an oleophilic coating to form the image area.

Digital platemaking. Most of the discussion over digital platemaking concerns thermal plates versus visible-light plates. Thermal plates are exposed by the component of a laser beam's energy that modulates too slowly to be seen; this spectrum is referred to as thermal energy (heat). This thermal energy can be generated in great quantities by the use of a YAG laser or in a less powerful fashion through the use of thermal laser diodes. Visible-light plates are exposed by laser beam energy that falls within the visible spectrum of light. Lasers used to expose visible-light plates include light-emitting diode, YAG, frequency-doubled YAG, argon-ion, and helium-neon.

The vast majority of digital plates adhere completely to the principle of offset lithography as a planographic reproduction process, meaning that the image carrier (printing plate) is a totally flat surface whose image and nonimage areas are defined by differing chemical properties. In this case, exposure from a laser beam (whether visible light or thermal) is used to initiate the change in selected areas from one chemical state to another. For a few new products, however, exposure from a thermal laser is used to actually ablate (vaporize) a portion of the plate's surface, revealing another layer below. A popular laser ablation plate, the Presstek PEARLgold, uses a thermal laser to ablate the top (hydrophilic) surface so that the bottom (oleophilic) surface can receive ink.

Although every vendor's product is unique, certain statements are typically true of thermal plates: thermal plates have the highest resolution of all digital plates (some products are able to resolve spots as small as 5 micrometers); thermal plates can be handled under yellow or ultraviolet-filtered white light; chemistry required for processing is minimal and nonhazardous (with the exception of bimetal plates); and thermal plates can be run with no special handling or fountain solution and are easy for a pressroom to adopt. Some of these plates (such as the KPG Thermal Plate/830) can also be exposed conventionally with an ultraviolet light source, allowing both digital and photomechanical imaging to occur on a single plate. This capability can allow a printshop using computer-to-plate to stock only a single type of plate, yet utilize both film- and laser-exposed work flows.

The higher cost and typically longer imaging times of thermal plates leave plenty of opportunity for use of visible-light plates. There is also the high expense of converting the existing base of prethermal platesetters to a new light source. Therefore, the computer-to-plate marketplace will undoubtedly continue to be a mixture of visible-light and thermal plates for some years.

For background information *see* BOOK MANUFACTURE; COMPUTER PERIPHERAL DEVICES; INK; LASER; PHOTOGRAPHY; PRINTING; TYPE (PRINTING) in the McGraw-Hill Encyclopedia of Science & Technology.

Hal Hinderliter

Bibliography. R. M. Adams and F. Romano, *Computer-to-Plate: Automating the Printing Industry*, 2d ed., 1999; M. Limburg, *The Essentials of Computer-to-Plate Technology*, 1995.

Computer virus

A computer virus is a computer program that is designed to replicate itself from file to file or from disk to disk. Viruses comprise the same basic instructions and program logic that constitute application programs, such as word processors, games, or spreadsheets. However, the computer virus is designed to spread itself without the user's knowledge, while normal applications are designed to provide a useful function. Computer viruses are written by people, and they must be intentionally designed and programmed to self-replicate. However, once a computer virus has been introduced into a system, it is capable of spreading without the aid of a human.

Prevalence. The Pakistani Brain virus, discovered by researchers in 1986, is widely believed to be the first computer virus. By 1990, roughly 50 viruses were known. Following a rapid increase during the latter 1990s, the number of distinct computer virus strains was estimated at over 20,000 in 1998. Luckily, very few viruses have found their way out of research laboratories and onto end-user computers. Based on statistics of the Wild List, an industry consortium, only 200–300 computer viruses are known to be in general circulation at any one time. (These viruses are called wild because they spread on their own outside of laboratories). While most infections are caused by a small set of viruses, many virus authors now post their creations on VX (Virus eXchange) sites on the World Wide Web, exposing the unwary Internet user to a potential hazard.

Damage. Contrary to widespread belief, most computer viruses are not intentionally damaging. Over two-thirds of them are designed with only self-replication in mind. In many cases, a user would be unaware that his or her computer was infected without checking it with antivirus software. Programs which are designed to cause damage to a computer system but do not replicate are called Trojan horse programs. Unfortunately, many computer viruses are poorly designed or "buggy" and may often cause significant unintended damage to computer programs and data. Viruses which do inflict intentional damage to computer systems are said to deliver a payload. Common virus payloads include formatting of the hard drive or deletion of computer files. These viruses typically have a "trigger" criterion and will wait until the criterion is met before delivering their payload. Common trigger criteria include waiting for the fiftieth boot-up since initial infection or waiting until a date such as March 6 (Michelangelo's birthday).

Design and programming. The typical computer virus is a very small program, several kilobytes or less in size. In general, designing and programming a computer virus is considered a fairly simple task. The typical virus writer can construct a simple new virus in a matter of hours or days. More complex viruses may take weeks or months of effort. Most computer viruses are written in assembly language, a low-level programming language which is often used to program operating systems or interact directly with the hardware of the computer.

Types. Three distinct classes of computer viruses are known: file viruses, boot viruses, and macro viruses.

File viruses. These viruses infect application files such as spreadsheets, computer games, or accounting software. When an infected application program is run on a computer, the virus embedded in the application gains control of the computer and searches for other application files to infect. Once the virus finds a new, uninfected application, it copies its own program logic and instructions into this file. Finally, the virus alters the new file such that it runs the virus's logic each time the application is used. If such an infected file is copied and run on another computer or shared on a computer network, the virus will soon spread to other applications on the new computer. Very few file viruses make their way outside research laboratories and affect end users.

Boot viruses. These viruses are designed to self-replicate from one disk to another. They infect special areas of floppy diskettes and hard drives known as boot records, which are used by the computer during system startup. If a computer is started with a boot-virus-infected diskette in the disk drive, the virus will gain control of the computer and copy itself to the system's hard drive. The virus will then spread from the hard drive to other floppy diskettes as they are inserted and used in the computer system. Since boot viruses can infect only floppy diskettes and hard drives, they cannot spread over computer networks or the Internet. Until recently, boot viruses were the most prevalent infections encountered by users.

Macro viruses. These viruses hide inside document and spreadsheet files used by popular word processing and spreadsheet applications. If a user loads an infected data file into the word processor, or merely views an infected document or spreadsheet attachment from an e-mail program, the virus will quickly gain control and infect the system. Any time the user edits subsequent documents or spreadsheets, the virus will copy itself into these data files, spreading the infection.

Of the three classes of computer viruses, macro viruses are by far the most prolific, based on their ability to rapidly spread and infect actual (nonresearch) end-user computers. The first macro virus was discovered in 1995, and by 1998 thousands of them had been cataloged. Computer-literate users share information at an ever-increasing rate. The documents and spreadsheets that contain this information also can contain viruses. The growing use of e-mail and other mechanisms for rapid information exchange enables macro viruses to flourish. An infected document or spreadsheet sent by e-mail can infect every user who receives that e-mail and who views or edits the infected file attachment.

Antivirus software. The rapid growth of the computer virus problem has spurred a large antivirus industry. Numerous computer software vendors produce software to protect against computer viruses.

Most antivirus products detect computer viruses by scanning each suspect file or disk for viral "fingerprints." Just as each person has a unique fingerprint, so does each computer virus. This fingerprint is a short sequence of 1's and 0's which can be found only in the virus and is not likely to be found in uninfected programs. When an antivirus researcher encounters a new virus, it is analyzed to extract its digital fingerprint. This fingerprint is added to a database which is regularly shipped to antivirus software customers.

When an antivirus product scans a computer for viruses, it examines every file or disk to see if it contains one of its thousands of viral fingerprints in its database. If the antivirus software finds one of its fingerprints in a file or disk, it reports this to the computer user, who can then repair the infection. Antivirus researchers must constantly analyze new computer viruses, extract their unique fingerprints, and add them to the antivirus database. This is a costly and tedious process, and many companies have started to employ specially designed automated analysis systems to perform much of this work.

Advanced viruses. Viruses constitute a tangible threat to end users and corporations. As antivirus vendors improve their protection, virus authors construct more insidious threats. For instance, many of the newer viruses are polymorphic, or many-formed. These viruses mutate each time they spread to a new file or disk, eliminating any consistent digital fingerprint and making antivirus detection much more difficult. These digital pathogens avoid detection in the same way that human immunodeficiency virus (HIV) and other viruses evade the human immune system. Other viruses, termed retroviruses, actively seek out and disable antivirus programs to avoid detection. This contest has been going on since the 1980s and will continue into the foreseeable future.

For background information *see* COMPUTER PROGRAMMING; COMPUTER STORAGE TECHNOLOGY; DIGITAL COMPUTER; ELECTRONIC MAIL; SOFTWARE in the McGraw-Hill Encyclopedia of Science & Technology.
Carey Nachenberg

Bibliography. D. Atkins et al., *Internet Security*, New Riders Publishing, 1996; F. B. Cohen, *A Short Course on Computer Viruses*, 2d ed., John Wiley, 1994.

Computerized searches

Digitizing and retrieving information with a computer is the latest development in the quest to organize and retrieve the intellectual and artistic products of humankind. Computers can search for single words, parts of words, or combinations of words. Those words may be subject headings, descriptors, or keywords. The search can burrow though indexes and catalogs, databases, or complete documents (fulltext).

One-word, one-computer searches. A simple search seeks a single word or part of a word on a single computer. Most word processors are able to search a directory for a particular string of letters. First the computer is instructed to list the files in the directory to be searched. Then it is commanded to "find" the desired string. The computer searches through all documents in that directory, listing only those that contain the specified characters.

Keywords versus descriptors. In addition to searching for words embedded in documents in its own files, a computer can search for keywords on other computers to which it is linked. Such a search is most efficient in databases that use descriptors or subject headings.

Descriptors (designed for computers) and subject headings (originally designed for library catalogs) are standardized words and phrases to identify the topics in particular documents. A descriptor is applied to all material on a topic, including items in which an author may have used a synonymous expression.

Library organizations and others have developed lists of discriptors. The Library of Congress maintains the *Library of Congress Subject Headings*, which is used by English-language libraries around the world. Similarly, the Educational Resources Information Center (ERIC) has published its thesaurus of descriptors for *Resources in Education* and *Current Index to Journals in Education*.

Librarians and indexers use such lists to catalog documents, audiovisual materials, maps, graphics, and so on. As a result, searchers who find a heading that matches their inquiry can search a database built upon that particular set of terms with confidence that the appropriate materials will be displayed.

Databases that lack descriptors must be searched with keywords. A successful keyword search depends upon a searcher's ability to think of all the synonyms that various authors might have used. In addition, the database may include only titles of documents, abstracts, or full text. The more text that is included in the database, the more likely the possibility of retrieving needed material with a keyword search.

Boolean combinations of terms. The most powerful capacity of computers is that of combining words and phrases. Derived from the work of George Boole,

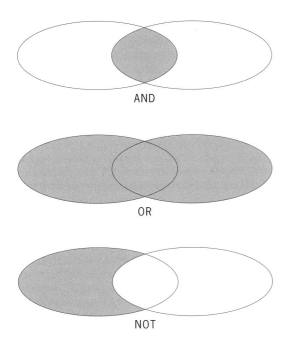

AND

OR

NOT

Diagrams illustrating actions of logical operators.

use of the logical operators AND, OR, and NOT in computer searches is known as boolean searching.

The operator AND instructs the computer to find items that contain two search terms (see **illus.**). Thus, the command "counseling AND discipline" retrieves only those items in which both "counseling" and "discipline" appear. The operator OR instructs the computer to find items containing either of the search terms. Thus, the command "counseling OR discipline" retrieves all items in which the word "counseling" appears as well as all those containing the word "discipline." The operator NOT directs the computer to find items that contain the first term, then toss out articles containing the second term. Thus, the command "counseling NOT discipline" retrieves items containing "counseling," then eliminates any that also contain "discipline."

To illustrate the impact of these commands, the numbers of documents obtained in a search of the ERIC databases are shown below. The count of documents retrieved with OR is the sum of those found with the other three commands.

counseling AND discipline	512
counseling OR discipline	50,566
counseling NOT discipline	37,222
discipline NOT counseling	12,832

Truncation and proximity operators. Computers offer other options to speed a search. For example, most search engines permit truncation. Usually coded with a question mark or other sign of punctuation, truncation has the effect of multiple OR commands. For example, "gramma?" has the effect of "grammar OR grammatical OR grammarian," and so on.

A proximity operator tells the computer to look for words only when they occur close to one another. For example, using the abbreviation NR for "near," "high NR2 school" directs the computer to retrieve items containing the words "high" and "school" only when they are within two words of one another. A similar effect is often available with a Web browser, where many search engines make it possible to specify that "high school" is to be searched "as a phrase."

Subject headings and descriptors. Subject headings and descriptors, where they are provided in an index or database, allow far more efficient searching than keywords. Procedures for locating these terms through the Internet will be illustrated.

Library of Congress subject headings. Perhaps the most widely used list of subject headings is that developed at the Library of Congress. A Web browser is used to enter the URL (Uniform Resource Locator, the address on the World Wide Web), where clicking with a mouse accesses library catalogs, browse search, and finally subjects. After entering the topic of interest in the box provided, the user clicks on "Browse." The screen then offers subdivisions of the subject and other search suggestions. Here are found related, narrower, and broader topics.

ERIC descriptors. The *ERIC Thesaurus of Descriptors* can also be consulted online via the Internet. The ERIC Wizard is accessed from its URL, Upon looking up a search term, the computer displays thesaurus results. Terms entered into the same search box are handled as if joined by OR; the boxes themselves are automatically joined with AND.

Searching the Internet. The Internet, comprising interconnected computer networks, has become a worldwide repository of information and resources from government, colleges and universities, commercial firms, and the general public. Its resources range from highly technical papers on scientific topics to advertising and pornography. These resources are accessed by directories and search engines.

Internet directories. Directories are categorical lists of Internet sites. A well-known commercial directory includes such categories as "arts and humanities," "entertainment," "news and media," and "society and culture," and also offers a directory for students with categories like "ask an expert" and "homework help." Another directory, the World Wide Web Virtual Library, is voluntarily updated by a worldwide team of librarians.

Clicking on a category in a directory produces subcategories until a list of actual Web sites is reached for immediate access. Because humans must analyze Web sites to determine appropriate categories, and because human indexers choose sites for interest and information, directories do not attempt total coverage of the World Wide Web.

Internet search engines. Search engines are much broader in scope than directories. During the early years of the Internet, both documents and software were retrieved with file transfer protocol (ftp). A search program, Archie, was created to search for files. Because ftp filenames were abbreviated,

Gopher was offered, a program that expanded filenames to readable menus, and another program, Veronica, to search Gopher menus. Archie and Veronica made it possible to connect to the desired file, wherever its home computer.

With the World Wide Web, however, Archie and Veronica were largely superseded by search engines using hypertext. Web technology has the same worldwide scope as Archie and Veronica, but the user simply clicks on a highlighted item to make a connection. Well over 100 search engines have been created.

Unlike directories, which are managed by human beings, search engines use automated software known as a robot, spider, or like term. This software examines the Web, collecting Web pages as it goes. Some engines collect only portions of a page; others, entire pages. Engines index pages differently. They cannot assign descriptors—that still requires a human being—but use words in the Web pages themselves. An engine may index either the first few lines or entire pages; it may or may not index the URL or descriptive tags that are invisible on the computer screen but are embedded in the hypertext.

Lacking descriptors, it is necessary to search with keywords. Successful searching, therefore, depends on thinking of likely synonyms. The engine searches its index, created from perhaps millions of Web pages, and presents items most closely matched to the search terms. Many engines use relevance ranking; that is, they choose Web pages containing the desired terms most often, or closest to the beginning, or perhaps in a title, so that the most relevant pages come to the top. Some engines allow a searcher to choose the most relevant pages and re-run the search to find the best matches. Many search engines feature the use of boolean and other operators; proximity searching; the ability to prioritize words in a search; and the ability to limit a search to title words, postings in Usenet groups, a selected language, the time the material was posted, and other features of Web pages. A few search engines automatically apply truncation; others do so with the proper command. Search engine developers continue to work with matching, truncation, synonyms, and other techniques to manage with automated software what is still done more precisely by human catalogers and indexers.

Broadening and narrowing a search. A searcher may need to broaden a search to obtain more items or to be certain of having found all the important documents. At other times, faced with more items than can possibly be perused, the searcher may wish to narrow a search.

Broadening. As shown above, the OR command is one way to broaden a search. This command is most often used with strings of synonyms in a keyword search. Truncation, since it is a variation of OR, also broadens a search.

Another broadening strategy involves searching additional databases or indexes, especially branching out to related fields. In a keyword search, synonyms expand the search. A broader term may also yield additional information. Thus, if nothing is retrieved about "ants," the searcher may try "insects," and if nothing results, then perhaps "invertebrates."

Surprisingly, using multiple narrower terms, *if* those terms are subcategories of the more general term, also broadens a search. For instance, given insufficient information on "insects," the searcher might try "ants OR butterflies OR moths OR beetles," and so on.

Narrowing. Computers provide powerful means of retrieving items that meet several criteria simultaneously. The boolean AND command effectively narrows a search, as discussed above. Proximity operators, including the ability to search for a phrase (that is, adjacent words in a specific order), provide an additional means of adding precision to a search.

The NOT command, too, focuses a search, but the searcher must avoid inadvertently discarding important material. Other limits—such as by language, date, field, type of document, publisher, and institution—also focus a search.

Of course, the reverse of several broadening strategies narrows the search. Thus, searching fewer indexes or databases reduces the number of items found in a search, as do narrower terms.

For background information *see* DATA COMMUNICATIONS; DATABASE MANAGEMENT SYSTEMS; DIGITAL COMPUTER; SOFTWARE in the McGraw-Hill Encyclopedia of Science & Technology.

Lillian Biermann Wehmeyer

Bibliography. P. A. Gilster, *Finding It on the Internet: The Internet Navigator's Guide to Search Tools & Techniques*, 2d ed., John Wiley, 1996; A. and E. Glossbrenner, *Search Engines for the World Wide Web*, Peachpit Press, 1998; L. B. Wehmeyer, *The Educator's Information Highway* (and updates), Technomic, 1995–1996.

Conifer

Conifers first appear in the fossil record during the Pennsylvanian Era, approximately 310 million years ago (Ma). The earliest conifers were probably small trees that had needle leaves, compound seed cones, and simple pollen cones. They looked similar to living conifers, but their reproductive biology was more like that of the most primitive seed plants, the seed ferns. For many years botanists thought that conifers originated from the free-sporing progymnosperms of the group Archaeopteridales and were not very closely related to seed ferns, cycads, gnetophytes, and the flowering plants. That theory assumed that conifers resulted from a separate origin of seed plants. Modern phylogenetic systematic methods have revealed that the seed plants all have a common ancestor and that conifers originated among the seed ferns. This suggests that some of the primary changes leading to the origin of conifers were a reduction and simplification of fernlike fronds to simple needle leaves, and the aggregation of fertile regions into seed cones and pollen cones. See TREE.

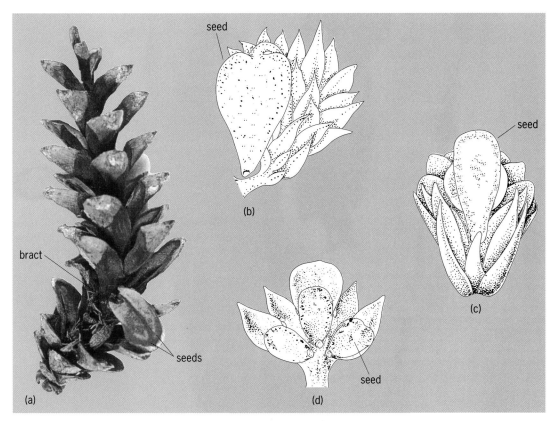

Fig. 1. Conifer seed cone structure and evolution. (*a*) Pine cone consisting of seed scales attached to the cone axis immediately above the bracts. One scale is bent back to reveal two seeds attached to the upper surface. (*b*) Fertile shoot from above a bract in an *Emporia* seed cone. (*c*) Fertile shoot from above a bract in a *Moyliostrobus* seed cone. (*d*) Fertile shoot from above a bract in a *Voltzia* seed cone. This highly modified shoot is the lobed equivalent of a lobed pine cone scale, as in *a*.

Distribution. The diversity and distribution of Pennsylvanian-age conifers were narrow, with all or most of the species inhabiting only relatively dry slopes near equatorial lowland plant communities. The oldest conifers conform to a small number of genera assignable to two or three extinct families, and are informally referred to as the walchian conifers. The best known of these is the genus *Emporia* from Kansas. *Emporia* was a small tree with needle leaves, simple pollen cones, and compound seed cones. Several other Permo-Carboniferous conifers such as *Utrechtia* (formerly referred to by the illegitimate name *Lebachia*) and *Ernestiodendron* are similar to *Emporia* in many features, but they vary in the shapes of the leaves and in features of the seed cones.

Both diversity and geographic distribution of conifers increased greatly throughout Permian time, 290–245 Ma. Conifers continued to be most abundant in the equatorial regions of the Euramerican continent and China, where the extinct families Utrechtiaceae (formerly Lebachiaceae), Majonicaceae, and Ullmanniaceae are found. In addition, there are several Permian genera such as *Ferugliocladus*, *Ugartecladus* (Ferugliocladaceae), and *Buriadia* (Buriadiaceae) from the Southern-Hemisphere Gondwana continent (Africa, South America, Antarctica, and India) and *Kungurodendron*, *Timanostrobus*, and *Concholepis* from the Northern-Hemisphere Angara realm (Russia and northernmost China) that reveal an increasing geographical distribution for conifers by the end of the Paleozoic.

At the beginning of the Mesozoic (245 Ma), the extinct family Voltziaceae was present in Euramerica and elsewhere. Other genera of uncertain affinities around the world suggest that the origins of modern conifer families occurred during the Mesozoic. One of the most widespread and diverse of these Mesozoic conifer families, the Cheirolepidiaceae, became extinct in the Cretaceous. Other conifer families diversified during the Mesozoic, and most are part of the living flora today. The first unequivocal evidence for modern families comprises the genera *Rissikia* and *Notophyllum*, which demonstrate that the family Podocarpaceae was established in the South African and Antarctic regions of the Gondwana continent by the Triassic (245–210 Ma). Evidence for the origin of other modern conifer families (Pinaceae, Araucariaceae, Taxodiaceae/ Cupressaceae, Taxaceae, Cephalotaxaceae) is less certain during the Triassic, but they all appear to be well established by the Jurassic Period (210–145 Ma).

Biological adaptations. There has been a large number of changes in conifer reproductive structures

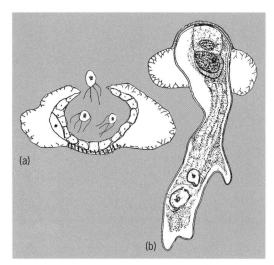

Fig. 2. Prepollen and pollen of ancient and modern conifers. (*a*) Prepollen of *Emporia* showing proximal germination to release swimming sperm, and the primitive cellular structure thought to characterize ancestral seed plants. Note the single vegetative cell and the jacket of cells surrounding the swimming sperm. (*b*) Pollen of pine showing the pollen tube germinating from the distal surface to carry the nonmotile sperm to the egg of the seed. Note that the pollen contains a very small number of cells.

and reproductive biology since conifers originated in the Pennsylvanian Era. The most well known is a series of changes in the forms and structures in the seed-bearing cones that allow understanding of the basic morphological similarities and taxonomic relationships of ancient and modern conifers. Seed cones of pine trees are representative of many living conifers (**Fig. 1**). They consist of a main axis that bears modified leaves called bracts, and directly above each bract is a flattened scale that has two seeds on the upper surface (Fig. 1*a*). This transformational series begins with the Pennsylvanian-age

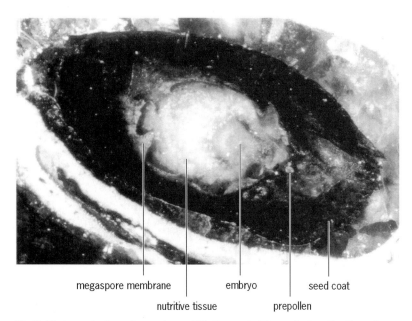

megaspore membrane   embryo   seed coat

nutritive tissue   prepollen

Fig. 3. Photograph of a mature embryo within the seed of the ancient conifer *Emporia*.

conifer *Emporia*, which has seed cones that contain a complete leafy shoot immediately above each bract (Fig. 1*b*). Most of *Emporia*'s scalelike cone leaves are small and simple, and one or more of these are fertile leaves, or sporophylls, each of which bears a seed which is bent back with its open end facing inward toward the cone axis. In the Permian-age *Moyliostrobus* (Fig. 1*c*), each fertile shoot is flattened, with a smaller number of scalelike leaves, and has only one ovule which is attached to such a short sporophyll that the ovule appears to arise directly from the surface of the fertile shoot. In the Triassic-age *Voltzia*, each axillary fertile shoot is highly flattened and has only about five scalelike leaves that are arranged in a single plane (Fig. 1*d*). Three of those leaves are sporophylls, each of which bears an ovule, and all five are partly fused to give the shoot the appearance of a lobed seed scale somewhat like that of pine. From this transformational series, botanists are able to interpret the seed scales of pine and all other conifers as highly modified fertile shoots.

Changes in the pollen cones of conifers through time are much less obvious than those in the seed cones. Like those of the most ancient conifers, the pollen cones of living conifers consist of an axis that bears simple leaflike sporophylls with microsporangia attached. In the earliest conifers there were several microsporangia on each sporophyll's upper surface, but in modern pine each sporophyll bears two microsporangia on its lower surface. Except for positions of the microsporangia, the pollen cones and sporophylls of the ancient and modern conifers look generally similar. However, the pollen has changed significantly. The pollen of many conifers has air bladders that aid in wind dispersal and pollination (**Fig. 2**). Pollen grains of the earliest conifers looked much like those of modern conifers, but they were very different in their reproductive biology. The earliest conifer pollen, or prepollen, probably produced swimming sperm that escaped from the upper, or proximal, surface of the pollen grain (Fig. 2*a*). Prepollen appears to consist of a single vegetative cell, adjacent to a jacket layer of cells, and swimming sperm (Fig. 2*a*), and is thought to be very similar to that of ancestral seed ferns. Over time there were additional significant biological changes such that in modern conifers there is a pollen tube that grows from the distal surface of the grain and carries nonmotile sperm (Fig. 2*b*) directly to the egg inside the seed.

Dormancy. One of the most important features for the success of conifers through time is the origin of seed dormancy. This biological innovation may have played an important role in the conifers' rapid rise to dominance at the end of the Paleozoic (245 Ma). Dormancy allows seeds to persist in the soil until conditions are favorable for germination and the successful establishment of a seedling. Mature seeds with fully developed embryos have been discovered in the seed cones of *Emporia* (**Fig. 3**). The mature embryos suggest that seed dormancy first evolved among the most primitive Pennsylvanian-age conifers.

Beginning in the Upper Pennsylvanian Era and continuing throughout the Permian, there was a fundamental change in the global landscape. By the beginning of the Mesozoic, wetland habitats had diminished and become isolated, and the topographic diversity of better-drained dryland habitats had increased dramatically. This environmental change was coupled with rapid diversification of conifer species and the establishment of conifer forests in drier, fire-prone habitats worldwide.

Additional biological adaptations, such as needle-like leaves and thick cuticles, and the establishment of mycorrhizal associations in the roots, further contributed to the ability of conifers to survive and thrive in water-stressed environments. Together with seed dormancy and improved sperm delivery, such specializations allowed conifers to become the first seed plants to extensively colonize habitats outside the wetlands. This success led to the emergence of modern conifer families that dominated the landscape until the rise of flowering plants approximately 80 Ma.

For background information *see* PALEOBOTANY; PINALES; STEM; TREE; TREE GROWTH in the McGraw-Hill Encyclopedia of Science & Technology.

Gene K. Mapes; Gar W. Rothwell

Bibliography. C. B. Beck (ed.), *Origin and Evolution of Gymnosperms*, Columbia University Press, 1988; R. Florin, Evolution of cordaites and conifers, *Acta Horti Bergiani*, 15(11):285–388, 1951; W. N. Stewart and G. W. Rothwell, *Paleobotany and the Evolution of Plants*, Cambridge University Press, 1993; T. N. Taylor and E. L. Taylor, *The Biology and Evolution of Fossil Plants*, Prentice Hall, 1993.

Conservation

To conserve an endangered species, an adequate number of populations must be maintained within a sufficient amount of protected habitat, with the goal of minimizing the likelihood of future extinction. In fact, of the 1154 species listed as threatened or endangered under the U.S. Endangered Species Act as of November 1998, the majority are declining due to loss of habitat (**Fig. 1**). The most direct approach for ensuring that a species will continue to exist is to conserve its environment. The primary elements that determine the overall effectiveness of a habitat conservation measure are (1) the number and type of factors causing species decline; (2) the condition of the habitat and the surrounding land matrix; and (3) the specific regulatory and management approaches used on the land, which may not always directly relate to species conservation. Although a variety of land protection strategies are available for conserving endangered species, the most common is federal land management. Government agencies often have direct responsibility for ownership and management of lands containing endangered species, but there are a growing number of private initiatives that can also be successful in habitat protection.

Endangered species. To preserve a species within its habitat, it is imperative that scientists first identify the habitat components critical for species persistence, in addition to identifying the specific causes of species decline. For example, if species decline is linked to an exotic predator, scientists should determine not only the environmental features necessary for feeding and breeding but also the requirements for protection from the predator. The California clapper rail is endangered because the bird's habitat has been converted for other uses, but recently the populations have been declining due to a second factor, the introduction of the nonnative red fox. Thus, to protect this species, it is imperative that large tracts of habitat be restored and the numbers of red foxes be controlled in local areas.

If the unique ecology of each species is not well documented or is particularly complicated, it is more difficult to protect the right type of habitat or combination of habitats. Although many species have broad habitat requirements, others may be specialists, depending on a particular food item or nesting only in a certain size of tree. These data are often time-consuming to acquire since many species can be secretive and their habits difficult to discern, or they may use particular habitats only under rare but critical circumstances such as fire or drought. Additionally, if the habitat is degraded by human activities, a larger area may need to be protected to compensate for the lowered quality. Due to human encroachment, many habitats are too small to support viable populations, and therefore restoration of formerly occupied lands is the only option. Thus, scientific studies linking species reproduction and survival to specific features of the environment, including human processes, are necessary before a proper land protection strategy can be implemented.

Habitat management. One key factor affecting the success of any habitat conservation approach is the

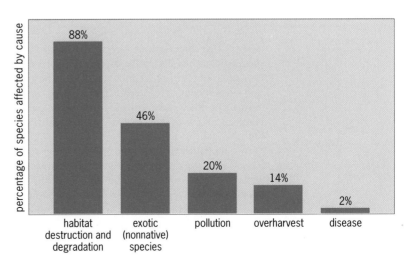

Fig. 1. Causes of endangerment of species in the United States. A species can have more than one cause of endangerment. The data were obtained from the Federal Register and cover all species listed or proposed for listing as of December 31, 1995. (*After R. F. Noss, M. A. O'Connell, and D. D. Murphy, The Science of Conservation Planning: Habitat Conservation Under the Endangered Species Act, Island Press, 1997*)

existence of a long-term plan for managing the populations. These plans should include clear obligations of all involved parties, such as providing funding, monitoring the status of the populations, and taking management actions to benefit the species. Because endangered species are already under severe threat of extinction, any multitude of factors could potentially create the conditions that finally drive the population to extinction. Therefore, the owners or managers of the land must be willing and able to monitor the status of the populations and quickly take action to halt population declines. Long-term management plans must include not only the species' currently known habitat needs but also the possibility of unforeseen circumstances that will impact the species. Obviously, many private landowners may not have this capability, so a partnership with a government or nonprofit group is required.

Protection on government lands. For conservation of endangered species, the common approach is to protect habitat through government ownership and management of the land. In the United States 50% of all endangered species are found solely on federal land. Federal agencies such as the Fish and Wildlife Service, Bureau of Land Management, Forest Service, and National Park Service are the primary managers of public lands and therefore are the defacto managers of habitat for endangered species. Management of these public lands for the benefit of endangered species is mandated by law, but it is often in conflict with other agency goals such as logging, cattle ranching, or human recreation. There is additional support at the state level, where numerous agencies manage public lands, some with direct responsibility for managing endangered species. Regional governments can also be active in habitat preservation. It is increasingly recognized that locally endemic species with restricted distributions can often be well managed by local interests. All levels of government are involved, in varying degrees, with species protection.

Direct ownership by a federal or state agency is clearly the most direct approach for conserving habitat, but other forms of land agreements are also used successfully. One alternative is to place a conservation easement on privately owned land. Conservation easements are legal agreements, often permanent, that place specific conditions on the landowner's use of the property, with incentives often provided by a nonprofit group or government agency. Both parties can gain through these agreements, with government costs reduced since the government is not managing the land, and with the landowner receiving direct incentives or continued land use. For example, The Nature Conservancy, a large nonprofit organization, directly owns or had under conservation easement over 1.6 million acres in the United States as of 1998. Other types of incentives for private landowners, particularly financial, have been proposed by the federal government but have not yet been widely employed. Such incentives range from tax relief for easements or donations of property, to credits or compensation for attracting or maintaining species on private lands. One problem with easements or financial incentives is that these strategies often lack secure funding for future management of the habitat. Without a clear source of funding, management generally fails to meet its objectives and the species suffers.

Protection on private lands. Because half of all endangered species have more than 80% of their habitats on nonfederal lands, protecting species on private property is an important conservation approach. Habitat conservation plans are one regulatory tool that is being used to resolve the conflict between private development interests and protection of endangered species populations. The plans have become a popular conservation technique only within the last 10 years. The number of habitat conservation plans escalated from 13 permits issued during 1982–1992 to 212 permits issued by 1997, with another 200 in the pipeline. The requirements of these plans hinge on the Endangered Species Act definition of "taking," which means harassment, harm, or killing of the species. The regulations state that any project actions authorized through the habitat conservation plan process must demonstrate that taking will not "appreciably reduce the likelihood of the survival and recovery of the species in the wild," and impacts of the taking will be "minimized and mitigated to the maximum extent practicable." The primary form of taking through the plan process is habitat destruction (**Fig. 2**), with the most common techniques for lessening the impact to the species being minimization, mitigation, and restoration of these losses. Thus, individuals of an endangered species can be taken by human activities, but only if the result of these actions is in compliance with the above requirements and does not cause the species to decline to extinction.

International agreements. Finally, for endangered species with habitat requirements that span more

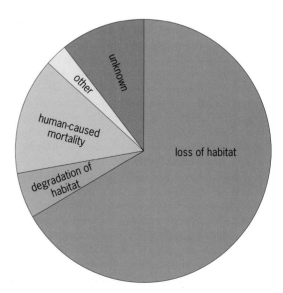

Fig. 2. Primary form of species "take" by project actions of habitat conservation plans.

than one continent, international agreements are often essential. Migratory species such as birds pose unique problems for conservation, since they use habitats in a variety of places and for entirely different functions. For instance, many shorebirds breed in northern Canada and North America but then migrate south in the winter to feed in habitats as far away as South America. In fact, the conservation of habitats for migratory birds has been a central concern over the last century and has spurred a number of international protection measures, including the North American Waterfowl Management Plan, which protects habitat for ducks and geese in the United States, Canada, and Mexico. International agreements are increasingly necessary to protect the variety of habitats used by wide-ranging species that are declining due to a variety of often unrelated causes.

For background information *see* CONSERVATION OF RESOURCES in the McGraw-Hill Encyclopedia of Science & Technology   Elaine K. Harding

Bibliography. L. C. Hood, *Frayed Safety Nets: Conservation Planning Under the Endangered Species Act*, Defenders of Wildlife, 1998; R. F. Noss, M. A. O'Connell, and D. D. Murphy, *The Science of Conservation Planning: Habitat Conservation Under the Endangered Species Act*, 1997; R. F. Noss and A. Y. Cooperrider, *Saving Nature's Legacy: Protecting and Restoring Biodiversity*, 1994.

Coral diseases

Coral is a generic term for a group of simple organisms in the phylum Cnidaria. Most often, it refers to reef-building corals of the order Scleractinia (class Anthozoa). These organisms form single-polyp or multipolyp colonies, and most are symbiotic with single-celled plants called dinoflagellates or zooxanthellae (see **illus.**). These plants live within the cells of corals and are essential to the host. Corals form the structural elements of the dominant ecosystem of shallow tropical oceans. Recent surveys worldwide have revealed that a wide variety of factors are degrading coral reefs on a large scale. One factor is an apparent increase in the incidence of disease. The causal agents for a large number of these diseases are currently unknown; however, most evidence points to a combination of physical (temperature, light) and biological agents. *See* CORAL REEF COMMUNITIES.

Mass bleaching. Mass bleaching has been the most conspicuous disease to strike coral reefs in the past 20 years. There have been six major periods of mass bleaching (1979–1980, 1982–1983, 1987, 1991, 1994, 1998). Before 1979 such a disease was not reported. Diseased corals lose their characteristic brown color over a period of several weeks and take on a brilliant white appearance. The loss of color is due to the reduction in the number of zooxanthellae, which are essential for the health of the coral. Thousands of square kilometers may be affected in any single event, which may stretch across several tropical oceans at once. Events last for 6–12 months, but the ramifications for coral reefs may last for years.

Individual corals may recover and regain their zooxanthellae. In many cases, however, corals suffer rates of mortality that may rise to 80–100%. Survivors may show reduced growth rates and reproductive capacity.

Global patterns. Coral reefs in almost every tropical ocean have been affected by mass bleaching. Mass bleaching events coincide with strong El Niño years. This association has been tracked to sensitivity to increases in temperature and light experienced by reef-building corals and other symbiotic invertebrates such as anemones and clams. In 1994, mass bleaching in the central Pacific was triggered by the arrival of water that was 0.8°C (1.4°F) above the summer maximum. Three weeks later, most coral reefs from French Polynesia to Fiji were 40–80% bleached. This pattern has been seen in most events over the past 20 years, with 80% of reports being associated with observations of warmer than normal conditions. The National Oceanic and Atmospheric Administration (NOAA) used satellite measurements of sea temperature to predict every incident of bleaching during the 1998 event by tracking 1°C (1.8°F) positive temperature anomalies. For example, a February 10, 1998, advisory statement from NOAA predicted the bleaching event that occurred on the Great Barrier Reef (Australia) 1 week later. Water that was 1°C above the normal summer maximum led to widespread bleaching on the Great Barrier Reef, affecting 80% of all reefs in the 2100-km-long (1304-mi) marine park.

Causal factors. While temperature is considered to be the primary variable associated with mass bleaching, several other factors also appear to be important. Changes in the salinity of the water and the quality of light accompanying temperature changes can have significant effects. Some evidence indicates that bleaching events are exacerbated by the doldrum conditions that usually accompany strong El Niño events. The doldrum conditions are typified by clear skies and no wind, leading to increased levels of both photosynthetically active radiation (PAR) and ultraviolet (UV) radiation. These forms of solar radiation appear to increase the bleaching of corals

White-striped red soft coral.

in response to small increases in sea temperature. A salient observation is that bleaching will occur faster if corals are exposed to either PAR or UV light and will not occur if corals are in the dark.

Mechanism. Mass bleaching is caused by a failure of the photosynthetic processes of the zooxanthellae that reside in corals. Recent work has shown that increases in temperature lead to a reduced capacity to process the excitation energy coming from the light reactions of photosynthesis, and the zooxanthellae of corals become more sensitive to photoinhibition. There is an overreduction of biochemical components within the light reactions, resulting from the fact that energy cannot be passed to the dark reactions of photosynthesis. Destruction of the chloroplast follows. Damaged zooxanthellae (and the host cell in which they reside) are quickly removed from the body of the coral polyp.

Pathogenic diseases. The growth of human populations worldwide has increased pressures on coral reefs. One important factor is the decline in water quality. Generally, corals grow best in warm, clear, low-nutrient waters of tropical oceans. Increased development of the coastal zone in many tropical countries has led to increases in nutrient, sediment, and heavy-metal concentrations, which have had a range of direct impacts on coral reef organisms. In addition, the incidence of pathogenic infections among corals has increased, especially in densely populated areas such as the Caribbean Sea. Similar increases appear to be occurring among other organisms inhabiting the reefs, with disease increasing among sea urchins, gorgonians, and coralline algae.

A number of these diseases have been identified since the 1970s. While the specific pathogen often remains undescribed, a diversity of host pathologies seems to indicate that a range of microorganisms are involved. These organisms are thought to be either opportunistic species that have capitalized on the reduced health of corals or pathogens that have been recently introduced into the marine environment. There are now over 20 different diseases described for corals, and several appear to be increasing in various parts of the world.

Black band disease. First recorded in 1974, this disease is characterized by a thick black band of tissue that advances rapidly across infected corals. Empty coral skeleton is left behind. The total destruction of coral colonies takes less than 60 days in many cases. The primary infection has been attributed to *Phormidium corallyticum*, which is blue-green alga, or cyanobacterium. A range of other organisms, including the blue-green alga *Spirulina* sp., various sulfur-oxidizing and sulfate-reducing bacteria, and other bacteria are associated with the band and cause the death of the underlying coral tissues through asphyxiation. Transmission of the disease is through contact between colonies. Red band disease and brown band disease are similar but appear to involve different pathogens.

White band disease. This disease is typified by a loss of tissue that is visible as a band of bare white skeleton. It has been reported on reefs in the Caribbean and Australian waters, and the Red Sea. No pathogen has been identified, and it is possible that this disease has multiple causes. The impact of this disease has been major in some areas. For example, up to 95% of the acroporid (staghorn) corals have died in some parts of the Caribbean Sea. This disease is thought to have important ramifications for the diversity of coral species and is considered a major problem in many parts of the Caribbean.

Rapid wasting syndrome. This disease was first observed in the Dutch Antilles in December 1996. By 1998, it had been reported from the Florida Keys and many other Caribbean locations (Venezuela, Grenada, Mexico, Grand Cayman). Rapid wasting syndrome is typified by a rapid loss of tissue and destruction of the underlying coral skeleton. It is thought to be associated with parrot fish predation and an unidentified fungus.

Tumors and growths. Coral colonies sometimes suffer abnormal growth forms. While rare, these abnormalities of either tissue or skeleton have attracted attention because of their similarity to human tumors and cancers. The intensity of ultraviolet radiation has been hypothesized as a causal agent, but no causal agent has actually been verified.

Vibrio. A form of coral bleaching has been associated with a biological agent. The bacterium in the genus *Vibrio* can cause symptoms that are similar to the coral bleaching caused by increased temperature. This disease has been reported among corals off the Mediterranean coast of Israel. *Vibrio* comprises aerobic rod-shaped and gram-negative bacteria, which include the causal agent responsible for cholera in humans (*V. cholerae*). The group also has extensive representation in marine ecosystems as both benign and symbiotic forms; for example, luminescent forms are symbiotic with a range of fishes and squids.

Role of human activity in disease. The rapid increase in disease among corals has implicated human activity. Humans have massively increased the sediment and nutrient loading of rivers, which is suspected to play a major role in the decline of the health of corals. These changes to the rivers that flow into coastal areas adjacent to coral reefs are linked to changes in land use. Global climate change leading to increased sea temperatures is yet another way that humans have been affecting the health of corals, and it has been implicated in mass bleaching and the increase in the incidence of disease. The reduced health of corals is suspected to be leading to an increase in the susceptibility to invasion by microorganisms. Changing land use has also led to the introduction of novel pathogens. For example, the fungus affecting gorgonians (sea fans) throughout the Caribbean Sea has been identified as *Aspergillus* sp., a terrestrial form that is unable to reproduce in seawater. Another hypothesis concerning the rise in disease in the Caribbean is the increasing amount of dust blowing across the Atlantic from North Africa. This dust may stimulate abnormal conditions under which corals then suffer, or may introduce a range of terrestrial bacteria, viruses, and fungal spores.

For background information *see* ALGAE; CORALLINALES; REEF; SCLERACTINIA in the McGraw-Hill Encyclopedia of Science & Technology.

Ove Hoegh-Guldberg

Bibliography. A. Antonius, Coral reef pathology: A review, *Proceedings of the 4th International Coral Reef Symposium*, Manila, 2:3–6, 1981; J. Bythell and C. Sheppard, Mass mortality of Caribbean shallow corals, *Mar. Pollut. Bull.*, 26:296–297, 1993; S. L. Coles and D. G. Seapy, Ultra-violet absorbing compounds and tumorous growths on acroporid corals from Bandar Khayran, Gulf of Oman, Indian Ocean, *Coral Reefs*, 17(2):195–198, 1998; W. B. Gladfelter, White-band disease in *Acropora palmata*: Implications for the structure and growth of shallow reefs, *Bull. Mar. Sci.*, 32:639–643, 1982; P. Glynn, Coral reef bleaching: Ecological perspectives, *Coral Reefs*, 12:1–17, 1993; O. Hoegh-Guldberg and G. J. Smith, The effect of sudden changes in temperature, irradiance and salinity on the population density and export of zooxanthellae from the reef corals *Stylophora pistillata* (Esper 1797) and *Seriatopora hystrix* (Dana 1846), *J. Exp. Mar. Biol. Ecol.*, 129:279–303, 1989; R. Jones et al., Temperature induced bleaching of corals begins with impairment of dark metabolism in zooxanthellae, *Plant Cell Environ.*, 21:1219–1230, 1998; A. Kushmaro et al., Bacterial infection and coral bleaching, *Nature*, 380:396, 1996; D. L. Santavy and E. C. Peters, Microbial pests: Coral disease in the western Atlantic, *Proceedings of the 8th International Coral Reef Symposium*, 1:607–612, 1997; G. Smith et al., Aspergillosis associated with Caribbean sea fan mortalities, *Nature*, 382:487, 1996.

Coral reef communities

Coral reefs are among the most diverse communities on Earth. This diversity occurs at low (species) and high (phylum) taxonomic levels and includes cryptofauna living within the substrate (for example, boring sponges, worms, and bivalves; sessile encrusting bryozoans, sponges, tunicates, and worms; motile worms, mollusks, echinoderms, and crustaceans), sessile epifauna on the substrate surface (for example, scleractinian corals, sponges, and coralline and fleshy algae), and suprabenthic fishes in the overlying water column. Approximately 93,000 species have been described globally on modern reefs, although 500,000 to 1 million species are estimated to inhabit reefs (in comparison with 20 million species in tropical rainforests). In addition to the tropical location of coral reefs, a number of other factors appear to be responsible for this high biodiversity, including elaborate biologically generated physical heterogeneity, sophisticated specializations, common sibling species (morphologically similar), and coevolved associations among species.

Several theories attempt to explain reef community structure. Some workers view reef communities as highly organized systems that are stable in composition, having reached equilibrium as a result of species interactions. Others interpret coral reef communities as unstable, nonequilibrium systems that vary in composition as a result of physical disturbances. A third group contends that community composition should be evaluated on a broader, regional (metapopulation) scale that considers dispersal among populations. The first theory predicts that community change is episodic, the second that it is gradual, and the third that it is regionally patchy and highly variable.

Reef biodiversity through geologic time. Like tropical rainforests, reef communities are currently declining rapidly in response to pollution, overexploitation, and habitat destruction by humans. Understanding the history of reef communities on a million-year time scale is therefore important for predicting the long-term consequences of this decline. However, tracing the history of reefs through the fossil record is complicated, because most reef dwellers are soft-bodied and not readily preserved. Moreover, because coral reefs are restricted to a narrow range of environmental conditions, they are patchily distributed through space and time. Long continuous sequences of reef deposits are rare, so historical interpretations must be made by correlating relatively short sequences in different locations.

Because of these difficulties, quantitative studies of the evolutionary dynamics of reef communities are best performed on carefully designed samples of well-preserved sessile dominants (that is, reef-building corals), which serve as proxies for the community as a whole. One such study involved 141 samples that were uniformly collected at closely spaced intervals within numerous scattered Caribbean sequences (Dominican Republic, Curaçao, Bahamas, Costa Rica, Panama, Jamaica) spanning the past 20 million years. The sequences consist of sedimentary (carbonate and siliciclastic) deposits in a range of exposed and protected reef environments. Geologic ages were determined for each sample using high-resolution chronostratigraphic methods that integrate microfossil, paleomagnetic, and strontium isotope data. A total of 47 genera and 189 species were identified using standardized sets of morphologic characters, established in part by comparison with molecular data.

Using the age data for each sample, the oldest and youngest ages were determined for each genus and species, and these age ranges were used to estimate the numbers of genera and species within the Caribbean over the past 20 million years. The results (**Fig. 1**) show that numbers of genera and species increased significantly between 5.5 and 10 million years ago (Ma) and remained high until about 1.5 Ma, when diversity dropped dramatically. Study of the oldest and youngest ages for each taxon (**Fig. 2**) suggests that a pulse of extinction occurred between 2 and 1.5 Ma. Species origination was slightly elevated (greater than 10%) for 3–5 million years preceding extinction. In contrast, genus formation remained low (less than 10%) over the past 15 million years.

Increased origination corresponded with changes in oceanographic circulation that took place

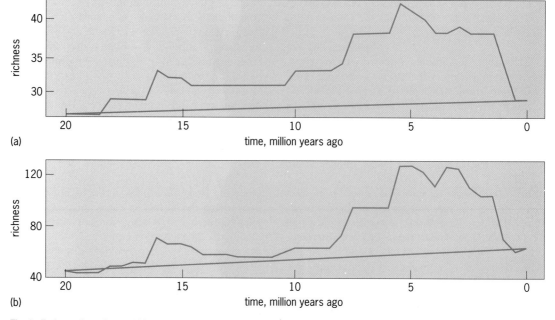

(a)

(b)

Fig. 1. Estimated numbers of (a) genera and (b) species of reef corals living in the Caribbean region over the past 20 million years. The estimates were made by counting the number of taxa within each half-million-year time slice. (*After A. F. Budd and K. G. Johnson, Origination preceding extinction during Late Cenozoic turnover of Caribbean reefs, Paleobiology, 25:188–200, 1999*)

between about 10 and 3 Ma in association with the emergence of the Central American Isthmus. However, the pulse of extinction corresponded with the onset of Northern Hemisphere glaciation at approximately 2-2.5 Ma. Rates of extinction were surprisingly low during the most intense climatic fluctuations later in Pleistocene time (0.1-1 Ma).

Dynamics of community change. During the five great mass extinctions in Earth history (Ordovician, Devonian, Permian, Triassic, Cretaceous), diversification followed extinction. Benthic marine ecosystems collapsed and took 1 to more than 3 million years to recover. In reef communities, recovery was unusually slow (5-20 million years) due to the high

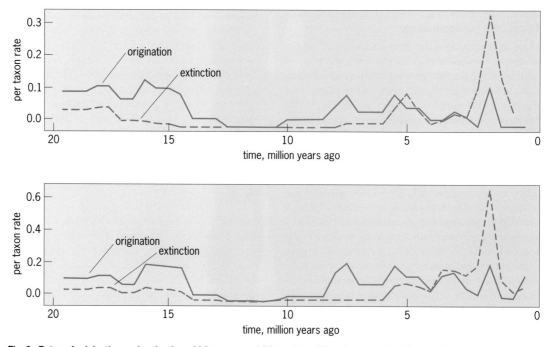

Fig. 2. Rates of origination and extinction of (a) genera and (b) species of Caribbean reef corals over the past 20 million years. The estimates were made by counting the number of oldest and youngest occurrences of taxa recorded within each half-million-year time slice and dividing by the total number of taxa within that time slice. (*After A. F. Budd and K. G. Johnson, Origination preceding extinction during Late Cenozoic turnover of Caribbean reefs, Paleobiology, 25:188–200, 1999*)

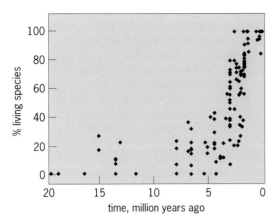

Fig. 3. Proportion of living species within 141 samples of Caribbean reef communities spanning the past 20 million years. Geologic ages are midpoints of the age range for each sample. *(After A. F. Budd and K. G. Johnson, Origination preceding extinction during Late Cenozoic turnover of Caribbean reefs, Paleobiology, 25:188–200, 1999)*

sensitivity of reef builders to environmental perturbations.

The smaller-scale, climate-mediated extinctions of the Cenozoic are unique in that origination preceded extinction, resulting in increased diversity during faunal change. Not only did preturnover (extinct) and postturnover (living) species simultaneously exist within the Caribbean region during faunal change, but they also coexisted within samples, suggesting that pre- and postturnover species lived together in the same communities. Study of the 141 previously described samples shows that, although maximum numbers of species within samples remained constant at 40–60 species between 4 and 1 Ma, percentages of living species range from less than 10 to 100% (**Fig. 3**). Within any half-million-year time slice between 4 and 1 Ma, samples vary by 75% in proportion of living species. This variability cannot be explained by geographic location but may be related to reef environment. Samples from exposed reef environments (with high wave energy) have higher percentages of living species.

These results support conflicting interpretations of the dynamics of reef community change. Asynchroneity in origination and extinction, and variable mixtures of extinct and living species within samples, suggest that communities may not have been tightly integrated during faunal change, and that they exhibit properties of unstable, nonequilibrium systems. However, constant numbers of species within samples indicate that reef community composition may have been constrained by biological factors, characteristic of equilibrium systems. Although more data are needed, preliminary results suggest that the conflict could be resolved if community change was influenced both by local species interactions and by migration of species between reefs across the Caribbean region. Faunal change may have been rapid at any one location, but patchy when viewed across the region as a whole. The summed result for the region would be gradual faunal change that took

place over a considerably longer period of time than occurred at any one place.

For background information *see* CORALLINALES; EXTINCTION; MARINE ECOLOGY; REEF in the McGraw-Hill Encyclopedia of Science & Technology.

Ann F. Budd

Bibliography. A. F. Budd and K. G. Johnson, Origination preceding extinction during Late Cenozoic turnover of Caribbean reefs, *Paleobiology*, 25:188–200, 1999; D. Jablonski et al. (eds.), *Evolutionary Paleontology*, 1996; J. B. C. Jackson et al. (eds.), *Evolution and Environment in Tropical America*, 1996; M. L. Reaka-Kudla et al. (eds.), *Biodiversity II: Understanding and Protecting Our Biological Resources*, 1997; J. E. N. Veron, *Corals in Time and Space*, 1995.

Coreceptors

Human immunodeficiency virus (HIV) enters target cells by a process involving direct fusion between the virus and cell membranes. This process begins when the HIV envelope glycoprotein (Env) binds to CD4, the primary receptor on the target cell surface. It has long been recognized that CD4 alone is not sufficient to allow HIV entry; an additional "coreceptor" on the target cell is required. Evidence has suggested that the tropism of different HIV strains for infection of various target cell types (for example, T-cell lines, macrophages, primary T cells) results from the ability of the corresponding Envs to use distinct coreceptors, which in turn are differentially expressed on various CD4-positive target cell types.

The identity of the coreceptors has been an elusive problem. They have now been identified as members of the chemokine receptor family, within the

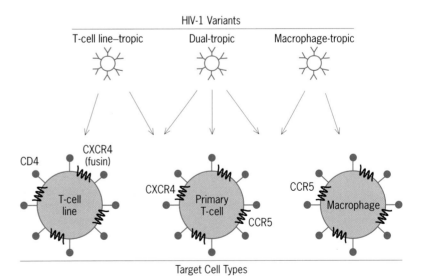

Fig. 1. Model for coreceptor usage and HIV-1 tropism. T-cell-line–tropic strains are specific for CXCR4 and can infect continuous T-cell lines and primary T cells, since these cells have CXCR4. Macrophage-tropic strains are specific for CCR5 and can infect primary macrophages and primary T cells, since these cells have CCR5. Dual-tropic strains can use both CXCR4 and CCR5, and can infect continuous T-cell lines, macrophages, and primary T cells.

Fig. 2. Model for coreceptor usage in HIV-1 entry. Upon binding to CD4, gp120 undergoes a conformational change that enables it to bind to the coreceptor. The interaction with the coreceptor triggers Env to undergo another conformational change, leading to extension of gp41 and insertion of the fusion peptide into the target cell membrane.

superfamily of G protein–coupled receptors. Two specific chemokine receptors account for the major features of HIV-1 tropism (**Fig. 1**): the CXCR4 coreceptor is preferentially used by T-cell-line tropic variants, and the CCR5 coreceptor is used by macrophage-tropic variants; both coreceptors can be used by dual-tropic variants. To date, the coreceptor repertoire has expanded to include a dozen members, mostly chemokine receptors and related orphans. The coreceptors have provided new molecular insights into the membrane fusion mechanism involved in HIV entry. In the most favored model (**Fig. 2**), CD4 binding to the gp120 subunit of the Env causes a conformational change that allows gp120 to bind to the coreceptor; the coreceptor interaction then triggers a conformational change in the gp41 subunit of Env, causing it to extend its fusion peptide so that it can insert into the membrane of the target cell. Fusion ensues, thereby enabling entry of the virus into the target cell.

The coreceptors have created entirely new perspectives on some of the most perplexing problems in HIV disease. It has long been known that the HIV-1 variants found shortly after infection are almost always macrophage-tropic. The awareness that the CCR5 receptor is preferentially used by macrophage-tropic variants prompted analysis of the possible role of CCR5 in HIV transmission. Remarkably, it has been found that some people are resistant to HIV infection because they contain two defective alleles of the CCR5 gene. Overproduction of chemokines that inhibit CCR5 might provide another basis for protection against infection. Individuals who are heterozygous for the defective CCR5 allele are not protected from infection, but disease progression proceeds less rapidly. Thus, CCR5 plays a critical role in HIV transmission, and progression to pathogenesis within the infected person.

The coreceptors also represent new targets for the development of novel approaches to treat HIV infection. Several strategies are under investigation, including blocking agents that directly bind to the coreceptors (chemokines and their derivatives, antibodies, small-molecular-weight inhibitors), modalities to modulate coreceptor expression, gene therapy approaches to downregulate coreceptor expression, and genetically engineered viruses containing CD4 plus coreceptor to selectively target (and kill) HIV-infected cells. These concepts are still at an early stage, but there is much hope that they will

lead to valuable components of future therapeutic regimens.

For background information *see* ACQUIRED IMMUNE DEFICIENCY SYNDROME; CELLULAR IMMUNOLOGY; CYTOKINE in the McGraw-Hill Encyclopedia of Science & Technology.   Edward A. Berger

Bibliography. G. Alkhatib et al., CC CKR5: A RANTES, MIP-1alpha, MIP-1beta receptor as a fusion cofactor for macrophage-tropic HIV-1, *Science*, 272:1955–1958, 1996; E. A. Berger, P. M. Murphy, and J. M. Farber, Chemokine receptors as HIV-1 coreceptors: Roles in viral entry, tropism, and disease, *Annu. Rev. Immunol.*, 17:657–700, 1999; Y. Feng et al., HIV-1 entry cofactor: Functional cDNA cloning of a seven-transmembrane, G-protein coupled receptor, *Science*, 272:872–877, 1996; R. Liu et al., Homozygous defect in HIV-1 coreceptor accounts for resistance of some multiply-exposed individuals to HIV-1 infection, *Cell*, 86:367–377, 1996.

Cosmological constant

A small fraction of all stars die violently, destroying themselves in a cataclysmic explosion that rivals the power of 5×10^9 suns. Such supernovae are beacons that can be seen hundreds of millions, or even billions, of light-years away. In addition to being incredibly luminous, these objects are extremely predictable; most of them output the same power to within 10%. Hence, if their apparent brightness can be measured, they can be used to accurately measure the distances of their host galaxies. In the past few years, astronomers have found that very distant supernovae appear unexpectedly faint, suggesting that they are even farther away than anticipated. Although other interpretations have not yet been definitively eliminated, a plausible conclusion is that the universe is expanding at an ever-increasing rate, pushing the supernovae and their host galaxies to enormous distances. This suggests the presence of a long-range repulsive effect, perhaps produced by quantum fluctuations in the vacuum.

Expansion of universe. In 1929, Edwin Hubble discovered that the universe is expanding: Other galaxies are moving away from the Milky Way Galaxy, with distant galaxies having greater speeds of recession than nearby galaxies. This discovery, together with other evidence such as the uniformly distributed cosmic microwave background radiation and the relative abundances of light elements, suggests that the universe began its existence billions of years ago in a hot, dense initial state, generally known as the big bang. Unlike a conventional explosion, where particles move through a preexisting space away from a common center, space itself is expanding between galaxies, and there is no unique center of expansion within the physically accessible dimensions. A useful two-dimensional analogy is the surface of a spherical, expanding balloon on which paper stickers have been placed: Each sticker sees the others receding from it due to the expansion of the rubber, but the center of expansion (the balloon's center) is not part of the hypothetical universe (the rubber itself).

Quantitatively, the expansion of the universe is given by Hubble's law, Eq. (1), where v is the radial

$$v = H_0 d \qquad (1)$$

velocity (recession speed) of a galaxy whose distance is d, and H_0 is the Hubble constant or Hubble parameter at the present time. (The factor of proportionality, H, is constant throughout the universe at a given time, but its value changes with time.) It is difficult to accurately determine H_0, but the best current estimate (uncertainty 10–15%) is 65 km/(s·Mpc), where a megaparsec (Mpc) is 3.26 million light-years. For example, a galaxy whose current distance from the Milky Way is 100 Mpc recedes with a speed of 6500 km/s, while one that is 200 Mpc away recedes at 13,000 km/s. The radial velocity of a galaxy is measured from the redshift z of features (absorption or emission lines of elements) in its spectrum: Photons are stretched by the expansion of space, making all of them longer in wavelength ("redder") as they travel through the universe. Though this lengthening is not really the Doppler effect, at low velocities the Doppler formula, Eq. (2), is valid, where λ

$$z = \frac{\Delta\lambda}{\lambda_0} = \frac{\lambda - \lambda_0}{\lambda_0} \approx \frac{v}{c} \qquad (2)$$

and λ_0 are the observed and rest-frame wavelengths of a feature, respectively.

Deceleration of universe. Ever since Hubble's discovery, astronomers have assumed that the mutual gravitational attraction of all the matter (and energy) would gradually brake the expansion of the universe. The situation is similar to that of a ball launched from the Earth's surface: Its upward speed continually decreases due to the pull of Earth's gravity, even if its initial speed exceeded the escape velocity. Of course, if the ball is thrown at less than the escape velocity, the upward motion eventually halts, and the ball falls back to the surface. When considering the entire universe, the relevant parameter is Ω, defined as the ratio of the average density (ρ) of matter to the critical density, which is given by Eq. (3), where G is

$$\rho_{\text{crit}} = \frac{3H_0^2}{8\pi G} \qquad (3)$$

the gravitational constant. In the absence of exotic forms of matter, Ω determines the ultimate fate of the universe. If Ω is greater than 1, the expansion decelerates so quickly that the universe eventually reaches a maximum size and subsequently collapses (the big crunch). However, if Ω is less than 1, the expansion easily continues forever, albeit progressively slowed by gravity. The dividing line, when the average density equals the critical density ($\Omega = 1$), corresponds to a universe that just barely expands forever: The recession speed of a given galaxy asymptotically approaches zero as time goes to infinity.

Thus, a major challenge in observational cosmology has been to determine the deceleration rate of the universe. A few measurements of the expansion rate when the universe was younger can be combined with the current expansion rate to yield the deceleration. Fortunately, the finite speed of light provides a method for looking back in time and effectively measuring the expansion rate as it was long ago: When galaxies are observed billions of light-years away, they are seen as they were billions of years ago, since light has taken that long to reach the Earth. Their measured distances depend on the degree to which the expansion of the universe has been decelerating: A galaxy with a given redshift should be closer for larger values of the deceleration rate (that is, larger values of Ω). If the universe is essentially empty ($\Omega \approx 0$) and therefore not decelerating, the derived distance should be its largest possible value for that redshift.

Type Ia supernovae. The key to using distances to gauge the ultimate fate of the universe is to find a class of objects whose luminosity is large (so that they can be seen at great distances) and predictable. Relative distances to such standard candles can be obtained by comparing the objects' apparent brightnesses: Objects become fainter by the square of their distance. One possibility is to use galaxies themselves, but these come in a variety of shapes and sizes that seem to change with time.

In recent years, type Ia supernovae (identified by the lack of hydrogen in their spectra, coupled with prominent features of silicon and iron) have emerged as the best prospect for cosmological distance determinations. These explosions are believed to arise from dense white dwarf stars, which consist of carbon and oxygen nuclei supported by the quantum-mechanical pressure of tightly packed "degenerate" electrons. Such stars are the cooling ember which forms when normal stars, like the Sun, exhaust their nuclear fuel supplies of hydrogen and helium. When a white dwarf is in a binary stellar system (unlike the Sun), it can gravitationally attract gas from the companion star, and under certain circumstances retain the stolen material, thereby increasing its mass. If the mass grows to a value close to 1.4 solar masses (called the Chandrasekhar limit), the white dwarf undergoes a sudden thermonuclear runaway, fusing carbon and oxygen into heavier nuclei such as silicon and sulfur, and releasing energy that disrupts the star. About half of the star's mass is converted to radioactive nickel, and its subsequent decay into radioactive cobalt and stable iron is responsible for the observed luminosity of the explosion. After brightening for 2–3 weeks, type Ia supernovae subsequently fade over the course of several months. Since the white dwarf always becomes unstable at about the Chandrasekhar limit, the explosion is theoretically expected to show little variation in peak luminosity, and this is indeed verified with type Ia supernovae in nearby galaxies of known distance.

Although type Ia supernovae are quite good standard candles, they are not all exactly alike, perhaps due to variations in the detailed properties of the white dwarf (for example, the quantity of trace heavy elements from previous generations of stars, or the ratio of carbon to oxygen nuclei). It has been shown, however, that the more luminous supernovae take a longer time to rise and decline in brightness than do less luminous ones; that is, their light curves are slower. Thus, the observed light curve of a given supernova can be used to determine a correction factor to its peak luminosity, thereby bringing all type Ia supernovae to a common standard luminosity. Moreover, the observed color of a supernova provides a measurement of the amount by which its light is extinguished by gas and dust along the line of sight, just as the setting Sun looks abnormally red because of the extra material through which its light passes when close to the horizon.

Careful observations of relatively nearby type Ia supernovae have proven their utility as cosmological distance indicators: The observed Hubble diagram of distance versus redshift shows that Eq. (1) is followed, with very little scatter. Indeed, type Ia supernovae provide distances with an accuracy of about 6–7% (**Fig. 1**).

High-redshift supernovae. In principle, the use of type Ia supernovae to determine the deceleration of the universe is straightforward: The procedure is to find them at high redshifts, measure their peak brightnesses, and derive the corresponding distances after making suitable corrections for colors and light-curve shapes. After measuring the redshift of the host galaxy of each supernova, the resulting Hubble diagram provides the deceleration of the universe: Supernovae at a given redshift should be closer if Ω is large.

But type Ia supernovae are rare: Depending on its type and size, one galaxy might host a supernova every 100–1000 years. To find many supernovae in a given night so that they may subsequently be monitored with assigned observing time on a suite of telescopes, tens or hundreds of thousands of distant galaxies must be surveyed. This is done by using large telescopes equipped with wide-angle cameras and sensitive charge-coupled-device (CCD) detectors. The same regions of sky are photographed about 3 weeks apart, and any new, stationary object in the second set of images is flagged as a possible supernova. (Moving objects are typically asteroids.) A spectrum is obtained shortly thereafter, typically with the Keck 10-m (400-in.) telescopes in Hawaii, to determine whether the object is a type Ia supernova and to measure its host-galaxy redshift. Light curves

and colors are obtained from the follow-up observations, with the most precise data coming from the Hubble Space Telescope (**Fig. 2**).

Accelerating universe. Over the past few years, two independent teams of astronomers have used this procedure to find several dozen high-redshift type Ia supernovae ($z = 0.3$–1.0). Their spectra are essentially indistinguishable from those of nearby supernovae, suggesting that the peak luminosities are also the same. The supernovae appear very faint for these redshifts, implying that they lie at great distances; it is therefore inferred that Ω is less than 1, and that the universe is unlikely to end in a big crunch. This answer is consistent with the results of other, independent techniques, which typically try to add up all the matter in the universe contained in clusters of galaxies.

More provocatively, however, the type Ia supernovae are so faint that their derived distances are about 10% larger than expected even in an empty (that is, nondecelerating) universe. The expansion rate of the universe seems to actually be accelerating with time, contrary to the effect produced by normal matter and energy. If the speed of a given galaxy is increasing with time, space itself must have a

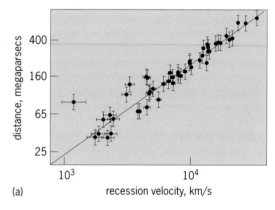

(a)

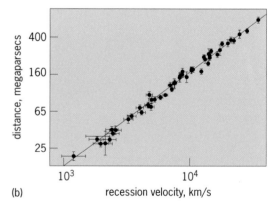

(b)

Fig. 1. Two Hubble diagrams derived from relatively nearby type Ia supernovae. Both distance and recession velocity are measured on logarithmic scales. (*a*) Diagram that is based on the assumption that all supernovae are perfect standard candles with uniform peak luminosity and that makes no correction for intervening gas and dust (extinction). (*b*) Diagram with corrections for intrinsic luminosity and extinction of each supernova; the scatter is much reduced.

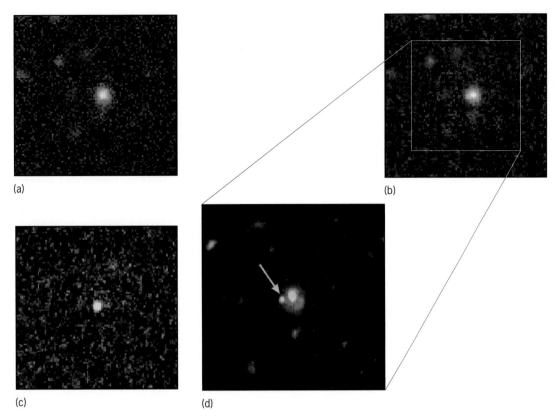

Fig. 2. Discovery of the supernova SN1997cj (redshift = 0.5). (*a*) Small portion of a charge-coupled-device frame obtained on April 7, 1997, with the Blanco 4-m (158-in.) reflector at Cerro Tololo Inter-American Observatory in Chile. (*b*) Similar image obtained on April 28, 1997. (*c*) Difference between these two images, revealing a supernova candidate that was subsequently confirmed to be a type Ia supernova with a spectrum obtained by the Keck 10-m (400-in.) telescope. (*d*) Image obtained on May 12, 1997, with the Hubble Space Telescope, showing the supernova well resolved from the center of the host galaxy.

long-range repulsive property, somewhat like a cosmic antigravity—though it probably has nothing to do with normal matter or antimatter. Albert Einstein initially introduced such an effect (called the cosmological constant) into his equations of general relativity to prevent the universe from gravitational collapse. After Hubble's discovery of universal expansion, however, he dismissed the idea as the biggest blunder of his career; its philosophical and esthetic motivation (a static universe) had vanished.

Although this discovery is tentative and must still be verified with completely independent techniques, its cosmological consequences are quite attractive. First, the derived expansion age of the universe is about 14×10^9 years [assuming a Hubble constant of 65 km/(s·Mpc)], larger than previous values and consistent with the revised ages of the oldest known objects in the universe, globular star clusters ($11-14 \times 10^9$ years). Second, the global spatial geometry of the universe appears to be nearly flat, in agreement with the predictions of popular inflation theories of the early universe. If the expansion of space continues to accelerate (that is, the cosmological constant is really constant with time), the universe will expand forever at an increasingly fast rate. Within about 10^{12} years, all of space beyond the Milky Way Galaxy and its nearest neighbors will have expanded beyond the ability of any inhabitants

to see it, leaving them in a very cold and dark universe.

The cause of this putative acceleration is not totally understood, but it might be related to a vast array of virtual particles allowed by the Heisenberg uncertainty principle. Unfortunately, simple estimates suggest that the cosmological constant, if nonzero, should have a value roughly 100 orders of magnitude larger than that implied by supernovae. To some theorists, this strongly suggests that the observational result, or its interpretation, is incorrect. One alternative is the possibility that dust partially obscures the light of distant supernovae, making them fainter than expected. Another is that the high-redshift supernovae themselves are less powerful than their nearby counterparts. However, at present there is no persuasive evidence for either of these (or other) possibilities.

For background information *see* BIG BANG THEORY; COSMOLOGY; DOPPLER EFFECT; HUBBLE CONSTANT; INFLATIONARY UNIVERSE COSMOLOGY; QUANTUM GRAVITATION; RELATIVITY; SUPERNOVA; UNCERTAINTY PRINCIPLE; UNIVERSE; WHITE DWARF STAR in the McGraw-Hill Encyclopedia of Science & Technology.   Alexei V. Filippenko

Bibliography. J. Glanz, Cosmic motion revealed, *Science*, 282:2156–2157, 1998; C. J. Hogan, R. P. Kirshner, and N. B. Suntzeff, Surveying space-time with supernovae, *Sci. Amer.*, 280(1):28–33, January 1999; R. Irion, Exploding stars tell all, *Astronomy*, 26:50–55, November 1998; A. G. Riess et al., Observational evidence from supernovae for an accelerating universe and a cosmological constant, *Astron. J.*, 116:1009–1038, 1998.

Data mining

Data mining is the automated process of turning raw data into useful information to enhance knowledge and facilitate better decisions. It allows intelligent computer systems to sift and sort through data, with little or no help from humans, to look for patterns or to predict trends. In addition to traditional analytical tools, data mining can use many artificial intelligence techniques.

With the proliferation of computerized data acquisition systems in retail markets and manufacturing enterprises, large amounts of data have been gathered and stored. Large, specialized databases called data warehouses have been constructed to store these data. In many cases, so much data has been collected that it becomes impossible for humans to analyze it effectively. For example, searching for purchase patterns of consumers using grocery store chains requires the analysis of millions of transactions, a transaction perhaps averaging 30–40 items. A data warehouse could hold hundreds of terabytes (trillions of bytes) of data.

Traditional databases were designed to store smaller amounts of data and to receive very specific instructions about which data to retrieve. An example can illustrate the difference between using a database query system and using data mining. The task is to determine ways to classify potential insurance customers according to risk so that high-risk customers can be identified. Using traditional databases and query techniques, the analyst has to first develop a hypothesis about which factors contribute to being a high-risk customer. Next, the analyst has to query the database to receive the required information. In modern systems this could amount to millions or billions of transactions to analyze. Finally, the analyst has to use an analytical approach to either prove or disprove the hypothesis. Modern reporting tools and on-line analytical processing (OLAP) systems have made this task easier. However, a human is still required to initiate and drive the information discovery process. With data mining techniques, the analyst can mine the data using more general instructions. The analyst can instruct the system to look at all the data and cluster the higher-risk customers into one or more groups. The data mining system then analyzes data from the millions of customers and claims to group them into different levels of risk. One or more analytical approaches can be used by the data mining system to determine the groupings. After the determination, the system can provide insight about common factors among the members of the group.

Applications. Data mining has many varied areas of application, including classification, estimation, segmentation, and association.

Classification. In classification, each item or record is placed into a group of similar members. An example occurs in the insurance risk analysis problem discussed above. The analyst may predefine the categories, providing the system with the categories and examples that have already been classified. The system will then determine rules or procedures to classify new records. The analyst may also allow the data mining system to define the number and type of categories. In this task (sometimes referred to as clustering), the analyst provides the system with a number of example records and allows the system to group them together.

Estimation. Also known as forecasting or prediction, estimation is the task of filling in missing values in a record as it is added to the system. The estimated value is normally numerical but can also be other data types, such as textual, categorical, or logical. For numerical data, statistical techniques such as regression or artificial intelligence techniques involving neural networks are often used. Examples of estimation are determining the likely amount that a person offered a low-cost trip to a ski resort will spend on ski lifts; and forecasting the telecommunications load on a network as a result of a new service that is offered. Often the task of estimation is combined with classification by using the estimated value to classify the object.

Segmentation. Segmentation is the task of breaking a population into subgroups before other processing is performed. It is a specific type of clustering task where the output must be further processed using other techniques to provide useful results. Segmentation is required in some cases because different patterns in subpopulations tend to cancel each other out when a single-population approach is followed.

Association. Association is the task of determining two or more events or characteristics that are related. For example, determining items that consumers usually purchase together in a grocery store may provide knowledge to create new marketing campaigns. Monitoring acoustical emissions of submarines allows identification of the type of submarine, by identifying certain patterns associated with distinct submarine types. A sequence is a specific type of association that occurs over time; that is, when one event occurs, another event will likely occur within a certain time period. An example is a manufacturing line using machine tools. By monitoring the vibrations of the cutting tool, certain patterns indicate that a tool failure will occur in the near future. Depending on the amount of machining left to perform, the operator can choose to continue machining until the part is finished or can stop the process and change the tool.

Process phases. The process of data mining is an iterative process that has three phases.

1. Data selection. The data warehouse is queried

to get representative data for the problem. In the first iteration the data extracted may not be the ideal data to solve the problem. As iterations proceed, the data selection method usually becomes more highly refined.

2. Data preparation. The data is transformed into the form required by the model that will be used or derived. Often there are missing pieces of data that must be estimated or predicted. Sometimes entire records are discarded because they lack the required information. Another common preparation step is to remove unwanted bias from the data.

3. Model generation phase. This is the heart of the data mining process. Many different types of models and modeling techniques can be used. The techniques depend on whether the data is quantitative or qualitative, predictive or descriptive, structured or unstructured. Many of these modeling techniques come from the field of artificial intelligence. The overall approach is iterative. As the model is developed, new data may be selected that is more representative of the features that the model is trying to exploit. Additional preparation may be developed to eliminate bias or estimate missing values.

Modeling techniques. There are many modeling techniques that can be used in data mining. Some of them are supervised, meaning the analyst helps to guide the model generation. Often the analyst uses graphical or pictorial representations to assist with supervised model generation. The human mind can still be a powerful tool to find abstract links in data. Traditional statistical methods are often used to help build the models. Rarely are statistical methods alone enough to perform data mining. Other common modeling techniques are rough sets, fuzzy sets, machine learning, neural networks, and rule-based systems.

The most common modeling techniques to date are probably neural networks and rule induction. Neural networks are based on models of the human brain where layers of neurons are linked to other layers in specialized patterns. When levels are high enough, neurons fire, sending to other layers in the network. Neural networks generally have one input and one or more output layers. Often there are multiple hidden layers between the input and output layers. Both supervised and unsupervised neural networks can be used in data mining.

Rule-based systems determine rules, also known as production rules, of the form IF-THEN, which can be used to process the data. Some of these systems arrange the rules as decision trees and then search the trees for branches that can be combined or eliminated. Other systems use induction to generate results that have certain confidence levels.

The relatively young field of data mining is sure to see increased activity in the future. Many companies are developing large data warehouses without an understanding of how they will process the data into usable knowledge. Data mining will likely be used to perform this transformation.

For background information *see* ARTIFICIAL INTEL-LIGENCE; COMPUTER; COMPUTER STORAGE TECHNOLOGY; DATABASE MANAGEMENT SYSTEMS; DECISION SUPPORT SYSTEM; DISTRIBUTED SYSTEMS (COMPUTERS); ESTIMATION THEORY; NEURAL NETWORKS; STATISTICS in the McGraw-Hill Encyclopedia of Science & Technology.   Gregory L. Tonkay

Bibliography. K. J. Cios, W. Pedrycz, and R. Swiniarski, *Data Mining Methods for Knowledge Discovery*, Kluwer, 1998; C. Westphal and T. Blaxton, *Data Mining Solutions: Methods and Tools for Solving Real-World Problems*, John Wiley, 1998; A. Zanasi et al., *Discovering Data Mining: From Concept to Implementation*, Prentice Hall, 1997.

Deep crustal microbes

The existence of life below the Earth's surface challenges the ecological "golden rule" of primary production based solely on photosynthesis. Yet, since early in the twentieth century, scientists have reported evidence of microbial life deep in the Earth's subsurface, where there is often no oxygen or sunlight. These deep-dwelling microorganisms derive energy by degrading organic materials contained within subsurface rock in an inhospitable environment where temperatures may be extreme and nutrients scarce. Initially it was believed that microorganisms detected in subsurface rock, sediment, and ground water were contaminants introduced from the Earth's surface during sampling. During the 1990s, however, microbiologists, together with geologists and geochemists, have developed techniques to track the possible introduction of microbes during sampling and have employed careful procedures to avoid contamination. These techniques have allowed definitive demonstration of the presence of microorganisms in the subsurface and have elucidated the mechanisms by which an impressive array of diverse subsurface microbial communities are maintained.

Recent interest in the microbiology of subsurface environments stems from investigations in which researchers looked for evidence of past or present microbial life within Martian meteorites. Mars's cold temperatures and high levels of solar radiation make life on its surface unlikely. However, life may exist below the surface of Mars and other planets, where water is liquid, temperatures are more favorable, and intense solar radiation cannot penetrate.

Other scientists are interested in the potential ability of deep crustal microbes to carry out metabolic processes that decontaminate sewage or degrade oil spills. Taking advantage of deep crustal microorganisms as bioremedial agents could have beneficial environmental and ecological effects.

Limits to growth. Considering the huge volume of sediment and rock contained within the Earth's crust, the extent of biomass colonizing the deep underground could exceed the level of microbial growth on the Earth's surface. However, certain factors keep the growth of deep crustal microbes in check. The

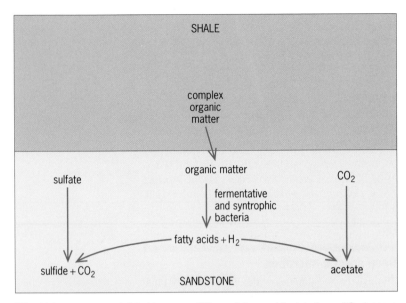

Microbial processes postulated to occur within sandstones at the interface of Cretaceous sandstones and shales.

growth and metabolism of organisms bound in subsurface rock may be limited by slow diffusion of nutrients or by a microorganism's ability to migrate through the narrow rock pores toward nutrients. Temperature also limits microbial growth below the surface of the Earth. If the assumption is made that $20°C$ ($68°F$) is the surface temperature, $110°C$ ($230°F$) is the temperature limit for microbial activity, and temperatures rise $25°C$ ($45°F$) for each kilometer (0.6 mi) of depth, then it can be conjectured that microbial life may exist at least to a depth of 3.5 km (2.2 mi). Nutrient availability has traditionally been thought to be extremely low at this level. Yet, growth of thermophilic, or heat-loving, microorganisms has been demonstrated on rock and sediment from several locations in the Earth's subsurface.

Although low nutrient availability imposes restrictions on growth, microbial communities have been found that are supported through novel mechanisms. These include harnessing geochemical energy produced by contact of ground water with reactive rock (lithotrophic microbial communities) and using rock- and sediment-bound organic materials previously thought to be unavailable (heterotrophic microbial communities). It seems likely that these communities of microbes will turn out to be quite common, thus raising the possibility that the subsurface is home to an abundant and diverse microbiota.

Heterotrophic microbial communities. In western regions of the United States, sands and muds were alternately laid down during the Cretaceous Period (90–93 million years ago). Over time, consolidation and other rock-forming processes resulted in alternating layers of sandstones and shales. Study of the anaerobic microbial communities living at the interface of these highly permeable sandstones and finer-grained, low-permeability, organic-rich shales show that they are able to thrive on this extremely old carbon. The nutrients originate from the organic-rich

rock and diffuse out to microorganisms living in adjacent, more permeable nutrient-depleted rock (see **illus.**).

Samples of these subsurface microbes have been obtained in the Cerro Negro region of western New Mexico within rock cores (cylindrical rock samples) brought to the surface by deep drilling with a hollow-core drill bit. The rock cores are immediately placed in an anaerobic glovebox to minimize exposure of potentially oxygen-sensitive microorganisms to air and to prevent oxidation of reduced chemical species. Much of the core material comes from sediments below the water table where ground water is devoid of oxygen but rich in sulfate. Microorganisms living within this subsurface environment respire sulfate in the same way that organisms on the surface breathe oxygen. The level of sulfate reduction (respiration) in the subsurface rock cores can be determined using a technique where sheets of silver foil are placed on intact core faces along with radiolabeled sulfate (containing sulfur-35) and incubated. During this incubation, radiolabeled sulfide formed by sulfate reduction activity reacts immediately with the silver and becomes fixed. The foils are then subjected to autoradiographic imaging to determine the location of sulfate-reducing activity within the cores.

When the total counts on each foil are determined as a function of depth, samples from the sandstone-shale interfaces within the sandstone and below the interface exhibit extremely high activity. The steepest gradients of organic carbon exist across the same interfaces where the greatest level of sulfate-reducing activity exists. It appears that organic matter trapped in shales during deposition diffuses out across the sandstone-shale interfaces. This activity in turn fuels discrete microbial communities at and near the interfaces within the adjacent sandstone. These microbial communities are thriving on organic matter that was laid down approximately 100 million years ago.

Several microbial communities living within unconsolidated materials have been observed obtaining nutrients in a manner similar to those colonizing sandstone-shale interfaces. These communities include those living at Cretaceous clay-sand interfaces, where the nutrients diffuse out of the clay layers toward microbes living within nearby sand. Buried lignite can provide a similar source of nutrients for adjacent sand-based microbial communities.

The idea of ancient organic matter fueling microbial growth is not new. For decades it has been postulated that indigenous microorganisms living near oil deposits can grow by catabolizing components of buried petroleum. In the late 1990s, several research groups isolated microorganisms from marine and continental oil reservoirs. Although oil-degrading strains have not yet been isolated from the indigenous populations, it is believed that microorganisms are living on petroleum or products of petroleum degradation. A restricted nutrient or water supply or a limiting pore size must restrict microbial growth in petroleum reservoirs; otherwise vast reservoirs of crude oil would not exist. Microorganisms living beneath the ocean floor appear to carry out metabolic processes similar to those of microorganisms living below the terrestrial surface. Sulfate reduction has been observed 500 m (1640 ft) below the sea floor where a huge microbial biomass appears to be present.

Lithotrophic microbial communities. The above mechanisms maintain subsurface microbial communities for extended periods, yet ultimately rely on the photosynthetically generated organic material trapped within shales, lignite, and clays during deposition. However, it has been hypothesized that certain microbial communities can survive strictly on available geochemical energy.

In deep aquifers located in the Columbia River Basalt group of the northwestern United States, an active anaerobic microbial community consisting of iron-reducing, sulfate-reducing, methanogenic, acetogenic, and fermentative bacteria has been identified. There is no evidence for migration of nutrients from the surface or from depth, nor is there an endogenous source of organic matter in basalts, which originate as molten rock from within the Earth. However, individual microbial populations are quite dense, often exceeding 10^4 cells per milliliter of ground water. There appears to be an external source of hydrogen for these microbial communities where the dissolved hydrogen concentrations vary between 20 nanomoles and 100 micromoles per liter. This level is considerably greater than in typical anaerobic ecosystems, where fermentation of organic matter is occurring, which rarely exceed 10 nanomoles per liter of dissolved hydrogen.

In controlled laboratory experiments, basalts incubated with buffered water produced hydrogen gas. This phenomenon may be the result of the interaction of the water with reduced iron [Fe(II)] within iron-containing minerals in the basalt. While hydrogen can be used as an energy source for many types of microorganisms, it is not generally thought of as a primary energy source for maintenance of an ecosystem. However, in this subsurface ecosystem the primary consumers of hydrogen may be hydrogenotrophic (H_2-oxidizing) bacteria. The organic materials provided by these microorganisms via excretion and lysis may support a range of microbial types in a fashion similar to that carried out by photosynthetic organisms in surface aquatic ecosystems. A comparable hypothesis has been proposed to explain the abundance of microbial life observed in deep aquifers within fractures in granitic bedrock.

Microbial remnants have also been observed within deep oceanic volcanic minerals located near basalts on the sea floor. This finding provides evidence for the existence of microbial ecosystems possibly fueled by geochemical energy where seawater penetrates deep oceanic basalts.

Survival versus growth. Microorganisms extensively colonize a variety of subsurface environments, using different support mechanisms. Some microbial populations are large and diverse in spite of scarce resources. This situation occurs at sites above the water table where water is scarce and in other regions where oxygen is abundant and extensive degradation of organic materials occurs. In the face of these hardships, microorganisms have developed survival mechanisms to maintain their population until conditions become more favorable for growth. The microorganisms enter a nongrowing or resting state to allow for long-term microbial survival. Often very little change will occur over geologic time in a given region of the subsurface, while other settings experience periodic rises and falls of the water table allowing for intermittent infiltration of nutrients and water. Organisms adapted to these less stable subsurface environments may be the most likely to endure times of nutrient deprivation.

Outlook. Scientists have explored only a small fraction of the vast space under the Earth's surface. During the 1990s, a variety of types of subsurface microbial ecosystems were discovered that are much more densely populated than expected. The deep crustal layer of the Earth harbors a diversity of microorganisms capable of carrying out almost any thermodynamically favorable reaction given the appropriate conditions.

The understanding of deep crustal microbial communities will be expanded through the study of the relationships among subsurface organisms, their natural energy sources, and recently introduced anthropogenic contaminants. These investigations will help predict the fate of contaminants in different subsurface regions and may also help in the design of microbially based systems for large-scale environmental decontamination.

For background information *see* BACTERIAL PHYSIOLOGY AND METABOLISM; BACTERIOLOGY; METHANOGENESIS (BACTERIA); MICROBIAL ECOLOGY; MICROBIOLOGY; SANDSTONE; SHALE; SULFUR in the McGraw-Hill Encyclopedia of Science & Technology.

Lee R. Krumholz; Cecelia M. Brown

Bibliography. J. K. Fredrickson et al., Pore-size constraints on the activity and survival of subsurface bacteria in a late Cretaceous shale-sandstone sequence, northwestern, New Mexico, *Geomicrobiol. J.*, 14: 183–202, 1997; L. R. Krumholz et al., Confined subsurface microbial communities in Cretaceous rock, *Nature*, 386:64–66, 1997; R. J. Parkes et al., Deep bacterial biosphere in Pacific Ocean sediments, *Nature*, 371:410–413, 1994; T. O. Stevens and J. P. McKinley, Lithotrophic microbial ecosystems in deep basalt aquifers, *Science*, 270:450–454, 1995.

Dengue fever

Dengue fever is a mosquito-borne viral infection throughout the tropical and subtropical zones. The virus exists in four antigenically distinct serotypes or subtypes. Its main natural reservoir is humans, and it is normally transmitted through the bite of infected female *Aedes* spp. mosquitoes.

Uncomplicated dengue fever can cause a tempo-

Medically important members of the genus _Flavivirus_*

Virus	H/E*	Vertebrate host	Vector†	Geographical distribution
Yellow fever virus (YFV)	H	Monkeys, humans	M: _Aedes, Haemagogus_	Tropical forest areas of Africa and South America
Dengue viruses 1–4 (DEN 1 to 4)	H	Humans, (monkeys)	M: _Aedes_	(Sub)tropical regions worldwide
Japanese encephalitis virus (B encephalitis)	E	Birds, pigs	M: _Culex_	Southeast Asia
Tick-borne encephalitis virus complex (TBE)‡	E	Rodents, ungulates	T: Ixodid ticks	Eurasia
West Nile fever virus	E	Birds	M: _Culex_	Africa, Middle East, South Europe, India
St. Louis encephalitis virus (SLE)	E	Birds, horses, (bats)	M: _Culex_	U.S. Caribbean
Murray Valley encephalitis	E	Birds	M. _Culex_	Australia, Papua-New Guinea
Omsk hemorrhagic fever virus	H	Rodents	T: Ixodid ticks	Siberia
Kyasanur forest disease virus	H	Rodents, monkeys	T: Ixodid ticks	India

*H = hemorrhagic fever, E = encephalitis.
†M = mosquito, T = tick.
‡Including Central European and Russian spring-summer encephalitis.

rarily crippling illness, but it is transient and carries a good prognosis. The severe manifestation of dengue virus infection, called dengue hemorrhagic fever, is life-threatening. First described in the 1950s in Southeast Asia, it has become a major cause of childhood morbidity and mortality there. Dengue hemorrhagic fever now also occurs in Central and South America and the Caribbean and Pacific islands, where it has caused epidemics.

Epidemiology and transmission. Dengue is a major public health problem in tropical and subtropical nations. It is the most important arthropod-borne arboviral disease of humans worldwide and occurs in more than 100 countries, with over 2.5 billion people at risk. The annual incidence is probably tens of millions of cases. About half a million individuals per year, mostly children, are hospitalized with dengue hemorrhagic fever, with an associated mortality rate of 5%.

Outbreaks occur most frequently in areas where environmental conditions are optimal (high temperature and high humidity) for the vector, mosquitoes of the genus _Aedes_, especially _A. aegypti_. This species is found around the globe, its range limited by a winter isotherm of 10°C (50°F).

Rapid expansion of urbanization has led to an increase of mosquito breeding sites, such as discarded tires and water storage containers. Humans experience increased exposure to dengue-carrying mosquitoes, which may have acquired insecticide resistance. The vertebrate reservoir species is humans. In some areas, monkeys may represent a jungle reservoir. Mosquitoes become infected when feeding on a viremic person. After the virus has a so-called extrinsic incubation period, replicating inside the insect, it is passed on to a different person by another bite. Transovarial transmission has been demonstrated in _Aedes_ mosquitoes, but its role under natural conditions is unclear. Direct person-to-person spread is extremely rare and has been described only through inoculation of blood from an acutely infected patient. Aerosols also present a hazard of infection

in the laboratory. The epidemiologically important, natural infection cycle is human-mosquito-human.

Biology. The dengue virus belongs to the family Flaviviridae (formerly group B arboviruses). Of the approximately 70 flavivirus species, most are arthropod-borne, and many have been associated with human disease (see **table**). The prototype of the family, yellow fever virus, was the first filterable agent shown to cause a human disease, and it was isolated in 1927.

The Flaviviridae possess a single-stranded positive-sense, ribonucleic acid (RNA) genome (approximately 11 kilobases in size). They consist of a spherical ribonucleoprotein core surrounded by a lipoprotein envelope. Flaviviruses are rapidly inactivated by high temperature (total inactivation of virus suspended in blood solutions occurs within 30 min at 56°C or 133°F).

The dengue virus particle (virion) contains three structural proteins: the nucleocapsid or core protein; the membrane-associated protein; and the envelope protein, which contains the antigenic determinants for inducing immunological responses in the infected host.

On the basis of antigenic differences, four dengue virus subtypes or serotypes can be distinguished, termed dengue-1 to -4 (DEN-1, and so on). They share 62–77% of their amino acid sequences, type 1 being most closely related to type 3 (77%), followed by types 2 (69%) and 4 (62%).

All four subtypes were formerly known to occur in Southeast Asia. Types 1 and 2 were originally endemic in West Africa, and type 2 in East Africa. Caribbean types were dengue-1 and -4, and in the Americas types 2 and 3 were endemic. Because of increased international travel activity, all four serotypes are found in both tropical and subtropical regions.

The different members of the genus _Flavivirus_ share common antigenic sites. This explains the marked antibody cross-reactivity not just between types but also between different virus species. For instance, there is a sequence homology of 46–53%

between Japanese encephalitis virus and dengue viruses.

Clinical features. Dengue virus infection is commonly asymptomatic. If it leads to clinical symptoms, these often present as a nonspecific febrile viral illness with fever, malaise, and so on, indistinguishable from other common viral infections.

Classical dengue fever disease starts abruptly after an incubation period of 2–7 days with high fever, severe headache, arthralgia (pain in the joints), and rash. The characteristic feature is severe generalized muscle, bone, and joint pain (thus the colloquial name, break-bone fever). The clinical features are age-dependent.

The fever may persist for 6–7 days and may take a biphasic course with two peaks (saddle-back fever). After the initial symptoms, generalized arthralgia, anorexia, and weakness follow. Respiratory symptoms such as a sore throat and rhinitis may occur, especially in children. The skin rash (exanthema) of dengue fever is not distinctive. A transient generalized macular rash may appear on the first or second day of illness. Later, a secondary rash may develop, first appearing on the trunk, and fever may rise again. Sometimes there is generalized lymphadenopathy. The white blood cell and the platelet counts are frequently depressed; rarely, hemorrhage occurs. A relative bradycardia (slow heart rate) is common during the febrile phase. In some cases, viscerotropic infection occurs and may result in hepatitis and jaundice. Classical dengue fever is normally a benign and self-limiting disease.

The severest manifestation of dengue infection is called dengue hemorrhagic fever, of which four grades of severity are recognized. Grades III and IV are known as dengue shock syndrome. As defined by the World Health Organization, dengue hemorrhagic fever is characterized by four major clinical manifestations: high continuous fever for 2–7 days; hemorrhagic diathesis (positive tourniquet test, petechiae, ecchymoses or purpura, bleeding); often, hepatomegaly; often, circulatory failure. The distinctive laboratory findings are moderate to severe thrombo-cytopenia (reduction in the number of platelets to less than 100,000 per microliter) and concurrent hemoconcentration (elevation of hematocrit by 20% or more above normal). The latter results from plasma leakage due to incresed vascular permeability. The absence of hemoconcentration allows differentiation of those occasional cases of classical dengue fever with some hemorrhagic manifestations from true dengue hemorrhagic fever. In the absence of spontaneous bleeding, a patient with dengue hemorrhagic fever is classified as grade I, otherwise grade II.

If, in addition to displaying the characteristic features of dengue hemorrhagic fever, a patient shows evidence of circulatory failure (rapid and weak pulse and narrow blood pressure, hypotension, cold skin, and so on), the illness is called dengue shock syndrome (equivalent to dengue hemorrhagic fever grade III or IV, the latter with profound shock and undetectable blood pressure or pulse).

The pathophysiological hallmarks of dengue hemorrhagic fever are capillary leakage and abnormal hemostasis. Hemostasis includes a rise in hematocrit, hypoproteinemia, and reduced plasma volume, which may lead to hypovolemic shock with renal failure. The disorder of hemostasis involves the major factors, thrombocytopenia, coagulopathy, and disseminated intravascular clotting which may lead to severe bleeding. Generalized hemorrhage and multisystem organ failure are the end results of dengue hemorrhagic fever, mimicking malaria or typhus.

All four dengue virus serotypes are capable of causing dengue hemorrhagic fever. The immune status and the age of the patient are important factors. Most patients with dengue hemorrhagic fever are children under 16 years of age experiencing their second dengue virus infection or infants with waning levels of maternal anti-dengue antibody.

Pathogenesis. After infection (through mosquito bite), the virus reaches the regional lymph nodes and disseminates into the reticuloendothelial system. The virus multiplies and then enters the bloodstream. In dengue hemorrhagic fever/dengue shock syndrome, the main abnormality occurs in and around small blood vessels and consists of endothelial swelling, perivascular edema, and infiltration of mononuclear cells.

Originally described in Southeast Asia in the 1950s, dengue hemorrhagic fever has also appeared in the Caribbean, Venezuela, Brazil, and the Pacific islands (Fiji), but it has not yet been documented from Africa. This observation of dengue hemorrhagic fever cases in regions experiencing their second wave of dengue infections, in the wake of the expansion of another, hitherto absent dengue virus subtype, led to the secondary infection theory.

The antibody-dependent immune enhancement hypothesis suggests that during the secondary infection with a different dengue virus serotype the pre-existing antibodies from the first infection, unable to neutralize the virus completely, enhance the virus uptake. The mononuclear cells so infected may become a target of an immune elimination mechanism, triggering the production of complement activators.

The possible role of antibody-dependent enhancement in the immunpathogenesis of dengue hemorrhagic fever has been demonstrated in vitro with a number of flaviviruses, including dengue and yellow fever virus. The enhancement of flavivirus replication, that is, the increased absorption of virus particles to host-cell plasma membrane mediated by subneutralizing antibodies, has been shown to increase viral replication in monocyte- or macrophage-like cells.

Further studies with monoclonal antibodies have revealed that there is no correlation between serological specificity and antibody-dependent enhancement; it can be mediated by broadly cross-reactive as well as type- and even subtype-specific antibodies. However, antibodies with different functional activities (for example, neutralizing, hemagglutination-inhibiting antibodies, and so on) can mediate antibody-dependent enhancement.

Studies in Thailand and Cuba demonstrated primarily the interval between the two different dengue serotype infections, and secondarily that the sequence of infecting dengue serotypes may be important in the severity of dengue hemorrhagic fever.

Diagnosis. Virological diagnosis can be achieved by virus isolation or through serological tests. Virus isolation can be attempted during the febrile phase of illness using live mosquitoes, or mosquito and mammalian cell lines with immunofluorescence staining.

Different antibody tests allow diagnosis of acute dengue virus infections. The detection of antidengue immunoglobulin M (IgM; the early antibodies) with commercial tests is now widely used to document primary infection. A positive IgM antibody result is described as a specific method to distinguish dengue from other flavivirus infections and to allow a definitive diagnosis from a single acute blood specimen. Otherwise, a fourfold increase in dengue IgG antibody titer in paired sera (taken during the acute and during the convalescent phase) using an enzyme immunoassay or a hemagglutination inhibition assay is required to confirm an infection.

Prevention. The control of dengue fever depends primarily on the control of the vector. *Aedes* spp. are peridomestic and often breed in human-made containers. The elimination of these breeding sites is the most effective control method, but it requires considerable effort which must involve the affected community and has to be sustained. During outbreaks, the use of larvicides can reduce transmission.

Dengue virus vaccines are being developed. One obstacle is the need to effectively immunize against all four serotypes; otherwise, vaccination might lead to the formation of enhancing antibodies and thereby increase the risk of dengue hemorrhagic fever. A tetravalent live attenuated dengue vaccine is among those in development.

For background information *see* ANIMAL VIRUS; ARBOVIRAL ENCEPHALITIDES; DENGUE FEVER in the McGraw-Hill Encyclopedia of Science & Technology.

Hans Wilhelm Doerr; Regina Allwinn; Wolfgang Preiser

Bibliography. S. W. Gollins and J. S. Porterfield, Flavivirus infection enhancement in macrophages: Radioactive and biological studies on the effect of antibody on viral fate, *J. Gen. Virol.*, 65:1261–1272, 1989; N. Gratz, The rise and spread of dengue, dengue haemorrhagic fever and its vectors: A historical review (up to 1995), *WHO Monogr.*, CTD/FIL(DEN) 96.7, 1996; S. B. Halstead, The pathogenesis of dengue: The Alexander D. Langmuir Lecture, *Amer. J. Trop. Med. Hyg.*, 114:632–648, 1981; B. L. Innis et al., An enzyme linked immunosorbent assay to characterize dengue infections where dengue and Japanese encephalitis co-circulated, *Amer. J. Trop. Med. Hyg.*, 40:418–427, 1989; M. Isaäcson and M. J. Hale, The viral haemorrhagic fevers, in W. Doerr and G. Seifert (eds.), *Tropical Pathology*, 2d ed. Springer, 1995; S. Nimmanitya, Dengue and dengue haemorrhagic fever, in G. C. Cook (ed.), *Manson's Tropical Diseases*, 20th ed., W. B. Saunders, 1996; W. F. Scherer et al., Laboratory safety for arboviruses and certain other viruses of vertebrates, *Amer. J. Trop. Med. Hyg.*, 29:1359–1381, 1980; B. de Wazières et al., Nosocomial transmission of dengue from a needlestick injury (Letter), *Lancet*, 351(9101):498, 1998.

Deoxyribonucleic acid (DNA)

Deoxyribonucleic acid is the genetic material which serves as the template or code for cellular processes in living organisms. DNA is transcribed into messenger ribonucleic acid (mRNA), which is then translated into amino acids, the subunits of the proteins required for life. DNA is constantly subjected to damage by both internal (endogenous) and external (exogenous) assaults. This damage can lead to blocks in transcription and replication or to permanent alterations in the genetic code, and can ultimately lead to disease and death of the organism. It has been reported that up to 90% of human cancers are associated with DNA damage. Because the survival of an organism depends upon precise transcription and replication of its genome, several mechanisms of DNA repair have evolved. In general, these repair systems are universal, correcting defects in the genomes of organisms as diverse as microbes and humans.

Sources of damage. Exogenous DNA damage, both chemical and physical, arises from exposure to ultraviolet radiation from the Sun and natural and synthetic chemicals. Internal sources of damage arise daily from DNA and cellular metabolic processes. For example, base mismatches due to replication fidelity errors are a major source of abnormal base pair formation. Loss of a purine base (depurination) is a common event in cells; it has been estimated at nearly 20,000 purines per day in human cells. Deamination of adenine, guanine, and cytosine (for example, cytosine is converted to a uracil) also occurs with a high frequency in mammalian cells. DNA strand breaks are induced by ionizing radiation, oxidative damage, or chemical agents. Finally, bulky adduct formation by monofunctional DNA damaging agents (for example, procarbazine) and induction of intrastrand and interstrand crosslinks by bifunctional agents (such as cisplatin) may prevent DNA metabolic functions such as replication and transcription. Overall, DNA is subject to many types of damage resulting from oxidative metabolism as well as exposure to physical and chemical mutagens in the environment.

Mechanisms of recognition. DNA is subjected to a wide variety of distinct lesions, yet this damage is recognized and corrected by a small set of repair proteins. It is unlikely that the repair molecules are able to recognize each unique type of lesion. Instead, repair proteins recognize a common feature of the lesions, such as helical distortions in the damaged DNA. There are only a few DNA repair systems known to play major roles in maintaining the integrity of the mammalian genome. Nucleotide excision repair (NER) is largely responsible for removing bulky lesions in the DNA; in fact, studies show that

Proteins involved in human nucleotide excison repair		
Protein	Activity	Repair function
XPA	DNA binding	Damage recognition
XPB	Helicase, ATPase	Dissociate oligomers
XPC	DNA binding	Repair of inactive regions
XPD	Helicase, ATPase	Dissociate oligomers
XPE	DNA binding	Damage recognition(?)
XPF	Nuclease	5′ incision
XPG	Nuclease	3′ incision
ERCC1	Interacts with XPF	Assists in 5′ incision
RPA	DNA binding	Damage recognition
TFIIH	Multicomponent transcription factor	Formation of preincision complex
PCNA	DNA binding	Polymerase clamp
RFC	ATPase	Molecular matchmaker
POL δ(ε)		Repair synthesis
Ligase		Ligation

nucleotide excision repair is capable of removing every DNA lesion tested to date. Clearly, DNA damage recognition is a complex process, and it is thought to be the rate-limiting step in DNA repair. Several mechanisms have been proposed to explain the broad and versatile damage recognition capacity. One mechanism depends on the degree of distortion in the DNA helix and on the single-strandedness induced by the damage. Supporting evidence is provided by studies demonstrating increased repair rates of lesions that induce large helical distortions as compared to those of lesions that are less distorting.

At least 14 proteins are involved in nucleotide excision repair (see **table**). Several proteins are involved in the damage recognition step, including xeroderma pigmentosum group A (XPA), which is thought to be the damage recognition protein in humans as evidenced by its ability to bind damaged DNA with a greater affinity than DNA in its standard form. The multiple protein-DNA and protein-protein interactions involved in nucleotide excision repair lead to another hypothesis of damage recognition termed selectivity cascades. Selectivity for sites of damage occurs via a series of sequential and overlapping steps of low selectivity resulting in high specificity due to the total number of interactions involved.

The flexible-fit hypothesis is another suggested mechanism of damage recognition. XPA contains a binding region for generalized binding to DNA and a flexible region containing a single-stranded DNA (ssDNA) binding region that interacts with another repair protein, RPA. This flexible region may insert itself into regions of ssDNA, allowing recognition of a variety of damaged sites.

Mechanisms of repair. A fundamental component of DNA repair is the redundancy of the genetic code (that is, one strand is the exact complement of the other). Therefore, when one strand of the duplex is damaged, the other strand is available as a template for repair synthesis. Four mechanisms of repair are discussed below.

Base excision or direct repair. The substrates for base excision repair generally include simple base pair modi-

fications such as those formed by endogenous oxidative events (for example, formation of uracil, thymine glycols, or 8-oxoguanine). This system of repair has a narrow substrate range with highly specific recognition. Unlike the other repair systems, the protein that recognizes the damage is catalytically involved in the removal of the damaged base.

Damaged bases are repaired in a two-step process: first, a damage-specific DNA glycosylase (for example, one recognizing a uracil, thymine glycol, methylpurine, or 8-oxoguanine) releases the base from the sugar backbone, leaving an apurinic or apyrimidinic (AP) site. Next, the abasic sugar is released by the actions of an AP lyase and an AP endonuclease (**Fig. 1**). The one nucleotide gap is then filled in and ligated.

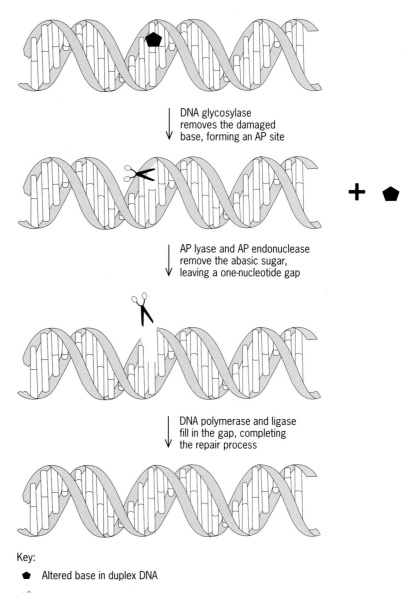

DNA glycosylase removes the damaged base, forming an AP site

AP lyase and AP endonuclease remove the abasic sugar, leaving a one-nucleotide gap

DNA polymerase and ligase fill in the gap, completing the repair process

Key:

⬟ Altered base in duplex DNA

✂ Cleavage sites

Fig. 1. Base excision repair pathway. The altered base is removed by the action of a specific DNA glycosylase, leaving only the sugar backbone or an apyrimidinic (AP) site. Next, the abasic sugar is cleaved and released by the actions of AP endonuclease and AP lyase. Finally, polymerase and ligase fill in and seal the gap.

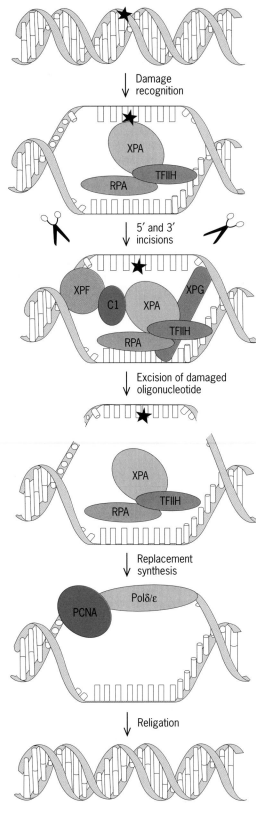

Key:

★ Lesion in DNA duplex

✂ Cleavage sites

Fig. 2. Nucleotide excision repair pathway. Graphic representations of proteins involved in nucleotide excision repair (NER) are not to scale.

Nucleotide excision repair. The major repair system for removing bulky lesions in the DNA, including ultraviolet photoproducts, is nucleotide excision. Defects in nucleotide excision repair can lead to devastating consequences as demonstrated by clinical disorders such as xeroderma pigmentosum, Cockayne syndrome, and trichothiodystrophy. Nucleotide excision repair removes an oligonucleotide which contains the damage. The patch size is specific to the organism. In eukaryotes the excised oligomer is generally 27–29 nucleotides, formed by incisions of 3–5 bases on the 3′ side of the lesion and 21–25 bases on the 5′ side.

This repair pathway has five steps (**Fig. 2**): damage recognition, 5′ and 3′ incisions, excision of the damaged DNA oligonucleotide, gap filling by replacement synthesis, and religation. First, XPA binds the damaged DNA and forms a complex with the repair protein, RPA, which then recruits the multicomponent transcription factor, TFIIH. This complex contains the helicases (enzymes capable of unwinding the DNA double helix) XPB and XPD, among other factors. Next, the excinuclease, XPG, binds RPA on the 3′ side of the lesion and cuts the strand containing the damage. A protein repair complex then binds XPA on the 5′ side of the lesion and incises the DNA. The oligomer containing the damage is displaced, leaving a single-stranded gap. This gap is filled in by DNA polymerase δ or ε in conjunction with DNA polymerase accessory proteins, using the underlying DNA as a template. Ligation completes the repair reaction.

Mismatch repair. Mismatch repair is responsible for correcting DNA replication errors and processing heteroduplex regions (double-stranded DNA in which the two strands show noncomplementary sections) in homologous recombination intermediates, rendering it crucial for genomic stability. Cells deficient in mismatch repair are hypermutable and predisposed to oncogenesis (cancerous growth). In fact, mutations in the mismatch repair proteins are implicated in the development of hereditary nonpolyposis colon cancer in humans.

Mismatches contain mispairing of otherwise normal bases, and therefore a signal is required to distinguish the newly synthesized DNA (containing the error) from the parental strand. This strand specificity is achieved through adenine methylation (that is, newly synthesized DNA is briefly unmethylated) at the sequence guanine-adenine-thymine-cytosine (GATC) in *Escherichia coli*. In humans the details of mismatch repair are not fully understood, but the suggested signals are cytosine hemimethylation at cytosine-guanine (CG) sequences and single-strand breaks.

The mechanism of mismatch repair as understood in *E. coli* involves the interaction of three repair proteins (MutH, MutL, and MutS). The repair process proceeds as follows (**Fig. 3**): (1) MutS binds the mismatch, (2) MutL binds MutS at the mismatch, which activates the GATC endonuclease from MutH, (3) MutH incises the unmodified strand, (4) MutU

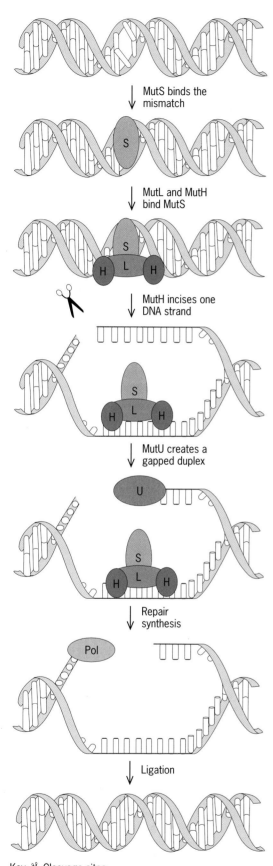

Key: ✂ Cleavage sites

Fig. 3. Mismatch repair pathway in *Escherichia coli*. The misalignment of the base pair represents the mismatch in the duplex DNA. Graphic representations of the proteins are not to scale.

and an exonuclease remove the mismatch and surrounding bases, creating a gap, and (5) the gap is filled in and ligated by repair synthesis.

Recombinational repair. In the repair systems discussed above, one strand of the DNA duplex serves as a template for repair of the complementary strand. One strategy that is used when both strands are damaged in the same location (a double-strand break) and therefore no template strand is available for repair is homologous recombination. Using the homologous chromosome (or sister chromatid) as the donor fragment, the double-strand break can be repaired via a crossover event.

For background information *see* DEOXYRIBONUCLEIC ACID (DNA); DEOXYRIBOSE; GENETIC ENGINEERING; MUTATION; NUCLEIC ACID; PROTEIN; PURINE; PYRIMIDINE in the McGraw-Hill Encyclopedia of Science & Technology.   Karen M. Vasquez; Peter M. Glazer

Bibliography. E. Bertrand-Burgraff et al., Identification of the different intermediates in the interaction of (A)BC excinuclease with its substrate by DNase I footprinting on two uniquely modified oligonucleotides, *J. Mol. Biol.*, 219:27–36, 1991; J. Cleaver and J. States, The DNA damage-recognition problem in human and other eukaryotic cells: The XPA damage binding protein, *Biochem. J.*, 328:1–12, 1997; R. Doll and R. Peto, The causes of cancer: Quantitative estimates of avoidable risks of cancer in the United States today, *J. Nat. Cancer Inst.*, 66:1192, 1981; C. Helene, Molecular mechanisms for the recognition of damaged DNA regions by peptides and proteins, *Adv. Biophys.*, 20:177–186, 1985; A. Sancar, DNA excision repair, *Annu. Rev. Biochem.*, 65:43–81, 1996; A. Sancar, Mechanisms of DNA excision repair, *Science*, 266:1954–1956, 1994; S. Thibodeau, G. Bren, and D. Schaid, Microsatellite instability in cancer of the proximal colon, *Science*, 260:816–819, 1993.

Deuterium

Rapid advances in astronomical telescopes, their instruments, and computers have increased quantitative precision in cosmology. For example, there is now only one main cosmological model, the big bang model, because all the others have been ruled out by data. The universe began in a hot dense state, with a period of exponential expansion, called inflation. Structure grew hierarchically, by the gravitational growth of tiny primordial fluctuations in density, with low-mass (10^3–10^6 solar masses) objects collapsing before high-mass objects, such as galaxies and clusters of galaxies. Accurate measurements have been obtained for several of the ten or so main parameters of the big bang model, which describe the geometry and expansion of the universe, its contents, and the tiny primordial density fluctuations. Observations of the amount of the deuterium in the universe have recently given a much more precise value for one parameter: the amount of ordinary matter. A high density is found, but only about 4% of that which would stop the universe from expanding.

Testing the big bang model. Four fundamental observations imply that the universe has evolved from a hot dense state, described by the big bang: (1) The universe is expanding, and was denser and hotter in the past. (2) Microwaves are observed coming from all directions in the sky. The number of microwaves with each wavelength follows an exceedingly simple distribution which can be explained only if the whole universe was a hot ionized gas. (3) There are tiny fluctuations of 1 part in 10^5 in the intensity of the cosmic microwave background in different directions. These fluctuations measure equivalent fluctuations in the density of matter when the universe was about 300,000 years old, and they have just the right size to grow to make galaxies and clusters of galaxies like those seen today. (4) The relative numbers of light elements in the universe can be explained only if the universe was hot and dense enough to undergo nuclear reactions in a process called big bang nucleosynthesis about 0.1 to 100 s after the big bang started.

Of the four observations, big bang nucleosynthesis provides the strongest test of the big bang model because it is the earliest cosmological process with well-established physics, and it makes precise predictions, which can be tested in several ways. It is sensitive to possible physics beyond that used in the standard model of particle physics, and it provides the best estimate of the density of ordinary matter.

Big bang nucleosynthesis. Deuterium is one of the five light nuclei created in big bang nucleosynthesis. In the early universe, all particles which could be made were moving around freely in a hot dense gas. There were no nuclei, because they would be split apart by collisions with particles, including photons, protons, neutrons, and electrons. As the universe expanded, it cooled, like an ordinary gas, and by about 0.1 s it reached $10^{10}\,°C$, low enough for the first simple nuclei. These nuclei consisted of one proton and one neutron bound together in a deuterium nucleus.

The nucleus of normal hydrogen contains only a single proton. Deuterium is a form, or isotope, of the element hydrogen, because it also has one proton, but it is heavier because of the neutron. Isotopes are denoted by their element abbreviation and by their mass. Ordinary hydrogen is ^1H, because the proton has one unit of mass, while deuterium is D, or ^2H, because neutrons have about the same mass as protons. These deuterium nuclei soon combined to form two isotopes of helium, ^3He and ^4He, and a lithium isotope, ^7Li. In the standard model of the big bang, using known physics, the relative amounts of these five nuclei depend on only one unknown parameter: $\eta = $ (density of radiation)/(density of baryons). Here, density is the number of particles per cubic centimeter, radiation includes the photons in all electromagnetic radiation, and baryons are ordinary protons and neutrons, which make up the nuclei of all atoms. If the ratio of any two of these nuclei can be measured, the theory of big bang nucleosynthesis makes it possible to calculate the value for η, and the relative amounts of the other three nuclei can be

used to test whether the theory contains all the relevant physics. The ratio of the number of deuterium nuclei to hydrogen nuclei, D/H, is the most sensitive to η.

Advantages of deuterium. The deuterium/hydrogen ratio should provide the most accurate measure of η for several reasons. First, the astrophysics of deuterium is simple: The big bang is the only known source, and all deuterium is completely destroyed inside stars. The gas which stars eject contains hydrogen and some other elements, which were made inside the star, but no deuterium. Hence, the deuterium/hydrogen ratio declines over time as the fraction of the gas in the universe that has been inside a star increases.

Another advantage of measuring the deuterium/hydrogen ratio is that both elements are relatively common nuclei: Hydrogen is the most abundant element in the universe, and deuterium is eighth, comparable to iron and magnesium. Both are readily observed because their neutral atoms make more than 20 strong absorption lines in spectra of near-ultraviolet light, between 91.2 and 121.6 nanometers. These lines can be observed from space or from the ground when they are redshifted to visible wavelengths. The lines of deuterium are displaced by about 0.1 nm from those of hydrogen, because the deuterium nucleus has twice the mass of hydrogen. This displacement is easily resolved by modern spectrographs, many of which can detect a shift ten times smaller.

Quasars and primordial D/H. It is advantageous to measure the deuterium/hydrogen ratio toward quasars because they show absorption from gas whose composition is similar to that produced by big bang nucleosynthesis, which is called the primordial composition. The Hubble Space Telescope has been used to accurately measure the deuterium/hydrogen ratio in the gas between local stars; the value obtained is $D/H = (1.6 \pm 0.2) \times 10^{-5}$. Unfortunately, this ratio cannot be used to get an accurate measure of η because the fraction of the gas which has been inside stars is unknown. This fraction depends on the distribution of stellar masses in the past, and on the infall of primordial gas into the Milky Way Galaxy.

Toward quasars, it is possible to observe absorption lines from gas which is believed to be nearly primordial because it contains only 1 or 0.1% of the heavy elements which are found near the Sun. These elements were released by stars, the same stars which would destroy deuterium. Since these elements are rare, the fraction of this gas which has been inside stars must be less than 1%, and hence more than 99% of the primordial deuterium should remain in the gas. The gas toward quasars samples large volumes of space, in different directions, at early times.

D/H measurement from quasar spectra. Gas will reveal absorption lines from deuterium only when two conditions are met. First, there must be enough gas to show the deuterium line, which happens in about one gas cloud toward every second quasar. Second, the hydrogen must be restricted to a narrow range of velocities. In about 98% of gas clouds, the hydrogen has a wide range of velocities. In such clouds, the Doppler effect causes some of the hydrogen absorption to shift into the position of deuterium, which it can hide. Early attempts to measure the deuterium/hydrogen ratio toward quasars were unsuccessful because many quasars must be examined to find one in which the deuterium is not completely hidden or strongly contaminated by hydrogen.

Two quasars have now been found in which deuterium and hydrogen are both clearly detected. Observations with the HIRES spectrograph on the W. M. Keck telescope in Hawaii give accurate values for the deuterium/hydrogen ratio in each case. The two measurements agree within their errors: $D/H = (3.4 \pm 0.4) \times 10^{-5}$, which corresponds to $\eta = (5.1 \pm 0.3) \times 10^{-10}$. The measurement of 411 photons per cubic centimeter in the cosmic microwave background can be used to convert η into the baryon density: $\rho_b = 3.6 \times 10^{-31}\,\mathrm{g \cdot cm^{-3}}$, or 0.21 atom of hydrogen per cubic meter. The critical density of matter which is just enough to stop the universe from collapsing is given by Eq. (1). Here, H_0 is the Hubble constant, which measures the expansion rate of the uni-

$$\rho_{\mathrm{crit}} = \frac{3H_0^2}{8\pi G} = 1.88h^2 \times 10^{-29}\,\mathrm{g \cdot cm^{-3}} \qquad (1)$$

verse, in units of $\mathrm{s^{-1}}$, G is the gravitational constant, and h is the Hubble constant in units of $100\,\mathrm{km \cdot s^{-1} \cdot Mpc^{-1}}$, in which it is commonly measured (1 megaparsec $= 3.1 \times 10^{19}\,\mathrm{km} = 1.9 \times 10^{19}\,\mathrm{mi}$). The fraction of the critical density contributed by baryons, Ω_b, is given by Eq. (2). For $h = 0.7$, $\Omega_b = 0.04$.

$$\Omega_b = \frac{\rho_b}{\rho_{\mathrm{crit}}} = \frac{0.019 \pm 0.001}{h^2} \qquad (2)$$

It is known from many other observations, including those of supernovae, the numbers of clusters of galaxies, the baryon contents of clusters, and their mass-to-light ratios, that the total density of gravitating matter in the universe is $\Omega_m = 0.2 \pm 0.1$. This means that most of the mass in the universe is not ordinary baryons. This dark matter is probably an exotic particle, of a type that has not yet been detected, such as the lightest supersymmetric particle.

The abundances of the five light nuclei allow tests of the physics in big bang nucleosynthesis. The η value from the deuterium/hydrogen ratio predicts the abundance of the other three nuclei. Statistically, observations agree with these predictions at about the two-sigma level, not a good agreement, and not a major disagreement. New physics could explain these slight differences, but this is not required because there are known difficulties with the measurements.

The baryon density from the deuterium/hydrogen ratio is in good agreement with other less accurate estimates, from clusters of galaxies, the amount of hydrogen in the intergalactic medium, and the amount needed to form galaxies. The primordial deuterium/ hydrogen ratio toward quasars is about two times the value observed in the gas near the Sun, which means that about half of the atoms in this gas have been inside stars. *See* X-RAY ASTRONOMY.

For background information *see* BIG BANG THEORY; COSMIC BACKGROUND RADIATION; COSMOLOGY; DEUTERIUM; NUCLEOSYNTHESIS; QUASAR; UNIVERSE in the McGraw-Hill Encyclopedia of Science & Technology. David Tytler

Bibliography. C. Copi, D. N. Schramm, and M. S. Turner, Big-bang nucleosynthesis and the baryon density of the universe, *Science*, 267:192–199, January 13, 1995; C. J. Hogan, Primordial deuterium and the big bang, *Sci. Amer.*, 275(6):68–73, December 1996; J. Silk, *The Big Bang*, W. H. Freeman, 1989; D. Tytler, X. Fan, and S. Burles, The cosmological baryon density derived from the deuterium abundance at redshift $z = 3.57$, *Nature*, 381:207–209, May 16, 1996.

Digital presses

Digital presses are used to reproduce documents through the process of toner-based electrophotography, a technology inherent in the common desktop laser printer. Digital press equipment varies greatly in terms of production speed, numbers of colors possible, quality levels obtainable, and sophistication of in-line finishing. In recent years there has been a rapid growth in the usage of this type of reproduction equipment, and in some cases digital presses are replacing the traditional offset lithographic printing press.

While each vendor employs slightly different techniques to produce an image, the process of electrophotography is used in the majority of digital presses. Electrophotography involves the electrical charging of a photoconductive imaging drum usually made of selenium. A positive electric charge is produced either by a reflective image of the original hard copy or by the operation of a laser or light-emitting-diode (LED). The image area has a positive charge, and the nonimage area remains in a negatively charged state. Negatively charged toner particles are then passed over the imaging drum, and the toner is attracted to the positively charged image areas of the drum. The toner that remains on the drum is then transferred to the paper, and heat is used to melt or fuse the toner to the paper (see **illus.**). With multicolor printers, additional imaging drums are used for each color.

Digital presses are the preferred method of reproduction when a document must be produced in a timely fashion or small quantity. Since digital presses do not use traditional image carriers or plates or require the sometimes elaborate setup process of common offset lithographic presses, print production times can be extremely fast. Most digital presses can act as traditional photocopiers, reproducing hard-copy original pages, or can act as laser printers, printing directly from a digital file created on

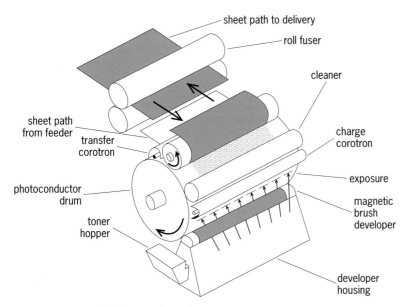

sheet path to delivery

roll fuser

cleaner

sheet path from feeder

transfer corotron

charge corotron

photoconductor drum

exposure

toner hopper

magnetic brush developer

developer housing

Interior of a typical digital press print engine.

a desktop computer. When printing directly from a digital file such as Adobe PostScript®, the files can be transmitted over a network or can be transported to the digital press by using removable computer media, such as a floppy disk. The prime advantage of using a digital press is that the print information can be saved and recalled for later reprinting. This allows a customer to spread out a print order over a longer period of time and to order reprints whenever they are needed, spawning the nickname "on-demand printing."

Types of equipment. Digital presses use the process of electrophotography, which is also known as xerography. A digital press can be broken down into three components: front end, print engine, and finishing.

Front end. This is the entry point of pages into the press. The front end can consist of a scanner, either off-line or incorporated into the press, which will digitize and store hard-copy originals, often consisting of text or photographs. A raster image processor (RIP) may also be included. A RIP is usually a high-power computer that accepts digital files via floppy disk or a network and creates a print-ready file. The digital press may also use the RIP computer station for archiving and manipulation of the job. Common digital press RIPs accept different print control languages, the most common being Adobe® PostScript. Other acceptable languages include Hewlett Packard Print Control Language (PCL), Tagged Image File Format (TIFF), and plain text.

Print engine. The print engine of a digital press is sometimes known as a marking engine. It is characterized by maximum number of colors possible, print speed, either cut sheet or roll fed, and its highest resolution. The resolution of most digital presses is 400–600 pixels per linear inch. This is a lower resolution than common lithographic printing, so digital presses do not deliver the same quality levels

as the offset lithographic press. Imaging speed can vary greatly between different models of printers. The fastest cut-sheet black and white printer images 180 8.5 × 11 in. (21.6 × 27.9 cm) pages per minute. Full-color printers, using cyan, magenta, yellow, and black to produce photographic quality, can range 8–70 8.5 × 11 in. pages per minute. Print engines can use either cut-sheet or roll-fed paper supplies; higher-speed print engines commonly use roll-fed paper supplies.

Finishing. Finishing capabilities can vary greatly from machine to machine. There are also a number of third-party attachments that can be incorporated. Finishing is considered to be in-line when integrated with the print engine and is automated. Black and white machines generally have many more finishing options than color machines; these options can range from basic stapling to hot-melt tape binding along the spine. Several optional devices can perform booklet making and soft cover perfect binding similar to that found in paperback novels. Because a digital press prints in a sequential fashion, starting with page one and ending with the last page, and can have in-line finishing, book production is a popular application.

Applications. Applications that are best suited for digital presses include short run quantities (fewer than 2000), multiple-page documents, and printing that is needed quickly. Additionally, a printing project that will undergo frequent changes to text or graphics is a prime candidate for a digital press.

The basic consideration when making a decision on using a digital press or a traditional lithographic press is quantity. The offset process is very efficient at long-run printing but can have a long setup time before producing good sheets. The cost of each printed piece using offset drops as the quantity increases and rises drastically on a short-run printing job. Digital printing has a relatively high cost per page that is constant regardless of print quantity. Because a digital press has no (or a very short) setup time, short-run jobs are much more economical to produce on the digital press versus the offset process.

A digital press also is preferred when the print job consists of multiple pages. For example, to produce one copy of an encyclopedia (containing many volumes) via offset, the cost would be astronomical because the offset process favors long print runs. The production of one copy via a digital press would be extremely easy because the digital press will print a job in sequential order, completing a book in its entirety and then starting the next copy.

Printing that is required in a short period of time also favors a digital press. Without the long setup times associated with offset lithography, a digital press can begin to image sheets quickly. The quick-printing industry has made extensive use of digital presses in their work flow, since most of their work is centered on the "while you wait" concept.

Because a digital press can store the print-ready file for later reprints, changes can be made easily. Substituting, adding, or eliminating pages or sections of pages can be accomplished readily because there is no plate to remake. A new digital page can simply be inserted electronically into the existing print-ready file. Customers having printing projects that will change frequently are one of the biggest users of digital presses.

An application gaining favor with digital presses is the growing process of personalized and database printing, commonly referred to as printing for an "audience of one." A digital press can vary the image or text on every page, unlike a conventional lithographic press that must print the exact same image. This feature of a digital press allows a printed piece to be customized for a particular recipient. The most common type of information that will be incorporated into a printed piece will be the recipient's past buying patterns, and this type of personalized direct mail will result in a much higher response rate from the recipient.

For background information *see* BOOK MANUFACTURE; COMPUTER PERIPHERAL DEVICES; ELECTROSTATICS; LASER; LIGHT-EMITTING DIODE; PHOTOCONDUCTIVITY; PHOTOCOPYING PROCESSES; PRINTING in the McGraw-Hill Encyclopedia of Science & Technology.
Frank Kanonik

Bibliography. H. Fenton and F. Romano, *On-Demand Printing: The Revolution in Digital and Customized Printing*, 2d ed., GATFPress, Sewickley, Pennsylvania, 1997; J. Hamilton, *The Print On Demand Opportunity*, CAP Ventures, Marshfield, Massachusetts, 1996.

Digital proofing

Digital proofing refers to processes that obtain hard-copy output directly from digital files for purposes of predicting the appearance or other aspects of printed matter. In practice, there are different purposes and, therefore, different requirements for digital proofs. The competing technologies in digital proofing include thermal transfer, electrophotographic, inkjet, dye sublimation, and photographic processes.

Historically, proofs have been used to check the accuracy and quality of prepared materials before the labor-intensive operations of readying the press for printing. Prior to the 1950s, special proofing presses were exclusively used. Photomechanical proofing, from high-contrast films, was the dominant method in the 1970s and 1980s. However, all-digital work flows do not require the exposure of film prior to imaging the printing plates, so photomechanical proofing is not applicable.

Purposes. The choice of a digital proofing method depends on the purpose of the proof. There are three categories of proofs: (1) design proofs aid in the visualization process when a graphic composition is being developed; (2) contract proofs provide an accurate predictor of the appearance expected from a given press system; and (3) imposition proofs confirm that the layout of graphic elements on a press form is proper. The requirements of design proofs and imposition proofs are less stringent than those of contract proofs. The design proof must accurately represent graphic and type elements with reasonably good color rendition. Typically, several design proofs are made as the final layout for the page is developed. Design proofs should be made quickly and relatively inexpensively. In fact, the first stages of design often utilize soft proofing, referring to using an image on a calibrated color monitor for judging the applicability of a design solution.

Contract proofs are made after the design process has been completed. They are intended to be an accurate prediction of the appearance of the printed job. These proofs are presented to the customer to approve the color reproduction and overall accuracy. Frequently, the customer is required to sign the contract proof to verify approval. The contract proof often acts as the color guide during subsequent production stages. The common understanding is that these proofs represent a "visual contract" with the customer. Failure to accurately match these proofs on press can lead to rejection of a printing job. Therefore, the digital proofing methods used to produce contract proofs come under the most severe scrutiny within the graphic communications industry.

Imposition involves laying out completed pages in a press form so that the pages will be in correct order after the signature (printed sheet) is folded. The imposition proof shows the layouts of completed press forms into signatures for platemaking. The sequence of folding and binding of signatures is generally optimized for the specific graphic product. In most cases, the imposition proof is made in one color since the color guides for the individual pages have already been made as contract proofs (see **illus.**).

Design proofs. Most graphic designs today are made with the aid of a personal computer and desktop publishing software. The design may start as a series of thumbnail sketches, but as it is developed into its final form, the computer is used. Typically, graphic design consists of type, graphic elements, and photographs. Page assembly programs, like QuarkXPress^{Tm} and PageMaker$^{©}$, are used to bring these elements together from individual pieces imported from specialized word processing, illustration, or image editing programs.

Soft proof. The simplest form of a digital proof is the soft proof, showing the image on a calibrated monitor to evaluate the graphic design. One drawback is that there is no hard-copy record of the approved image. Soft proofing is the primary digital proofing method used during image capture when digital photography is employed. It is also convenient when deadlines are short and quality expectations are not too high, such as in newspaper printing. Soft proofing is the most readily accessible form of remote proofing. If the client and the graphic designer are in different locations, individual color monitors can be calibrated with color management software to give like color appearance, and the soft proof can be displayed on both monitors simultaneously.

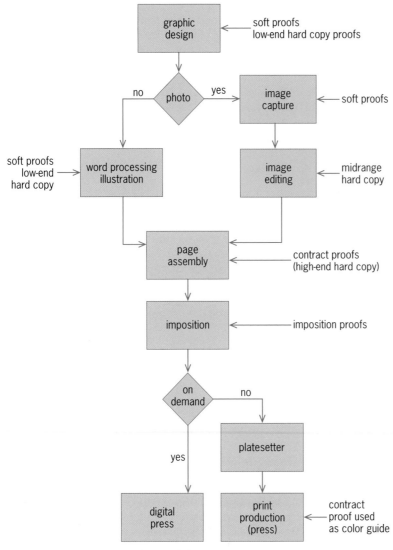

Proofing requirements of the digital work flow.

the finished job. The proofing system must mimic the ink hues, dot gain, gloss, paper color, and transfer characteristics of the specific press system in use. The contract proof should be presented at the same size and resolution that will be inherent in the final printing.

Indication of dot shape and screen ruling. The contract proof should consist of halftone dots that are of the same size, shape, frequency, and orientation as will be used for the printed piece. This is necessary to predict any optical interference patterns (commonly called moiré) that will affect the printing. Most of the digital proofing systems fail in at least one of these criteria. The constraints of cost and time, and a realistic appraisal of the quality requirements, guide practitioners in their choice of proofing systems.

Few digital proofing systems provide true halftone dot rendition. Kodak[Tm] Approval is one system that uses a laser imaging system to transfer pigments from donor sheets onto specially prepared substrates. The costs of the proofing engine and consumables are high. Furthermore, the throughput time is fairly slow (25 min for the first proof). However, the Kodak Approval system is well established due to its consistency and excellent match with press conditions. Recently, a larger four-page format was introduced.

Thermal transfer and photographic technologies have been developed that can accurately render halftone dot patterns. One of their advantages is that they use the same raster image processor (RIP) and marking engine that will eventually be used to write printing plates or color-separated films. Another feature of the thermal solutions is that they are able to use the same pigments that are commonly incorporated into printing inks. These systems are also able to make proofs up to the full size of press forms, which is often 25×38 in. Overall, the photographic solution has not been well received by the market because it involves chemical processing impacting environmental concerns and because it is inherently inconsistent.

Digital proofing systems that do not offer halftone dot rendition are popular due to their low cost, good repeatability, and accurate color fidelity. This market is dominated by dye sublimation machines, with some competition from electrophotographic and inkjet machines. The color rendition from some of these dye sublimation proofers has been judged acceptable in a series of digital proofing studies carried out by the Graphic Arts Technical Foundation. The absence of accurate halftone dot simulation precludes their use for the most critical jobs, but for general commercial work they are often deemed acceptable for contract proofs. Dye sublimation proofers are a popular choice for remote proofing operations where a hard-copy output is required.

Inkjet proofing systems are often chosen for large-format contract proofs. The most widely used system is the Scitex© Iris proofer, which utilizes continuous-stream inkjet imaging. Iris printers are available in a range of sizes offering high resolution, good color fidelity, and good repeatability.

Hard proof. Hard-copy design proofs are often made with color laser printers or desktop drop-on-demand inkjet printers. These systems have replaced thermal wax transfer printers. These proofs are relatively inexpensive and quickly made, but they lack the high resolution that is sometimes required to evaluate fine type and hairline rules. They also provide nonrepresentative halftone dot structures and very coarse screen rulings when compared to the eventual output that will be used for printing. They lack the color accuracy that is required for a contract proof.

Contract proofs. Contract proofs are the most demanding digital proofs since customers rely on them as visual predictors of the appearance of the final printing. Some of the requirements for an ideal contract proof are described below.

Consistency and repeatability. A proofing system should be repeatable from one proof to the next, and it should be consistent over time.

Accuracy of color rendition, size, and resolution. The contract proof should match the color appearance of

Contract proofs are sometimes made on digital presses, which utilize electrophotographic technologies. The digital presses cost in excess of $300,000 for 12 × 18 in. format, but the consumable costs per print are negligible. Print times are on the order of 90 s, and the color rendition is often acceptable for general commercial applications.

Imposition proofs. Imposition proofers are usually large-format machines since they must be the full size of the press form. These proofs do not need to be in full color because their sole purpose is to confirm that all the parts for a press form are properly positioned. Large-format pen plotters are sometimes used for these proofs, but the more popular solution is to use electrostatic or drop-on-demand inkjet plotters.

For background information *see* BOOK MANUFACTURE; COMPUTER GRAPHICS; COMPUTER PERIPHERAL DEVICES; DIGITAL COMPUTER; IMAGE PROCESSING; INK; PHOTOGRAPHY; PRINTING; TYPE (PRINTING) in the McGraw-Hill Encyclopedia of Science & Technology.   Anthony Stanton

Bibliography. E. T. Chrusciel and K. Sepos, The new direct digital proofing: Imagesetter and proofer in one, *Adv. Imaging*, March 1994; P. Green, *Understanding Digital Color*, Graphic Arts Technical Foundation, 1995; P. Hutton, G. Leyda, and K. Williamson, *GATF Digital Proofing Study*, RTR No. 4, Graphic Arts Technical Foundation, 1997; H. Tolliver-Nigro, Choosing a digital proofer, *Electr. Pub.*, pp. 24–31, December 1998.

Disease

Infectious diseases are caused by pathogens (such as bacteria, viruses, or fungi) or parasites (for example, protozoa or helminths). Not all infectious diseases are contagious, that is, pass directly from one person to another. Not all infections become diseases. An infection is defined by the presence of a pathogen or parasite in or on the host's body, whereas a disease can be a state of compromised health without an infection. Noninfectious diseases are caused by preexisting genetic conditions or the effects of abiotic environmental conditions, or by a combination of the two. For instance, heart disease is usually a noninfectious disease that has a strong genetic component (that is, it is more likely to occur in people who have a family history of heart disease), and it can be exacerbated by a diet high in saturated fats and by lack of exercise.

Infectious versus noninfectious. Research is beginning to reveal that many diseases formerly thought to be noninfectious are in fact caused by infectious agents. Some gastric ulcers are caused by the bacterium *Helicobacter pyloris*. Viruses have been implicated in causing certain types of lymphomas, leukemias, and other types of cancer, as well as some cases of multiple sclerosis. It is likely that many diseases are stimulated by the body's immune response to infectious agents, and that the response itself is affected by both heredity and abiotic environmental factors. Therefore, the distinction between infectious and noninfectious diseases may be fuzzy.

Infectious diseases can be classified by the source of the infectious agent and by the means of transmission to the victim. When pathogens and parasites are maintained largely in human populations, the diseases they cause are considered anthroponotic. When the disease agents are maintained largely in other animals, the disease is called zoonotic. Both anthroponotic and zoonotic diseases can be transmitted to the victim through direct contact, through aerosols, via drinking or cooking water, or by animal vectors. An example of direct contact is cold viruses moving from one's nose to one's hand, then to another person via a handshake, and finally to that person's nose. Cold viruses may also be transmitted through aerosols launched by a sneeze or cough. Many diarrheal diseases are transmitted when people with the disease defecate at a river or lake that provides drinking water for others. Many diseases are transmitted by arthropods, such as mosquitoes, flies, ticks, or fleas, that acquire the pathogen or parasite from an animal (sometimes a human) and later pass it on to a human host during a blood meal.

New and emerging or reemerging. The rate of discovery of new infectious diseases increased dramatically in the late twentieth century. Some diseases that are thought to be new are actually quite old in that the pathogen has existed and caused disease in humans for millennia. Some recently recognized diseases are not new and are termed emerging or reemerging diseases. In some cases, old diseases are thought to be new problems simply due to recent advances in methods of diagnosis, disease surveillance, and disease reporting by health agencies. In other cases, diseases emerge or reemerge due to changing ecological conditions that affect the disease agent, its animal hosts or vectors, or human behavior or population density.

Climate change. Changes in global or local climate, especially in temperature and precipitation, are frequently claimed to influence the occurrence of diseases, but strong evidence to support this allegation exists for only a few diseases. Some examples of the influence of global climate change on disease risk are found in diseases which are transmitted by mosquitoes of the genera *Anopheles* (malaria) and *Aedes* (dengue and yellow fever). These disease-bearing mosquitoes are sensitive to cold temperatures, and the northern limits of their geographic ranges in the Northern Hemisphere are determined largely by temperature patterns. Evidence suggests that as the minimum temperature of some regions has increased due to the greenhouse effect, so too has the geographic range of these disease vectors. The direct effects of global climate change on the distribution of mosquito vectors, and thus on the incidence of mosquito-borne diseases, are complicated by coincident evolution of pesticide resistance in mosquitoes and drug resistance in the malarial parasite (*Plasmodium* spp.).

Perhaps the best evidence for the effect of climate change on disease dynamics concerns cholera, a diarrheal disease caused by the bacterium *Vibrio cholerae*. Vibrios can be passed from person to person through direct contact or via local contamination of water supplies. Epidemics in southern Asia and western South America have been caused by massive contamination of drinking water and food, particularly shellfish, in coastal areas. Vibrios survive within the tissues of planktonic animals, particularly copepods, in fresh-water, brackish-water, and marine habitats. When weather conditions are particularly warm and rainy, coastal waters increase in both temperature and nutrient concentration (via runoff). Such conditions cause blooms of phytoplankton, which in turn elicit explosions in zooplankton biomass and associated vibrios. Local weather patterns and global climate change can thus be linked to local or regional outbreaks of cholera. This situation is exacerbated by the global transport of vibrios and other pathogens in the ballast water of oceangoing ships. Outbreaks of cholera have been predicted by monitoring sea surface temperatures and plankton blooms near population centers.

Many viruses and bacteria that cause human disease are maintained as benign infections within populations of wild mammals, especially rodents. One such pathogen, a hantavirus, was reported as a new disease afflicting people in the southwestern United States beginning in 1993, although subsequent analyses indicated that the virus has persisted within its rodent host for thousands to millions of years. In fact, it appears that many species of wild mice and voles maintain within their bodies a specific strain of hantavirus with which they have coevolved for millennia. In the case of the hantavirus pulmonary syndrome, a particular hantavirus that causes no disease in its main host, the deer mouse (*Peromyscus maniculatus*), suddenly began to appear in humans. It was particularly virulent, causing death in about half the people infected. The virus migrates to the capillaries lining the lungs, where it stimulates massive leakage of fluids into alveolar cavities, in essence causing the victim to drown. The reason that this virus, which historically was an exceedingly rare cause of human disease, suddenly emerged as a public health problem lies with the conditions that favor high population density of deer mice. Patients with hantavirus pulmonary syndrome frequently have deer mouse infestations in their homes or cars. The virus is shed in the urine and feces of mice and then can be inhaled in aerosol form by humans. Apparently, dense mouse populations cause airborne contamination rates to increase. Scientists who have been monitoring populations of rodents in the southwestern United States, for reasons unrelated to hantaviruses, found that deer mice showed an unusual increase in numbers in the winter and spring of 1993. The suspected cause was higher than normal rainfall in the deserts of the American west, arising from an El Niõ event (a warming of surface waters in the central Pacific Ocean). It is possible that this heavy rainfall increased food supplies (piõn nuts, fruits, and insects) for the mice, allowing their populations to increase, and promoting an outbreak of this zoonotic disease. El Niõ events, which typically persist for one to several years, represent a case of global climate change. However, a link to human activities, such as greenhouse gas emissions, has not been established.

Land-use patterns. Another zoonotic disease in which mice play a key role is Lyme disease. It involves ticks of the genus *Ixodes*, which transmit the spirochete bacterium *Borrelia burgdorferi* to humans. The bacterium causes both the skin rashes in early stages and the arthritis in late stages of the disease. In the eastern and central United States, ticks become abundant only in areas supporting plentiful white-tailed deer (*Odocoileus virginianus*), because the adult female ticks require a blood meal from a deer in order to lay eggs. However, deer are incapable of transmitting Lyme disease bacteria to feeding ticks. Instead, the bacteria are known to reside within the bodies of wild mammals, especially white-footed mice (*Peromyscus leucopus*), in which they do not cause disease. The only way for the tick vector to acquire the bacteria is to feed on a reservoir host, which efficiently transmits the pathogen to the vector. Infected ticks maintain the bacteria for life and may transmit them to people while sucking their blood. Therefore, optimal conditions for the maintenance of Lyme disease include plentiful deer and white-footed mice.

Deoxyribonucleic acid (DNA) and other evidence suggest that Lyme disease has persisted in North America and Europe for at least centuries and probably much longer. It is likely that changes in land use by Americans strongly influenced the epidemiology of this disease. American colonists cleared vast expanses of native forests along the eastern seaboard for agriculture, forestry, urbanization, tanneries, and fuel production. This deforestation decimated populations of deer and relegated some rodent species to habitat fragments. After the massive reductions in agriculture and other industries requiring deforestation, forests of the northeastern United States began to regenerate, particularly in the twentieth century. Deer populations rebounded, and rodents expanded their ranges to occupy regrowing forests. By the late twentieth century, suburbanization, which creates landscapes consisting of an ideal mixture of forest, ornamental plantings, and open fields, together with the extirpation of natural predators such as wolves (*Canis lupus*) and mountain lions (*Felis concolor*), appears to have created optimal conditions for the proliferation of deer. White-footed mice also perform particularly well in these human-impacted landscapes. These changes in land use set the stage for ticks to proliferate and maintain spirochete infections.

Outbreaks of several hemorrhagic fevers may be promoted by other types of changes in land use by humans. Argentine and Bolivian hemorrhagic fevers are transmitted to humans through inhalation of aerosols containing excreta of mice of the genus *Calomys*. The Lassa fever virus in central and western

Africa has a similar route of transmission from mice of the genus *Mastomys*. Similar to the situation for deer mice and hantavirus pulmonary syndrome, outbreaks of these diseases are associated with eruptions of the rodent reservoirs. Both *Calomys* and *Mastomys* are agricultural pests that attack cereal crops both in the field and in storage, and mouse populations may be prone to erupting in landscapes containing a heavy food supply (crop fields) near safe refugia (intact forest or human habitation). In addition, both types of rodent reservoirs appear to be suppressed by the presence of other, dominant rodents and by predators, whose populations may be reduced in and around human settlements.

Effects of climate change, pollution, and land-use patterns on disease agents and their reservoirs have been addressed only recently, and evidence is not yet strong. Future research in this field will be critical to anticipating and ameliorating emerging infectious diseases.

For background information *see* CANCER (MEDICINE); CHOLERA; CLINICAL MICROBIOLOGY; EPIDEMIC; INFECTIOUS DISEASE; LEUKEMIA; LYME DISEASE; MALARIA; YELLOW FEVER in the McGraw-Hill Encyclopedia of Science & Technology.

Richard S. Ostfeld

Bibliography. R. R. Colwell, Global climate and infectious disease: The cholera paradigm, *Science*, 274: 2025–2031, 1996; L. Garrett, *The Coming Plague: Newly Emerging Diseases in a World out of Balance*, Farrar, Straus and Giroux, 1994; J. N. Mills and J. E. Childs, Ecologic studies of rodent reservoirs: Their relevance for human health, *Emerging Infect. Dis.*, 4:529–538, 1998; R. S. Ostfeld, The ecology of Lyme-disease risk, *Amer. Scient.*, 85:338–346, 1997.

Distance education

Since the late 1980s, satellites have been used to provide distance education to large numbers of students and adults. Teachers can extend their reach to remote parts of the world, providing students with access to advanced courses that are unavailable locally. Until recently, most of these courses have been delivered in the traditional educational style with students watching the teacher on video. However, emerging technologies are reshaping distance education. The ability of these technologies to handle video, data, and voice gives designers the opportunity to take maximum advantage of a rich distributed-learning environment. Individuals can be offered a chance to engage in learning at their convenience and in a variety of modes, and they can have access to experts as well as current resources.

One-way versus interactive technologies. Distance learning can use old or new technologies or combinations of both. Correspondence courses were an early form of distance learning. The availability of satellites for telecommunications has resulted in several other forms, ranging from one-way broadcast of video (instructional television) that is designed for either real-time viewing (synchronous) or videotaping for later use (asynchronous) to fully interactive courses that utilize multimedia.

When videoconferencing first appeared, it combined one-way video with a return audio via regular phone lines to allow participants to interact with presenters. The distant stations, called television receive only stations (TVROs), required only satellite dishes, receivers, and a standard phone line. This point-to-multipoint transmission via satellite is still frequently used because it is more cost-effective than full two-way videoconferencing between sites. The program can be produced in a studio or on location and uplinked to the satellite at a fixed ground station or by a mobile unit.

Until the mid-1990s, broadcast video signals were analog. The introduction of digital technology resulted in more efficient use of satellite transponders and the ability to use one transponder to handle both video and data signals. The development of direct broadcast satellite (DBS) technology, which uses small, inexpensive, fixed satellite dishes at receive sites, also provided for more efficient use of bandwidth. Several companies use this technology to broadcast distance learning programming to hundreds of sites across the United States. These networks use one-way video point-to-multipoint designs. Students interact with instructors by phoning in questions and comments or by using e-mail, bulletin boards, or chat rooms.

Full two-way videoconferencing via satellite has been relatively expensive, requiring an investment in full-duplex earth stations at each site, videoconferencing hardware and software, and a control center (bridge) to coordinate the activities of each location. Compressed video systems are being introduced that provide a less expensive alternative to full broadcast-quality video and use substantially less bandwidth. However, the cost of this equipment is still high compared with a television-receive-only installation, and this design also requires a bridge to coordinate interaction between sites. The video usually operates at a maximum of 30 frames per second compared to 60 for full broadcast-quality video. However, advances in video compression technology now allow for the scaling of image quality from medium-bit-rate (384 kilobits per second) to high-bit-rate Motion Picture Expert Group (MPEG) (3 megabits per second) compression, which provides higher-quality video transmission.

Internet Protocol over satellite. The Internet Protocol (IP) for the transmission of both data and video provides the ability to deliver just-in-time distance education (that is, asynchronous courses that can be accessed by students whenever they need them) throughout the world. It also allows quality video programming to be combined with World Wide Web content for delivery via satellite. Very small aperture terminals (VSATs) can be used to deliver data over both one-way and two-way links, although the cost of full-duplex (transmit and receive) sites is still relatively high.

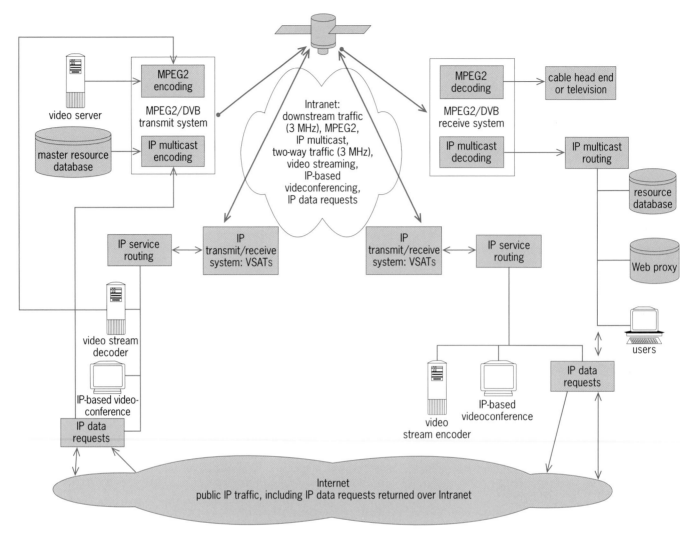

Pacific Region Satellite Network, showing the technology used to provide distance learning from a head end with a video server or master resource database. The Intranet is the Internet Protocol network connecting all the sites in this network via the satellite, and it connects at the head end with the Internet.

Internet Protocol over satellite provides the ability to deliver high bandwidth to remote locations that do not have access to terrestrial fiber. Using digital technology, throughput on a single transponder can reach 58 megabits per second, significantly reducing costs. Speeds of 512 kilobits to 3 megabits per second are now affordable and provide for high-speed Internet access. For most distance learning applications, bandwidth between 1.5 and 3 megabits per second is sufficient.

Internet Protocol over satellite can also take advantage of the asymmetric nature of Internet traffic by providing higher bandwidth for the forward channel and lower bandwidth for the back channel. In distance learning applications, this design allows for high-bandwidth applications to be delivered on the forward channel to students while their low-bandwidth requests and responses are handled on the back channel.

Digital satellite technology allows a transport stream to carry a variety of applications, including video and data services. Management of these ser-

vices can be accomplished at the transmission site, resulting in complete control of remote receivers. All of these services are integrated into seamless networks using MPEG, Digital Video Broadcast (DVB), and TCP/IP (Transfer Control Protocol/Internet Protocol) standard protocols. Data streams and other information can be multiplexed into a composite data stream which is then connected to a DVB-compatible satellite modulator for modulation to an intermediate frequency for satellite transmission. On the receiving end, PC (personal computer) cards have been developed to break out video and data from the multiplexed signal, allowing for users to view both on their personal computers. This development provides an opportunity to use effectively video streaming to deliver video and other resources to students on demand.

A number of problems are inherent in the delivery of Internet Protocol over satellite. The delay resulting from the time for a signal to be transmitted to a satellite 22,000 mi (36,000 km) above the Earth and then back to the receive site can cause problems with bit error rates. Since Internet Protocol uses forward error correction, any significant delay in the receipt of Internet Protocol packets can degrade IP-over-satellite performance. However, several companies have developed new products to deliver Internet Protocol over satellite. In addition, some satellite networks use frame relay to allocate dynamically bandwidth to support voice, data, and compressed video.

Several VSAT networks are currently being deployed or are in operation. In the Pacific region, a VSAT network is being built to connect students spread over 4.9 million square miles (12.7 million square kilometers) with a combination of full-duplex and receive-only sites. Full-duplex sites will transmit and receive broadcast-quality video and provide for two-way Internet connectivity. Some receive-only sites will receive video and data via satellite with the back channel to the Internet via a local Internet service provider. Remote receive-only sites will receive video and access to information on the Internet through the use of Internet Protocol multicasting and proxy servers. This architecture increases the already asymmetric nature of Internet traffic and allows for the use of a slow-speed back channel with a high-speed broadcast. The network will use 8 MHz of C-band transponder space to deliver both video and data and create an Intranet for high-speed delivery of data (see **illus.**).

Internet connectivity to classrooms to support distance learning is also being provided in the United States using high-speed satellite download systems direct to personal computers at speeds of up to 3 megabits per second. This architecture includes a proxy server and a return path via a local Internet service provider. Currently, this service is being provided to a number of remote rural schools whose only previous access to the Internet was through a modem at 28.8 kilobits per second.

Distributed learning environments. The availability of multimedia technologies now allows for truly interactive distance learning via satellite. Internet-based distance learning is being delivered via satellite throughout the world, especially to remote locations that previously had little or no telecommunications access. This type of distance learning allows the student to engage in synchronous or asynchronous interaction with teachers, other students, and experts in the field. These technologies enable the creation of a distributed learning environment that provides individualized experiences for the learners.

Current distance learning applications use a variety of media to create a distributed learning environment. Such an application might include print, video, voice, and software components. The print component might consist of course text, readings, and syllabi delivered via regular mail or posted at a Web site on the Internet. The video component might include either two-way interactive videoconferencing or video accessed through video streaming on the Web or by videotape. The voice component might be through an audio conference or through a record-

ing accessed through the Web. Finally, software components might include the use of collaborative software that allows learners to interact with other learners and the instructors both synchronously or asynchronously through bulletin boards, chat rooms, e-mail, multiuser dungeons (MUDs), multiuser object-oriented environments (MOOs), or threaded discussions. [Multiuser dungeons are based on synchronous, text-base, computer-mediated communication between a group of people who create their own virtual reality or world. Players congregate in different "rooms" in order to "chat" with other players. Multiuser object-oriented environments are a category of multiuser dungeons that use an object-oriented programming language. Threaded discussions are conducted through an on-line forum where users can post messages and read previous postings, which are connected together (threaded) by either date or subject.]

For background information *see* COMMUNICATIONS SATELLITE; DATA COMMUNICATIONS; DIRECT BROADCASTING SATELLITE SYSTEMS; MULTIMEDIA TECHNOLOGY; TELECONFERENCING; TELEPHONE SERVICE; VIDEO TELEPHONE in the McGraw-Hill Encyclopedia of Science & Technology.

Steve Baxendale

Bibliography. H. Berman, Telehealth and satellites: A prescription for the 21st century, *Via Satellite*, 13(9):72–80, 1998; P. J. Brown, Interactive distance learning: Satellites move to the head of the class, *Via Satellite*, 13(11):22–36, 1998; A. E. Hancock, The evolving terrain of distance learning, *Sat. Commun.*, 23(3):24–28, 1999; C. Kunz, Earth stations: Our link to the stars, *Via Satellite*, 13(5):24–36, 1998.

Distributed dynamic systems

Distributed dynamic systems for querying the Web that depend on dynamic run-time planning have the advantage that flexible and extensible queries can be specified. Use of metadata and agents allows the system to unify data from heterogeneous sources. User efficiency can be greatly improved.

Web-based queries. As the World Wide Web is increasingly used to publish and acquire information, there is a growing need to improve both the efficiency of finding information of interest and the way this information is disseminated. Most searches for information on the Web (for example, to answer a specific user question) are handled through the use of search engines. However, search engines are not capable of handling the growth in the size of the Web and are being augmented by alternative approaches that unify information sources, permitting query strategies within particular knowledge domains to help search for an appropriate answer.

More sophisticated query mechanisms abound on the Web in particular domains. For example, virtual catalogs aggregate merchandise over multiple online stores; airlines offer route planners for their customers; a major credit card offers geographical information as to the nearest cash machine; and on-line brokerages permit many sophisticated financial queries.

A more general technique is to add one or more programs (or layers, as they are usually called in software engineering) between the data sources on the Web and the actual query to mediate between them. This query layer breaks the query into subqueries against each information source and then aggregates (gathers together) the results for presentation to the user. The query layer typically presents the user with a user interface form that constrains and controls the type of queries that can be made.

The implementation of the query layer depends on describing the capability of information sources in such a way as to make it possible to dynamically decide a strategy for answering a query. In other words, depending on what the query layer knows about the state of the information sources at the time of the query, it decides how best the query can be handled, including whether to solicit more information from the user.

Web browsing. The Web is a protocol that enables computers to communicate with one another and that people use to publish information from one computer system (called a Web site) to another (called a Web browser, or browser for short). Web sites are made up of collections of blocks of text and graphics, known as Web pages, and increasingly, programs that dynamically generate Web pages based on input from a browser.

Search engines. Search engines are the most common query mechanism on the Web. The search engine is itself a separate program that takes structured input from a browser, typically one or more words or text fragments. The result of the query is a dynamically generated list of Web pages or Web sites containing those words or text fragments. Public search engines work by automatically finding as many Web pages as they can, reading through their text contents, and building tables which link the words in the text back to the original Web page. Search engines make it possible to find all pages that might mention the word "cancer" or "gasoline" or "antique furniture." *See* COMPUTERIZED SEARCHES.

Increasingly, organizations (such as companies) are using internal Web sites to publish to users within the organization. These Web sites are growing larger and more complex, and a limited search engine is often included in the Web site so that users can find information there.

The growth of the Web has made search engines more difficult to use. For example, in 1998 a search for the word "cancer" with a popular search engine found nearly 3 million references to Web pages, whereas a similar search a few years earlier might have found 50–100 references.

Web publishing. Accessing content published on the Web is only half the task of information retrieval; publishing is the other half. Most information in electronic format is not in text form but in spreadsheets, databases, or files with their own format. If informa-tion in a Web page is published simply by extracting the information and putting in a text-based Web page, it is difficult to update that information if the original information changes. It is also not possible for a person using a browser to see the original information in another format. Because of these drawbacks, publishing on the Web is not done in this way. Instead, based on input from the browser, a program uses the information in its original format to dynamically generate output as a Web page.

A simple example involves the organizational structure of a company, which is kept in a company database. This structure can be extracted into a Web page and published so that anyone who looks at that Web page sees the organization chart. Alternatively, the Web page can be a program so that when the browser looks at the Web page the program executes and builds the organization chart that would be seen in the browser based on accessing the company database. The Web page is built on demand instead of being prepackaged. Search engines cannot help to locate such dynamically built Web pages since they distinguish only two types of data sources: those that contain text and those that do not. Those that do not contain text, they ignore.

Metadata. Intelligence can be added to the query layer by giving a description of the data source. This description is known as metadata to distinguish it from the actual data in the source. For example, one data source for purchasing goods may measure prices in dollars, while another may measure prices in German marks. The data are numbers that represent the actual prices; the metadata are the information that in one data source the data are price in dollars, and in the other data source, in German marks.

Metadata allow the query layer to decide at the time of the query how best to execute the query. Metadata descriptions are used to break queries into subqueries, choose which data sources to query, unify heterogeneous data representations into one data model, apply rules to validate query results, and unify the results of subqueries into the query result.

Metadata for each information source are used to specify a program or agent that can execute subqueries on the data source directly. This agent also makes the metadata for the data source available to the query layer, which at run-time can ask the agent what its capabilities are.

An example of these concepts involves a search for flights between San Francisco and Los Angeles at a certain fare range and in a range of dates. The query layer is to use the Web to ask each airline what flights it has available and what the fares are. This information is presented differently by different airlines, and the fare structure may have to come from a different database.

The query layer must deal with several issues: (1) Information sources on the Web are frequently modified, new ones added, and old ones removed. Existing sources may be unavailable at the time of the query. (2) Information must be unified across different sources. For a given airline, fare structures are available in one database, but flight availability is in another. (3) Similar information from different sources may need to be combined. For example, a travel agency may publish special fares on the Web which are not available by using the airline's Web services.

Knowledge and query planning. The collection of all the metadata—information about the available data sources—is often called knowledge. The query layer uses its knowledge of the metadata to plan how to run the query. To make this architecture work well, metadata must be expressed in a metadata language which is used by the query layer, and it must be expressive (to flexibly capture capabilities) and unambiguous (so it can be processed by a program). Examples of fairly well worked out metadata languages and agent communication protocols are KIF (Knowledge Interchange Format) and KQML (Knowledge Query and Manipulation Language).

For background information *see* ARTIFICIAL INTELLIGENCE; HUMAN-COMPUTER INTERACTION in the McGraw-Hill Encyclopedia of Science & Technology.
James Veitch

Bibliography. V. Jagannatham, R. Dodhiawala, and L. S. Baum (eds.), *Blackboard Architectures and Applications*, Academic Press, 1989; Y. Labrou and T. Finin, A semantic approach for KQML: A general purpose communication language for software agents, in *Proceedings of the 3d International Conference on Information and Knowledge Management*, 1994; S. Russell and P. Norvig, *Artificial Intelligence: A Modern Approach*, Prentice Hall, 1995; G. Wiederhold and M. Genesereth, The conceptual basis for mediation services, *IEEE Expert, Intelligent Systems and Their Applications*, vol. 12, no. 5, September-October, 1997.

Echolocation

Echolocation refers to an animal's use of sound reflections to localize objects and to orient in the environment. Echolocating animals transmit acoustic signals and process information contained in the reflected signals, allowing the detection, localization, and identification of objects by sound. Echolocation has been documented in bats, marine mammals, some species of nocturnal birds, and to a limited extent in blind or blindfolded humans. Specialized perceptual and neural processes for echolocation have been detailed only for bats and dolphins.

Sonar signals. The acoustic signals used for echolocation by bats and marine mammals are primarily in the ultrasonic range, above 20 kHz and above the upper limit of human hearing. These ultrasonic signals have very short wavelengths and therefore reflect well from small objects in the environment (**Fig. 1**). All bat species of the suborder Microchiroptera use echolocation, producing sounds through the open mouth or through the nostrils, depending on the species. The signal types used by different bat species vary greatly, but all contain some

Fig. 1. Emitted pulse of a bat and the frequency modulations in the echo due to the wing movements of a nearby moth.

frequency-modulated (FM) components, whose frequency varies over the duration of the signal. Such components are well suited to carry information about the arrival time of target echoes. Constant-frequency (CF) components are sometimes combined with frequency-modulated components, and these tonal sounds are well suited to carry information about target movement through Doppler shifts in the returning echoes (**Fig. 2**). A Doppler shift is an apparent change in sound frequency that is proportional to the relative velocity of the source (in this case, sonar target) relative to the listener (in this case, the bat). If the bat approaches the target, the apparent echo frequency will increase, denoted as a positive Doppler shift. There is evidence that bat species using both frequency-modulated and constant-frequency signals show individual variations in signal structure that may facilitate identification of self-produced echoes. One echolocating bat species of the suborder Megachiroptera, *Rosettus aegyptiacus*, produces clicklike sounds with the tongue. The most widely studied echolocating marine mammal, the bottlenose dolphin (*Tursiops truncatus*), produces brief clicks, typically less than 50 microseconds in duration, with spectral energy from 20 kHz to well over 150 kHz. The temporal and

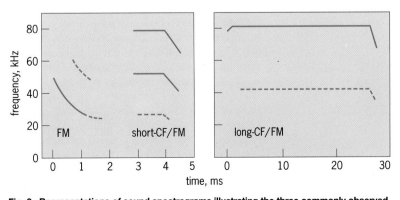

Fig. 2. Representations of sound spectrograms illustrating the three commonly observed types of echolocation sounds used by bats. Constant-frequency (CF) and frequency-modulated (FM) signals are the basic elements. There may be several harmonics in a single sound; solid lines indicate strong harmonics, and broken lines show weak harmonics.

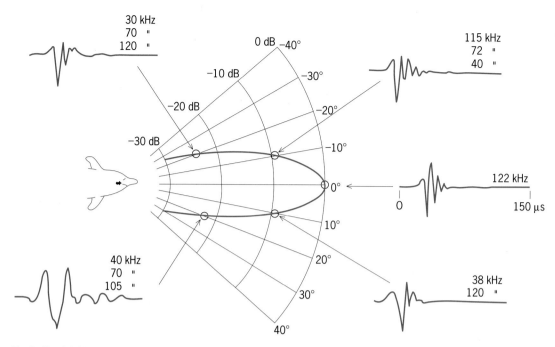

Fig. 3. For dolphin-produced clicks, examples of average waveforms at different azimuths in the horizontal plane. The peak frequency of each waveform is shown to the right. For frequency spectra with multiple peaks, the frequencies of the peaks are listed in descending order according to amplitude.

spectral characteristics of the dolphin clicks depend on the angle at which the sounds are recorded and the acoustic environment in which the sounds are produced (**Fig. 3**).

Perception. In echolocating animals, detection of sonar targets depends on the strength of the returning echoes. Large sonar targets reflect strong echoes and are detected at greater distances than small sonar targets. Behavioral studies of echo detection in bats and dolphins indicate that performance depends strongly on the acoustic environment. The presence of echoes from objects surrounding the target can interfere with its detection. Objects in front of the target produce echoes that arrive earlier in time, giving rise to forward masking of the target echo; and objects behind the target produce echoes that arrive later in time, giving rise to backward masking of the target echo. In addition, background noise level and reverberation (sound bouncing around in the environment) can influence sonar target detection.

Once an animal using echolocation detects a target, it uses information carried by the features of the signal to localize the object in three-dimensional space. In bats, the horizontal location of a target influences the features of the echo at the two ears, and these interaural cues permit calculation of a target's horizontal position in space. Laboratory studies of target tracking along the horizontal axis in bats suggest an accuracy of about 1°. The vertical location of a target influences the travel path of an echo into the bat's external ear, producing spectral changes in the returning sound that can be used to code target elevation. Accuracy of vertical localization in bats is approximately 3°. The third dimension, target distance, is measured from the time delay between the outgoing sound and returning echo. Behavioral studies

of distance discrimination in bats using frequency-modulated echolocation signals report thresholds of about 1 cm, corresponding to a difference in echo arrival time of approximately 60 μs. Experiments that require a bat to detect changes in the distance (echo delay) of a jittering target report thresholds of less than 0.1 mm, corresponding to a temporal jitter in echo arrival time of less than 1 μs. Successful interception of insect prey by bats requires accuracy of only 1–2 cm.

In marine mammals, psychophysical data show that the dolphin's horizontal angular discrimination depends on sound frequency; for pure tones, horizontal discrimination is 2.1° at 10 kHz, 3.6° at 6 kHz, and 3.8° at 100 kHz. In the vertical plane, discrimination increases from 2.3° at 20 kHz to 3.5° at 100 kHz. Localization accuracy in dolphins improves with broadband click stimuli, with thresholds in both the horizontal and vertical planes of less than 1°. Along the distance axis, dolphins can discriminate a target range difference of approximately 1 cm—performance similar to that of the echolocating bat.

Many bat species that use constant-frequency components in their signals are specialized to detect and process frequency and amplitude modulations in the returning echoes that are produced by the moving wings of flying insect prey. The constant-frequency signal components are relatively long in duration (up to 100 ms), sufficient to encode echo changes that occur with the movement of a fluttering insect over one or more wing-beat cycles. For example, the greater horseshoe bat (*Rhinolophus ferrumequinum*), using constant-frequency sonar signals, can discriminate frequency modulations in the returning echo of approximately 30 Hz (less than 0.5% of the bat's 83-kHz constant-frequency signal component). In addition, this species can discriminate fluttering insects with different echo signatures. Several bat species that use constant-frequency signals for echolocation adjust the frequency of sonar transmissions to offset Doppler shifts in the returning echoes, the magnitude of which depends on the bat's flight velocity. These adjustments in the frequency of constant-frequency echolocation calls, referred to as Doppler shift compensation, allow the bat to isolate small amplitude and frequency modulations in sonar echoes that are produced by fluttering insects.

More complex aspects of perception by sonar have been examined in some bat species. Free-flying bats can discriminate between mealworms and disks tossed into the air. Both mealworms and disks present changing surface areas as they tumble through the air. Research findings indicate that bats use changes in the features of returning echoes to discriminate target shape. Three-dimensional recognition of fluttering insects from novel perspectives has been reported in the greater horseshoe bat, a species that uses a constant-frequency signal for echolocation.

Dolphins can use echolocation to discriminate between objects made of different materials and shapes. Target dimensions and materials influence the intensity, spectrum, and temporal characteristics of the returning echoes, providing a number of potential cues for complex target discrimination by dolphins.

Neural mechanisms. Successful echolocation relies on specializations in the auditory receiver to detect and process echoes of the transmitted sonar signals. Central to the conclusive demonstration of echolocation in bats were data on hearing sensitivity in the ultrasonic range of the biological sonar sounds. Subsequent research has detailed many specializations for the processing of sonar echoes in the auditory receiver of bats. In dolphins, studies of the central auditory system have been limited, but early work clearly documents hearing in the ultrasonic range of echolocation calls.

Some bat species using constant-frequency components in their echolocation signals show specializations in the peripheral (the inner ear) and central auditory (the brain) systems for processing echoes in the constant-frequency range of the sonar sound. The greater horseshoe bat, for example, adjusts the frequency of its sonar vocalizations to receive echoes at a reference frequency of approximately 83 kHz. Its auditory system shows a large proportion of neurons devoted to processing this reference frequency, and this expanded representation of 83 kHz can be traced to mechanical specializations of this bat's inner ear.

There are other specializations in the bat central auditory system for processing echolocation signals that may play a role in the perception of target distance. In several bat species, researchers have identified neurons in the midbrain, thalamus, and cortex that respond selectively to pairs of frequency-modulated sounds, separated by a particular delay. The pairs of frequency-modulated sounds simulate the bat's sonar transmissions and the returning echoes, and the time delay separating the two signals corresponds to a particular target distance. The pulse-echo delay that elicits the largest facilitated response, referred to as the best delay, is topographically organized in some bat species using constant-frequency signal components. Researchers report that most neural best delays are in the range 2–40 ms, corresponding to target distances of approximately 34–690 cm (13–272 in.). These best delays represent a biologically relevant range for localizing prey items. Such topography has not been demonstrated in bat species using only frequency-modulated signals.

For background information *see* DOPPLER SHIFT; ECHOLOCATION; FREQUENCY MODULATION; PHONO-RECEPTION; SONAR in the McGraw-Hill Encyclopedia of Science & Technology.   Cynthia F. Moss

Bibliography. W. L. Au, *The Sonar of Dolphins*, Springer-Verlag, 1993; R. J. Busnel and J. F. Fish (eds.), *Animal Sonar Systems*, Plenum Press, 1980; R. R. Fay and A. N. Popper (eds.), *Springer Handbook of Auditory Research: Hearing by Bats*, Springer-Verlag, 1995; P. E. Nachtigall and P. W. B. Moore, *Animal Sonar: Processes and Performance*, Plenum Press, 1988; G. D. Pollak and J. H. Casseday, *The Neural Basis of Echolocation in Bats*, Springer-Verglag, 1989.

Ecosystem dynamics

Structurally modern terrestrial ecosystems originated approximately 380–340 million years ago (Ma) from the Middle Devonian Period through the earliest Carboniferous. Although on different temporal and spatial scales, the evolutionary assembly of these ecosystems parallels the assembly of modern ecosystems following major disturbances or recovery from widespread physical and climatic disruptions, such as the end of the last glaciation. The deep evolutionary history of terrestrial ecosystems documents the emergence of constraints that would affect all subsequent ecological dynamics and dictate the nature of smaller-scale landscape and successional patterns. The fossil record is the sole source of information about the establishment of these basic patterns of species interaction and resource partitioning.

Early ecosystems. Plant communities appeared on land during the Silurian Period, possibly earlier, more than 420 Ma. Well into the Middle Devonian, these communities were dominated by small, largely rootless plants with space-filling growth forms, living almost entirely on wet soils. For the most part, the dominant plants were opportunists, exploiting landscape disturbance from fires, floods, and other physical events. Plant assemblages were of low diversity and often consisted of a mix of large patches, each patch dominated by one plant species. There is little evidence of the extensive tiering and dense canopies found in later ecosystems. Although many plant species have been described, variation in their ecological strategies was relatively small. Arthropods appeared early in association with plants, as detritivores, living off dead plant material, and perhaps as herbivores. Vertebrates, however, do not appear to have had a direct impact on plants until the Carboniferous or later.

Such systems were open to evolutionary invasion by organisms with more complex life histories. Not only did the early plants occupy a small part of the land surface, but also they had not yet reached the levels of structural complexity permitted by their basic tissues and developmental systems, which were evolving gradually. The fossil record shows that plants were increasing in structural complexity through the Early and Middle Devonian, adding vegetative and reproductive components one upon the other. In the later Middle Devonian, plants in several lineages appear to have reached a threshold of structural complexity, following which major changes in body plan appeared in a relatively short interval of geological time. Coupled with the origin of new body plans, plants radiated into a wide array of new habitats.

Spatial and structural complexity. Evolutionary events in plants appear to have been the driving force behind the development of complex landscapes. During the Middle to Late Devonian and perhaps into the earliest Carboniferous, all modern vascular-plant body plans evolved. These body plans are characteristic of distinctive evolutionary lineages and are generally accorded the taxonomic rank of class. Included are the familiar ferns and seed plants, as well as the sphenopsids (horsetails) and some now extinct groups, all of which evolved from a common ancestral stock, the trimerophytes. At the same time, the lycopsids, a relatively obscure group today, also were radiating. They include the lycopods (clubmosses), the selaginellas, and the isoetoids (today represented by the tiny quillworts, the relatives of the giant scale trees of the Carboniferous). The lycopsids evolved from zosterophyll ancestors. The zosterophylls and trimerophytes had separated in the Early Devonian or earlier.

As these groups evolved, they began to partition the unutilized or underutilized ecological resources uniquely available at that historical time. Studies relating physical habitats and plant distribution indicate that isoetoid lycopsid trees evolved in and became major elements of wetland habitats, including swamps and mires. This establishment was made possible by unique root systems and reproductive organs that functioned optimally in wetlands. Seed plants and their immediate ancestors, the progymnosperms, originated near wetlands but radiated into drier, terra firma habitats, a consequence of deeply penetrating root systems and, ultimately, seeds, which permitted them to escape the need for free water to reproduce. Sphenopsids became most common and often dominant in streamside and swampy habitats where repeated burial by flood-borne sediments was common. As the only group of trees with rhizomes (underground stems), they were able to recover rapidly from disturbances in such habitats. Ferns initially were opportunistic weeds, occurring in a wide range of disturbed environments from

swamps to volcanigenic landscapes—a life history consistent with their massive reproductive output and highly dispersible spores.

Taxonomic partitioning. The landscape partitioning created during the class-level evolutionary radiation was complete by the early part of the Carboniferous and persisted, with some modifications, into the Early Permian. Each of the major classes had a distinct ecological center, with some overlap among groups. Thus, particular habitats of the Early Carboniferous remained relatively low in diversity, averaging three to five species in a typical paleobotanical census. Speciation within lineages tended to conform to the basic ecological centers defined in the class-level radiation so that, through time, species filled in the resource space by refining resource partitioning. Species turnover also was strongly constrained by evolutionary lineage; as species went extinct, they were replaced by descendants of similar morphology and ecological tolerance.

This evolutionary pattern of defining the ecological envelope followed by filling in the resource space has been inferred on smaller spatiotemporal scales for several groups of plants and animals that radiated recently. A taxonomic group of higher rank, through its included species, exercises incumbency, meaning occupation and utilization of resources to the exclusion of other species. Incumbents present a challenge to species that would invade such resource space. Assemblages of species, commonly called communities, create a fabric of resource use that tends to keep stable the dynamic partitioning of the larger landscape. Large-scale Paleozoic landscape assembly thus had consequences for the dynamics of community assembly at smaller spatiotemporal scales. Strongly expressed ecological centers of the major class-level lineages tended to restrict narrowly species replacement patterns in time and space. Assembly rules inferred for extant plant communities are very similar in kind, if not in scale, to those inferred for the Paleozoic.

Animal component. The fossil record of arthropods and of arthropod damage to plants indicates that detritivory was a major strategy for arthropods and was probably the major way that plant primary productivity entered animal food webs, at least through the later Carboniferous. However, herbivory (the consumption of plants) recently has been convincingly documented as a significant part of swamp and terra firma habitats as early as the Late Carboniferous. Current findings indicate all tetrapod vertebrates were carnivorous until the Late Carboniferous, from which time there is tenuous evidence of herbivores. Herbivory certainly was established by the Early Permian, when animals with dental wear and body architectures typical of herbivores became more common. At this time, no clearly defined structural or dynamic changes have been identified in plant assemblages as a consequence of the appearance of herbivores. Although largely conjectural, vertebrate herbivory, and possibly much insect herbivory, may have originated in areas of seasonally dry vegetation that is generally represented poorly in the Paleozoic fossil record. Appearance of drought-tolerant ecosystems may coincide with the drying of the tropical lowlands in the Permian as polar glaciers decreased.

Upland ecosystems. The fossil record of terrestrial organisms is strongly biased toward basinal lowlands. Through much of Earth history, especially in the Carboniferous tropics, these basinal lowlands were also wetlands. Such lowlands certainly experienced extended intervals of seasonal dryness, inferred from a variety of geological information. Anomalous fossil floras and faunas representing such time periods occasionally may be found, often rich in xeromorphic seed plants (those plants able to survive in dry environments); they show evidence of complex plant-animal interactions unseen in the primeval wetlands. The rarity of such xeric assemblages in the Carboniferous strongly suggests that extrabasinal areas, often categorized as uplands, were colonized by vegetation and faunas tolerant of seasonal dryness. Such areas appear to have been sites of the most active evolution, where major morphological and ecological innovations were taking place. As the tropics and basinal lowlands in other latitudes became more strongly seasonal, many of these biological innovations appeared suddenly in the Late Carboniferous and Permian geological record. In all probability, seed plants and smaller numbers of other groups had begun their movement into extrabasinal regions much earlier, during the Devonian class-level radiation.

Implications. Ecological partitioning and landscape dominance by four classes and numerous orders of vascular plants is a uniquely Paleozoic pattern. The degree to which the inferred dynamics can be found in a modern world dominated by flowering plants must be questioned. Nonetheless, there is little evidence to suggest differently at the level of ecological dynamics, which appear to be basically the same today as in the Paleozoic. Due to lower species diversity and the high-level taxonomic and structural differences among the dominant plants, the starkness of the patterns in the deep past may be more readily discerned and less subtle than at later times in Earth history. As today, the roles of incumbent advantage and long-term persistence of resource partitioning should be considered carefully as potential factors stabilizing ecosystem responses in the face of apparent small-scale, high-amplitude fluctuations in spatial and temporal patterns.

For background information *see* ECOLOGY; ECOSYSTEM; PALEOBOTANY; PALEOZOIC; PLANT EVOLUTION in the McGraw-Hill Encyclopedia of Science & Technology.   William A. DiMichele

Bibliography. T. J. Algeo et al., Late Devonian oceanic anoxic events and biotic crises: "Rooted" in the evolution of vascular plants?, *GSA Today*, 5:45, 64–66, 1995; R. M. Bateman et al., Early evolution of land plants: Phylogeny, physiology, and ecology of the primary terrestrial radiation, *Annu. Rev. Ecol. Systemat.*, 29:263–292, 1998; A. K. Behrensmeyer et al. (eds.), *Terrestrial Ecosystems Through Time*, 1992; W. A. DiMichele and R. M. Bateman, Plant paleoecology and evolutionary inference: Two examples from the Paleozoic, *Rev. Palaeobot. Palynol.*, 90:223-247, 1996; R. A. Gastaldo, W. A. DiMichele, and H. W. Pfefferkorn, Out of the icehouse and into the greenhouse: A late Paleozoic analog for modern global vegetational change, *GSA Today*, 6:1-7, 1996; N. Hotton, E. C. Olson, and R. Beerbower, Amniote origins and the discovery of herbivory, in S. Sumida and K. L. M. Martin (eds.), *Amniote Origins*, pp. 207-264, 1996; C. C. Labandeira, Early history of arthropod and vascular plant associations, *Annu. Rev. Earth Planet. Sci.*, 26:329-377, 1998.

El Niño

El Niño is a warming of the tropical Pacific Ocean that occurs roughly every 3 to 7 years. It develops in association with swings in atmospheric pressure known as the Southern Oscillation. During El Niño, the tradewinds weaken along the Equator as atmospheric pressure rises in the western Pacific and falls in the eastern Pacific. This condition allows warm water, normally confined to the far western Pacific, to migrate eastward. Upwelling, a process which brings nutrient-rich cold water to the surface along the coast of South America and along the Equator, is shut down, and sea surface temperatures warm in the central and eastern Pacific. Deep cumulus clouds and heavy rains, normally occurring in the western Pacific over the warmest water, migrate eastward in response to these surface temperature changes. These changes leave the western Pacific dry but bring torrential rains to the islands of the central Pacific and the west coast of South America.

Changing air currents. Tropical rainfall also releases heat into the upper troposphere, providing a source of energy to drive global wind fields. Shifts in these precipitation patterns cause changes in the atmospheric circulation that carry the influence of El Niño to parts of the globe remote from the tropical Pacific. The jet streams in both hemispheres of the Pacific intensify and shift equatorward during El Niño, steering wintertime storms into southern California and northern Chile. Northward deflection of air currents at higher latitudes over the North Pacific during El Niño years also brings warmer winter temperatures to parts of Alaska, Canada, and the northern tier of the United States.

ENSO. La Niña is characterized by stronger than normal tradewinds, colder tropical Pacific sea surface temperatures, and a shift in heavy rainfall to the far western tropical Pacific. It often produces effects on global weather patterns opposite to those of El Niño. As a result, El Niño, La Niña, and Southern Oscillation are often referred to collectively as ENSO, a cycle which oscillates between warm, cold, and neutral states in the tropical Pacific.

Drastic weather changes. In 1997-1998 El Niño brought torrential rainfalls and flooding to parts of California, the southeastern United States, equatorial east Africa, and Chile. It was also responsible for severe droughts in Mexico, Indonesia, and northeast Brazil. It virtually shut down the Atlantic hurricane season in 1997, yet spawned deadly swarms of tornadoes in nine southeastern states in the spring of 1998. Parts of the Midwest and the Great Lakes region experienced their mildest winter in over 100 years, as temperatures soared to record highs between November 1997 and February 1998.

Marine ecological shifts. Effects of the 1997–1998 El Niño on Pacific marine ecosystems were dramatic. The anchovy fishery collapsed off the coast of Peru, and thousands of marine mammals and seabirds perished for lack of food off the coast of California. Sportfishing along the west coast of the United States enjoyed a banner year as exotic tropical fish species migrated northward with El Niño warmed waters.

Economic and environmental losses. One preliminary estimate has put the total cost of the 1997–1998 El Niño at $33 billion due to crop failures, damaged infrastructure (such as roads, bridges, and buildings), reduced energy production and industrial output, and other economic losses. It has also been estimated that over 23,000 lives were lost worldwide as a result of weather-related disasters, and millions more were affected by the damage left in the wake of El Niño. El Niño was an environmental disaster in places such

as Indonesia, northeast Brazil, and Mexico, where forest fires raged out of control for months. Among its side effects was the spread of infectious diseases such as malaria, dengue fever, and cholera in Southeast Asia, South America, and Africa.

Observing and predicting trends. The 1997–1998 El Niño was, by some measures, the strongest on record, surpassing the record 1982–1983 occurrence (**Fig. 1**). These two climate events delimit a remarkable chapter in climate research. The 1982–1983 El Niño was neither predicted nor even detected until nearly at its peak. This failure shocked the scientific community which had been planning a major decade-long international program to study the ENSO cycle. The 1982–1983 El Niño made it starkly clear that both observational and forecasting capabilities were woefully inadequate, so that developing such capabilities became a central theme of the Tropical Ocean-Global Atmosphere (TOGA) research program, which took place from 1985 to 1994.

ENSO observing system. Within the context of TOGA, a new ENSO observing system was developed. It consists of arrays of moored and drifting buoys, shipboard measurements, and a network of island and coastal sea-level measurement stations (**Fig. 2**). It required financial support from many nations to complete, and was not finished until the final month of the TOGA program (December 1994).

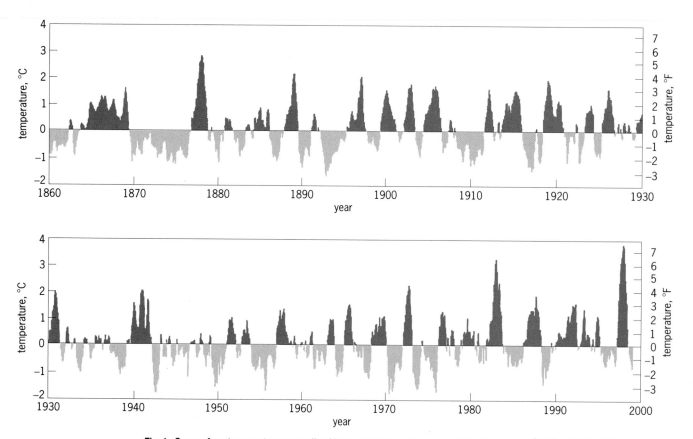

Fig. 1. Sea surface temperature anomalies (that is, deviations from normal) for the region 5°N–5°S, 90°W–150°W. Positive anomalies greater than about 0.5°C indicate El Niño events. Negative anomalies less than about −0.5°C indicate La Niña events.

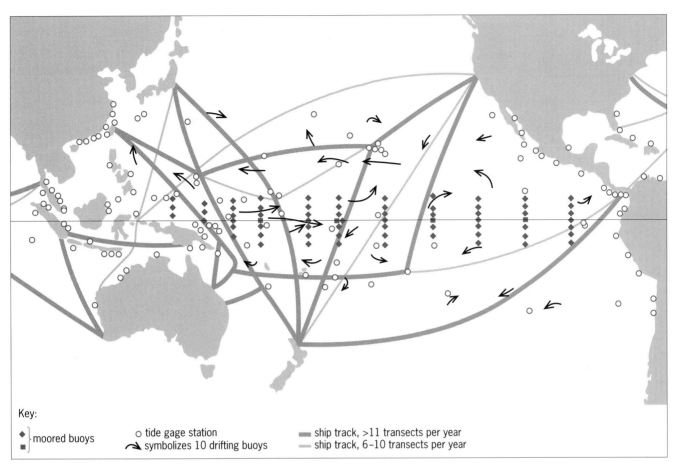

Key:

♦} moored buoys
■}

○ tide gage station
↗ symbolizes 10 drifting buoys

━━ ship track, >11 transects per year
‒‒ ship track, 6–10 transects per year

Fig. 2. **ENSO observing system developed under the auspices of the TOGA program. The four major elements are (1) a volunteer observing ship program for surface marine meteorological observations and ocean temperature profiles; (2) an island and coastal tide gauge network for sea-level measurements; (3) a drifting buoy network for sea surface temperatures and ocean currents; and (4) a moored buoy array for wind, temperature and, at some locations, current measurements.**

A key feature of the observing system is the fast delivery of data via satellite relay, often within hours of collection. These data are used for monitoring evolving climatic conditions, scientific analyses, and ENSO forecasting. Complementing this suite of ocean-based measurements is a constellation of space satellites measuring key environmental parameters.

Mapping 1997–1998 El Niño. This El Niño was the first for which the ENSO observing system was in place from start to finish, so that this event was not only the strongest on record but also the best documented. Though its development was similar in many respects to that of previous El Niño events, enhanced definition and fast delivery of the data from the observing system provided crucial information on its rapid evolution. This El Niño developed so explosively that from June to December 1997 each month set a new record high for sea surface temperatures in the eastern equatorial Pacific. By December 1997, most of the equatorial Pacific was covered with water at 28–29°C (82–84°F), which is near the maximum sustainable temperatures possible in the open ocean (**Fig. 3**). The global impacts of this El Niño were equally spectacular in keeping with

the extreme conditions observed in the equatorial Pacific. Then, even more suddenly than it developed, El Niño ended with an unprecedented drop in sea surface temperatures in the eastern and central Pacific, falling at some locations nearly 8°C (14°F) in 30 days. The climate system shifted from the strongest El Niño on record to cold La Niña conditions in the span of a month.

ENSO forecasting. Great strides have been made in ENSO forecasting since the 1982-1983 El Niño. A wide variety of forecasting approaches have been developed, ranging from statistical models based on the average behavior of previous ENSO events, to complex dynamical models that try to represent the physical processes at work in the coupled ocean-atmosphere system. Many of these models had success in predicting, at least one to three seasons in advance, that 1997 would be unusually warm in the tropical Pacific. Many of them also predicted that El Niño would give way to La Niña in 1998. Long-range weather forecasting schemes that included information about tropical Pacific sea temperature conditions were successful in predicting temperature and precipitation patterns in widely disparate parts of

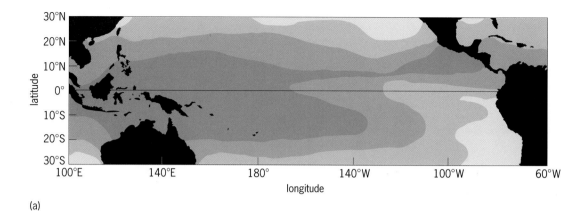

(a)

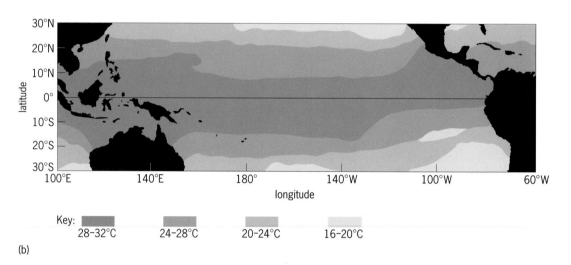

Key:

| 28–32°C | 24–28°C | 20–24°C | 16–20°C |

(b)

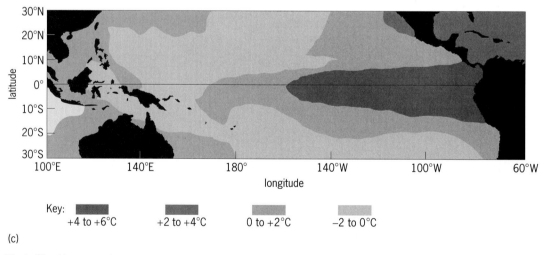

Key:

| +4 to +6°C | +2 to +4°C | 0 to +2°C | −2 to 0°C |

(c)

Fig. 3. Monthly averaged sea surface temperature for (*a*) December 1996 and (*b*) December 1997. (*c*) Monthly averaged sea surface temperature anomaly for December 1997.

the globe many months in advance. For example, the forecast for wintertime precipitation issued by the National Centers for Environmental Prediction in the fall of 1997 was the most accurate ever for the continental United States.

Damage control. Successful forecasts, unprecedented high-definition ocean measurements, and record warmth in the tropical Pacific all combined to cap-

ture the public attention in 1997–1998. Media coverage was so intense that El Niño became a household word all over the world. As a result, many individuals, municipalities, businesses, and in some cases national governments mobilized resources in an effort to prepare for El Niño's onslaught. It is likely that without the advance warning the toll from El Niño would have been much higher.

Forecasting difficulties. However, there were some forecasting failures related to the 1997–1998 El Niño. None of the forecast models predicted the rapid development or intensity of the El Niño before its onset, and none predicted the suddenness of its demise. Expectations of severe droughts in Australia and Zimbabwe and reduced summer monsoon rainfall over India failed to materialize. The reasons for these failures have yet to be fully determined. Factors that can influence the ENSO cycle and its global consequences include chaotic or random processes in the climate system that might enhance or obscure ENSO-related variations. Ocean-atmosphere interactions originating in regions outside the tropical Pacific may also be important, such as in the Indian Ocean where there has been a warming trend in tropical sea temperatures for the past 20 years. In addition, temperatures have been elevated in the tropical Pacific since the mid-1970s in association with a naturally occurring basin-scale phenomenon with a period of several decades. This Pacific Decadal Oscillation affects the background conditions on which ENSO events develop, potentially altering ENSO's character. Finally, 1998 and 1997 were, in that order, the warmest years on record. Occurrence of the 1997–1998 El Niño contributed in part to these extremes, since it is known that global temperatures rise a few tenths of a degrees Celsius following the peak El Niño warming in the tropical Pacific. However, aside from record warmth in 1997–1998, there has been a century-long trend of rising global temperatures, which may be due to anthropogenic greenhouse gas warming. *See* GLOBAL WARMING.

Climatic factors. Exactly how global warming, influences from outside the tropical Pacific, decadal time-scale variations, and random and chaotic elements of the climate system interact with one another and with ENSO is not entirely clear. It is clear, though, that there have been more El Niños than La Niñas since the mid-1970s, the early 1990s was a period of extended warmth in the tropical Pacific, and the extremely strong 1997–1998 El Niño followed by only 15 years the record-setting El Niño of 1982–1983. Further research is required to better understand the interactions between El Niño and other climate phenomena and to translate that understanding into improved forecasting capabilities.

For background information *see* CLIMATE MODELING; CLIMATIC PREDICTION; EL NIÑO; EQUATORIAL CURRENTS; PACIFIC ISLANDS; TROPICAL METEOROLOGY; UPWELLING; WEATHER FORECASTING AND PREDICTION in the McGraw-Hill Encyclopedia of Science & Technology.   Michael J. McPhaden

Bibliography. M. J. McPhaden, Genesis and evolution of the 1997–1998 El Niño, *Science*, 283:950–954, 1999; National Research Council, *Learning to Predict Climate Variations Associated with El Niño and the Southern Oscillation: Accomplishments and Legacies of the TOGA Program*, National Academy Press, 1996; S. G. H. Philander, *El Niño, La Niña, and the Southern Oscillation*, Academic Press, 1990.

Electric vehicle

In real-world driving conditions, the electric vehicle has yet to provide—at reasonable cost—range and performance equal to a conventionally powered automobile. One reason is that the power source for the electric vehicle has been the lead-acid battery with its inherent limitations. New developments, especially in batteries and fuel cells, should help overcome those obstacles and result in an electric vehicle that is consumer-acceptable.

An electric vehicle generally is defined as a self-propelled battery-powered vehicle having two to four wheels, powered by one or more electric motors, and used primarily for personal transportation. In a typical electric vehicle, the propulsion system converts electrical energy stored chemically in a battery into mechanical energy to move the vehicle; this is classified as a battery-only powered electric vehicle. Another major class is the hybrid electric vehicle, which has more than one power source, such as a thermal engine and an electric motor. The auxiliary power can be supplied from electrical energy sources such as a battery, fuel cell, or solar cell; an electrical storage device such as an ultracapacitor; or mechanical energy stored in a flywheel (kinetic-energy storage) or pressure accumulator.

Interest in the electric vehicle was renewed by various environmental and societal initiatives during the 1990s. In 1990 the California Air Resources Board mandated (but later rescinded) that 2% of each automakers' sales in the state should have zero emissions during the 1998 model year. Although the exhaust gas from some automotive engines was clean enough to meet the standard for a low-emission vehicle (LEV) and production of an ultra-low-emission vehicle (ULEV) was feasible, the demand for a zero-emission vehicle (ZEV) could be met only by the electric vehicle. This focused research on the weaknesses of the lead-acid battery, which include high weight for the amount of energy stored (low energy-storage density), quick-charge limitations, and reduced capacity in cold weather. However, the lead-acid battery has some advantages: it utilizes familiar technology and is readily available, compatible with existing infrastructure, cost-effective, and completely recyclable.

Environmental benefits of the electric vehicle include quiet operation, no waste from engine or gearbox oil changes, and no tailpipe or exhaust emissions. While cleaner air and lessening dependence on petroleum have pushed electric-vehicle research and development, so have military planners. Vehicles powered by an internal combustion engine have an easily detectable thermal signature that can serve as an aiming point for heat-seeking missiles. Cold combustion or other alternative power sources could prevent such targeting.

Advanced batteries. To help develop a better battery for electric vehicles, the US Advanced Battery Consortium was formed 1991. The purpose of this partnership among United States automakers, the electric utility industry, and the Department of Energy is to develop advanced batteries capable of providing future generations of electric vehicles with significantly increased range and performance.

Advanced batteries are generally classified as improved lead-acid (for example, using gel instead of liquid electrolyte), nickel-based, sodium-based, and lithium-based. The Advanced Battery Consortium has concentrated on nickel-metal hydride, lithium-polymer, and lithium-ion batteries. Nickel-metal hydride (NiMH) batteries have the potential to effectively double the range and performance provided by lead-acid batteries, while lithium-polymer and lithium-ion batteries could provide even more energy and power at lower cost. Lithium-based batteries would then make the range and performance of electric vehicles comparable to vehicles powered by gasoline engines.

Battery-only power. In late 1996, General Motors began limited marketing of the EV1, a two-seat electric vehicle powered by lead-acid batteries (**Fig. 1**). The EV1 was the first specifically designed electric

car produced by a major automaker since before World War II. The assembly of 26 lead-acid batteries forms a T-shaped battery pack that runs down the center and across the rear of the car. The 12-volt, 48-ampere-hour batteries are wired in series, with the battery pack providing a nominal voltage of 312 V.

The EV1 uses a valve-regulated, lead-acid (VRLA) battery. "Valve-regulated" means the battery is sealed but has a one-way valve through which a small quantity of gas can vent if pressure builds up inside the battery. Use of gas-recombinant technology eliminates the need for replenishing the electrolyte, making the valve-regulated, lead-acid battery maintenance-free. The electrolyte is stored in absorbent pads between the battery plates to minimize leakage in a collision or rollover.

Operationally, valve-regulated, lead-acid batteries differ from conventional 12-V automotive batteries primarily by having an increased duty cycle, delivering up to 85% of their charge without damage to the batteries or shortening their useful life. This could provide the EV1 with a useful driving range

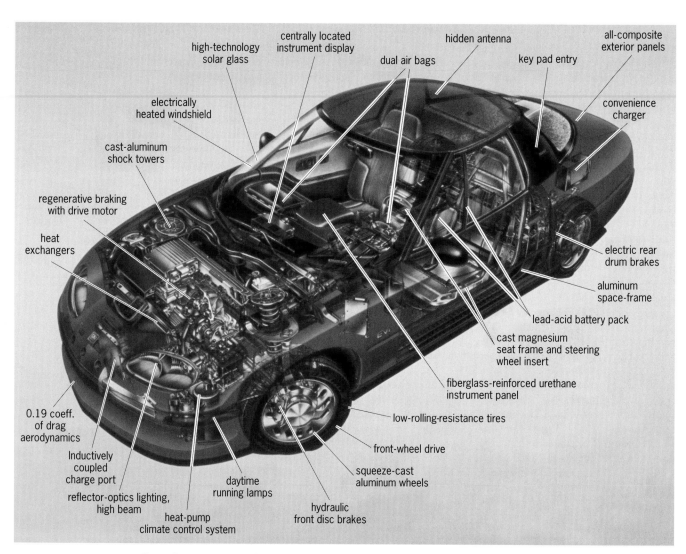

Fig. 1. General Motors' EV1, a two-seat electric vehicle powered by lead-acid batteries (*General Motors Corp.*)

per charge of 70–90 mi (113–145 km), based on the Environmental Protection Agency (EPA) city/highway driving schedule. A top speed of 80 mi (129 km) per hour is possible but at a sacrifice in range, which may also be shortened by topography, cold weather, accessory use, and driving style.

The propulsion system of the EV1 is a three-phase alternating-current induction motor integrated into a drive unit with a two-stage gear reduction and differential (**Fig. 2**). The assembly weighs 150 lb (68 kg), is filled-for-life with 56 fluid ounces (1656 milliliters) of oil, and has a peak power rating of 102 kW. This enables the vehicle to accelerate from 0 to 60 mi (0 to 97 km) per hour in less than 9 s. During braking, the drive motor is changed into a generator that reclaims energy from the motion of the vehicle as the vehicle's speed decreases. This provides a charging current to the batteries while helping the friction brakes slow the vehicle. The regenerative braking, which is computer-controlled and the only onboard charging, can increase vehicle range by up to 20% in stop-and-go driving.

To charge the batteries in the EV1, an inductive charge coupler is used. A lightweight, weatherproof plastic-covered paddle is inserted into the vehicle's charge port (**Fig. 3**). The paddle has no direct electrical contact with the vehicle. Charging current is transferred from the paddle to the vehicle by a magnetic field. Recharging the batteries at 220 V/30 A takes approximately 3 h. Some electric vehicles use a contact-type conductive charge coupler.

The charging system can be programmed to activate a heat pump in the vehicle's climate control system 15 min before the next scheduled vehicle usage. Heat-pump operation then heats or cools the passenger compartment, while helping to extend driving range because power to run the heat pump is supplied by the charger instead of the vehicle batteries.

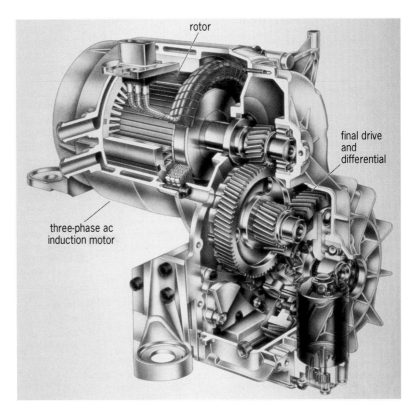

Fig. 2. Electric-vehicle propulsion system. (*General Motors Corp.*)

The heat pump runs as needed to keep the batteries cooler in summer and warmer in winter to improve their performance.

Some automobile manufacturers are adapting electric vehicles to use nickel-metal hydride batteries (**Fig. 4**). In this cell, the cathode, or positive electrode, is a nickel compound. The anode, or negative electrode, consists of vanadium, titanium, zirconium,

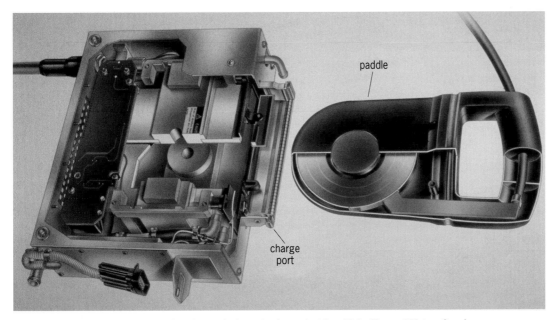

Fig. 3. Inductive charge coupler used to charge the batteries in an electric vehicle. (*General Motors Corp.*)

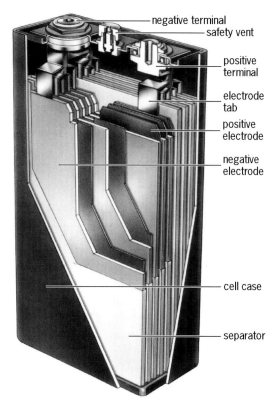

Fig. 4. Construction of a nickel-metal hydride cell which produces a nominal 1.2 V. (*General Motors Corp.*)

cle, as well as its manufacturing, safety, and maintenance problems. While not a common propulsion system for automotive vehicles, hybrid power has long been used in diesel-electric locomotives and submarines.

Two types of hybrid power trains are classified as series systems and parallel systems. In a series system, power flows through only one path (Fig. 5a). For example, a small engine drives a generator to charge the battery that powers an electric motor which drives the wheels. The engine may be a Stirling, gas-turbine, diesel, or spark-ignition engine running at constant speed to reduce fuel consumption and exhaust emissions. Use of an alternative fuel, instead of gasoline, could further lower exhaust emissions.

In a parallel hybrid vehicle, power flows along two paths such that both the engine and the electric motor can drive the wheels (Fig. 5b). This system allows the engine to charge the batteries while powering the vehicle. Under high-load conditions, both power plants operate simultaneously, but the system can choose to power the vehicle with either the engine or the electric motor, depending on speed and driving conditions. Regenerative braking provides battery charging during deceleration and braking, and the engine automatically shuts off when the vehicle stops moving. Toyota calls this arrangement parallel series hybrid. The result is improved fuel

nickel, and chromium alloyed to create a metal hydride. The electrodes are immersed in an electrolyte which is a strong alkaline solution of potassium hydroxide (KOH) and water, all sealed within a stainless steel case. Hydrogen ions are stored in the hydride during charging, and released during discharging to provide electron flow through the external circuit.

Each nickel-metal hydride cell produces a nominal 1.2 V. Eleven cells are connected in series to form 13.2-V batteries. The batteries are sealed, maintenance-free, and recyclable. They can accept high-rate charging of 50 kW for 15 or 20 min, which can boost the state of charge to the 80% level. However, the cost of nickel-metal hydride batteries is much higher than lead-acid batteries.

Hybrid power. A hybrid electric vehicle has two power sources which work in combination to propel the vehicle (**Fig. 5**). These usually are a thermal engine and an electric motor which uses electricity supplied from energy storage devices such as batteries, flywheels, and ultracapacitors, or from other energy sources such as fuel cells and solar cells. Like batteries, flywheels and ultracapacitors can be used both to store energy and to provide power.

Although power losses occur each time energy is converted from one form to another, hybrid power can be more efficient than today's automotive engine. Each power source is selected to maximize its advantages and offset the disadvantages of the other. However, having two power sources increases the complexity, cost, and weight of a hybrid vehi-

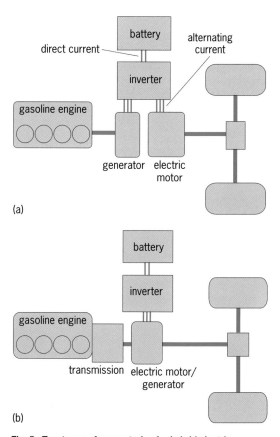

Fig. 5. Two types of power trains for hybrid electric vehicles: (*a*) series hybrid power train; (*b*) parallel hybrid power train. (*Toyota Motor Sales, U.S.A., Inc.*)

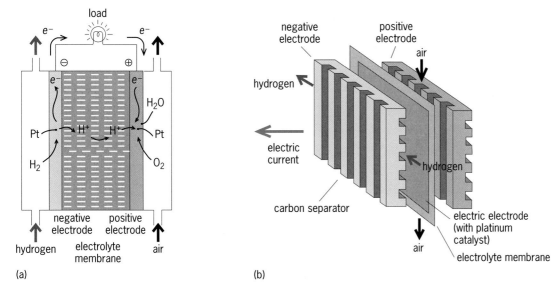

Fig. 6. Fuel cell. (a) Chemical action. (b) Construction. (Toyota Motor Sales, U.S.A., Inc.)

economy, reduced emissions, smooth performance, and increased vehicle range. Reportedly, engine exhaust emissions of hydrocarbons (HC), carbon monoxide (CO), and oxides of nitrogen (NO$_x$) are about one-tenth that of a conventional gasoline-engine vehicle, while fuel efficiency is doubled.

Fuel cell. A fuel cell is an electrochemical device in which the reaction between a fuel, usually hydrogen, and an oxidant, usually oxygen from the air, converts the chemical energy of the fuel directly and continuously into electrical energy (**Fig. 6**). As long as hydrogen is supplied to one electrode and oxygen to the other, the fuel cell produces a voltage between the two electrodes. Since fuel-cell voltage typically is less than 1.0 V, stacks of fuel cells are connected in series to provide the needed electrical energy. By using hydrogen as the fuel, water vapor (H$_2$O) and electricity are the only products. There are no significant amounts of unwanted emissions. The overall electrochemical reaction in a fuel cell is 2H$_2$ + O$_2$ → 2H$_2$O + electricity.

When fuel cells are the primary power source in a hybrid electric vehicle, batteries provide auxiliary power (**Fig. 7**). Fuel cells do not provide immediate output during a cold start. Until the fuel cells reach operating temperature, which may take a few minutes, batteries supply power for initial vehicle startup and movement.

Fuel cells may operate on direct hydrogen or on reformed hydrogen extracted from a hydrocarbon or fossil fuel. Hydrogen can be stored directly onboard the vehicle as a compressed gas, cryogenic liquid, or metal hydride. Vehicle range is determined by tank or hydride storage capacity. However, widespread use of hydrogen stored in these forms as a near-term vehicle fuel is unlikely because, among other problems, no infrastructure of public-accessible refueling stations exists.

Hydrogen can also be obtained on the vehicle by filling its fuel tank with a liquid hydrocarbon fuel such as gasoline or methanol. The liquid fuel undergoes an onboard conversion process called reforming which vaporizes the fuel, converting it into hydrogen and carbon monoxide. Then the carbon monoxide is converted into carbon dioxide (CO$_2$) and additional hydrogen. Use of the reformer

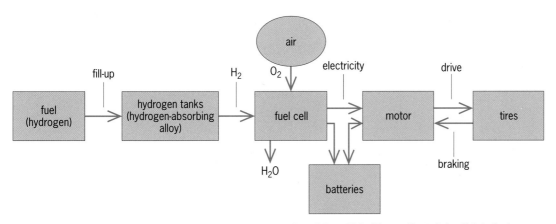

Fig. 7. Fuel-cell electric vehicle using direct-hydrogen storage onboard the vehicle. (Toyota Motor Sales, U.S.A., Inc.)

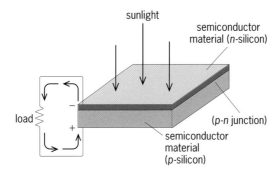

Fig. 8. A solar cell produces a voltage when exposed to sunlight. (ATW)

to provide hydrogen avoids refueling problems and makes the hybrid electric vehicle compatible with existing infrastructure. Motorists could refuel their vehicles in the conventional manner at local service stations.

The ability to fuel the vehicle with gasoline could move fuel-cell technology years closer to production while improving vehicle fuel efficiency by 50% and reducing emissions by 90%. When compared to a direct-hydrogen fuel cell, the reformer lowers overall vehicle efficiency and creates unwanted emissions of carbon dioxide.

Solar cell. A solar cell is a semiconductor device that produces a small dc voltage (typically 0.5–1.2 V) when exposed to sunlight (**Fig. 8**). An electric current will flow when the solar cell is connected to an external load. Solar power is attractive because it is a free and inexhaustible energy source that is environmentally benign.

Panels or arrays of solar cells placed on top the vehicle can be used for battery charging and additional electric power when needed. Some automotive vehicles have roof-top solar panels that power small ventilating fans. When needed, these fans are automatically switched to cool the interior of a parked vehicle. When the fans are not needed, the output from the solar cells is sent to the vehicle battery. This charges the battery and keeps it charged even if the vehicle remains parked for a long time. Separate solar-powered battery chargers are also available.

General Motors built an experimental vehicle to evaluate solar-powered vehicles (**Fig. 9**). The curved solar array covers about 86 ft^2 (8 m^2). The 8800 solar cells, each approximately the size of a postage stamp and as thick as a credit card, are grouped in 20 strings of 440 cells connected in series. This can produce 150 V and a current of up to 10 A to run the motor and charge the batteries. No solar-powered automotive vehicles are considered to be near production by major automotive manufacturers.

Flywheel. The flywheel is a mechanical device that stores energy by its motion. Theoretically, a spinning flywheel will continue to rotate at the same speed unless energy is either added to increase its speed or removed to decrease its speed. The flywheel is used in submarines and satellites, and can both store and deliver energy at high rates.

Various concepts for flywheel power in automotive vehicles have been studied and tested, including the use of a heavy steel flywheel and a high-speed, lightweight, vacuum-housed composite flywheel (**Fig. 10**). The flywheel is attached to a motor/ generator which includes regenerative braking capability. However, new and potentially costly technology is needed to suspend the flywheel in a vacuum with very little friction. Also, special materials for the flywheel are needed to contain it if it breaks apart.

Ultracapacitor. A capacitor, found in nearly all electronic devices, stores energy by allowing a charge to build up between two conductive plates separated by an insulator. An ultracapacitor stores energy between two electrodes immersed in an electrolyte. With further development in materials science, an ultracapacitor may be able to store hundreds of times

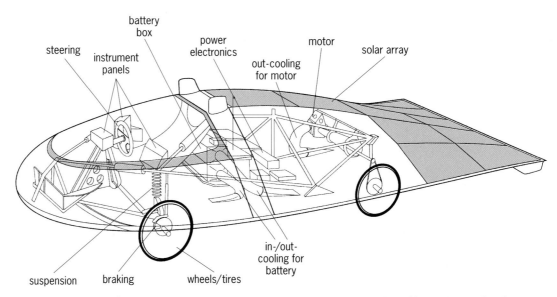

Fig. 9. General Motors' Sunraycer, used to evaluate the potential of solar-powered vehicles. (General Motors Corp.)

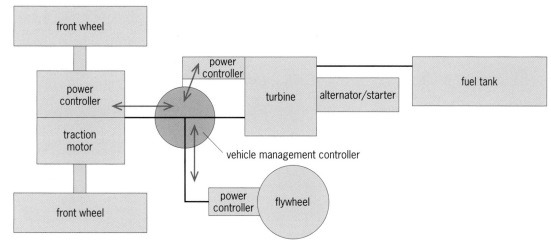

Fig. 10. Hybrid power train that uses a flywheel. (*Chrysler Corp*.)

as much energy as today's capacitors. This would allow the ultracapacitor to deliver the high power required for acceleration of a hybrid electric vehicle. However, safety and environmental concerns may also have to be overcome.

For background information *see* AUTOMOTIVE ENGINE; BATTERY; CAPACITOR; FLYWHEEL; FUEL CELL; GASOLINE; HEAT PUMP; HYDROGEN; METAL HYDRIDES; MOTOR VEHICLE; SOLAR CELL; STIRLING ENGINE; STORAGE BATTERY in the McGraw-Hill Encyclopedia of Science & Technology.   Donald L. Anglin

Bibliography. Robert Bosch GmbH, *Automotive Handbook*, 4th ed., 1996; Society of Automotive Engineers, *SAE Handbook*, 3 vols., annually; Society of Automotive Engineers, *Technology for Electric and Hybrid Vehicles*, SP-1331, 1998.

Electron transfer

Electron transfer between the electronically excited state of a dye and a semiconductor was discovered in 1887 by James Moser at the University of Vienna. He observed that the photoelectric effect detected on silver plates treated with iodide or bromide was greatly enhanced in the presence of erythrosine. A few years later, H. Vogel in Berlin exploited this phenomenon to render silver halide microcrystals sensitive to visible light, a process which eventually found wide application in color photography. Apart from the fundamental importance for photography and xerography, current investigations are fueled by the recent discovery of dye-sensitized injection solar cells. These cells have already reached an impressive conversion efficiency of 11% under full sunlight, rendering them competitive with commercial silicon-based devices.

Light harvesting. To obtain such a high yield in the conversion of sunlight to electric power, most of the visible and near-infrared part of the solar radiation must be collected. Light is absorbed by a dye attached to the surface of the large-band-gap oxide which assumes a role analogous to chlorophyll in green plants. On a flat surface a monolayer of dye would capture only a very small fraction of the incoming sunlight. A newly designed solar cell overcomes this problem by using mesoporous oxide films to support the dye. Such films are constituted by nanocrystalline particles ($\sim$20 nanometers in diameter) of titanium oxide (TiO_2), providing a substrate material characterized by a very large effective surface area. The roughness factor for a 10-micrometer-thick layer can easily reach 1000, implying that the real surface is 1000 times larger than the apparent one. Light penetrating the nanocrystalline film crosses hundreds of adsorbed dye monolayers and hence is collected in an efficient manner. The mesoporous structure thus fulfills a function similar to the thylakoid vesicles in green leaves which are stacked in order to enhance light harvesting.

Light-induced-charge separation. The use of mesoporous oxide films as a substrate to anchor the dye molecules allows sunlight to be harvested over a broad spectral range in the visible region. In addition, for efficient photoconversion, both charge injection and collection must occur with a yield close to unity. Rate constants for electron transfer from excited dyes into titanium oxide have been determined by time-resolved laser photolysis experiments

Some rate constants for electron transfer from excited dyes into titanium oxide*

Sensitizers	k_{inj}, s^{-1}	$|V|$, cm^{-1}	t_f, ns	Φ_{inj}
Ru^{II}(bpy)$_3$	2 ¥ 10^5	0.04	600	0.1
Ru^{II}L$_3$(H$_2$O)	3 ¥ 10^7	0.3	600	0.6
Eosin-Y	9 ¥ 10^8	2	1	0.4
Ru^{II}L$_3$ (EtOH)	4 ¥ 10^{12}	90	600	1
Coumarin-343	5 ¥ 10^{12}	100	10	1
Ru^{II}L$_2$(NCS)$_2$	10^{13}	130	50	1
Ti$_s^{IV}$-Alizarin	>10^{13}	5 ¥ 10^3	—	1

*Electron injection rate constants k_{inj} and electronic coupling matrix elements $|V|$ are measured by nanosecond and femtosecond laser flash photolysis for various sensitizers adsorbed onto nanocrystalline TiO_2. t_f and Φ_{inj} are the excited-state lifetime and the injection quantum yield, respectively. In the sensitizers, L stands for the 2,2'-bipyridyl-4,4'-dicarboxy ligand.

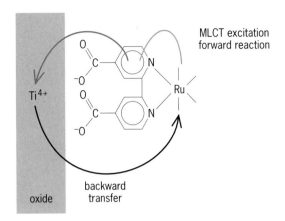

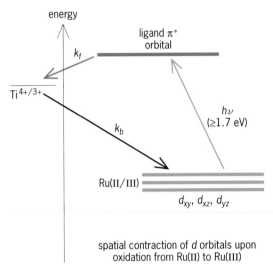

Fig. 1. Dye-sensitized electron transfer at the surface of an oxide semiconductor. k_b and k_f are forward and backward rate constants.

(see **table**). The rates vary over eight orders of magnitude depending on the type of dye employed, the fastest injection times being in the femtosecond domain. The key to obtaining such rapid electron trans-

fer is to endow the dye with a suitable anchoring group such as a carboxylate or phosphonate substituent or a catechol moiety through which the sensitizer is firmly grafted onto the surface. The role of these groups is to provide good overlap between the dye molecule's excited state orbital and the empty acceptor levels; that is, the Ti(IV) has a $3d$ orbital manifold forming the conduction band of titanium oxide. Alizarin is a special case, as it produces a strongly colored complex by reacting directly with titanium ions present at the surface of titanium oxide. The extent of coupling is expressed by the electronic coupling matrix element (V) which is related to the rate constant for charge injection (k_{inj}) by Eq. (1). Here

$$k_{inj} = (2\pi/b)|V|^2\rho \qquad (1)$$

b is Planck's constant, and ρ is the density of electronic acceptor states in the conduction band of the semiconductor. Equation (1) assumes that there is no activation energy for the charge transfer from the excited dye molecules into the semiconductor. In such a case, the reaction proceeds along a channel where the driving force for electron injection compensates the reorganization energy, that is, the free energy required to rearrange the nuclear coordinates of the dye during the charge transfer process. This has been experimentally confirmed for several dyes where the rate was found to be temperature-independent.

Photovoltaic applications. The best photovoltaic performance, both in terms of conversion yield and long-term stability, has been achieved with polypyridyl complexes of ruthenium. In particular, the complex RuL$_2$(NCS)$_2$ [L is 2,2′-bipyridyl-4,4′-dicarboxylate] has attracted worldwide attention due to its outstanding properties which make it a likely candidate for use in the first practical photovoltaic devices being commercialized. This dye is distinguished by two broad absorption bands in the visible region of the spectrum, allowing for the harvesting of a large fraction of incident sunlight. The

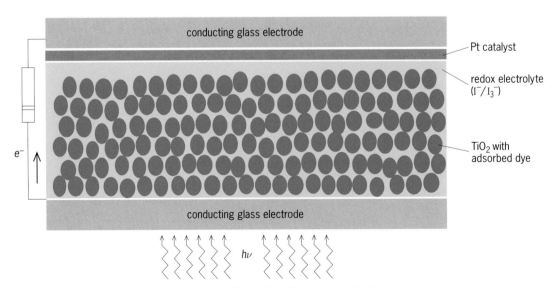

Fig. 2. Cross section of a dye-sensitized nanocrystalline solar cell to show its principle of operation.

optical transition has metal-to-ligand charge-transfer (MLCT) character. Excitation of the dye promotes an electron from the ruthenium to the π^* orbital of the bipyridyl ligand. The latter is coupled through the carboxyl substituents to the titanium ions present at the substrate surface, allowing for the electron to be injected into the titanium oxide within 50 femtoseconds and with unit quantum yield (**Fig. 1**).

The recapture of the electrons by the oxidized ruthenium complex is many orders of magnitude slower than charge injection. One reason for this behavior is that the molecular orbitals involved in the back reaction overlap less favorably with the wavefunction of the conduction band electron than the orbitals involved in the forward process. Thus, while the injecting orbital is the π^* wavefunction of the carboxylated bipyridyl ligand, the recapture of the electron by the oxidized dye involves a d-orbital localized on the ruthenium metal whose electronic overlap with the titanium oxide conduction band is small. The spatial contraction of the wavefunction upon oxidation of the Ru(II) to the Ru(III) state further reduces this electronic coupling.

In the nanocrystalline solar cell the back reaction is intercepted by transfer of positive charge from the oxidized dye to iodide ions contained in an electrolyte and thence to the counter electrode (**Fig. 2**). The circuit is closed via this last charge transfer, which returns the mediator to its reduced state. The system converts light into electricity without permanent chemical transformation. The maximum voltage that such a device could deliver corresponds to the difference between the redox potential of the mediator and the conduction band position of the semiconductor.

The monochromatic current output of the cell as a function of the wavelength of the incident light is called the photocurrent action spectrum (**Fig. 3**). Very high efficiencies of current generation, exceeding 75%, are obtained. The yields are practically 100% after correction, for the inevitable reflection and absorption losses in the conducting glass serving to support the nanocrystalline film. Historically, RuL$_3$ was the first efficient and stable charge-transfer sensitizer to be used in conjunction with high-surface-area titanium oxide films. However, its visible light absorption is insufficient for solar light conversion. A significant enhancement of the light harvesting was achieved in 1990 with the trimeric complex of ruthenium in which two peripheral ruthenium moieties were designed to serve as antennas. The advent of RuL$_2$(NCS)$_2$ in 1991 marked a further improvement, since it extended the light absorption over a broad range in the visible. Its performance was superseded only recently by the discovery of a new black dye, RuL'(SCN)$_3$, having a spectral onset at 900 nm. This wavelength is optimal for the conversion of AM 1.5 solar radiation (which has a path length 1.5 times longer than at normal incidence to the Earth) to electric power in a single-junction photovoltaic cell.

Several industrial organizations are engaged in bringing the nanocrystalline injection solar cell to

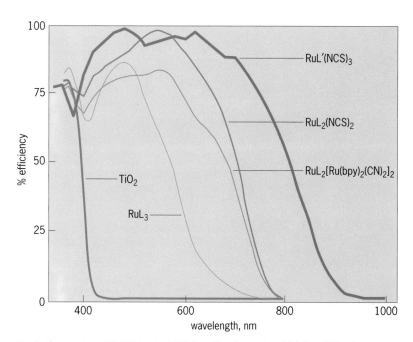

Fig. 3. Conversion of light into electricity by ruthenium polypyridyl dyes. This photocurrent action spectrum was obtained with a nanocrystalline injection solar cell sensitized by different ruthenium polypyridyl complexes. The incident photon-to-current conversion efficiency is plotted as a function of excitation wavelength. L is 4,4'-COOH-2,2'-bipyridine, and L' is 4,4',4''-COOH-2,2': 6',2''-terpyridine.

market. The near-term applications will be mainly in the low power range. One example is a solar watch with the photovoltaic cell incorporated into the watch glass. Considerable progress has also been made in the development of in-series interconnected modules and the production of solar tiles in Australia.

For background information *see* DYE; ELECTRON-TRANSFER REACTION; ENERGY CONVERSION; INORGANIC PHOTOCHEMISTRY; LIGHT; PHOTOCHEMISTRY; PHOTOELECTRIC DEVICES; PHOTOVOLTAIC CELL; PHOTOVOLTAIC EFFECT; SEMICONDUCTOR; SOLAR CELL; SOLAR ENERGY in the McGraw-Hill Encyclopedia of Science & Technology.   Michael Grätzel

Bibliography. U. Bach et al., Solid-state dye-sensitized mesoporous TiO$_2$ solar cells with high photon-to-electron conversion efficiencies, *Nature*, 395: 583–585, 1998; A. Hagfeldt and M. Grätzel, *Chem. Rev.*, 95:49, 1995; B. O'Regan and M. Grätzel, A low cost, high efficiency solar cell, *Nature*, vol. 353, 1991.

Electronic commerce

The rapid development of computer power and information technology has greatly affected the way that business is conducted, and has opened up a new means, electronic commerce (e-commerce or EC). In particular, the commercialization and privatization of the Internet and its swift rise in popularity are fueling the growth of electronic commerce. Electronic commerce requires methodologies and systems that support the creation of information sources; effective and efficient interactions among

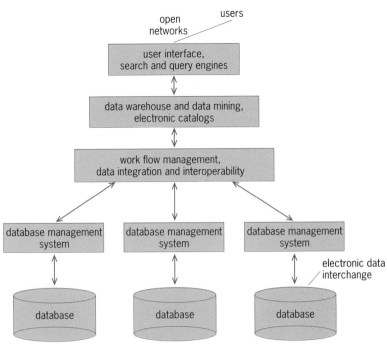

Architecture of electronic commerce technology.

sellers, consumers, intermediaries, and producers; and the movement of information across global networks (see **illus.**).

By adopting electronic commerce, businesses expect to be able to achieve such goals as the streamlining of procurement processes, the shortening of production cycles, closer customer and vendor relationships, and the effective conduct of business with distant partners—while cutting costs. For example, it is estimated that traditional distribution channels add 135% to the cost of bringing an item from the manufacturer to the consumer, whereas direct distribution from manufacturer to consumer will add only 10%. An electronic airline ticket is estimated to cost around one-eighth as much to process as a traditional airline ticket.

Electronic commerce activities can be placed in three classes: business-to-business, business-to-consumer, and business-to-government. The U.S. Department of Commerce estimates that 90% of electronic commerce is business-to-business. Business-to-business electronic commerce is typically built on established trust relationships, makes use of shared computer/telecommunication infrastructures, and is able to achieve efficiency through large volume of transactions. Business-to-consumer electronic commerce is not built on mutual trust, has a small volume of transactions, and requires a ubiquitous, inexpensive infrastructure providing for personalization and customization. The number of small businesses with Web sites is expected to increase from 900,000 in 1997 to 2,000,000 in 2000. Business-to-government electronic commerce is more restrictive due to government regulations. The Federal Acquisition Streamlining Act (FASA) has mandated that all U.S. government agencies conduct bidding by electronic data interchange (EDI) by late 1999.

Electronic commerce systems have to deal increasingly with multimedia data, such as video, images, and audio, as well as text. In addition, they must deal with information from heterogeneous, autonomous sources and disseminate information to, and deal with, users having a wide range of specialties, interests, capabilities, and characteristics. Although electronic commerce systems use various communication networks, including value-added networks (VANs), the use of the Internet as an infrastructure is becoming commonplace. This presents a security challenge.

Security. Security assumes paramount importance in successful electronic commerce and has many requirements. Privacy is the ability to keep the transactions as well as the identities of the parties involved confidential. User authentication is the ability of the authorized parties in a communication session to ascertain each other's identity. Integrity is concerned with ascertaining that the exchanged information has not been subjected to unauthorized modification, deletion, or addition. Nonrepudiation is the guarantee that neither party can deny the existence or content of a communication session. Secure payment mechanisms with low overhead are needed, as are mechanisms to audit access of electronic commerce objects while upholding confidentiality, as well as copyright protection and copy prevention.

Cryptography is a popular tool for dealing with these security requirements. Cryptosystems typically use either secret (symmetric) key encryption or public key encryption. With the former, a single secret key is shared between two parties and used for both encryption and decryption. The Data Encryption Standard (DES) is the approved standard for exchanging unclassified information. DES uses a 56-bit key, which is one of its most controversial aspects, especially since a key was broken in 5 months. Although secret key encryption is computationally less expensive than public key, it suffers from two major limitations: It requires a secure transmission channel to distribute the secret key, and it requires a large number of keys [$n(n-1)$ keys among n customers].

Public key cryptosystems solve these problems. They use a public key and a private key known only to the party that owns it. To ensure confidentiality of a message from A to B, the message has to be encrypted with B's public key and decrypted with B's own private key. Since no one else other than B knows B's private key, only B can decrypt the message. However, A must obtain B's public key from a reliable channel so that no one can impersonate B. This is carried out through the Public Key Infrastructure (PKI). The PKI is a system of digital certificates, certificate authorities, registration authorities, certificate management services, and directory services that verifies the identity and authority of each party involved in any transaction over the Internet. The PKI issues public key certificates, revokes them as needed, establishes governing policies regarding the issuance and revocation of certificates, and archives information needed to later validate certificates. RSA (Rivest-Shamir-Adleman) is one of the widely used public key cryptographic protocols.

While the requirements of ensuring confidentiality and integrity of messages, authentication, and nonrepudiation are fairly well understood, and established solutions exist, ensuring copyright protection and establishing a secure, low-overhead payment system are still major challenges in electronic commerce.

Electronic payment systems. Central to global electronic commerce are electronic payment systems that provide the financial infrastructure needed in an electronic marketplace. Digital currency relies on information technology and high-speed communications networks to store, transmit, and receive representations of value. Furthermore, it relies on cryptography to provide security in an open network environment and strives to reduce costs through economies of scale and technological advances. One form of digital currency is the stored-value smart cards (debit and credit cards). These cards may be relationship-based, where an accounting settlement at the end of a billing cycle is required, or e-purses and debit cards, where a given dollar amount is stored and each purchase deducted (thus no accounting settlement is required). Another form of digital currency is the e-token or e-money, which should be bank-certified, exchangeable with other forms of payments such as certified checks, tamper-resistant, and remotely accessible. E-payment schemes can be classified into bearer certificate (similar to cash—whoever holds the certificate holds the value) or national, where third parties keep track of transactions.

The Secure Socket Layer (SSL) protocol, widely deployed on the Internet, is implemented in most major Web browsers used by consumers as well as in merchant server software. SSL provides a secure electronic pipe between the consumer and the merchant for exchanging encrypted payment information. Hundreds of millions of dollars are already changing hands as cyber shoppers enter their credit card numbers on Web pages secured with SSL technology. However, SSL must have strong server authentication—otherwise, it could be susceptible to attack by intermediaries.

The Secure Electronic Transaction (SET) protocol is designed to provide a mechanism for secure electronic payment by credit card over an otherwise insecure public Internet. With SET, the bank does not have to know the terms of the transaction between the merchant and the cardholder, and the merchant does not have to know the account number of the cardholder. Authentication is preformed by digital certificates issued by the PKI. Messages are encrypted with 56-bit DES keys, which are then transmitted between the two parties using a 1024-bit RSA public key envelope.

Bidding and negotiation are integral parts of commerce; a good electronic commerce solution should be able to mimic traditional bidding and negotiation protocols such as closed-bid auctions (in which all bidders submit their bids by a deadline, and the bids are evaluated at the deadline), and English auctions (in which those interested in an item bid against one another until all but the highest bidder are eliminated). Intelligent agents are needed that can act on behalf of their human counterparts. These agents should be robust enough to withstand attacks by malicious servers or intruders, and should make decisions by traveling via insecure networks.

Electronic catalogs. Electronic catalogs are considered a gateway to a digital version of the business itself. They enjoy several advantages over printed catalogs: (1) They provide two-way, real-time communication between businesses and consumers. (2) They provide an opportunity for personalization and customization, thus allowing businesses to have new and closer relationships with their customers. (3) They enable companies to quickly adapt to market conditions by adjusting prices, repackaging, and rechanneling, and to instantaneously add value everywhere their catalogs reach.

Realizing these advantages requires addressing several issues regarding infrastructure support such as the ability of such systems to track potential customers worldwide and provide businesses with dynamic updating capabilities; and user interfaces with computer interactivities that address presentation, spatial arrangement of the catalog, and multimedia capabilities.

In an electronic commerce environment, customers need to be able to search for products using catalogs from multiple suppliers. This requires efficient mechanisms for seamless access to heterogeneous sources whose data are related but may be incompatible due to, for example, syntactic or semantic differences.

Buyers and sellers need to exchange requests for quotes, price lists, purchase orders, invoices, shipping notices, and other documents. While today's World Wide Web sites publish information for people, tomorrow's Web sites need to publish information for computers. This exchange calls for automated interactions between buyers and sellers. Electronic commerce applications need to extract information from Web resources such as comparison shopping and tracking services, which cannot be handled by HTML (Hypertext Markup Language, the document format commonly used on the World Wide Web). XML (Extensible Markup Language) is emerging as a replacement that will make it possible for information to be available to machines and will enable Web publication of documents that computers can understand.

Major challenges are found in content management. Buyers should be able to find desired items by means of a graphical user interface and navigational environment that are consistent across suppliers. Shopping carts used by the requisition/order management system should be compatible with the general ledger coding practices used by the buying organization. Suppliers need to easily review, correct, and update their product information. Procurement officials should be able to review the product information for compliance with all regulations and contractual agreements. Controlled access to the procurement system by authorized and authenticated users must be provided. Privileges and preferences that have been established for each authorized user must be administered.

Data warehousing and mining. Electronic commerce is an information-intensive environment in which businesses typically collect enormous amounts of information about consumers. Retailers and suppliers need to customize their shops and services. On-line consumers have shown interest in making quicker, better-informed decisions rather than just demanding the lowest price. All these requirements call for data warehousing (the process of extracting, cleaning, augmenting, and organizing operational data) and data mining (the extraction of structure from the warehoused data). Data mining tools for clustering, associating, and performing pattern analysis on the data generated by electronic commerce enable usage tracking.

Work flow management. Business processes such as financial transactions, invoicing and payment, and acquisition of resources are set up by organizations to define precisely how they are supposed to work. Work flows define the steps followed in a process, who carries out each step, how these steps are coordinated, and the inputs and outputs of each step. Work flows in an electronic commerce environment are dynamic and distributed. This requires integration of business processes, real-time information sharing and planning, and dynamic adaptation of work flows to reflect rapid changes and developments. Business processes and their information support must be constantly reengineered and optimized to ensure that distributed work flow is executed without requiring centralized control.

Search and resource discovery. The interest of on-line consumers in making better-informed decisions necessitates the development of powerful search engines as well as catalog integration and cross-catalog search. Users need to effectively retrieve information using familiar terms. In addition, electronic commerce providers need to cater to users with diverse backgrounds by offering broader, more general ontologies that are interlinked to cover many domains.

The challenge is to provide retrieval tools for searching images, video, and text by contents and concepts. Concepts must be identified and extracted from objects, and the objects classified and indexed based on extracted concepts. This activity should be automatic (or at least semiautomatic), as manual classification and indexing of objects is resource-intensive and economically infeasible.

Electronic data interchange (EDI). This is a means for electronic exchange of business documents, for example, purchase orders and invoices among trading partners. This kind of interoperability required the development of certain standards providing a framework for formatting a given EDI message, and the adherence of trading partners to such standards. Two types of standards exist: ANSI X12 is the United States national standard and was developed by the Accredited Standards Committee. EDIFACT is a worldwide standard developed by a United Nations committee. In general, these standards provide the format and establish the data contents of EDI messages. Sending an EDI transaction involves mapping, in which elements in the database that are needed to create an EDI transaction are identified (done once per new transaction); extraction, where predefined data are extracted from the database; translation, where extracted data are translated into EDI standards using translation software; and transmission of the EDI message, typically over a value-added network.

The relatively high cost of investment needed for EDI has restricted its use to large companies. There is a move toward replacing the value-added network with the Internet. When the Internet is used as the infrastructure supporting EDI, the previously mentioned issues, such as message integrity, nonrepudiation, and security, must be addressed.

For background information *see* COMPUTER SECURITY; CRYPTOGRAPHY; DATABASE MANAGEMENT SYSTEMS; HUMAN-COMPUTER INTERACTION; MULTIMEDIA TECHNOLOGY; WIDE-AREA NETWORKS in the McGraw-Hill Encyclopedia of Science & Technology.

Nabil R. Adam; Vijayalakshmi Atluri; Yelena Yesha

Bibliography. N. R. Adam et al., *Electronic Commerce: Technical, Business and Legal Issues*, Prentice Hall, 1999; N. R. Adam and Y. Yesha, Electronic commerce and digital libraries: Towards a digital agora, *ACM Comput. Surv.*, 28(4):818–835, December 1996; W. Ford and M. S. Baum, *Secure Electronic Commerce*, Prentice Hall, 1997; R. Kalakota and A. B. Whinston, *Frontiers of Electronic Commerce*, Addison-Wesley, 1996.

Environmental engineering

For thousands of years the oceans have provided humankind with food and trade routes, and more recently with petroleum and minerals to fuel society. Since the 1960s, there has been a dramatic change in the general view of the oceans and the human impact on their sustaining ecosystems. In past times, little regard was given to disposal of wastes from ships. As greater understanding of the marine environment developed, it became clear that human behavior must change to ensure the health of that environment.

Much has changed with enactment of regulations on the international, national, state, and local levels to control pollution from ships and dumping of sewage sludge and garbage from land sites. Environmental compliance at sea is typically achieved through a combination of changes in operational procedures, education of personnel, source reduction and pollution prevention, and installation of

TABLE 1. Status of annexes to MARPOL 73/78[*]

Annex	Topic	Entry into force in United States
I	Oil	October 2, 1983
II	Noxious liquids carried in bulk	April 6, 1987
III	Harmful substances carried in package form	July 1, 1992
IV	Sewage	Not in force[†]
V	Garbage	December 31, 1988
VI	Air pollution	Not in force[†]
VII	Ballast water exchange	Under development[†]

[*]Regulations for the Control of Pollution from Ships, International Convention for Control of Pollution from Ships.
[†]As of December 1998.

shipboard pollution control equipment. At great expense, most existing ships have been modified to be environmentally compliant, and future ones are being designed and built to meet this criterion.

Regulations. International regulations affecting ship operations on the high seas have been developed by the International Maritime Organization. These rules are found in the International Convention for the Prevention of Pollution from Ships (MARPOL 73/78) and its annexes (**Table 1**). Depending upon language used by the U.S. Congress in ratification, these annexes become federal law and are applicable to U.S. commercial flagged and public vessels (vessels of the Navy, Coast Guard, and so forth).

At the federal level, statutes (laws), regulations of executive agencies and departments of the federal government, and executive orders affect operations at sea. State and local regulations vary widely between locations. The challenge facing the ship designer and operator is to consider the most restrictive scenario within the expected operating area, anticipate changes in the regulatory environment over the life of the class of ship, and design accordingly.

Commercial ships versus warships. Regulations governing operations of commercial ships and warships are similar, but the challenges for implementing technological solutions on naval warships are more difficult due to their unique characteristics. These ships are designed for extended periods at sea in a variety of sea states, weather, and climate conditions. Surface warships are typically volume-limited, with weapon systems designed in the smallest package to conserve energy requirements and minimize visual and electronic detection. The majority of equipment built for ashore or commercial marine application is not designed for the unique naval ship operating environment. The primary differences include space and weight limitations, shock and vibration requirements, electromagnetic interference hardening, ship motions in six degrees of freedom, and most importantly the average skill level of the typical naval operator. In most cases, differences in ship operating profiles (frequency of port calls, area of operations, logistic support constraints, and so forth), crew size, and training and certification levels of

the crew make this transition difficult. Commercial cruise ships make frequent port calls allowing opportunities for offload of solid and liquid wastes; ships engaged in commercial shipping are crewed at minimal levels resulting in minimal waste generation; and most dedicated operators and maintainers of commercial ships are licensed via an appropriate regulating body. Naval ships make infrequent port calls over 6-month deployments, have a much greater crew density, and have a greater number of people (frequently with high turnover rates) maintaining and operating equipment, thus compounding the training and certification challenge.

Technological solutions. The U.S. Navy is in the process of retrofitting all ships in the fleet with new technologies and procedures for managing sewage and wastewater, solid waste, oily waste, and excess hazardous materials. Hazardous materials are managed through minimization and holding used or excess materials for shore disposal.

Solid waste. In 1987 the U.S. Congress ratified Annex V to MARPOL 73/78 as the Marine Plastic Pollution Research and Control Act, which amended the Act to Prevent Pollution from Ships (APPS). Annex V established a worldwide prohibition on the discharge of plastic waste at sea, and established eight special areas where the discharge of other solid waste was deemed to be detrimental to the marine environment. Approximately 50% of U.S. Navy deployed ships operate in these areas (**Fig. 1**). Currently only the Baltic Sea, the North Sea, and the Antarctic Region special areas are in effect. The other areas will go into force following certification of adequate shore-based solid-waste management facilities.

In developing the U.S. Navy solid-waste management program, shipboard generation studies were performed to determine equipment design parameters. Aircraft carriers produce 10 tons and destroyer-size ships produce about 1000 pounds of solid waste every day (**Table 2**). To manage these wastes, the U.S. Navy developed a suite of specifically designed equipment. Plastic waste processors enable a 30:1 reduction in volume by shredding, compressing, and melting plastics into 20-in.-diameter (50-cm) disks weighing about 10 lb (4.5 kg) each (**Fig. 2**). These disks are stored on-board until they can be offloaded for shore disposal.

TABLE 2. Solid-waste generation rates

Solid-waste category	Generation rate by weight, lb/person-day (kg/person-day)	Generation rate by volume, ft³/person-day (liters/person-day)
Food	1.21 (0.55)	0.03 (0.8)
Paper/cardboard	1.11 (0.50)	0.19 (5.4)
Metal/glass	0.54 (0.24)	0.05 (1.4)
Plastic	0.20 (0.09)	0.15 (4.2)
Wood	0.01 (0.005)	<0.01 (0.3)
Textiles	0.12 (0.05)	0.01 (0.3)
TOTAL	3.19 (1.45)	0.43 (12.2)

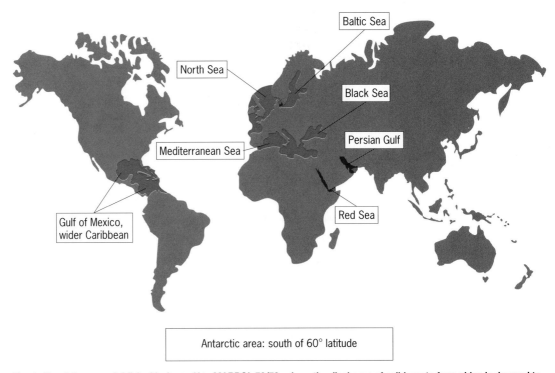

Fig. 1. Special areas established in Annex V to MARPOL 73/78, where the discharge of solid waste from ships is deemed to be detrimental to the marine environment.

Paper, food, and cardboard are processed in pulpers. These devices reduce the material to a slurry which is discharged directly overboard when operating farther than 3 nautical miles (5.5 km) from land. Metal-glass shredders crush glass and shred cans into strips which are then placed in burlap bags

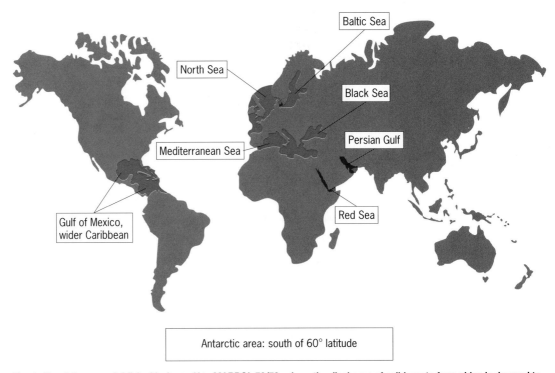

Fig. 2. Plastic waste processor aboard USS *John C. Stennis.*

for overboard discharge when greater than 12 nmi (22 km) from land. Some ships burn textiles, cardboard, and paper wastes in incinerators. Other solutions used by navies of other countries include compaction and refrigeration of all solid wastes, and frequent off-load of unprocessed waste to shore facilities or supply ships. The U.S. Navy is developing advanced shipboard thermal destruction technologies that will afford complete destruction of organic material and solidification of inorganic material into glasslike bricks for shore-based disposal.

Most commercial ships rely on incineration to dispose of paper, food, textiles, and cardboard waste. Some burn plastics, metal, and glass; others have onboard recycling programs for offload of these materials.

Sewage and wastewater. In the United States, the Clean Water Act prevents overboard discharge of unprocessed sewage within 3 nmi (5.5 km) of land. Annex IV to MARPOL 73/78, not yet ratified by the United States, prohibits this discharge within 4 nmi (7.4 km) of land. Current U.S. Navy sewage management strategy is to hold sewage when within 3 nmi of land, and discharge it overboard when outside this limit. When in port, sewage is pumped to shore receiving stations. Wastewater (graywater) collected from showers, sinks, and the galley is currently pumped overboard when away from the pier, and pumped ashore when moored. Future regulations will likely increase the unprocessed sewage discharge limit to 12 nmi (22 km) and regulate wastewater discharge as well.

Various marine sanitation device designs are available to process sewage and wastewater. These systems use technology similar to land-based sewage treatment plants that rely on physical, chemical, or biologic processes to break down sewage into discharge-grade effluent and concentrated sludge for shore disposal. These systems are found on most oceangoing and Great Lakes commercial vessels. Current requirements have not compelled use of these systems aboard U.S. Navy ships, but anticipated regulatory changes and rising sewage disposal costs overseas have combined to spur development of an integrated liquid disposal system. This system uses membrane and ultraviolet-light technology to filter and sterilize graywater for overboard discharge. The remaining material (permeate) is then combined with sewage and waste oil and burned in a thermal destruction device.

Oily waste. Oily wastewater is generated primarily from machinery space bilges and from compensated fuel-seawater ballasting systems. Annex I to MARPOL, the APPS, and the Clean Water Act regulate discharge of oily wastewater. Current restrictions generally limit overboard discharge of oily water to 15 parts per million (ppm) oil or to no sheen on the water. Some state and local municipalities have even stricter standards preventing overboard discharge of any oil-bearing water. Depending on the ship design, age, and type of equipment, generation of bilge water can be significant. Typical sources are condensation from pipes and the ship hull; leaking valves, fittings, and seals; and leaking equipment and machinery. Equipment to separate oil and water reduces this waste to a manageable volume by allowing overboard discharge of the clean effluent. Designs of oil-water separators are usually based on centrifuge technology, use of coalescing plates in tanks, or use of absorbent material to remove oil from water. Oil content monitors installed in the system are designed to prevent inadvertent discharge of water with greater than 15 ppm oil. The concentrated oil is then directed to a waste-oil holding tank for shore disposal.

U.S. Navy oil-water separators are predominantly coalescing plate designs. Sludge buildup and use of detergents and other agents that emulsify oil, making separation difficult, significantly degrade system performance. Compounding these problems is the likelihood of future reduction in allowable discharge from 15 to 5 ppm oil. To meet these challenges, the U.S. Navy is testing an ultrafiltration ceramic membrane polishing system.

As ships burn fuel or offload cargo, they become lighter and therefore less stable. Water is brought aboard to offset this weight loss. Clean ballast systems use separate tanks for water and fuel; compensated fuel systems use the same tanks. Compensated systems have become routine in warship design because they allow fewer tanks, reducing overall ship size and cost. However, these systems are more likely to result in spillage of small amounts of fuel during fueling operations. Commercial ships predominantly use clean ballast systems, and future warships will likely use these designs as well.

Air-pollution control. Emission standards from ship propulsion plants and incinerators have been primarily concerned with smoke production. Recent amendments to the Clean Air Act, proposed new regulations of the U.S. Environmental Protection Agency and the State of California, and the expected ratification of Annex VI to MARPOL 73/78 will affect vessel operations. Limits for oxides of nitrogen (NO_x) for new diesel engines, limits on sulfur content of fuel [the control strategy for oxides of sulfur (SO_x)], and emission standards for incinerators are expected. These new requirements will force diesel engine design changes, use of higher-quality fuel, possible installation of exhaust gas management systems, and changes in incinerator operations.

The U.S. Navy has worked with industry to develop new refrigerants to replace ozone-depleting substances (that is, chlorofluorocarbons). Retrofits to replace refrigeration and air-conditioning systems are in progress. These conversions are in response to the Montreal Protocol and the amendment to the Clean Air Act banning production of compounds suspected in stratospheric ozone layer reduction.

Guidelines published by the Environmental Protection Agency limit the amount of volatile organic compounds that can be emitted from various industrial processes, in order to limit smog and tropospheric ozone production. These regulations have forced the development of new paints and preservation systems that significantly reduce the amount of volatile organic compounds produced. The new high-solids-content paints meet federal air-quality standards and provide longer service life.

Ballast water management. A consequence of taking on ballast water is the introduction of marine life into the ballasting systems. Annex VII to MARPOL 73/78 is being developed to control the spread of nonindigenous species from one locale to another. Zebra mussels were introduced into North America by a ship that took on ballast water in a foreign port and discharged it in the Great Lakes. By exchanging ballast water taken close to shore for water from the deep ocean, organisms that may have inadvertently entered the system can be flushed out. Most ships now perform two ballast water exchanges at sea under voluntary U.S. Coast Guard rules.

For background information *see* AIR POLLUTION; ENVIRONMENTAL ENGINEERING; MARINE MACHINERY; NAVAL SHIP; WATER POLLUTION in the McGraw-Hill Encyclopedia of Science & Technology.

Stephan P. Markle

Bibliography. American Society of Naval Engineers, *Proceedings of the ASNE Environmental Symposium "Environmental Stewardship: Ships and Shorelines,"* 1997; International Maritime Organization, *International Convention for the Prevention of Pollution from Ships (MARPOL 73/78)*, 1991; S. P. Markle, S. E. Gill, and P. S. McGraw, Engineering the environmentally sound warship for the 21st century, *Proceedings of the American Society of Naval Engineers, ASNE Day '99*, May 1999; U.S. Department of the Navy, *Final Environmental Impact*

Statement: Disposal of U.S. Navy Shipboard Solid Waste, 1996.

Exotic atoms

An exotic atom is a system in which either the nucleus of a hydrogen atom is replaced by an exotic particle (such as a muon, to form muonium, or a positron, to form positronium), or one electron in an ordinary atom is replaced by an exotic particle (such as a muon, pion, or antiproton), or even both substitutions are made (as in antihydrogen). This article deals with those exotic atoms in which a normal nucleus is orbited by electrons and one exotic particle. Atoms of this kind are formed by stopping exotic particles, usually produced in particle accelerators, in matter. The stopped particle replaces an electron in an ordinary atom (**Fig. 1**). The first orbit of the exotic particle after capture is very similar in size to that of the electron before ejection. Afterward, it cascades down the ladder of exotic-atom states by x-ray and Auger transitions (that is, ejection of electrons). If the exotic particle is a muon, it reaches the lowest energy level, $1s$. The muon experiences only the Coulomb interaction with the protons in the nucleus and the weak interaction with all the nucleons. In the case of the hadrons (such as the pion, kaon, or antiproton) the cascade ends earlier for all exotic atoms except those with atomic number 1 or 2, due to nuclear absorption or annihilation of the particle by the short-range strong interaction. Since exotic particles are all much heavier than the electron, they are more strongly bound to the nucleus than electrons, and their transitions during the deexcitation are much more energetic than those of electrons. In addition, exotic particles may come much closer to the nucleus than the electrons in an ordinary atom.

Strong interaction. This ability of exotic particles to closely approach the nucleus makes possible some interesting experiments. As an exotic hadronic particle comes close to the nucleus, it experiences the short-range strong interaction with the nucleons. Thus, it may serve to explore both the nature of this interaction and the extension of the neutron cloud in the nucleus; since neutrons carry no charge, strong interaction is their only interaction with the hadrons.

The energy of the last observable exotic x-rays emitted from pionic, kaonic, and antiprotonic hydrogen is shifted by the strong interaction of the exotic particle with the nucleus from the value that would follow from pure electromagnetic forces. At the same time the line is broadened by the absorption of the hadron, which shortens the lifetime of the state appreciably. Indeed, according to Heisenberg's uncertainty principle, which states that $\Delta E \cdot \Delta t \gtrsim h/2\pi$ (where h is Planck's constant), the uncertainty in energy, ΔE, becomes observably large (up to values in the kiloelectronvolt region) as the lifetime, Δt, becomes small. In experiments with protium, the hydrogen isotope with only one proton and no neutrons in the nucleus, the strong interaction was investigated for the simplest possible system, whereas by a comparison of the results from protium and deuterium the influence of the added neutron could be disentangled. The results are important for the understanding of the hadron-nucleon interaction at rest.

The antiproton-nucleus interaction in heavy antiprotonic atoms may be used to probe the neutron distribution at the very periphery of the nucleus. Although the exotic particle comes much closer to the nucleus than the electron, it only grazes the nuclear periphery before it disappears. Recent experiments with elements of medium and high atomic number established that the neutron density at a distance about 2–3 femtometers outside the half-density radius (the radius at which the neutron density is half that at the center of the nucleus) is strongly increased compared to the value that would be expected simply from the ratio of neutron to proton number in the nucleus. This result implies the existence of a neutron halo. Knowledge about such a halo is important, especially for neutron-rich nuclei, because the most promising way to produce superheavy elements is to shoot neutron-rich (radioactive) projectiles of medium atomic number at heavy nuclei such as uranium or thorium. A neutron halo would let the strong-interaction potential cancel in part the electrostatic repulsion between the nuclei and would, hence, strongly enhance the production probability for such superheavies.

Capture and cascade. Exotic hydrogen atoms, the simplest systems of their kind, are best suited to study the atomic capture and cascade of exotic particles. Capture is expected to take place after the particle has been slowed down to kinetic energies which roughly correspond to the ionization energy of the capturing atom. The atom gains a kinetic energy

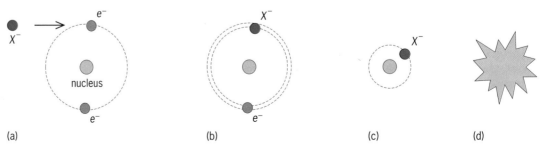

(a)   (b)   (c)   (d)

Fig. 1. History of an exotic atom. (a) Slowing down of exotic particle, X^-. (b) Atomic capture. (c) Cascade. (d) Absorption.

during capture which is determined, due to momentum conservation, by the particle kinetic energy before capture and by the binding energies of the exotic particle after its capture and the electron before its ejection. During the cascade the system may be accelerated even more by the Coulomb deexcitation. In this process the energy which is gained in a transition of the exotic particle to a lower level in the atom, induced by collisions with other atoms in the target, is used to increase the kinetic energy of both the exotic atom and its collision partner. As recent experiments with muons and pions have shown, the exotic atom may gain energies in the 10-electronvolt region. Knowledge about this energy is important for a number of basic experiments on quantum electrodynamics and the weak interaction.

Transfer. The muonic hydrogen atom is a special system in still another respect. The muon is very efficiently transferred from this system to other atoms or even to heavier hydrogen isotopes (that is, from protium to deuterium or tritium). In view of the large proton-muon binding energy, this phenomenon is astonishing; it is enabled by the fact that the system is small and neutral. It travels through the electron cloud of the struck atom until, close to the nucleus, transfer becomes possible. The process is of considerable interest for the development of calculational tools in atomic and molecular physics. It plays an important role in processes such as muon-catalyzed deuterium-tritium fusion, where the muon binds the two hydrogen nuclei so closely together that they fuse very rapidly. The process has been considered as a source of megaelectronvolt neutrons, since the muon is able to catalyze hundreds of such fusions and one neutron is generated during each fusion.

Spectroscopy of antiprotonic helium. Highly excited, Rydberg-like antiprotonic helium atoms (**Fig. 2**) were successfully exploited to investigate the three-body interaction in the system (helium nucleus + antiproton + electron). Such exotic atoms were found to be produced in states with principal quantum numbers, n, around 38, and in about 3% of the cases to be long-lived compared to ordinary exotic atoms (which means that the lifetime of these metastable states is measured in microseconds instead of picoseconds). The metastable atom lives long enough to be irradiated with laser light. If the frequency of this laser light corresponds to the energy of a transition to a short-lived state of the system, the antiprotonic helium atom is resonantly deexcited to this state and cascades down very rapidly to a level from which it annihilates. The products of this annihilation, several charged or neutral pions, may be easily detected. Hence, at the moment of laser irradiation, a burst of annihilation products would be expected, and has indeed been observed. As the wavelength of the laser light can be measured with part-per-million (ppm) precision, the transition energy in the antiprotonic helium atom may be determined with very small uncertainty. This type of measurement makes it possible to test the theory for this three-body system by experiment, and theoreticians have also improved their calculations by orders of magnitude. As theory and experiment now agree on the sub-part-per-million level under the assumption that the antiproton mass is equal to the proton mass,

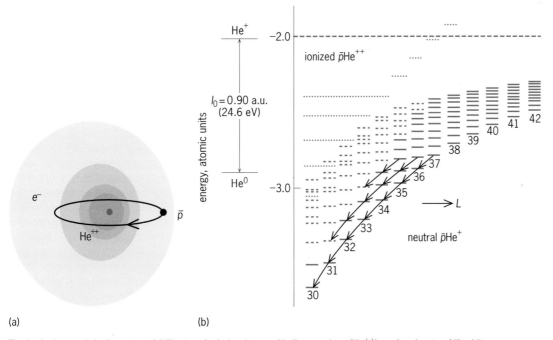

(a)   (b)

Fig. 2. Antiprotonic helium atom. (a) Electron (e^-) cloud around helium nucleus (He^{++}), and antiproton ($\bar{p}$) orbit. (b) Energy-level scheme. Numbers below energy levels give angular momentum quantum number (L). Solid lines and broken lines represent, respectively, metastable levels and short-lived states in the antiprotonic helium atom (neutral $\bar{p}He^+$; helium nucleus + antiproton + electron). Dotted lines represent levels of the antiprotonic helium ion (ionized $\bar{p}He^{++}$; helium nucleus + antiproton). Ground-state energy levels of neutral and ionized helium (He^0 and He^+) and helium ionization potential (I_0) are also shown. Energies are expressed in atomic units (1 a.u. = 27.2 eV).

it may be concluded that these masses are equal at the same level of accuracy. This comparison constitutes a test for the CPT theorem, a basic principle of physics, which, among other things, postulates that the proton and antiproton masses are equal.

Prospects. Additional experiments should improve the determination of strong-interaction widths and shifts in kaonic hydrogen and, thus, increase knowledge about the strong interaction of the kaon with nuclear matter. The very precise determination of the energy difference between the 2s and 2p states in muonic hydrogen by a laser-resonance method should shed additional light on quantum electrodynamics (QED), the quantum theory of the electromagnetic interaction, and should enable determination of the radius of the proton (a system of three quarks) very precisely. The uncertainty of this radius limits the precision of tests of quantum electrodynamics with ordinary hydrogen atoms. Finally, a precision measurement of the lifetime of the muon in the 1s state of muonic hydrogen, which is influenced by the weak interaction between muon and proton, should lead to a more accurate determination of the parameters of the weak interaction between leptons and nucleons. Leptons are a class of pointlike particles which experience only weak (and electromagnetic) interactions with the nucleon. The best-known representative of this class is the electron.

A new generation of antiprotonic-atom experiments is in preparation at the Antiproton Decelerator at CERN, near Geneva, Switzerland. The laser resonance technique applied to antiprotonic helium is to be refined and supplemented by microwave experiments in order to measure the interaction of the antiproton magnetic moment with the electron spin. These measurements should enable a precision comparison between the magnetic moments of the proton and its antiparticle, another test of the CPT theorem. The proposed experiments also aim at producing a beam of antiprotons with kinetic energies in the electronvolt region in order to directly study atomic capture and cascade of exotic particles. Another goal is the laser spectroscopy of antiprotonic hydrogen, the simplest antiprotonic atom. The inherent high precision would allow a more thorough test of the equality of proton and antiproton masses.

For background information *see* ATOMIC STRUCTURE AND SPECTRA; CPT THEOREM; HADRONIC ATOM; LASER SPECTROSCOPY; NUCLEAR FUSION; NUCLEAR STRUCTURE; QUANTUM ELECTRODYNAMICS; STRONG NUCLEAR INTERACTIONS; TRANSURANIUM ELEMENTS in the McGraw-Hill Encyclopedia of Science & Technology.　　　　　　　Joachim Hartmann

Bibliography. D. Horváth, Progress in the physics of exotic atoms at CERN, *Nucl. Instrum. Meth. B*, 124: 314–319, 1997; C. Petitjean and L. Schaller (eds.), Proceedings of the Monte Veritá Workshop on Exotic Atoms, Molecules and Muon-Catalyzed Fusion, *Hyperfine Interactions*, 1999; L. M. Simons, D. Horváth, and G. Torelli (eds.), *Electromagnetic Cascade and Chemistry of Exotic Atoms*, Plenum Press, 1990.

Exotic nuclei

The study of atomic masses has significantly contributed to understanding the basic properties of matter. The development of new and precise experimental methods in this field has led to an extension of knowledge far into the region of exotic nuclei and toward the limits of nuclear stability. This experimental progress is also reflected by a revolution in precision mass measurements by frequency analysis of stored and cooled exotic nuclei.

Nuclear mass measurements. The nucleus is a bound system of protons (of number Z), which carry positive charge, and neutrons (of number N), which carry no charge. In an atomic system the positively charged nucleus is surrounded by an extended cloud of Z negatively charged electrons. The protons and neutrons are collectively called nucleons. The mass of a nucleus (M_{nuc}) is the sum of the proton masses ($Z \times m_p$) and the neutron masses ($N \times m_n$) reduced by the total binding energy B, as given in Eq. (1), where B is given in mass units according

$$M_{nuc} = Z \times m_p + N \times m_n - \frac{B}{c^2} \qquad (1)$$

to Einstein's energy-mass equivalence and c is the velocity of light. For a neutral atom the rest mass of Z electrons, reduced by their binding energy, is only a small correction to the nuclear mass, which represents 99.98% of the total atomic mass.

The nuclear binding energy, which results from all interactions of the nucleons via the nuclear and Coulomb forces, is responsible for the structure and stability of a nuclide and is the origin of the existence of all stable matter. The mass number A of a nucleus is the sum of neutrons and protons, $Z + N$. The average binding energy per nucleon, B/A, is about 8 MeV, and is less than 10^{-2} of the rest mass of a nucleon. This relation demonstrates that a mass determination has to be very precise to deduce significant information about the nuclear binding. Systematic studies of mass differences provide knowledge about the binding energy of single nucleons or nucleon clusters, closures of shells and subshells in the nuclear structure, and changes in nuclear shapes.

Mass measurements naturally started with stable nuclei. The masses of all stable nuclei are now known to high precision and are used to determine the values of sets of parameters used in theoretical predictions. Progress in understanding and crucial tests of the models can be obtained by mass measurements of nuclei which are far from stability. Since such species are unstable and hence cannot be found in nature, they must be produced in nuclear reactions.

Production and separation. In recent years a new era of nuclear physics studies has been opened up by powerful heavy-ion accelerators, novel separation techniques, and efficient detector systems, which allow investigations of nuclei with extreme ratios of protons to neutrons compared to stable nuclei. Fusion, fission, nuclear transfer, proton-induced spallation, and fragmentation of relativistic heavy ions are effective tools for creating exotic species in the laboratory. Then the reaction products must be separated and identified prior to investigation. This task can be performed by on-line isotope separators (ISOL), where the exotic nuclei are slowed down, implanted in an ion source, and reaccelerated for separation. Alternatively, in-flight separators can be used, where the nuclear-reaction products are separated with the full kinetic energy from the production process.

The on-line isotope separator facilities use very thick production targets and provide intense beams of exotic nuclei with kinetic energies of 30–100 keV extracted from an ion source. This method is limited to nuclei with half-lives longer than 10–100 milliseconds. The in-flight technique uses thin targets and takes advantage of the reaction kinematics. It can be applied to exotic nuclei with half-lives as short as microseconds, the only limitation being the time of flight of the nuclear species through the separator.

Mass measurement methods. Traditionally, mass spectroscopy and measurements of nuclear reaction energies (Q-values) have been the main experimental principles used to study nuclear masses. In mass spectroscopy the ionized atoms are analyzed in high-resolution electromagnetic spectrometers. This technique is confined to stable and long-lived nuclei. The second method is based on energy measurements and uses the conversion of mass to energy in a nuclear reaction.

Since the pioneering experiments in mass spectroscopy in the first half of the twentieth century, much progress has been made. Fundamental results of the early mass studies were the discovery of the existence of isotopes, the explanation of basic properties of nuclear binding by the liquid-drop model of the nuclei, and the first quantitative proof of the stabilization of nuclei by shell effects.

Precision measurements of stored ions. Experimental techniques have been further developed to increase sensitivity and allow mass measurement of shorter-lived nuclei far from stability. Important developments have improved magnetic rigidity and time-of-flight spectrometers.

Recently the ultimate precision in direct mass measurements was achieved with stored ions. The disadvantages of the relatively large velocity spread (broad range of velocities) and large angular divergence of the nuclear reaction products, which makes precision measurements impossible, can be overcome by cooling the stored exotic nuclei in ion traps and storage rings.

Novel experimental setups have been developed for direct mass measurements of both low- and high-energy exotic ions (**Fig. 1**). At the low energies typical of on-line isotope separation techniques, the exotic nuclei are efficiently cooled in a trap filled with a buffer gas. For high-energy ions, the atomic interaction of the stored ions with a merged cold electron beam has been applied for the preparation of high-precision experiments.

Storage rings and ion traps follow similar ion optical principles. They confine the motion of the injected ions with electromagnetic fields under ultrahigh vacuum conditions. In an ion trap, precise manipulation of the particle motion is achieved by static and radio-frequency electromagnetic fields applied to special electrode structures. The most commonly used types are the Paul and the Penning traps. The Penning trap, which employs a weak electrostatic quadrupole field with its electrode structure placed in a strong homogeneous magnetic dipole field, is particularly well suited for precision mass measurements. The mass M of the stored ions, which has a known charge state q, is determined in a Penning trap by measurement of the cyclotron frequency, given by Eq. (2), of an ion confined in a magnetic field B.

$$\omega = \frac{qB}{M} \qquad (2)$$

There are several trap systems under construction. Presently only the ISOLTRAP, at CERN (European Center for Nuclear Research), near Geneva, Switzerland, has been operated as a mass spectrometer for exotic nuclei. In the first stage a bunching system (Paul trap or quadrupole mass filter) and a cooler trap separate a monoisotopic beam of a few ions, which is then injected into the precision trap, where the mass measurement is performed with a resolving power of about 10^7 (Fig. 1a). The ISOLTRAP can measure masses of exotic nuclei with half-lives down to a few seconds.

A complex facility for mass measurements of stored relativistic nuclei has been used at GSI (Gesellschaft für Schwerionenforschung) in Darmstadt, Germany. The storage-cooler ring (ESR) with a circumference of 108 m (355 ft) acts as a precision trap, while the fragment separator (FRS) separates the nuclei of interest before they are injected into the ring (Fig. 1b). The mass measurement in the ESR is based on precise measurements of the revolution frequency of the stored ions. The velocity spread of the nuclear reaction products imposes a serious limitation on the resolving power of these measurements. This problem can be solved by two different experimental approaches. First, by applying electron cooling, the velocity spread of the stored ions can be reduced below 10^{-6} within less than a minute. Second, operation of the ion-optical system of the ESR in the isochronous mode shapes the trajectories of the ions in such a way that a larger velocity for a specific isotope is exactly compensated by a longer path length of the orbiting ions.

Both principles can be understood by a relation (3) between the mass M, the revolution frequency f,

$$\frac{\Delta M}{M} = -\gamma_t^2 \frac{\Delta f}{f} + \left(\gamma_t^2 - \gamma^2\right) \frac{\Delta v}{v} \qquad (3)$$

and the velocity v of the ions, where γ_t is an ion-optical parameter and γ is the Lorentz factor, given

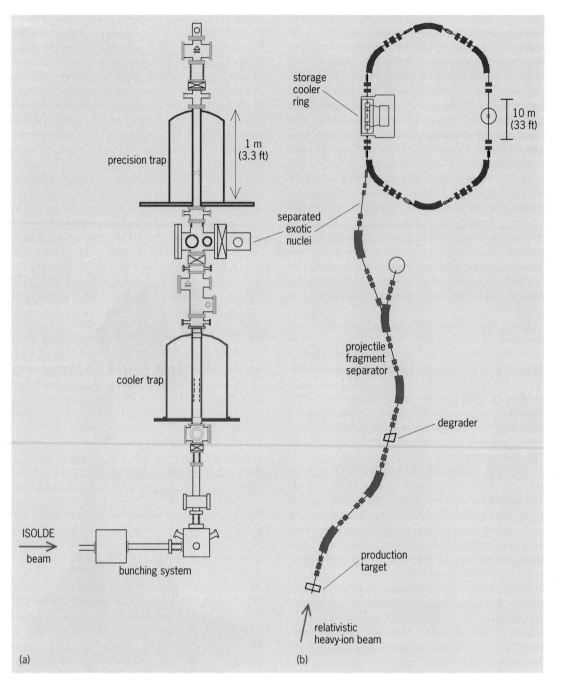

Fig. 1. Facilities for direct precision mass measurements of stored exotic nuclei. (*a*) ISOLTRAP, a low-energy facility. (*b*) FRS-ESR, a high-energy facility.

by Eq. (4). The symbol Δ indicates the deviation of a

$$\gamma = \left(1 - \frac{v^2}{c^2}\right)^{-1/2} \qquad (4)$$

quantity from its average value, and $\Delta f/f$ is the relative accuracy for the determination of the revolution frequency. In the case of electron cooling, $\Delta v/v \approx 0$, while in the case of isochronous measurement, $\gamma_t^2 \approx \gamma^2$. Equation (3) illustrates that with stored circulating ions the accuracy in mass determination is limited by the velocity spread, since the frequency measurements can be very precise.

The mass measurement is performed by Schottky analysis. The induced signals of the circulating ions are recorded in capacity probes and Fourier-transformed. The difference in the mass-over-charge ratio determines the difference in the revolution frequency. More than 100 new masses in the range of $54 \leq Z \leq 84$ have been determined using this method, with an accuracy of about 100 keV. While the cooling method is limited to nuclei with half-lives longer than about 1 second, the isochronous method, which has been successfully tested, permits the study of short-lived nuclei down to microsecond half-lives.

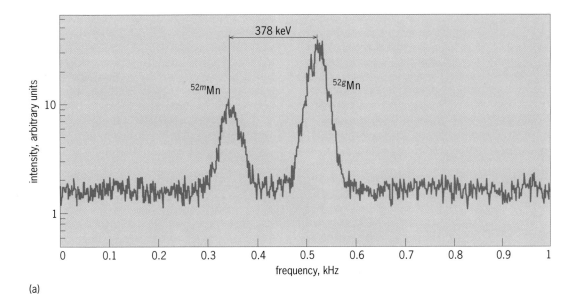

(a)

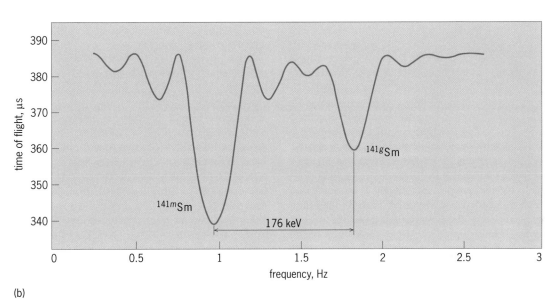

(b)

Fig. 2. Mass-resolved isomer spectra from ISOLTRAP and FRS-ESR experiments. The indices *g* and *m* indicate the ground and isomeric states, respectively. (*a*) Spectrum of manganese-52 (^{52}Mn) from FRS-ESR. (*b*) Spectrum of samarium-141 (^{141}Sm) from ISOLTRAP.

The excellent resolving powers of the ISOLTRAP and the FRS-ESR experiments allow resolution of the mass difference of ground and isomeric states of a nucleus (**Fig. 2**). Such mass spectra are a direct proof of Einstein's mass-energy equivalence. Both facilities have made a large contribution to the field of mass measurements (**Fig. 3**).

Measurements of the heaviest elements. All elements heavier than uranium ($Z = 92$) must be produced in nuclear reactions since they do not occur naturally on Earth. Based on extrapolations of the known nuclear shell structure, the existence of an island of superheavy elements at the magic proton number of 114 has been predicted. The heaviest elements, presently known up to $Z = 112$, have been created via complete fusion reactions at low excitation energies using the velocity filter at GSI. They are charac-

terized by very small production cross sections and are produced in quantities of single atoms. They predominantly decay by alpha emission with half-lives as short as a few hundred microseconds.

The masses of these nuclei are obtained from the kinetic energies of the alpha particles (Q_α-values) by linking them to known daughter nuclei. Observations of long alpha-decay chains down to nuclei with known masses enable their masses to be determined. With this method a new shell region located at hassium (element 108) has been discovered.

Measured masses compared with theory. Various approaches have been used to describe nuclear masses, ranging from pure systematics to macroscopic-microscopic models, refinements of the Weizsäcker mass formula with the inclusion of nuclear structure effects from shell-model calculations, and, finally,

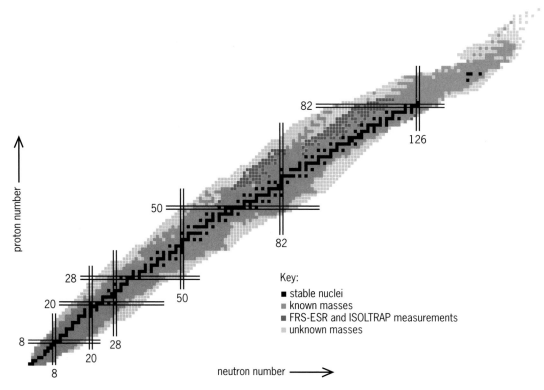

Fig. 3. Nuclei with known and unknown masses. The ISOLTRAP and FRS-ESR direct mass measurements are indicated.

self-consistent microscopic theories, which are the most fundamental descriptions. From that point of view it is expected that the predictive power of the self-consistent microscopic models is more reliable near the neutron and proton driplines though they presently show larger deviations from the experimental results.

A representative comparison of the experimental and theoretical mass values for lead isotopes (**Fig. 4**) reflects the present status. A macroscopic-microscopic model and two microscopic models are compared to the measured data. The microscopic mod-

els are based on self-consistent calculations using an effective nuclear force and a relativistic mean field approach. The high quality of the experimental data is a challenge to further improve the microscopic theories.

For background information *see* ATOMIC STRUCTURE AND SPECTRA; EXOTIC NUCLEI; ION SOURCES; MASS SPECTROSCOPE; NUCLEAR BINDING ENERGY; NUCLEAR REACTION; NUCLEAR STRUCTURE; PARTICLE ACCELERATOR; PARTICLE TRAP; TRANSURANIUM ELEMENTS in the McGraw-Hill Encyclopedia of Science & Technology.   Hans Geissel; Gottfried Münzenberg

Bibliography. I. Bergström, C. Carlberg, and R. Schuch (eds.), *Trapped Charged Particles and Related Fundamental Physics*, Physica Scripta, 1995; F. Bosch and P. Egelhof, Nuclear physics at storage rings, *Nucl. Phys.*, A626:1–636, 1997; H. Geissel, G. Münzenberg, and K. Riisager, Secondary exotic nuclear beams, *Annu. Rev. Nucl. Part. Sci.*, 45:163–205, 1995; W. Mittig, A. Lepine-Szily, and N. A. Orr, Mass measurement far from stability, *Annu. Rev. Nucl. Part. Sci.*, 47:27–66, 1997.

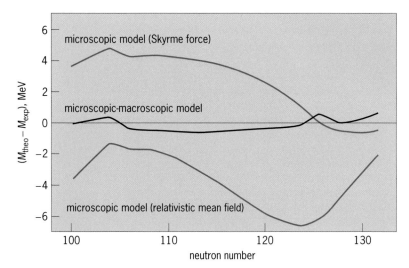

Fig. 4. Various theoretical predictions for mass values of lead isotopes with even neutron number (M_{theo}) compared with experimental mass values (M_{exp}).

Explosives detection

Over the last two decades, there has been a large increase in the use of explosives by terrorists, manifested in deadly bombings of such targets as passenger airlines, public buildings, and embassies. In response to this threat, existing detection technologies have been improved and new ones are being developed.

Detection of explosives, whether concealed on a human body, in passenger luggage, or inside cars and trucks, is a very difficult task. It must always be conducted under very unfavorable conditions, that is, within a short time; with minimum disturbance and delay to people, flights, or flow of commerce; with no limit on size or content of inspected objects; and with significant financial and space constraints.

Detectable features of explosives. Detection of explosives is based on certain of their attributes that distinguish them from the surrounding benign material.

Shape. Shape was a distinguishing feature years ago, when a typical bomb was made of several sticks of dynamite with an alarm clock as a timing device, and thus could be readily identified by an x-ray screener. Today explosives can be made in any shape or molded into the side of a briefcase, making the standard x-ray system of limited value in examining the cluttered contents of typical suitcases, trucks, or shipping containers.

Density. Another distinguishing feature of explosives is their physical density. All powerful explosives are very dense compared to other organic materials. This is especially true for military explosives (such as C4, Semtex, PETN, and TNT), with densities in excess of 1.5 g/cm^3, while most of the benign organic materials have densities below 1.2 g/cm^3 and many are below 1 g/cm^3 (water). Unfortunately, industrial explosives have a lower density than military ones, approaching that of water and other common organic substances. In order to uncover a small, dense, organic substance in a sea of slightly lower-density benign substances, the density has to be measured in small volumes all over the object under inspection. This can be achieved by x-ray computed tomography.

Crystalline structure. Another distinguishing feature of some explosives, such as with C4 and PETN, is their crystalline structure. This structure manifests itself via coherent (small-angle) scattering of a low-energy beam of x-rays off the probed object. Since only a small fraction of the x-rays are coherently scattered in this way, and both the probing and scattered rays are highly attenuated by the materials present, very high intensity x-ray sources are required. Another manifestation of the crystalline structure of some explosives is the molecular-crystal interaction which affects the nitrogen atom within the molecules of the explosive. This interaction can be detected by the absorption and emission of specific (resonance) electromagnetic radio-frequency fields. This technique is called nuclear quadrupole resonance (NQR). Different resonance frequencies are needed for the different crystalline explosives. The radio-frequency transmission is greatly affected by the presence of metallic parts and items in the scanned object, which may cause some missed detections and false alarms.

Molecular composition. The most specific method of distinguishing explosives from benign substances is based on their molecular composition. In fact, the molecular composition of explosives is very different from that of all other materials, including all organic ones.

There are many sensitive chemical-analytical and chemical-physical techniques, such as mass spectroscopy, gas chromatography, and ion mobility spectroscopy, that can detect and identify trace amounts of materials when these materials are accessible to the detection system. Unfortunately, the explosives are usually well concealed inside the object, and no chemical technique can determine their chemical composition without penetrating the object and taking a sample. Generally this cannot be done.

For trace techniques to be applicable, several conditions should be fulfilled simultaneously. (1) There should be a contamination (traces) of explosives on the outside of the item, for example, the suitcase, which relates to the presence of explosives inside. (2) There must be means, such as manual swiping of surfaces (for example, handlebars of suitcases), which are effective in removing and collecting the traces. (3) The chemical or analytical device must be able to detect the presence of an explosive from the traces, and does not generate false alarms. Most trace detectors detect minute amounts of explosives if properly presented. The first and second conditions are much less predictable, as they depend on many unknown factors, including the skill of the terrorist, the type of explosive, and the type of material constituting the container. Thus the performance of these chemical "sniffers" is highly unpredictable.

In addition to chemical and analytical trace detectors, there exist sensitive biological sensors, such as the olfactory capabilities of dogs. Dogs can be trained to detect the low levels of vapor emitted from most explosives. The basic limitations are having enough explosive vapors outside the container (some explosives have a very low vapor pressure) and the attention span of the trained dogs.

Elemental composition. The elemental composition of compounds is close to molecular structure in its ability to provide a technique for distinguishing one substance from another. This is particularly true for

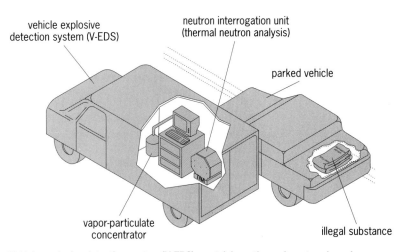

Vehicle explosive detection system (V-EDS), containing a thermal-neutron-based sensor and a secondary vapor-particulate sensor, scanning a parked car for large amounts of explosives hidden in the trunk. The V-EDS is stationary or moving at about 1 mi/h (0.5 m/s).

Explosive detection systems

Detected property	Specificity	Technique type	Principle	Key assumptions and problems	Automatic	Probability of detection	Probability of false alarm	Speed	Applicability to large objects (trucks and containers)
Molecular composition and crystalline structure	High	Trace detection	Detects particulates/traces of explosives	Depends on existence of contamination related to explosives and ability to collect it; unpredictable	Semi	Difficult to predict	Highly variable	Slow	Marginal
		Electro-magnetic	Nuclear quadrupole resonance (NQR)	Only crystalline explosives, such as C4 (RDX), are detected; presence of metals renders technique problematic	Yes	High	Medium	Fast	No
		Low-energy x-ray	Coherent scattering	Requires crystalline structure; in/out beam is highly attenuated by content	Yes	High	Low (expected)	Very slow	No
Shape (two-dimensional)	Very low	X-ray (single or dual energy)	X-ray attenuation	Explosives are assumed to be generally denser than other organic materials; screener looks for anomalies in image; explosives have no shape	No	Very low	Indeterminate	Fast	Yes (with higher energy)
		Enhanced dual energy x-ray	X-ray attenuation with image enhancement	Same as above; however, image processing helps focus screener's attention on limited area	Semi	Low	Indeterminate	Fast	No
Density (three-dimensional)	Medium	X-ray computed tomography (CT)	Three-dimensional CT density of explosives	Explosive density assumed to be significantly higher than benign material; good for military explosives	Yes	High	Medium-high	Medium	No
Elemental composition	High	Thermal neutron analysis (TNA)	Neutron-capture gamma rays	Detects explosives characterized by high nitrogen density; applies to all nitrogenous explosives, including liquids	Yes	Medium	Medium	Medium-fast	Yes (for large explosives)
		Pulsed fast neutron analysis (PFNA)	Elemental densities through neutron inelastic scattering	Detects nitrogen, oxygen, carbon (and hydrogen) density over volume of the size of explosive; unequivocal decision; applies to all explosives	Yes	Very high	Very low	Very fast	Yes

explosives, whose elemental composition is very different from most benign materials. Explosives are characterized, in general, by a relatively high concentration of oxygen and nitrogen and a low concentration of carbon and hydrogen. Thus the ability to determine elemental composition and features related to it can provide an effective means for explosive detection.

Neutron-based inspection techniques. The presence of chemical elements within an object can be determined nonintrusively by probing it with neutrons from the outside. The neutrons, being neutral particles, can penetrate through the various materials before they are scattered many times and absorbed. During these nuclear processes, gamma rays characteristic to the nuclei that interact with the neutrons are emitted and are detected by gamma-ray detectors outside the inspected object. The energy of the emitted gamma rays unequivocally determines the presence of the specific elements, while the intensity relates to their amount.

Thermal neutrons, the main probing tool of the thermal neutron analysis (TNA) technique, determine the presence of explosives primarily by detecting nitrogen and hydrogen. A variety of thermal-neutron-based interrogation systems for explosives and other materials have been developed. They have different embodiments, depending on their objectives. An example is a thermal-neutron-based inspection system for explosives hidden in vehicles (car bombs) called the vehicle-explosive detection system (V-EDS; see **illus.**).

Fast neutrons are more penetrating than thermal neutrons, and can detect many elements, including the key organic ones, oxygen and carbon, which thermal neutron analysis cannot. Furthermore, if fast neutrons with the same energy are produced in very brief pulses (lasting, for example, 10^{-9} s) and their time of interaction is recorded by timing the emission of the resulting gamma rays, the spatial distribution of the various chemical elements can be determined by employing the time-of-flight technique. This is the principle of the pulsed fast neutron analysis (PFNA) technique, which provides a three-dimensional map of the material in the inspected container and thus determines and locates most contraband: explosives, drugs, hazardous chemicals, dutiable goods, and other materials. It can be applied to luggage to inspect for small explosives which are dangerous to airplanes. It is uniquely suitable to inspect large trucks and shipping containers, the preferred conveyance of large amounts of drugs smuggled into the United States.

Evaluation and outlook. The techniques described above (see **table**) cover a wide variety of explosive attributes, and vary greatly in performance, as measured by the probability of detection and the probability of false alarm. Most of the techniques apply only to small objects, such as passenger suitcases. A few techniques apply also to the detection of explosives, drugs, and other materials in large objects such as cars, trucks, and shipping containers.

Most of the systems currently employed for aviation security are standard x-ray and trace detector (popularly called chemical-sniffer) units, together with an increased number of computerized tomography x-ray systems. Several higher-energy x-ray systems are used for truck and container inspection (mostly for drugs). In the next several years, the more advanced techniques, such as thermal neutron analysis and pulsed fast neutron analysis, probably will be combined with the existing ones to achieve higher performances and inspection regimes that are more cost-effective in providing overall security.

For background information *see* ACTIVATION ANALYSIS; COMPUTERIZED TOMOGRAPHY; GAMMA-RAY DETECTORS; GAMMA RAYS; NEUTRON; NONDESTRUCTIVE TESTING; NUCLEAR QUADRUPOLE RESONANCE; NUCLEAR SPECTRA; RADIOGRAPHY; TRACE ANALYSIS; X-RAYS in the McGraw-Hill Encyclopedia of Science & Technology.   Tsahi Gozani

Bibliography. S. M. Khan (ed.), *Proceedings of the 1st International Symposium on Explosive Detection Technology*, FAA Technical Center, Atlantic City, February 1995.

Field-emission displays

High-resolution flat-panel displays aim to achieve the visual and optical qualities of cathode-ray tubes, but with a significant reduction in overall size, weight, and power consumption. Among flat-panel displays, the benchmark for performance is the active-matrix liquid-crystal display. Field-emission displays are now emerging as a leading alternative for low-power, flat-panel multimedia applications and high-brightness automotive and avionic applications. Improvements in field-emission display technology have resulted from technological advances in vacuum microelectronics involving large-area lithography and thin-film processing, improved stability of field-emitter arrays, new emitter structures and materials, and improvements in low-voltage phosphor technology.

The primary reason for the growing interest in field-emission displays is that they offer significant advantages over liquid-crystal displays. Field-emission displays combine the best performance features of cathode-ray tubes, but in a thin and light package. Like the cathode-ray tubes, field-emission displays are emissive displays and can be seen from virtually any angle. Liquid-crystal displays are nonemissive light valves requiring a backlight for their operation, with a loss of clarity, contrast, and color purity when the display is viewed from different angles. In addition, they have a slow response time and are not suitable for operation over a wide temperature range.

Structure. Field-emission displays use the cold field emission of electrons and the cathodoluminescent generation of light to produce video images, similar to a cathode-ray tube (**Fig. 1**). A cathode-ray tube uses an electron gun that sequentially scans an electron beam across the pixels of a phosphor-coated screen. Due to the electron optics, the electron gun must be located far from the anode, at a distance similar to the width of the display area. A field-emission display operates from a matrix addressable array, with hundreds of individual electron emitters per pixel, thus removing the need for electron beam scanning. Images are created simply by turning the individual electron-emitting microtips on or off in sequence behind each phosphor pixel. As a result, field-emission displays need a spacing of a millimeter or less between the emitters and the phosphor screen, allowing for a thin and flat display.

Structurally, a field-emission display is a hermetically sealed vacuum package with the emitter array in the back glass sheet and the phosphor screen on the front glass sheet, separated by insulating spacer posts that are invisible to the viewer. On the emitter side, the addressable x-y emitter array layout eliminates the nonlinearity and pincushion effects associated with standard cathode-ray tube images. The anode plate hosts standard cathode-ray-tube phosphors, resulting in a display comparable to a cathode-ray tube in every respect.

Cathode arrays. Field emission is the quantum-mechanical tunneling of electrons from a surface into

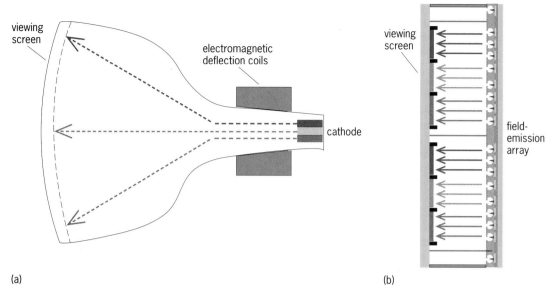

(a)

(b)

Fig. 1. Cross sections of (a) cathode-ray tube and (b) field-emission display.

vacuum under the influence of a very high electric field. These very high electric fields, when applied to a conducting surface, make it possible for some electrons to spontaneously tunnel through the vacuum-metal barrier. For many metals, appreciable electron tunneling current is observed at electric fields greater than 3×10^7 V/cm.

However, it is difficult to achieve such high fields under practical conditions without some form of field enhancement at the surface. For this reason, micrometer-size structures with a high height-to-width ratio are desirable. A process to produce self-aligned tip and gate structures using semiconductor process technology initiated the most important technical advance leading to the current state of vacuum microelectronics. This tip deposition process is now called Spindt tip deposition, and the structures thus prepared, Spindt tips (**Fig. 2**). A major problem is that a simple field-emission cathode is prone to failures due to uncontrolled emission current. An important advance in field-emission arrays is the inclusion of a ballast resistor, which has a current-limiting effect and minimizes uncontrolled emission current. This also results in a more uniform and stable cathode array, which is necessary for display devices.

The Spindt tip process is based on the fabrication of well-defined emitter cones in micrometer-size wells (**Fig. 3**). It is very effective as a self-aligned, self-limiting process for producing millions of submicrometer structures over a large area. A difficulty with the process is that the shape and height of the emitter cone are dependent on several variables such as size and shape of the emitter hole, dielectric thickness, and thickness and composition of the gate metal. Refinements to the basic concept are under development to address this challenge.

While Spindt technology has received considerable attention, there is continuing research on al-

ternative emitter structures, encompassing a variety of procedures and materials. Planar emitters such as diamondlike carbon films and surface-emitting structures, aimed at reducing process complexity and manufacturing costs, are being explored. One

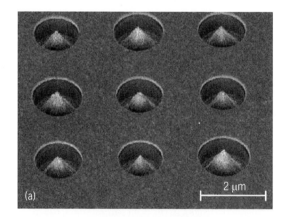

Fig. 2. Electron micrographs of a field-emission cathode array. (a) A 3 × 3 array of emitters. (b) Cross section of a single emitter cone.

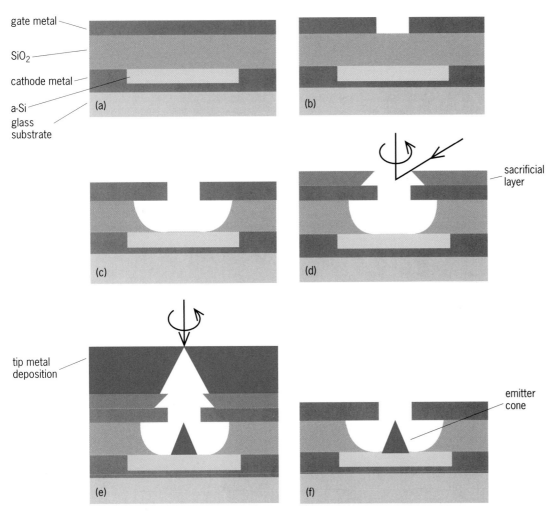

Fig. 3. Spindt-tip deposition method for fabricating field-emission cathode arrays. (*a*) Electrode and dielectric layer deposition on glass. (*b*) Definition of emitter well using photolithography. (*c*) Formation of emitter well. (*d*) Deposition of sacrificial liftoff layer, typically aluminum, by angle evaporation. Rotation of glass substrate during deposition is indicated by arrow. (*e*) Tip deposition by electron beam evaporation. Tip materials are usually refractory metals such as molybdenum or niobium. Rotation of substrate is indicated by arrow. (*f*) Removal of sacrificial and closure layers.

surface-emitting structure is based on field emission across nanometer-size tunneling barriers. Nanometer-width cracks are created in thin-film materials containing nanocrystalline particles. When a potential is applied across this small gap, electrons can tunnel across the gap. Some of these electrons scatter at the edge of the positive electrode and are attracted to the phosphor screen. These structures can be created with simple printing techniques such as inkjet deposition, with minimal use of expensive semiconductor process equipment.

Fabrication. Field-emission displays integrate two generations of electronics manufacturing technology. Field-emission cathode arrays and anodes are fabricated with sophisticated microelectronic manufacturing techniques. The emitter-array glass sheet and phosphor screen are integrated using glass-to-glass vacuum packaging techniques similar to those used in electron tube manufacturing. Engineering of field-emission display systems requires expertise in semiconductor processing, vacuum tube packaging and methods to maintain high vacuum, and

high-voltage electronics and electron beam technologies. A great amount of compatibility is needed at the materials, packaging, and component design levels to successfully manufacture field-emission displays.

A standard fabrication process for field-emission displays includes (1) the microfabrication of the emitter structure, (2) the production of the phosphor screen, (3) the placement of spacers and vacuum gettering material between the emitter and phosphor glass plates, and (4) the vacuum sealing of the assembly (**Fig. 4**). For color displays, the alignment of the individual subpixel elements with the corresponding electron emitter array is essential to produce saturated colors. Therefore, the sealing method must accurately maintain alignment between the two pieces of glass. As in liquid-crystal displays, the device drivers and the addressing electronics are integrated along the edges of the package.

Phosphor screen fabrication. The phosphor screen of the field-emission display operates in a manner analogous to the cathode-ray tube, and hence the basic fabrication processes are similar. However, the details

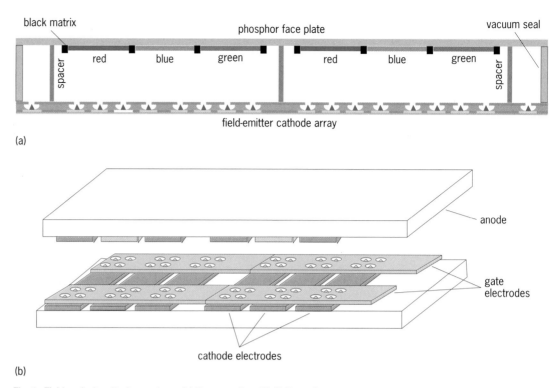

black matrix   phosphor face plate   vacuum seal

spacer   red   blue   green   red   blue   green   spacer

field-emitter cathode array

(a)

anode

gate electrodes

cathode electrodes

(b)

Fig. 4. Field-emission display package. (*a*) Cross section. (*b*) Oblique view.

vary somewhat because of operational differences such as applied voltage. For example, cathode-ray tubes operate with a typical anode voltage of 15,000–30,000 V, whereas field-emission displays operate in the 200–5000-V range. For high-voltage operation, above 4 kV, the anode screen is generally aluminized to reduce charge buildup, to improve the optical efficiency, and to maintain the phosphor. Operation at less than 1000 V minimizes high-voltage effects such as arcing. Though the system design in this case is simpler, phosphor efficiency is severely reduced, thus diminishing the color quality of the display. There is considerable research to develop efficient low-voltage phosphors that address this problem.

Spacers and vacuum packaging. In order to reduce the package weight and provide thin profiles, field-emission displays are built with thin flat glass. The typical thickness of the cathode and anode glass is 0.7–1.5 mm. Because the display is a vacuum package, these glass plates are required to withstand a differential pressure of at least 1 atmosphere. Under this force, thin flat glass undergoes bowing, changing the optical uniformity of the display. Severe bowing can lead to breakage. Therefore, insulating spacers are placed between the two glass plates to prevent the package from collapsing due to this pressure differential. These spacers are required to withstand electric fields as high as 5 V per micrometer. Furthermore, the electron beam can also interact with the spacer materials, allowing charge buildup on the spacer. If this charge becomes too large, it can lead to catastrophic breakdown between the anode and the cathode. Thus field-emission display spacers should have not only good mechanical strength and stability

and vacuum compatibility, but also a high breakdown voltage and low secondary emission characteristics, so as not to become visible to the viewer.

Vacuum pumping in sealed displays. Field-emission displays are vacuum tubes, and a good vacuum is essential for their operation. Field-emission sources are more sensitive than cathode-ray-tube thermionic emitters to residual gases because of the sensitivity of the height of the electron tunneling barrier to contamination. Poor vacuum in the display envelope also increases the likelihood of sputter damage, arcing between the emitter array and the high-voltage phosphor screen, and phosphor degradation. Also, the small dimensions of the panel create poor conductance for both initial vacuum pump-down and continued vacuum operation. Because of these effects, most field-emission displays use getter vacuum pumps to maintain a low vacuum during their operation. After the placement of spacers and getters at suitable locations, the two glass plates are aligned and sealed together using low-melting-point sealing glasses. Since field-emission displays require a high vacuum in the panel, good vacuum sealing methods are essential for long device life.

Options. For field-emission displays to be commercially viable, they must be at least comparable in performance to competing technologies, such as the active-matrix liquid-crystal display. While the full potential of field-emission display performance has not been realized, there are several examples of working prototypes. There are two distinct design strategies based on the operational voltage of the anode. Low voltage is generally defined as less than 800 V for the anode acceleration. A typical low-voltage design operates in the range of 200–400 V and a cathode-to-anode separation of about 200 μm. In this configuration, proximity focusing is achieved without focus grids. Another advantage to low voltage is that the switched anode provides excellent color saturation. Also, with only a few hundred volts across the spacers, problems due to electrical breakdowns and vacuum arcs are reduced. The main disadvantage to the low-voltage approach is the lack of red, green, and blue phosphors with good efficiency at low voltage.

While high-voltage operation offers the potential for high brightness and high contrast more similar to cathode-ray tubes, there are significant technical challenges to the high-voltage approach. Even though higher anode voltage improves beam collimation, this advantage is offset by the beam divergence over the longer path length that is required to maintain the voltage difference between the anode and the cathode. Successful long-term operation of high-voltage field-emission displays requires the development of suitable spacer materials that can withstand several thousand volts across a millimeter, and the elimination of the potential for catastrophic arcs between the anode and cathode due to electrical breakdown. The major hurdles are in vacuum sealing and processing. Since maintenance of a good vacuum is essential for achieving long display life, this is a critical area of research.

For background information *see* CATHODE-RAY TUBE; ELECTRONIC DISPLAY; FIELD EMISSION; INTEGRATED CIRCUITS; LIQUID CRYSTALS; VACUUM TUBE in the McGraw-Hill Encyclopedia of Science & Technology.　　　　Babu R. Chalamala; Robert H. Reuss

Bibliography. I. Brodie and C. A. Spindt, Vacuum microelectronic devices, *Adv. Electr. Electron Phys.*, 83:1–106, 1992; D. Cathey, Jr., Field-emission displays, *Inform. Display*, 11:16–20, 1995; B. R. Chalamala, Y. Wei, and B. E. Gnade, FEDup with fat tubes, *IEEE Spectrum*, 36:41–50, 1998; H. F. Gray, The field-emitter display, *Inform. Display*, 9:9–14, 1993.

Filamentous fungi

From the beginning of recombinant fungal biotechnology, the yeast *Saccharomyces cerevisiae* has been the most commonly used host organism for the production of heterologous proteins. Most developments of fungal biotechnology were initiated by work with this organism. In recent years, other yeast species, such as *Kluyveromyces lactis*, *Hansenula polymorpha*, and *Pichia pastoris*, have become accessible for applications with recombinant proteins. Interestingly, in some instances, yeasts are unable to express eukaryotic genes which encode proteins that require posttranscriptional modifications (such as glycosylations). In contrast, filamentous fungi are able to synthesize some of these proteins in the fully functional form. Simultaneous with the development of molecular genetic techniques and gene transfer systems in yeast, comparable systems were established for filamentous fungi, making strain improvements feasible. Filamentous fungi have been used for decades as major producers in the pharmaceutical, food, and food-processing industries, leading to a high technical standard in fermentation processes with large-scale fermenters. This, together with the fact that many filamentous fungi possess the GRAS ("generally recognized as safe," in terminology of the U.S. Food and Drug Administration) status, makes them ideal organisms for the production of recombinant proteins. Many filamentous fungi with the GRAS status are standard organisms for large-scale fermentation, able to secrete large amounts of proteins, and therefore are ideal host organisms for the production of recombinant proteins. Included in this category are *Aspergillus niger, Aspergillus oryzae, Acremonium chrysogenum, Penicillium chrysogenum*, and *Trichoderma reesei*, among others.

Recombinant fungal strains. One prerequisite for the expression of heterologous genes in fungi is the availability of an established deoxyribonucleic acid (DNA) transformation system (used to transfer and incorporate foreign DNA into a cell's genome). The first successful DNA-mediated transformations were reported in the late 1970s, and since then the list of transformable fungal species has been growing constantly. In most species, the transforming DNA is not stably maintained as a self-replicating DNA molecule within the fungal cell, but integrates into the fungal genome. While in yeast the integration of transforming DNA is entirely homologous, in other eukaryotes, including filamentous fungi, ectopic integration is more common. Therefore, every single fungal transformant has to be analyzed individually whether or not optimal gene expression of the heterologous gene is guaranteed.

The general approach for the production of heterologous proteins in filamentous fungi by genetic engineering consists of two steps: (1) Construction of an expression vector which usually contains a strong (inducible) promoter sequence from fungal sources in front of the heterologous gene. Signals for transcription termination are located downstream of the gene of interest. Termination signals are usually derived from fungal genes; however, it has been shown that even termination signals from higher eukaryotes are functional in filamentous fungi. (2) Development of an optimal DNA-mediated transformation system. In general, fungal cells are treated by chitin-degrading enzymes to obtain spheroplasts or protoplasts for the polyethyleneglycol-mediated uptake of transforming DNA. Alternatively, electroporation procedures or particle guns have been successfully used for the delivery of recombinant DNA into fungal cells. However, this technique is applied only to those organisms which cannot be transformed by conventional techniques.

The first generation of overproducing fungal strains carried multiple copies of the gene of interest which were stably integrated into the fungal genome. Generally, gene expression is controlled by a strong

Examples of the application of filamentous fungi for heterologous expression of fungal and other eukaryotic genes *

Fungal host	Promoter/secretion signal sequence	Heterologous protein (origin of gene)
Acremonium chrysogenum	Alkaline protease (F. sp.)	Alkaline protease (Fusarium sp.)
		Lysozyme (human)
	pcbC (A.c.)	Hirudin (Hirudo medicinalis)
	pgK (A.c.)	Cephalosporin-C-acylase (Pseudomonas sp.)
		Thrombomodulin (human)
Aspergillus awamori	exl A (A.a.)	Cutinase (Fusarium solanipisi)
	glaA (A.n.)	Aspartic proteinase (Mucor miehei)
		Chymosin (Kalb)
	gpd (A.s.)	Glucoamylase (Aspergillus niger)
Aspergillus nidulans	aphA (A.n.)	Interferon α2 (human)
	alcA (A.n.)	Interferon α2 (human)
	glaA (A.n.)	Interleukin-6 (human)
	gpdA, alcA	Glucoamylase (Aspergillus niger)
	α-L-arabinofuranosidase A (A.n.)	α-L-arabinofuranosidase A (Aspergillus niger)
Aspergillus niger	glaA (A.n.)	scFv antibody fragments
		Insulin (human)
		Pancreatic phospholipase (pig)
	gpdA (A.n.)	Lysozyme (hen egg white)
	gpdA (A.n.)	Alpha-amylase (Aspergillus niger)
	Xylanase (A.a.)	Xylanase (Aspergillus awamori)
Aspergillus oryzae	glaA (A.o)	Chymosin (Kalb)
	glaA (A.a)	Lactoferrin (human)
	Taka-amylase (A.o.)	Manganese-peroxidase (Phanerochaete chrysosporium)
		Aspartic proteinase (Mucor pusillus)
Fusarium graminearum	try (F.o.)	Cellulase (Scytalidium thermophilum)
Fusarium oxysporum	try (F.o.)	Lipase (Thermomyces lanuginosus)
Neurospora crassa	glaA (N.c.)	Zeamatin (Zea mays)
Penicillium chrysogenum	phoA (P.c)	Tear lipocalin (human)
		Xylanase (fungal)
Trichoderma reesei	cbh1 (T.r.)	Glucoamylase P (Hormoconis resinae)
		Fab antibody fragments
		Acid phosphatase (Aspergillus niger)

*alcA alcohol dehydrogenase gene
aphA alkaline phosphatase A gene
cbh1 cellubiohydrolase 1 gene
glaA glucoamylase gene
gpdA glyceraldehyde-3-phosphatide hydrogenase gene

exlA endoxylanase A gene
try trypsin gene
pcbC isopenicillin N synthetase
A.a. Aspergillus awamori
A.c. Acremonium chrysogenum
A.n. Aspergillus nidulans

A.o. Aspergillus oryzae
F.o. Fusarium oxysporum
F.sp. Fusarium species
N.c. Neurospora crassa
P.c. Penicillium chrysogenum
T.r. Trichoderma reesei

fungal promoter sequence (a region on DNA to which an enzyme binds prior to transcription of an operon) which often functions in different hosts. An example is the constitutive *gpdA* promoter, which was first derived from the *Aspergillus nidulans* gene encoding the enzyme glyceraldehyde-3-phosphate dehydrogenase. This approach led to the overproduction of several extracellular or intracellular proteins (see **table**). However, in cases where heterologous gene products are part of a complex biosynthetic pathway, the increased production rate often does not result in a higher yield of the end product.

The second generation of fungal hosts, showing a major increase in production yield, were generated by DNA-mediated transformations of gene fusions. The corresponding expression vectors contain cassettes (regions containing nucleotide sequences that may be substituted for one another) which allow the fusion of nonfungal genes to a highly expressed fungal gene. The corresponding gene product is then secreted into the culture medium, allowing for its easy and convenient harvest. Characteristic examples for this application are the glucoamylase gene from *Aspergillus niger* or *A. awarmori* and the cellobiohydroxylase I gene from *Trichoderma reesei*. As a result, the secretion of human and nonhuman products into the culture medium became feasible (see table).

Transcriptional gene regulation. Many experiments with recombinant filamentous fungi have demonstrated that the increase in copy number of recombinant genes results in elevated amounts of the corresponding transcript. Similarly, strong fungal promoter sequences such as the constitutive *gpdA* promoter from *Aspergillus nidulans* or the inducible *glaA* (glucoamylase) promoter from *A. niger* allow a significant increase in the level of recombinant transcripts.

However, these approaches are less suitable to exogenously regulate expression of recombinant genes. In industry, the control over the timing and level of gene expression is highly preferred when recombinant products are synthesized in large-scale fermenters. Therefore, several attempts have been undertaken to use chimeric promoters that respond to otherwise inactive chemicals. An example is chemically regulating the expression of the *alc* regulon from *A. nidulans*. The chemically induced *alcA* promoter from *A. nidulans*, which is controlled by the *alcR* transcription factor, was successfully used for regulated gene expression not only in filamentous fungi but also in higher plants. Nontoxic amounts of ethanol can be used as chemical inducers for rapid initiation of the expression of recombinant genes.

Other promising attempts will try to control protein synthesis through fungal transcription factors, which bind specifically to fungal promoter sequences. The best-characterized positive-acting factors are AREA, or PACC (the transcription factor mediating pH regulation), from *A. nidulans* or the CREA repressor from the same species. In future experiments, emphasis will be placed on controlling gene expression through the coexpression of genes for transcription factors which control promoter-directed expression of recombinant genes.

Posttranslational protein modifications. At the posttranslational level, protein folding and glycosylation processes substantially influence structure and function of heterologous proteins. One strategy to synthesize functional heterologous proteins is the cosynthesis of chaperones or foldases, which are involved in the translocation of proteins into the endoplasmic reticulum. Their presence in the fungal cell should lead to improved levels of protein secretion.

Another strategy attempts to modify the glycosylation pathway. Targeted modification of glycosylation enzymes such as glycosyltransferase may result in heterologous proteins with comparable activity found in the natural host. It has been shown that fungal *N*-glycanes can serve as acceptor substrates for mammalian glycosyl transferases to produce glycans of the high-mannose type found in higher eukaryotes. Therefore, it seems desirable to introduce mammalian genes for glycosyl transferases into the fungal host genome to obtain the synthesis of mammalian proteins. This will become more relevant when an increased number of therapeutic proteins are produced for which product authenticity is a prerequisite before commercialization.

Another problem arises when therapeutic proteins are produced. Filamentous fungi usually secrete many different proteases (enzymes that digest proteins), which reduce the efficient synthesis of heterologous proteins. In *Aspergillus niger*, nine genes for different proteases have already been characterized at the molecular level, making gene disruption experiments feasible. Gene inactivations have resulted in the construction of *Aspergillus* strains, which show a deficiency of major proteases, thus making them preferable to serve as heterologous hosts.

Pathway (metabolic) engineering. In recent years, attention has been given to improve industrially important filamentous fungi by pathway or metabolic engineering. This technique is defined as the purposeful modification of intermediary metabolism using recombinant DNA techniques. It allows the in vivo production of metabolites which can be chemically synthesized only with great effort. The first example concerns the beta-lactam antibiotics, which are synthesized by filamentous fungi (such as *Acremonium chrysogenum* and *Penicillium chrysogenum*) as well as by prokaryotic organisms (such as *Streptomyces clavuligerus*). Metabolic engineering allows the removal of bottlenecks in the beta-lactam biosynthetic pathway and the combinatorial synthesis of new molecules. For example, prokaryotic beta-lactam genes not found in eukaryotic microorganisms can be expressed in fungal hosts in order to generate novel antibiotics by modifying biochemical pathways. Similarly, metabolic engineering of another class of secondary metabolites, the polyketides (such as tetracyclines), has been reported. Polyketides are used in human and animal health for a wide variety of effects, including immunosuppression (rapamycin), antimicrobial activity, and lowering blood cholesterol levels (savalestatin). New polycyclic compounds can be generated by recombining genes from prokaryotic and eukaryotic sources in a fungal host.

Finally, the manipulation of the isoprenoid pathway, which was already used for metabolic engineering in *Saccharomyces cerevisiae* and other yeasts to design novel carotenoids, seems to be suitable for further manipulation of hosts such as phycomycetous fungi.

These examples indicate that future generations of fungal strains will be genetically manipulated to metabolize novel compounds which broaden the spectrum of biotechnically important secondary metabolites.

Future prospects. There are many examples demonstrating that filamentous fungi are suitable host organisms for the production of heterologous proteins. In future experiments, fungal strains will be constructed that produce posttranslational modification enzymes together with the heterologous protein in order to guarantee a functional product. Another advance will come from experiments in which the manipulation of fungal transcription factors will improve the controlled overexpression of functional therapeutic compounds. Finally, combinatorial synthesis of new molecules by metabolic engineering will strengthen the position of filamentous fungi as potential producers in fundamental and applied microbiology.

For background information *see* FUNGI; GENETIC ENGINEERING; MOLECULAR BIOLOGY; YEAST in the McGraw-Hill Encyclopedia of Science & Technology.

U. Kück

Bibliography. R. J. Gouka, P. J. Punt, and C. A. van den Hondel, Efficient production of secreted proteins by *Aspergillus*: Progress, limitations and prospects, *Appl. Environ. Microbiol.*, 47:1–11, 1997; R. Radzio and U. Kück, Synthesis of biotechnologically relevant heterologous proteins in filamentous fungi, *Process Biochem.*, 32:529–539, 1997; J. P. T. W. van den Hombergh et al., *Aspergillus* as a host for heterologous protein production: The problem of proteases, *Tibtech*, 15:256–263, 1997.

Food-borne disease

The vast majority of reported cases of food-borne illness of known etiology are of bacterial origin. Among the bacterial pathogens are *Campylobacter jejuni*, *Salmonella*, *Staphylococcus aureus*, *Escherichia coli* O157:H7, *Shigella*, *Bacillus cereus*, *Clostridium perfringens*, *Listeria monocytogenes*, *Yersinia enterocolitica*, and *Clostridium botulinum*. These pathogens account annually for more than 5 million cases (estimated) of food-borne illness in the United States.

Illness can result from a bacterial infection or intoxication. Food-borne infection occurs when a pathogen grows in the human body. Food-borne intoxication results when a pathogen grows in food and releases toxins that cause illness when ingested in sufficient amounts. Infectious pathogens include *C. jejuni*, *Salmonella*, *E. coli* O157:H7, *Shigella*, and *Y. enterocolitica*. Toxin-producing pathogens are *C. botulinum* and *Sta. aureus*.

Campylobacter jejuni. *Campylobacter jejuni* is the leading cause of acute bacterial diarrheal illness in the United States, responsible for more than 2 million cases (estimated) of gastroenteritis annually. The disease is inexplicably most prevalent in California, where the incidence is up to four times greater than in other states. Poultry is the primary vehicle of infection, being associated with 50–75% of cases of Campylobacter enteritis. Approximately 50–80% of retail poultry is contaminated by *C. jejuni*, often at levels of 10^4–10^6 cells per gram.

Campylobacter jejuni also has been associated with Guillain-Barré syndrome, the leading cause of acute neuromuscular paralysis in the United States. Guillain-Barré syndrome is an autoimmune disease that is triggered by several factors, including an acute infectious illness of the gastrointestinal tract. It is estimated that 25–50% of cases of the syndrome are precipitated by *C. jejuni* infections. Hence, *C. jejuni* accounts for up to 2100 cases annually.

Salmonella species. Another leading cause of food-borne illness is *Salmonella*, which is responsible for 2 million cases (estimated) of gastroenteritis per year. *Salmonella enteritidis* and *Sal. typhimurium* are the two leading serotypes causing human food-borne illness. Each serotype is responsible for about 20–25% of salmonellosis cases. *Salmonella enteritidis* is largely transmitted by eggs. This pathogen colonizes the ovarian tissue of laying hens and is present within the egg contents. Present estimates suggest that 1 in 20,000 eggs in the United States is internally contaminated with *Sal. enteritidis*. Properly refrigerating and thoroughly cooking eggs are the best preventive measures to avoid egg-borne salmonellosis.

An increasing public health concern is the development of strains of *Salmonella* that are multiresistant to antibiotics used for human therapy. *Salmonella typhimurium* phage type DT 104, which is resistant to five commonly used antibiotics, including chloramphenicol, has emerged as a major cause of salmonellosis in humans. In 1979–1980, 0.6% of cases of *Sal. typhimurium* infection in the United States were attributed to phage type DT 104. In 1996, DT 104 was responsible for 34% of such cases. Interestingly, the total number of annual cases of salmonellosis caused by *Sal. typhimurium* has not changed substantially in recent years, but DT 104 is rapidly becoming the dominant phage type among the strains of *Sal. typhimurium* responsible for gastroenteritis. Only one effective antibiotic, fluoroquinolone, remains available for use in the treatment of *Sal. typhimurium* DT 104 in human infections. Reports from the United Kingdom indicate that some strains of DT 104 have also acquired resistance to fluoroquinolone. Strains having this antibiotic-resistance profile may not be treatable in life-threatening cases of salmonellosis.

Escherichia coli O157:H7. Enterohemorrhagic *E. coli* O157:H7 (EC O157) is an unusually virulent bacterium that has led food microbiologists to rewrite the rule book on food safety. This pathogen has a very low infectious dose (it is estimated that ingestion of less than 10 cells can cause illness), and produces severe illness that affects all age groups. It has unusual tolerance to acid, and an inexplicable association with ruminants. The symptoms of *E. coli* O157:H7 infection are often severe and can be life-threatening. They include hemorrhagic colitis (overtly bloody diarrhea), hemolytic uremic syndrome (acute kidney failure), and thrombotic thrombocytopenic purpura (diminished platelet count, prolonged bleeding time). The last resembles hemolytic uremic syndrome but causes less renal failure and has neurological involvement that includes seizures and strokes. *Escherichia coli* O157:H7 can survive for several days to weeks in acidic foods such as apple juice, fermented sausage, and yogurt that contain sufficient amounts of acid to kill most food-borne pathogens. Cattle are the primary reservoir of *E. coli* O157:H7. The bacteria persist in the animal's forestomachs and lower gastrointestinal tract. From 1 to 3% of cattle carry *E. coli* O157:H7 with no apparent symptoms of sickness. Eating undercooked ground beef and handling animals (especially cattle) on farms are leading risk factors associated with *E. coli* O157:H7 infection. An estimated 10,000–20,000 cases of such infection occur annually in the United States.

Trends. Many changes have occurred during the past generation that likely contribute to increases in food-borne illness. These trends include changes in diet, increasing use of commercial food services, new methods of producing and distributing foods, new and reemerging food-borne microbial pathogens, and an increased number of immunocompromised people who are at risk of severe food-borne illness.

During the past two decades, consumption patterns of many foods have changed substantially. Efforts to promote a heart-healthy diet have led to increased consumption of fruits and vegetables, which has led to an increase in importation of produce. Fresh produce has been increasingly associated with food-borne disease, accounting for one-third of all outbreaks reported in Minnesota from 1990 to 1996. During the last 20 years, imported produce has been associated with many outbreaks of food-borne illness in the United States. In addition, beef consumption has decreased in part because of public concern about *E. coli* O157:H7 contamination, whereas poultry consumption has almost doubled in 20 years, greatly increasing the risk of acquiring *Campylobacter* infection.

Cultural changes have also affected where people eat and how their food is prepared. The time-pressured life-styles of Americans lead to reduced time available for shopping and preparing meals. These pressures can affect food selection and safety, with more reliance on leftovers, increased purchase of prepared or convenience foods, and frequent eating away from home.

There has been a dramatic change in consumption of foods away from home. About one-third of food dollars was spent eating away from home in 1970, whereas in 1996, 46% of food dollars went to meals and snacks prepared outside the home. These changes have led to a substantial increase in the number of people handling food. Recent studies in Minnesota have revealed an increased risk of transmission of *Sal. typhimurium* associated with food handlers.

Major changes have occurred in the processing and distribution of foods. Today, large volumes of foods are processed in very large manufacturing plants. A few decades ago there were many smaller plants. Mass preparation and packaging of food in large processing facilities increases the potential of very widespread outbreaks if contamination occurs. Low-level contamination of a ready-to-eat product can result in thousands of cases of illness because of the large quantity of product consumed. The largest reported outbreak of salmonellosis in the United States was associated with ice cream that was sporadically contaminated with low levels of *Salmonella* and then widely distributed.

Emerging and reemerging food-borne pathogens, which include new, recurring, or drug-resistant pathogens, have caused an increase in the incidence of food-borne illness in the last two decades. Examples include *E. coli* O157:H7, *Cyclospora* (a coccidia), *Sal. enteritidis* in eggs, and *Sal. typhimurium* DT 104. Many of these cause more severe symptoms than the typical mild gastrointestinal ailments frequently associated with food-borne illness. Overt

bloody diarrhea, kidney failure, arthritis, and paralysis are some of the pronounced symptoms attributed to the emerging food-borne pathogens.

Host susceptibility is an important factor associated with the risk of acquiring food-borne illness. The competence of an individual's immune system to resist infection is critical to reducing the risk of food-borne infection. Elderly persons with weakened immune systems and people with cancer and chronic diseases that result in an immunosuppressed state are highly susceptible to food-borne microbial pathogens. The number of elderly in the United States is increasing rapidly, with an estimated one-fifth of the population being over age 65 by 2030. There is also an increasing number of people with immunosuppressive diseases such as human immunodeficiency virus.

Food-borne disease of microbial origin is a major issue in the United States. With the emergence of previously unrecognized food-borne pathogens and new influences that facilitate the dissemination of pathogens or the acquisition of food-borne illness, even greater concerns regarding food-borne pathogens are likely.

For background information *see* BACTERIA; ESCHERICHIA; FOOD MICROBIOLOGY; FOOD POISONING; FOOD SCIENCE; GASTROINTESTINAL TRACT DISORDERS; KIDNEY DISORDERS; SALMONELLA in the McGraw-Hill Encyclopedia of Science & Technology.

Michael P. Doyle

Bibliography. R. L. Buchanan and M. P. Doyle, Food-borne disease significance of *Escherichia coli* O157:H7 and other enterohemorrhagic *E. coli*, *Food Technol.*, 51(10):69–76, October 1997; Council for Agricultural Science and Technology, *Food-borne Pathogens: Risks and Consequences*, CAST Task Force Report 122, 1994; National Academy of Sciences, *Ensuring Safe Food from Production to Consumption*, 1998; P. I. Tarr and M. P. Doyle, Food safety: Everyone's responsibility, *Pediat. Basics*, 78: 1–13, Fall 1996.

Food irradiation

The treatment of fresh or processed foods with ionizing radiation renders them safe to consume and extends their storage lifetime. Irradiation is effective in ridding foods of insects and disease- and spoilage-causing microorganisms. These benefits are achieved without appreciably raising the temperature of the food and without causing significant physical or chemical effects. Consequently, food irradiation only minimally affects nutritional value and food quality. Over 50 years of intensive research has shown this technology to be safe, efficient, and acceptable.

Biological contaminants. Foods are often contaminated with insects, molds, parasites, or bacteria that cause spoilage or lead to serious illness. *Escherichia coli* O157:H7, found in raw beef, poultry, and other foods, causes hemorrhagic diarrhea and has been responsible for several serious outbreaks and some fatalities. *Listeria monocytogenes, Campylobacter jejuni*, various strains of *Salmonella*, and *Staphylococcus aureus* are pathogenic, leading to illnesses characterized by fever, vomiting, diarrhea, and severe discomfort. While refrigeration generally can control the growth of contaminants, it is not effective for *L. monocytogenes*. In the United States it has been estimated that food-borne contaminants cause about 33 million cases of illness each year and as many as 9000 deaths result. There is an associated economic cost estimated at $6 billion due to hospitalization, lost time, and decreased productivity. There is a clear need to make food safe for the consumer.

Treatment effectiveness. The interaction of ionizing radiation with the vital components within the contaminants effectively causes their inactivation. Minor changes in bacterial deoxyribonucleic acid (DNA) lead to an inability to replicate and ultimately to destruction. The susceptibility of bacteria to inactivation, defined in terms of the dose required to reduce the population by 90%, varies by species and also depends upon the ability to form spores. Values for the susceptibilities of pathogens found in beef are listed in **Table 1**. Accordingly, if a product is treated to a dose of 1.8 kilograys, there would be six such reductions of *E. coli* O157:H7 and the original population would be reduced by over 99.9999%. The reduction in *Campylobacter jejuni* would be at least 99.9999999%. The spore-forming *Clostridium botulinum* is considerably less susceptible to inactivation. The effect of irradiation on insects is especially significant, eliminating the adult insects and their eggs and pupae. At quite low doses, irradiation can make the adult male sexually sterile—the basis of the sterile insect release technique for controlling insects in the environment.

Irradiation processing. Foods are irradiated using either radioactive sources emitting gamma rays or machine sources producing high-energy electrons or x-rays. Approved sources include cobalt-60, cesium-137, electrons with energies less than 10 MeV, and x-rays with less than 5 MeV. Gamma rays and x-rays are highly penetrating, but the charged electrons have a low penetration power of about 1 cm (0.4 in.) in water.

Packaged foods are exposed to these radiations as they are moved by conveyor through a shielded facility. By carefully controlling the speed of the conveyor or the residence time in the facility, the processor can control the total radiation dose absorbed, defined in terms of the gray (Gy) rather than the former rad [100 rad = 1 Gy and 1000 Gy = 1 kilogray (kGy)]. There are standardized procedures for measuring the absorbed dose. Very little energy is actually absorbed as the radiation passes through the food. There is only a 0.24°C (0.43°F) increase in temperature for a dose of 1 kGy, which justifies describing irradiation as a cold process.

The same techniques used in irradiating foods have been used for many years in the irradiation curing of paints and in irradiation sterilization of medical supplies, including bandages, sutures, surgical instruments, and needles. In contrast to sterilization processes involving autoclaving or ethylene oxide, the irradiation process is highly uniform, highly repeatable, and easy to control.

Safety. Extensive research into the physical, microbiological, chemical, nutritional, and toxicological aspects of irradiated foods has produced a wealth of evidence that such foods are "wholesome." This term is used by international experts to mean that foods handled according to good manufacturing practices and irradiated in accordance with standardized procedures are safe to consume and nutritionally adequate.

Theoretically, foods treated with the approved sources cannot become radioactive. Since the gamma rays from cobalt-60 and cesium-137 are not energetic enough to produce radioactivity and since these sources never come in contact with the food, no activity above natural background levels can be detected. The energies allowed for the electrons and x-rays are also lower than would be required to produce radioactivity in the food, so no detectable activity above background is found in treated foods.

The microbiological data demonstrate that no new radiation-resistant strains are produced and that, under proper conditions, surviving microorganisms do not pose any threat. Generally speaking, a mutant with high radiation resistance, if produced, would have to compete against great odds to grow. Moreover, microorganisms, such as spore-forming bacteria that are not destroyed in foods treated to low, pasteurizing doses, are unlikely to compete under the temperature conditions of refrigerated storage. Generally, the low temperature and the presence of oxygen would prevent the outgrowth of all microorganisms except type E *Clostridium botulinum*.

The interaction of ionizing radiation with the chemical constituents in the food could cause some minimal but predictable changes. Major constituents of foods include proteins, fats, and carbohydrates; minor constituents include vitamins and any approved additives. Changes in the major constituents lead to the formation of other larger or smaller molecular weight substances, similar to what is seen in the canning or cooking of foods. Changes in fats lead

TABLE 1. Radiation dose required to reduce the population of food-borne pathogens by 90% in beef at 5°C (41°F)*

Pathogen	Dose, kGy
Bacillus cereus endospore	2.45 ±0.31
Campylobacter jejuni	0.16 –0.20
Clostridium botulinum endospore	3.43 at −30°C
Escherichia coli O157:H7	0.30 ±0.02
Listeria monocytogenes	0.45 ±0.03
Salmonella species[†]	0.70 ±0.40
Staphylococcus aureus	0.48 ±0.02

*Adapted with permission from Council for Agricultural Science and Technology, *Radiation Pasteurization of Food*, CAST Issue Pap. 7, 1996.

[†] *Salmonella dublin, S. enteritidis, S. newport, S. senftenberg*, and *S. typhimurium*.

to the formation of free fatty acids and, to a lesser extent, of volatile hydrocarbons, such as pentane. The amount formed depends on the nature of the food, the presence or absence of oxygen, whether the food is frozen, and the total absorbed dose. Indeed, highly sophisticated analytical techniques are needed to determine if a food has been irradiated.

Because the effects of irradiation on the constituents are minimal, the nutritional data demonstrate that irradiated foods adequately retain all their essential nutrients. Proteins and amino acids, essential fatty acids and oils, and energy-producing carbohydrates are not compromised. Some vitamins, though they are minor constituents, do have a discernible radiation susceptibility. Vitamin B_1, thiamin, is an example. The amount lost in pork samples treated to pasteurizing doses is considered tolerable. Moreover, the sensitivity of vitamin B_1 to heat is considerable: more thiamin is lost in a pork product sterilized by heat than one sterilized by irradiation.

The supporting data on the nutritional adequacy of irradiated foods come from animal feeding studies. In well-designed studies with properly irradiated foods, all of the indices of growth, longevity, and specific nutrient utilization were found to be the same for the irradiated and control groups.

Toxicological data demonstrate the absence of harmful effects due to irradiation of foods. No carcinogens, mutagens, or teratogens have been found in carefully designed studies with appropriately irradiated foods. Multigenerational studies using rats, mice, beagle dogs, and other animals have shown no sign of toxicity. Shorter-duration studies with humans in the United States and in China have shown no evidence of deleterious consequences of consuming many different kinds of irradiated foods. Studies using cell cultures or mutation-susceptible *Drosophila* flies were negative for genotoxicity. In some cases in which adverse effects were reported, subsequent attempts to reproduce these effects failed. While a compound with a toxic potential might possibly be formed, scientists at the U.S. Food and Drug Administration (FDA) and international experts believe that the level of the compound in the food would be too small to be of any toxicological significance.

Applications. The food irradiation process has been approved in approximately 40 countries. Approvals in the United States are listed in **Table 2**. Recent changes and additions include the generic approval for tropical fruits and for meat and meat products, both chilled and frozen. Approval for fresh eggs is expected soon.

Fruit and vegetables. Irradiation can be used for fruits and vegetables to prevent overripening or sprouting as well as to eliminate infesting insects and microorganisms. In Japan, which bans chemical sprout inhibitors, potatoes are irradiated on a commercial scale to ensure availability throughout the year. In the United States, fruits from Hawaii can now be marketed in other states provided that they are irradiated to disinfest them of fruit flies, including the Medfly. An especially successful application is the irradiation of strawberries to prevent mold formation and to extend the refrigerated storage time to about 3 weeks. Mushrooms have also been irradiated.

Wheat and flour. Irradiation can be used to ensure that grain products are free of adult insects and their eggs and pupae. Although the treatment is not used in the United States, it has been used in Odessa, Ukraine, to disinfest imported wheat. Two machines producing electrons were installed to treat as much as 400,000 tons per year.

Poultry and meat pasteurization. Technically, irradiation is ideal for ensuring that poultry and meat products reach consumers free of contaminating pathogens. Although steps are taken in the handling and processing of the foods to minimize contamination, those pathogens that unavoidably are introduced are destroyed by irradiation to low doses.

Chicken is being irradiated within the approved dose range and sold in selected markets. Consumers

TABLE 2. United States approvals for irradiated foods*

Product	Agency	Date	Dose, kGy	Purpose
Wheat, wheat flour	FDA	1963	0.2–0.5	Insect disinfestation
White potatoes	FDA	1964	0.05–0.15	Sprout inhibition
Spice and vegetable seasonings	FDA	1983	max. 10	Microbial decontamination
Pork	FDA	1986	0.3–1.0	Trichina inactivation
Fruits and vegetables	FDA	1986	max. 1.0	Insect disinfestation; maturation delay
Papaya, carambola, litchi	USDA	1997	0.15–0.25	Insect disinfestation
Herbs, spices, and dry vegetable seasonings	FDA	1986	max. 30	Insect disinfestation; microbial decontamination
Dehydrated enzymes	FDA	1986	max. 10	Microbial decontamination
Animal and pet foods	FDA	1986	max. 25	Microbial decontamination
Poultry	FDA	1990	max. 3.0	Microbial decontamination
	USDA	1992	1.5–3.0	Microbial decontamination
Red meats, nonfrozen	FDA	1997	max. 4.5	Microbial decontamination
Red meats, frozen	FDA	1997	max. 7.0	Microbial decontamination
Whole shell eggs	FDA	Pending	0.7–1.7	Microbial decontamination

*Adapted with permission from Council for Agricultural Science and Technology, *Radiation Pasteurization of Food*, CAST Issue Pap. 7, 1996.

have accepted it, especially when provided with information on the treatment and its benefits.

Meat and meat products can be irradiated. Actual sales of irradiated meats will have to await the publication of a U.S. Department of Agriculture (USDA) final implementation rule. Such products have been tested in controlled studies to determine the packaging and conditions of treatment needed to assure acceptance by consumers. Red meats, because of the pigment myoglobin and the unsaturated fats, could be slightly affected in an adverse way if not optimally treated. The potential for meeting the demands of the public for safety, particularly in ground beef, is considerable.

Shelf-stable, prepared foods. Irradiation can be used advantageously to produce ready-to-eat meals or meal components that are completely free of spore-forming bacteria and, consequently, safe to store at room temperature. Though not yet approved for general use in the United States, the meals and their components are permitted by the FDA for use by astronauts. They are approved by South Africa for use by the military and for sale to outdoor enthusiasts, campers, and sailboat racers. They are also approved for use in many countries as sterile meals for hospital patients whose immune functions have been impaired or suppressed.

The technology for producing radiation-sterilized products was developed in the 1960s by the U.S. Army. The overall process focuses on eliminating three causative agents of food deterioration: enzymes, oxygen, and bacteria. The enzymes that would cause the food to break down during storage are inactivated by precooking. The oxygen is removed by vacuum-sealing the precooked food in a container, usually a flexible pouch with an aluminum middle layer. The pathogenic, spore-forming bacterium *C. botulinum* is destroyed by irradiating the food in the container while frozen at $-30 \pm 10°C$ ($-22 \pm 18°F$) to a dose corresponding to 12 times the 90% reduction dose. For cured meats, this dose is 25 kGy; for noncured meats, 44 kGy. Since the food is irradiated while frozen, the quality attributes and vitamins are only slightly affected. Once processed, the product is allowed to thaw and then can be stored for an extended period. Samples of irradiated ham prepared for the 1977 Apollo-Soyuz space flight are still intact.

Spices. The most extensive use of food irradiation is to disinfest and sterilize spices, herbs, and dry vegetable seasonings. These food ingredients are known to be contaminated with insects and high levels of fungi and bacteria. Currently used fumigants, such as methylbromide and ethylene oxide, are toxic and environmentally problematic. Irradiation is safe and efficient, and so the FDA allows food processors to use doses of up to 30 kGy. Whether used directly by the homemaker or put into products as an ingredient, these irradiated items decrease the potential for cross-contaminating other foods.

Future prospects. Irradiated foods, because of their high safety assurance, their extended shelf-lives, and their equivalent or superior quality attributes relative to other preservation technologies, will eventually make up a large proportion of the foods people consume. Consumers will have the option to choose them from among foods preserved in other ways. Foods must have a label indicating the irradiation treatment, but the label may also indicate the beneficial effects associated with the treatment.

For background information *see* FOOD IRRADIATION; FOOD MICROBIOLOGY; FOOD PRESERVATION; FOOD SCIENCE in the McGraw-Hill Encyclopedia of Science & Technology.

Edward S. Josephson; Irwin A. Taub

Bibliography. Council for Agricultural Science and Technology, *Radiation Pasteurization of Food*, CAST Issue Pap. 7, 1996; J. F. Diehl, *Safety of Irradiated Foods*, 2d ed., Marcel Dekker, 1994; J. F. Diehl and E. S. Josephson, Assessment of wholesomeness of irradiated food (a review), *Acta Alimentaria*, 23(2):195–214, 1994; International Atomic Energy Agency, *Food Irrad. Newsl.*, 20(2), Suppl. A and B: Clearance of Item by Country, and Clearance of Item by Name, 1996.

Up-to-date information on food irradiation and its approvals and applications can be accessed at http://www.iaea.org/icgfi/: International Consultative Group on Food Irradiation.

Forest certification

Until recently consumers of paper, lumber, and other wood products have had little influence over the type of management practices applied in growing and harvesting the wood. Sustainable forest management, however, has been an enduring, high-priority goal of environmental organizations, governments, foresters, and other concerned parties over the past several decades. Worldwide, efforts to achieve sustainable forest management have been primarily guided through regulatory approaches and, more specifically, through established criteria and indicators developed through extensive negotiations. With the failure of regulatory-based mechanisms to adequately address the increasingly severe problems of deforestation and unsustainable forest management practices, voluntary market-based initiatives have been developed.

In recent years, certification in forestry and in the wood products industries has emerged as the most promising market-based initiative in the promotion of sustainable forest management. On the demand side, certification can be a valuable and informative tool for the conscientious consumer or retailer of wood products. On the supply side, certification can be valuable for gaining access to markets where consumers are environmentally sensitive, achieving a stronger position within an established market through product differentiation, and improving forest management systems and practices, based on voluntary participation.

The two dominant approaches are system-based certification (ISO 14000 environmental management system series) and performance-based certification (Forest Stewardship Council).

Assessments. Broadly, certification can be defined as an evaluation process which results in a written statement (a certificate) attesting to the origin of wood raw material, and its status or qualifications following validation by a qualified auditor. The assessment and any resulting certificate may be applied to a forest area, a management system, or an end-use product. Performance-based assessments generally result in the certification of a forest area or end-use product, while systems-based (also known as process-based) assessments result in the certification of the management system.

The assessment may be classified as one of three types: first-party assessment, conducted internally by the organization managing the forests; second-party assessment, conducted by a customer of the supplier; or third-party assessment, conducted by an independent institution. First-party audits are largely limited to assessments for internal purposes, while third-party audits are primarily used to communicate the performance quality or quality of management practices (environmental, social, and otherwise) to consumers and other external audiences. Independent third-party audits generally are more costly for the company or forest manager, but they provide the benefit of greater credibility in the eyes of customers.

Regardless of significant differences between systems-based approaches (ISO 14000) and performance-based approaches (Forest Stewardship Council), many basic features are shared. Among the fundamental elements of any certification initiative, there must be a standard setting body and a set of basic principles and criteria for management and performance (**Fig. 1**). Additionally, there must be an accrediting body (often the same as the standard setting body) responsible for the certifying bodies. The certifying bodies, in turn, are responsible for conducting assessments of forest areas, management systems, or chains of custody, based on national or regional standards, and granting or refusing certificates.

ISO 14000. The systems-based approach, which is predominantly associated with ISO 14000, evaluates the environmental management system of a given organization. It does not evaluate the actual performance within the management system. The ISO system does not include any chain-of-custody elements or product labeling elements. The International Standards Organization (ISO) itself acts as the standard setting body and the accrediting body.

ISO is a globally recognized federation of national standards bodies that is best known for the ISO 9000 series of quality standards for internationally traded goods and services. The ISO 14000 series was developed in the aftermath of the 1992 United Nations Conference on Environment and Development in order to provide a uniform set of standards for all types of industries and companies; there are currently no ISO standards specifically designed for forestry. ISO standards, like those of the Forest Stewardship Council, are global in nature and require national or local interpretation to fit within a given set of national or local laws and regulations. The ISO 14000 series, launched in 1996, is still in development with only ISO 14001 and 14004 published in final form while various other standards are working documents (**Fig. 2**).

The underlying theme of any environmental management system is the promotion of continuous improvement in management system and practices. The basic idea is for companies to set incremental goals for improvement, to work toward and meet these targets, and to subsequently set more ambitious

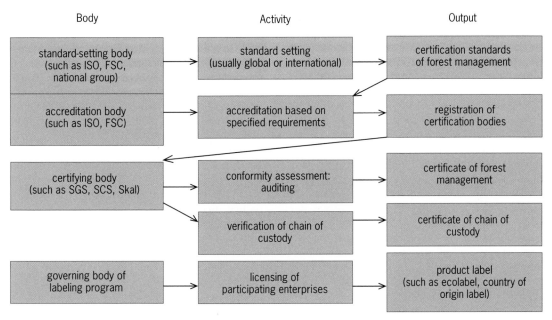

Fig. 1. Elements of certification and labeling initiatives.

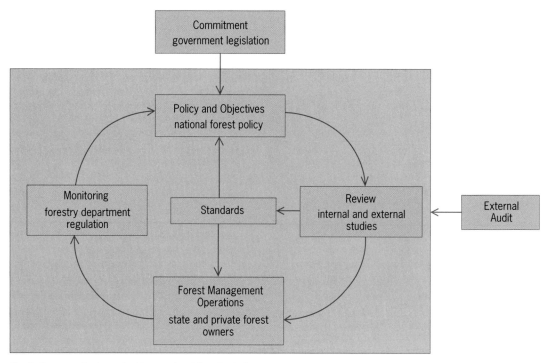

Fig. 2. Framework of a typical International Standards Organization environmental management system.

targets. A typical certification process involves an initial review of a company in which its key environmental impacts and pertinent regulations are identified. The certifying body then conducts an on-site audit, including a review of the company's policy and procedures manuals and its environmental protection objective. No forest area surveys are conducted. After successful certification, followup surveillance visits are conducted periodically. This same auditing process can also be conducted by the company itself for internal management purposes. Third-party certification is voluntary and is useful for external credibility. However, the certification is primarily useful in business-to-business communication; it is less useful in business-to-consumer communication since the end product has no label or certificate.

Performance-based approach. The performance-based approach requires that forests be managed in accordance with a set of defined principles and criteria. It is the forest area itself that becomes certified, not the forest manager or the system of management. Management performance is evaluated according to environmental, economic, and social criteria.

The Forest Stewardship Council, formally established in 1994, is currently the only internationally recognized program of independent third-party, performance-based assessments that results in an end-use product label of certification. It has established a global set of 10 principles and criteria for forest stewardship and uses this set as a basis upon which it accredits certifiers (entities that actually conduct the audits and assessments).

In order to be certified, a company must meet the following criteria:

1. Meet all applicable laws
2. Have legally established rights to harvest
3. Respect indigenous rights
4. Maintain community well-being
5. Conserve economic resources
6. Protect biological diversity
7. Have a written management plan
8. Engage in regular monitoring
9. Maintain high-conservation-value forests
10. Manage plantations to alleviate pressures on natural forests

To date, there are only six Council-accredited certifying organizations. Only Sweden has a national set of standards endorsed by the Council that will be used for all certifications in the country. Elsewhere, certifying bodies, such as the SGS of the United Kingdom or the SCS of the United States, have developed their own specific standards according to national and local conditions and in adherence to the Council's global principles and criteria. More than 15 countries currently have Council working groups that are in the process of developing national standards. Although governments may provide an enabling regulatory environment and other incentives for the generation of such standards, only independent entities representing a mixture of social, environmental, and economic interests can fulfill this function.

The approach of the Forest Stewardship Council offers two forms of certification: forest management certification and chain-of-custody certification. Forest management certification involves an on-the-ground assessment of a landowner's forestry practices in accordance with regional and national standards. Chain-of-custody certification involves an audit of each step of the manufacturing process, from harvest to retail store, and results in the Council label for products.

Such labels for forest products—predominantly a Forest Stewardship Council output—can be achieved only by assessing the entire chain of custody (for example, from forest to mill to manufacturer to retailer). Although monitoring the chain of custody can be difficult, it is the only method through which the average consumer can be assured that the product has been produced according to the specifications.

Summary. The certification approaches described above date back only to the early to mid 1990s, yet both have grown rapidly throughout many regions of the world. Regardless of which approach is best for which parties, voluntary certification of forest management is a novel and authentic approach in the efforts to reduce environmental destruction and to promote forestry management practices that are environmentally, socially, and economically sustainable. As both types of approaches mature, the prospects for sustainable forest management are becoming more optimistic.

For background information *see* FOREST AND FORESTRY; FOREST MANAGEMENT; FOREST RESOURCES; WOOD PRODUCTS in the McGraw-Hill Encyclopedia of Science & Technology.   Bruce Cabarle

Bibliography. B. H. Ghazali and M. Simula, *Timber Certification: Progress and Issues*, Report for the International Tropical Timber Organization, Yokohama, 1997; C. Upton and S. Bass, *The Forest Certification Handbook*, Earthscan Publications, London, 1995; V. M. Viana et al., *Certification of Forest Products: Issues and Perspectives*, Island Press, Washington, D.C., 1996.

Forest ecosystem

Trees evolved approximately 100 million years after the supposed origin of land plants and 50 million years after the rise of vascular plants. Better access to light and more effective spore dispersal are the selective advantages often invoked for their evolution. Medium-sized trees appeared in the Late Givetian, about 376 million years ago, at a time of the Devonian when plants experienced maximal diversity and a high rate of new genera formation. This innovation predated the appearance of the seed habit. The earliest trees were essentially free-sporing with a pteridophytic (seedless) life cycle that was largely restricted to wet habitats. Earliest forests had a worldwide distribution and were dominated by the genus *Archaeopteris*, of which the closest relatives today are the gymnosperms. Extinction of *Archaeopteris*, which coincides with the diversification of early seed plants in the latest Devonian and Early Carboniferous, marks the end of these earliest forests which

may have played a major role in changes that affected the biosphere at that time.

Earliest trees. Included in current definitions of trees are the self-supporting habit (upright structure) and a maximum height of 4 m (13 ft) or more. Direct evidence for high stature is generally lacking in the fossil record, and evidence for this structure is usually assessed from the diameter of the available stem fragments. Reconstructions of the earliest trees rely on the connection, but most often on the association, of isolated plant parts. They are based on two sets of evidence: fossil impressions that account mostly for the external morphology of the tree parts, and anatomically preserved specimens that provide information on the cellular structure of individual organs. The tree structure arose independently in four systematic groups of Devonian plants.

Trees of the extinct group Cladoxylopsida (for example, *Pseudosporochnus*) were especially prosperous in the Givetian (Middle Devonian). They are reconstructed as small individuals comprising a swollen base covered with tuft roots and a stout trunk 2–4 m (7–13 ft) high bearing a crown of acutely inserted, three-dimensional branches. The lower branches characteristically divide in a fingerlike (digitate) manner, then dichotomously (into two parts). Cladoxylopsid trees were devoid of leaflike flat organs. The smallest terminal branches, instead, consisted of little, irregularly arranged, three-dimensional branched systems. Structurally, cladoxylopsid axes comprise a dissected vascular system of interconnecting strands which probably provided both support and conduction.

Plants of tree size affiliated to the Lycopsida also first appeared in the Givetian. They were diverse and some had a worldwide distribution (for example, *Lepidosigillaria, Cyclostigma*). Trunks typically grew unbranched for an extended period of time and produced a crown of erect, dichotomizing branches at a height that rarely exceeded 10 m (33 ft). Leaf-type organs consisted of microphylls (small leaves) that were undivided, densely inserted on all axes, moderate in size, and supplied by a single vessel. The central structure of these trees consisted of a xylem strand of limited width devoted to conduction of fluid, and a wide cortical zone with external fiber cells that may have provided support. The base of *Cyclostigma*, one of the best-known genera of the latest Devonian, is described as a four-lobed root-bearing structure.

The natural group that includes the extinct progymnosperms and all living gymnosperms and angiosperms evolved the most successful trees. Its major innovation is the evolution of a bifacial cambium layer that produced secondary xylem (wood) internally and secondary phloem externally. *Archaeopteris*, the major arborescent genus of progymnosperms that evolved in the Devonian, is reported worldwide and had an extensive paleolatitudinal distribution, from the northern to the equatorial. Its earliest representatives are also Late Givetian and include the tallest trees of the Devonian. The most abundant remains of *Archaeopteris* are represented by frondlike lateral branch systems bearing multiveined, laminate leaves, and structurally preserved portions of large axes that may have lacked leaves (for example, that bore only branches). The latter typically possess an extensive cylinder of secondary xylem that may have provided both support and water conduction, a strategy comparable to that of existing conifers. Popular reconstructions show a tree that spreads out into lateral branch systems which are regularly shed during growth. The branches are almost horizontal and arranged in a spiral fashion (helically) on a trunk that may have exceeded 30 m (98 ft) in height. *Archaeopteris* is the earliest plant that produced adventitious (abnormally positioned) branches, an innovation discovered recently, which may have extended its survival in case of damage to the trunk apex. *Archaeopteris* was also the first among early trees to develop an extensive root system which comprised both superficial axes and roots penetrating to an estimated depth of 1 m (3 ft).

Trees of the group Sphenopsida arose later in the Devonian. They were little diversified (*Pseudobornia*, *Archaeocalamites*) and distributed mainly in northern paleocontinents. The largest, *Pseudobornia*, might have exceeded 15 m (49 ft) in height, and its trunk produced three orders of branches at distinctive nodes. Leaves were multisegmented and borne in whorls on terminal branches. These plants are suspected to have had hollow stems, a strategy favorable to fast-growing individuals inhabiting swampy habitats. Underground parts are unknown.

Structure and occurrence. The few paleoecological studies undertaken on Middle and Late Devonian plant assemblages are mostly based on deposits from the northeastern part of the United States. Evidence for forest communities is often indirect and relies principally on the abundance and limited transportation of isolated tree parts in fluvial sediments. Such evidence gives limited yet significant information on the occurrence and original composition of these ecosystems.

The earliest assemblage suggestive of a forest consists of massive deposits of erect stumps from the Late Givetian (late Middle Devonian) locality of Gilboa, New York. These fossils, referred to as *Eospermatopteris*, have been fossilized on their growth site. Their affinities with the Cladoxylopsida, the Lycopsida, or the arborescent gymnosperms are controversial. From the same locality are reported impressions of sterile branches possibly referrable to the progymnosperms, and claimed by some to be part of the same plant. Others favor the possibility that these remains belonged to a different, smaller, shrubby plant related to the progymnosperm genus *Aneurophyton* or to the prefern *Ibyka*. Other remains found at Gilboa include a variety of lycopsids that encompass both arborescent and ground-cover forms.

The Frasnian sequence of plant assemblages from New York demonstrates a modification of the early Late Devonian landscapes, from shrub-dominated communities comprising a diverse flora of aneurophytalean progymnosperms, cladoxylopsids, and related forms in the Early Frasnian, to extensive shaded forests in the Middle and Late Frasnian. These forests formed low-diversity communities composed of an arborescent strata dominated by *Archaeopteris*, where arborescent lycopsids were less common, and of an eventual ground cover of herbaceous lycopsids. They grew from coastal habitats to uplands, along streamsides and on wet floodplains. Changes that followed the biotic crisis of the Frasnian–Famennian transition showed their decline. *Archaeopteris* communities became restricted to floodplains, where they occasionally cohabited with shrubby fernlike *Rhacophyton*. They were excluded from the earliest peat-forming swamps, which were occupied by *Rhacophyton* and surrounded by communities of treelike lycopsids. Subsequent changes that started at the Devonian–Carboniferous boundary involved the extinction of *Archaeopteris* and *Rhacophyton*, increasing colonization of drier habitats by seed plants, and domination of wet environments by medium-sized treelike lycopsids.

Impact. The period of approximately 15 million years that spanned the Late Givetian to latest Devonian was characterized by major changes that affected the Earth on a global scale. The transformation of atmospheric composition involved a steady increase in oxygen and a dramatic depletion in carbon dioxide. The Earth also experienced a global cooling that resulted in a brief Late Devonian glaciation. In the marine realm, the Frasnian–Famennian transition was marked by rapid sea-level fluctuations, extended bottomwater anoxia (lack of oxygen), and the extinction of 70–80% of species of the tropical fauna.

Models have been provided that link this evolution to two important features related to the evolution of the tree structure and extension of the earliest forests: (1) the increase in global terrestrial biomass and accumulation of large amount of litter, and (2) the spread of deep-rooted vascular plants. They predict that the large available quantity of litter is responsible for a significant increase in marine primary productivity through food webs involving organisms from riparian and coastal habitats. Development of extensive root systems greatly enhances rock weathering and soil formation. This created conditions for eutrophication (nutrient enrichment stimulating algal blooms) in epicontinental seas and development of the anoxic horizons linked to massive extinctions of marine organisms. This process resulted in removal of atmospheric carbon dioxide and further global cooling.

Charcoal remains of *Cyclostigma* and *Archaeopteris* around the Devonian and Carboniferous boundary provide evidence that, for the first time in the history of Earth, conditions were amenable for the ignition of extensive fires. These conditions include an appropriate level of atmospheric oxygen and sufficient fuel, corresponding to the large amount of plant debris accumulated from the new forested ecosystems. Formation of inert charcoal that resists weathering and biodegradation may have also contributed to the trapping of terrestrial organic matter and the resulting global decrease of atmospheric carbon dioxide. The double role of fires, in providing open spaces for colonization and acting as a selective force, may have further resulted in the evolution of new biological forms and the emergence of new ecosystems such as the fern and pteridosperm-dominated drier landscapes of the Early Carboniferous.

For background information *see* PALEOBOTANY; PLANT EVOLUTION; TREE in the McGraw-Hill Encyclopedia of Science & Technology.

Brigitte Meyer-Berthaud

Bibliography. T. J. Algeo and S. E. Scheckler, Terrestrial-marine teleconnections in the Devonian: Links between the evolution of land plants, weathering processes, and marine anoxic events, *Phil. Trans. Roy. Soc. London B*, 353:113–130, 1998; A. K. Behrensmeyer et al. (eds.), *Terrestrial Ecosystems Through Time*, University of Chicago Press, 1992; S. E. Scheckler, Geology, floristics and paleoecology of Late Devonian coal swamps from Appalachian Laurentia (U.S.A.), *Ann. Soc. Geol. Belg.*, 109:209–222, 1986; T. N. Taylor and E. L. Taylor, *The Biology and Evolution of Fossil Plants*, Prentice Hall, 1993.

Forest genetics

Cloning, which occurs naturally in many tree species, is assuming a rapidly expanding role in commercial forestry. Cloning is the production of genetically identical copies (ramets) from an original, selected individual (ortet). Direct evidence of humans cloning woody plants dates to circa 1000 BCE. Indeed, such cloning probably occurred much earlier when some prehistoric person noticed that fresh branches stuck in the ground would sometimes take root and produce a new stem. Cloning has long been the standard practice for producing many tree food crops, such as apples and bananas. It is also a main propagation technique in the woody ornamental field, to faithfully reproduce tree shapes, foliage color, lack of seed production, and other unique traits. Cloning specifically for producing improved plantations dates back at least to the 1800s with sugi (*Cryptomeria japonica*) in Japan and hybrid poplars (*Populus* spp.) in Europe. With the rapid growth of genetic improvement efforts in the 1950s, cloning became important as the basis of most seed orchards (plantations of trees established for the purpose of producing genetically improved seed). Until recently, the establishment of large plantations of cloned trees was limited to a few species that could be rooted inexpensively from stem cuttings. Now, all that is changing due to tissue culture and other biotechnology methods. The ability to clone trees is one of the most important tools that will be used in constructing the highly productive, very specialized tree plantations the world will need to meet its future requirements for wood products, renewable energy, and some biological chemicals. However, indiscriminate cloning also carries significant risks for the long-term stability of plantations.

Natural cloning. The natural ability to clonally reproduce trees is an important characteristic of many species. It appears to have evolved under the selection pressures of fire, flooding, and other environmental disturbances that may kill or damage the aboveground portion of a tree. The ability to sprout after major disturbances provides a self-replacement strategy. Sprouting from roots or pieces of broken stem to produce multiple offspring provides a colonization strategy that is conveniently timed with the opening up of sites. Natural cloning also provides a way to deal with the fitness versus flexibility compromise that selection forces on all organisms. Because most tree species grow under a wide variety of conditions over space and time, they need to generate a large pool of genetically diverse offspring to have the flexibility to deal with environmental variation. The diversity in the genetic makeup of seedling populations means many individual trees will not have optimal fitness for the particular set of environments being colonized. Many of the less fit genotypes will die out as stands of trees develop to maturity. Tree species that can propagate by both sexual and vegetative means can colonize new environments with a diverse seedling population, while replicating fit genotypes for stand replacement or for spread into similar environments. The Yellowstone National Park fires of 1988 have provided a large-scale demonstration of this phenomenon as aspen (*P. tremuloides*) seedlings colonize newly available soils and root sprouts regenerate and expand aspen stands where the aboveground portions were killed. In the Rocky Mountains, single aspen clones cover 30 or more acres (12+ hectares).

Almost all angiosperm tree species will sprout from the lower stem, at the root collar where the stem transitions into the roots, or from adventitious buds that develop in wound tissue. This ability declines with the age or diameter of the individual tree in many species. A restricted set of species can send up sprouts from their root systems. The best temperate zone examples are the aspens (*Populus* section Populus) in forests and the tree-of-heaven (*Ailanthus altissima*) in cities. In contrast, natural cloning is relatively rare in the gymnosperms. The coast redwood (*Sequoia sempervirens*) is one of the few that can stump-sprout. A few other species that grow in moist environments can propagate by layering, which involves the formation of roots on a branch that is in contact with the ground and the subsequent growth of the branch tip into a new stem. The rarest form of natural cloning is found in a few species such as alders (*Alnus* spp.), where seed embryos can be produced from maternal cells that have not undergone meiosis (apogamy).

Advantages of cloning. Sexually produced offspring inherit only the average genetic qualities of the two or more parents involved in producing a family (narrow-sense heritability). Clonal offspring inherit all the genetic variability of the original selections (broad-sense heritability). The difference is due to the reassortment of chromosomes and genes during meiosis and the new gene combinations from the union of sexual cells (gametes) from the parents. Furthermore, sexual offspring of most trees are highly variable due to the many gene combinations that are possible. Clones are quite uniform, with only environmental variation and a rare mutation causing differences between ramets. The most efficient way to genetically improve trees is to combine the controlled breeding of highly selected parents with the clonal magnification of the very best offspring from each cross. Even if a cross is difficult to make, it takes only one offspring to be multiplied into plantations. Inbreeding, or the crossing of related individuals, usually is not recommended in tree improvement because it produces many poor genotypes. They have the same deleterious message (allele) on both members of some chromosome pairs (homozygous). The relatively rare, very good individuals that have not inherited homozygous deleterious alleles can be placed into commercial service through cloning. As biotechnology provides new tools for inserting desired genes into trees, it is much more efficient to use clonally propagated lines where all trees will carry the new gene inserted in an ortet that is preselected for other important traits. Cloning also offers an efficient way to conserve genotypes, potentially forever, that would otherwise be lost with the death of the individual tree and then reassorted and recombined in its sexual offspring.

Commercial advantages are centered on the relative uniformity of clonal forests. There can be very close matching of clones with particular soil and cultural environments that optimize overall performance. Plantation spacings do not have to allow for the fact that many trees will not have high fitness to a particular set of conditions as would be the case in seedling plantations. As with food crops, there are substantial commercial advantages to having other tree crops where all the trees reach harvestable potential at essentially the same time and with the same anatomical and chemical characteristics. If a harvested plantation is regenerated by stump or root sprouts, there is no additional investment needed in expensive, genetically improved stock. The plantation will grow even faster than it did the first time because the sprouts inherit mature-size root systems.

Artificial cloning techniques. Some cloning techniques are just mimics of natural processes, while newer techniques require more manipulations of tree anatomy and biochemistry. Some tree species, such as the poplars and willows, are very easy to propagate by stem cuttings (stecklings). A few of the gymnosperms, such as spruces (*Picea* spp.) and Monterey pine (*Pinus radiata*), also can be rooted from cuttings. Multiple one-year-old shoots are harvested during the dormant season from stool beds or hedges that are specifically maintained for this purpose. The cuttings are then planted in nursery beds

or directly in plantations where roots form along the lower portion of the stem and a new top is produced from an upper bud. Treatment of the stem bases with rooting hormones, such as indole butyric acid, is needed in some cases. European poplar culture is based on a special form of stem cuttings. Stem cuttings are allowed to grow for 1–2 years in a wide nursery spacing, then the resulting 3–5-m-tall (10–16-ft) stems (barbatels) are cut at the ground line with little or no roots attached and replanted in 1-m-deep (3-ft) holes. The result is a near-instant forest where the trees already dominate over their surroundings.

In those species that root-sprout in nature, it is possible to develop plantations by planting root segments from which new shoots arise (rootlings). With enough study, all trees can be cloned by grafting branches (scions) or even single buds from one selected tree onto the stem or root collar of other trees. This technique has been used extensively in establishing seed orchards, germplasm banks, and high-valued food or ornamental tree crops, but it has not been generally considered an economically feasible approach for most forestry applications.

The production of apomitic (asexual) seed from in vivo plants has not been used, but substantial effort has been invested in developing in vitro techniques for many tree species. Many angiosperm trees can be propagated in tissue culture from bud, internode, root, or leaf segments (plantlets). Following extensive research, it became possible to propagate gymnosperms through tissue culture, first from immature stages of seed development and then, with continuing research, from older and older tissues. The first approach was quickly adopted to multiply scarce seed supplies from tree improvement programs. The full-scale application of clonal forestry in any species is dependent on being able to use older ortets that have already been screened for their genetic values. Whenever possible, tissue culture techniques have been sought that will induce single or small groups of somatic cells (emblings) to develop along the same pathways as sexually produced embryos. This is the most efficient way to produce trees that have been genetically transformed with selected gene constructs at the single-cell stage, and it also is leading to the development of clonal, artificial seed.

Examples of clonal forestry. The most widespread use of clonal plantings has been with the willow family (Salicaceae), involving willow (*Salix*) and poplar (*Populus*) species and hybrids. Paper companies in Ontario, British Columbia, Oregon, Washington, Mississippi, Kentucky, Minnesota, and elsewhere have begun planting clonal forests of poplars to supply special fiber qualities for their products and to replace wood resources that have been taken out of production to protect old growth and other values of natural forests. In the process, they can produce wood at least five times faster than it is grown in natural forests (**Fig. 1**). As paper companies develop techniques for cloning gymnosperms, they are able to focus almost all their plantation forestry on clones (**Fig. 2**). Clonal plantations of sugi have been grown

Fig. 1. Harvest of 8-year-old clonal plantation of poplars for paper manufacture.

in Japan since the 1400s to provide an alternative to the use of natural forests. Modern techniques would allow the same thing with coast redwood which comes from the same tree family (Taxodiaceae); this tree is relatively easy to clone experimentally.

The clonal production of *Eucalyptus* species and hybrids in South America may revolutionize markets for paper raw materials. Very ambitious clonal planting programs are in progress in China to support that country's growing resource and environmental conservation needs. In the Po River valley and many other parts of Europe, barbatel poplar plantations are extensively planted to support the areas' solid wood industries. Sweden has led the efforts, now developing in many other countries, to grow willow or other species in clonal plantations to produce biomass for energy. Most energy plantations are

Fig. 2. Large-scale clonal planting of loblolly pine (*Pinus taeda*).

now aimed at producing solid fuels for use in electric power plants. The research groundwork is being done to grow thousands of clonal tree plantations in the United States to serve as a feedstock for making ethanol fuels from cellulose. Neem (*Azadirachta indica*), a producer of natural insecticides in southern Asian and African plantations, is being cloned through tissue culture.

Combinations of trees (usually as clones) with agricultural practices have been pursued in many tropical regions since ancient times. Agroforestry concepts have spread slowly into the temperate zones, and they have become an important component of stream and soil protection in parts of the United States. The U.S. Department of Agriculture has set a goal of establishing 2 million miles (3.2 million kilometers) of conservation buffers in agricultural landscapes by the year 2002. The resulting stream side buffer strips, dynamic windbreaks, and alley cropping schemes are likely to make extensive use of poplar and willow clones. Besides their soil and water protection functions, these plantings should provide improved wildlife habitat and additional raw materials from periodic harvests and clonal regeneration.

Minimizing risks in clonal forestry. Stecklings and rootlings can become infected with microorganisms that cause problems. An example is the typical decline in health of many Lombardy poplars before they reach 20 years of age. This problem can be reduced by good sanitation in propagating cuttings, and can be eliminated in sterile tissue culture.

Just as uniformity can be very beneficial to the ways humans want to use trees, it also can promote the success of insects and pathogens. Epidemics can rapidly spread through plantations and cause serious economic and environmental damage. Genetic resistance to these pests may be built into clonal plantations, but large plantings of single clones are likely to lead to selection of new strains of insects or pathogens that overcome that resistance. To minimize these risks, a number of superior clones must be developed for each species and each set of environmental conditions. Theoretical analyses suggest that a minimum of 7 and no more than 40 unrelated clones may be needed to have enough genetic diversity in the tree population to minimize the risks from pest organisms.

If the tree species is tolerant enough of shade, or can be planted at wide enough spacings, or can be harvested at young enough age to avoid serious shading from adjacent trees, then maximum uniformity is not needed in harvested material and all of the superior clones can be mixed together in each planting. Such plantings may be even more genetically diverse and, therefore, safer than natural stands of trees.

Where product uniformity standards are high or where shading would lead to dominance by the tallest clones, an alternative strategy of planting mosaics of pure clone blocks must be pursued. The safe size for these pure clone blocks has not yet been determined, but current operational practice seems to be based on the hope that if clones of aspen up to 30 acres (12 ha) in size have worked in natural landscapes, then similar sizes of clonal blocks can be used for plantations.

A future alternative may be to produce and plant mixtures of multiple sublines of a clone that carry a wide variety of different inserted genes for different possible pest-resistance mechanisms. This may result in plantations with uniform growth rates and product characteristics, but diverse reactions to and selection pressures on pests.

As clonal forestry becomes more common, it will be important to have a much better understanding of all these relationships. The temptation will always exist to plant only the best single clone. Some governments (for example, Germany) have established regulations on the number of clones that must be used and the size limits on pure clone blocks.

For background information *see* FOREST GENETICS AND BREEDING; TREE; TREE DISEASES in the McGraw-Hill Encyclopedia of Science & Technology.

Richard B. Hall

Bibliography. M. R. Ahuja, W. Boerjan, and D. B. Neale (eds.), *Somatic Cell Genetics and Molecular Genetics of Trees*, 1996; M. R. Ahuja and W. J. Libby (eds.), *Clonal Forestry*, vols. 1 and 2, 1993; H. H. Hartmann et al., *Plant Propagation: Principles and Practices*, 1997; B. Zobel and J. Talbert, *Applied Forest Tree Improvement*, 1984.

Fossil

The fossil record is incomplete. A particular species may not have been preserved for a variety of reasons. It then becomes difficult to interpret many patterns in the fossil record because they may represent real evolutionary change or they may simply reflect the incompleteness of the fossil record. Paleontologists have long been able to qualitatively document many causes of incompleteness of the fossil record. Now, they are able to use computers to quantitatively simulate processes of evolution, ecology, fossilization, and sediment accumulation as a means of more confidently interpreting the actual fossil record. These models have revealed that incomplete preservation not only can mask patterns of evolution but can create the impression of evolutionary patterns where none exist. Paleontologists can also use these models to devise new ways of avoiding problems of incompleteness when interpreting the fossil record.

Incompleteness of record. Many factors contribute to the incompleteness of the fossil record. Methods of sampling may cause paleontologists to fail to collect fossils that are present. Organisms may have been rare during life, causing their fossils to be correspondingly scarce. Only hard parts of organisms are commonly preserved, and breakage, abrasion, and dissolution destroy many of these forms. Organisms may live in certain habitats, causing their fossils to be found only in specific facies, which are bodies of rock formed from sediment accumulated in those environments. Sediments may not have accumulated where the organism lived, or the rocks or sediment in which fossils are preserved may be removed by subsequent erosion.

Computer simulations. Marine settings, in which much of the fossil record occurs, may be simulated on a computer. Any reasonable computer simulation of the fossil record must account for several natural processes. Simulated species must originate and become extinct. Sediments must accumulate in a wide range of sedimentary environments across a region. Species must be limited, to some degree, to specific environments; that is, they must have ecological limits like those recognized among species today. Certain environment-specific processes can prevent the burial and preservation of fossils. Finally, the rocks must be sampled for fossils as a paleontologist would do.

These models produce simulated outcrops across a sedimentary basin. For each outcrop or rock exposure, the sedimentary facies is recorded for each level, as are the fossils found at that level. Paleontologists can then use these as bases for simulating a variety of evolutionary phenomena as well as processes that prevent fossilization and bias their impression of the fossil record.

Blurring of fossil record. These computer models accurately reproduce the Signor-Lipps effect, notorious for causing abrupt evolutionary events such as mass extinctions to appear in the fossil record as if they took place gradually over a long span of time. Because fossilization is a rare process, not all species that were alive at any given time are preserved as fossils in sedimentary rocks. In the case of a sudden and catastrophic mass extinction, only a fraction of the species alive prior to the extinction event will be found as fossils in the sediments laid down at the time of extinction. However, many of the species not preserved in this layer might well be found in older layers of rock. The result of this lack of preservation is that the fossil record appears to record the gradual elimination of species prior to the event, rather than the abrupt elimination of all of these species at the time of the mass extinction. This apparent blurring of evolutionary events is known as the Signor-Lipps effect.

The computer models also reveal that the Signor-Lipps effect is much more intense than previously recognized. In fact, the models predict that it is unusual for the first appearance of a fossil to coincide with the time of origination of that species, or for the last appearance to coincide with the time of extinction. If speciation events are rapid, as many evolutionary theories argue, computer modeling suggests that the fossil record will rarely preserve the morphological changes that occur during speciation. Instead, paleontologists should normally expect to find one species prior to the speciation event and a new daughter species, fully formed, following the speciation event with no detailed record of the transition. The abrupt appearance of many species in the fossil record probably does reflect the incompleteness of the fossil record, as Darwin himself argued. The Signor-Lipps effect therefore not only can blur large-scale evolutionary patterns but can also cause individual speciation events to appear more rapidly than they actually occurred.

Artifacts of preservation. Computer models show that the processes by which sediments accumulate can also create patterns in fossil occurrences that mimic evolutionary patterns. In much of the stratigraphic record, facies gradually pass laterally and vertically into other facies. Because many fossils are specific to certain facies, the fossils found in one part of an outcrop gradually pass into a different suite of fossils higher in the outcrop. These changes are not evolutionary but merely reflect gradual changes in the sedimentary environment over time.

However, not all facies changes are gradual. The field of sequence stratigraphy recognizes two important types of surfaces or horizons across which facies change abruptly, not gradually. Consequently, the fossils found in an outcrop may also change abruptly (see **illus.**).

The first of these surfaces is a flooding surface, which records an abrupt switch from a facies deposited in a shallow-water marine environment to one deposited in a deep-water marine environment. In this case, fossils found in the outcrop will also display an abrupt switch from shallow-marine to deep-marine. If deep marine conditions persist above the flooding surface, sufficient time may elapse for many of the shallow-water species now living elsewhere to become extinct. In this outcrop, these shallow-water species will have their last occurrence at the flooding surface, though they may have actually become extinct much later. Conversely, if shallow-water conditions were present for a long time before the flooding event, deep-water species may have actually originated in a deep-water environment elsewhere at a time predating the event. Their first occurrences will then lie immediately above the flooding surface. Computer models show that these clusters of first occurrences of deep-water species and last occurrences of shallow-water species can be quite pronounced. Without knowledge of the facies changes, this pattern of faunal change at the flooding surface could easily be misinterpreted as a major extinction episode followed by the migration or speciation of many new species. Similar patterns have been described for trilobite extinctions during the Cambrian Period, and further work is needed to distinguish the artifacts imposed by these flooding surfaces from any actual evolutionary changes. *See* TRILOBITA.

The second of the abrupt surfaces is a sequence boundary, which can be expressed in several ways. Some sequence boundaries mark abrupt shifts from deep-water facies to shallow-water facies—that is, the opposite of a flooding surface. In these cases, the surface will record the last occurrences of many deep-water species and the first occurrences of many shallow-water species. Like a flooding surface, this type of sequence boundary could be misinterpreted as a major extinction episode followed by an immigration or diversification of many new species. Other sequence boundaries simply record a long period of nondeposition or erosion of sediment and may be overlain or underlain by many kinds of facies. The abrupt appearances and disappearances of many species will also characterize these sequence boundaries. However, the species will not be exclusively from shallow-water or deep-water facies as for the other surfaces.

Avoidance of problems. Computer models also provide paleontologists with guidelines for avoiding these biases and interpreting the fossil record more accurately. Sequence-stratigraphic study of outcrops can indicate the locations of potential artifacts in fossil occurrences, even before fossil collection has begun. Paleontologists can then be aware of horizons at which sudden and marked changes in fossil content are expected as a result of rapid changes in depositional environment or long periods of nondeposition of sediments. If fossil collections reveal that major changes in fossil content do occur at these horizons, these changes may be artifacts of fossil preservation and sediment accumulation. Alternatively, they may be real evolutionary events that just happen to coincide with these surfaces. The computer models suggest that these two possibilities can be distinguished if more fossil collections are made at

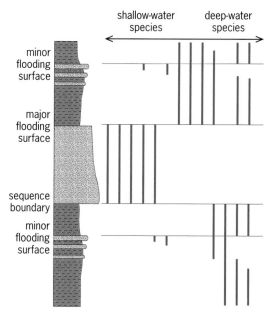

Key:

▨ rock type indicative of
deep water, such as a shale

▨ rock type indicative of
shallow water, such as a sandstone

Vertical section through a rock outcrop, showing changes in sedimentary rock types at flooding surfaces and sequence boundaries, and the parallel changes in fossil content. Species that disappear and reappear are still present elsewhere in the sedimentary basin. Likewise, most first and last occurrences of species do not correspond to the times at which those species originate and become extinct.

the same horizon but in other outcrops farther away. If the species appear and disappear at the same positions in the outcrop, despite changes in facies that occur laterally, the changes in fossil content probably do reflect evolutionary change. However, if the fossils can be shown to track the facies in which they occur, the abrupt change is an artifact of stratigraphic accumulation and not an evolutionary event.

It may not always be possible to collect from multiple outcrops to overcome these biases. In these cases, computer models suggest that the only way to mitigate the effects is to decrease the resolution of the study and focus on evolutionary studies over longer time scales. In these studies, stratigraphic biases impose a limit to the resolution of these evolutionary patterns.

For background information *see* FACIES (GEOLOGY); FOSSIL; STRATIGRAPHY in the McGraw-Hill Encyclopedia of Science & Technology.　　Steven M. Holland

Bibliography. C. E. Brett, Sequence stratigraphy, biostratigraphy, and taphonomy in shallow marine environments, *Palaios*, 10:597–616, 1995; S. M. Holland, The stratigraphic distribution of fossils, *Paleobiology*, 21:92–109, 1995.

Frequency and time standards

Frequency is the most accurately measured physical quantity. It is used to measure the time intervals from which time scales are generated. Accuracy is important in navigation, communications, space exploration, and scientific research, particularly in relativity. Recent developments in physics and communications have brought about advances in the accuracy of frequency measurement and in the accessibility of time scales.

International time scales. The International Bureau of Weights and Measures (BIPM) maintains the world's time scale, receiving contributions from clocks in a large number of laboratories worldwide and computing the free atomic time scale (EAL). EAL is more regular than individual contributions, but for long-term stability it is corrected to produce Coordinated Universal Time (UTC) by accurate measurements of the hyperfine splitting in the ground state of the cesium-133 atom. These measurements, done on primary standards in a very few laboratories, realize the International System (SI) second, which is defined as the time taken by 9,192,631,770 cycles of the radiation whose frequency corresponds to this energy difference.

Cesium fountain. The cesium fountain exploits some of the techniques that are widely used in the laser cooling of atoms. The use of cold atoms facilitates improvement in both the stability of a frequency standard and the accuracy with which it relates to the unperturbed atom. The cesium D line, the energy difference or transition between the ground state and the higher-energy P state, falls at an optical frequency in the near infrared. A laser can be tuned to the transition between one ground-state hyperfine level and the P state. The cesium atoms in the device acquire momentum from the light emitted by a laser that has been tuned in this manner, and are slowed from thermal velocities (root-mean-square velocities of about 100 m/s) to a root-mean-square velocity of about 9 cm/s or 120 microkelvin temperature by laser (Doppler) cooling. In a typical configuration (**Fig. 1**), six laser beams are arranged in counterpropagating pairs along three orthogonal axes, and are tuned to a frequency a few megahertz below the energy gap. Atoms absorb the below-resonant light, brought into resonance by the Doppler shift due to their motion relative to the beams, and emit light on resonance. They lose energy because they emit higher-energy photons than they absorb. The cloud of cesium atoms is then thrown upward at a few meters per second by shifting the frequency of the vertical lasers. The atoms are then cooled further by sub-Doppler techniques to about 2.5 μK, equivalent to 1.2 cm/s root-mean-square velocity and prepared in one of the hyperfine states. The atoms then pass through a microwave field region before falling under gravity into a detection region. The process is repeated until the microwave frequency is found that transfers the maximum number of atoms to the other hyperfine state.

Laser cooling enables the atoms to exist as a cloud of centimeter dimension during the typical 1-second period of ballistic flight. The cesium fountain has advantages over cesium-beam devices due to two basic differences: the longer interaction time and the reversal of the path traveled by the atoms during the interaction with the microwave field.

Interaction time. The precision with which a frequency can be measured is fundamentally limited by the Q, the ratio of the linewidth of the radiation whose frequency is measured to the frequency itself. In the cesium fountain the Q is increased by lengthening the measurement time, which reduces the linewidth. Both beam devices and fountains use the separated-oscillatory-fields method to identify the optimum microwave frequency with higher resolution. The atoms are exposed to the field twice, with some time separating the exposures. Multiple maxima in the transferred population are produced, analogous to fringes seen due to interference of light in the Young's slit experiment. The largest fringe peak indicates the maximum in the number of atoms transferred, and the width is inversely proportional to the time between interactions. The width is about 100 times narrower in the fountain than in beam devices.

Microwave field. The largest uncertainty in the beam device is usually the cavity phase difference. This uncertainty is due to the phase of the microwave field being different in the two separate regions through which the beam of cesium atoms passes. Its size is about 6 parts in 10^{15} in the best beam devices but is greatly reduced in the fountain because the atoms pass twice through the same microwave region.

Limitations. There are two sources of uncertainty in the cesium fountain that increase relative to beam standards. The first is the pulsed nature of the experiment. Atoms are cooled and travel in ballistic flight, and their states are detected. A measurement is made about once per second. In the beam devices, the atoms pass continuously through the system and there is a continuous error signal. The discrete nature of the fountain signal introduces noise in the measurement.

The use of cold atoms introduces a higher probability that the atoms will collide during the time between the microwave interactions. This is a consequence of the Heisenberg uncertainty principle: Since the atom's momentum is known more accurately, this principle requires that the spatial extent of the quantum-mechanical probability density function for the location of its center of mass have a greater range. Much work has been done to estimate the collisional shift and reduce the associated uncertainty.

Currently, one laboratory has operated a fountain as a frequency standard and estimates the uncertainty at 2×10^{-15}, which compares favorably with 7×10^{-15} for a "classical" cesium-beam device now in operation. Several laboratories are constructing fountain standards, and a number of beam devices are under development. The possibility of further extending the interaction time in cold-atom clocks has been recently explored in a micro-gravity experiment that was carried out on an airplane flying so as to produce near-weightless conditions.

Ion traps. Unlike neutral atoms, the measurement time of ions may be increased by confining them for long periods in a trap using electromagnetic fields (**Fig. 2**). There is much interest in the use of cold ions for frequency standards, and a number of different ions and transitions in the optical, infrared, and microwave regions are being studied.

Particularly promising for frequency standards are transitions at optical rather than microwave frequencies, since the Q may be 10^4 times larger due to the higher frequency. For example, optical-clock Q's of about 10^{15} are possible for weak electric quadrupole transitions, and experimental Q's in the region of 10^{13} have already been achieved. Recently, an extremely weak electric octupole transition in a single cold ytterbium ion, ^{172}Yb$^+$, at 467 nanometers was observed. This is possibly the narrowest atomic transition available at optical frequencies, having a theoretical lifetime of 10 years and a corresponding intrinsic Q of 10^{24}. In practice, the octupole transition may enable a small number of orders of magnitude increase in Q over quadrupole-transition optical frequency standards. An ion of another ytterbium isotope, ^{171}Yb$^+$, has nuclear spin $1/2$ and some energy transitions which are independent of magnetic field to first order, and so makes a very good candidate for a frequency standard.

Ion-trap standards have recently progressed from pure scientific study. A 40-GHz microwave-frequency clock based on a mercury ion trap operates at each of the three deep-space tracking stations of the National Aeronautics and Space Administration (NASA).

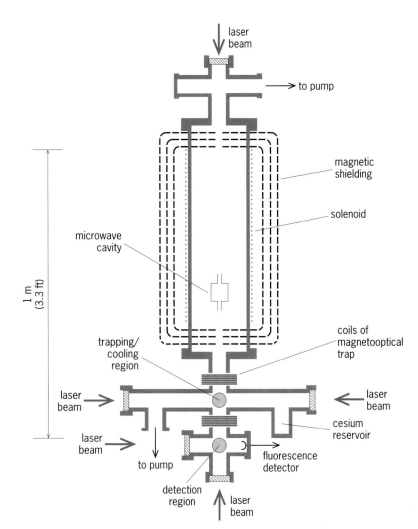

Fig. 1. Side view of atomic fountain frequency standard.

A more precise frequency standard at this frequency has recently been constructed using laser-cooled mercury ions with both a frequency stability and an accuracy comparable to the cesium fountain. In 1997 the International Committee for Weights and Measures adopted the optical quadrupole transition of a single cold ^{88}Sr$^+$ ion at 674 nm as one of the recommended standard wavelengths for the realization of the meter. Such standard wavelengths are used as references in the measurement of length by exploiting the interference of light. Since the meter is defined in terms of the distance that light travels in a given time, it is necessary to relate these standard wavelengths, or optical frequencies, to the 9.2-GHz microwave frequency in the SI definition of the second. This requires the construction of frequency chains by which optical and microwave frequencies can be compared.

Time transfer. Comparisons of time and frequency between timing laboratories require that the time taken to transfer the information is precisely known. Radio transmission via satellite is the established mechanism for this determination, though there are also experimental optical satellite links. The

Fig. 2. End-cap ion trap. (*National Physical Laboratory, Teddington, Middlesex, England***)**

mechanisms are two-way satellite time and frequency transfer (TWSTFT) and various techniques involving the NAVSTAR Global Positioning System (GPS) satellites or the similar Global Navigation Satellite System (GLONASS) constellation. The GPS consists of 21 operational and 3 spare (in-orbit) satellites orbiting in six fixed planes inclined at 55° to the Equator. Each vehicle repeats its ground track each 23 h 56 min, and there are usually four or more above the horizon simultaneously at a given location. Each carries four atomic clocks referenced to the clock at the U.S. Naval Observatory and transmits time and position information on two frequencies. Civilian utilization of the timing information has been restricted to the standard positioning service (SPS) by coding on the precise positioning service (PPS). This coding has an effect similar to random noise on the arrival of the time signal. Both this coding and the variability of the propagation delays due to the ionosphere have been significant impediments to the use of the GPS for precise time transfer.

GPS developments. Techniques are emerging whereby the time transfer capability of the GPS may be more widely exploited, assisted in part by the commercial availability of inexpensive receiving apparatus driven by the large market for navigation.

Common view. Both the dither coding of the standard positioning service and the calibration of the satellite clock can be removed from the time-transfer process if observations of the satellite signals are made at the two locations simultaneously. Tracking schedules are published by the BIPM for various geographical locations of pairs of observers (for example, the United States and Europe). These schedules provide times that each GPS satellite is well above the horizons of both observers simultaneously. By making observations at such times, it is possible to reduce the effect of the signal being transmitted to the two observers with different ionospheric delays.

At the expense of greater computational complexity, it is now possible to measure simultaneously the signals from several vehicles relative to a local clock. If the two observers compare their local clocks relative to each vehicle before they compute an average, then noise, such as the effect of ionospheric delay, is reduced to a level similar to that achieved by following the tracking schedule. The advantage of this technique is that the intercomparison can be continuous. Both techniques achieve an accuracy of approximately 4×10^{-9} s, in the multichannel case for a 5-min comparison.

Carrier phase. Rather than using the comparatively noisy (coded) timing pulses alone, it is possible to combine the information from these pulses with information from the measurement of the phase of the radio carrier wave at the two frequencies transmitted by GPS satellites. The timing pulses overcome the ambiguity with regard to which cycle of the carrier is timed. The ionospheric delay can be estimated from the difference between the phases of the two frequencies. Besides time transfer, this technique is used in survey work.

Augmentation. Many commercial applications of the GPS such as in data network synchronization and aircraft landing systems require that integrity of the GPS information together with corrections to the individual satellite clocks be available in real time. Various augmentation systems are in operation or planned. The U.S. Coast Guard operates a positioning system for coastal and inland waters enabling 10-m accuracy using some 50 surface transmitters. In Europe an overlay system for the GPS, EGNOS, is under development. The GNSS1 geostationary satellite system is expected to begin tests during 1999.

For background information *see* ATOMIC CLOCK; ATOMIC TIME; FREQUENCY MEASUREMENT; LASER COOLING; LINEWIDTH; PARTICLE TRAP; Q (ELECTRICITY); SATELLITE NAVIGATION SYSTEMS; TIME; UNCERTAINTY PRINCIPLE; WAVELENGTH STANDARDS in the McGraw-Hill Encyclopedia of Science & Technology.

Dale Henderson

Bibliography. J. C Bergquist (ed.), *Proceedings on the 5th Symposium on Frequency Standards and Metrology*, World Scientific, 1996; S. Chu, Laser trapping of neutral particles, *Sci. Amer.*, 266(2):71–76, February 1992; W. Itano and N. Ramsey, Accurate measurement of time, *Sci. Amer.*, 269(1):56–63, July 1993; J. Vanier and C. Audoin, *The Quantum Physics of Atomic Frequency Standards*, Adam Hilger, 1989.

Friction

The modern study of friction may be said to have begun in the fifteenth century, when Leonardo da Vinci deduced the laws governing the motion of a rectangular block sliding over a planar surface. However, Guillaume Amontons is credited with the first published account of the classic friction laws, which in 1699 described his observations of solid surfaces in sliding contact.

Amontons concluded that (1) the friction force that resists sliding at an interface is proportional to the normal load, or force, which presses the surfaces together; and (2) contrary to intuition, the friction force is independent of the apparent area of contact. Thus, a small block experiences as much friction as does a large block of the same material so long as their weights are equal. A third law, frequently included with these two, is attributed to Charles Augustin de Coulomb in the eighteenth century: The friction force is independent of velocity for ordinary sliding speeds. Amontons' law is written as $F = \mu N$, where the coefficient of sliding friction, μ, is defined as the ratio of the frictional force F to the load N, or force normal to the interface. The coefficient is independent of the apparent contact area A, the loading force N, and the sliding speed v. Considering its simplicity, Amontons' law is remarkably well obeyed for a wide range of materials such as wood, ceramics, and metals.

Origin. Amontons' and Coulomb's classical friction laws have far outlived a variety of attempts to explain them on a fundamental basis. Surface roughness was ruled out definitively in the mid-1950s as a possible mechanism for most friction. Molecular adhesion, though, remained a strong possibility, due in large part to studies which found that friction, although independent of apparent macroscopic contact area as Amontons had stated, is in fact proportional to the true contact area. That is, the microscopic irregularities of the surfaces touch and push into one another, and the area of these contacting regions is directly proportional to the friction force. Subsequent experiments explored the possibility that friction arose from sufficiently strong bonding at the true contact points so as to produce continual tearing away of tiny fragments of material. This explanation failed, however, to predict experimental observation. Indeed, it was proved incorrect in the 1970s, when a "surface force apparatus" developed for atomic-scale friction measurements found clear evidence for friction in the total absence of wear. The surface force apparatus consists of two perfectly flat, cleaved mica surfaces which are separated by lubricant films that can be as thin as a few molecules. It is employed to measure how such lubricant films affect the sliding friction associated with relative motion of the mica surfaces.

In the 1990s it was established that friction did not correlate directly with the strength of the adhesive bond itself. Instead, it is associated with adhesive "irreversibility," or how surfaces behave differently while being stuck together compared to being unstuck. The fundamental atomic-scale origins of the friction forces nonetheless remained unclear.

Phonon friction. One notion concerning the origins of friction, which dates back to 1929, suggests that vibrations of atomic lattices will result in wear-free friction. Friction arising from such atomic-lattice vibrations occurs when atoms close to one surface are set into motion by the sliding action of atoms in the opposing surface. The lattice vibrations, which are called phonons, are produced when mechanical energy needed to slide one surface over the other is converted to sound energy, which is eventually transformed into heat. Hence, to maintain the sliding motion, more mechanical energy must be added, and a larger force is required.

Atomic force microscope results. The concept of phonon friction inspired measurements of nanometer-scale friction in the mid-1980s. A new invention, the atomic force microscope, was adapted for measurements of lateral forces, leading to the first observations of friction, measured atom by atom. While no additional evidence could be adduced for or against the case of phonon-induced friction, the instrument revolutionized studies of friction at atomic length scales.

An atomic force microscope consists of a sharp tip mounted at the end of a compliant cantilever. As the tip is scanned over a sample surface, forces that act on the tip deflect the cantilever. Various electrical and optical means (such as capacitance and interference) quantify the horizontal and vertical deflections. In the early 1990s, a friction force microscope was set up in an ultrahigh vacuum, allowing studies of the sliding of a diamond tip over a crystalline diamond surface with a contact area estimated to be less than 20 atoms in extent. These measurements yielded a friction force that exhibited no dependence on the normal load. According to the classical friction laws, this result would have implied zero friction. Not only was friction evident, but the shear stress, or force per area required to maintain the sliding, was enormous: 10^9 newtons per square meter, or 150,000 pounds per square inch. That force is large enough to shear high-quality steel.

Quartz microbalance results. Shear stresses more than 11 orders of magnitude lower have been observed for molecularly thin films sliding on metal substrates. In this work, friction is measured with a quartz crystal microbalance, a device that has been used for decades to weigh samples with masses as small as nanograms. The basic component of a quartz crystal microbalance (see **illus.**) is a single quartz crystal that has very little internal dissipation (or friction). As a result, it oscillates at an extremely sharp resonance frequency (usually 5–10 MHz) that is determined by its elastic constants and mass. The oscillations are driven by applying a voltage to thin metal electrodes that are deposited on the surface of the quartz in a manner which produces a crystalline texture, generally (111) in nature. Atomically thin films of a different material (solid or liquid) are then

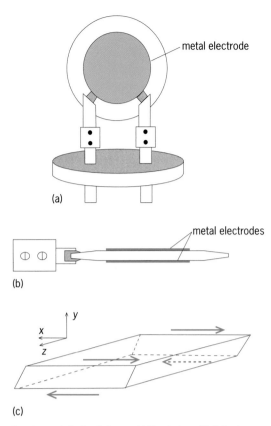

Quartz crystal microbalance. (*a*) Front view. (*b*) Side view. (*c*) Transverse shear oscillatory motion of the quartz crystal. Electrodes are parallel to the *x*-*z* plane.

adsorbed onto the electrodes. The extra mass of the adsorbed layer lowers the resonance frequency of the microbalance, and the resonance is broadened by any frictional dissipation due to relative motion of the adsorbed layer and the microbalance. By simultaneously measuring the shift in frequency and the broadening of the resonance (as evidenced by a decrease in the amplitude of vibration of the microbalance), the sliding friction of the layer with respect to the metal substrate can be deduced. The experiment is analogous to pulling a tablecloth out from under a table setting, whereby the degree of slippage is determined by the friction at the interface between the dishes (that is, the adsorbed film material) and the tablecloth (that is, the surface of the electrode).

The friction can be measured only if it is sufficiently low to result in significant sliding, which will be accompanied by a measurable broadening of the resonance. For this reason, quartz crystal microbalance measurements of sliding friction tend to be carried out on systems exhibiting very low friction, such as rare-gas solids adsorbed on noble metals. For many other systems that exhibit higher friction (such as chemically bonded layers), the slippage on the surface of the quartz crystal microbalance is too small to produce a measurable broadening.

The quartz microbalance operates on a time scale short enough to be able to detect phonons, and indeed, its measurements have provided definitive experimental confirmation of the existence of this mechanism for frictional energy dissipation. Measurements have also been performed of nitrogen sliding on lead in its normal and superconducting states, and changes have been observed in the friction at the superconducting transition. Such changes indicate that electronic mechanisms may also give rise to friction.

Computer simulations. The comparison of the quartz crystal microbalance data with theories of phononic friction is dependent on comparisons to computer simulations. Such simulations have also demonstrated that static friction, the friction which initially resists all motion, is likely to be related to adhesive forces of thin adsorbed films (water, hydrocarbons, and so forth), which are known to be present on most surfaces. The simulations indicate that these films are capable of locking surfaces together when in static contact.

Stick-slip friction. Another phenomenon associated with friction is stick-slip motion, in which surfaces gliding past one another momentarily cling and then let go. Examples include screeching train brakes and fingernails on blackboards. Roughness was thought to cause the random nature of the sticking and the slipping. But stick-slip friction was recently observed in lubricated contacts between nominally perfect mica surfaces. Millions of repetitive cycles of a sinusoidal force were applied to confined liquids without wear, and the observed results suggest that randomness (specifically, so-called 1/f noise) may be intrinsic to the stick-slip friction itself.

For background information *see* FRICTION; LATTICE VIBRATIONS; MOLECULAR ADHESION; PHONON; SCANNING TUNNELING MICROSCOPY; WEAR in the McGraw-Hill Encyclopedia of Science & Technology.

Fereydoon Family; Jacqueline Krim

Bibliography. B. Bhushan (ed.), *Handbook of Micro/Nanotribology*, CRC Press, 1995; J. Krim, Friction at the atomic scale, *Sci. Amer.*, 275(4):74–80, October 1996; B. N. J. Persson, *Sliding Friction: Physical Principles and Applications*, Springer-Verlag, 1998; B. N. J. Persson and E. Tosatti (eds.), *Physics of Sliding Friction*, Kluwer, Dordrecht, 1996; I. L. Singer and H. M. Pollock (eds.), *Fundamentals of Friction: Macroscopic and Microscopic Processes*, Kluwer, Dordrecht, 1992.

Fungal enzyme

Enzymes are proteins found in all living cells. They catalyze biochemical processes but are not consumed in the process. As proteins, they are biodegradable and operate at specific pH values and at temperatures found in nature. Enzymes are fairly specific, and there is one reaction that is optimal for each. Fungal enzymes are used in various ways, and new applications are constantly being discovered. Enzymes can replace chemical catalysts in some industrial processes.

Detergents. The detergent industry is the largest single market for enzymes. Proteases (enzymes that digest proteins) of bacterial origin dominate the market, but fungal lipases (enzymes which catalyze the hydrolysis of fats) and cellulases (enzymes which catalyze the hydrolysis of cellulose) are also widely used. The lipase used in detergents to remove lipid-containing stains from clothes originates from *Thermomyces lanuginosus*. Cellulases have a specific application in washing. Cellulose exists in a very complex, highly crystalline state and requires the combined action of different types of cellulases to become fully degraded. The cellulase used in commercial detergents is one of many natural cellulases and therefore does not degrade the cotton fibers. Its specific use is removal of microfibers from the surface of cotton fibers. The microfibers result from mechanical abrasion during washing and wearing. Their removal restores the smooth appearance of the cotton fibers. Thus, the fabric remains soft and retains its original color much longer than when washed without cellulase.

A peroxidase preparation from the inkcap (*Coprinus* sp.) is under development. Since it decolorizes solubilized dyes, its application in laundry will make it possible to wash white and colored clothes together.

Starch conversion. A fungal glucoamylase produced by *Aspergillus niger* is used together with two bacterial enzymes in the conversion of starch to fructose. First, the starch is liquefied by a bacterial enzyme. Then, the fungal glucoamylase catalyzes the saccharification of the liquefied starch to glucose, and another bacterial enzyme isomerizes the glucose to fructose. The fructose syrups are commercially used in soft drinks and other products.

Textiles. Fungal cellulases from *Humicola* sp. are replacing stone washing of denims. This application solves wastewater problems and uses much less energy during dyeing and bleaching processes. A fungal catalase is also used to remove surplus hydrogen peroxide during the bleaching of cotton, reducing the number of washing steps.

Paper and pulp. Cellulases are used to decolorize recycled paper (deinking), and lipases to prevent accumulation of pitch on the large rollers used in paper manufacturing.

Food production. Fungal enzymes dominate the food industry. A milk-coagulating enzyme produced by *Rhizomucor miehei* has been used in cheese production, instead of rennin from suckling calves. Pectinases and other plant cell wall–degrading enzymes from *Aspergillus* sp. not only improve the yield of oil from seeds and juice from fruits but also make filtering easier. Dextranases produced by *Penicillium* spp. remove the unwanted dextrans from sugarcane juices before the manufacture of sugar.

Flavor, texture, appearance, and stability of the final product can also be improved by enzymes. A fungal xylanase is used in baking to produce a volume increase and a more stable dough; the exact way it works on the dough is unknown but is being studied. The gluten network of the dough is influenced by a variety of chemical treatments, and the current effort to develop enzymes primarily seeks subsitutes for some less acceptable chemical additives such as potassium bromate, presently used for gluten strengthening.

Animal feed. Increasing the digestibility of feeds and thus the growth rate of animals, is the primary aim of research here. Xylanase, cellulase, and phytase are of great value as the use of antibiotic growth promoters becomes more restricted. This market for fungal enzymes will probably grow and diversify.

Phytase hydrolyzes inorganic phosphate from phytate. Phytate is a common source of stored phosphate in cereals and oil seeds. This phosphate is unavailable to monogastric animals, such as chickens and pigs, and inorganic phosphate must be added to their feed. Treatment of the feed with phytase increases the available phosphate, aids in its uptake, and lessens the risks associated with phosphate runoff from feedlots.

Industrial production. Fungal enzymes have been produced on an industrial scale for 100 years. In 1896, Japan made Takadiastase, which is a mixture of hydrolytic enzymes produced by *A. oryzae* acting on wheat bran. Such surface cultivation is not suitable for large-scale production, and technicians may be exposed to large numbers of spores which can be hazardous to their health. Not much development took place until the 1960s, when enzymes were added to detergents. Since then, submerged cultivation in stirred and closed fermentors has replaced surface cultivation. Fermentors, of 0.5–150,000-liter capacity, are easily monitored to maintain the desired pH, temperature, and nutrient level. They have made liquid fermentation safer to handle, and have become the preferred production method. Normally fungi produce complex enzyme mixtures at low yield. Large-scale production of the desired enzyme is very costly, and purification and characterization are almost impossible. The production of fungal enzymes was made easier by the introduction of gene technology in the 1980s. The producing fungus might have weaknesses, such as slow growth rate, low production levels, toxicity, and pathogenicity. These weaknesses can be overcome by transferring the gene encoding the desired enzyme from the original species into a host organism. *Aspergillus oryzae* is often chosen as the host because it grows quickly on cheap media and is regarded as safe since it has been used for the production of human food for over a century.

Gene transfer is made possible through a relatively new process called expression cloning. This is a faster and more efficient method for cloning genes. For the expression of a cloning system, synthesis of complementary deoxyribonuclic acid (cDNA), reliable and sensitive enzyme assays, and expression of the enzyme in yeast are necessary. Yeast is not as productive an organism as the filamentous fungi. Therefore, the clones are transformed into *A. oryzae*, and large-scale fermentation of almost pure monocomponent enzymes can be carried out. Good yields have been achieved, and the purification and characterization have become much easier.

For background information *see* DETERGENT; ENZYME; FOOD MICROBIOLOGY; FUNGI; GENETIC ENGINEERING; STARCH in the McGraw-Hill Encyclopedia of Science & Technology.   Ruby I. Nielsen

Bibliography. H. Dalbøge, Expression cloning of fungal enzyme genes: A novel approach for efficient isolation of enzyme genes of industrial relevance, *FEMS Microbiol. Rev.*, 21:29–42, 1997; H. Dalbøge and H. Heldt-Hansen, A novel method for efficient expression cloning of fungal genes, *Mol. Gen. Genet.*, 243:253–260, 1994.

Gas hydrate

Gas hydrate has been known to chemists since the early 1800s, but in the 1960s geologists were the first to recognize it as naturally occurring. Thereafter, interest in natural gas hydrate increased dramatically, leading to speculation regarding its energy resource potential, its effect on global climate change, and its role as a geohazard.

Definition. Gas hydrate, also called gas clathrate or clathrate hydrate, is a naturally occurring solid composed of water molecules forming a rigid lattice of cages (a clathrate), with most of the cages containing a molecule of natural gas, mainly methane (**Fig. 1**). Gas hydrate is essentially a water clathrate of natural gas in which water crystallizes in the isometric crystallographic system, rather than the hexagonal system of normal water-ice. Three structures, I, II, and H, of the isometric (cubic) lattice are found in nature. Gas hydrate containing mainly methane is most common, forming structure I, wherein the cages are arranged in body-centered packing. These cages are large enough to include methane, ethane, and other gas molecules of similar molecular diameters, such as carbon dioxide, nitrogen, and hydrogen sulfide.

The maximum amount of gas that occurs in gas hydrate is fixed by the clathrate geometry. In a fully saturated structure I methane hydrate, for example, one

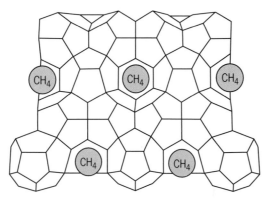

Fig. 1. Gas hydrate structure. The rigid cages are composed of hydrogen-bonded water molecules, and each cage, both exposed and covered, contains a molecule of methane (CH_4).

methane molecule is present for every 5.75 molecules of water. Thus, when appropriate expansion factors are considered, 1 m^3 (35.3 ft^3) of methane hydrate can contain up to about 170 m^3 (6000 ft^3) of methane gas at standard conditions.

Occurrence. The occurrence of gas hydrate in nature is controlled by an interrelation among the geological factors of temperature, pressure, amount and composition of gas, and ionic content of pore water. As a result of the specific requirements of these factors (mainly cold temperatures and high pressures), gas hydrate is restricted geographically to two regions: polar continental and continental shelves, and deep-ocean outer continental margins in all oceans, including polar oceans (**Fig. 2**). In polar regions, gas hydrate can be expected to occur within a depth range of 150 to perhaps 2000 m (490 to 6560 ft) below the continental and continental shelf surface. In outer continental margin settings (the region seaward of the continental shelf), gas hydrate can be expected where water depths are greater than about 300 m (980 ft), and occurs from the sea floor to depths of about 1100 m (3610 ft). At least 19 locations are known worldwide where gas hydrate samples have been recovered; the occurrence of gas hydrate has also been inferred from well logs and from marine seismic surveys over outer continental margins.

Origin. Methane that forms gas hydrate results from the decomposition of organic materials following biochemical and thermal pathways, with biochemical pathways involving methanogenesis being the most important. Methane generation for gas hydrate formation occurs by carbonate reduction [reaction (1)], acetate (methyl) fermentation [reaction (2)], and thermal cracking [reaction (3)]. In

$$CO_2 + 4H_2 \longrightarrow CH_4 + 2H_2O \qquad (1)$$

$$CH_3COOH \longrightarrow CH_4 + CO_2 \qquad (2)$$

$$CH_3COOH + heat \longrightarrow CH_4 + CO_2 \qquad (3)$$

reaction (1), CO_2 derived from organic matter is reduced microbially by methanogens in mainly marine depositional settings. In reaction (2), methyl fermentation by methanogens occurs mainly in fresh-water settings. In reaction (3), methane results from the thermal cracking of organic chemical bonds, in contrast to reaction (2) where chemical bonds are broken biochemically. Because most gas hydrate occurs in marine depositional settings, the biochemical process illustrated by reaction (1) is the dominant one for methane generation in gas hydrate formation. Given the appropriate conditions of pressure and temperature and an adequate supply of methane and water, gas hydrate is the mode in which methane can be expected to occur in nature.

The amount of methane in gas hydrate is likely enormous, with current estimates at about 10^{19} g (2.2×10^{16} lb) of methane carbon. The amount of methane carbon in gas hydrate may be about twice the carbon present in all known fossil fuel deposits of coal, crude oil, and natural gas (**Fig. 3**). Therefore,

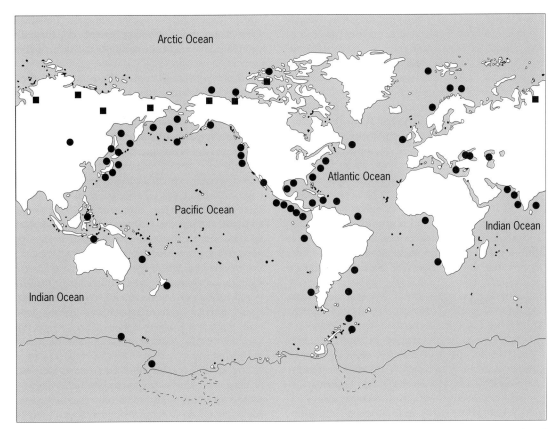

Fig. 2. Locations of known and inferred gas hydrate in aquatic (mostly oceanic) sediments [circles] and in continental (polar) regions [squares].

naturally occurring methane hydrate is a significant carbon source and sink in the shallow geosphere.

Societal issues. Given the huge amount of methane that is sequestered in gas hydrate at relatively shallow depths in the geosphere, methane hydrate might be regarded as a potential energy resource. After all, methane is the principal energy component of conventional natural gas. But the technological problems for gas hydrate exploitation are formidable and expensive, especially for gas hydrate occurrence under the ocean in outer continental margins. Production of methane from gas hydrate will have to await demonstrated and urgent energy needs and technological breakthroughs.

Because gas hydrate occurrence is controlled by pressure and temperature, changes in these factors will affect gas hydrate stability. When gas hydrate becomes unstable because of increased temperature or decreased pressure, methane gas and water are released. Methane is radiatively active and acts as a greenhouse gas in the atmosphere. However, the amount of methane from gas hydrate reaching the atmosphere now or in the future is very uncertain. Methane oxidation is a ubiquitous geochemical process that exerts a powerful influence on the global methane inventory. For the immediate future, the effects of methane from destabilized gas hydrate should not be a major societal concern.

Of immediate concern, however, is gas hydrate as a geohazard. Gas hydrate is a proven geohazard

to drilling in polar regions, and it becomes increasingly important as exploration for conventional energy resources moves to ever-increasing depths in

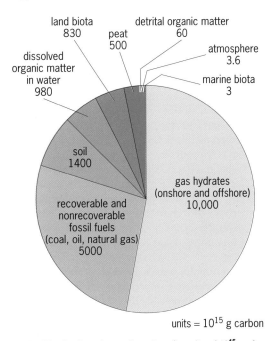

units $= 10^{15}$ g carbon

Fig. 3. Distribution of organic carbon (in units of 10^{15} g of carbon, or gigatons) in the Earth (excluding dispersed organic carbon such as kerogen and bitumen, which equals nearly 1000 times the amount of organic carbon in the diagram).

the oceans. There is compelling evidence that gas hydrate has been involved in submarine landslides. Gas hydrate's role as a geohazard is now recognized, and future technological advances will be needed to engineer for the physical consequences of gas hydrate instability.

Gas hydrate has been a nuisance for petroleum engineers, who discovered in the 1930s that gas hydrate formation in pipelines can block the transport of natural gas. Technological solutions have been found, however, to inhibit such formation, and there is a growing interest in other industrial applications of gas hydrate research. For example, consideration is being given to the making of methane hydrate for use in energy storage and transport, refrigeration, and desalination. Because carbon dioxide and water form a gas hydrate, the sequestration of fossil fuel by-products in carbon dioxide hydrate is under investigation.

For background information *see* CARBON DIOXIDE; CLATHRATE COMPOUNDS; CRYSTAL STRUCTURE; ENERGY SOURCES; FERMENTATION; FOSSIL FUEL; HYDRATE; METHANE; METHANOGENESIS (BACTERIA); NATURAL GAS in the McGraw-Hill Encyclopedia of Science & Technology.   Keith A. Kvenvolden

Bibliography. K. A. Kvenvolden, Gas hydrates: Geological perspective and global change, *Rev. Geophys.*, 31:173–187, 1993; E. D. Sloan, Jr., *Clathrate Hydrates of Natural Gases*, 2d ed., Marcel Dekker, 1998.

Genetic engineering

Adding deoxyribonucleic acid (DNA) to an organism by artificial means does not in itself pose an ecological risk. Most populations of organisms vary considerably among individuals in the complement of genes carried in the DNA. The difference between genetically engineered organisms and their nonengineered counterparts is usually only one gene or a handful of genes. Genetically engineered organisms have new genes, placed into their DNA using molecular technology, that they would not have by traditional breeding methods. Because such a small fraction of genes is typically altered by genetic engineering, the possible risks or ecological effects of this new technology lie with the particular effects these novel genes have on the physical and physiological characteristics (the phenotypes) of the newly transgenic carriers. Though only a small amount of DNA may be involved, small changes in DNA can have profound effects.

For example, genetically engineered bacteria can metabolize petroleum products, making them useful in the cleanup of oil spills. Plants have been engineered for a variety of functions, from the uptake of heavy metals (for use in the cleanup of toxic mine tailings), to the resistance of diseases and herbivores (for use, instead of susceptible plants, as crops), to the production of modified food products (high-laurate canola oil and delayed-ripening tomatoes). Even mammals have been genetically engineered to produce in their milk life-saving drugs such as human insulin.

Ecological risks. Changing the way organisms interact with each other and their environment results in a change in their ecology. Ecological risks associated with the release of genetically engineered organisms include unexpected consequences of new physical and physiological characteristics. These organisms have altered ecologies by design, and some differences are expected and desired. The risks of unexpected changes in ecology are of two sorts. First, there are the risks that genetically engineered organisms will disrupt the way that individuals within a species interact, the way that the genetically engineered organisms interact with other species, and the way that genetically engineered organisms influence the environment. Second, there are the risks that the new gene in genetically engineered organisms will move into other species that are not genetically engineered, and that those other species, now armed with these new genes, will incur the hazards of the first sort.

Ecological risk assessment, then, is the evaluation of these two kinds of risks. In particular, assessors of ecological risk question the different ways genetically engineered organisms interact with individuals of the same species, different species, and the environment. This altered interaction may threaten the stability of communities of organisms that rely on the complicated interactions among species for survival. In addition, new genes may move from the organism into which they were engineered into organisms where they were never intended to be. Newly modified organisms may interact in new and hazardous ways with each other, other species, and the environment.

As a result of transgenes, there is almost always some change in the interactions of genetically engineered organisms with conspecifics, with other species, or with the environment. After all, typically genetically engineered organisms are designed to be changed in ecologically meaningful ways. The risk comes from the possibility of those changes leading to a chain reaction of unexpected interactions throughout food webs. For instance, it is plausible that a newly insect-resistant plant could drive the herbivore that usually eats it to local extinction. Another herbivore, previously unimportant because of its competitive inferiority to the first insect, could become a serious problem to these plants. Thus, engineering the plant to resist one herbivore could result in no long-term protection of the plant, just a change in the problem. In fact, the cascade of possible consequences of a change in one species to the rest of a community is as diverse as the web of complicated linkages that tie the community together. These ecological risks remain largely unexplored, mainly because they are ungeneralizable: knowing that such a risk exists for one genetically engineered organism in one ecological community does not help in understanding whether the same challenge will exist for other genetically engineered organisms or in other communities. These risks are unpredictable.

Gene escape. The focus of most ecological risk assessment of genetically engineered organisms has been in evaluating the second kind of ecological challenge, the more predictable risk of escape of a transgene from the genetically engineered organism to another population or another species. To understand this risk, it is useful to consider two broad classes of genetically engineered organisms. Engineered organisms are created for two reasons: (1) to improve ability to survive and reproduce under harsh conditions (examples include squash plants that can survive attack by viruses, corn plants that can withstand herbicide spray, and cotton plants that can fend off insect attack), and (2) to produce products that should not influence the ability of the organism to survive and reproduce (such as changes in the oil quality of canola seeds). The reasons differ in an important way: it is by improved survival and reproduction that genes can spread.

There is a great deal of information to determine the risk of a transgene moving from a transgenic plant to a nontransgenic plant. The risk has three levels. The first is the risk that pollen from a transgenic plant will reach a nontransgenic plant of the same species. This risk is substantial, particularly in plants that depend on pollinators such as bees. It has been established that this risk is a function of pollinator density, distance between the transgenic plant and the nontransgenic target, and the presence of intervening plants. The intervening plants can serve as places where pollinators can stop to drop the transgenic pollen before the pollen reaches the target. If the pollen is distributed by wind, distance and wind direction influence this level of risk. In plants that self-pollinate (that is, plants that reproduce asexually), this first level of risk is small.

The second level of risk associated with the movement of transgenes from one plant to another is that the gene can move from species to species. This risk is necessarily low, since species are defined as reproductively isolated units, but it is not absent. In fact, species definitions are less clear when plants are considered. Many related plant species, such as the *Brassica* species, which include cultivated broccoli, cabbage, and cauliflower as well as roadside weeds, are capable of some very limited gene exchange, despite being considered different species. This second level of risk depends on the probability that two species are not perfectly reproductively isolated, and it is higher when the species in question are very closely related.

The third level of risk associated with the movement of transgenes from one plant to another is that the gene will persist and spread once it arrives on the nontransgenic plant. Genes spread through populations when carriers survive and reproduce more successfully than individuals that do not carry the genes.

This process, called evolution by natural selection, is no different than evolution by natural selection acting on the natural variation that occurs in species. The risk depends almost entirely on what the gene does to the recipient organism in the environment in which it lives. If, for example, the gene increases plant disease resistance in a region where plants suffer from disease, the risk of spread of this gene is high. Plants with the new gene will gradually replace plants without it. If the gene increases petroleum metabolism in a region where there is no petroleum available, the risk is low.

Thus the risk of spread of transgenes from one species to another is the combined risks of the gene moving to a nontransgenic organism, successfully reproducing in that organism if it is not the same species as the genetically engineered organism, and persisting and spreading in that new species. In plants, all of these risks exist, though the combined risk is small and, under many conditions, unlikely. By contrast, ecologists still have no way to assess how risky genetically engineered organisms are to the stability of natural communities. Once these risks are assessed, they can be counterbalanced with assessments of the benefits of genetic engineering, and that balance can permit informed decisions about whether or not to use genetically engineered organisms.

For background information *see* BIOTECHNOLOGY; ECOLOGY; GENETIC ENGINEERING in the McGraw-Hill Encyclopedia of Science & Technology.

Juliette Winterer

Bibliography. J. Bergelson, J. Winterer, and C. Purrington, The ecological impact of transgenic crops, in V. Malik (ed.), *Biotechnology and Genetic Engineering of Plants*, Oxford University Press, 1998; M. A. Levin (ed.), *Engineered Organisms in Environmental Settings: Biotechnological and Agricultural Applications*, CRC Press, 1996; P. M. Kareiva and I. Parker, *Environmental Risks of Genetically Engineered Organisms and Key Regulatory Issues*, Greenpeace International, 1994; H. A. Mooney and G. Bernardi (eds.), *Introduction of Genetically-Modified Organisms into the Environment*, Scientific Committee on Problems of the Environment 44, John Wiley, 1990; J. Rissler and M. Mellon, *The Ecological Risks of Engineered Crops*, MIT Press, 1996; H. J. Rogers and H. C. Parkes, Transgenic plants and the environment, *J. Exp. Bot.*, 46(286):467–488, 1995.

Geographic information systems

A geographic information system (GIS) is an automated system for the capture, storage, retrieval, analysis, and display of spatial data. These systems have been in use since the 1970s to make the information contained in maps legible to the computer, so that the information can be linked to ordinary databases. For example, since 1790 the U.S. Bureau of the Census has conducted the census of population. Starting with experiments for the 1970 census, the Bureau has now completed a digital street map of the entire United States that can be linked to the census information. Users can make their own maps of any of the attributes recorded in the census tables and use these maps in making decisions.

Innovations in computing, in databases, and in graphics technology have recently led the GIS field into global scales for data handling. Now scientists can look at the global distribution of population and model its future patterns.

Improvements in GIS have been partly technical, including faster computers, more user-friendly GIS software, better storage, and the increasing sophistication of GIS use, now often called a science by itself (geographic information science). A powerful technological improvement has been the link between GIS, the Global Positioning System (GPS), and remote sensing. The Global Positioning System is a positioning and navigation system of satellites placed in orbit by the U.S. Air Force. The 24 satellites in 20,200-km (12,552-mi) orbits transmit highly accurate time signals that can be received and decoded into three-dimensional Earth positions (latitude, longitude, and elevation). While the system normally is used in a coarse acquisition mode, giving an accuracy of about 75 m (246 ft) on the ground, several methods can now improve this accuracy to less than a meter. This technology has allowed rapid creation and update of GIS databases, even in unmapped, remote, or rapidly changing areas. Software has allowed the data from the Global Positioning System to be directly input into GIS. For example, with a Global Positioning System antenna on a car roof, the user can drive along a road and simultaneously watch the road appear on a GIS map displayed on a portable computer. With the Global Positioning System operational worldwide 24 h a day and in all weather (although with some constraints due to vegetation and mountains blocking the signal), the late 1990s have seen the system collect and validate an astonishing amount of new Earth-related map data.

Satellite mapping and aerial photography have allowed an increase both in the amount of world mapping taking place and in the spectral and spatial details with which the Earth can be imaged. Remote sensing instruments can image at night, through cloud and smoke, and in parts of the electromagnetic spectrum invisible to the human eye. New sensors and the release of previously secret intelligence imagery have decreased the spatial resolution—the ground distance associated with a single picture element or pixel on the Earth—to less than 10 m (33 ft) and in some cases to approximately 1 m (3 ft). The U.S. Geological Survey's digital orthophoto maps, for example, include 1-m (3-ft) ground coverage at a scale of 1:12,000 for much of the United States.

The convergence of GIS with these new data sources has led to its broad-scale use for handling global data sets. The demand for global-scale work has grown from many needs, including the increased modeling of global systems such as weather, ocean circulation, and climatic variation. The GIS has become a major provider and integrator of data for weather prediction, for study of the El Niño-Southern Oscillation and its effects, and for experimentation with problems such as global warming, sea-level rise, and ozone depletion. Recent additions to this global-scale work have been the study of deforestation, especially in the tropical rainforests, biodiversity and habitat loss, and the land-use changes associated with human settlement and urbanization. *See* OCEAN CURRENTS.

For example, at the Earth Resources Data Center in Sioux Falls, South Dakota, researchers have devised a method for repeated assessment of the global condition of growing vegetation, using a numerical measurement called the Normalized Difference Vegetation Index, computed from satellite imagery collected by the National Oceanic and Atmospheric Administration's advanced very high resolution radiometer. Choosing sections of hourly radiometer images aggregated over periods of 2 weeks, the center is able to eliminate the effect of clouds and other atmospheric disturbances and provide images of the growing conditions for vegetation and agricultural crops on a global scale. One highly successful application of these data has been by the U.S. Agency for International Development in the estimation of the condition of the African food harvest by the Famine Early Warning System (FEWS). The FEWS project uses this, and other sources, to anticipate and prevent major famine events in the area.

Many efforts are under way to generate global maps of geographical variables. An initial effort by the U.S. Defense Mapping Agency (now the National Imagery and Mapping Agency) produced the Digital Chart of the World, with 14 data layers at the equivalent of the 1:1M map scale, and including cities, place names, transportation, and topography. Global mapping of topography has improved from the older 3-arcsecond data set from the National Geophysical Data Center to a new 1-km (3281-ft) data set being assembled at the Earth Resources Data Center. Other new global-scale data sets include hydrography (rivers, glaciers, and lakes); land characteristics, including land use; vegetation according to the Matthews classification; soils using the Zobler classification; population from the Global Demography project at the National Center for Geographic Information and Analysis; Chris Elvidge's Nighttime Lights of the World imagery; carbon dioxide emissions from fossil fuel burning; United Nations Environment Programme data on desertification; and major ecosystem complexes. Most of these data are available over the World Wide Web.

Another theoretical concern over data availability is the choice of a map projection. Most global data sets are made available with relatively little concern about how the three-dimensional surface of the Earth is to be depicted on a two-dimensional map or within a GIS database. Some map projections preserve the areas of features, an important consideration when global GIS data are to be used in global systems modeling; but others distort area to preserve shape, direction, or some other map property. While GIS software is capable of transforming between map projections so that, for example, two of the above map layers can be compared, little is known about how the inherent distortion in the projections affects the modeling and subsequent use of the data. Often these errors are maximal at places that are also of maximal geographic interest, such as over the Poles. While classical cartography is good at describing what patterns of global distortion are inevitable, what actually happens inside a GIS, where data are grid cells and where computational errors such as rounding and averaging are common, remains the subject of cartographic research.

The GIS has a unique role when global data are concerned. Most global models, human or physical, require vast amounts of data to be brought together in a common geometric reference frame. At this stage, the GIS can be used to conduct retrieval operations across layers and then to perform analyses of the results. For example, the data described above could be used to extract those areas of the world above 1000 m (3281 ft) in elevation and that have large concentrations of population on fertile soil close to rivers. This is called an overlay operation. The GIS also can do analysis. Examples are computing distances from major roads, adding Strahler stream order to a river network, or selecting a site for waste storage. Increasingly, the GIS is used to assemble data for modeling. These models can be of urban growth, land-use change, global air circulation, contaminant transport in the ocean, flooding, earthquakes, or other phenomena. While most models use the data outside the GIS software, the results are often passed back into the software to perform further analysis. Another important role for GIS is in supporting data visualization and mapping, using methods such as three-dimensional representation, animation, and virtual imaging. These powerful methods offer some value in data exploration, allowing the scientist who can visualize the global data sets to perceive systematic relationships and data connections that would be impossible to detect if the data were examined one theme at a time.

For background information *see* COORDINATE SYSTEMS; DATABASE MANAGEMENT SYSTEMS; DECISION SUPPORT SYSTEM; GEOGRAPHIC INFORMATION SYSTEM; REMOTE SENSING; SATELLITE NAVIGATION SYSTEMS; TERRESTRIAL COORDINATE SYSTEM in the McGraw-Hill Encyclopedia of Science & Technology.

Keith C. Clarke

Bibliography. K. C. Clarke, *Getting Started with Geographic Information Systems*, 2d ed., Prentice Hall, 1999; M. F. Goodchild, L. T. Steyaert, and B. O. Parks, *GIS and Environmental Modeling: Progress and Research Issues*, GIS World Books, 1996; D. R. Steinwand, J. A. Hutchinson, and J. P. Snyder, Map projections for global and continental data sets and an analysis of pixel distortion caused by reprojection, *Photogrammetric Eng. Remote Sens.*, 61(12):1487–1497, 1995.

Geological time scale

The history of the Earth can be displayed in the form of a calendar which is based on the observation of rocks formed through time. Depending on the location, the formation of rocks has recorded different (regional) histories describing the magmatic, tectonic, hydrospheric, or biospheric evolutions, and thus different calendars have been generated. Geologists (stratigraphers) are attempting to unify these different calendars. Stratigraphers define rock units bracketed between two boundaries that can be correlated worldwide. The succession of these modern units is the global geological time scale.

Stratigraphical units. In general, the evolution of the Earth is continuous, so fixing the location of the unit boundaries can be achieved only through convention (decisions are made under the aegis of the International Commission on Stratigraphy). Stratigraphers use a variety of tools for characterizing the age of a rock, including physical ages, fossils, magnetism, and chemical properties, all of which have evolved through time. There are three intervals in the history of the Earth (Archean-Proterozoic, Phanerozoic, Plio-Quaternary) with each one showing distinct material available for characterizing rocks; thus, there

Eon	Era	Period	Ma
PHANEROZOIC	PALEOZOIC	CAMBRIAN	
			540
	NEO-PROTEROZOIC	NEOPROTER. III	
			650
		CRYOGENIAN	
			850
		TONIAN	
			1,000
	MESO-PROTEROZOIC	STENIAN	
			1,200
		ECTASIAN	
			1,400
		CALYMMIAN	
			1,600
PROTEROZOIC	PALEO-PROTEROZOIC	STATHERIAN	
			1,800
		OROSIRIAN	
			2,050
		RHYACIAN	
			2,300
		SIDERIAN	
			2,500
	NEOARCHEAN		
			2,800
	MESOARCHEAN		
ARCHEAN			3,200
	PALEOARCHEAN		
			3,600
	EOARCHEAN		–4,550 +20/–80

Fig. 1. Agreed-upon geological time scale of the pre-Phanerozoic history of the Earth. The beginning of the Earth is estimated to be 4,470 to 4,570 Ma, with a preferred age at 4,550 Ma.

are three successive kinds of geological time scale. In the earliest time scale, during Archean and Protero-zoic eons, fossils are rare or absent in most rocks, and the major tool is the physical dating process based on naturally unstable isotope decay used by geochronologists. For this interval of time, the bound-aries of the calendar units are defined by selected nu-merical ages (**Fig. 1**). These ages are conventionally abbreviated Ma (Mega anna, or million years, follow-ing recommendation of the Subcommission on Geo-chronology).

During the Phanerozoic Eon certain animals and plants elaborated skeletons or exoskeletons that, when preserved as fossils, provide an additional means of time correlation (**Fig. 2**). Up to about 5 Ma, these fossils become the major tool for determin-ing the relative age of a rock. Together with other physicochemical characteristics, fossils allow a defi-nition of stages which cover an average duration of about 5 Ma each. For the last 5 Ma, Plio-Quaternary time, there is evidence that the evolution of the envi-ronment on Earth is diversified and well represented in the geological record.

Most stratigraphical tools can be used continu-ously in this young interval, each tool providing a specific time scale. Because several tools can often be used to characterize (that is, to date) a single rock, it is easy to connect all these time scales. A variety of easily interchangeable time scales are useful. However, the absolute time dimension for all time scales can be calibrated only by using geochronol-ogy. The progress toward a common terminology and geological time scale benefits from three fac-tors: better definition of the conventional units, bet-ter geochronological calibration, and a potential for extrapolating between calibrated ages.

Geochronological calibration. Progress in geochro-nological calibration benefits from both new tech-nology and new geochronological information. Dur-ing the 1990s, the precision and reproducibility of the measurements obtained using mass spectrome-try have increased significantly. This improvement offers the possibility of dating minute quantities of certain material with remarkable precision. Using the uranium-lead (U-Pb) dating method, the age of a single crystal of zircon, 200 micrometers in length and 10 μm in diameter, can be measured. More sig-nificantly, several portions of this same crystal can be dated separately, allowing for verification of the internal consistency of the data obtained from the single crystal. For the potassium-argon (K-Ar) dating method, developments in the irradiation techniques which transform the original potassium into argon allow single biotite mica flakes 500 μm in diameter and 10 μm thick to be dated. Laser heating can also give several ages obtained from different points of a single crystal.

Explosive volcanic eruptions producing ash clouds blown over large areas are common. The abil-ity to date small quantities of material has led to the search for minor volcanic events within the strati-graphical record. This advance is important because the same volcanic material covers both marine and continental areas, sometimes at the scale of a conti-nent, allowing correlation of distant deposits. In ad-dition, there is great interest in this method since ex-plosive volcanism often scatters a variety of minerals, including uranium-bearing and potassium-bearing ones. Thus, the geochronologist can perform mea-surements on independent isotopic systems in order to make a reciprocal check of the validity of the cal-culated ages. The calibration of the geological time scale is mostly realized through the study of crystal-bearing volcanic dust discovered in sediments. Pre-cise dating can also be achieved through a variety of other datable materials. For example, a few tens of milligrams of microtektite particles (scattered in wide areas when an asteroid collides with the Earth) can be dated. Another example is calcite crystal-lized in paleosols during cyclic deposition in shal-low basins. Because this calcite is associated with organic matter which favors uranium enrichment, and is formed essentially at the time of deposition, calcite crystals become a potentially datable material using uranium-lead methods.

Two examples of refinement. One example of re-cent refinement concerns the dating of the Eocene-Oligocene boundary. That boundary has long been known as the *Grande Coupure* (great break) in the history of European land-mammal evolution. There have been a few European studies which docu-mented an age at about 34 Ma. But the age of the boundary was assumed to be about 37–38 Ma by some North American geologists. Paleontologists did not like the latter age because, in North America, the 38-Ma-old mammals were dated using contempora-neous volcanic flows and seemed less evolved than those known to be at the stratigraphical boundary in Europe. The problem was solved due to a better boundary stratotype discovered near Ancona in east-central Italy. Minerals sampled from several layers of volcanic dust interbedded in the marine deposits of the stratotype were dated. The results indicated an age of about 33.7 ± 0.5 Ma for the boundary. This result confirmed the later of the two previous pro-posals demonstrating that the evolution of mam-mals in Europe and North America was synchro-nous.

Another significant example of the beneficial com-bination of improved stratigraphical definition and modern geochronological dating is given by the Precambrian-Cambrian boundary. The base of the Cambrian (and of the Phanerozoic Eon) had long been placed at the first occurrence of skeletalized fossils including trilobites (arthropods). Later, a Tom-motian pretrilobitic stage was added below it in view of the presence of older faunal remains, such as ar-chaeocythids (calcitized spongelike forms), which have been well documented on the Siberian Plat-form. Before 1980, the earliest skeletalized faunas were estimated to be between 570 and 590 Ma. How-ever, independent geochronological data were gath-ered from northern France, southern Britain, Mo-rocco, and Israel in the early 1980s. These data were

Fig. 2. Simplified geological time scale of the Phanerozoic Eon.

Left table

Era	Period	Epoch	Stages	Ma	+/-
CENOZOIC	NEOGENE	QUATERNARY		1.75	0.05
		PLIOCENE		5.3	0.15
		MIOCENE	MESSINIAN	7.3	0.15
			TORTONIAN	11.0	0.3
			SERRAVALLIAN (+LANGHIAN)	15.8	0.2
			BURDIGALIAN	20.3	0.4
			AQUITANIAN	23.5	1
	PALEOGENE	OLIGOCENE	CHATTIAN	28	1
			RUPELIAN	33.7	0.5
		EOCENE	PRIABONIAN	37.0	1/0.5
			BARTONIAN	40	1
			LUTETIAN	46.0	1/0.5
			YPRESIAN	53	1
		PALEOCENE	THANETIAN	59	2
			DANIAN	65.0	0.5
MESOZOIC	CRETACEOUS	LATE	MAASTRICHTIAN	71.5	0.5
			CAMPANIAN	83	1
			SANTONIAN	87	1
			CONIACIAN	88	2
			TURONIAN	92	2
			CENOMANIAN	96	2
		EARLY	ALBIAN	108	3/1
			APTIAN	113	3
			BARREMIAN	117	5/2
			HAUTERIVIAN	123	6/2
			VALANGINIAN	131	4
			BERRIASIAN	135	5
	JURASSIC	LATE	TITHONIAN	141	?/5
			KIMMERIDGIAN	146	—
			OXFORDIAN	154	5
		MIDDLE	CALLOVIAN	160	2
			BATHONIAN	164	2
			BAJOCIAN	170	4/3
			AALENIAN	175	3
		EARLY	TOARCIAN	184	3
			PLIENSBACHIAN	191	—
			SINEMURIAN	200	4/?
			HETTANGIAN	203	3
	TRIASSIC	LATE	RHETIAN	—	—
			NORIAN	220	—
			CARNIAN	230	6
		MIDDLE	LADINIAN	233	5
			ANISIAN	240	5
		EARLY	OLENEKIAN	—	—
			INDUSIAN	250	3

Right table

Era	Period	Epoch	Stages	Ma	+/-
	PERMIAN	LATE	(TATARIAN)	255	
			(KAZANIAN)	258	
		EARLY	(KUNGURIAN)	265	
			(ARTINSKIAN)	275	
			(SAKMARIAN)	285	
			(ASSELIAN)	295	5
	CARBONIFEROUS	LATE (SILESIAN)	(GZHELIAN)	—	—
			(KASIMOVIAN)		
			(MOSCOVIAN)		
			(BASHKIRIAN)		
			(SERPUKHOVIAN)	325	5
PALEOZOIC		EARLY (DINANTIAN)	(VISÉAN)	345	—
			(TOURNAISIAN)	355	5
	DEVONIAN	LATE	FAMENNIAN	370	5
			FRASNIAN	—	—
		MIDDLE	GIVETIAN	380	
			EIFELIAN	390	5
		EARLY	EMSIAN	400	—
			PRAGIAN	—	—
			LOCHKOVIAN	410	8/5
	SILURIAN	PRIDOLI		415	—
		LUDLOW		425	5
		WENLOCK		430	6
		LLANDOVERY		435	6/4
	ORDOVICIAN	ASHGILL		445	4
		CARADOC		455	5
		LLANVIRN		465	5
		ARENIG		480	5
		TREMADOC		500	
	CAMBRIAN	LATE		—	—
		MIDDLE		520	—
		EARLY	(LENIAN)	525	5
			(ATDABANIAN)	530	5
			(TOMMOTIAN)		
			("PALEOCAMBRIAN")	540	5

PRECAMBRIAN

Fig. 2. Simplified geological time scale of the Phanerozoic Eon. From about 120 possible stages, only one-fourth are formally defined in modern terms. Among others, many are mostly used regionally (in parentheses) with less precise boundaries and content compared to formal ones. Stages are not shown for some epochs for which no realistic subdivision can be recommended today. Ages given in two columns are shown without error bar when they are obtained from an interpolation procedure. For some boundaries, the error bar is asymmetrical; for example, the Aptian-Albian boundary has a preferred age at 108 Ma, but this age is constrained only between 111 and 107 Ma. (*After G. S. Odin, Geological Time Scale, C. R. Acad. Sci., 318:59–71, 1994*)

obtained from levels located below the first occurrence of trilobites in the different countries. The data showed that trilobites were younger than 530 ($\pm$10) Ma. In the following years, older faunas known as small shelly fossils contemporaneous with trace fossil assemblages were discovered in China, Australia, the Siberian Platform, and Canada. A modern Precambrian-Cambrian conventional boundary was definitely fixed in 1992 at the base of this fauna in Canada. From new geochronological information obtained from volcanic zircon sampled from the above locations, an age of 540 ($\pm$5) Ma was documented for that boundary. This has been of great consequence, considering that the end of the Cambrian is about 500 Ma. The apparently extraordinary radiation of skeletalized metazoans observed within the Cambrian took only a few tens of millions of years (instead of 100 Ma as thought in the mid-1980s). This extraordinary radiation must be compared to the evolution observed over the next 500 Ma during which no new important phyla were created. Two examples that help provide better understanding of geological phenomena connected to the precise dating of geological strata are the short duration of the important biological cuts occurring at the Permian-Triassic (Paleozoic-Mesozoic) boundary and the Cretaceous-Palaeogene (Mesozoic-Cenozoic) boundary.

Extrapolation procedures. The direct geochronological calibration method will never allow for the continuous calibration of every point of geological history, since datable material is much too scarce in rocks. However, continuous dating can be refined through the use of interpolation procedures between geochronologically calibrated points. This principle consists in combining those tie points with a continuous geological phenomenon. Commonly used phenomena are rhythmic sedimentation and the oceanic record of past magnetic fields. When the rhythmic deposition of sediments can be related to the orbital (Milankovitch) parameters of the Earth, the time scales of which are reasonably well understood, the duration of deposition can be estimated when combined with nearby measured ages.

Another procedure considers the aperiodic change (reversal) of the direction of the Earth's magnetic field that is recorded in the oceanic plates being continuously formed at midocean ridges (separating two tectonic plates). For a given plate, the distance between two magnetic reversals is proportional to the time durations between reversals and spreading rate (which can be calculated from two geochronologically dated points). Thus, geological ages can be calculated for each point of the record, though with some degree of uncertainty.

The geological time scale is gradually becoming unified through an internationally agreed upon scale which is replacing a variety of regional scales. It is ironic that such an important improvement in calibrating this vast expanse of time is linked to the discovery of volcanic dust interbedded in sedimentary rocks.

For background information *see* ARCHEOLOGICAL CHRONOLOGY; DATING METHODS; FOSSIL; GEOCHRONOMETRY; GEOLOGICAL TIME SCALE; INDEX FOSSIL; RADIOCARBON DATING; ROCK AGE DETERMINATION; SEDIMENTARY ROCKS; STRATIGRAPHIC NOMENCLATURE; STRATIGRAPHY in the McGraw-Hill Encyclopedia of Science & Technology.   G. S. Odin

Bibliography. H. Blatt, W. B. N. Berry, and S. Brande, *Principles of Stratigraphic Analysis*, 1991; J. P. Grotzinger et al., Biostratigraphic and geochronologic constraints on early animal evolution, *Science*, 270:598–604, 1995; G. S. Odin, Geological time scale, *C. R. Acad. Sci.*, 318:59–71, 1994; G. S. Odin et al., Numerical dating of Precambrian-Cambrian boundary, *Nature*, 301:21–23, 1983; G. S. Odin and A. Montanari, Radio-isotopic age and stratotype of the Eocene/Oligocene boundary, *C. R. Acad. Sci. Paris*, 309:1939–1945, 1989.

Global Positioning System (GPS)

The Global Positioning System is a satellite-based system providing worldwide continuous position, velocity, time, and related data to civil and military users. It has a growing number of applications in the fields of marine, land, and aerospace navigation and precise time and time transfer, as in surveying, geodesy, and mapping; precision farming; air-traffic control; asset location and tracking; and timing of communication systems and power grids. Since the 1960s, GPS has grown from a navigation concept to an operational system of about 24 spacecraft (**Fig. 1**) serving millions of users. Over a million GPS receivers a year were produced during 1997–1999.

GPS has performed extremely well and has generally exceeded expectations. However, a number of deficiencies have been identified, and some significant improvements are needed that could be implemented with the new GPS replenishment spacecraft.

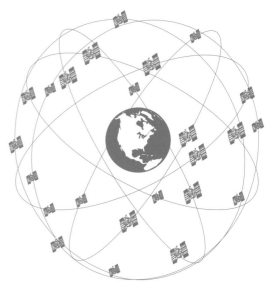

Fig. 1. Constellation of operational GPS spacecraft.

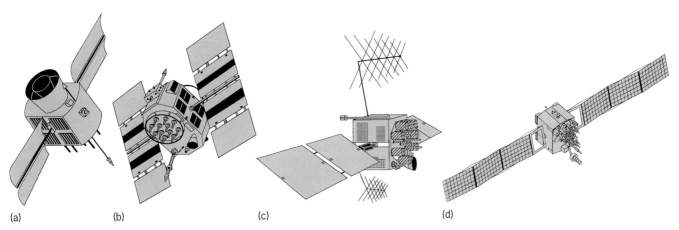

Fig. 2. Generations of GPS spacecraft. (*a*) Block I. (*b*) Block II-IIA. (*c*) Block IIR. (*d*) Block IIF.

Satellites. The satellites' limited lifetime in orbit, about 7.5 years for the current operational Block II and IIA spacecraft (**Fig. 2**), establishes the need and deployment schedule for their replacements. Twenty-one third-generation (Block IIR) replenishment spacecraft have been ordered by the U.S. Department of Defense to continue the GPS constellation to 2010 and possibly beyond. In July 1997, the first of these was launched to replace the Block II and IIA spacecraft that will phase out by about 2005.

The Department of Defense has also contracted for 6 of a planned 30 fourth-generation, follow-on (IIF) spacecraft. These are to replace the IIR spacecraft and will carry the GPS constellation well be-

yond 2010. The Delta 2 launch vehicle (**Fig. 3**) carries these spacecraft to their medium-altitude orbits (20,180 km or 10,898 nautical miles above the Earth).

Modernization activities. A number of committees have investigated the needs and deficiencies of the GPS in order to determine what capabilities and features should be incorporated into a future system to satisfy both military and civil users. The modernization will include a new frequency, new signals, higher signal power levels, more extensive ground tracking, and more frequent spacecraft position updates, all of which will dramatically improve accuracy, integrity, and other aspects of performance. The management of GPS has also changed and now involves coordinated civil and military funding and oversight. These factors, combined with the increasing worldwide importance of navigation systems and services, provide a basis for integrating GPS into an international Global Navigation Satellite System (GNSS) consisting of a number of independent but coordinated elements. The modernized GPS will continue to play a central role in providing position, velocity, attitude, and time services in an economical manner.

Selective availability. Removal of selective availability (SA) is scheduled between 2000 and 2006, in accordance with the Presidential Decision Directive on GPS of March 29, 1996. This removal will provide undegraded accuracy of the signals for civil users. This modification, together with the additional civil signal frequencies (which include means for correcting ionospheric delay errors), will improve civil GPS performance by an order of magnitude or more, to a position determination accuracy of 5 m (15 ft) or better, by 2010 (**Fig. 4**).

Civil signals. New civil signals are planned, including one centered at frequency L2 (1227.6 MHz), which has heretofore been used exclusively by the military (**Fig. 5**). The long-standing civil signal centered at L1 (1575.42 MHz) will be retained. On January 25, 1999, Vice President Gore announced that a third civil frequency, L5, had been selected at 1176.45 MHz in

Fig. 3. Delta 2 ready to launch a Block II spacecraft.

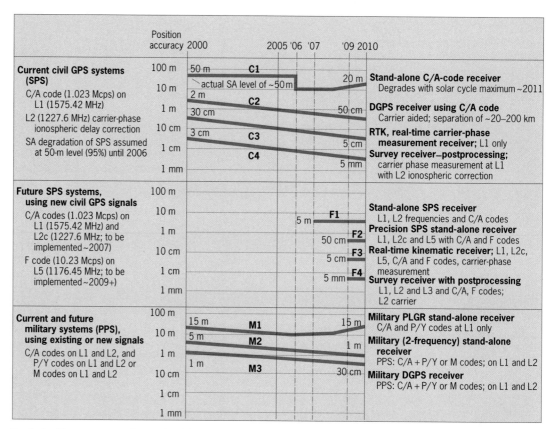

Fig. 4. Position accuracy estimates for civil and military GPS receivers in various modes of operation for the 2000–2010 period. SPS is Standard Positioning Service; PPS, Precise Positioning Service; SA, selective availability; M codes, new military codes; DGPS, differential GPS; RTK, real-time kinematic; PLGR, precise lightweight GPS receiver. Position accuracy means horizontal position accuracy at 95% confidence level. 1 km = 0.6 mi; 1 m = 3.3 ft; 1 cm = 0.4 in; 1 mm = 0.04 in.

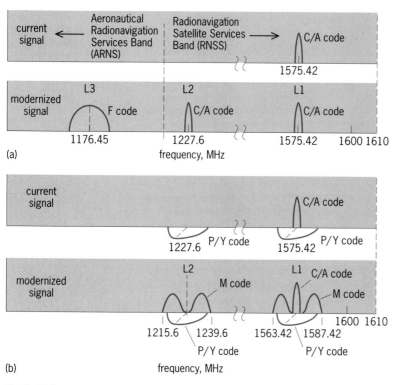

Fig. 5. GPS current and modernized signals. (*a*) Civil signal spectrum. (*b*) Military signal spectrum.

the Aeronautical Radionavigation Services band. This selection was intended principally to satisfy aviation safety concerns but also to benefit applications requiring real-time kinematic measurements. (Basically, these are precision measurements of carrier phases for a number of GPS signals that are measured between two differential receivers. They are done in real time, or near real time, and provide almost survey-quality position information, currently about 5–20 cm or 2–8 in.) The new arrangement provides to civil users capabilities for correction of errors caused by ionospheric delays, increased signal robustness, and improved techniques for resolving the cycle ambiguities associated with precision carrier-phase measurements.

Signal structures. The new civil signal at L5 is planned at a code rate ten times that of the Coarse/Acquisition (C/A) code, which is now available to civilian users, and with a 1-millisecond period. These specifications will improve measurement accuracy, reduce noise, and provide improved mitigation of multipath errors. New military signals named M codes will provide improved measurement accuracy, a desirable power distribution in the spectrum, and a capability for direct access. (In order to acquire the P/Y code, currently employed in military applications, authorized users must normally first access the civil C/A code, which contains information about the timing of the military code. No such procedure is needed to access the M code.) The civil signals have most of their power in the center of their bands, while the military signals have most of their power in the outer regions of their bands (Fig. 5). The existing civil and military signals will remain available throughout the decade 2000–2010, while the new military (M-code) signals and the new civil signals at L2 and L5 are to be introduced during the latter part of the decade.

Receiver system. Improved receivers such as narrow correlator types, which provide considerably lower noise and other desirable performance characteristics, will become generally available and commonly employed. Also, the use of carrier-phase (real-time kinematic) measurements to obtain precision position, velocity, attitude, and time determinations will become commonplace. For example, position precision at the 2–10-cm (1–4 in.) level will become available in moderate-cost receivers using phase and wide-lane measurements of the three civil frequencies. (Wide-lane measurements are based on the "lanes" formed by the wavelength corresponding to the difference in frequency between signals. For example, the wide lane formed by the L1 and L2 frequencies, separated by 347.72 MHz, is about 86 cm.)

Spacecraft. The GPS spacecraft will offer increased signal availability and power, and also have greater reliability and longer lifetimes. Power in the new civil L2 (C/A-code) signal is to be consistent with the L1 civil signal, which may be increased in power level by about 6 dB for greater system robustness. Power in the military M-code signals is to be substantially greater (by about 6–10 dB or more at times) than the current P/Y military code signal power levels. While an increased number of spacecraft (30–36) in the GPS constellation cannot be assured, there is strong interest in this expansion.

Error reduction. Systematic errors will be reduced not only by the removal of selective availability and by ionospheric error correction but also by substantial improvements in GPS receivers and in the control segment. The GPS ground control system will be expanded by the addition of six or more tracking stations, principally by incorporating those of the National Imagery and Mapping Agency. More frequent uploads to the GPS spacecraft are also planned. Control segment determinations of spacecraft position and prediction errors will improve from about 2 m (80 in.) to 10–50 cm (4–20 in.).

Augmentations. Augmentations supporting improved performance will become available worldwide before 2010. Augmentations include the U.S. Coast Guard Differential Network (available now), the U.S. Federal Aviation Administration's Wide Area Augmentation System (WAAS) and Local Area Augmentation System (LAAS), the European Geostationary Navigation Overlay System (EGNOS), and the Japanese Mobile Satellite Augmentation System (MSAS). Also, a large number of other differential GPS systems are in use or will become available that

can provide highly precise position, velocity, attitude, and time measurements.

International implications. The GPS has become the de facto standard for navigation satellite system operations, but there have been long-standing concerns internationally because of the United States military origin and control of the system. However, system and institutional changes have occurred such that GPS now has a joint civil-military management structure and provides independent civil and military capabilities, both of which are being considerably improved. Additionally, GPS has an important role as a resource worldwide. The management of GPS appears ready to take a significant role in an international Global Navigation Satellite System. On February 10, 1999, the European Commission requested the governments of the 15 states in the European Union to support the development of an advanced system called the Galileo project. Proposals have been made for the GPS frequencies and civil signal structure to be integrated into the Galileo system and possibly into the MSAS as well. The Europeans indicate a strong desire for Galileo to support their launch vehicle, spacecraft, and ground control system industries. There is also the potential for a coordinated global navigation satellite system capability involving GPS as a principal element. The transition to an international system is likely to occur before 2010.

For background information *see* SATELLITE NAVIGATION SYSTEMS in the McGraw-Hill Encyclopedia of Science & Technology.   Keith D. McDonald

Bibliography. K. D. McDonald, The GPS modernization dilemma and some topics for resolution, *Quart. Newsl. Inst. Navig.*, vol. 8, no. 2, summer 1998; K. D. McDonald, The modernization mantra, *GPS World*, Directions '99, 9(12):46, December 1998; K. D. McDonald, Technology, implementation and policy issues for the modernization of GPS and its role in a GNSS, *J. Navig.*, vol. 51, no. 3, Royal Institute of Navigation, September 1998.

Global warming

Carbon dioxide (CO_2) is a major greenhouse gas, and its level in the atmosphere has been increased by human activities. This increase should bring about global warming. Anthropogenically produced atmospheric carbon dioxide becomes involved in the global carbon cycle and may be stored in the oceans, soil, and living things. The fluxes between the atmosphere and various reservoirs are poorly known but must be understood in order to predict future rises in atmospheric carbon dioxide and global warming. There are also a number of other greenhouse gases whose effect on global warming is nearly comparable to that of carbon dioxide.

Atmospheric carbon dioxide. Carbon dioxide is the fourth most abundant gas in the atmosphere and is important in maintaining the Earth's surface temperature. Along with water vapor, it absorbs infrared

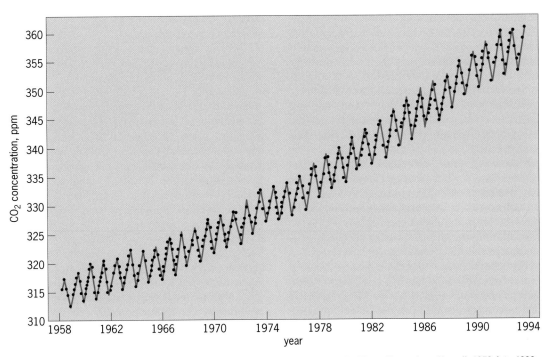

Fig. 1. Mean monthly concentration (parts per million) of atmospheric carbon dioxide at Mauna Loa, Hawaii, 1958–late 1993, as measured by C. D. Keeling and associates. The yearly oscillation is explained mainly by the annual cycle of photosynthesis and respiration of plants in the Northern Hemisphere. (1 ppm $CO_2 = 2.12$ Gt C $= 2.33 \times 10^9$ tons C.) (*After E. Berner and R. Berner, Global Environment: Water, Air and Geochemical Cycles, Prentice Hall, 1996*)

(long-wave) radiation given off by the Earth and reradiates this energy back to the Earth's surface, keeping the Earth much warmer ($33\,°C$ or $59\,°F$) than it would be otherwise (the so-called greenhouse effect).

The concentration of atmospheric carbon dioxide has been measured on Mauna Loa in Hawaii since 1958 (**Fig. 1**), and there has been a sharp increase from 315 parts per million in 1958 to 361 ppm in 1995. This carbon dioxide increase is anthropogenic and is primarily due to the burning of fossil fuels (oil, natural gas, and coal) which contain carbon (and to a minor extent the increase is due to cement production). An additional source of atmospheric carbon dioxide is deforestation and burning of trees, which also contain carbon.

Measurements of carbon dioxide concentrations in bubbles trapped in Greenland ice suggest that there has been a rise in atmospheric carbon dioxide beginning in the 1850s and starting from about 280 ppm, referred to as the preindustrial concentration. Because of the greenhouse effect due to carbon dioxide, this rise from 280 ppm to that at present, along with predicted further increases in the future, is expected to cause global warming.

Carbon cycle. Carbon dioxide contains carbon, the dominant element in life and in biogeochemical cycles on Earth. The concentration of atmospheric carbon dioxide has a conspicuous natural yearly oscillation (Fig. 1) because of its uptake due to excess photosynthesis and storage of carbon by plants during the growing season, and carbon dioxide release to the atmosphere in the fall and winter when plant decay through respiration dominates. Plants plus the soil are an important reservoir of carbon on the Earth's surface (**Fig. 2**). In fact, they are a much larger reservoir than atmospheric carbon dioxide. The largest reservoir of carbon is the oceans, where atmospheric carbon dioxide is dissolved and converted to inorganic bicarbonate (HCO_3^-) which is stored in the oceans. The surface ocean (top 100 m or 328 ft) exchanges carbon dioxide with the atmosphere and mixes slowly with the deeper ocean (over about 1000 years), which allows long-term carbon storage at greater depths.

Anthropogenic carbon budget. The average yearly amount of carbon in the form of carbon dioxide produced by fossil fuel burning and cement production (for 1990) was estimated as 5.9 Gt C/y (gigatons of carbon per year; 1 Gt $= 10^9$ metric tons $= 1.1 \times 10^9$ tons). An additional source of carbon dioxide from net tropical deforestation (the excess of deforestation over growth in tropical forests) is poorly known, with estimates varying from 0.2 to 1.6 Gt C/y. The total anthropogenic carbon dioxide emissions are thus estimated as 6.1–7.1 Gt C/y.

Of this amount, only about half (2.8–3.3 Gt C/y) occurs as an increase in the concentration of atmospheric carbon dioxide. The rest is taken up at the Earth's surface. The oceans are thought to store 2.0 to 2.3 Gt C/y. There is also storage in the terrestrial biosphere. Evidence for increased activity by the terrestrial biosphere in northern latitudes comes from a longer growing season and a greater amplitude of the annual cycle of carbon dioxide caused by terrestrial vegetation. Northern forests which were heavily deforested in the nineteenth and early twentieth centuries and abandoned farmland are regrowing and are believed to be storing carbon. North America is a large terrestrial sink.

In addition to forest regrowth, anthropogenic

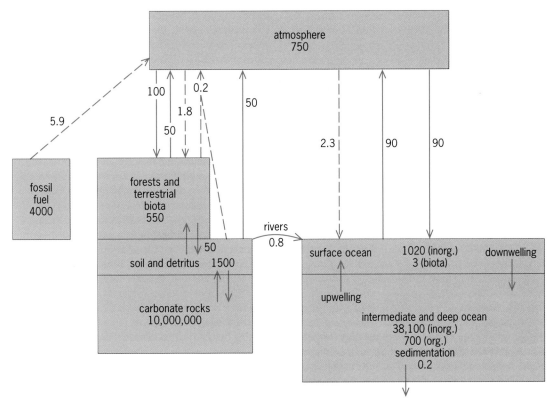

Fig. 2. Carbon cycle. Reservoirs are in Gt C, fluxes in Gt C/y. Broken-line fluxes are due to human activity; solid fluxes are natural. (*After E. Berner and R. Berner, Global Environment: Water, Air and Geochemical Cycles, Prentice Hall, 1996; also after the data in the table in this article*)

nitrogen deposition from fossil fuel burning enhances terrestrial carbon uptake, acting as a fertilizer (0.2–2.0 Gt C/y); much of this is in the Northern Hemisphere. This process is referred to as nitrogen fertilization.

Furthermore, global warming, if it has occurred, might enhance growth and carbon uptake, mostly in the North Temperate Zone. However, there is the possibility that global warming might have the opposite effect as well, by increasing the decomposition of soil organic matter to carbon dioxide. In addition, increased carbon dioxide concentrations themselves are thought to stimulate more plant growth, essentially causing carbon dioxide fertilization (0.5–2.0 Gt C/y); this is believed to be most important in the tropics, and would reduce net emissions from tropical deforestation.

The sum of these sinks for anthropogenic carbon dioxide is 6.9–7.1 Gt C/y, depending upon which estimates are chosen. The sources and sinks of carbon dioxide actually should balance. There has been progress recently in improving estimates, but clearly more information is needed. In order to predict the rise in atmospheric carbon dioxide concentration expected from further increases in carbon dioxide released by fossil fuel burning, it is necessary to know how much of the carbon dioxide stays in the atmosphere, where it actually contributes to global warming.

The term radiative forcing is often used to express the changes due to carbon dioxide and other gases and aerosols that cause global warming or cooling.

Radiative forcing is the change in average radiation at the top of the troposphere (lower atmosphere) because of a change in either solar or infrared (earth) radiation. The doubling of carbon dioxide alone from its preindustrial value accounts for a mean radiative

Average anthropogenic carbon budget*	From IPCC	From Fan
Carbon dioxide sources		
Fossil fuel combustion and cement production	5.5 ± 0.5[†]	5.9 (1990)
Net emissions from tropical deforestation	1.6 ± 1.0	0.2 ± 0.9
Total anthropogenic emissions	7.1 ± 1.1	6.1
Carbon dioxide sinks		
Atmospheric storage	3.3 ± 0.2	2.8 (1990)
Ocean uptake	2.0 ± 0.8	2.3
Northern forest regrowth	0.5 ± 0.5	
Other terrestrial sinks (CO_2 fertilization, N fertilization, climate effects)	1.3 ± 1.5	
Total Northern Hemisphere terrestrial uptake		1.8 ± 0.5
Total sinks	7.1	6.9

*In gigatons (10^9 tons) of carbon per year (Gt C/y). 1 Gt C/y = 10^9 metric tons = 1.1×10^9 tons.
[†]Average value for 1980–1989.
Source: Data from Intergovernmental Panel on Climate Change, *Climate Change, 1995: The Science of Climate Change*, Cambridge University Press, 1996; S. Fan et al., A large terrestrial carbon sink in North America..., *Science*, 282(5388):442–446, 1998.

forcing increase of 4 W/m². Taking internal feedbacks involving such things as water vapor and ice into account, the increase in global mean surface temperature for a doubling of carbon dioxide has been calculated to be 1.5–4.5°C (2.7–8.1°F); the best estimate of the temperature change is an increase of 2.5°C (4.5°F). The predicted temperature increases are greatest at high latitudes in late autumn and winter.

Other greenhouse gases. There are a number of other anthropogenically produced greenhouse gases whose concentration has increased and which contribute to global warming. Methane (CH_4) occurs naturally, but its atmospheric concentration has been increased by various human activities such as fossil fuel production and use, rice paddies, raising animals, landfills, and biomass burning. Methane is the second largest contributor to radiative forcing at 0.5 W/m², resulting from the increase in its concentration since preindustrial times.

Nitrous oxide (N_2O) occurs naturally, but its atmospheric concentration has also been increased since preindustrial times by humans. The various sources of nitrous oxide include agriculture, biomass burning, and industry. The increase in radiative forcing from nitrous oxide is 0.1 W/m². Chlorfluorocarbons (CFCs) are another major greenhouse gas source with an anthropogenic atmospheric increase from the use of spray cans and refrigerants, among other sources. Chlorfluorocarbons have produced increased radiative forcing of 0.3 W/m². Also chlorfluorocarbons have been implicated in ozone destruction in the stratosphere (and this process also

represents indirect negative forcing). Because of ozone destruction, there has been international agreement on reducing the production of chlorfluorocarbons. The total effect of the anthropogenic increase in methane, nitrous oxide, and halogens above their preindustrial concentrations in producing global warming is about 60% of that of carbon dioxide (**Fig. 3**).

In addition, there are effects on radiative forcing from changes in ozone and atmospheric aerosols. Ozone, another greenhouse gas, occurs in the stratosphere and troposphere. In the stratosphere, ozone concentrations have been reduced due to reactions between ozone and chlorine and bromine compounds from halogens, particularly over the Antarctic during Southern Hemisphere winter, producing the so-called ozone hole. There have also been Northern Hemisphere ozone losses at high and middle latitudes. Drops in stratospheric ozone cause a globally averaged negative forcing of −0.1 W/m², but since stratospheric ozone loss is concentrated at high latitudes the negative radiative forcing is not uniform. Tropospheric ozone, associated with photochemical smog from automobiles, has increased particularly in the Northern Hemisphere. This causes positive radiative forcing of 0.2–0.6 W/m².

Anthropogenic aerosols. Anthropogenic aerosols which affect global warming include sulfate aerosols, and organic and elemental carbon aerosols from biomass burning. Aerosols cause a direct cooling effect by absorbing and scattering solar radiation back to space. Sulfate aerosols are produced from sulfur dioxide (SO_2) released by fossil fuel burning. They cause a direct negative forcing of −0.25 to −0.9 W/m². Sulfate aerosols are most important over North America, Europe, China, and Japan. Biomass burning, particularly in the tropics, produces organic and elemental carbon aerosols which are thought to produce direct radiative forcing of −0.05 to −0.6 W/m².

In addition, aerosols have indirect effects because they serve as cloud droplet nuclei and affect cloud formation and cloud brightness, thus increasing the reflection of solar radiation. This indirect radiative effect of aerosols would also cause negative radiative forcing which may be similar in size to direct forcing by aerosols, but it is less well known.

For background information *see* ATMOSPHERIC CHEMISTRY; CARBON DIOXIDE; CLIMATOLOGY; FOREST ECOSYSTEM; GREENHOUSE EFFECT; HALOGENATED HYDROCARBON; METEOROLOGY; WEATHER MODIFICATION in the McGraw-Hill Encyclopedia of Science & Technology.　　　　Elizabeth Kay Berner

Bibliography. E. Berner and R. Berner, *Global Environment: Water, Air and Geochemical Cycles*, Prentice Hall, 1996; S. Fan et al., A large terrestrial carbon sink in North America implied by atmospheric and oceanic carbon dioxide data and models, *Science*, 282(5388):442–446, 1998; Intergovernmental Panel on Climate Change, *Climate Change, 1994: Radiative Forcing of Climate Change and an Evaluation of the IPCC 1992 Emission Scenarios*, Cambridge University Press, 1995; Intergovernmental Panel on

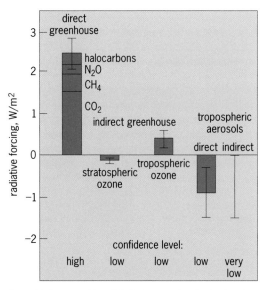

Fig. 3. Estimates of the globally averaged radiative forcing due to changes in greenhouse gases and aerosols from preindustrial times to the present. The height of the bar indicates a midrange estimate of the forcing, while the lines show the possible range of values. An indication of the relative confidence in the estimates is given below each bar. (*After Intergovernmental Panel on Climate Change, Climate Change, 1994: Radiative Forcing of Climate Change and an Evaluation of the IPCC 1992 Emission Scenarios, Cambridge University Press, 1995*)

Climate Change, *Climate Change, 1995: The Science of Climate Change*, Cambridge University Press, 1996.

Globular clusters

Globular clusters are gravitationally bound stellar systems containing, on average, about 100,000 stars. With one significant exception (omega Centauri), it appears that all stars in any one cluster formed at the same time out of a parental gas cloud, and all the stars began their lives with the same chemical composition. Not all globular clusters are alike, however. Some have very low abundances of elements heavier than helium, so that Z, the heavy-element mass fraction, may be over 100 times lower than that found in the Sun's atmosphere (where Z is about 0.018). Other clusters have relatively high Z values, reaching almost as high as that of the Sun. Most of the globular clusters in the Milky Way Galaxy lie within a few kiloparsecs of the galactic center, but some lie as far out as 100 kiloparsecs (1 kiloparsec equals 3260 light-years, 3.1×10^{16} km, or 1.9×10^{19} mi; for reference, the Sun lies about 8 kiloparsecs from the center).

Relative ages. Globular clusters are the best probes of the Galaxy's early stages of evolution, when the gas out of which it formed had not yet been enriched in heavy elements and before it had taken final form as a rapidly rotating disk. Measurement of the relative ages of globular clusters helps clarify the sequence and timing of the most significant events in the Galaxy's formation and, in principle, in the formation of other disk-shaped galaxies as well.

As a globular cluster ages, more and more of its stars exhaust the hydrogen fuel in their cores and move away from the main sequence on a temperature-luminosity (Hertzsprung-Russell) diagram (see **illus.**). They first evolve up the red giant branch to its high-luminosity end and then undergo a sudden change in nuclear energy generation, appearing on the horizontal branch. The stars spend most of their remaining lives burning helium in the core and hydrogen in a shell near their zero-age horizontal-branch positions before evolving up the asymptotic giant branch and ultimately becoming stellar remnants. If luminosity (magnitude) and temperature (color) information for stars of a particular cluster were obtained now and then obtained again 1 or 2 billion years from now, stellar evolution calculations predict that the intrinsic luminosity of those stars that were exhausting their hydrogen fuel and beginning to move away from the main sequence would be smaller in the later observation. These stars are identified with the main-sequence turn-off, and the relative present ages of clusters may be estimated by comparing the intrinsic luminosities of their turn-offs, after adjusting for secondary effects due to differences in Z.

The measurement of intrinsic luminosities includes measurement of the apparent luminosities,

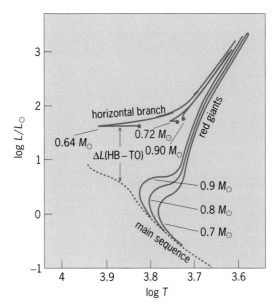

Calculated evolutionary tracks of two sets of three stars, all with heavy-element mass fraction $Z = 0.001$. The stars' luminosities (more precisely, $\log L/L_\odot$, where L is the star's luminosity and $L_\odot$ is the solar luminosity) are plotted against their approximate surface temperatures ($\log T$, where T is the absolute temperature in kelvins). Three stars of 0.9, 0.8, and 0.7 solar mass ($M_\odot$) are shown evolving from positions on the zero-age main sequence. They reach their turn-off positions ($\log L/L_\odot = +0.5, +0.3,$ and $+0.1$) in 8.4, 13.7, and 23.5 billion years. Three other tracks represent later evolution of stars of 0.64, 0.72, and 0.90 solar mass from their zero-age horizontal-branch positions, indicated by dots. (*Sukyoung Lee*)

plus knowledge of the distances and extinction due to intervening interstellar dust. The problem of determining distances and extinction disappears if the intrinsic luminosity of the horizontal-branch stars, $L(\text{HB})$, is known. The difference between the apparent luminosity of the horizontal branch and that of the main-sequence turn-off also equals the difference in their intrinsic luminosities since all the stars in a cluster are at the same distance. When this difference, $\Delta L\,(\text{TO} - \text{HB})$, is known, the intrinsic luminosity of the turn-off, $L\,(\text{TO})$, can be derived. Since the intrinsic luminosity of the horizontal branch is believed to be a simple function of Z, the relative ages of clusters with similar chemical abundances may be compared by simply superposing their temperature-luminosity diagrams such that the apparent luminosities of the clusters' horizontal branches are equal. Relative ages are then judged easily by measuring any differences in relative turn-off luminosities.

Another possible relative-age discriminant is the difference in temperature or color between the turn-off and the red giant branch. As clusters age, the difference diminishes, and is also independent of distance and interstellar obscuration. Unfortunately, the difference has a larger and less well understood sensitivity to heavy-element abundances, and therefore it is most reliable when applied to globular clusters with similar Z values.

Early star formation. Intriguing results have been obtained regarding the relative ages of globular clusters using these two methods, with the help of the world's largest ground-based telescopes and especially the Hubble Space Telescope. When the Milky Way Galaxy first formed, the Z value of the gas was approximately 0. Supernovae gradually enriched the gas, but clusters with the smallest Z values probably formed first. It appears that these clusters all have the same age, at least to within about 1 billion years, the limit of the methods' age resolution. Such results have been found for four globular clusters lying between 6 and 16 kiloparsecs from the galactic center. The low-Z but very distant (90 kiloparsecs) cluster NGC 2419 also shares the same age. Further, very deep imaging of low-Z globular clusters in the dwarf galaxy in Fornax and of stars in the low-Z dwarf galaxy in Draco shows that all the low-Z globular clusters in the Milky Way Galaxy and beyond have the same ages. At least in the Local Group of galaxies, the lights seem to have come on all at once.

A further subject of investigation is the speed at which star formation, having begun in a burst, then proceeded throughout the Milky Way Galaxy. The globular cluster Palomar 14, which has an intermediate Z value ($Z = 0.0004$) and lies about 60 kiloparsecs from the galactic center, has been found to be about 2 billion years younger than clusters with comparable Z values closer to the galactic center. Similar results have been found for three other very distant globular clusters with Z values of 0.0005 to 0.0009. Thus, star formation appears to have proceeded more slowly in the outer parts of the Galaxy.

Absolute ages. With the continuing progress in the study of the expansion rate of the universe, the importance of the absolute ages of globular clusters grows. It is critical to determine whether the estimated ages of the clusters agree with the age of the universe estimated from its current and past expansion rates.

Various errors may arise in the calculations and from the observations. The key problem is the luminosity of the horizontal branch stars, L(HB), and, unfortunately, this is not yet resolved. The *Hipparcos* satellite measured accurate distances for many stars near the Sun. While these "field" stars are not members of globular clusters, they are nonetheless the same types as found in globular clusters, including main-sequence and horizontal-branch stars. A large sample of low-Z main-sequence stars has been used to determine distances to globular clusters, assuming that the intrinsic luminosities of the field stars and cluster main-sequence stars are the same if their Z values and temperatures (or colors) are the same. For $Z \cong 0.0002$, the intrinsic luminosity of the clusters' horizontal branch stars in the visual band is found to be 61 times greater than that of the Sun, with a formal error of $\pm 7\%$. Unfortunately, this result does not agree with the luminosities of horizontal-branch stars in the field obtained with *Hipparcos* observations. Direct trigonometric parallax data indicate that the horizontal-branch luminosities are only 50 times that of the Sun (with an error of about $\pm 9\%$). Statistical analyses of the motions of field horizontal-branch

stars have also been used to determine their luminosities, finding that they are only 45 times that of the Sun (with an error of 15%). The range in measured L(HB) is thus from 45 to 61 times that of the Sun, a factor of 36%, much greater than could be explained by the estimated errors. The implications for the absolute cluster ages are considerable. The age of M92, which has $Z = 0.0001$, has been estimated to be about 14 billion years assuming that horizontal-branch luminosities are 61 times that of the Sun. The age estimate will be much larger if the true value of L(HB) is lower. *See* ASTROMETRY.

Internal mixing in red giants. The burning that transmutes hydrogen into helium with the release of energy in main-sequence stars takes place only in the hottest and densest regions: the cores. As stars exhaust their core hydrogen, the core contracts, heats up, and ignites hydrogen burning in a thin shell surrounding the core. As the star's outer layers expand and cool, turning the stars into red giants, convective turbulence extends from the surface deep into the stars, enabling them to mix some of the nuclear burning products into the stars' atmospheres, where it may be studied spectroscopically. Such probing of stellar interior processes has yielded two surprising results.

The first result concerns how deeply the mixing penetrates. During the red giant stage, hydrogen fusion is accomplished using carbon-12 nuclei as catalysts. Through a sequence of proton and positron captures, four hydrogen nuclei are transformed into helium, using isotopes of carbon, nitrogen, and oxygen in the CNO cycle. In equilibrium, the cycle acts to deplete the carbon abundance and enhance the nitrogen abundance (because the proton capture onto nitrogen-14 is the bottleneck in the flow around the cycle). Anticorrelations between nitrogen and carbon abundances in the atmospheres of globular-cluster red giants have been known for a long time, so mixing does occur. There can also be leakage out of the main cycle into other cycles of reactions. As a result, anticorrelations between nitrogen and oxygen and between oxygen and sodium may occur, and these have also been observed. The biggest surprise has to do with the anticorrelation of magnesium and aluminum abundances observed in the globular cluster M13. Proton captures onto the 12-proton magnesium nuclei require very high temperatures, so mixing must reach very deep layers, far deeper than predicted.

The second surprise is that the degree of anticorrelation between oxygen and sodium is much greater in globular-cluster red giants than in field red giants with similar Z values. This result suggests that the high-stellar-density environment in globular clusters may enhance mixing, perhaps by interactions that increase stellar rotational velocities. This mixing, which would also have to include hydrogen and helium, could mean that field and cluster red giants are different. If so, the luminosities of the following evolutionary stage, the horizontal branch, could depend on whether a star is in the field or in a globular cluster. The confirmation of such a dependence could help to resolve the remaining problems of the absolute ages of globular clusters, as well as achieve a better understanding of the internal processes in their stars.

For background information *see* CARBON-NITRO-GEN-OXYGEN CYCLES; HERTZSPRUNG-RUSSELL DIAGRAM; MILKY WAY GALAXY; NUCLEOSYNTHESIS; STAR; STAR CLUSTERS; STELLAR EVOLUTION; STELLAR POPULATION in the McGraw-Hill Encyclopedia of Science & Technology.   Bruce W. Carney

Bibliography. C. J. Hansen and S. D. Kawaler, *Stellar Interiors: Physical Principles, Structure, and Evolution*, Springer-Verlag, 1994; R. Kippenhahn and A. Weigert, *Stellar Structure and Evolution*, Springer-Verlag, 1990; B. E. J. Pagel, *Nucleosynthesis and the Chemical Evolution of Galaxies*, Cambridge University Press, 1997; R. J. Tayler, *The Stars: Their Structure and Evolution*, Cambridge University Press, 1994.

Hearing (insect)

Many insects, like other terrestrial animals, use their sense of hearing to detect and locate potential predators or prey, and mates or rivals. While the number of insects with a highly developed sense of hearing is less than that with a strong visual sense, the number that can hear is greater than might be apparent from a casual acquaintance with singing insects, namely the conspicuously noisy crickets, katydids, grasshoppers, and cicadas. Just because an insect does not produce sounds does not mean it lacks hearing. For example, many large night-flying moths (including noctuids, geometrids, arctiids, and notodontids) possess well-developed tympanal hearing organs (ears) that are tuned to the ultrasonic emissions of insectivorous bats that use biosonar to detect and locate the moths. A sense of hearing has also been reported in some praying mantises, cockroaches, beetles, flies, and green lacewings. The sensitivity of some insect ears is quite remarkable and in some species far exceeds human capabilities. For example, humans can hear frequencies from about 20 Hz to 15 kHz, but katydids and crickets are sensitive to ultrasound exceeding 50 kHz, and some moths can hear ultrasound exceeding 100 kHz.

Ears. The hearing organs of insects are called tympanal organs and are usually identified externally by the presence of a tympanal membrane (eardrum). Tympana are localized regions of cuticle that are thinner (often being transparent) than the surrounding cuticle. The tympanum is backed internally by an air-filled sac or cavity, and the inner face of the tympanum or the air sac is innervated by a chordotonal sensory organ. This sensory organ may be made from just one bipolar sensory neuron (in some moths), to a hundred neurons (in crickets), to nearly a thousand (in cicadas); for reference, the human inner ear contains about 15,000 sensory cells. Sound waves from a source propagate through the air and impinge upon the external surface of the tympanum, setting this membrane into vibration. In some animals, the sound wave may also be conducted internally, from ear to ear, through specialized pathways to excite the inner surface of the tympanum, and in such cases provides a mechanism for localizing the sound source. The vibrations of the eardrum are transferred to the sensory neurons of the chordotonal organ, resulting in the transmission of signals along the auditory nerve into the central nervous system.

Location. Like vertebrate ears, insect ears always come in pairs located on opposite sides of the body. Unlike vertebrate ears, always on the head, the insect ears may be found virtually anywhere in the body: in mouthparts (some moths), legs (crickets and katydids), wings (green lacewings), thorax (some moths, praying mantises), abdomen (some moths, beetles, grasshoppers), and neck (some beetles). The difference is a matter of developmental and evolutionary origins.

Origin. The internal hearing organ of all vertebrates develops in the embryonic head and is within the otic capsule of the skull, where the acoustic receptor cells are located. Hearing appears to have evolved just once in the vertebrates. However, hearing organs have evolved many times and independently in different orders of insects. The insect tympanal hearing organ is derived developmentally from sensory organs (chordotonal organs) that are sensitive to movement (such as stretch receptors) and are distributed throughout an insect's body. Such receptors are commonly associated with the joints of appendages or are within and between body segments. Thus, the evolutionary and developmental precursors for hearing organs are much more widely distributed anatomically in insects than in vertebrates.

Acoustic signals. The sense of hearing serves an insect in the same way it does a vertebrate, including mammals. Some insects have evolved long-range mate-detection systems based on acoustic signals, analogous to the mate calling systems of frogs, birds, and other vertebrates. While mating calls (usually produced by males of the species) may be emitted to attract prospective reproductive partners, they may also serve to warn or repel prospective rival males. Thus, for the most part, a sense of hearing is found in both sexes. Moreover, the sense of hearing alerts many insects to the presence of potential prey or predators; indeed, this may have been the primitive role for hearing in many insects. The ability to send long-range signals without relying on vision is obviously important in nocturnally active insects. The calls of crickets and katydids are most conspicuous at night. In contrast, grasshoppers and cicadas are diurnally active, and the penetrating calls of the latter are among the loudest of all insects. The distance over which one insect can hear another depends on the species involved as well as on environmental and climatological conditions, and ranges from a few meters up to a kilometer (0.6 mi).

The most far-reaching insect sounds are in a range

from about 2 to 5 kHz; very high frequencies (for example, ultrasounds over 20 kHz) are not transmitted over distance efficiently, and are attenuated by a number of physical and environmental factors which become more telling in their effect as sound frequency increases. Insect hearing, as defined here, refers to hearing in the far field of sound, and is accomplished by the tympanal chordotonal hearing organ. While low-frequency sounds would be propagated over longer distances than high frequencies, the small sizes of insects prevents them from efficiently producing loud, low-frequency signals, in the hundreds of hertz range. Indeed, insects that employ low-frequency signals confine their range of communication to short range, sometimes just a few body lengths in distance. In any case, tympanal hearing organs of the type discussed here would be insensitive to low-frequency, low-amplitude signals. A different kind of receptor, based on the movements of hairlike receptors, does permit insects to detect very low frequency sounds, and examples are found on the cerci (posterior feelers) of cockroaches and crickets or on the antennae of mosquitoes. The frequency sensitivity of these near-field auditory organs is in the range of tens to a few hundred hertz, and detection range is mostly limited to centimeters.

When acoustic signals are produced for communication, as in mate calling, the sensitivities of the auditory systems in these insect species are usually well matched to the acoustic features of the call. The auditory system of a field cricket is well tuned to the 5-kHz frequency of its calling songs, as well as to the 15 kHz of its courtship songs. In crickets and katydids, however, it is not sufficient that the male's calling song contain the species-specific spectral frequency in each sound pulse; the pattern of song pulses over time is also species-specific, and a singing male will not attract a female unless his song is delivered in the right tempo or rhythm. The temporal pattern of the calling song for a given species is not only stereotyped within an individual but is species-specific within the genus; it is under firm genetic control, just as the spectral specificity is genetically conferred. The spectral and temporal specificity of the auditory system is also under genetic control and, for a given species, is usually matched to the characteristics of the song. There is little evidence that learning plays any role in perfecting the sender-receiver match in the calling-song systems of insects.

Many crickets and katydids are sensitive to frequencies in the range of 20 to 60 kHz. This characteristic may reflect tuning to the biosonar signals of predatory bats that hunt by echolocation. In other insects, such as many moths, the auditory system seems to be sensitive only in the range of 20 to 80 kHz and may reflect specialized tuning to predatory bats; most moths do not produce sounds themselves. However, arctiid moths produce ultrasound-containing clicking, and in some contexts these sounds appear to be used in courtship, but more often they are emitted in the presence of actively hunting bats. Presumably, such signals from the moth could be used to jam the biosonar system of the bat, to startle the bat, or to warn the bat that the moth is distasteful or poisonous. This warning is an example of aposometic signaling, as in the visual example of warning coloration in many insects. In any event, bats appear to be repelled by the clicks of such moths. It seems clear that any night-flying insect is at risk for predation by insectivorous bats and that many such insects have evolved ultrasound-sensitive ears to enable them to evade hunting bats.

This discussion restricts hearing to the realm of airborne sounds. However, many insects produce highly structured, stereotyped, species-specific vibrational signals that are propagated through their supporting substrate, such as leaves and branches. In terms of behavioral context and sensory organs for perception, there are similarities between airborne and substrate signaling. If such signals were to be considered within the realm of hearing, the number of insects that could be claimed to hear would be tremendously expanded beyond the current list.

For background information see PERCEPTION; PHONORECEPTION; PSYCHOACOUSTICS; SOUND in the McGraw-Hill Encyclopedia of Science & Technology.

Ronald Hoy

Bibliography. A. W. Ewing, *Arthropod Bioacoustics*, Cornell University Press, 1989; R. R. Hoy, A. N. Popper, and R. R. Fay (eds.), *Comparative Hearing: Insects*, Springer-Verlag, 1998; R. R. Hoy and D. Robert, Tympanal hearing in insects, *Annu. Rev. Entomol.*, 41:433–450, 1996; F. Huber, T. E. Moore, and W. Loher, *Cricket Neurobiology and Behavior*, Cornell University Press, 1989.

Hearing impairment

Approximately 1 out of every 2000 infants is born profoundly deaf, and more than one-third of people born with normal hearing will have a significant hearing loss by their 60s. Hearing loss has a variety of causes, including trauma (such as a head injury), illness (such as rubella or meningitis), the environment (notably, long-term exposure to loud noise), drugs (including some powerful chemotherapy agents and certain antibiotics), genetics, or some combination of these factors. Genetic hearing loss has not been well understood until recently, when several of the responsible genes were identified. The hearing loss that results from inherited defects in these genes can be divided into syndromic and nonsyndromic.

Syndromic hearing loss. Syndromic hearing loss occurs in the presence of one or more other symptoms (see **table**). The responsible genes usually fall into three categories. First are those genes that make a structural protein that is needed both in the ear and in one or more other tissues (collagen genes and *MYO7A*). For example, mutations in three different type IV collagen genes have been found in patients with Alport syndrome (hearing loss and kidney defects). These genes make proteins that are important

Hearing loss syndromes

Syndrome	Additional symptoms	Mutated genes
Alport syndrome	Kidney defects	COL4A3, COL4A4, COL4A5
Branchio-oto-renal syndrome	Kidney defects, branchial cleft fistulas	EYA1
Jervell and Lange-Nielsen syndrome	Heart defects	KVLQT1, KCNE1
Norrie disease	Eye defects	NDP
Pendred syndrome	Goiter	PDS
Stickler syndrome	Vision loss, joint defects	COL2A1, COL11A1, COL11A2
Treacher Collins syndrome	Eye and facial malformations	TCOF1
Usher syndrome	Retinitis pigmentosa	MYO7A, USH2A
Waardenburg syndrome	Reduced pigmentation in the eyes, skin, and hair	PAX3, MITF, EDNRB, EDN2, SOX10

in maintaining basement membranes in the ear and in the kidney. Basement membranes form barriers between different tissues. When defective, they leak, possibly causing damage to the tissues on either side of the membrane.

The second category of syndromic hearing loss genes includes those genes that code for transcription factors, proteins that bind to deoxyribonucleic acid (DNA) and turn on or off genes (*PAX3*, *MITE*, *SOX10*). When a gene for a transcription factor is mutated, many other genes in the cell may not be regulated properly. For example, patients with Waardenburg syndrome type 1 have mutations in the *Pax3* transcription factor gene. The Pax3 protein turns on other genes that ensure that certain types of cells migrate to appropriate locations during development. When the *Pax3* gene is mutated, these cells do not migrate properly to the ears, eyes, skin, and hair, possibly leading to hearing loss and reduced pigmentation in the eyes, skin, and hair.

Third, the genes responsible for Jervell and Lange-Nielsen syndrome, Pendred syndrome, and Treacher Collins syndrome do not code for structural proteins or transcription factors. These genes synthesize proteins that are probably required for the transport of specific molecules or ions both in the ear and in other organs.

Nonsyndromic hearing loss. In nonsyndromic hearing loss, the individual has no symptoms except hearing loss. There are two main reasons why genes for nonsyndromic hearing loss have been more difficult to identify than their syndromic hearing loss counterparts. First, mutations in different hearing loss genes can often cause the same type of hearing loss. (With syndromic hearing loss, this is usually not a problem, as a particular syndrome may be diagnosed based on other symptoms.) This makes it difficult to know the exact number of hearing loss genes. Also, deaf people often intermarry, making the genetic analysis even more complicated.

The second reason that nonsyndromic hearing loss genes can be difficult to identify is that many forms of such loss affect individuals as they age, causing a progressive decline in hearing. This hearing loss is often attributed simply to old age and is not recognized as a genetic problem. Nonsyndromic hearing loss genes can be divided into three categories based on the pattern of inheritance: autosomal dominant, autosomal recessive, and X-linked.

Autosomal dominant genes. Autosomal dominant hearing loss genes typically (but not always) cause progressive hearing loss. (Hearing may have been normal initially, but it worsens with age.) Autosomal dominant indicates that every person has two copies of each of these genes (one copy from each parent), but inheritance of a single defective copy from either parent will cause hearing loss (**Fig. 1**). Thus, the hearing loss is usually passed on directly from parent to child, and large families usually have several affected individuals.

Eighteen of these autosomal dominant genes are known to exist, named *DFNA1* through *DFNA19* (numbers 8 and 12 were shown to be the same). The identities of only six of these genes are known. The existence of a deafness gene does not reveal the nature of the protein for which it codes. Deafness genes are identified through genetic mapping, in which the inheritance of specific chromosomes or segments of chromosomes is tracked with the transmission of hearing loss. Thus, it can be established that a certain deafness gene, such as *DFNA10*, lies on a small portion of chromosome 6, without having identified the actual gene.

The five identified genes are *GJB2*, which forms gap junctions between cells in the inner ear; *DIAPH1*, which is thought to play a role in the polymerization of the protein actin into long filaments; *MYO7A*, which can bind to actin filaments and move along them; *POU4F3*, a transcription factor that operates in the inner ear; and *TECTA*, a structural

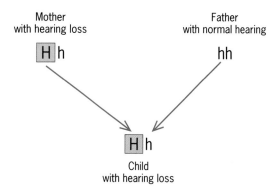

Fig. 1. Autosomal dominant hearing loss. Hearing loss results when one of an individual's two copies of the autosomal dominant hearing loss gene is mutated (boxed). A child gets one copy of the gene from each parent, so an affected child must have at least one affected parent.

protein in the extracellular matrix in the inner ear.

Different mutations in *MYO7A* can cause syndromic hearing loss (Usher syndrome), nonsyndromic autosomal dominant hearing loss, or nonsyndromic autosomal recessive hearing loss. Mutations in *GJB2* can cause either autosomal dominant or recessive nonsyndromic hearing loss. In the case of these two genes, the type of hearing loss depends on exactly where and how the gene is mutated.

Autosomal recessive genes. Autosomal recessive genes usually cause congenital hearing loss, meaning that the hearing loss is present at birth. These genes often cause more severe hearing loss than the autosomal dominant genes, up to and including complete deafness. Autosomal recessive means that a person must inherit mutant copies of the gene from both parents in order for hearing to be damaged. This inheritance usually occurs when each of the affected individual's parents has one mutant copy of the gene, and is thus a carrier with normal hearing. In order for hearing loss to occur, each parent must pass on a mutated copy of the gene to the child (**Fig. 2**).

There are many more autosomal recessive hearing loss genes that remain to be identified. Twenty (*DFNB1-DFNB20*) locations are known, but only four genes have been identified. The four include the genes *MYO7A* and *GJB2*; the gene that makes a sulfate transporter protein (*PDS*) [also defective in Pendred syndrome]; and *MYO15*, which, like its cousin *MYO7A*, can bind to and travel along filaments of actin.

X-linked genes. X-linked hearing loss genes are found on the X chromosome, and they have a very special pattern of inheritance. Since females have two X chromosomes (XX), they have two copies of each X-linked hearing loss gene. For a female to have hearing loss, both of her copies of a particular deafness gene must be mutated, just as with the autosomal recessive genes. Males have one X and one Y chromosome (XY), and thus they have only one copy of each X-linked hearing loss gene. If this single copy is mutated, a male will have hearing loss. Thus, X-linked deafness is usually seen when a woman with normal hearing (a carrier) transmits her one mutant

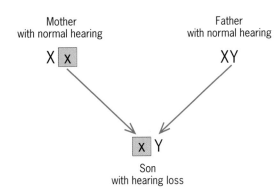

Fig. 3. X-linked hearing loss. X-linked hearing loss genes are located on the X chromosome. Since males are XY, they have only one X chromosome. If a male receives a mutated copy of an X-linked hearing loss gene from his mother (boxed), he will have hearing loss. The mother, who has two X chromosomes, is usually unaffected.

copy of an X-linked hearing loss gene to her son (**Fig. 3**).

There are now seven reported X-linked hearing loss genes, *DFN2-DFN8*. Only *DFN3* has been identified. It is *POU3F4*, a transcription factor, and it appears to play a critical role in directing the development of the ear. In males with mutations in this gene, the ear does not form normally, resulting in abnormal hearing.

For background information *see* EAR; HEARING (HUMAN); HUMAN GENETICS; PHONORECEPTION; SOUND in the McGraw-Hill Encyclopedia of Science & Technology.   Frank J. Probst; Sally A. Camper

Bibliography. C. T. Baldwin et al., An exonic mutation in the *HuP2* paired domain gene causes Waardenburg's syndrome, *Nature*, 355:637-638, 1992; D. P. Kelsell et al., Connexin 26 mutations in hereditary non-syndromic sensorineural deafness, *Nature*, 387:80-83, 1997; M. Tassabehji et al., Waardenburg's syndrome patients have mutations in the human homologue of the *Pax-3* paired box gene, *Nature*, 355: 635-636, 1992; D. Weil et al., Defective myosin VIIA gene responsible for Usher syndrome type 1B, *Nature*, 374:60-61, 1995.

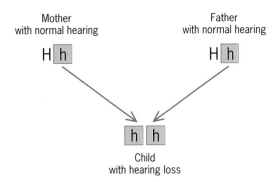

Fig. 2. Autosomal recessive hearing loss. A child will have hearing loss only if both copies of an autosomal recessive hearing loss gene are mutated (boxed). Each parent usually has one mutant copy of the gene, so each is said to be a carrier of the mutation and has normal hearing.

Hypercarbon chemistry

One of the foundations of organic chemistry is A. Kekulé's concept of the tetravalency of carbon (also suggested by A. S. Couper). This led to the octet rule, proposed by G. N. Lewis in 1916, that only a maximum of four ligands can be bound to carbon by two-electron two-center sigma bonds, for a total of eight electrons around carbon. Violations of the octet rule in the compounds of the second and higher rows of the periodic table are recognized, but carbon, being a first-row element, cannot extend its valence shell. Through quantum theory, it is understood that the idea of bonding pairs of electrons always being between two atomic centers is, at best, only an approximation. Many organic molecules require a description that involves the electrons in the bond being distributed over two or more atoms. The majority of these molecules are conjugated unsaturated π systems that are described in terms of delocalized bonds, as in benzene [structure (**1**),

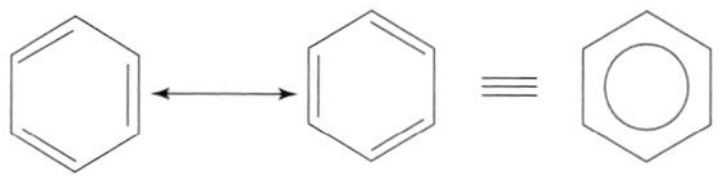

(**1**)

with resonance structures]. Multicenter delocalized bonding is equally possible in a sigma-bond framework. One example of such bonding is diborane (**2**)

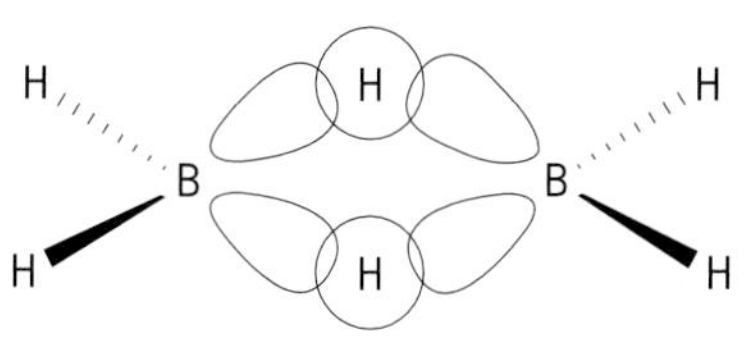

(**2**)

with the three-center two-electron (3c-2e) bonds resulting from the overlap of two sp^3-like boron orbitals and a hydrogen *1s* orbital.

Multicenter sigma bonding is also common in the chemistry of carbon compounds—thus the terms hypercarbon compounds and hypercarbon chemistry. The coordination number of an atom represents the number of nearest neighboring atoms around the central atom. Valence generally represents the number of electron pairs around a central atom. It is important to distinguish between coordination number and valence. Even though a hypercarbon molecule may have a carbon with more than four nearest neighbors, at no time does the carbon atom have more than four pairs of bonding electrons (thus not violating the octet rule). Although references were made in the literature to possible violations of the octet rule in organic chemstry, subsequent studies have shown them to be incorrect.

Carboranes. Electron-deficient boron cage compounds containing one or more carbon atoms in their framework have formed a major area of interest in inorganic and organometallic chemistry. Such boron cage compounds are called carboranes and include both open-cage and closed polyhedral structures. Carboranes can be viewed as derivatives of boron hydrides, where B^- and BH groups have been replaced by isoelectronic carbon atoms. The skeletal bonds of the carboranes, in general, are partial and consist of fewer than two electrons each. The atoms involved in the cage may have coordination numbers greater than four, with the greater number of bonds compensating for the weaker nature of individual bonds. Two representative polyhedral *closo*-carboranes containing five and six coordinate carbons are *closo*-1,2-$C_2B_{10}H_{12}$ (ortho) and *closo*-1,2-$C_2B_4H_6$ [structures (**3**) and (**4**)], respectively.

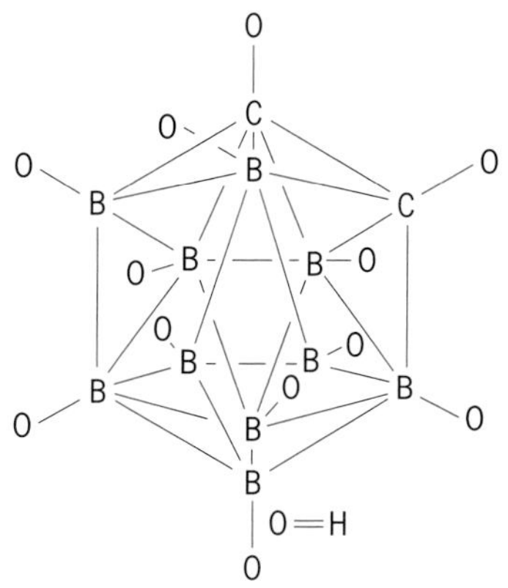

(**3**)

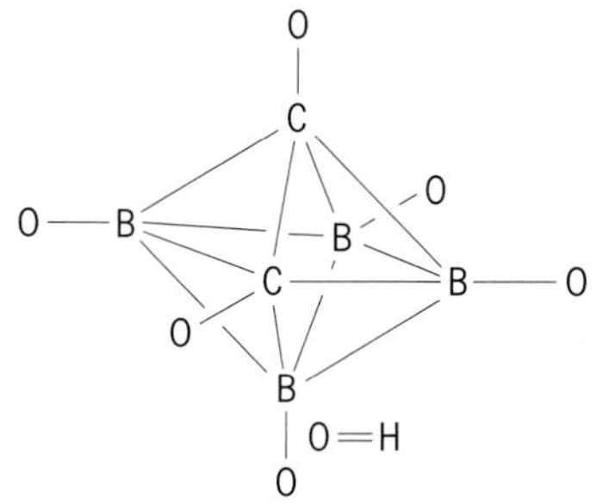

(**4**)

Alkylated metals. The alkyl derivative of certain electron-deficient elements of groups I through III (in the periodic table) of the form $[MR_n]_x$ (where M = Li, Be, Cu, Mg, and Al and where R = —CH_3, —C_2H_5, —C_3H_7, and so on) are associated through bridging alkyl groups. Thus, the electron deficiency of the electropositive metals is decreased by association. Such association, or aggregation, can be broken by adding electron-donating ligands. One example of such an association is methyllithium, which exists as a tetramer (**5**). In the tetramer, the carbon is hexacoordinate and the lithium is tricoordinate. There are some higher aggregates of methyllithium in which carbon can even be considered heptacoordinate. In the methyl-bridged organo-aluminum derivative (**6**),

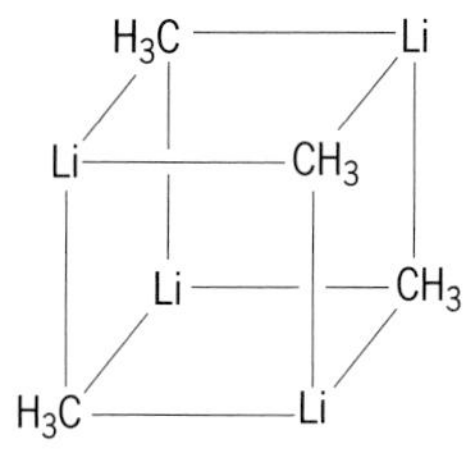

(**5**)

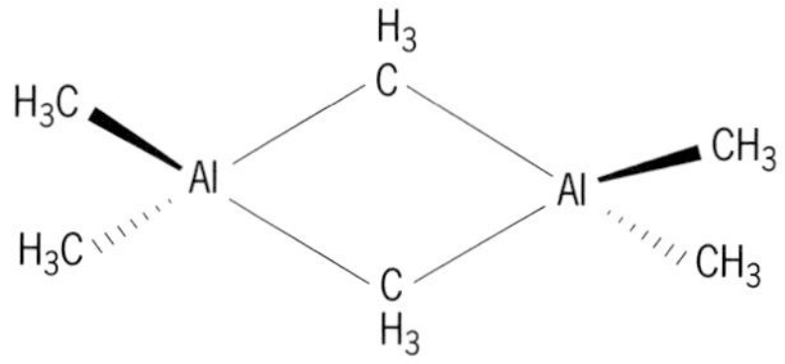

(**6**)

the carbon is pentacoordinate. Similar pentacoordinate carbons are observed with copper, beryllium, and magnesium analogs, as well as with some lanthanum compounds.

Transition-metal carbides. Another interesting class of hypercarbon compounds, in which carbon exhibits coordination numbers of five, six, and even eight, is the transition-metal carbides, or carbide carbonyl clusters. One of the earliest carbide carbonyl clusters (containing pentacoordinate carbon) to be reported was $CFe_5(CO)_{15}$ (pentairon pentadecacarbonyl carbide; **7**). It is a *nido* cluster, wherein

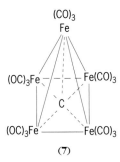

(7)

the central carbide carbon atom is situated among five iron atoms [as $Fe(CO)_3$ units] with a C_{4v} symmetry group arrangement. In this structure, the carbon atom is slightly below the plane defined by the four iron atoms. There is a basal C—Fe bond distance of 0.189 nanometer (1.89 angstroms) and an apical Fe—C bond distance of 0.196 nm (1.96 Å). Hexacoordinate carbon atoms are also found in $CFe_6(CO)_{16}^{2-}$, $CRu_6(CO)_{17}$, and $CRh_6(CO)_{15}^{2-}$. In the first two cases (**8**), the carbon atom resides approximately at the

M
M—|—M
C
M——M
M
M = Fe, Ru

(8)

center of an M_6 octahedron. In the third case the carbon atom is surrounded by six rhodium atoms in a trigonal prismatic arrangement with D_{3h} symmetry.

Heptacoordinate (as M_7C) carbon clusters have not been observed, although heptacoordinate carbon may be present in the higher aggregates of methyllithium and its homologs. There are a few examples of octacoordinate carbon, such as $CCo_8(CO)_{18}^{2-}$ (octacobalt octadecacarbonyl carbide

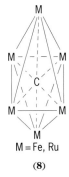

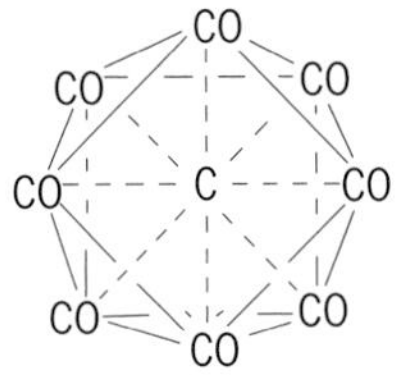

(9)

dianion; **9**, partial structure shown). Here the carbon is situated in the center of a quadratic antiprism arrangement, with four cobalt atoms at the top and four at the bottom.

Unsaturated transition-metal compounds. The other category of hypercarbon compounds that has seen rapid growth is the coordinatively unsaturated transition-metal compounds, wherein the metal atoms, with fewer electrons in their valence shells than can be accommodated, form strong bonding interactions with the neighboring C-H groups, resulting in (3c-2e) CHM bonds (where M is the metal atom). The metal atom in such bonds tends to distort the coordination sphere of the carbon atom involved, converting it from a normal carbon atom to a hypercarbon atom and drawing the C-H unit close to itself; such bonding is described as agostic (derived from Greek meaning "to hold or clasp to oneself, as of a shield"). A representative example is structure (**10**). Such compounds play a major role

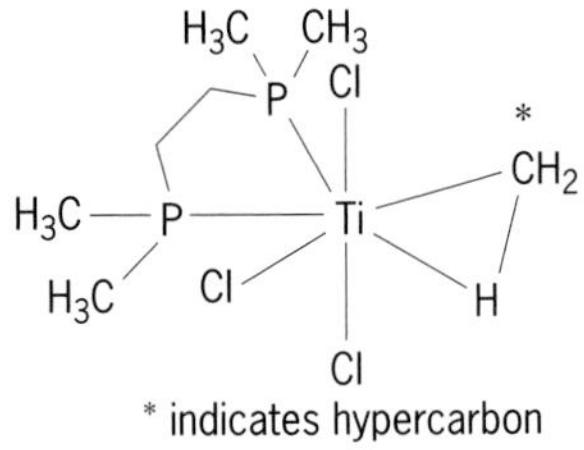

H₃C   CH₃
   \   /
    P
H₃C—P   Cl   *
   | \   CH₂
H₃C   Cl—Ti—|
           | H
           Cl
* indicates hypercarbon

(10)

in C-H activation of saturated hydrocarbons.

Carbocations. In addition to the well-established area of neutral hypercarbon compounds containing carbon atoms with coordination numbers of five to eight, hypercoordinate carbocations have been recognized. They demonstrate the ability of carbon atoms to participate in two-electron multicenter bonding in electron-deficient hydrocarbons ions. A number of methods to study carbocation intermediates under superacidic conditions have been developed.

Protonated alkanes. Protonated alkanes $(C_nH_{2n+3})^+$ play a pivotal role in the acid-catalyzed cracking and isomerization of hydrocarbons. The methonium ion (**11**) is the parent member of this class of compounds,

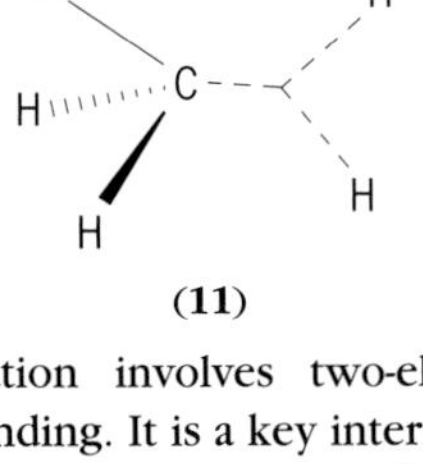

H   H⁺
 \   /
  H····C - - ⟨
 /   \
H   H

(11)

and its formation involves two-electron three-center HCH bonding. It is a key intermediate in the superacid-induced condensation of methane into gasoline-range hydrocarbons. Protonated methane is well known in the gas phase, especially through mass spectrometric studies of methane at high source pressures. In the condensed state, evidence for CH_5^+ comes mainly from the hydrogen-deuterium scrambling of methane in deuterated superacids. Based on experiment as well as theory, the structure of protonated methane has the C_s symmetry with the front side protonation. Polycharged CH_6^{2+}, CH_7^{3+} containing hexa and heptacoordianted carbon atoms were computed to be minima on the potential energy surface by ab initio calculations. Utilizing the (triphenylphosphine)gold(I) ligand (LAu^+) as the isolobal (number of electrons and shape of the orbitals to the central carbon are similar) substitute for H^+, monopositively charged gold complex (**12**) has been

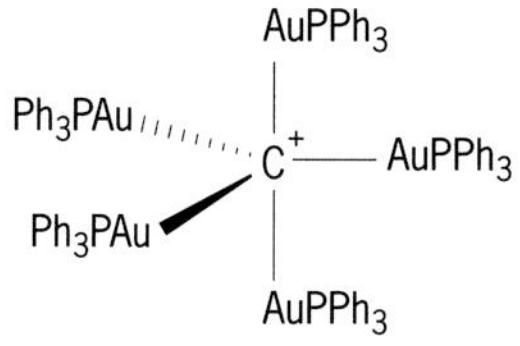

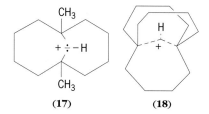

(**12**)

prepared and characterized as an analog of (**11**) containing a pentacoordinate carbon. A gold analog of CH_6^{2+} has also been synthesized.

The long-controversial 2-norbornyl cation is an analog of protonated methane. The symmetrical, sigma-bridged structure (**13**) is preferred over the

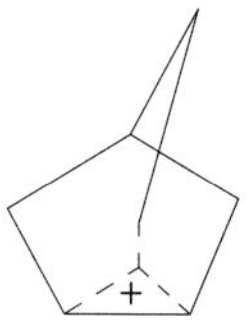

(**13**)

two classically represented degenerate structures (**14**). Similarly, a large number of sigma-bridged cat-

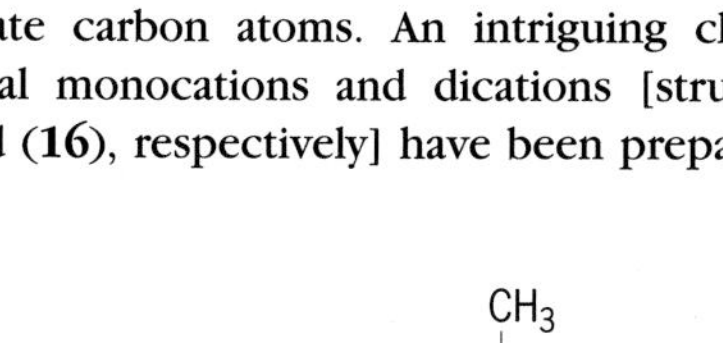

(**14**)

ions are known, all containing well-defined pentacoordinate carbon atoms. An intriguing class of pyramidal monocations and dications [structures (**15**) and (**16**), respectively] have been prepared in

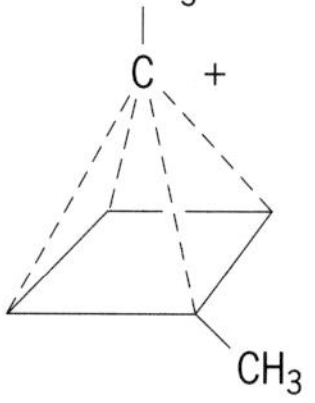

(**15**) 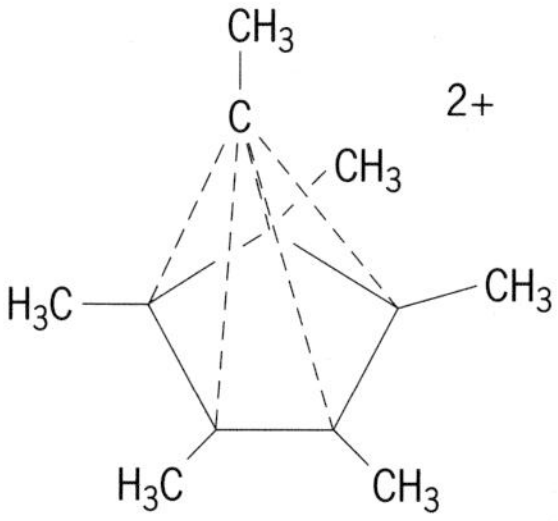   (**16**)

the superacid media. The dication possesses a unique hexacoordinate carbon atom. Even μ-hydrido-

bridged cations (**17** and **18**) have been characterized

(**17**)   (**18**)

(involving 3c-2e CHC bonds). Remarkably, (**18**) is stable at room temperature in trifluoroacetic acid solution.

Hypercarbon intermediates. Electrophilic reactions of saturated hypercarbons (practiced in large scale in the petrochemical industry) proceed via hypercarbon intermediates. Acid-catalyzed isomerization of alkanes involves initial abstraction of hydrogen (with an electron pair) to form a trivalent alkyl cation which undergoes alkyl and hydride shifts to form a highly branched tertiary carbocation. The intermolecular transfer of hydrogen completes the isomerization of one molecule while forming a new cation to continue the process. It was initially presumed that the transition state for intermolecular hydrogen transfer was linear; later it was suggested that two-electron three-center interactions in carbocations tend to be triangular in nature. However, in sterically congested systems the transition state may be more linear. The nonlinearity of the transition state for the interaction of a tertiary alkane with a tertiary alkyl cation is shown by the reaction of isobutane with the *t*-butyl cation giving highly crowded 2,2,3,3-tetramethylbutane. Hypercoordinate carbocations are also involved in the degenerate rearrangement of many carbocationic systems. Other reactions involving hypercarbon intermediates are the electrophilic substitutions of hydrocarbons such as halogenation, nitration, and superacid-catalyzed oxyfunctionalization using hydrogen peroxide or ozone; the electrophilic reactions of alkanes, alkenes, alkynes, and aromatics; and transition-metal-catalyzed reactions of alkanes and alkenes in which metals obtain a stable electron configuration by drawing electron density from neighboring sigma and pi bonds to form a variety of stable bridged compounds. Hypercarbon chemistry increasingly plays a key role in a wide variety of chemical transformations.

For background information *see* BONDING; CARBON; CARBORANE; CHEMICAL BONDING; COORDINATION CHEMISTRY; LIGAND; ORGANIC CHEMISTRY in the McGraw-Hill Encyclopedia of Science & Technology.   George A. Olah; G. K. Surya Prakash

Bibliography. G. N. Lewis, The atom and the molecule, *J. Amer Chem. Soc.*, 38:762–785, 1916; G. A. Olah et al., *Hypercarbon Chemistry*, John Wiley, 1987; G. A. Olah, G. K. S. Prakash, and J. Sommer, *Superacids*, John Wiley, 1985; G. A. Olah and G. Rasul, From Kekulé's tetravalent methane to five-, six-, and seven-coordinate protonated methanes, *Acc. Chem. Res.*, 30:245–250, 1997; F. Scherbaum et al., "Aurophilicity" as a conseqence of relativistic effects: The hexakis(triphenyl-phophineaurio)methane dication

[(Ph$_3$PAu)$_6$C]$^{2+}$, *Angew. Chem. Int. Ed. Engl.*, 27: 1544–1546, 1988; F. Scherbaum et al., Synthesis, structure, and bonding of the cation [{(C$_6$H$_5$)$_3$PAu}$_5$C]$^+$, *Angew. Chem. Int. Ed. Engl.*, 28: 463–465, 1989.

Immune system

The ability of the immune system to protect the body from various insults declines with age. Changes in both innate and acquired immunity appear to be involved in this phenomenon, and they can be described at the level of the entire organism or the cell. For example, the innate response of mounting a fever with infection occurs less frequently in aged persons; 20% of adults over age 65 with serious bacterial infections do not have fevers. This may be due to decreased sensitivity of the central nervous system to immune signals, rather than actual changes in the ability to generate such stimuli. Most changes in immune function associated with advancing age have been observed at the cellular level and involve acquired immunity.

T and B lymphocytes. T lymphocytes are distinguished by markers that indicate whether they have (memory cell) or have not (naive or virgin cell) yet encountered antigen. Quantitative changes in lymphocytes with aging include declining numbers of virgin T cells and increasing numbers of memory T cells. The overall number of B cells does not change appreciably, but there is a decrease in bone marrow B-cell precursors, and structural changes in B-cell membranes. Qualitatively, the proliferative response of lymphocytes from old adults is significantly less than the response of lymphocytes from young adults. The ability to increase the number of lymphocytes is a key factor in mounting cellular and humoral immune responses. In old individuals, fewer cells respond to proliferative stimuli and the vigor of the response is decreased. These changes have been seen in helper cells, memory cells, cytotoxic cells, suppressor cells, and B cells.

The decreased proliferative response of lymphocytes in aging is due in part to impaired membrane signal transduction. Activation of the protein kinase families in cell membranes leads to increased transcription and translation of genes coding for the T-cell growth factor, interleukin-2 (IL-2), and its receptors. IL-2 stimulates proliferation of cells bearing the IL-2 receptor. Accessory cells present antigen that occupies and cross-links T-cell receptors and enhances the production of and the response to IL-2. These accessory cells secrete interleukin-1 (IL-1) and other cytokines that provide additional signals necessary for complete cell activation. The level of protein kinases involved in the cascade of cell activation is reduced with age, resulting in decreased intracellular calcium release and lower levels of other cell-activating factors.

Once cells are activated and proliferate, they participate in cellular immune responses or in humoral immune responses (production of antibody). Cytotoxic lymphocytes from aged mice are less able to bind targets, though they appear to be equally effective in destroying their targets. The production of antibody also changes with increasing age. There is an increase in the prevalence of autoantibodies (antibodies formed against components of the self) especially after age 70. This may be due to the loss of suppressor cell inhibition of autoantibody formation. Such declines in T suppressor cell function with age have been associated with the development of oral tolerance. There is also evidence that B-cell repertoires shift with age, predisposing to autoantibody production.

The ability to respond to a particular antigenic challenge with specific antibody production lessens with aging. This has been described in studies of both the primary (first time) and secondary (memory) antibody responses. The duration of the response is shorter than in young adults, and fewer cells participate in antibody production. The accumulation of antibodies directed against other antibodies (anti-idiotypes) may also interfere with the efficiency of the humoral response. Although changes in antibody production are mostly the result of declines in T-lymphocyte function, there is evidence for a decline in intrinsic B-cell function.

Macrophage function. Macrophages and monocytes are accessory cells that have a primary role in presenting antigen to T and B lymphocytes and in producing cytokines that activate the cells. There may be fewer macrophage precursors with advancing age, and there appear to be defects in macrophage-T cell communication. Monocytes from old adults display less cytotoxicity against certain tumor cell lines, and they produce lower levels of reactive oxygen intermediates (H$_2$O$_2$ and NO$_2$) than monocytes from young adults.

Natural killer cells. Natural killer (NK) cells are cytotoxic cells that differ from cytotoxic T cells by the ability to lyse targets without the need for antigenic sensitization. Lymphokine-activated killer cells are able to lyse certain cell lines which are resistant to natural killer cells. In aging humans, the majority of studies have shown no change in natural killer cell cytotoxic ability, although murine studies suggest an age-related decline in ability to lyse target cells. Natural killer cells from young individuals may respond to lower levels of activating substances, and the activity of lymphokine-activated killer cells does appear to be reduced in aged as compared to young humans.

Prostaglandins. Prostaglandins are metabolites of cell membranes which are implicated in age-related changes in immunity. Prostaglandin E2 is a feedback inhibitor of T-cell proliferation, and T cells from adults over age 70 are more sensitive to inhibition by prostaglandin E2. Thus, prostaglandin E2 may interfere with the expansion of antigen-specific T-cell clones. There is also some evidence that prostaglandin E2 production increases with age.

Interleukins. Accessory cells secrete cytokines that provide signals necessary for complete activation of

T cells. In aging, there is decreased expression of messenger ribonucleic acid (mRNA) for IL-2, decreased production of IL-2, decreased density of IL-2 receptors, and decreased proliferation in response to IL-2. Newer studies suggest that some of the observed differences may be due to experimental conditions.

Once activated, T cells produce a variety of growth and differentiation factors. There are age-related declines in the production and response to several interleukins (IL-1, IL-4, IL-8), gamma-interferon, and tumor necrosis factor, while tonic IL-6 production is greater. As with IL-2, production of these cytokines with aging varies depending on experimental conditions. Some of the decrease in antibody production may be due to delays in secretion of growth factors such that they inhibit rather than stimulate antibody production.

Declining immune function. A number of general mechanisms have been postulated to account for the age-related decline in immunity. An obvious change with aging is the involution of the thymus. Exposing lymphocytes of old individuals to thymic hormones results in at least a partial restoration of immunity on a temporary basis. Declines in other hormones (such as growth hormone and melatonin) have also been associated with changes in immunity. Increased sensitivity to prostaglandins with age may be a major cause of immune decline, especially since prostaglandins are produced by most tissues in the body. Suppressing prostaglandin production stimulates immune responses.

Nutrition across the life-span has been associated with age-related immune changes. Dietary manipulations have an effect on immunity. Nutritional deprivation of experimental animals at a young age results in preservation of normal immune function into old age. Vitamin and caloric supplementation has been associated with enhanced immune responses and fewer days of infectious illness in both nutritionally deficient and healthy elderly adults.

Complex and direct links exist between the central nervous system and the immune system. The neurohumorally mediated effects of psychological stress on immunity have been demonstrated in controlled experiments in animals. Studies in humans demonstrate similar effects, though it is impossible to achieve the same degree of control. Associations have been observed between depression, bereavement, poor social support, and suppressed immunity. Old age is a time when many stresses and losses are experienced, and this may contribute to immune decline.

The theory that a genetically programmed biological clock may account for the immune decline with aging is countered by observations that changes in different physiological systems with age do not appear to be synchronized. Varying levels of decline are seen in different systems within a given individual. Immunomodulating substances may affect some systems and not others. No global mechanism has yet been found to account for immune senescence.

Clinical implications. There is little direct causal evidence linking specific changes in immunity to specific clinical illness or mortality. A few studies have reported correlations between the suppression of immune parameters and an increased incidence of infection or mortality. Autoantibodies and circulating immune complexes have been linked to the development of atherosclerosis by contributing to inflammation of the vascular lining, thrombosis, and free-radical generation. In animals, these associations are more pronounced when combined with a high fat diet.

With age, there is an increase in the incidence and severity of infections and a decrease in the response to protective vaccines. Influenza becomes clinically important when it occurs in the elderly, who account for over 80% of deaths related to epidemics. When old adults receive the influenza vaccine, the humoral and cytotoxic responses are lower than in young adults. The decreased response is mediated by reduced IL-2 production and diminished T-cell activation.

A similar pattern is seen with other infections for which vaccines are available. Streptococcal pneumonia is more likely to cause morbidity and mortality in the old adult than in the young. The antibody produced in response to pneumococcal vaccination has diminished affinity for its target and is less effective in preventing infection than antibody produced by young individuals. Serum levels of antibody also fade more rapidly in old adults.

Relatively rare in developed countries, tetanus occurs in persons who do not have protective levels of antitetanus toxoid antibody (more than 70% of adults over age 70). The risk of contracting the illness is more than 10 times higher for adults over age 80 than for adults in their 20s. When old people are immunized with tetanus toxoid, lower serum antibody levels are achieved, fewer cells participate in the production of antibody, and less antibody is produced on average per cell. In spite of decreased responses to all these vaccines, epidemiological evidence suggests that immunizations reduce morbidity and mortality in the elderly. This justifies the clinical practice of targeting old adults for vaccination.

Age-related waning of cellular immunity is related to the increased incidence of tuberculosis and herpes zoster infection. T cells from old mice are defective in their ability to fight tubercular infection and to prevent the spread from the lung to other organs. Most cases of herpes zoster occur in adults over age 75, paralleling a decline in the cellular immune response to the virus.

For background information *see* AUTOIMMUNITY; CELLULAR IMMUNOLOGY; CYTOKINE; IMMUNOLOGIC CYTOTOXICITY; IMMUNOLOGICAL AGING; INTERLEUKIN in the McGraw-Hill Encyclopedia of Science & Technology.   Edith A. Burns

Bibliography. E. A. Burns and J. S. Goodwin, Immunodeficiency disorders of aging, *Drugs & Aging*, 11(5):374–397, 1997; C. Caruso et al., Cytokine production pathway in the elderly, *Immunol. Res.*, 15:

84–90, 1996; R. Chandra, Nutrition and immunity in the elderly: Clinical significance, *Nutrit. Rev.*, 53: S80–S85, 1995; G. Doria et al., Age restriction in antigen-specific immunosuppression, *J. Immunol.*, 139:1419–1425, 1987; N. Fabris, A neuroendocrine-immune theory of aging, *Int. J. Neurosci.*, 51:373–375, 1990; J. S. Goodwin, Decreased immunity and increased morbidity in the elderly, *Nutrit. Rev.*, 53: S41–S46, 1995; J. S. Goodwin, R. P. Searles, and K. S. K. Tung, Immunological responses of a healthy elderly population, *Clin. Exp. Immunol.*, 48:403–410, 1982; D. R. Jackola, J. K. Ruger, and R. A. Miller, Age-associated changes in human T cell phenotype and function, *Aging*, 6:25–34, 1994; R. A. Miller, Aging and immune function, *Int. Rev. Cytol.*, 124:187–215, 1991; R. A. Miller, Calcium signals in T lymphocytes from old mice, *Life Sci.*, 59:469–475, 1996; L. Nagelkerken et al., Age-related changes in lymphokine production related to a decreased number of CD45RBhi CD4+T cells, *Eur. J. Immunol.*, 21: 273–281, 1991; C. Nicoletti, Antibody protection in aging: Influence of idiotypic repertoire and antibody binding activity to a bacterial antigen, *Exp. Molec. Pathol.*, 62:99–108, 1995.

Immunosuppression

Recognition of foreign antigens by host T cells is an essential defense mechanism to protect the body from pathogens such as viruses and bacteria. In the case of transplantation, this system serves to work against the desired outcome by attacking the donor organ, which it correctly perceives as foreign. The main focus of this attack is the products of the major histocompatibility complex (MHC), a set of genes that encode cell surface molecules. In recent years great advances have been made in understanding the interaction of the T cell with antigen-presenting cells (APCs), hence the mechanisms of allorecognition by which foreign (allo)-MHC is recognized. MHC molecules are structured such that they form a groove in which a peptide is bound. It is the three-dimensional structure formed by an MHC–peptide complex that is recognized by the T-cell receptor (TCR).

There are two known pathways of allorecognition: the direct pathway in which recipient T cells recognize donor MHC as intact molecules on the surface of donor antigen-presenting cells, and the indirect pathway whereby the donor antigen is endocytosed (taken in) by recipient antigen-presenting cells, then broken down into the form of peptides which are subsequently presented on the cell surface bound to recipient MHC molecules. In both of these pathways, the peptide plays a pivotal role in the interaction between the MHC molecule and the T-cell receptor with the subsequent response. Minor changes in the sequence of the peptide affecting critical residues involved in the interaction with the T-cell receptor may alter the nature of the T-cell response, resulting in T-cell activation, a state of partial activation or anergy, or the failure of recognition.

The realization that peptides play a central role in the immune response has stimulated research on their ability to function as immunoregulators. It is not just at this site that peptides have the potential to exert an effect: as the complexities of the immune response are unraveled, it has become apparent that there are many critical protein–protein interactions which may be disrupted by their use. While peptides offer the significant advantage of being nontoxic, the issue of bioavailability is a significant one, since peptides are rapidly broken down by plasma proteases.

The two major groups of peptides which have been found to effectively influence the immune response are MHC peptides derived from either conserved or nonpolymorphic areas of the MHC and those that are derived from variable or polymorphic regions of the MHC. These peptides act at various sites (see **illus.**).

Polymorphic peptides. Synthetic peptides derived from regions of the MHC which are highly variable have been used to investigate the mechanisms of allorecognition. These experiments have established the indirect pathway as a significant contributor to the rejection process, and they emphasize its role in chronic rejection. Those peptides which are recognized by T cells primed by immunization or transplanation are termed immunogenic peptides. Class I and II MHC polymorphic (having many shapes) peptides corresponding to segments of these polymorphic regions have been shown to induce tolerance in several small animal transplant models; however, their ability to do so is dependent on the route of administration. Intrathymic (within the thymus) injection of either class I or II MHC immunogenic allopeptides prior to engraftment has resulted in long-term survival of the transplanted organ. Induction of anergy, and peripheral deletion or development of a state of immune deviation, in which the cytokine profile is altered to the anti-inflammatory T-helper 2 (interleukin-4, interleukin-10) pattern, have been suggested as underlying mechanisms. Orally administered synthetic class II MHC allopeptides also induce tolerogenic effects and can inhibit delayed-type hypersensitivity, a T-cell-mediated reaction to antigen. Only immunodominant peptides are tolerogenic, and the unresponsive state is associated with a state of immune deviation.

The induction of tolerance to allografts has been achieved by the development of allochimeric proteins in which polymorphic regions of donor class I MHC are incorporated into a backbone of the recipient class I MHC. Following intraportal administration of these proteins, long-term engraftment, which is donor-specific, may be achieved. The precise underlying mechanisms are currently under investigation. Other peptides have been investigated based on the knowledge that variation in the nature of the peptide bound within the groove of the MHC molecule may alter the immune response. This may be achieved in several different ways. There may be MHC blockade

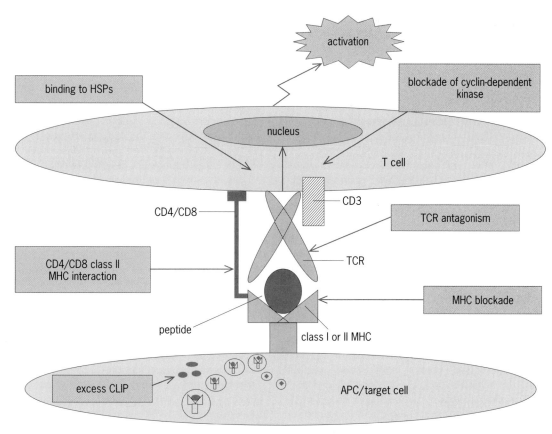

T-cell receptor and antigen-presenting-cell interaction, showing sites of action of immunoregulatory peptides.

in which an excess of a peptide from a relatively non-polymorphic region of a particular allele competes for presentation with an immunogenic peptide of the same allele. Second, there may be T-cell-receptor antagonism with an inhibition of the T-cell response that occurs through the use of antigen analogs in which the major T-cell contact residues have been modified to produce a powerful antagonist. Finally, there can be high-zone tolerance, when high concentrations of antigen stimulate initial T-cell activation followed by cell death through activation-induced apoptosis. These principles have been used successfully in vitro in an allogeneic system, and in vivo in experimental autoimmune models.

Nonpolymorphic peptides. Peptides derived from a conserved region of the alpha$_1$ domain of the human class I MHC, corresponding to residues 75–84, have been shown to inhibit class I restricted immune responses in an allele-nonspecific manner. These HLA-B7.75-84 and HLA-B2702.75-84 peptides prevented the differentation of precursors into effector cytotoxic T lymphocytes in vitro, inhibited lysis by established cytotoxic T lymphocytes, and inhibited natural killer cell-mediated cytotoxicity. These peptides have immunomodulatory effects in vivo, resulting in long-term graft survival which was donor-specific in several animal transplant models when given with or without low-dose cyclosporine. The peptides appear to mediate their effect through binding to heat shock proteins (HSPs). Based on these encouraging results

in animal models, there was a randomized, double-blind placebo controlled study of the safety and the pharmacokinetics of HLA-B2702.75-84 in human recipients of a first renal allograft. The initial report demonstrated a significant reduction in natural killer cytotoxicity noted in patients from day 15 through 2 months following the end of treatment. No toxic effects were noted. The rational design of analogs of this first-generation therapeutic is now in progress.

Since it had been documented that peptides bound within the groove of the MHC molecule were predominantly derived from MHC molecules themselves, the possibility existed that they served not only to stabilize the heterodimer but might also influence the binding of antigenic peptides or the interaction with the T-cell receptor. Three peptides, HLA-DQAI*0101 (62-77), RT1.Bα (62-78), and RT1.Dα (61-75), derived from the alpha chains of either human or rat class II MHC, have been demonstrated to have an immunosuppressant effect. HLA-DQAI*0101 (62-77) was found to be the most efficacious, suppressing the mixed lymphocyte response in several species. In addition, it inhibits the generation of cytokine T lymphocytes, proliferation to antigen, cytokine production, and superantigen, but not mitogen. The mechanism of action mediating this effect is the induction of apoptosis in T cells upon activation. A peptide derived from a similar region of the alpha chain of a different allele, DQA03011 (65-79), has also been shown to downregulate the immune response. Interestingly, it functions in a different manner from HLA-DQAI*0101 (62-77), through inhibition of cell-cycle-dependent kinase 2, thus preventing cell cycle progression. These peptides offer exciting potential for clinical application either through their use or through the understanding of their mechanisms of action.

Other immunomodulatory peptides. Proteins important to cell-cell interactions have been targeted as potential ways of manipulating the immune response. Peptide inhibitors of MHC class II–CD4/CD8 coreceptor interactions have been particularly successful in vivo. Structure-based design and nuclear magnetic resonance (NMR) spectroscopy techniques have been used to develop compounds which either mimic the CD4 interaction site with class II MHC or serve to block the interaction site. These reagents, a peptide analog and an organic compound respectively, have been shown to significantly prolong both skin and bone marrow engraftment and to prevent graft versus host disease. Furthermore, they were effective in preventing the development of experimental autoimmune encephalomeylitis in a mouse model. The precise mechanism by which they mediate their effect is yet to be established. Clinical trials are being initiated.

The pathway involved in loading of class II MHC and its subsequent presentation on the cell surface has been well documented. MHC class II–associated invariant chain peptide (CLIP) is essential for proper loading of exogenous peptide on the MHC. It stabilizes the heterodimer and permits presentation of MHC class II with its associated antigenic peptide on the cell surface. A synthetic peptide with a sequence derived from CLIP was synthesized and shown to inhibit antigen-specific T-cell responses in vitro and in vivo following immunization. It is presumed to exert its effect through inhibition of loading of MHC with antigenic peptide. This demonstrates that there are multiple points at which peptides may be used to manipulate the immune response, from the disruption of antigen presentation to the prevention of cell-cell contacts, the alteration of T-cell signaling, and finally, the inhibition of proliferation.

For background information *see* AUTOIMMUNITY; CELLULAR IMMUNOLOGY; HISTOCOMPATIBILITY; IMMUNOSUPPRESSION; LYMPHOKINES; PEPTIDE in the McGraw-Hill Encyclopedia of Science & Technology.

Barbara Murphy

Bibliography. C. Clayberger, Immunosuppressive peptides corresponding to MHC class I sequences, *Curr. Opin. Immunol.*, 7:644-648, 1995; A. M. Krensky, HLA-derived peptides as novel immunosuppressives, *Nephrol. Dial. Transplant.*, 12:865-868, 1997; C. C. Magee, Peptide-mediated immunosuppression, *Curr. Opin. Immunol.*, 9:669-675, 1997; B. Murphy, Immunomodulatory function of MHC-derived peptides, *Curr. Opin. Nephrol. Hyper.*, 5: 262-268, 1996; H. L. Weiner, Oral tolerance: Immunologic mechanisms and treatment of murine and human organ specific autoimmune diseases by oral administration of autoantigens, *Annu. Rev. Immunol.*, 12:809-837, 1994.

Infrared astronomy

The observation of celestial objects at wavelengths ranging from 1 micrometer to several hundred micrometers is known as infrared astronomy. Infrared radiation is that part of the electromagnetic spectrum located between radio waves and visible light. The major contribution of infrared astronomy to the field of astrophysics is the observation of dusty regions and solid bodies too cold to emit visible radiation. Dusty regions not only are invisible to the eye but also absorb the optical light originating behind them; the absorbed light warms up the dust and makes it glow in infrared light. Dusty regions are seen as dark patches in many optical images. Infrared photons are very weakly absorbed by dust, and thus infrared observations reveal what happens inside and behind these obscured regions (**Fig. 1**).

Recent improvements in the sensitivity and in the number of active elements (pixels) of detecting devices, coupled to the ability to carry out observations from above the Earth's atmosphere with airborne telescopes, balloons, and satellites, have opened up the field of high-resolution spectral studies throughout the infrared range. Many ions and molecules have very closely spaced energy levels so that transitions between these levels result in the emission of infrared photons at well-determined wavelengths. Infrared astronomers are now capable of observing the spectral signature of a large number of molecules and solids, thus contributing to the very active field of astrochemistry. Some spectral features are still unidentified, and much laboratory work is being carried out with the aim of identifying the source of these lines.

Infrared Space Observatory. The most sensitive infrared observations were obtained by the *Infrared Space Observatory* (*ISO*) of the European Space Agency (ESA). On April 8, 1998, *ISO* completed a 29-month mission (it was launched on November 17, 1995). It was placed in a highly elliptical orbit by an Ariane 44P launcher. The orbital period was 24 h, with perigee at 1000 km (620 mi) and apogee at 70,000 km (43,000 mi). Two other national space agencies, the National Aeronautics and Space Administration (NASA) and Japan's ISAS, collaborated with ESA to ensure 24-h telemetry coverage of the spacecraft. *ISO* was a major technical and scientific success. Its reserve of liquid helium, which was used to cool the telescope and the infrared detectors to a temperature of 2.7 K ($-454.8°$F), lasted 29 months whereas the nominal duration was 18 months. Its pointing accuracy and stability were measured to be several times better than the nominal specifications.

Results. During its lifetime *ISO* performed more than 25,000 observations on behalf of over 500 astronomers from Europe, the United States, and Japan. Owing to the large number of operational modes built into *ISO*'s four scientific instruments, astronomers could obtain, at their choice, spectral, photometry, or imaging information on selected areas of the sky, over a wavelength range spanning 2–240 μm.

glow discovered by the *Infrared Astronomical Satellite* (*IRAS*, launched in 1983 for a 10-month mission) in the Milky Way results from a series of spectral bands at 11.3 and 12.7 μm. *IRAS* observations lacked the spectral resolution to confirm this hypothesis. Most astronomers believe that these features, as well as other bands seen by *ISO* at 3.5, 6.2, 7.7, and 8.6 μm, result from stretching and bending oscillating modes of hydrocarbons. *ISO* has observed these bands in many other galaxies, so they appear to be universal tracers of the presence of interstellar matter.

Planetary nebulae. Stars are born inside dense interstellar clouds. Planetary nebulae are the final stage of the death of small stars such as the Sun. When the star's nuclear fuel is exhausted, it blows off the outer layers of its atmosphere; this is one of the mechanisms whereby stellar material is fed back into the interstellar medium, ready to become a next-generation star. Surprisingly, *ISO* observations of the Helix planetary nebula (**Fig. 2**) found no traces of hydrocarbon bands. Instead, the spectrum is dominated by emission lines due to molecular hydrogen. Hydrocarbon molecules may have been destroyed by the powerful ultraviolet emission of the central star.

Water vapor. Water vapor is so prevalent in the Earth's atmosphere that before *ISO* it was virtually impossible to measure its content in cosmic bodies. *ISO* has found water vapor in the outer planets, Titan (Saturn's satellite), the Milky Way, and distant galaxies. The highest density of water molecules has been observed in the Orion Nebula. The concentration of any chemical species is the resulting balance between creation and destruction processes. The water production rate in Orion has been estimated to create enough molecules to fill the Earth's oceans 60 times per day. A fraction of these water molecules will sooner or later become ice crystals that will stick onto dust grains in interstellar clouds. Similar

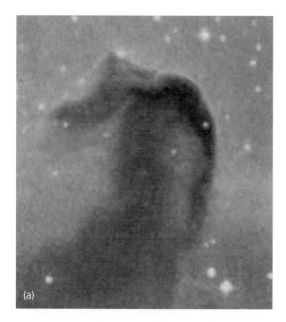

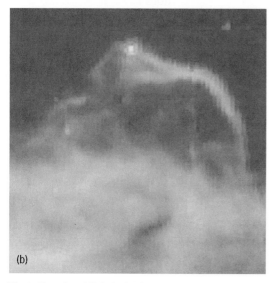

Fig. 1. Horsehead Nebula. (*a*) Optical image (© *1980 Royal Observatory, Edinburgh, and Anglo-Australian Telescope Board*). (*b*) Infrared image obtained by *ISO*'s camera, ISOCAM, a composite of observations at wavelengths of 7 and 15 μm; a small glowing filament is visible inside the Horsehead (*ESA/ISO/ISOCAM; L. Nordh, Stockholm Observatory, et al.*).

Spectral information is necessary to characterize the chemical composition of the emitting material; imaging information is required to study the morphology of infrared sources, as well as to establish their energy balance by relating the measured flux to the size of the emitting region.

Major astronomy fields addressed by *ISO*'s observations include planets, comets, zodiacal dust, interstellar medium, star formation and young stars, supernova remnants, planetary nebulae, galaxies, and clusters of galaxies. Several deep surveys were conducted, two of them in parallel with deep fields obtained by the Hubble Space Telescope, with the aim of understanding galaxy evolution.

Hydrocarbon emission. The spectroscopic capabilities of *ISO* have confirmed that the ubiquitous 12-μm

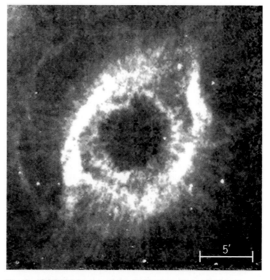

Fig. 2. ISOCAM image of the Helix planetary nebula at a wavelength of 7 μm. The ring of matter ejected by the collapsed star is clearly visible, but the star has become too hot to be visible in infrared light. (*ESA/ISO; ISOCAM; P. Cox, Institut d'Astrophysique Spatiale, et al.*)

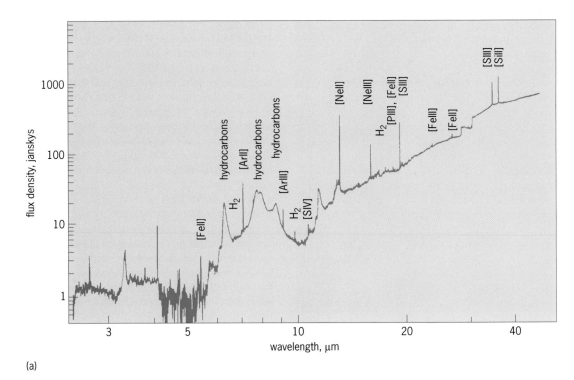

(a)

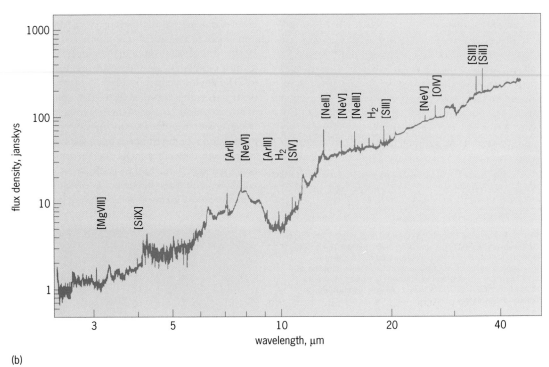

(b)

Fig. 3. Spectra of ultraluminous infrared galaxies (ULIRGs) obtained by the Short Wave Spectrometer on board *ISO*. 1 jansky = 10^{-26} W m^{-2} Hz^{-1}. (*a*) M82, a starburst galaxy, known to be powered by star formation. Emission bands of hydrocarbons are present at wavelengths of 6.2, 7.7, and 8.6 μm. (b) Galaxy in the constellation Circinus, known to have a black hole in its center. Emission from highly ionized species is present. (*ESA/ISO, SWS/MPE, D. Lutz et al.*)

ice-coated particles may have been present within the gas cloud from which the solar system was formed, explaining why water is present on the Earth.

Ultraluminous galaxies. One of the main results of the *IRAS* mission was the identification of a class of bright galaxies radiating most of their energy at far-

infrared wavelengths. These galaxies are known as ultraluminous infrared galaxies (ULIRGs). It has since been shown by high-resolution optical studies that ULIRGs result from the collision and merging of galaxies. The gravitational equilibrium in the galaxies is disrupted by the collisions, and huge clouds of gas and dust fall toward the galactic nuclei. It has been proposed that the enormous surge of infrared emission is due to a dramatic enhancement of the rate of star formation in these dusty clouds. A second scenario is that they are attracted by a central black hole and heat up as they fall. However, the dust at the origin of the observed emission prevented optical astronomers from seeing the center of the galaxies and distinguishing between the two mechanisms. The Short Wave Spectrometer on board *ISO* was finally able to solve the problem since it is able to see through the dust and examine the spectral signatures predicted by one or the other scenario. Astronomers are now able to tell for each observed ULIRG which mechanism is at work. Both starburst- and black-hole-driven ULIRGs have been observed (**Fig. 3**). Highly ionized species are present in the black-hole galaxies, while the starburst galaxies display emission from hydrocarbons.

Star formation. Observations of nearby colliding galaxies with ISOCAM, the camera on *ISO*, showed that sites of star formation are invisible at optical wavelengths but brightly lit in infrared light. To perform statistical studies of star formation in more distant galaxies, deep surveys, that is, long exposures of a field pushing the instrument sensitivity to its limits, were undertaken by ISOCAM. The deep surveys have shown that indeed most sites of star formation are inside dusty regions and thus invisible to all but infrared observations. For example, the star formation rate when the universe was one-third of its present age was measured to be four times larger than previously deduced from optical and ultraviolet observations alone.

A deep survey conducted using ISOPHOT, *ISO*'s photometer, resulted in the discovery of almost 200 very distant galaxies which were still evolving to their final shape at the time the emission seen by *ISO* left the objects, some 12×10^9 years ago. These observations will aid in research on galaxy formation in the early ages of the universe.

Postmission archive. All the European teams that participated to the construction of *ISO*, together with the Infrared Processing and Analysis Center of the Jet Propulsion Laboratory, are preparing the legacy archive of *ISO*'s observations. The interim archived material is available to any person interested in working with data obtained during the mission and will be accessible via any Internet browser. The archive contains raw data as well as data processed by an automated pipeline in a manner to make it suitable to non-*ISO* experts. Advanced interactive data-analysis packages will be made available to those who would rather analyze the raw data.

For background information *see* GALAXY, EXTERNAL; INFRARED ASTONOMY; INTERSTELLAR MATTER; MILKY WAY GALAXY; PLANETARY NEBULA; STARBURST GALAXY; STELLAR EVOLUTION in the McGraw-Hill Encyclopedia of Science & Technology.

Catherine Cesarsky; Diego A. Cesarsky
Bibliography. *The Universe as Seen by ISO*, European Space Agency, 1999.

Integrated electric propulsion

Many ship types experience a wide variation in propulsion power demand during normal operating conditions. These variable load demands are coupled with small percentages of operating time when full propulsion power is required. Therefore an integrated electric plant providing both ship service and propulsion power can be very effective in reducing operating costs. Variable operating profiles in conjunction with stable ship service load demands commonly occur in such diverse ship types as oceanographic research vessels, cruise ships, icebreakers, product oil tankers, automobile ferries, and combatant ships with multiple warfare roles.

The use of integrated electric propulsion systems in these ship types results in greatly reduced annual fuel oil costs, due to the ability of such systems to automatically match the number of operating generator sets to the specific electric load demand at any given time, and each individual prime mover can operate a reduced number of hours each year, with significant savings in maintenance and overhaul costs. These savings are made possible by new solid-state power electronics technology, which permits variable voltage and frequency outputs to be derived from a unit with fixed voltage and frequency inputs. The applications have resulted in a tremendous resurgence of orders for ships with alternating-current electric propulsion, approaching the levels before World War II.

Direct-current propulsion. Electric motors powered by direct-current generators were first used for ship propulsion in the late 1800s, when it became apparent that separately excited dc motors exhibited the same characteristics of high output torque at very low speed as did steam reciprocating engines. The high reliability of these machines coupled with extremely responsive reversing and maneuvering capability led to many applications in icebreakers, tugboats, and ferries. This technology was also used in naval ships in many countries, culminating in construction of hundreds of diesel-electric destroyer escorts, submarines, minesweepers, fleet tugs, and tankers in the 1930s and 1940s.

Alternating-current propulsion. As propulsion motor horsepower requirements in larger vessels increased beyond the practical limits of dc motor applications due to the inherent limitations of brushes and commutators to transmit the high current levels required, ac propulsion motors came into general use in the 1910s. Variable-speed steam turbines were applied to ac propulsion generators as a means of varying their output frequency, and ac propulsion motors with as many as 30 poles and a maximum speed of 240 revolutions per minute (rpm) provided output power to the propeller shafts. Separate fixed-frequency generators were still required for ship service power. This type of propulsion plant remained popular for special ship applications such as the liner *Normandie*, built in 1935, on into the 1960s, when the liner *Canberra* went into service.

In addition to commercial tankers and colliers, battleships, cruisers, seaplane tenders, and the first U.S. Navy aircraft carriers used this means of propulsion. The advantages were the same as for the dc propulsion plants: rapid maneuvering and the ability to operate only the minimum number of generator sets required to meet the load demand during each ship operating condition. A little-known application of steam turbine-electric propulsion plants was the emergency supply of electric power to land-based utilities in the wake of natural disasters which disabled generating stations. A notable example occurred in the 1930s when the turbine generator sets of a moored aircraft carrier provided electric power to Tacoma, Washington, for several weeks, including the power for the street railway system. U.S. Navy ships were also deployed to individual land-based generating plants to supply power to assist in the recovery from the Northeast power blackout in 1965.

Demise of conventional systems. The requirements for larger, separate, ship-service generator sets, the improved efficiency of steam turbines and gears, and the increasing first cost of turbine-electric propulsion plants brought about the end of these applications after World War II. Many ships of that era still remain in service, however, attesting to the reliability of this type of machinery. Of the hundreds of T-2 tankers built in the 1940s, many remain in service worldwide with the original turbo-electric propulsion plants powering modified midbodies containing everything from petroleum products, chemicals, containers, dry cargo, and orange juice to high-powered radar sets for missile tracking. The U.S. Military Sealift Command continued to operate the range tracking ships *Redstone* and *Vanguard* until 1995, with electric propulsion plants over 50 years old. Nine other turbine-electric vessels are still operating in the U.S. flag commercial marine service. An alternative to the ac steam turbine-electric propulsion plant was the ac diesel-electric generating plant. These ships utilized multiple variable-speed propulsion diesel generator sets which powered the propulsion motors at sea and then operated at a fixed frequency when in port or at anchor to supply auxiliary loads. Examples were the *USS Hunley* and *USS Holland*, naval submarine tenders which were commissioned in 1963.

Technology. Solid-state propulsion-motor drive converters make it possible to utilize fixed-frequency ac generator sets supplying a common bus to feed both propulsion and ship service loads simultaneously. Supplying both propulsion and ship service loads from a common system permits the use of a central power station concept whereby the on-line generating capacity can be efficiently matched to the actual load demand.

The solid-state power electronic drive converter modules, utilizing programmable microprocessor-based digital control such as silicon controlled rectifiers, thyristors, and isolated gate bipolar transistors (IGBT), receive fixed-voltage-and-frequency ac

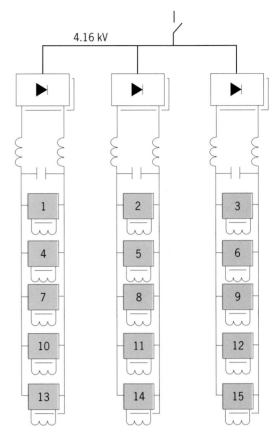

Fig. 1. Fifteen-phase isolated gate bipolar transistor (IGBT) stack. Transistors are numbered 1 through 15. (*Alstom-Cegelec USA*)

power from the main switchboards. The modules convert this ac input power to dc power, then convert it back to the variable-voltage-and-frequency ac power needed by the propulsion motors to accurately respond to the remote-control speed and direction signals received from the navigating bridge by the microprocessors.

Compact IGBT drive converter modules offer simple robust construction with greatly reduced harmonic distortion and motor noise. The water-cooled converters utilize modular power stacks, with the power transistors arranged in an H-bridge configuration for each individual phase supply (**Fig. 1**).

A saturable reactor is provided for each bridge arm to suppress high-frequency pulses and reduce switching losses. Fast-response digital regulators provide precise propulsion motor control and have extensive built-in diagnostic capabilities.

Applications. The application of ac/dc integrated propulsion motor drives with fixed-frequency ac input and variable-voltage dc output became common in icebreakers, tugboats, and offshore supply vessels in the 1970s. In the early 1980s, ac/ac propulsion motor drives with fixed-frequency ac input and variable-frequency ac output began to be applied in icebreakers, ferries, and large cruise vessels. This type of plant has been increasingly popular in commercial and naval ships for both new construction and conversions. A notable propulsion plant conversion was the

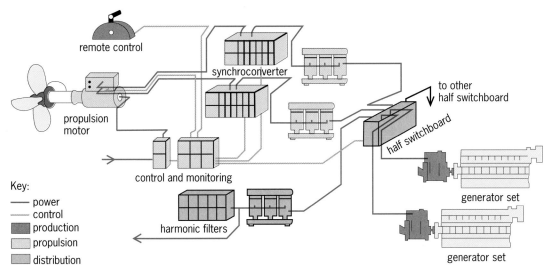

Fig. 2. Marine integrated alternating-current propulsion plant. (*Alstom-Cegelec USA*)

1987 repowering of the *Queen Elizabeth 2* from geared steam-turbine propulsion to ac/ac diesel electric propulsion utilizing two 45-megawatt (60,000-horsepower) propulsion motors. Major cruise lines are ordering increasingly larger ships with integrated ac/ac electric propulsion plants even as the application of these systems to military vessels, such as the 22.5-MW (30,000-hp) U.S. Coast Guard icebreaker *Healy*, continues. The British Royal Navy Type 23 frigates have a combined gas turbine and diesel electric (CODELAG) propulsion plant in which the electric propulsion motors are installed directly into the gas turbine gearbox output shaftline. The motors are energized from the ship service generator sets when the gas turbines are shut down, to provide an economical cruise mode when full power is not required. The U.S. Navy has acquired a prototype 19-MW (25,500-hp) integrated electric propulsion plant utilizing IGBT technology for installation in a land-based test site designed for development testing of electric propulsion plants for new combatant ships.

Plant configuration. An integrated ac electric propulsion plant for a typical large ship (**Fig. 2**) includes a power generation element, a power distribution element, and propulsion motors and drives.

The power generation element could consist of as many as six 6-MW (8000-hp) diesel or gas turbine generator sets. This quantity gives the capability to provide power at the most efficient generator set utilization level, from anchor to the full-power operating condition. The sets would be installed in two separate generator rooms to provide redundancy in the event of a casualty. Each generator set would have an associated generator control panel for local control and monitoring and could be automatically or manually started both locally and remotely at an electric plant control console in the central control station. Automatic paralleling and load-sharing capability would be provided for each set.

The power distribution element would consist of two main switchboards, located in separate compartments, interconnected in a ring bus configuration. Each main switchboard would feed a propulsion motor drive and two or more ship service transformers. The secondary of each ship service transformer feeds a ship service switchboard, which supplies power to loads in the electrical zones throughout the vessel. Main switchboard and ship service switchboard power circuit breakers would have both local manual control and remote automatic control from the electric plant control console in the central control station.

Ship propulsion could be achieved by the application of two 15-MW (20,000-hp), 210-rpm, synchronous motors, with dual windings for system redundancy. Each motor would receive power from two IGBT propulsion drives with 60-Hz fixed-voltage-and-frequency input, and ac variable-voltage-and-frequency output. These drives would be provided with an emergency local propulsion motor control panel and with remote control capability from propulsion control consoles on the navigating bridge and in the central control station.

For background information *see* GENERATOR; ICEBREAKER; MARINE ENGINE; MARINE MACHINERY; MOTOR; NAVAL SHIP; POWER INTEGRATED CIRCUITS; SEMICONDUCTOR RECTIFIER in the McGraw-Hill Encyclopedia of Science & Technology.   Robert J. Whitfield

Bibliography. Institute of Electrical and Electronics Engineers, *IEEE Recommended Practice for Electric Installations on Shipboard*, IEEE Std 45-1998, Institute of Marine Engineers, London, *Proceedings of the 1998 All Electric Ship Conference*, 1998.

Intelligent collaborative agents

Human agents, such as travel agents, insurance agents, and real-estate agents, are common. The Agency Theory in economics proposes that complex organizations must employ representatives (agents) to supervise and manage the organization's operations. Such agents operate on behalf of their organizations and interact with clients to accomplish specific economic goals, for instance, planning travel, issuing insurance policies, or executing realty transactions. With the increasing influence of information technologies on economic activities, artificial agents have begun to assume a similar role. Artificial, computer-based agents also operate as intermediaries in complex organizations and perform their functions with specific goals. They are called intelligent because they are part of artificial intelligence and their intelligence can be designed by their programmer.

Early agents were defined for Distributed Artificial Intelligence, which comprises distributed problem solving (DPS) and multi-agent systems (MAS). Distributed problem solving focuses on centrally designed systems solving global problems and applying built-in cooperation strategies. Multi-agent systems deal with heterogeneous agents whose goal is to plan their utility-maximizing coexistence. Examples of distributed problem solving are mobile robots exploring uncertain terrain, and task scheduling in manufacturing facilities. Both can be operated with centralized programs, but in relatively more distributed environments they are usually more effective with autonomous programs, or agents. Examples of multi-agent systems are collaborative product design, and group behavior of several mobile robots.

Definition of agents. An agent is a computing hardware- or software-based system with the following properties. (1) Autonomy: agents operate without the direct intervention of humans or other agents, and have certain control over their actions and internal state. (2) Interaction ability (also called social ability): agents interact with other agents, and possibly humans, via some agent-communication language. (3) Reflexivity: agents perceive their environment, which may be the physical world, a user via a graphical user interface, a collection of other agents, the Internet, or perhaps all these combined, and react in a timely fashion to changes that occur in this environment. (4) Proactiveness: agents do not simply act in response to their environment, but are able to exhibit goal-directed behavior by taking the initiative.

An agent, such as a mobile robot or a Web-search software agent, can autonomously interact with other agents and with its environment, act, and respond reflexively to the impacts from the environment, all in accordance with its given goals. The importance of agents is their ability to perform complex intelligent activities in a highly distributed, global environment with the advantages of synergy and in parallel to their human users. Since individual agents are rarely able to fulfill their system's goals, most agent applications are termed multi-agent systems.

Interaction is fundamental to agents. The paradigm shift from algorithms to interactions, and from procedure-oriented to object-based and distributed processing, captures the technology shift from mainframes to workstations and networks. Interactions are thus viewed as more powerful than algorithms. Agents can interact as needed and when needed, based on current conditions and not on predefined logic which may already be unsuitable.

An agent may seek collaboration through communication with other agents. The communication is regulated by protocols, or structures of dialogs, that are defined to enhance the effectiveness and efficiency of the numerous information exchanges.

Collaboration. Agents are often applied to integrate information and knowledge from their distributed environment and to achieve integrated actions or results. Because of the increasingly distributed nature of systems, the agents must obtain information, knowledge, and support from other cooperative agents and systems. Collaboration, cooperation, and integration depend on open communication, and are interrelated terms. Collaboration is the active participation and work of the subsystems toward integration. Cooperation is the willingness and readiness of subsystems to share or combine their tasks and resources. Integration is a process by which subsystems share or combine physical or logical tasks and resources so that the whole system can produce better (synergistic) results.

Based on these definitions, collaboration is impossible without cooperation, and the quality of integration will be determined directly by the nature of the collaboration. Therefore, agents in benevolent applications would usually be designed to be collaborative (with certain rational restrictions and defenses). Computer virus agents, obviously nonbenevolent, must be initially collaborative, but eventually become destructive.

Classification of agents. Frequently, agents are defined indirectly by their characteristics. Such definitions may be confusing because many of these characteristics are also common to computer programs. Indeed, any agent is also a computer program and can include hardware, as in a mobile robot, but not every program is an agent. Usually, a program does not have the four properties of an agent: autonomy, interaction ability, reflexivity, and proactiveness.

Based on recent research and development of agents, agents can be classified by their nature or by their application, as outlined below. Future applications are as far-reaching as human ingenuity.

I. Classification by nature
  A. Biological
  B. Robotic
    a. Mobile robot
      +Service
      +Manufacturing
      +Other
    b. Robotic tool
    c. Smart sensor
    d. Other robot
  C. Computational
    a. Artificial life

b. Software
  +Task-specific
    Search ("Bot")
    Operating system helper
    Controller
    Other software task
  +Entertainment
    Computer game
    Other
  +Virus
II. Classification by application
  A. Industrial
    a. Process control
    b. Design
    c. Scheduling
    d. Manufacturing
    e. Logistics
  B. Commercial/service
    a. Information management
    b. Electronic commerce and banking
    c. Business process management
    d. Environmental monitoring
  C. Medical
    a. Patient monitoring
    b. Health care
  D. Military
    a. Intelligence gathering
    b. Robotics
  E. Entertainment
    a. Computer games
    b. Interactive theater and cinema

Characteristics of agents. Researchers have variously described agents' characteristics, sometimes even proposing conflicting definitions. Some characteristics are considered optional for certain applications. The following list contains a sample survey of the most typical agent characteristics.

Autonomous: Exercises control over its own action/Without direct intervention of humans or other agents

Interactive, Communicative, Sociable: Communicates with other agents/Communicates with agents and with humans

Reactive: Responds in a timely manner to changes in the environment

Reflexive: Responds immediately to certain changes in its environment without any planning

Proactive, Goal-oriented: Acts in pursuit of goals, not only in response to the environment

Temporally continuous: Is a continually running process

Communal: Lives in a "world" with other agents

Reproductive: Capable of producing other agents when necessary

Mobile: Capable of migrating from one machine or system to another

Flexible: Its actions are not scripted

Believable: Can assume a believable personality and emotional state

Competitive: Optimizes its state and functions locally

Benevolent: Not lying, nondestructive

Adaptive: Capable of learning/Capable of improving over time

Resourceful: Able to handle unknown situations

Agent networks and organizational communication. Usually, an agent does not function alone but as a member of a group of agents, or an agent network. The interaction and communication among agents can be explained by the analogy of organizational communication. An organization is an identifiable social entity pursuing multiple objectives through the coordinated activities and relations among members and objects. Such a social system is open-ended and dependent for its effectiveness and survival on other individuals and subsystems in the larger society. Following this analogy, three characteristics of an organization, and of an agent network, can be observed: (1) entities and organization; (2) goals and coordinated activities; and (3) adaptability and survivability of the organization.

There are also five motivations observed for organizational communication and for agent network communication: (1) generate and obtain information; (2) process and integrate information; (3) share information needed for the coordination of interdependent organizational tasks; (4) disseminate decisions; and (5) reinforce a group's perspective or consensus. These five motivations can serve as a checklist for developing interaction protocols. One of the most influential factors affecting interpersonal or interagent communication patterns among group members involves the characteristics of the task on which they are working. As task certainty increases, the group coordinates more through formal rules and plans than through individualized (personal) communication modes. Therefore, the interacting behaviors and information exchanges among agents have to follow interaction and communication protocols.

For any agent-based collaborative problem solving, three reasons for uncertainty exist: (1) lack of information necessary to make a decision; (2) lack of the outcome of a decision, or of the perceived risk associated with a decision; and (3) lack of understanding the root causes of a problem. For each application, the design of individual agents and of the agent networks has to follow a strategy to reduce the uncertainty to the point of fulfilling the intended goals.

Agent design. Although different agent applications require different agent design, five general areas have to be addressed: (1) goal identification and task assignment; (2) distribution of knowledge; (3) organization of the agents; (4) coordination mechanisms and protocols; and (5) learning schemes.

There are two basic approaches to the design of agent systems. The first is to allow as many agents as necessary to be formed and reproduced. The second advocates self-controls that would destroy or remove agents that have fulfilled their functions and are no longer needed. The latter approach is preferred when there is a concern for the overburdening of systems with software and possibly hardware of unnecessary agents.

Trends. Intelligent collaborative agent research is highly active and in its preliminary stages. While the most known applications have been in Internet search and remote mobile robot navigation, emerging examples combine agents through computer networks with remote monitoring for security, diagnostics, maintenance, and repair, and remote manipulation of robotic equipment. Emerging applications will soon revolutionize computer and communication usefulness. Interaction and communication with and among intelligent tools, home appliances, entertainment systems, and highly reliable mobile service robots will change the nature of manufacturing, services, health care, food delivery, transportation, and virtually all equipment-related activities.

For background information *see* ARTIFICIAL INTELLIGENCE; CONTROL SYSTEMS; CYBERNETICS; EXPERT SYSTEMS; HUMAN-COMPUTER INTERACTION; HUMANFACTORS ENGINEERING; HUMAN-MACHINE SYSTEMS; INTELLIGENT MACHINE; NEURAL NETWORK; ROBOTICS; VIRTUAL REALITY in the McGraw-Hill Encyclopedia of Science & Technology.   Shimon Y. Nof

Bibliography. C. Y. Huang and S. Y. Nof, Formation of autonomous agent networks for manufacturing systems, *Int. J. Prod. Res.*, 1999; N. R. Jennings and M. J. Wooldridge (eds.), *Agents Technology Foundation, Applications, and Markets*, Springer-Verlag, 1998; P. Maes, Modeling adaptive, autonomous agents, *Artif. Life*, 1(1/2):135–162, 1994; S. Y. Nof (ed.), *Handbook of Industrial Robotics*, 2d ed., John Wiley, 1999; L. Steels, When are robots intelligent, autonomous agents?, *Robotics Autonomous Sys.*, 15:3–9, 1995.

Intelligent transportation systems

While travel on roads and streets continues to increase, the available capacity of the roadway system is rapidly being exhausted. Between 1976 and 1996 the number of annual vehicle miles of travel in the United States increased by 77%, but the road and street miles increased only 2%. In 1996, 54% of peak-period vehicle miles of travel occurred under congested conditions on urban interstates. As a result of congestion, drivers spend more time than ever on the road—an average of approximately 1.2 hours per day. Congestion adversely affects commercial vehicles, causing delays in transporting goods and services and thus adding to their costs. Congestion also increases the risk of traffic jams and air-quality degradation.

Environmental and financial constraints on land use make it difficult to expand the highway system appreciably in the United States. Consequently, alternative approaches must be explored to add to the capacity of highway systems without physical

expansion. The concept of intelligent transportation systems (ITS) is to enhance traffic capacity, safety, and environmental quality through efficient management of the operation of existing facilities using emerging information and communication technologies. Although this concept and attendant research were initiated in the United States as early as the 1960s, a serious effort to plan and deploy these systems was not made until the early 1990s. The Intermodal Surface Transportation Efficiency Act (ISTEA) of 1991 first authorized federal funds for the planning and operational tests of these systems. The Transportation Equity Act for the 21st Century (TEA 21) of 1998 provided continued funding for these systems, and many states and local areas are currently involved in implementing one or more of the possible applications of ITS technologies.

User services. An intelligent transportation system involves the integrated application of advanced information processing and communications, serving, display, and control technologies to surface transportation—both in the vehicle and on the highway. ITS technologies are considered in terms of a collection of interrelated user services for application to the surface transportation problems. To date, 30 user services have been identified, bundled into six categories (**Table 1**). Individual user services are building blocks that may be combined for implementation in a variety of ways depending upon local priorities, needs, and market forces. These services may require multiple technological elements or functions that are common for several services. For example, a single user service will usually require several technologies, such as advanced communications, mapping, and surveillance, which may be shared with other services. At present, user services are in various stages of development. Many of these services can be deployed in rural and suburban as well as urban settings.

Travel and traffic management. User services collect and process information about the surface transportation system and provide commands to various traffic control devices. For example, real-time information about traffic conditions can be provided to drivers through highway advisory radios (HAR), through variable message signs (VMS), or directly in the vehicles. Other information may include incidents, construction work zones, transit schedules, and weather conditions. This information, if pretrip, can assist travelers in planning the timing or location of trips, and if en route, in selecting another route or shifting to another mode. Traveler services information can also provide a business directory, or "yellow pages," of service information such as the location, operating hours, and availability of food, lodging, parking, vehicle repair, hospitals, and public facilities. This information can be accessible in the house, office, or other public locations as well as en route. The incident management services use advanced sensors, data processing, and communications to promptly identify incidents and implement a response to minimize their effects on traffic. This group of user services is also aimed at policies and strategies for

TABLE 1. Intelligent transportation systems user services

Bundle	User services
Travel and traffic management	Pretrip travel information Enroute driver information Route guidance Ride matching and reservation Traveler service information Traffic control Incident management Travel demand management Emissions testing and mitigation Highway-rail intersection
Public transportation management	Public transportation management Enroute transit information Personalized public transit Public travel security
Electronic payment	Electronic payment services
Commercial vehicle operations	Commercial vehicle electronic clearance Automated roadside safety inspection On-board safety monitoring Commercial vehicle administrative processes Hazardous material incident response Commercial fleet management
Emergency management	Emergency notification and personal security Emergency vehicle management
Advanced vehicle safety systems	Longitudinal collision avoidance Lateral collision avoidance Intersection collision avoidance Vision enhancement for crash avoidance Safety readiness Precrash restraint deployment Automated vehicle operation

reductions in travel demand, particularly of single-occupancy vehicles. For example, pretrip travel information or real-time ride matching information and reservations can significantly affect the timing, mode, and location of trips.

An example of a traffic management system is the one implemented along 26 mi (42 km) of freeway in San Antonio, Texas, to speed up the detection and response to traffic incidents. The system uses a battery of detectors and video cameras to detect problems from the traffic management center and quickly dispatch appropriate emergency vehicles. Additionally, the data are sent to the area's traffic signal computer, which draws on over 34,000 pre-programmed responses to adjust signal timing on arterials, lane control signals, and message signs for commuters along the roadway.

Public transportation management. These services can involve the automation of operations, planning, and management functions of public transit systems, providing enroute real-time transit information to travelers as well as efficient demand-responsive transit services. For example, in Kansas City, Missouri, 160 transponder-equipped signposts were installed along 38 bus routes to communicate with the transit management center and detect when a bus passes. A display appears on the bus indicating if it is running early or late. Additionally, buses are equipped with communications equipment that allows them to respond quickly to emergencies by immediately notifying the transit management center.

Electronic payment. A common electronic payment medium is envisioned for all transportation modes

and functions, including tolls, transit fares, and parking, using "smart cards" or other technologies. The flexibility offered by such a system can also facilitate travel demand management and application of congestion pricing policies.

For example, the Go Card used by the Washington Metropolitan Area Transit Authority enables a commuter to pay for bus rides, trips on the subway, and parking fees. In New York City the MetroCard (similar to a credit card) is used by the Metropolitan Transit Authority with a variety of choices, such as Unlimited Ride (7 days or 30 days), Fun Pass (unlimited subway and bus rides per day), and Pay-Per-Ride. Also in New York City, various toll bridges have incorporated E-ZPass, utilizing a sticker inside the windshield near the rear-view mirror. As the vehicle enters the toll lane, the E-ZPass system recognizes the individual code on the sticker. The appropriate discounted toll is automatically deducted from the E-ZPass account, and the gate goes up to pass the vehicle. Recognition-activation requires about 2 seconds.

Commercial vehicle operations. These services are geared to improve the safety and productivity of the motor carrier industry. Most of the applications have been to provide (1) safety assurance through automated inspections and reviews; (2) credential administration which allows filing credential and permit information through a single transaction; (3) electronic screening which automates weight, safety, and credential screening at roadside weigh stations and facilities; (4) seamless travel through state and international borders; and (5) carrier fleet management through automated vehicle location (AVL) and optimal route planning. An example of these services is the PREPASS program currently being implemented on the West Coast and in southwestern states, allowing trucks preclearance in each state. A participating truck has a transponder on its dashboard that provides information electronically to weigh station facilities. As a truck on the highway proceeds at a normal speed, a reader embedded in or above the roadway identifies the truck and a computer in the weigh station verifies the truck's credentials. Additionally, a Weigh-in-Motion (WIM) device embedded in the road checks the vehicle weight. If the truck passes inspection, it is given a green light to continue. If not, a light flashes to indicate that the truck must pull over at the upcoming weigh station. This system allows a truck to clear a weigh station without slowing, thus saving truckers time, keeping highway traffic moving smoothly, and reducing emissions.

Emergency management. Police, fire, ambulance, and other emergency operations can improve their management of and response to emergency situations through ITS technologies. These services can particularly improve safety in rural areas through emergency management systems, helping drivers by immediately notifying police and other emergency operations of accidents or hazardous conditions. The emergency management system in Chicago includes an electronic map of the city with a high level of detailed information, including 20,000 street segments, over 20,000 alleys, the location of hydrants, and "footprints" of the buildings. Emergency vehicles are outfitted with automatic vehicle locators, allowing a central computer system to locate the vehicle nearest the scene of an emergency and dispatch it with instructions and routing. This system has the potential to link up to a database which can provide information on the medical conditions of people who request assistance.

Advanced vehicle control and safety systems. These services are primarily directed to improve vehicle safety through collision avoidance and warning systems, including vision enhancement, safety readiness, and precrash restraint deployment. In addition, automated highway systems (AHS) are a long-term ITS goal which would provide a fully automated, hands-off operating environment. To improve the safety of rail-highway and rail-pedestrian crossings of its Light Rail System, Los Angeles County has implemented video surveillance of crossings and intrusion detection devices. In addition, the system uses photo enforcement for red-light or grade crossing violations. The system also equips vehicles such as buses and hazardous-materials trucks with alert devices that can warn upcoming trains of these vehicles on the tracks.

Architecture. ITS architecture defines the framework around which multiple design approaches can be developed, each one specifically tailored to meet the needs of the individual user. The architecture defines the functions (for example, gathering traffic information or requesting a route) that must be performed to implement a given user service, the physical entities or subsystems where these functions reside (for example, the roadside or the vehicle), the interfaces or information flows between the physical subsystems, and the communication requirements for the information flows (for example, wireline or wireless). The system architecture also identifies and delineates types of standards needed to accommodate national and regional interoperability, as well as product standards needed to support economy-of-scale considerations in deployment.

Logical architecture. The logical architecture provides a model of ITS functions that can be depicted using data flow diagrams. Logical architecture includes electronic payment, driver, traveler, and emergency services. It also incorporates commercial vehicle operations, planning and deployment, vehicle and monitoring control, and transit and traffic management (**Fig. 1**).

Physical architecture. The physical architecture partitions the functions defined by the logical architecture into systems and subsystems (**Table 2**). Traveler subsystems represent platforms for ITS functions of interest to travelers or carriers in support of multimodal traveling. Center subsystems refer to functions that can be performed at control centers. Roadside subsystems include functions that require convenient access to a roadside location for the deployment of sensors, signals, programmable signs, or other interfaces with travelers and vehicles. Vehicle subsystems represent functions (for example, navigation and tolls) that may be common across all types of vehicles.

Communication. ITS architecture provides the framework that ties the transportation and telecommunication fields together. Four communication media types are identified to support the communications requirements between the subsystems: wireline (fixed-to-fixed), wide-area wireless (fixed-to-mobile), dedicated short-range communications (fixed-to-mobile), and vehicle-to-vehicle (mobile-to-mobile). Center subsystems can be linked together over a

TABLE 2. Physical architecture of systems and subsystems

System	Subsystem
Traveler	Personal information access
	Remote traveler support
Center	Traffic management
	Emergency management
	Emission management
	Commercial vehicle administration
	Planning
	Transit management
	Information service provider
	Toll administration
	Fleet and freight management
Roadside	Roadway
	Toll collection
	Parking management
	Commercial vehicle check
Vehicle	Private
	Transit (public)
	Commercial
	Emergency

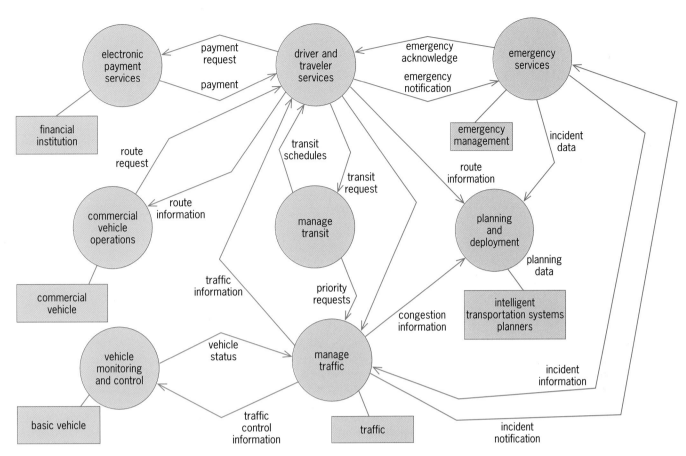

Fig. 1. Simplified top-level data flow diagram. Circles represent functions that are broken down into lower levels of detail, and rectangles represent external entities.

wireline network. There are two distinct categories of wireless communications based on range and area of coverage: wide-area and short-range. Wide-area wireless communications are further differentiated based on whether they are one- or two-way. Short-range wireless communications include dedicated short-range communications (DSRC) and vehicle-to-vehicle communications.

Technologies. ITS technologies can be primarily grouped into surveillance, communications, traveler interface, control strategies, navigation/guidance, and data processing. Basic devices for surveillance are sensors and detectors. Sensors are sensitive to light, temperature, or radiation level and transmit a signal to a measuring or control instrument. Detectors are used to note the existence or presence of something. Communication technologies can be wireline or wireless. Common media include fiberoptic wirelines, traditional telephone lines, commercial radios, cellular phones, satellites, video image processing, and infrared and microwave transmission. Traveler interfaces may involve variable message signs, touch screen, keypad/keyboard, voice recognition, voice output, visual display (on-board computer), and head-up display in vehicles. Control strategies can be realized by using ramp metering, high occupancy vehicle (HOV) restriction, sign control, parking restriction, ramp/lane closures, road use pricing, and reversible lanes. Navigation/guidance

refers to pretrip or en route advisory/instruction on trip choice. The measures may include position display, map database, guidance display, and dead reckoning. One application for navigation/guidance purpose is the Global Positioning System (GPS) technology. GPS components for civilian use comprise a constellation of satellites, a network of base stations, radio communication, receivers, a database (for example, map datum and application databases), and application software. Other mechanisms may include in-vehicle navigation systems (IVNS), variable message signs, highway advisory radios, electronic mass media, and kiosks. Technologies used for data processing mainly include dynamic traffic assignment (DTA), route selection algorithms, driver/vehicle/cargo scheduling, and real-time traffic prediction.

Implementation. Indiana's Borman Expressway is one of the most heavily traveled highways in the United States (**Fig. 2**). With an average of more than 120,000 vehicles per day, including 20–30% trucks, traffic management is crucial. An Advanced Traffic Management System (ATMS) is being implemented by the Indiana Department of Transportation along the I-80/94 freeway corridor in northwest Indiana with the purpose of rapidly detecting and alleviating congestion-causing problems.

The Borman Advanced Traffic Management System has four components: traffic management,

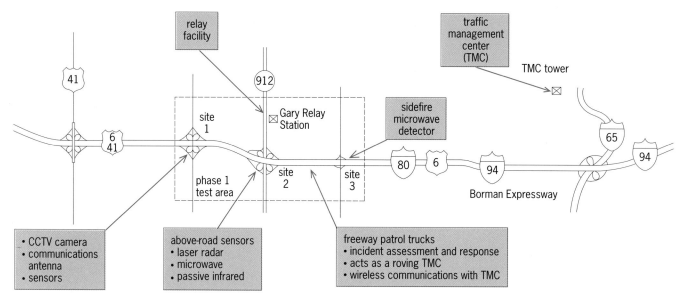

Fig. 2. Borman Advanced Traffic Management System.

motorist information, incident response, and advanced communications. For traffic management, expert systems software, hosted by a graphics workstation, is used to monitor the flow of information on corridor traffic flow from various sources to the traffic management center. Any unusual activity triggers an alarm and shows a video image of the problem area to the center operators, who then undertake appropriate traffic management strategies, including relaying advice to the freeway patrol trucks. The traffic management center makes use of the information gathered by the Advanced System to keep motorists advised of the Borman's current status. Communication with motorists is accomplished primarily by variable message signs and highway advisory radio. In order to clear incidents as expeditiously as possible, freeway patrol trucks continuously monitor the corridor. Equipped with special communications equipment, a video camera, and a video monitor, these trucks can transmit images at the scene to the center. This visibility enables the center to make informed decisions more quickly. The Borman Advanced System makes use of a special spread-spectrum network radio, which provides the critical link between the center freeway patrol trucks, and the various electronic devices along the highway. These network radios provide robust, secure communications and have an automatic rerouting capability. If one radio fails, others pick up the signal so that no information is lost.

It can be expected that ITS projects will be an integral component of transportation plans and programs. Various issues, however, can affect the success of ITS, including institutional coordination and cooperation, system interoperability along with standards for compatibility of technologies, establishment of human resources for the implementation and operation, and partnership with the private sector.

For background information *see* HIGHWAY ENGINEERING; SATELLITE NAVIGATION SYSTEMS; TRAFFIC-CONTROL SYSTEMS; TRANSPORTATION ENGINEERING in the McGraw-Hill Encyclopedia of Science & Technology.   Kumares C. Sinha; Zongzhi Li

Bibliography. American Society of Civil Engineers, *Proceedings of the 5th International Conference on Applications of Advanced Technologies in Transportation Engineering*, Irvine, California, 1997; U.S. Department of Transportation, *Intelligent Transportation Systems: Real World Benefits*, Federal Highway Administration, 1998; U.S. Department of Transportation, *National ITS Architecture*, Federal Highway Administration, 1998.

Invasion ecology

Biological invasion is the process by which species (or genetically distinct populations), with no historical record in an area, breach biogeographic barriers and extend their range. Invasions occur as a natural ecological process. For example, after North and South America collided 3 million years ago, numerous mammals (deer, rodents, cats, skunks, weasels, llamas) extended their ranges south, and a few (armadillo, porcupine, oppossum) extended their ranges north. However, over the course of human history, the pace of biological invasions has accelerated because, through their travels and trade, humans tend to move other species. The immediate consequence of biological invasion is a local increase in the number of introduced species, also called alien, invasive, exotic, or nonindigenous species.

Human causes. The Hawaiian Islands provide a dramatic case study of the causes and consequences of biological invasions. Because of their distance from any other landmass, the Hawaiian Islands initially experienced extremely low natural invasion rates, and 90% of island biodiversity evolved from the few species that arrived by drifting or being blown on-shore. People arrived by boat from Polynesia some 1500 years ago, bringing about 30 species of cultivated plants and several domesticated animals. In addition, there were stowaways, including weeds, rats, and lice. Today, the causes of Hawaiian invasions have changed only slightly. Humans still introduce species on purpose, for utility or esthetic appeal. Some of these purposeful introductions are imports that are not intended to escape. For example, about 260 alien fern species are cultivated in gardens and may at some point increase the 30 species that have established in Hawaii. Invasions continue to occur accidentally, as new species "hitchhike" on other species (including human travelers), trade goods, or containers.

Numbers of invasions. At a minimum, there are 4500 alien species in the United States, plus 100 that have been introduced beyond their native ranges. In Hawaii alone, successful invaders include 70 birds, 2500 insects, and 900 plants. In fact, introduced species now outnumber natives there. For large-bodied taxa that tend to be purposefully introduced, the number of exotic species has stabilized (for example, few new birds have established since 1920). For other taxa, numbers continue to accumulate, probably because their pathways for introduction remain open (for example, over 1000 new insect species have been recognized in Hawaii since 1948).

Impacts. Some of the most devastating invasions have occurred on islands. In Hawaii, where 42 native bird species are listed as threatened or endangered, 81% suffer in part because nonindigenous pigs create mud holes which become breeding places for introduced mosquitoes, which then transfer introduced malaria among birds. The malaria is deadly to native birds but tolerated by many introduced species. Overall in the United States, introduced species are a contributing cause of endangerment for 49% of 1880 imperiled species. Assessments of economic impacts of introduced species have concluded that, on an annual basis in the United States, alien weeds cost $4.6–6.4 billion, foreign insect crop pests cost over $1 billion, and, for several years in the 1980s, fresh-water invasions that fouled waterways and utility plants cost over $1 billion. These costs include lost revenues, attempts to control the invasion, and indirect health costs of some sorts of control such as pesticides.

Steps of an invasion. Each invasion goes through a sequence of steps: arrival, establishment, population growth, population expansion, and impact (see **illus.**). Each step is a filter; that is, not all species that arrive will become established. Considering an invasion as a sequence of steps leads to the practical separation of the factors influencing each step, and consequently of the strategies for control. There are instructive departures from a simple expectation that species will progress regularly through these steps. Species can arrive multiple times before establishing. Once established, a population may persist at low abundances for decades before it suddenly grows and spreads. Particularly abundant species sometimes decline later, presumably because they exhaust their resources or are limited by species at higher trophic levels.

Because invasions are idiosyncratic combinations of new and existing species in an area, perfect

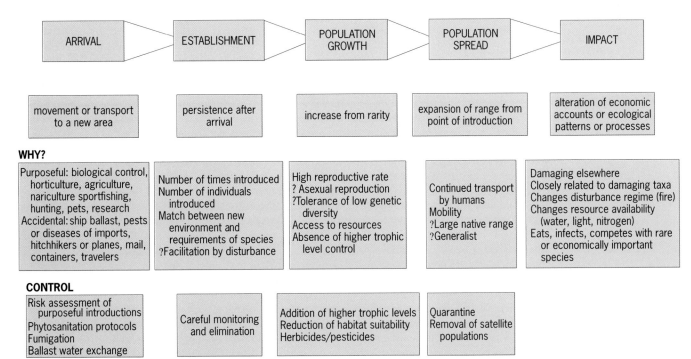

Steps of a biological invasion. Below each step is a brief description, possible causes, and strategies for control. A question mark means uncertainty because evidence is lacking or contradictory.

prediction of invasion fate is likely to remain elusive. However, invasion ecologists remain interested in determining whether a species will invade and the possible impacts.

Species invasion. Determining if a species will invade increasingly appears to be a function of autecology (whether the species' requirements are matched by the new environment) and opportunity. Successful establishment becomes more likely as a species is introduced more times or at higher abundances, in cases as disparate as insects introduced for biological control in Canada and birds introduced purposefully to New Zealand. Opportunity for invasion of coastal marine systems has recently increased for organisms carried in ballast water. For stability, ships often pump in water in one port and release it in another, transferring up to 3000 species around the globe daily. As ships become larger and faster, more organisms are likely to survive this transfer; an example is the change in establishment of new species in San Francisco Bay from less than one per year prior to 1960 to almost four per year now. Given sufficient opportunity, whatever can get in, will get in. Invasion may also be facilitated by human-caused disturbance for a number of reasons: many invaders are early-successional species; new resources may be available because native species have been disrupted; and humans can simultaneously be agents of disturbance and vectors of nonindigenous species.

Problems. Introduced species become problematic ecologically when they influence native species directly (for example, by eating them) or indirectly by altering resource availability, habitat structure, or ecosystem functioning. Direct effects are epitomized by the introduction of generalist predators to oceanic islands, where species are naive and have no evolutionary history of avoiding predators. Results are almost always disastrous: the demise of 12 bird species on Guam due to the nonindigenous brown tree snake is such a case. Examples of introduced species with substantial indirect effects include a nitrogen-fixing tree that has altered the nutrient properties of Hawaiian soil, plants such as tamarisk in the United States southwest and Brazilian pepper tree in the Florida Everglades that have lowered water tables, and cheatgrass that has invaded western North American prairies and shifted the fire cycle from once per century to once every 3 years. High impact depends in part on an invader reaching high abundance, which can occur when the new habitat lacks factors that limit the species in its native range (for example, predators, disease, and food resources with chemical or behavioral defenses).

Introduced species become problematic economically when their direct or indirect effects and high abundances occur in human-dominated or exploited landscapes. Of course, other introduced species, from livestock to garden varieties, provide economic benefits under the same circumstances.

In the short term, introduced species can provoke highly contentious disagreements about the balance between risks and benefits. In the long term, they seem to be accommodated and to be considered by most nonspecialists as part of the natural environment. Nevertheless, on a global basis, the inevitable result of biological invasions is an increasing homogenization of previously distinct regional biotas.

For background information *see* ECOLOGICAL METHODS; ECOLOGICAL SUCCESSION; ECOLOGY; ISLAND BIOGEOGRAPHY in the McGraw-Hill Encyclopedia of Science & Technology.   Jennifer Ruesink

Bibliography. J. A. Drake et al., *Biological Invasions: A Global Perspective*, John Wiley, 1989; C. S. Elton, *The Ecology of Invasions by Animals and Plants*, Chapman and Hall, 1958; R. H. Groves and J. J. Burdon, *Ecology of Biological Invasions: An Australian Perspective*, Australian Academy of Science, 1986; R. Hengeveld, *Dynamics of Biological Invasions*, Cambridge University Press, 1989; H. A. Mooney and J. A. Drake, *Ecology of Biological Invasions of North America and Hawaii*, Springer-Verlag, 1986; Office of Technology Assessment, *Harmful Non-Indigenous Species in the United States*, U.S. Government Printing Office, 1993; M. Williamson, *Biological Invasions*, Chapman and Hall, 1996.

Jupiter

When the *Galileo* spacecraft arrived at Jupiter on December 7, 1995, it first dropped a probe through the planet's atmospheric clouds. Data were relayed to Earth via the *Galileo* Orbiter, sweeping past Jupiter before going into orbit. When the *Galileo* Orbiter flew just 835 km (519 mi) above Ganymede's surface on June 27, 1996, it began devoting much of its resources (its on-board tape recorder and its radio downlink to Earth) to studying Jupiter's four large galilean satellites, discovered by Galileo Galilei with his primitive telescope nearly four centuries ago. There were three fly-bys of Europa, the second satellite out from Jupiter and the smallest of the four. The largest satellite, Ganymede, was encountered four times, and outermost Callisto three times. Meanwhile, *Galileo* studied innermost Io from afar, monitoring its continually erupting volcanoes while avoiding the intense radiation near that body.

Galileo's prime mission concluded at the end of 1997. However, the project was extended, as the *Galileo Europa Mission* (*GEM*), in order to follow up discoveries about Europa and to attempt the closeup investigation of Io. A year and a half after its October 1989 launch, *Galileo* had suffered a debilitating blow when its primary antenna failed to deploy. Therefore, a small, backup antenna has been used, permitting only a tiny fraction of the originally planned data to be beamed back to Earth. Nevertheless, by selecting the best of carefully planned observations for transmission to Earth, often using sophisticated data-compression techniques, the return has been richly rewarding. *Galileo* has provided a new window onto these remarkable worlds, which were only briefly glimpsed from afar during the swift fly-bys of the two *Voyager* spacecraft in 1979.

This article summarizes the major findings about these satellites obtained during *Galileo*'s prime mission and during the Europa-intensive phase of *GEM*. (Meanwhile, other studies have continued concerning Jupiter's atmosphere, its immense magnetosphere, its tenuous ring system, and several of its small moons.) *GEM* will officially conclude after *Galileo* dives closer to Jupiter to study Io in late 1999. Should the spacecraft survive the deadly radiation, there may be more studies of several galilean satellites before *Galileo* participates with *Cassini* (swinging by Jupiter en route to Saturn) in joint studies of Jupiter's magnetosphere in December 2000. *Galileo*'s quarter-century history has involved several generations of scientists and engineers; it will leave a legacy of unique observations of three planets, two asteroids, countless satellites, and exceptional events, like the crash of Comet Shoemaker-Levy 9 into Jupiter in 1994.

Io. This small, colorful satellite of Jupiter is by far the solar system's most geologically active body. Immense plumes, rising hundreds of kilometers above Io's surface, and localized hot regions centered on volcanic craters reflect Io's response to the immense wrenchings of its interior by Jupiter's tidal forces combined with the gravitational orbital locks between Io, Europa, and Ganymede. The volcanism, predicted just before the first *Voyager* encounter, was at first interpreted to be driven by the cosmically abundant, volatile element sulfur, which appeared responsible for the diversity of hues on Io's surface.

Monitoring by *Galileo*'s camera plus infrared spectroscopy and thermal studies by *Galileo*'s near-infrared mapping spectrometer have changed the current perspectives about Ionian volcanism. Remarkably, temperatures of several volcanoes turn out to be much higher than the 500 K (440°F) maximum expected for sulfur volcanism. Indeed, temperatures approaching 1800 K (2780°F) of magmas in the Pillan caldera surpass usual temperatures for silicate volcanism on Earth. *Galileo* images show that Io has about 100 high mountains; most are not old volcanoes, but their origin remains mysterious. It is surprising that the surface of this furiously active satellite—which is presumably molten at depth—can sustain mountains as high as 10 km (6 mi); evidently Io has a strong, solid crust at least 30 km (19 mi) thick.

Visually fascinating are the radical changes in Io's surface colors resulting from the active volcanism. Strangely, some changes observed during the few years of *Galileo*'s monitoring of Io are more dramatic than changes between *Voyager*'s fly-bys in 1979 and *Galileo*'s arrival. Evidently, the alterations, due to the dozen or so volcanic plumes that are active at any one time, are superficial and transient rather than permanent and cumulative. Other new *Galileo* images, taken while Io was in solar eclipse, have revealed faint, multihued glows in addition to bright spots of hot lava. Some are understood as volcanic emanations, while others reflect auroralike interac-

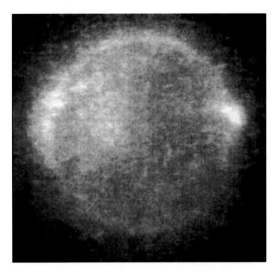

Fig. 1. View of Io while it was in eclipse (in Jupiter's shadow), taken by *Galileo* on May 31, 1998, from a range of 1.3×10^6 km (8×10^5 mi). Some of the glows are due to interaction of Io's tenuous atmospheric gases with charged particles trapped in Jupiter's magnetic field. The brighter glows are associated with volcanic plumes. (*Jet Propulsion Laboratory*)

tions between Io's thin, transient atmosphere and charged particles trapped in Jupiter's magnetosphere (**Fig. 1**).

Io is losing heat at the remarkable rate of 10^{14} watts, more than twice as fast as the much larger Earth. At times a single volcano, Loki, has been responsible for as much as a quarter of the total power radiating away from Io. However, the volcanoes vary in their activity; for example, a period of extreme activity of Loki in 1998 ended abruptly in early 1999. It is possible that Io as a whole responds to the tidal tug-of-war by undergoing episodes of high activity (like the present) with intervening epochs of quiescence.

Europa. None of *Galileo*'s discoveries have provoked so much interest as its pictures of the bright, icy surface of Europa. Europa is crisscrossed by a remarkable pattern of ridges, cracks, and families of linear features, all less than 200 m (600 ft) high. The ridges, often double, sometimes change direction abruptly. Some form curving, scalloped patterns. Apparently, the ridges begin to develop when Europa's brittle surface ice cracks; repeated jostling, perhaps driven by the daily tides, squeezes up dark-colored material below. However, the extreme lengths of some ridges (stretching far around Europa's globe) and the variety of intricate patterns have so far defied an agreed-upon, universal explanation (**Fig. 2**).

The uncommon terrains, where ridges are rare or absent, have caused the most interest. In certain "chaotic" regions, the crisscrossed ice apparently has broken apart and drifted about on a more mobile substance. Analogies to the splitting and foundering of ice rafts in terrestrial arctic zones do not really explain *Galileo*'s closeup views, which show that the would-be refrozen ocean in which the plates appear to float is actually very rough. Yet there are other places on Europa where the surface is extremely

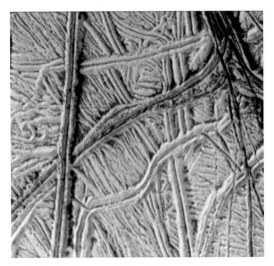

Fig. 2. Intricately textured ridged plains on Europa, photographed by *Galileo* on December 16, 1997, from a distance of 1300 km (800 mi). The region covered is about 20 km (12 mi) on a side, with a resolution of only 26 m (85 ft) per picture element. Note the multiple generations of overlapping ridges, many of them double. (*Jet Propulsion Laboratory*)

smooth, suggesting that an ocean not far beneath the surface has "soaked through."

Various pits, spots, domes, and moats—typically 5–12 km (3–7.5 mi) across—form when thermal instabilities push up more buoyant, warmer materials from the bottom of the ice crust, which may average less than 10 km (6 mi) thick. A depth of 10 km may mark the transition to the top of a 100-km-thick (60-mi) ocean of liquid water or brine. The practicality of future exploration of Europa's putative ocean may hinge on the possibility that the icy crust may be much thinner than 10 km in certain places. These interpretations overturn conventional wisdom that liquid water could not exist so far from the Sun.

Just as for Io (although less so), tidal heating warms Europa much more than the Sun's radiant heat. Just possibly, a European ocean could provide an abode for life, past or present.

Conceivably, the images are deceptive, and Europa's surface presents a frozen tableau of once-active arctic geology that happened long ago rather than a currently active world. However, enough comets randomly pass Jupiter (like Comet Shoemaker-Levy 9) that any old surface should have been battered by impact craters. Instead, Europa's nearly crater-free surface must have been renewed by later generations of ridges within the last 20 million years or so, probably even more recently in certain localities. (What had been tentatively interpreted as more numerous 10-km-diameter craters in images from *Voyager* and early *Galileo* orbits, implying older ages, have turned out to be pits and moats formed by internal processes.) Since Europa's surface is currently active, the presumed subsurface ocean still exists. The possibility that it harbors life arouses interest in future missions to Europa.

Ganymede and Callisto. Callisto is a large, dark-colored, cold world battered with large overlapping craters. Since *Voyager*'s early reconnaissance, it was thought that Callisto was made of a mixture of ices and carbon-bearing rocks that had never thoroughly segregated into crust, mantle, and core due to lack of long-term warmth. As a geologically inactive remnant from accretionary epochs of solar system history, Callisto was expected to be a dead, cratered relict, useful only as a contrast to the other, more geologically active satellites. Indeed, Ganymede, the largest satellite in the solar system (larger even than two of the nine planets), was thought to be intermediate between Callisto and Europa: Large, dark patches of cratered terrain on Ganymede resemble Callisto but have been broken apart by intervening,

Fig. 3. Perspective view of Callisto's dark surface, looking across a high scarp (cliff) which casts a long shadow. The scarp is part of the giant Valhalla multi-ring impact structure, about 4000 km (2500 mi) in diameter. This view of a portion only 33 km (20 mi) across, taken on November 4, 1996, shows the relatively crater-free, smooth surface of Callisto and icy knobs left over from the disintegration of crater ramparts. (*Jet Propulsion Laboratory*)

bright-colored, grooved terrains, thought to have been formed more recently (by processes reminiscent of what is still occurring on Europa) before Ganymede's internal heat finally decayed away and the world froze into a state of inactivity.

Galileo has presented a dramatically different picture of both of these worlds. Measurements of Callisto's gravity field show that its mass is more centrally concentrated than had been supposed, indicating at least some differentiation and segregation of materials (heavier ones sinking, lighter ones floating toward the surface). Moreover, the way that Callisto interacts with Jupiter's magnetic field suggests that it sustains a subsurface ocean. *Galileo* images show Callisto's surface, far from being pockmarked by generations of overlapping, smaller impact craters, to have few small craters. Instead, the surface is blanketed by dust still eroding from large-crater ramparts as their icy structural components sublimate away (**Fig. 3**).

Ganymede is also perplexing. Its dark terrains, like the circular Galileo Regio, hardly resemble Callisto at all. They have been heavily deformed by tectonic faulting, and the dark material seems to be a surficial coating that has oozed downhill into lowlands. It is not clear if the bright, grooved terrains (which turn out to be much more thoroughly faulted than previously suspected) are always younger than the dark terrains. In any case, the age of Ganymede's surface remains indeterminate, although Ganymede displays an unexpectedly strong magnetic field, hinting at continuing internal activity within this not-so-dead planet-sized body.

For background information *see* JUPITER; PLANETARY PHYSICS; SPACE PROBE in the McGraw-Hill Encyclopedia of Science & Technology.   Clark R. Chapman

Bibliography. J. K. Beatty, C. C. Petersen, and A. Chaikin (eds.), *The New Solar System*, 4th ed., Sky Publishing Corp. and Cambridge University Press, 1999; Special issue on *Galileo* remote sensing, *Icarus*, vol. 135, no. 1, September 1998.

Kilogram

Physical measurements depend on an evolving structure of base units, algorithms for their realization, and recommended procedures that constitute the International System of Units (SI). Until the second half of the twentieth century, most such base units were defined in terms of specific physical artifacts. These units entered the measurement process by means of replicas of the primary artifacts. Much work has been done to replace these artifact standards by invariant alternatives, such as atomic transition frequencies and fundamental physical constants. Only the base unit of mass remains defined by an artifact. Efforts involving atom counting and electromagnetic force generation may provide an invariant replacement.

Unit of mass. The kilogram is the unit of mass in the International System. It was defined as the mass of the International Prototype Kilogram (IPK) by the General Conference on Weights and Measures (CGPM) in 1901. A group of platinum-iridium (Pt-Ir) artifact kilograms had been prepared by the Johnson Matthey Company in 1878 in the form of right circular cylinders with both height and diameter of 39 millimeters. One of these cylinders was chosen as "*Le Grand K*," the International Prototype Kilogram. Other cylinders were distributed to the signatory countries of the Treaty of the Meter (signed in 1875) except for six "official copies" that are kept with the IPK. The U.S. national prototype kilogram, numbered K20, was among the initial group of national prototypes; it is maintained at the National Institute of Standards and Technology (NIST) in Gaithersburg, Maryland. The IPK is stored in ambient air in a triple bell jar at the International Bureau of Weights and Measures (BIPM) in Sevres, France. In this atmospheric environment, the surface of the IPK is exposed to the possibility of adsorption or absorption of whatever contaminants are present in the ambient atmosphere of this Paris suburb.

During the twentieth century a number of the national prototype kilograms have been returned to Sevres for comparison with the IPK. The first general verification was between 1899 and 1910 (without participation of the United States). The next general "reunion" took place in 1946 and showed a spread in values of about 10^{-7} kg, with most results clustered in about half this range. A subsequent international "periodic verification" in 1989 showed a larger variation with a spread approaching 2×10^{-7} kg and most values within a third of this range. An interpretation was then appended to the kilogram definition so that it currently reads: "The Kilogram is equal to the mass of the International Prototype of the Kilogram, after cleaning and washing using the BIPM method." The BIPM method involves "rubbing the artifact with a chamois cloth soaked in an ether/alcohol mixture, followed by washing in a jet of steam."

Alternatives to the artifact kilogram. Aside from a general preference for base units defined in terms of either atomic properties or fundamental constants, hierarchical schemes have uncertainties that increase with distance from the prototype standard. When this consideration is added to the historical situation described above, there is ample motivation for efforts to replace this last artifact standard by an invariant definition. Two alternatives to the artifact definition of the kilogram are being developed. One approach aims to connect the macroscopic (kilogram) mass scale with the microscopic (atomic mass unit) mass scale, thereby allowing the possible replacement of the kilogram by an amplified replica of the unit of the atomic mass scale. The second alternative balances the force generated by the Earth's gravitational field acting on a kilogram mass standard with the force on a current-carrying coil in a strong magnetic field.

Atom-counting alternatives. The direct approach to an atom-based kilogram is as appealing in its conceptual simplicity as it is daunting in its realization. The procedure involves production of a mass-resolved ion beam whose current is measured with respect to modern electrical standards; these standards are already well connected to fundamental atomic constants. To give some idea of the ion-counting method, scientists working on this procedure propose using a current of 6 milliamperes of gold ions for a period of 10 days to reach the desired 10 gram mass. This means maintaining or measuring the current with an accuracy of 10^{-8} during this 10-day period. The corresponding mass using gold as the accumulated substance can be compared with that of the IPK.

A second atom-based approach exploits the geometric regularity of certain crystalline forms to "amplify" the atomic mass content of a crystal's unit cell to a level at which direct comparison with the IPK is possible. Industrial need for single-crystal silicon has established a global manufacturing capability for large silicon ingots with high perfection, although the remaining imperfection continues to be a pressing problem. The key idea underlying this approach involves two ways of expressing the macroscopic mass content of a crystalline unit cell. Consider a cubic unit cell with edge dimension a that contains n atoms with average atomic weight (or mean molar mass) M. For definiteness, take the material to be silicon in the form of a spherical 1 kg object. Measurement of the mass, m, and volume of this spheroid provides a numerical value for its (macroscopic) density, ρ. This allows the two mass scales to be connected by the appropriate conversion factor, namely the Avogadro constant, N_A. This formal expression, Eq. (1), describes the x-ray/crystal-density approach

$$\rho \times a^3 = (n \times M) \times N_A^{-1} \qquad (1)$$

to the Avogadro constant. Its possible role in the kilogram replacement process will be significant only if its uncertainty can be reduced from a near-term limit near 10^{-7} to a value closer to current limits of the artifact-based scale (approximately 10^{-8}). Should such progress take place, the base unit would be related to the atomic mass unit by the Avogadro constant, N_A, to which would be assigned a definite numerical value.

Electrical generation of large forces. Although electrical current-balance technology, such as the Ampère balance, has been available since the nineteenth century, its possible role in the kilogram replacement process has been limited by several factors. To be effective, such electromagnetic force generation must be able to counteract the force exerted on a kilogram mass by the Earth's gravitational attraction. This force, 9.8 newtons, is easily generated with modest currents in large magnetic fields. However, there are large uncertainties associated with measurement of the field and of the coil dimensions. These uncertainties would effectively exclude such an approach to the kilogram problem were it not for an elegant and ingenious procedure proposed by Brian Kibble of the National Physical Laboratory (NPL) in Teddington, England, in 1978. The resulting system, now called a Watt balance, evades the need for coil geometry and

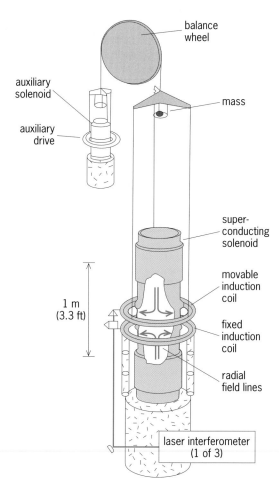

NIST Watt balance. (*After E. R. Williams et al., Accurate measurement of the Planck constant. Phys. Rev. Lett., 81:2404–2407, 1998*)

magnetic field measurements as is described below. This approach, subsequently pursued by groups at NPL and at NIST, led to initial results in 1988. Newer results are currently emerging from these two laboratories while related efforts are under way at standards laboratories in Switzerland and Japan.

Although the key idea of the Watt balance is very simple, realization of the full potential has required the utmost in metrological sophistication (see **illus.**). First, it is necessary to produce a strong, radial magnetic field over an extended vertical region. This field is generated by a pair of superconducting solenoids producing opposing fields; the field configuration is enhanced by additional windings. The next item is a circular coil hung from a balance wheel (this is a wheel, not an arm). On the other side of the balance wheel are disposed a countermass pan and a linear drive motor. This configuration has two modes of operation. In the first mode, an electric current is established in the hanging coil so as to counter the effect of gravity on a 1 kg mass added or removed from the mass pan located above the coil. The local value of the acceleration of gravity, g, is continuously monitored so that at any time the force, $F = m \times g$, is known. In the second mode, the linear motor is used to drive the hanging coil through the magnetic field at a constant speed, V (velocity, measured by laser interferometers) so as to generate a voltage, E, equal to that of a quantum standard Josephson effect reference array.

In the force mode, balance equilibrium is realized by setting a current, I, so that the force, F, equals $1 \text{ kg} \times g$. In velocity mode, a velocity is chosen so that the required voltage, E, is generated. The relevant conditions for a current, I, and velocity, V, are such that $F/I =$ (geometrical factor) $\times$ (magnetic field) and $E/V =$ (geometrical factor) $\times$ (magnetic field). Eliminating the difficult-to-measure (geometry factor) $\times$ (magnetic field) between these two expressions gives Eq. (2). Thus the experimental result is in the

$$F \times V = E \times I \qquad (2)$$

form "mechanical power = electrical power," hence the name Watt balance. The NIST group obtained results with an uncertainty below 10^{-7}. This result gives new determination of the Planck constant, h, since the voltage, E, is measured in terms of the Josephson constant, $2e/h$, where e is the magnitude of the charge of an electron, while current, I, is determined by voltage and the von Klitzing resistance, h/e^2. The success and further promise of this work has already prompted a detailed proposal for the structure of a possible kilogram redefinition.

For background information *see* CURRENT MEASUREMENT; ELECTRICAL MEASUREMENTS; ELECTRICAL UNITS AND STANDARDS; FUNDAMENTAL CONSTANTS; JOSEPHSON EFFECT; METRIC SYSTEM; PHYSICAL MEASUREMENT; PLANCK'S CONSTANT; UNITS OF MEASUREMENT in the McGraw-Hill Encyclopedia of Science & Technology.   Richard D. Deslattes

Bibliography. M. Gläser, Proposal for a novel method of precisely determining the atomic mass by the accumulation of ions, *Rev. Sci. Instrum.*, 62:2493–2494, 1991; D. Kestenbaum, Recipe for a kilogram, *Science*, 280:823–824, 1998; B. P. Kibble, A measurement of the gyromagnetic ratio of the proton by the strong field method, in J. H. Sanders and A. H. Wapstra (eds.), *Atomic Masses and Fundamental Constants*, pp. 545–551, Plenum, 1976; B. Petley, The Fundamental Physical Constants and the Frontier of Measurement, Adam Hilger, 1988; E. R. Williams et al., Accurate measurement of the Planck constant, *Phys. Rev. Lett.*, 81:2404–2407, 1998.

Magnetic levitation

The Inductrack is a magnetic levitation (maglev) concept for trains and other moving objects. A small working model of an Inductrack has demonstrated stable levitation over a 20-m (65-ft) track. As yet, no full-scale demonstration of the concept has been made.

Existing maglev systems employ superconductors at cryogenic temperatures or electronically controlled electromagnets to generate their levitating magnetic fields. The Inductrack should be simpler and potentially less expensive than these systems, since it utilizes special arrays of permanent magnets to achieve comparable levitation forces. Moreover, the Inductrack system is inherently stable, in contrast to those types of maglev systems which require electronic feedback systems to achieve a stable levitation state. In the latter type of maglev train, test tracks of which have been built in Germany, the train cars are levitated by the use of iron-pole electromagnets that are energized so as to produce a lifting force against the undersurface of specially designed tracks composed of iron plates. In such situations, stability theory (in particular, Earnshaw's theorem, published in 1839) dictates that the levitated system (the train car) will be unstable against vertical displacements from its equilibrium height. Thus, in these systems it is necessary to employ sensors and electronic control systems in order to maintain the spacing (a few millimeters) between the poles of the electromagnets and the lower surface of the iron plates of the track, independent of speed and load. Failure of these control systems could lead to a serious accident, so that high reliability is required. Energizing of the electromagnets also requires an on-board source of electrical power of high reliability, an additional complication.

Inductrack operating principles. Using only permanent magnets on the moving car, the Inductrack is a passive levitation system, relying only on the motion of the train to produce its levitating force. However, achievement of practical levels of levitating force using only permanent magnets requires innovation in the design of both the permanent magnet arrays and the track itself. The two critical elements of the Inductrack system are therefore (1) special arrays of permanent magnets (Halbach arrays) on the moving object, producing a strong and concentrated spatially periodic magnetic field below the array; and (2) a "track" made of a close-packed array of "shorted" inductively loaded circuits embedded in the track surface (**Fig. 1**).

At rest ("in the station") no levitation occurs, and the train car relies on auxiliary wheels to carry its weight. However, as soon as it is in motion at an appreciable speed (typically, walking speeds), the moving magnet array induces currents in the

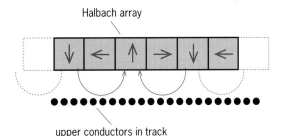

Fig. 1. Elements of the Inductrack system. An end view of the Halbach array on a moving car is shown above the upper conductors of the shorted levitation circuits in the track. Straight arrows show magnetization of permanent magnets in the Halbach array. Curved arrows show the periodic magnetic field below the array.

conductor array and thereby levitates the train. Owing to inductive effects that shift the phase of the induced currents, the drag power at high speeds is greatly reduced, being typically a small fraction of the power required to overcome aerodynamic friction. As long as the train is in motion, it is levitated, and the levitation mechanism is stable against vertical displacements, independent of speed or load (up to the maximum permitted load of the car). If the driving power fails, the train simply slows down and comes to rest on its auxiliary wheels at a low speed. No on-board power or control circuitry is required, and permanent-magnet arrays have a high degree of reliability.

Halbach arrays (Fig. 1), invented by Klaus Halbach for use in particle accelerators, turn out to be ideally suited for use in the Inductrack magnetic levitation system. Highly efficient in their use of permanent magnet material, they produce a strong, sinusoidally varying, periodic magnetic field just below the array, while canceling the fringing magnetic field above the array. Therefore, shielding the passengers of the train car from the levitating magnetic fields is not necessary, as would be the case for some of the previous magnetic levitation approaches.

Lift-to-drag ratio. From the theory of the Inductrack concept, a simple expression for the lift-to-drag ratio of the system has been derived. This expression, Eq. (1), shows that at high speeds the drag

$$\frac{\text{Lift}}{\text{Drag}} = \frac{\omega L}{R} \qquad (1)$$

force varies inversely with increasing speed, by contrast with aerodynamic drag or bearing-friction drag on a conventional train, both of which increase with speed. Here L is the inductance (self-inductance plus mutual inductance), in henries, of an individual circuit in the track; R is the resistance of that circuit in ohms; and ω is the angular frequency of the exciting wave. This frequency is in turn determined by Eq. (2) from the spatial wavelength, λ (in meters),

$$\omega = \frac{2\pi v}{\lambda} \qquad \text{radians per second} \qquad (2)$$

of the permanent magnet array; and the speed, v (in meters per second), of the train over the track.

Circuit design. It is straightforward to design the L/R ratio of the circuits so that the lift-to-drag ratio is large compared to unity at all but the lowest speeds, corresponding to a low power requirement per kilogram levitated. For example, one form that the track circuits might take would resemble an extended stack of elongated window frames, lying side by side down the track, with one of their long legs on the top surface of the track, and with the lower one directly underneath. Each circuit would be fabricated of conducting material (copper or aluminum) so as to make shorted loops. Thus the circuits could take the form of shorted coils of stranded wire, or they could be fabricated from thin sheets of copper or aluminum, punched out and laminated to form a track comprising nested shorted loops. A still simpler form, useful in some applications, would consist of a horizontally lying stack of thin copper or aluminum sheets, insulated from each other by coatings or anodizing. The sheets would be slotted transversely over most of their width, leaving their slot-free outer edges to complete the circuit.

Typical values. Typical values for L and for R for a full-scale Inductrack of the stacked window-frame type (assuming no additional inductive loading of the circuits) are $L = 2.5\ \mu\text{H}$, $R = 12.5\ \mu\Omega$. If, further, the wavelength of the Halbach array is assumed to be 0.5 m (1.6 ft), then Eqs. (1) and (2) yield Eq. (3) for the lift-to-drag ratio.

$$\frac{\text{Lift}}{\text{Drag}} = 2.5v = Kv \qquad (3)$$

At a maglev train speed of 500 km/h (140 m/s or 310 mi/h), the lift-to-drag ratio is therefore equal to 350, assuming no additional inductive loading of the circuits. With inductive loading, the lift-to-drag ratio can be increased, but with a corresponding reduction in the peak levitating force of a given Halbach array. Thus, inductive loading could be employed in order to drop the useful operating speed range of an Inductrack system to urban train speeds.

Variation with speed. **Figure 2** illustrates the variation of the lift-to-drag ratio with speed for three different cases: (1) a typical Inductrack with no extra inductive loading; (2) an Inductrack with extra inductive loading, for which $K = 3.0$; and (3) the lift-to-drag ratio that would be obtained if the circuits of the Inductrack were to be replaced by a copper sheet (conducting plate) within which repelling eddy currents would be generated by the moving Halbach arrays. In the third case, the lift-to-drag ratio would be much lower, and would increase with speed only as the square-root, rather than linearly. Some maglev systems have been proposed in the past that would employ eddy currents in conducting surfaces

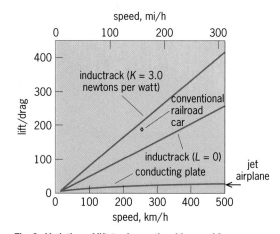

Fig. 2. Variation of lift-to-drag ratio with speed for an Inductrack with no extra inductive loading ($L = 0$), an Inductrack with extra inductive loading ($K = 3.0$ newtons per watt), and a maglev system with Inductrack circuits replaced by a conducting plate. Wavelength of Halbach array is 1.0 m (3.3 ft). Also shown are lift-to-drag ratios of a jet airplane and a conventional high-speed railroad car.

to produce their lift. As can be seen, such systems would pay a penalty for deriving their lift in such a way. Also shown for comparison purposes are the lift-to-drag ratio for a typical jet airplane wing, and the lift-to-drag ratio calculated from the frictional drag of mechanical origin (that is, excluding aerodynamic losses) of a conventional four-axle steel-wheel–steel-rail passenger car at 250 km/h (155 mi/h, half the maglev operating speed). It is generally accepted that 250 km/h is approaching the upper limit of speed for wheeled railroad systems.

Levitating power requirement. A lift-to-drag ratio of 350 at a speed of 500 km/h (310 mi/h) corresponds to a levitating power requirement (derived solely from the motion of the car relative to the track) of about 200 kilowatts for a train car weighing 50,000 kg (110,000 lb). Such a car would require about 8.0 megawatts to overcome aerodynamic losses at that operating speed. The calculated levitating power is thus about 2.5% of the required drive power at full speed.

Transition speed. The transition speed, v_t (measured in meters per second), defined as the speed at which the levitation force rises to 50% of its final value, is also equal to the speed at which the lift-to-drag ratio equals 1.0; that is, v_t is given by Eq. (4). For the example given above, where $K = 2.5$,

$$v_t = \frac{1}{K} \qquad (4)$$

the transition speed is 0.4 m/s, or about 1.5 km/h (0.9 mi/h), a very slow walking speed.

Levitating force. From the theory of the Inductrack it can be shown that, using modern high-field permanent-magnet material (neodymium-iron-boron), for which the remanent fields can exceed 1.4 tesla, it is practical to achieve a levitating force of 40 metric tons per square meter (4 tons per square foot) of Halbach array, thereby levitating in typical cases some 50 times the weight of the magnet arrays themselves.

Drive mechanisms. Being a passively levitated system, the inductrack is open to a variety of drive means. In those situations where cost or other considerations rule out electrical drive systems embedded in the track, a shrouded turbofan could be used to drive the car. Where electrical drive is feasible, either a linear induction motor drive (as is used in present-day maglev systems) or synchronously excited drive coils interleaved with the levitation coils could be used. The currents in these coils would be timed to interact with the vertical components of the Halbach-array magnetic fields, producing forward driving forces (or braking forces).

Application to rocket launching. In addition to its possible future use for high-speed maglev trains, the Inductrack has other potential uses. For example, another model Inductrack, now under construction, will attain a speed of Mach 0.5 (600 km/h or 375 mi/h) on a track 150 m (500 ft) in length. The model is intended to demonstrate critical aspects of technology that would be required to launch large (million-kilogram or thousand-ton) rockets by first accelerating them in a magnetically levitated and electrically propelled "cradle."

For background information *see* EDDY CURRENT; ELECTROMAGNETIC INDUCTION; INDUCTION MOTOR; MAGNET; MAGNETIC LEVITATION; MAGNETIC MATERIALS; RAILROAD ENGINEERING in the McGraw-Hill Encyclopedia of Science & Technology.   Richard F. Post

Bibliography. Electrical motor technology poised to catapult space travel into the 21st century, *Elect. Line* (Canadian), pp. 28–34, January/February 1999; S. R. Gourley, Track to the future, *Pop. Mech.* pp. 68–70, May 1998; Halbach arrays enter the maglev race, *Ind. Physicist*, pp. 2–13, December 1998; A new approach to magnetically levitating trains—and rockets, *Sci. Technol. Rev.* (Lawrence Livermore National Laboratory), pp. 20–24, June 1998; Prepare for lift off, *New Scientist*, p. 7, August 8, 1998.

Magnetic resonance imaging

In less than 25 years, magnetic resonance imaging (MRI) has revolutionized the practice of modern medicine and has become the imaging modality of choice in human diagnosis. The images are obtained by placing the subject within strong magnetic fields. Typical fields for clinical magnetic resonance imaging are in the 1.5-tesla (15,000-gauss) range. Under these conditions, nuclei that have inherent spin, such as the proton within water and fat molecules, precess about the magnetic field. Images can be obtained by disrupting this precession through the application of appropriate radio-frequency and gradient magnetic fields. Magnetic resonance imaging is noninvasive and can provide the clinician with spectacular anatomical, physiological, and biochemical images. These features are shared by no other single imaging modality. This technique is also capable of diffusion and perfusion imaging. These modalities enable the clinician to visualize blood flow through the arteries and veins, a technique known as MRI angiography.

With proton-based magnetic resonance imaging, the clinician has a variety of methods for obtaining contrast in the images. Image contrast can be achieved through the T1 (longitudinal) and T2 (transverse) spin relaxation mechanisms. In addition, proton density and externally injected contrast agents can be utilized to alter image contrast. The T1 relaxation times tend to increase with increasing field strengths, while the T2 values tend to decrease. Most importantly, the amount of detectable spin increases substantially with field strength. This increase results in the images having a higher signal-to-noise ratio, and subsequently in superior image quality, manifested either in improved resolution or in decreased scan time. As a result, since the establishment of the standard 1.5-T human MRI scanners in the early 1980s, there has been a constant push toward human imaging at higher fields. This has occurred despite early predictions that the human beings could not be imaged at frequencies above 10 MHz (corresponding to a magnetic field strength of approximately 0.25 T). Indeed, there are now more than 10,000 human scanners operating at magnetic field strengths of 1.5 T and at least 25 scanners operating at fields of 3 and 4 T.

Human imaging at 8 T. Recently, a human MRI instrument was constructed which operates at a field strength of 8 T, or 160,000 times the Earth's magnetic field at sea level. At the heart of this scanner is an 8-T, 80-cm (32-in.) solenoidal magnet which is manufactured from 414 km (257 mi) of copper-imbedded niobium-titanium superconducting wire wound on four aluminum formers. This results in a total inductance of 4155 henries. The four aluminum formers are enclosed in a helium dewar which can hold 1600 liters (423 gallons) of liquid helium. The helium dewar is positioned within a vacuum jacket and an aluminum magnet casing. The assembled magnet weighs approximately 35,000 kg (77,000 lb), operates at 4 K ($-452°$F), and carries 198 amperes of current once fully energized. As such, it contains 81.5 megajoules of stored persistent magnetic energy. It is believed that this currently represents the most stored magnetic energy in the world for all types of magnets. The inner windings of this magnet are designed to operate at 145 megapascals (21,000 lb/in.^{2}) of force. The magnet is characterized with a homogeneity of 5 parts per million over a 40-cm (16-in.) central volume. In order to contain the large stray magnetic field, the magnet is housed in a magnetically shielded room. The walls of this room are made from 30.5-cm-thick (12-in.) 1030 steel plate, and the ceilings and floor are made from 15.2-cm (6-in.) steel plate, resulting in a total shield weight of more than 200 metric tons (nearly 500,000 lb). In addition to the magnet itself, the 8-T magnetic resonance imager comprises radio-frequency amplifiers which are able to deliver a total of up to 10 kilowatts of power at frequencies up to 340 MHz, the operational proton frequency of the scanner.

Concerns for UHF MRI. Prior to the construction of the 8-T system, there were significant technological hurdles to overcome. Clearly, the construction of a whole-body magnet that could operate at such high magnetic fields was the foremost technological concern. Moreover, the ability to obtain a human image in the ultrahigh-frequency (UHF) range remained uncertain. Indeed, the frequency of 340 MHz approaches the microwave region of the electromagnetic spectrum. There was the possibility that the radio-frequency energy required to excite the spin would not penetrate to the center of the human head. In addition, even if this energy could penetrate, there were concerns that dielectric resonances (much like standing waves) would be produced inside the head which would deteriorate the quality of the resulting images. There were also human safety concerns. Foremost was the theoretical prediction that a tremendous amount of radio-frequency power would be required to obtain the image. There was the possibility that radio-frequency burns could be produced in the subject. The high radio-frequency power requirements at 340 MHz could also act to raise the internal temperature, much like heating in a microwave oven. Thus, the anticipated high radio-frequency power requirements became the key concern in advancing to the ultrahigh-frequency region of the spectrum.

In addition, humans had yet to be exposed to such high static magnetic field strengths. Previous experience at 4 T indicated that exposure to high magnetic fields could produce transient dizziness in the subject following rapid head motion in the scanner. This presumably occurs due to the interaction of the magnetic field with ions contained within the inner ear. It is assumed that eddy currents are set up during rapid motion within these canals. In turn, these currents cause a local osmotic pressure disruption, much like what is experienced when large quantities of alcohol are consumed. There was a natural fear of the unknown. Yet, a thorough review of the scientific literature pointed to the relative safety of high magnetic fields. Nonetheless, it would be naive to assert that everything is known about the interaction of static magnetic fields with a system which remains filled with mystery, the human body.

Advent of UHF MRI. An axial ultrahigh-frequency magnetic resonance image acquired from the human brain at 8 T (see **illus.**) not only represents a technological achievement in medical physics but also answers important scientific questions. There appear to be no dielectric resonances in the human head at 8 T. Other than the expected dizziness during rapid head motion in the scanner, which proved to be significant, no physiological effects were detected on the human body at this field strength. Cognitive and cardiac functions proved to be normal both during and following exposure to the static magnetic field for periods of up to 2 h. Thus, it appears that human

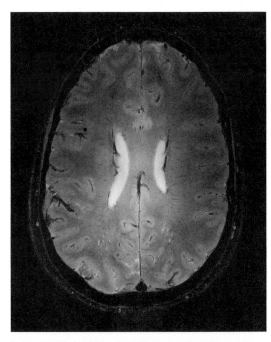

Representative axial image of the human brain by an ultrahigh-frequency magnetic resonance imaging scanner at a magnetic field strength of 8 tesla and frequency of 340 MHz.

imaging at this field strength will now move forward without significant short-term safety concerns.

Most importantly, human head images could be acquired with only a small fraction of the expected radio-frequency power requirements. Indeed, while 1–2 kilowatts of power had been expected, only about 100 watts was required. The signal-to-noise ratio was excellent, as expected at this field strength. There were no problems with radio-frequency penetration.

Prospects. The most promising aspect of imaging at 8 T remains the impact that such a device will have on functional imaging of the human brain. Magnetic resonance imaging can map regions of the brain involved in performing simple cognitive and motor tasks, such as visual perception and finger motion. It is already well recognized that 4-T scanners provide significant improvements in image contrast in this area.

The 8-T instrument also promises to have a significant impact in spectroscopy in vivo. This scanner should facilitate characterizing cellular bioenergetics within the human body under in vivo conditions. Spectroscopic methods should also help to better characterize tumors and may well assist in the differentiation between benign and malignant tissue.

The 8-T magnetic resonance imager should provide additional signal for the low-sensitivity nuclei that resonate at frequencies below 200 MHz (potassium-39, boron-10, boron-11, lithium-7, oxygen-17, nitrogen-15, sodium-23, carbon-13, and phosphorus-31). In addition, the 8-T instrument should offer a unique tool for the measurement of enzymatic rates under in vivo conditions. It also represents an opportunity for advances in spectrometers, radio-frequency coils, hardware, and software. Nonetheless, it is the promise of an 8-T scanner to help diagnose human diseases which will continue to drive the development of this technology.

For background information *see* BIOELECTROMAGNETICS; MAGNETIC FIELD; MAGNETIC RESONANCE; MEDICAL IMAGING; NUCLEAR MAGNETIC RESONANCE (NMR); SUPERCONDUCTING DEVICES in the McGraw-Hill Encyclopedia of Science & Technology.

Pierre-Marie Robitaille

Bibliography. Z. Fisk et al., *Physical Phenomena at High Magnetic Fields, III*, World Scientific, 1999.

Magnetoresistance

Magnetoresistance is the change in the electrical resistance of a material when it is subject to the application of a magnetic field. [That is, the magnetoresistance is the quantity $(R_{H=0} - R_H)/R_H$, where R_H is the electrical resistance when the material is subject to the magnetic field H, and $R_{H=0}$ is the resistance when the magnetic field is absent.] This property has widespread application in sensors and magnetic read heads (which read the signals encoded on a magnetic disk or tape). Its importance increases with the recent discoveries of giant magnetoresistance, junction or tunneling magnetoresis-

tance, and colossal magnetoresistance, due to various physical mechanisms.

Anisotropic magnetoresistance. All metallic magnetic materials exhibit anisotropic magnetoresistance (AMR). This effect comes about because conduction electrons have more frequent collisions with atoms when they move parallel to the magnetization in the material than when they move perpendicular to it. The higher rate of collisions means that the material's resistance is higher. Thus, passing a current down a stripe of magnetic material results in a change in the direction of magnetization of the material. Many materials have been investigated since the effect was discovered in 1857 by Lord Kelvin, and in even the best materials (nickel-iron alloys) the change of resistance is no higher than about 6%.

Giant magnetoresistance. Starting in the late 1980s, researchers found that structures consisting of alternating layers of ferromagnetic and nonferromagnetic metals with thicknesses in the nanometer range can have a magnetoresistance much larger than the anisotropic magnetoresistance effect. The change in resistance can be more than 110% at room temperature. This effect is also seen in bulk material consisting of disordered heterogeneous mixtures of ferromagnetic and nonferromagnetic metals, and is known as giant magnetoresistance (GMR). The giant magnetoresistance effect is due to a decrease in the scattering of current carrying electrons when the ferromagnetic planes have their magnetic moments aligned by a field compared to when they are not aligned (**Fig. 1a**). The largest magnetoresistance is thus obtained when the ferromagnetic planes are antiferromagnetically coupled in the absence of an applied field, that is, when their magnetic moments are antiparallel in zero field and can be made parallel by application of a field. This sort of antiferromagnetic coupling can be accomplished by tuning the thickness of the nonferromagnetic spacer layer separating the ferromagnetic layers, since the nature of the coupling depends critically on the thickness of the spacer layer.

Antiferromagnetic coupling through the nonferromagnetic layer is not exploited in a so-called spin valve, which consists of two ferromagnetic layers and a single spacer layer that is fairly thick so that it decouples the two ferromagnetic layers. In this configuration, the magnetization direction of one of the ferromagnetic layers is pinned in a particular orientation; that is, it is made less susceptible to switching by an applied magnetic field. In this way the resistance of the structure can be switched between the high-resistance state and the low-resistance state by application of a small field which flips the softer of the two layers but does not flip the harder one. The structure can be in either the low- or the high-resistance state in zero applied field, depending on how the field had been applied previously.

The actual value of magnetoresistance obtained in a giant magnetoresistance structure depends strongly on the details of its fabrication, the applied field, and the temperature at which the measurement is made, with the largest magnetoresistance occurring

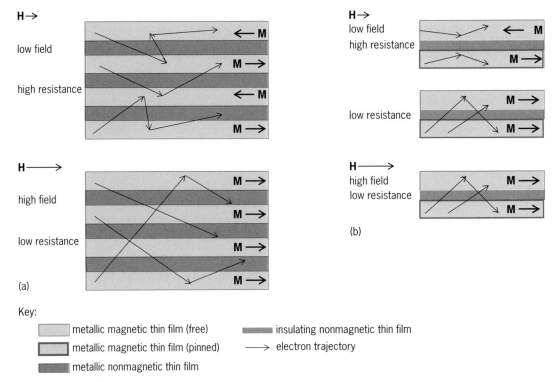

Key:

metallic magnetic thin film (free)

metallic magnetic thin film (pinned)

metallic nonmagnetic thin film

insulating nonmagnetic thin film

→ electron trajectory

Fig. 1. Low-resistance and high-resistance states in (a) giant magnetoresistance structure and (b) junction magneto-resistance structure. At low magnetic field (H), the magnetizations (M) of adjacent layers can be opposite in direction, leading to increased scattering of electrons in a or decreased tunneling probability in b. At high field, the magnetizations are parallel, leading to lower scattering or increased tunneling probability.

at cryogenic temperatures and in applied fields well over 1 tesla. Giant magnetoresistance structures can be made in which the low-field room-temperature magnetoresistance is larger than would be possible in bulk materials.

Junction or tunneling magnetoresistance. This term refers to the behavior of a trilayer thin-film structure consisting of two metallic ferromagnetic thin films sandwiching an insulating film. When the insulating film is thin enough (less than about 2 nanometers), electrons can pass from one ferromagnetic film to the other via quantum-mechanical tunneling. The ease with which this can happen depends on the relative alignment of the magnetizations in the two magnetic films: when they are parallel, the probability for electrons to jump from one metal film to the other is highest, and the resistance is lowest (Fig. 1b). This effect is strongest for metals that have the highest degree of polarization of conduction electrons, such as the manganite compounds (described below) and other relatively exotic metals.

The sensitivity of junction magnetoresistance devices depends on the choice of ferromagnetic films, the growth conditions, and the choice of insulating barrier material. The best results have been reported on devices using aluminum oxide (Al_2O_3) as the barrier material, since it can be formed as an extremely thin film with excellent insulating properties and no pinholes. Metals such as nickel-iron alloys (Permalloy) and cobalt are usually used for the ferromagnetic films, and junction magnetoresistance values of up to 37% at room temperature have been reported (**Fig. 2**). As in the case of the giant magnetoresistance

spin valves, a junction magnetoresistance structure can have one of its layers pinned, and in zero applied field it can be in either the high- or low-resistance state depending on the history of how fields were applied. Thus, these devices exhibit a memory effect, which has great potential for technological applications.

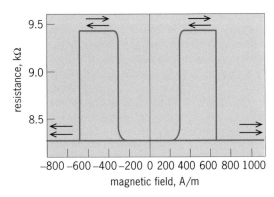

Fig. 2. Resistance versus applied field for a $Co/Al_2O_3/Ni_{0.8}Fe_{0.2}$ tunnel junction at room temperature, giving a junction magnetoresistance of 14%. Arrows indicate direction of magnetization of the two films at various fields; the bottom film allows magnetization to reverse at a lower field than the top film. In this experiment the applied magnetic field started at $H = 0$ (with parallel magnetization), was increased to $H > 1200$ A/m, then decreased to $H < -1200$ A/m, and then returned to $H = 0$. Increasing $|H|$ resulted in antiparallel magnetization of the two layers, but on decreasing $|H|$ parallel magnetization was retained. However, the high-resistance state is maintained if H is returned to zero (rather than increased to 800 A/m), showing a memory effect. (*J. S. Moodera, Massachusetts Institute of Technology*)

Colossal magnetoresistance. Colossal magnetoresistance (CMR) is the large magnetoresistance associated with magnetic phase transitions in certain homogeneous materials (as opposed to the intrinsically heterogeneous structures of giant magnetoresistance and junction magnetoresistance). While colossal magnetoresistance has come to refer to such effects in a variety of materials, the term typically is used to describe a class of perovskite manganites with the composition $R_{1-x}A_xMnO_3$, where R^{+3} is a rare-earth ion such as lanthanum, praseodymium, or reodymium, and A^{+2} is a divalent ion such as calcium, barium, strontium, or lead. The magnetoresistance in these materials can be several orders of magnitude, but such large magnetoresistance occurs only at low temperatures and requires fields of several tesla.

The very large magnetoresistance in these materials arises from a strong correlation between ferromagnetism and electrical conductivity which is understood in terms of the double-exchange mechanism, which involves the exchange of an electron between Mn^{+3} and Mn^{+4} ions and is strongly suppressed if the core spins of the manganese ions are not aligned. The resistivity at high temperatures is thermally activated, as is expected for a semiconductor or for small polaron-hopping conductivity. However, for certain cation concentrations these manganites undergo a phase transition into a ferromagnetic state at a transition temperature, T_c. At T_c the resistivity drops sharply as the spins align, and the sample behaves as a conductor at temperatures less than T_c (**Fig. 3**). Application of a magnetic field increases T_c and correspondingly increases the temperature at which the resistivity drops sharply. Thus, at a temperature near T_c, the material can be tuned from the high-resistivity phase to the low-resistivity phase by applying a magnetic field, leading to an extraordinarily large magnetoresistance peaked sharply near T_c.

In addition to the large magnetoresistance, these compounds display a rich variety of physical behavior, including metal-insulator transitions as functions of field, temperature, and composition; real-space charge ordering; spin-glass-like behavior; and a variety of unusual ferromagnetic and antiferromagnetic ordered magnetic phases. Recent experimental and theoretical work has demonstrated the need to consider effects such as the Jahn-Teller distortions associated with the Mn^{+3} ions in addition to double exchange to quantitatively understand the observed phenomena.

Large magnetoresistance associated with a phase transition is observed in many other classes of materials, including pyrochlore manganites and layered manganites with the composition $R_{2-x}A_{1+x}Mn_2O_7$. In these materials, conducting ferromagnetic planes are separated by layers that are more insulating, and a large magnetoresistance that resembles junction magnetoresistance and is not associated with a phase transition has been observed in the direction perpendicular to the planes. Similar temperature-independent magnetoresistance has been seen in polycrys-

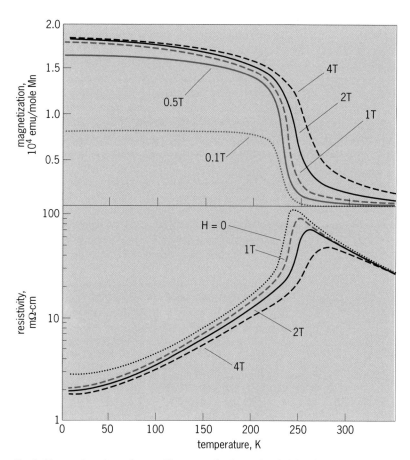

Fig. 3. Temperature dependence of the magnetization and resistivity of polycrystalline $La_{0.75}Ca_{0.25}MnO_3$ in different magnetic fields (*H*). The large change in the resistivity with applied field at temperatures near the transition temperature, T_c, and the temperature-independent change in resistance at low temperatures are attributable to intergrain tunneling. (*After P. Schiffer et al., Low temperature magnetoresistance and magnetic phase diagram of $La_{1-x}Ca_xMnO_3$, Phys. Rev. Lett., 75:3336–3339, 1975*)

talline samples of the perovskite materials and has been associated with intergrain tunneling.

Read heads. The data in a hard disk drive or floppy disk are stored as magnetic domains in a thin film on the disk. Changes in the direction of the magnetization define 1 and 0 bits. In order to read the data, a very sensitive magnetic detector must be used. As the density of bits increases, their size must shrink, and the problem of detection is a critical bottleneck. Until about 1990 an inductive-loop pickup was universally used because of its simplicity and low cost, but the need for greater sensitivity demanded a new technology, the magnetoresistive (MR) head. Initially, these heads used the conventional anisotropic magnetoresistance effect. In 1998 the giant magnetoresistance head was introduced—a design that actually uses a spin-valve configuration—and modified versions are likely to be used for the foreseeable future. Eventually, read heads may employ the junction magnetoresistance effect.

Magnetoresistive memory. Magnetoresistive memory is a random-access memory that uses the magnetic state of small ferromagnetic regions to store data and magnetoresistive devices to read the data, all integrated with silicon integrated-circuit electronics. The present technology uses the anisotropic magnetoresistance effect, but as with read heads, future generations will use spin-valve devices to improve density and access times. The junction magnetoresistance devices, with their relatively high resistance and simple structure, are well suited for this application. Many large semiconductor companies are actively investigating the potential of this technology, which may ultimately be competitive with standard dynamic random-access memory (DRAM) because of the relatively simple processing needed as well as the advantage of nonvolatility.

For background information *see* COMPUTER STORAGE TECHNOLOGY; FERROMAGNETISM; FREE-ELECTRON THEORY OF METALS; MAGNETIC MATERIALS; MAGNETIC RECORDING; MAGNETORESISTANCE; PEROVSKITE; POLARON; TUNNELING IN SOLIDS in the McGraw-Hill Encyclopedia of Science & Technology.

P. Schiffer; R. B. van Dover

Bibliography. K. Inomata, Giant magnetoresistance and its sensor applications, *J. Electroceramics*, 2: 283–293, 1998; S. S. P. Parkin, Giant magnetoresistance in magnetic nanostructures, *Annu. Rev. Mat. Sci.*, 25:357–388, 1995; Special issue on magnetoresistances, *IBM J. Res. Dev.*, vol. 42, no. 1, January 1998; Y. Tokura (ed.), *Colossal Magnetoresistive Oxides*, Gordon and Beach Scientific, 1998.

Magnetotellurics

Magnetotellurics is a geophysical exploration technique that uses surface measurements of natural electromagnetic fields to image subsurface electrical resistivity. First proposed by L. Cagniard and A. N. Tikhonov in the 1950s, it has developed into a powerful tool for geophysical exploration on a wide range of scales.

Physical basis. The electrical resistivity of the Earth's interior provides information about composition and structure. Naturally occurring rocks and minerals exhibit a very broad range of electrical resistivities. Crystalline igneous rocks typically have resistivities greater than 1000 Ωm, sedimentary rocks have intermediate resistivities, and fluid-saturated rocks and orebodies can exhibit resistivities below 1 Ωm. These resistivity values are strongly influenced by interconnected aqueous or magmatic fluids in the pores of the rock. Thus, remotely sensing the electrical resistivity of the subsurface can reveal its physical state and fluid content.

Diverse geophysical techniques can be used to map subsurface resistivities from depths of a few meters to hundreds of kilometers. Many techniques require electromagnetic waves to be generated by a transmitter. The magnetotelluric method utilizes naturally occurring electromagnetic waves to probe the Earth. This avoids the cost of generating signals, and the strength of the natural signals is such that magnetotellurics is the only electromagnetic exploration technique capable of imaging to depths greater than 10 km (6 mi). Magnetotelluric exploration maps the variation of resistivity with depth by means of the

skin depth effect. The amplitude of an electromagnetic wave decays with a characteristic skin depth, as in Eq. (1), where ρ is the resistivity of the ground

$$d = 503\sqrt{\frac{\rho}{f}} \qquad (1)$$

(Ωm), f is the frequency (Hz), and d is the skin depth (m). This attenuation occurs as electromagnetic energy is converted to heat in the Earth through ohmic losses. Equation (1) shows that low-frequency electromagnetic waves have a larger skin depth and thus penetrate deeper into the Earth than high-frequency waves (**Fig. 1**). High-frequency naturally occurring electromagnetic waves typically originate in worldwide lightning activity, and their small skin depth gives information about resistivity structure close to the surface. As the frequency is reduced, the electromagnetic wave penetrates deeper into the Earth and senses resistivity at greater depths. These low-frequency electromagnetic waves originate in magnetospheric oscillations caused by the solar wind.

By considering the impedance of a plane electromagnetic wave incident on the Earth, it can be shown that the apparent resistivity of the Earth is given by Eq. (2), where E and H are mutually orthogonal

$$\rho_a = \frac{1}{2\pi f \mu_0}\left|\frac{E}{H}\right|^2 \qquad (2)$$

electric and magnetic fields at the Earth's surface, and μ_0 is the magnetic permeability of free space. When the Earth has a uniform electrical resistivity, apparent resistivity is equal to true resistivity. However, in more complex Earth structures the apparent resistivity represents an average resistivity within a skin depth of the surface. For example, a two-layer resistivity model shows a variation of apparent resistivity with frequency that would be measured on the Earth's surface by a magnetotelluric instrument (**Fig. 2**). At frequencies above 10 Hz, the apparent resistivity is equal to 10 Ωm, since no signals

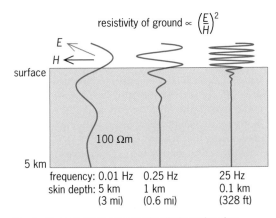

Fig. 1. Physical basis of magnetotelluric exploration. Electromagnetic waves penetrate a distance into the Earth that is inversely proportional to frequency. At each frequency, the ratio of the electric field (*E*) to the magnetic field (*H*) determines the Earth's average resistivity over a skin depth. By combining many frequencies, a depth sounding is obtained.

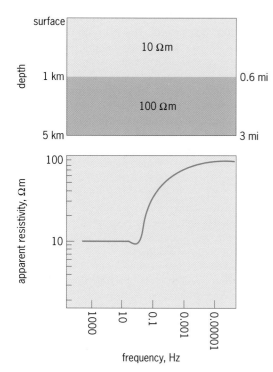

Fig. 2. Apparent resistivity as a function of frequency for a two-layer Earth model.

penetrate the lower layer. As the frequency is reduced, electromagnetic energy penetrates the more resistive lower layer and the apparent resistivity rises. At a frequency of 0.0001 Hz, the apparent resistivity is close to 100 Ω·m, the true resistivity of the lower half-space.

Data collection and processing. Modern magnetotelluric instruments record the time variation of horizontal electric and magnetic fields in two orthogonal directions. The vertical magnetic field is often recorded in addition. Electric fields are measured across approximately 100-m (330-ft) dipoles that are electrically coupled to the Earth with porous pot electrodes. The weak magnetic fields are measured with induction coils or fluxgate magnetometers. The time variations of the electric and magnetic fields are then Fourier-transformed, and apparent resistivity is computed at a range of frequencies. Depending on the frequency band of interest, there are two distinct types of magnetotelluric instrumentation available. Broadband instruments record data with a frequency content of 1000–0.001 Hz, while long-period (low-frequency) instruments sample 0.1–0.00001 Hz. The natural electromagnetic fields that are used in magnetotellurics are easily swamped by human-generated noise, so data are synchronously recorded on two instruments to allow this noise to be eliminated. In a typical survey, data are collected at locations spaced 1–10 km (0.6–6 mi) apart along a profile, or at a grid of points on the Earth's surface. Magnetotelluric data interpretation seeks to convert apparent resistivity as a function of frequency into true resistivity as a function of depth. Analysis techniques developed in the 1970s considered one-dimensional Earth models.

However, the subsurface electrical structure is often two- or three-dimensional in nature. Apparent resistivity values calculated from electric fields parallel and perpendicular to the structure will be different, and represent two independent modes of electric current flow. Magnetotelluric data are routinely converted into a two-dimensional resistivity model of the subsurface, and three-dimensional interpretation is becoming more widely used as digital computers become more powerful.

Applications. Magnetotellurics has been used in many aspects of pure and applied geosciences to image subsurface structure. Commercial applications routinely include mineral and geothermal exploration. The targets are orebodies or zones of hot rock and water that have an anomalously low electrical resistivity. Magnetotellurics has been used to map subsurface aquifers and plumes of contaminants with length and depth scales less than a kilometer. The aquifers and plumes can be either more or less resistive than the surrounding rocks. Hydrocarbon exploration has used the magnetotelluric method in contexts where seismic exploration is difficult or the data interpretation ambiguous (for example, imaging beneath volcanic layers or salt sheets). The shallow nature of these targets enables broadband systems and relatively high frequencies to be used. Broadband magnetotellurics has been used by academic investigators to study tectonic problems such as the internal structure of fault zones, crustal fluid flow, and volcanic processes at active calderas.

Imaging the lower crust and upper mantle requires lower frequencies (long periods) and recording times in excess of a week at each site. These surveys have enhanced understanding of the tectonic processes occurring at plate boundaries, primarily through their ability to map fluids in situ. In 1988 P. W. Wannamaker and coworkers showed how water is transported into the Earth's crust by the subducting Juan de Fuca plate beneath Oregon. In studies of continent-continent collision zones such as the Himalaya and Tibetan Plateau, magnetotellurics has shown the presence of partial melting in the crust. These results help explain the observed deformation patterns by providing a mechanically weak layer in the crust. Away from plate boundaries, magnetotellurics has improved understanding of the composition of the Earth's continental crust. Low-resistivity zones in the continental lower crust have been attributed to the presence of aqueous fluids or elemental carbon.

For background information *see* ELECTRIC FIELD; ELECTROMAGNETIC WAVE; GEOELECTRICITY; GEOMAGNETISM; GEOPHYSICAL EXPLORATION; MAGNETIC FIELD; MAGNETOMETER; RESISTANCE MEASUREMENT; ROCK, ELECTRICAL PROPERTIES OF in the McGraw-Hill Encyclopedia of Science & Technology.

Martyn Unsworth

Bibliography. L. Cagniard, Basic theory of the magnetotelluric method of geophysical prospecting, *Geophysics*, 18:605–635, 1953; M. J. Unsworth, G. D. Egbert, and J. R. Booker, High resolution electromagnetic imaging of the San Andreas Fault in Central California, 104:1131–1150, 1999; K. Vozoff, The magnetotelluric method, in M. N. Nabighian (ed.), *Electromagnetic Methods in Applied Geophysics*, vol. 2, chap. 8, Society of Exploration Geophysicists, 1991; P. W. Wannamaker et al., Resistivity cross section through the Juan de Fuca subduction system and its tectonic implications, *J. Geophys. Res.*, 94:14127–14144, 1989.

Magnetovision

Modern numerical methods enable the calculation of extremely complex magnetic fields. However, there are still many limitations on the exact determination of magnetic field distributions, especially in cases where all the parameters of magnetic materials (such as hysteresis, eddy currents, losses, material heterogeneity, and domain or crystal structure) must be considered. Moreover, computations of three-dimensional magnetic fields are very time consuming. In these cases, experimental determination of the magnetic field distribution is very useful.

The simplest way to measure a magnetic field distribution is to scan the magnetic field in the investigated area or space. Present scanning systems use either a small Hall sensor or a thin-film Permalloy magnetoresistive (MR) sensor to measure the magnetic field. Both types of scanning system have been used to investigate electrical steels.

In the magnetovision method, scanning results from a magnetoresistive sensor are processed numerically and presented in the form of a color map on a video display unit. This map can be printed in color or stored as a graphic or numeric file. The name of this approach derives from its similarities to the thermovision method.

Operating principles. In a typical magnetovision system designed for electrical steel testing (**Fig. 1**), the magnetic field strength distribution over the surface of a magnetized steel sheet is determined. The magnetic circuit is closed by a double C-yoke system (not shown). Two computer-controlled stepper motors move the magnetoresistive sensor along a meandering path over the sheet. The primary signal from the sensor is converted into a digital voltage signal. The signal from the sensor is sampled at a fixed rate, and every measurement point is characterized by a series of digitized readings.

Data analysis. The raw measurement data are converted into a more easily readable floating-number format. Various forms of mathematical manipulation may then be applied, for example, computation of root-mean-square (rms) values or fast Fourier transform (FFT) analysis. The results of the calculations are stored in a numeric file, and each measurement point corresponds to a single numerical value. The results from the numeric file are transformed into a color map or a graphic file.

Magnetoresistive sensor. The Permalloy magnetoresistive sensor has many advantages over other sensor types. It is small and much more sensitive than the Hall sensor. Magnetoresistive sensors with sensitivities of $20~\mu\mathrm{VA^{-1}m^{-1}}$ and dimensions of 1 by 1 mm are typical, but smaller sensors with dimensions of 1 by 0.04 mm and sensitivities of $1~\mu\mathrm{VA^{-1}m^{-1}}$, for example, are commercially available. The output signal of the sensor is proportional to the magnetic field strength H (not to its rate of change dH/dt, as in the case of the coil sensor). Both direct and alternating magnetic fields can be measured. Typical commercial sensors measure alternating magnetic fields up to a frequency of several megahertz.

Sensor arrays. The main disadvantage of the scanning system discussed above is the relatively long time required for the analysis of the magnetic field distribution. For example, testing an area of 5 cm² (0.8-in.²) with a 0.5-mm (0.02-in.) step from one measurement point (node) to the next takes about 30 min. Therefore, only a static (not varying in time) magnetic field can be analyzed.

In order to reduce the measurement time, the sensor positioning system can be replaced with a static array of sensors. Progress in the design of magnetoresistive read heads (particularly in the field of digital information processing) has enabled the manufacture of sensors dimensioned in micrometers. The process for manufacturing magnetoresistors is similar to that used for integrated circuits. This kind of technology allows the production of multisensors composed of several hundred elementary magnetoresisitve elements on an area not larger than several square millimeters.

There are some difficulties with the assembly of densely packed sensors. The sensitivity of a

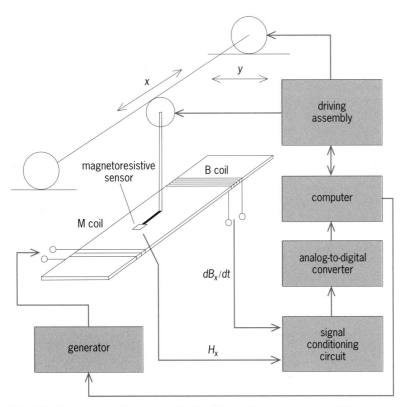

Fig. 1. Typical magnetovision system. The B coil is used for measurements of the magnetic flux density, and the M coil for magnetizing the sheet.

magnetoresistive sensor depends directly on the sensor area, and problems can arise in multiplexing the low-level signals received from microscopic sensors. The solution to this problem is to combine the scanning and multiplexing methods. Instead of one sensor, a line of many sensors may be moved. A system that tests 40,000 measurement points over a 10-cm² (1.6-in.²) area in minutes has been reported.

Application to material testing. A magnetovision system is quite versatile and may be used in all types of investigations of magnetic field distribution. So far, most of the interesting magnetovision results have been achieved in the testing of materials, especially electrical steel.

In the case of electrical steel, it can be assumed that the tangential component of the magnetic field over the sheet surface is a good approximation of the magnetic field inside a sheet. Therefore, the sensor should be located close to the sheet surface. (A thin-film magnetoresistive sensor measures the field component in the film plane.) For a given constant magnetic flux density in the sheet, the magnetic field strength distribution may be used to investigate the properties of the material.

Grain structure visualization. The grain structure of the steel can be analyzed by means of the magnetoresistive sensor. A very good correlation has been achieved between the map of magnetic field strength above the sheet and the grain structure. This correspondence enables the visualization of the grain structure even in the case of a coated steel sheet, for which other methods of analysis are useless. By mapping another component of the magnetic field strength, it is possible to make images of the domain structure of the sheet.

Quality assessment. For users of electrical steels, the specific power loss is the most important parameter. Unfortunately, standard methods of power loss measurement practically destroy the testing material. Moreover, complicated procedures are required to prepare the samples and carry out the measurements. Therefore, a nondestructive method for the fast assessment of electrical steel quality would be very useful. Testing results obtained from many samples of electrical steels have shown that the magnetovision method may be used for this purpose.

Four maps of magnetic field strength distribution, measured for the same value of the flux density, 1 tesla, are shown in **Fig. 2.** These maps have been determined for 30-by-30-mm areas of sheets with different power losses (estimated by conventional methods). The sheets with higher power losses exhibit a higher mean value of the field strength and a larger range of field strength values. In a full-color map, cool colors (shades of blue and green) are used to represent areas of low magnetic field strength, and warm colors (yellow and red) indicate areas of higher field strength. This color scale resembles closely those used in geographical maps. The appearance of shades of red in a magnetovision map may be a signal that some part of the mapped material is deteriorating. Thus, the map on the screen enables a quick visual assessment of steel quality.

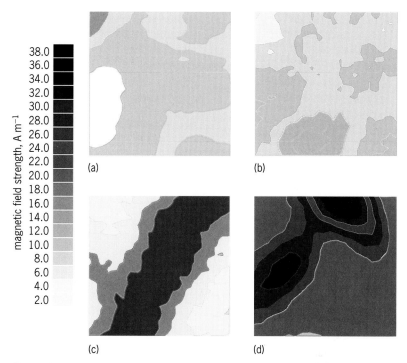

Fig. 2. Magnetovision maps of electrical steel samples with various power losses: (*a*) 0.31 W kg⁻¹; (*b*) 0.37 W kg⁻¹; (*c*) 0.45 W kg⁻¹; (*d*) 0.51 W kg⁻¹.

In electrical steel manufacturing, the analysis of material heterogeneity is useful in determining the causes of quality deterioration. Sheet areas with distinctly inferior properties can be detected and then tested with other methods. Since such deficiencies result most frequently from disturbances in the crystal structure, magnetovision helps in finding ways to improve the crystalline texture of the material.

Application to nondestructive testing. Ferromagnetic materials are very sensitive to a variety of stresses. Therefore, even small strains can be detected by means of the magnetovision method. Magnetoresistive sensors have been used in fatigue investigations of ferromagnetic materials. The reverse magnetostriction (Villari) effect has enabled the effects of cyclic mechanical loading to be measured without any external magnetic field excitation. This technique makes it possible to analyze the influence of material discontinuities on the fatigue characteristics of the sample. Because the changes in the sensor signal are alternating, a "sample-and-hold" system has been used to obtain a quasicontinuous (pulsating) magnetic image.

Many users of electrical steels underestimate the influence of stress on the properties of magnetic materials. A magnetovision picture of the tested sample gives information about the actual state of the material. For example, it is possible to observe the effect of the aging process or of applying stress. **Figure 3** presents magnetovision maps of a material before and after the sheet is cut into strips. Without annealing, this material was completely destroyed by the cutting device.

Even small changes in the shape of a magnetic material result in changes in the magnetic field around

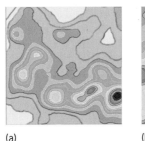 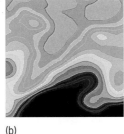

(a)   (b)

Fig. 3. Magnetovision map of electrical steel sample (*a*) before and (*b*) after the sheet containing the sample is cut into strips. The scale of magnetic field strength is the same as in Fig. 2.

the sample. This sensitivity makes it possible to detect microscopic cracks in a smooth specimen.

It is also possible to observe the influence of working conditions on the magnetic properties of a sample. By scanning a sufficient number of maps, it is possible to generate an animated presentation of the magnetizing process. In this way it is possible to observe, for example, the influence of the change in direction of excitation by controlling the rotation of the magnetizing field.

For background information *see* DOMAIN (ELECTRICITY AND MAGNETISM); MAGNETIC FIELD; MAGNETIC MATERIALS ; MAGNETORESISTANCE; MAGNETOSTRICTION; METAL, MECHANICAL PROPERTIES OF; NONDESTRUCTIVE TESTING; STEEL in the McGraw-Hill Encyclopedia of Science & Technology.

Slawomir Tumanski

Bibliography. J. Kaleta and J. Zebracki, Application of the Villari effect in a fatigue examination of nickel, *Fatigue Fracture Eng. Mater. Struc.*, 19:1435-1443, 1996; B. B. Mohd Ali and A. J. Moses, A grain detection system for grain-oriented electrical steels, *IEEE Trans. Magnetism*, 25:4421-4426, 1989; H. Pfützner, Computer mapping of grain structure in coated silicon iron, *J. Magnetism Magnet. Mater.*, 19:27-30, 1980; S. Tumanski and M. Stabrowski, The magnetovision method as a tool to investigate the quality of electrical steel, *Meas. Sci. Technol.*, 9:488-495, 1998.

Marine conservation

The precipitous loss of the Earth's species and ecosystems is among the greatest environmental challenges facing humankind. Loss and degradation of habitats and species is almost universal. Only recently have conservation biologists turned their attention to marine environments. The marine realm covers about 71% of the Earth's surface to an average depth of about 4 km (2.5 mi) and has the planet's tallest mountains, longest mountain range, and deepest canyons. The sea's permanently inhabited volume is at least 200 times that of land, and its vastness and topographic complexity accommodate levels of biological diversity far richer than on land—nearly half of all animal taxa are exclusively marine. It has been estimated that there are 10 million undescribed species in the deep sea. In general, basic understanding of marine biodiversity trails that of the terrestrial realm, particularly in relation to life histories. These knowledge gaps limit understanding of ecosystem function and hamper the ability to predict how human alterations may impact marine communities. Additionally, the vast differences in trophic dynamics and species life histories between marine and terrestrial ecosystems suggest that the ocean may respond to human perturbations in a fundamentally different way than terrestrial systems.

Damage. Primary threats to the sea include overexploitation, physical ecosystem alteration, pollution, introduction of alien species, and global climate change. For example, overexploitation in marine fisheries began to alarm scientists in the late 1800s when efficient methods of fishing were first developed. Even today, management often remains ineffective, as evidenced by the crashes of important fisheries—for example, North Atlantic cod (*Gadus morhua*), North Sea herring (*Clupea harengus*), and Alaska king crab (*Paralithodes camtschatica*). This ineffectiveness is due in part to intrinsic variability in recruitment dynamics of marine fisheries and uncertainty about that variability.

Humankind relies on the world's oceans for essential products and ecosystem services. The world's oceans provide a substantial fraction of animal protein—at least half of the protein in countries such as Ghana and Japan. Coral reefs, seagrass, mangrove, and salt marsh communities protect shorelines from storm damage. Primary producers in the ocean are critical in planetary gas exchange. Such organisms work as a biological pump by absorbing atmospheric carbon dioxide and conveying it into the deep sea, where it resides for centuries and slows the atmospheric buildup of this dominant anthropogenic greenhouse gas, thereby delaying global warming. Yet, from the mangroves fringing tropical coasts to the hydrothermal vent communities beneath the surface, no part of the sea has escaped the degradation associated with human activities.

Recovery zone. In the face of this global marine crisis, one approach is to establish no-take zones so that populations of overexploited marine species may recover or some portion of the marine realm can be maintained in a pristine state. Unlike standard fisheries regulations, which apply to entire stocks, the marine reserve approach is similar in spirit to terrestrial reserves. However, terrestrial reserves typically do not harbor heavily harvested species or species that regularly cross park boundaries (because park boundaries often represent sharp contrasts in habitat). As such, the terrestrial literature concerning the question of "how big is enough" cannot be easily applied to marine situations.

The design of a given marine reserve should depend upon the case-specific goals. A marine reserve might be created for conservation, fishery management, recreation, esthetics, intrinsic values, research, or education. For example, the justification for establishing a reserve for marine mammals might be to provide refuge from undesirable human activities such as shooting, entanglement, oil spills, waste dumping, fishing, noise disturbance, and collision with ships. In other cases, a reserve for a declining population of a commercially harvested fish might be designed to promote recovery such that harvest could be sustained. Ideally, reserves should include a comprehensive and representative distribution of habitats and allow sufficient area to support viable populations.

An interesting application of the no-take approach to dealing with overexploitation was recently adopted for managing small-scale benthic fisheries in Chile. In particular, joint management areas involving resource users and government, known as Management and Exploitation Areas, were established with the goal of achieving sustainable fishing. For inshore ocean bottom areas, small-scale fisheries may harvest those benthic resources that are not designated as fully exploited. Information about natural restocking of benthic resources, such as the muricid gastropod *Concholepas concholepas*, obtained by removal experiments from coastal preserves, was the primary scientific information considered by the Chilean government in embracing this approach. Comparison of the average size harvested and catch per unit effort between Management and Exploitation Areas and open-access fishing areas showed a clear increase in size and abundance for several benthic species. While knowledge about the long-term status of such species is slowly accumulating, the development of similar approaches for other small-scale inshore invertebrate fisheries provides a promising method for accumulating knowledge about the complexity and variability of marine systems.

Objectives. Marine reserves are widely advocated in marine conservation, and the number of newly designed reserves has increased dramatically. How to go about setting up marine protected areas and how to assess their efficacy remain to be determined. Clearly, marine protected areas provide unique research opportunities for addressing biodiversity research questions in controlled and protected settings. Some key questions include: What are the physical effects of humans on coastal biodiversity? How large an area is required to protect nursery populations that are fished elsewhere, and over how large an area might nurseries be effective? How well can degraded environments and communities be restored, and how long does it take? The answers will come with an increased understanding of basic patterns and processes of marine biological diversity (for example, ecology, taxonomy, and oceanography), and of changes in diversity resulting from human activities.

For most existing reserves, data are inadequate to determine whether they actually cause positive changes in species abundance and ecosystem integrity. The ability to detect the effectiveness of marine reserves depends upon species life history traits, level of depletion, adult distribution and larval dispersal mode, recruitment variability, and the time since the establishment of the reserve. Ongoing experiments test questions about ideal sizes for reserves and the activities that should be allowed

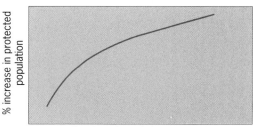

Relationship between percent of a population included in a marine reserve and the efficacy of that reserve.

within reserves. Clearly, the design and efficacy of marine reserves can be dramatically improved by better use of scientific information. To increase the likelihood of attaining conservation objectives, the number and size of marine reserves should be increased, and a scientifically based framework for reserve design should be developed (see **illus.**).

Leah R. Gerber

Bibliography. G. W. Allison, J. Lubchenco, and M. H. Carr, Marine reserves are necessary but not sufficient for marine conservation, *Ecol. Appl.*, 8(1):79–92, 1998 ; J. C. Castilla and M. Fernandez, Small-scale benthic fisheries in Chile: On comanagement and sustainable use of benthic invertebrates, *Ecol. Appl.*, 8:S124–S132, 1998; T. Lauck et al., Implementing the precautionary principles in fisheries management through marine reserves, *Ecol. Appl.*, 8(1):S72–S78, 1998; R. A. Myers and G. Mertz, The limits of exploitation: A precautionary approach, *Ecol. Appl.*, 8(1):S165–S169, 1997; M. H. Ruckelshaus and C. G. Hays, Conservation and management of species in the sea, in P. L. Feidler and P. M. Kareiva (eds.), *Conservation Biology for the Coming Decade*, 2d ed., Chapman and Hall, 1998.

Marine forensics

The discovery and exploration of famous shipwrecks such as the passenger ships *Titanic, Britannic* (**Fig. 1**), and *Lusitania* (**Fig. 2**) and the German battleship *Bismarck* have aroused public interest. While marine forensics has been practiced for many years, there are now new and more powerful investigative techniques and technologies, including apparatus to locate and explore wrecks in the deep ocean; improved means of relaying, transmitting, and processing information; and faster, more sophisticated computer models and analyses.

Marine forensics is a specialized branch of naval architecture and marine engineering that uses comprehensive analysis and reverse engineering to determine how and why a ship sank. All of the traditional methodologies of the naval architect and marine engineer, as well as the instruments and knowledge of the ship operator, are employed to perform the forensics analysis. Knowledge of structure, seakeeping, propulsion, hydrodynamics, navigation, mechanics, and piping systems is as important in the analysis of a maritime accident as they are in the basic ship

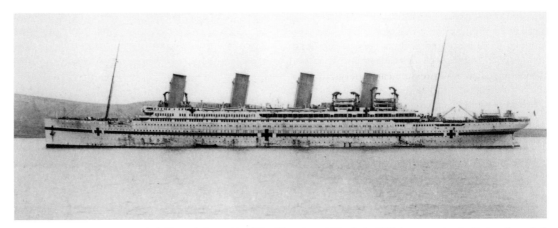

Fig. 1. HMHS *Britannic*, a hospital ship and sister ship of the *Olympic* and *Titanic*, in 1915. Large gantry davits are abreast of the forward and aft stacks. Completed with all the improvements recommended by the Mersey (*Titanic*) Inquiry, the ship sank in 55 min after the bow struck a mine. Preliminary forensic analysis indicates that massive flooding caused by open watertight doors and portholes may have accelerated the sinking despite improvements to the watertight subdivision. (*Harland and Wolff*)

design. A more recent factor is the development of crewed and crewless submersibles that can dive to depths of 20,000 ft (6000 m) or more.

Submersibles. The loss of the submarine USS *Thresher* in 1963 made the Navy aware of the difficulty of finding and exploring a deep-ocean wreck site with the equipment available. The specially equipped bathyscaph *Trieste* was used in the reconnoitering of the seabed where the *Thresher* was believed lost. This incident showed the need for deep-diving submersibles, and both crewed and crewless submersibles were rapidly developed.

Crewed submersibles must be designed to resist large pressures in order to work at extreme depths. Their significant diving and recovery times allow only a limited period to explore the deep-ocean environment. The occupants have limited vision through viewing ports of thick glass and small diameter. The quarters are austere due to the electrical and operating gear within the titanium sphere.

Modern electronics along with the development of fiber-optic cable has enabled the development of the crewless submersible, also known as the remotely operated vehicle (ROV), that can operate over the wreck site for an unlimited amount of time. Equipped with still and video cameras and sonar, such craft can dive to 20,000 ft (6000 m). They are self-propelled and are operated from a mother ship on the ocean surface through a console controlling movement of the craft and operation of the cameras. With the sonar on the crewless submersible, the operator can detect objects on the ocean bottom.

Physical phenomena. Certain occurrences during a ship's sinking are important in later analysis.

Implosion-explosion. This phenomenon has been observed on deep-ocean wrecks and makes the assessment of a ship sinking more difficult. While the damage that caused the sinking can be masked by many events during the process, the implosion-explosion phenomenon provides the greatest challenge to an investigator.

Implosion-explosion results from a conflict between the desire to make a ship as watertight as possible and the way that increasing water depth affects the ship as it sinks to the seabed. The water pressure compresses the structure until the weakest portion fails. Then, the water pressure very rapidly compresses the air, which had been at atmospheric

Fig. 2. RMS *Lusitania* in 1915, later sunk by a single torpedo from a German U-boat. Preliminary forensic analysis indicates that a mysterious secondary explosion was steam-induced, caused by water hammer in the steam piping. (*Mariners' Museum, Newport News, Virginia*)

pressure inside the structure. This sudden pressurization (implosion) causes both the water and air to overshoot like a spring and the air bubble expands (explodes) and pulsates outward. These large, rapid, alternating changes in pressure cause devastation to any structure or equipment in their path. This damage is very similar to that caused by an underwater explosion, which behaves according to the same principles. Assessment of wrecks has shown that the greater the amount of flooding at the surface, the less the amount of damage caused by this phenomenon.

Terminal velocity. As a ship sinks through the water column, it reaches a certain velocity, depending on the amount of superstructure. This velocity may be as low as 20 knots (10 m/s), as in the *Titanic*, or as much as 30–35 knots (15–18 m/s), as in the *Thresher*. If the ship has much implosion-explosion damage, it can be seriously damaged upon impact with the seabed, such as the stern of the *Titanic*. However, if the ship has time to progressively flood at the surface, such as the *Bismarck* or the American carrier *Yorktown*, the damage upon impact is more localized to the points of impact with the sea floor.

Biological phenomena. The action of deep-ocean organisms, which is still imprecisely known, makes the study of shipwrecks more difficult. For example, while the pine wood decks of the *Titanic* were devoured by tubeworms, the steel hull was covered by microbe-based structures termed rusticles. They are rust-colored, suggesting an oxidized form of iron, and hang down like icicles, making it difficult to observe the features of the plating underneath. The rusticles are bioconcretions, that is, hardened porous concreted masses that have been formed by growths of bacteria and fungi which are also found in deep oil and water wells. These microbes leach out the iron from the steel and recycle the steel components back into the environment.

Locating the wreck. A wreck site can be difficult to locate due to poor position plotting or the absence of communication about the sinking location. The *Titanic* wreck is 19 mi (30 km) from its position reported in its wireless messages. The wreck of the *Bismarck* was difficult to find because the reported sinking position was plotted from dead reckoning by exhausted British crews intent on fighting a battle. The reported position of the *Britannic*, sister ship of the *Titanic*, was inaccurate by a distance of 8 mi (13 km).

Search strategies. Most often, it is necessary to map out the entire area around a suspected wreck site. Careful search patterns must be plotted so that the searching ship will pass over the suspected area in a regular overlapping pattern known as "mowing the lawn." Such maneuvers require sophisticated navigation and sonar detection gear. Modern navigational systems based on the Digital Global Positioning System provide the precise location of the ship at a given time.

Sonar detection. Typical sonar systems, which use a sound signal beamed to the seabed and reflected back to the mother ship, are not sophisticated enough to distinguish a shipwreck of interest from the natural features in the ocean floor or other nearby uncharted shipwrecks. Advances in electronics in the late 1990s, however, provided sharper images and the ability to peer through sediments to determine features of a hidden wreck. Once a wreck is located, the forensic investigator can view it and begin an analysis of how the ship arrived there.

In the exploration of the *Titanic*, a chirp profiler was used to peer through the sediments along the side of the ship to detect openings in the bow. The profiler was mounted on the front of the submersible *Nautile* and oriented to look down through the mud mounds against the hull. A processor collected, processed, and displayed the data, highlighting six narrow separations in the bow covering six compartments of the ship. This information was invaluable in determining the nature of the damage from the iceberg collision.

Inspection of the wreck. The wreck can be explored by several means. As noted above, the submersibles which can dive to 20,000 ft (6000 m) have limited time on the bottom and a small viewing area. Remotely operated vehicles also can dive to such depths and take pictures as well as send back live television images from the ocean bottom. The vehicle is tethered to the mother ship by an umbilical as much as 4–5 mi (6–8 km) in length. Support of the vehicle requires a crane with adequate lifting capacity, plus deck area to store a number of vans that control and service the craft. With the proper light sources, the vehicle can be even more valuable than a crewed submersible at less cost and more coverage time on the wreck site. Ocean currents can play an important part in this operation. Crewed submersibles can be launched and recovered only in low sea states, whereas the remotely operated vehicle has a little more latitude. However, storms and high seas terminate operations and limit what can be covered in the wreck site investigation. The mother ship needs dynamic position devices that can maintain its position while the submersibles, with or without crews, are deployed below.

Entering the wreck. The wreck can be entered by divers up to depths of 750 ft (225 m). There is always danger in entering a shipwreck unless the ship sank under controlled conditions, such as is done in the artificial reef program. The danger arises from jagged structures due to the implosion-explosion phenomenon as well as from the damage to the structure that caused the sinking. The irregularities can snag the umbilical of remotely operated vehicles and the breathing devices of divers.

Photography. High-resolution video and still cameras are the most widely used equipment for surveying a shipwreck site, particularly in the deep ocean. Vital to successful photography is proper lighting to show features of the wreck. The lighting must be oriented to avoid shadows which can deceive a viewer of the two-dimensional image into perceiving a feature where none exists. A multilight source is best. To aid in the photographic investigation, the viewer should have an up-to-date set of ship plans in order to determine what features of the wreck are being viewed. The camera is important because of its capability to record everything visible. A recent advance in the use of such records is the creation of a photomosaic, combining many photographs of the wreck site.

Analysis of the sinking. Several techniques can help unravel the circumstances under which a vessel was lost. The most important starting point, however, is survivor testimony.

Survivor testimony. Useful information on how a ship foundered can be gleaned from careful and methodical reviews of survivors' testimony from interrogations or their statements made at inquiries. The expert in the loss is the ship officer who is specially trained in the ship's operation. This officer should be able to describe what critical problems brought about the sinking, as well as the lists and trims the vessel assumed during the sinking process and any special measures that were taken to countermand them. Other testimony can vary in degree of accuracy depending on the witness's familiarity with ships and the witness's background. An excellent witness is a trained engineer, like naval architect Thomas Andrews on the *Titanic*, who was able to tell Captain Edward Smith the dangers he was facing and the need for an evacuation of passengers as quickly as possible.

Other techniques. Finite element analysis has been increasingly valuable in determining inherent stresses in the ship's structure during the sinking process.

A damage stability analysis, a methodical reproduction of the ship's flooding process, gives an indication of the angles of the ship's trim and list; these data are analyzed; and with the use of a computer animation, the ship sinking can be replicated. It may also be necessary to recover steel from a wreck, if brittle fracture is suspected, metallurgical analysis determines the grain and chemical composition of the material, as well as any flaw in its microstructure that could have led to structural failure.

All of the foregoing techniques may be combined into a computerized simulation that provides a useful graphical analysis for assessors.

For background information *see* FINITE ELEMENT METHOD; SHIP DESIGN; SONAR; UNDERSEA VEHICLES; UNDERWATER PHOTOGRAPHY in the McGraw-Hill Encyclopedia of Science & Technology.

William H. Garzke, Jr.; Richard Silloway

Bibliography. W. Garzke et al., Deep ocean exploration vehicles: Their past, present, and future, *Trans. Soc. Naval Arch. Mar. Eng.*, 101:485–536, 1993; W. Garzke, Jr., et al., *Titanic*: The anatomy of a disaster, *Trans. Soc. Naval Arch. Mar. Eng.*, 105:3–59, 1997; S. Mills et al., The saga of HMHS *Britannic*: A preliminary marine forensics analysis, in *From Research to Reality in Ship Systems Engineering Symposium*, American Society of Naval Engineers and Society of Naval Architects and Marine Engineers, pp. 77–110, September 1998.

Marine natural products

The marine environment has proven to be a rich source of both biological and chemical diversity and has, therefore, become the focus of a major research effort in natural products drug discovery. A natural product is a chemical that is produced by a plant, animal, or microorganism. Terrestrial plants and microorganisms have traditionally been an important source of natural products for the development of new drugs. Over 80% of the medicines most commonly prescribed in hospitals have their roots in natural products. These include compounds such as aspirin, which is a stable derivative of a compound present in the bark of the willow tree; numerous antibacterial drugs, such as penicillin, streptomycin, and erythromycin, which are produced by microorganisms; and the anticancer drug Taxol®, derived from the bark of the Pacific yew tree.

The oceans cover approximately 70% of the Earth's surface and are estimated to contain at least 300,000 invertebrate and algal species. From the relatively small number of species that have been studied to date, thousands of chemical compounds have been isolated, yet only a small percentage of these chemicals has been studied for their potential as useful products. The oceans represent a virtually untapped resource for discovery of novel chemicals with potential as pharmaceuticals, nutritional supplements, cosmetics, agrichemicals, molecular probes, enzymes, and fine chemicals. Several marine-derived products are currently on the market (see **table**).

Chemical ecology. The diversity of chemical compounds produced by marine organisms is thought to be due to the extreme competition between organisms for space and resources in most marine habitats. Because many of these plants and animals live in densely populated habitats, are nonmotile as adults, and have only primitive immune systems, they have evolved a variety of chemical compounds to help them compete for resources. Some possible roles for the compounds are defense against predators and infection, attraction or inhibition of larval settlement, prevention of overgrowth by other organisms, and use as pheromones to synchronize reproduction among organisms that expel eggs and sperm into the water. The molecular mechanisms by which these chemicals prevent infection, encroachment, or predation may be through interactions with the same or similar enzymes and receptors as those involved in human disease processes. For example, many natural products have been identified that inhibit cell division, the primary target of many anticancer drugs. In most cases, there is a greater understanding of the effect of the natural product on human disease processes than of the role in the marine organism from which it was isolated.

Marine sponges are among the most prolific sources of diverse chemical compounds with therapeutic potential. Of the more than 9000 chemical compounds derived from marine organisms, more than 30% have been isolated from sponges. Sponges occur in every marine environment, from intertidal

Examples of commercially available products based upon marine-derived compounds		
Product	Application	Original source
Ara-A	Antiviral drug	Marine sponge, *Tethya crypta*
Ara-C	Anticancer drug	Marine sponge, *Tethya crypta*
Okadaic acid	Molecular probe: phosphatase inhibitor	Dinoflagellate
Manoalide	Molecular probe: phospholipase A_2 inhibitor	Marine sponge, *Luffariella variabilis*
Vent™ DNA polymerase	Polymerase chain reaction enzyme	Deep-sea hydrothermal vent bacterium
Formulaid® (Martek Biosciences, Columbia, MD)	Fatty acids used as additive in infant formula nutritional supplement	Marine microalga
Aequorin	Bioluminescent calcium indicator	Bioluminescent jellyfish, *Aequora victoria*
Green fluorescent protein (GFP)	Reporter gene	Bioluminescent jellyfish, *Aequora victoria*
Phycoerythrin	Conjugated antibodies	Red algae
Resilience® (Estée Lauder)	"Marine extract" additive in skin creams	Caribbean gorgonian, *Pseudopterogorgia elisabethae*

to abyssal regions, in all the oceans, and they produce a greater diversity of chemical structures than any other group of marine invertebrates. Other productive sources of bioactive molecules with therapeutic potential are the bryozoans (sea mosses), ascidians (sea squirts), mollusks, cnidarians (jellyfish), and algae.

Drug discovery teams. The discovery of marine natural products with potential as therapeutic agents requires a multidisciplinary team approach, with the collaboration of scientists from a number of disciplines: marine biologists and microbiologists to collect, document, and identify the organisms to be studied; immunologists, tumor biologists, microbiologists, virologists, and biochemists to evaluate potential drug targets and to design appropriate tests (assays) for evaluating extracts of the organisms collected; and natural products chemists to purify and determine the structures of the chemicals present in the marine organisms.

Sample collection. The sample collection team consists of marine biologists, microbiologists, and taxonomists who are responsible for the collection, isolation, and identification of macroorganisms and microorganisms. Samples are collected from shallow water by wading, snorkeling, and scuba diving. Manned submersibles, such as Harbor Branch Oceanographic Institution's Johnson-Sea-Link (JSL) submersibles, enable scientists to access unusual habitats, such as vent communities and deep-sea bottom (benthic) habitats. Submersibles also allow for collections from previously inaccessible marine habitats, such as rugged, vertical walls which are difficult to sample with surface gear such as trawls or dredges. The JSL submersibles are equipped with multifunctional manipulator arms to collect samples, video and still cameras for photodocumentation, and a data recorder which logs temperature, salinity, and oxygen during the dive. Some submersible systems are also equipped with specialized tools and chambers that allow samples to be maintained under ambient conditions, that is, high pressure and low temperature. Collections made by submersible and scuba are very selective compared to trawling or dredging

methods, and are of minimal impact to the environment. *See* UNDERWATER VEHICLES.

Permits and research clearances are generally required for collection and exportation of marine samples, so the collectors must work closely with government agencies to obtain the necessary documents.

Marine microbiology. Marine microorganisms are an important source of novel chemical compounds. In addition, there is evidence that some natural products derived from marine invertebrates may actually be produced by associated or symbiotic microorganisms. These microorganisms are obtained from macroorganisms (such as sponges and algae) as well as deep-sea sediment samples by growing them on agar media. The various microorganisms are separated into pure (axenic) cultures, and then they are fermented in liquid culture under a variety of conditions to optimize production of the natural products.

Extracts. An extract of the macroorganism or microorganism sample is prepared by grinding or steeping the material in a solvent, usually an alcohol. In some cases, successive extraction with solvents of varying chemical polarity is used to produce a series of extracts which contain different suites of compounds. The extract is filtered to remove pieces of the organism, and its concentration is calculated. This crude extract, which contains multiple chemical compounds, is tested to determine its biological activity.

Biological evaluation. Many biological evaluation programs use a random screening approach to discover natural products with therapeutic potential. In this process, a large number of extracts (from thousands to millions per year) are put through various biochemical and cell-based tests, termed screens or assays. These screens are designed to be highly selective in detection of natural products which have the highest probability of yielding marketable drugs. In this respect, the selection of targets to screen is possibly the most important aspect of a natural products drug discovery program.

The three commonly used approaches to screening are in vivo, or testing in animals; in vitro, or testing in cell cultures; and target-directed. For a number of years the U.S. National Cancer Institute ran a screen for anticancer agents, using an experimental leukemia model in mice. This approach, although possibly the best for indicating the effectiveness of a potential drug, is very expensive, requires large amounts of material for testing, is time-consuming, and can have significant numbers of false negatives due to the presence of multiple classes of compounds which interfere with the assay. In the cell culture approach, materials are tested for their ability to affect cells. An example is the A549 human lung tumor cell line assay: an extract or a purified natural product is added to the lung cancer cells, and its ability to inhibit cell proliferation is determined with respect to an untreated cell culture, or control. The major disadvantage to a cell culture–based drug discovery program is that a significant amount of follow-up work must be performed to determine the actual mechanism by which cell proliferation is affected. For example, a compound which stops a cell from dividing could be affecting deoxyribonucleic acid (DNA) replication, tubulin polymerization, protein synthesis, or many other cellular functions. It is important to know the mechanism by which a drug works in order to predict its specificity, toxicity, and hence its utility as a therapeutic agent.

Mechanism-based, or target-directed, screening involves selection of a molecular target, such as an enzyme or a receptor involved in a human disease process. The natural products are screened for compounds which specifically interact with this target molecule. For example, certain enzymes are known to be necessary for activating the T cells of the immune system. Natural products which inhibit the activity of such an enzyme would therefore be expected to suppress the immune system. A major disadvantage of a purely molecular-based discovery approach is that compounds which inhibit a purified enzyme may not be active in whole-cell systems due to their inability to penetrate the cell. One way to overcome this problem is to use recombinant techniques to engineer a cell line in which the molecular target is overexpressed, and to design an assay so that interaction of the drug with the target results in a measurable event such as a color change or luminescence. These assays ensure that the drug can penetrate the cell and that it interacts at the specified target.

The best approach to drug discovery combines all three approaches. Extracts of marine organisms are screened through molecular target–based assays, the activity is confirmed in cultured cells, and finally, pure compounds are tested in appropriate animal models.

Natural products chemistry. Biologically active compounds display a wide range of complexity and structural types. An extract of a marine organism typically contains many different compounds, only one of which is generally responsible for the bioactivity. The active compounds are purified from the mixture using a combination of methods.

Molecular structure determination of the purified compounds relies heavily on contemporary spectral methods, including nuclear magnetic resonance spectroscopy, infrared spectroscopy, ultraviolet spectroscopy, and mass spectroscopy. If a crystal of the parent compound or a derivative can be obtained, structures can be determined by x-ray crystallography.

Drugs under evaluation. Several marine-derived chemicals are being evaluated for their potential as pharmaceuticals. Bryostatin-1, isolated from the bryozoan *Bugula neritina*, is a polyketide with both anticancer and immune modulating activity. Ecteinascidin-743 is a complex alkaloid derived from the ascidian *Ecteinascidia turbinata* and is in human clinical trials for the treatment of certain cancers. Dolastatin-10 is an antibiotic peptide isolated from the sea hare *Dollabella auricularia* which is being evaluated for its utility against cancer. Discodermolide, a polyketide isolated from deep-water sponges of the genus *Discodermia*, is a potent antitumor agent which inhibits the proliferation of cancer cells by interfering with the cell's microtubule network; the compound is in advanced preclinical testing. Another promising sponge metabolite in advanced preclinical trials is halichondrin B, derived from a New Zealand deep-water sponge, *Lissodendoryx* sp. The pseudopterosins are terpene-glycosides derived from the Caribbean soft coral *Pseudopterogorgia elisabethae* and are in advanced preclinical trials as anti-inflammatory and analgesic drugs.

Sustainable use of marine resources. After the discovery phase, compounds are developed with the cooperation of an industrial partner or with the National Institutes of Health (in the United States). To fully evaluate the clinical usefulness of the compounds, large quantities are usually required. A critical issue in drug development for any natural product is ensuring adequate supply of the compound while protecting the source organism and its habitat from overexploitation. As a result of the United Nations Convention on Biological Diversity, legislators, biomedical researchers, and environmental resource managers are working together to address issues regarding sustainable use of resources, protection of a region's genetic resources, and equitable sharing of technologies and revenues which result from the development of natural resources.

Some options for sustainable use of marine resources are chemical synthesis, controlled harvesting, aquaculture of the source organism, in vitro production through cell culture of the macroorganism or microorganism source, and genetic engineering techniques (such as transgenic production). Each option has its advantages and limitations. Not all methods will be applicable to supply each marine bioproduct, and most of the biological supply methods are still under development. The approach to be used will be based on a number of factors: the abundance of the organism in nature will determine if harvesting is feasible; the chemical complexity of the molecule will determine if it can be synthesized at an industrial scale; the source of the compound (macroorganism or microorganism) will determine if biological supply options, such as fermentation, cell culture, or aquaculture, are possible; and if the biosynthetic pathway for the compound is known, it may be possible to identify, isolate, and clone the genes responsible for production of the compound.

Research to address each of these approaches for sustainable use is in progress. Methods for aquaculture of biomedically important marine species (including sponges, tunicates, and bryozoans) are under way at laboratories around the world. For example, aquaculture is envisioned for large-scale production of bryostatin-1 and ecteinascidin-743. It has been demonstrated that cell cultures of some bioactive sponges will continue to produce their antitumor compounds, indicating the feasibility of this approach. Many marine natural products have been successfully produced through chemical synthesis. For example, sufficient supplies of dolastatin-10 have been produced for clinical investigation. Perhaps the most exciting new approach is the potential for transgenic production; however, this is not a trivial process, and a significant amount of research is still required for this option to be feasible.

Outlook. Marine natural products research has resulted in the discovery of a significant number of novel chemical compounds with pharmaceutical potential. The collaboration of marine biologists, pharmacologists, cell biologists, molecular biologists, biochemists, and engineers will lead to the development of new approaches to sample the marine environment, new assays to test marine extracts, and new methods to produce the compounds for clinical development into useful drugs.

For background information *see* BIOASSAY; BIOCHEMICAL ENGINEERING; CHROMATOGRAPHY; OCEANOGRAPHY; PHARMACEUTICALS TESTING; PHARMACOGNOSY; PHARMACOLOGY; SPECTROSCOPY in the McGraw-Hill Encyclopedia of Science & Technology.

Shirley A. Pomponi; Amy E. Wright;
John K. Reed; Peter J. McCarthy

Bibliography. D. H. Attaway and O. R. Zaborsky (eds.), *Marine Biotechnology*, vol. 1: *Pharmaceutical and Bioactive Natural Products*, Plenum Press, 1993; V. P. Gullo (ed.), *The Discovery of Natural Products with Therapeutic Potential*, Butterworth-Heinemann, 1993; National Research Council, *From Monsoons to Microbes: Understanding the Ocean's Role in Human Health*, National Academy Press, 1999; National Research Council, *Understanding Marine Biodiversity: A Research Agenda for the Nation*, National Academy Press, 1995.

Micro-electro-mechanical systems (MEMS)

In the late 1980s, micromachining technology, which utilizes techniques similar to those employed in integrated-circuit manufacturing processes, became available to fabricate micrometer-size mechanical parts. One of the first micromachines with moving parts was a micromotor (**Fig. 1**) whose rotor could turn at 1000 Hz or higher. A comb structure derived from the micromotor concept eventually evolved into the airbag sensor, which has been installed in all automobiles manufactured in the United States since 1995. Micromachining technology opens new domains for both fundamental research and applications in all engineering disciplines. Such tiny devices are studied in the new field of micro-electro-mechanical systems.

Microscopic transport phenomena. Transporting mass, momentum, and energy efficiently is the main purpose of most mechanical systems. Efficiency must be designed into the system, so the governing physical mechanisms, which have different emphases in the microscopic and macroscopic worlds, must be well understood. For example, the surface-to-volume ratio of a device is inversely proportional to its length scale and, in the case of a length scale in the micrometer range, the large surface-to-volume ratio accentuates surface effects. Therefore, it is necessary to consider certain surface forces that can be neglected in larger systems, and reexamine the constitutive relations as well as the boundary conditions in the mass, momentum, and energy transport equations. An example of a constitutive relation that is altered in a microscopic configuration involves thermal conductivity, whose value in a thin-layered material can be as little as 1–10% of the bulk material value, depending on the size of the layer and its surface roughness.

The flow of fluid through a microchannel is used in numerous mechanical, aerospace, and biomedical engineering applications. A microchannel (with a cross section of $1\ \mu\text{m} \times 40\ \mu\text{m}$) with integrated microscopic pressure sensors distributed along the streamwise direction can be fabricated to investigate gas flows. When helium is used as the working medium, the mean free path of molecules in the gas is about $0.1\ \mu\text{m}$. In other words, flow becomes rarefied in this microchannel. The slip boundary condition, for which the flow velocity is not zero at the channel walls, is clearly observed through the pressure

Fig. 1. Micromotor, with a strand of human hair, approximately 100 μm in diameter, as a reference for size. (*From L. S. Fan, Y. C. Tai, and R. S. Muller, IC-processed electrostatic motors, in Technical Digest International Electronics Development Meeting, San Francisco, pp. 666–669, 1988*)

distribution measured by the microsensor arrays.

Liquid flow through a microchannel is even more interesting than gas flow. Experimental evidence indicates that the viscosity of a simple fluid is a function of the device size. This result means that the constitutive relation that governs the flow depends on the size of the test facility. Furthermore, although surface forces usually decay quickly from the surface, the thickness of the surface force layer can still be significant compared with the dimensions of the bulk flow. The governing momentum equation again becomes size-dependent. Size effects raise challenging and intriguing issues in the study of microscopic transport phenomena.

Microtransducers. Although the technologies for fabricating micromachines and integrated circuits are similar, the difference is that in micromachine fabrication some mechanical parts need to be able to stand above the substrate or move freely over it. Bulk micromachining, surface micromachining, LIGA (an acronym for the German phrase *Lithographie, Galvanoformung, und Abformung*, or lithographys, electroforming, and molding), and more recently deep reactive ion etching (DRIE) have been developed to manufacture various types of micromachines. The basic process in all these technologies is to expose a patterned photoresist and then selectively etch away the unwanted portion of the deposited layer or substrate.

Many varieties of microsensors and microactuators with attractive characteristics have been developed based on micromachine technology. The large decrease in the physical size of the transducers manufactured by micromachine technology has greatly improved the temporal and spatial resolutions of these devices. Because of the drastic reduction in inertia resulting from these smaller sizes, a micrometer-size hot-wire anemometer for velocity sensing can have a frequency response of 1 MHz instead of the 20 kHz of conventional hot-wires. The integrated-circuit batch-processing technique can produce transducers in a large quantities to achieve low unit cost. In addition, these microtransducers can be integrated with microelectronics to form a micro-electro-mechanical system. The electronic circuit can provide on-board biasing and signal processing capabilities. Several types of microscopic fluid sensors will be discussed to illustrate the enhancement of transducer functionality by MEMS technology.

Surface shear stress is an important quantity to measure since it can be used to determine the separation point of the boundary layer in a two-dimensional steady flow, and since the integrated value of the surface shear stress is the viscous drag. However, measuring surface shear stress in the air is difficult, especially with a thermal sensor, which suffers from a low signal-to-noise ratio. A typical thermal shear-stress sensor is a heating element placed at the surface of a substrate. The heat transfer to the air is proportional to the surface shear stress, but most of the heat is conducted into the substrate, resulting in low sensitivity. By using micromachine technology, a micrometer-thick vacuum chamber can be placed under the heating element to prevent substrate heat conduction, which increases the sensitivity by two orders of magnitude. More than 100 surface shear-stress sensors, each of which has a vacuum chamber underneath the heating element, have been integrated on a single, flexible, 80-μm-thick substrate.

Pressure is another important flow property. Pressure sensors are employed in many fields ranging from biomedical usage to space vehicles, and microscopic pressure sensors constitute a major portion of all MEMS devices currently manufactured. Although many detection principles are available, the piezoresistive type of sensor is the most common, but it suffers from low sensitivity, whereas the capacitive type is highly sensitive. Recently, a MEMS-based microphone has been developed that is sensitive enough to detect low-level sound waves.

Most of the pressure and surface shear-stress sensors have sizes of the order of 100 μm in order to achieve reasonable signal-to-noise ratio. Temperature sensors, thermocouple or thermistor, can be about 1 μm in size and still be sensitive enough to resolve temperature differences of 0.05°C (0.09°F).

Laser Doppler velocimetry and particle image velocimetry are common optical means of measuring velocity in flows, and fairly complex optical systems are used for beam steering. With MEMS processing, a micro-optical table with lenses and mirrors can be reduced to millimeter size. This type of micro-optical system provides numerous opportunities for applications in fluids engineering, graphical display, data storage, and communication systems.

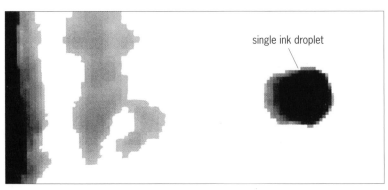

(a)

(b)

Fig. 2. Output of inkjet printers. (a) Primary droplet and stream of satellite droplets ejected from commercial inkjet with 60-μm-diameter nozzle. **(b)** Single droplet ejected from micromachined inkjet with 10-mμ-diameter nozzle. (*From F. G. Tseng, C. J. Kim, and C. M. Ho, Microinjector with a novel virtual chamber neck, IEEE MEMS Workshop, Heidelberg, pp. 57–62, 1998*)

Fluid controls. Microsensors integrated with integrated-circuit-based logic circuits and actuators are able to perform combined sensing, decision, and actuation functions. This capability enables the control of natural phenomena in real-time and in a distributed manner. Examples of the utilization of MEMS for fluid controls range from maneuvering a meter-size aircraft to manipulating nanometer-size deoxyribonucleic acid (DNA). Almost all engineering disciplines have been influenced by MEMS technology, which is still in its early stages. Only a few representative examples will be discussed.

Aircraft and microtransducers have a large disparity in size, with the kinetic energy of the aircraft being astronomical compared with the energy provided by the actuators. However, it has been demonstrated that dramatic maneuvering patterns of a scaled aircraft can be achieved by millimeter-size actuators. The key is to use minute amounts of perturbations to trigger a flow instability at the proper location on the wing. The properly applied perturbations take advantage of the flow amplification that can occur to achieve global control. This type of control is possible for a delta-wing aircraft flying at a high angle of attack. Two symmetric leading-edge separation vortices form and trail such an aircraft. The surface suction caused by the two large leading-edge vortices contributes 40% of the lift. If the symmetry of the vortices is broken, a rolling torque is expected, which can be used to maneuver the wing. From past experience with flow control, it has been learned that the evolution of the separated free shear layer is very sensitive to the perturbations at its origin. By using miniature actuators to move the separation point either upstream or downstream from the unperturbed location, the direction of the separating free shear layer is changed. The vortices are, therefore, shifted either in- or out-board from the midchord, such that a large rolling, pitching, or yawing torque is generated to maneuver the aircraft. Flight testing with a miniature mechanical actuator was carried out in 1997, and the first successful flight test with a MEMS actuator was accomplished in January 1999. *See* AIRPLANE TRAILING VORTICES.

Inkjet printers are commonly used for computer output. The thermally driven printer head uses a microheater to produce a small vapor bubble of ink, which functions as a pump to eject an ink droplet for printing. Formerly, however, commercial inkjets always generated a stream of droplets, rather than a single one, producing a blurred printout (**Fig. 2***a*). By employing a properly micromachined heater, a well-controlled unsteady pressure field is generated inside the printing head, eliminating the long ink steam so that only one droplet forms (Fig. 2*b*). Furthermore, because the surface micromachined nozzle can be 10 μm or less in diameter, the small ink droplets increase the number of dots per inch to the point that printing of photographic quality can be achieved with inkjet printers.

Identification of biological agents (such as bacteria) is regularly needed in biomedical applications.

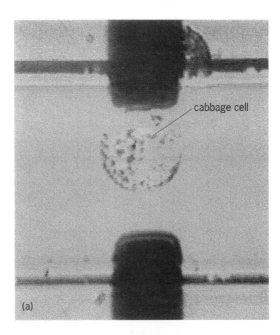

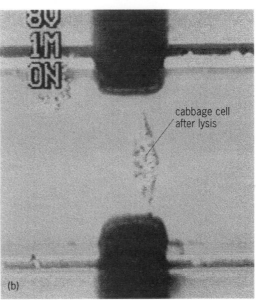

Fig. 3. Cell lysis by microelectrodes. (*a*) Cabbage cell between lysis electrodes separated by 5 μm, comparable to the size of the cell. (*b*) Cabbage cell after lysis, which was accomplished by applying a voltage to the electrodes. The important parameter in determining the efficiency of this procedure is the electrode-gap/cell-diameter ratio.

Use of the DNA sequence of the targeted agent as a sensor to check against the DNA of the bacteria in the collected sample offers the specificity needed for biological agent detection. After bacteria are collected, the sample is prepared by extracting DNA from the cell. The sample preparation process involves cell lysis and separation of DNA and ribonucleic acid (RNA) from other cellular materials; sometimes concentration of the DNA and RNA may also be required. The procedures involved include mixing of reagents, separation, and washing. By using MEMS-based cell-lysis electrodes, micropumps, and microvalves, the amount of fluids required and the processing time can be significantly decreased. The size matching between the MEMS devices and the cells can significantly facilitate the DNA detection procedures (**Fig. 3**).

For background information *see* BOUNDARY-LAYER FLOW; COMPUTER PERIPHERAL DEVICES; DEOXYRIBO-NUCLEIC ACID (DNA); INTEGRATED CIRCUITS; PRESSURE TRANSDUCER; RAREFIED GAS FLOW; TRANSDUCER; VISCOSITY in the McGraw-Hill Encyclopedia of Science & Technology.   Chih-Ming Ho

Bibliography. C. M. Ho and Y. C. Tai, Micro-electromechanical system (MEMS) and fluid flows, *Annu. Rev. Fluid Mech.*, 30:579–612, 1998; M. Madou, *Fundamentals of Microfabrication*, CRC, 1997; C. L. Tien, A. Majumdar, and F. M. Gerner, *Microscale Energy Transport*, Taylor & Francis, 1998; M. C. Wu, Micromachining for optical and optoelectronic systems, *Proc. IEEE*, 85(11):1833–1856, 1997.

Microstructures

Miniaturized devices require both structural and functional components that have dimensions in the micrometer range. These components may, in many cases, be three-dimensional (3D). For example,

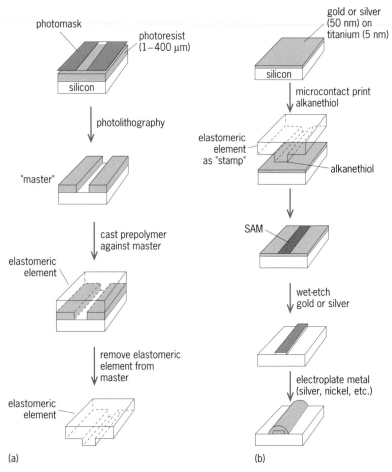

micro-electro-mechanical systems (MEMS) require the microscopic analogs of motors, gears, valves, and pumps; biomedical systems make use of microscopic metallic structures for stents (slotted expandable tubes that are mounted on a balloon catheter) and prostheses (artificial body parts); light, small air vehicles and microsensors are important for defense systems; portable consumer products need small power supplies. *See* NEUROTECHNOLOGY.

At the macroscale (dimensions from centimeters to meters), fabrication methods for producing and assembling components into 3D structures are well developed. Most of these techniques do not, however, scale down well to the micrometer range. In this regime of small sizes, the best-established methods for fabrication—photolithography and related techniques—were developed for the production of microelectronic circuitry. In photolithography, light is projected through a rigid mask that has the pattern of interest represented in transparent and opaque (chrome) regions. The projected light induces a chemical change in the illuminated regions of a thin polymeric layer (the photoresist). The exposed regions of resist become either more (a positive resist) or less (a negative resist) soluble in the developing solution. This technique can produce features as small as 0.18 μm.

Several methods can be used to transform these patterns in photoresist into metallic microstructures. If features with high-aspect ratios (that is, height: width ratios much greater than 1) are required, the lithographic step is performed on a conductive substrate. When the substrate is used as the cathode in an electroplating bath, the patterned photoresist serves as a mold that determines where metal can deposit, and also determines its cross-sectional profile. The complete process for forming metallic microstructures (from photolithography to completed structure) is referred to either as through-mask electroplating or, if x-rays from a collimated synchrotron source are used in the pattern transfer step, as LIGA (*Lithographie, Galvanoformung, Abformung*).

These techniques can produce structures with high aspect ratios (as high as 20:1) and small (submicrometer) feature sizes. They fail, however, if the substrate of interest is not flat. High-resolution pattern transfer during photolithography is possible only when the substrate is in contact with the mask (contact-mode lithography) or in the image plane (projection lithography) of the mask. If the substrate is rough or nonplanar, the image transferred to the substrate during photolithography will be distorted. Fully 3D structures, therefore, must be constructed layer by layer by sequential photolithographic steps.

Most alternatives to through-mask plating use a serial technique to produce 3D microstructures either by "carving" material from a solid object or by "writing" material using a localized deposition process. Soft lithography offers a low-cost, parallel approach to the problem of microfabrication in three dimensions that does not require the use of light in the pattern transfer step. When coupled with

Fig. 1. Elastomeric element (*a*) as formed for soft lithography and (*b*) as used for microcontact printing.

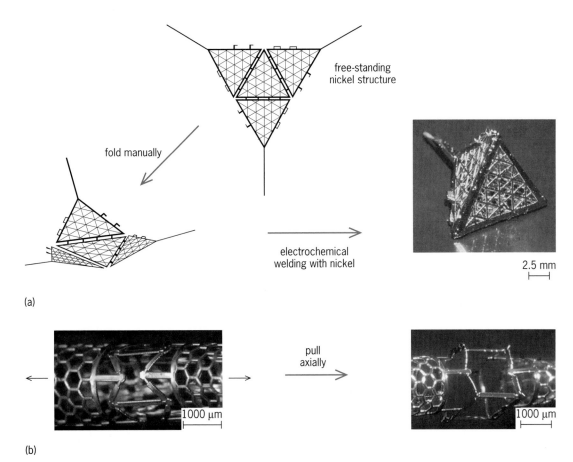

free-standing
nickel structure

fold manually

electrochemical
welding with nickel

2.5 mm

(a)

pull
axially

1000 µm

1000 µm

(b)

Fig. 2. Fabrication of 3D microstructures. (*a*) From planar patterns; for example, a tetrahedron is formed by folding a free-standing planar projection of the structure up manually and then electrochemically welding it together with nickel. (*b*) From cylindrical patterns; for example, microcubes are formed by axial deformation of a free-standing cylindrical mesh. (*After R. J. Jackman et al., Three-dimensional metallic microstructures fabricated by soft lithography and microelectrodeposition, Langmuir, 15:826–836, 1999*)

microelectrodeposition, this approach can produce 3D microstructures.

Soft lithography. Soft lithography comprises a set of techniques (such as microcontact printing, micromolding in capillaries, and microtransfer molding) that use an elastomeric or flexible element as the mask or mold to transfer patterns instead of a rigid chrome mask. This flexible element permits the formation of high-resolution features in metal or photoresist (on a conductive substrate) that can then be used to form free-standing, metallic 3D structures. The fabrication of an elastomeric element is accomplished by casting and curing a liquid prepolymer against a "master." This master usually consists of features in photoresist defined by photolithography on a silicon wafer, but can be any structure having 3D relief (**Fig. 1***a*).

Microcontact printing (µCP) uses a molded elastomer as a "stamp" in the pattern transfer step (Fig. 1*b*). In most cases, the "ink" used on the stamp is an ethanolic solution of an alkanethiol. When the alkanethiol is printed onto the surface of a gold or silver film, a self-assembled monolayer (SAM)—a one-layer-thick, semicrystalline film of the organic molecule—forms in the regions where the stamp contacts the substrate. As the sulfur groups react

to form a covalent bond with the gold or silver, the alkane chains extend from the substrate at a tilt angle that maximizes their van der Waals interactions with one another. This monolayer acts as a nanometer resist or protective coating that prevents removal of metal when the substrate is immersed in a wet chemical etching bath. After microcontact printing and etching, the patterns of metal replicate the pattern of the stamp that was used to form the self-assembled monolayer.

This technique has been used to produce features as small as 0.2 µm on planar substrates. The use of an elastomer rather than light to transfer a pattern relaxes the requirement that the substrate be planar. The elastomeric nature of the stamp allows it to come into conformal contact with a rough surface and enables the transfer of material to a curved substrate (for example, a capillary, lens, or fiber) by simply rolling the material across the surface of the stamp.

Microelectrodeposition is a tool that can take thin patterns of metal such as those formed by soft lithography and transform them into functional metallic microstructures. It increases the rigidity of microstructures formed by microcontact printing, reduces their electrical resistance, allows them to be stretched

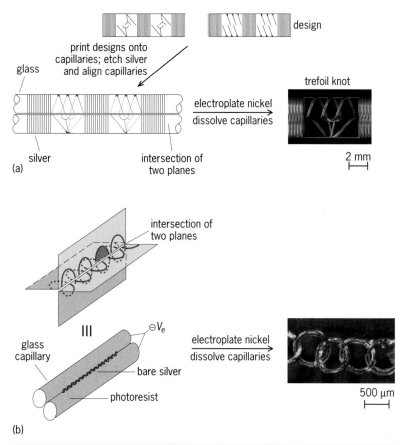

Fig. 3. Three-dimensional microstructures with complex topographies. (*a*) Trefoil knot. (*b*) Interlinked chain formed by soft lithography and using electrodeposition to weld the structure together. (*After R. J. Jackman et al., Design and fabrication of topologically complex, three-dimensional structures, Science, 280:2089–2091, 1998*)

or bent sharply, and welds separate components together. For example, metallic microstructures on cylindrical supports formed by microcontact printing followed by electrodeposition may find uses in a variety of applications. Microcoils (widths of 25–150 μm) on cylinders (diameters of 150 μm-2 mm) can serve as detection or receiver coils for performing nuclear magnetic resonance experiments on nanoliter volumes of material, and they can be used as the electrically conducting coils in microelectromagnets and microtransformers. These types of structures could be incorporated into microfluidic systems for performing on-line chemical analyses or into microelectromechanical systems for actuation or voltage conversion. Bands (50–150 μm wide) printed onto optical fibers can function as integrated photomasks for the formation of in-fiber gratings that are important in optical communications, for example, as wavelength-selective mirrors.

While all of these types of microstructures on cylindrical substrates are difficult to fabricate using conventional techniques, they are straightforward to produce using soft lithography and electrodeposition. Cylindrical structures are, however, still only pseudo 3D (having no variation in the z dimension). Further complexity and full three-dimensionality can be added by using electrochemistry to strengthen and weld structures together.

3D microstructures from planar patterns. Fabrication of such structures can be viewed as micro-origami. Many simple, fully 3D structures—for example, a tetrahedron—can be projected or unfolded onto a plane. Conversely, an appropriate design in two dimensions can be folded to produce a 3D structure. On the microscopic scale, this approach can be realized using microcontact printing and electrodeposition (**Fig. 2***a*). Microcontact printing forms a two-dimensional pattern of metal on a planar substrate in the shape of a projected tetrahedron. By electrodeposition of metal, the structure is made more rigid; it can subsequently be released from the underlying substrate. Folding the structure using tweezers produces the tetrahedron. The electrodeposited metal makes the structure self-supporting, it allows high strain deformation (folding), and in a final step, further electroplating welds the completed structure together. A series of tabs and slots, arranged around the sides of the tetrahedron, hold the sides of the folded structure in the correct locations during electroplating. This type of open, lightweight microstructure, if constructed from appropriate materials, could be a structural component in a microairplane wing or in a small space vehicle.

3D microstructures from cylindrical patterns. At smaller scales, different strategies for assembling the components that make up these 3D structures will be needed because it will not be possible to manipulate them using tweezers: the structures will need to fold themselves. One approach to 3D microfabrication that may scale down to micrometer–sized structures is to design a structure with small, compliant components. When a macroscopic force is applied to this structure, it is distributed (by design) among each of the microscopic components, and the movements that result give rise to a new 3D structure. This scheme allows the transformation of cylindrical structures into fully 3D microstructures. Application of a macroscopic, axial force to a free-standing, cylindrical, metallic mesh with hinges incorporated into its design (formed using soft lithography and electroplating; Fig. 2*b*) can produce structures with noncylindrical symmetry. The electroplating step prior to the deformation ensures that the structures are self-supporting and will yield, but not break, when they are pulled. After pulling, another electrodeposition step can weld any cracks, small tears, or stress-thinned regions that have formed and can largely restore the mechanical properties of the structure.

The ability of electroplating to function as a micrometer-scale welding tool furthers the complexity of microstructures that can be made: it can weld discrete structures in previously inaccessible geometries. Complex topographies, which are impossible in a single-layer, planar geometry, can be fabricated by welding structures formed at the intersection of two planes. Examples of these structures are interlinked chains and knots (**Fig. 3**). A trefoil knot can be formed by printing patterns onto cylinders. Electrodeposition onto the metallic structures on the two cylinders that are held in the correct orientation welds the structure together. This methodology should make it possible to produce knots of arbitrary complexity with just two cylinders.

For background information *see* ELECTROPLATING OF METALS; INTEGRATED CIRCUITS; MICROSENSOR; MONOMOLECULAR FILM; PRINTED CIRCUIT in the McGraw-Hill Encyclopedia of Science & Technology.

Rebecca J. Jackman; Scott T. Brittain; George M. Whitesides

Bibliography. R. J. Jackman et al., Design and fabrication of topologically complex, three-dimensional microstructures, *Science*, 280:2089–2091, 1998; R. J. Jackman et al., Three-dimensional metallic microstructures fabricated by soft lithography and microelectrodeposition, *Langmuir*, 15:826–836, 1999; Y. Xia and G. M. Whitesides, Soft lithography, *Angew. Chem. Int. Ed. Eng.*, 37:550–575, 1998.

Mid-ocean ridge structures

Oceanic crust generated at mid-ocean ridges covers two-thirds of the Earth's surface, yet geoscientists have had only a limited understanding of the processes that create the ocean's floor. As two tectonic plates separate at a mid-ocean ridge spreading center, the mantle upwells to fill in the gap, producing a small fraction of pressure-release melting. This melt migrates through the mantle to feed a narrow (a few kilometers wide) axial volcanic region and form the nascent oceanic crust. The Mantle Electromagnetic and Tomography (MELT) experiment was designed to obtain geophysical information to determine more detailed characteristics of mantle flow and magma generation beneath mid-ocean ridges. Passive arrays of seismometers, electrometers, and magnetometers were placed on the sea floor across the East Pacific Rise to record seismic waves from earthquakes and coupled variations in the electric and magnetic fields.

A principal goal of the MELT experiment was to distinguish between two different types of models depicting mantle upwelling and melting. In passive flow models, a broad zone, approximately 100 km (60 mi) wide, of upwelling and melt is present, and melt must migrate horizontally to the narrow axial volcanic region. In dynamic flow models, upwelling and melting occurs in a narrow zone, perhaps only a few kilometers wide, and melt migration is

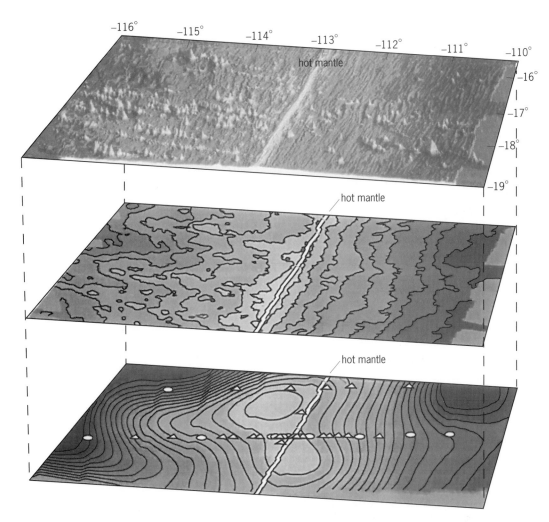

Fig. 1. Sea-floor topography (top), gravity anomalies (middle), and seismic Rayleigh-wave velocities (bottom) measured in the MELT experiment. The positions of ocean-bottom seismometers are indicated by triangles and circles in the bottom panel. Light tones indicate shallow sea floor, low densities, and low phase velocities (top, middle, and bottom, respectively). (*From D. S. Scheirer, based on data from the MELT experiment*)

predominantly vertical. Additional questions addressed by the MELT experiment are the depth extent of melting, such as whether melting primarily occurs at regions above 100 km depth or extends much deeper due to the presence of water, and whether there is a link between mid-ocean ridges and the structure of the lower mantle, such as if upwelling beneath ridges were an active part of a whole-mantle convection system rather than a passive response to plate spreading.

The MELT experiment site is on the longest, straightest section of the global spreading system and at one of the fastest-spreading mid-ocean ridge regions. Geophysical surveys show that there is an asymmetry to the ridge in this area. The sea floor is subsiding more slowly on the Pacific plate to the west than on the Nazca plate to the east. Mantle Bouguer gravity anomalies, which can be considered a proxy for mantle densities, are lower to the west, and the west side has a greater population of small seamounts (**Fig. 1**). These asymmetries likely indicate an asymmetry in the thermal structure and melt production of the underlying mantle, with hotter temperatures to the west. The absolute and relative plate motions are also asymmetric. The Pacific plate is moving almost twice as fast to the west as the Nazca plate is moving to the east in the hot-spot coordinate frame, causing the axis to migrate to the west, and rapid rift propagation has been transferring sea floor from the Pacific plate to the Nazca plate, so that the relative half-spreading rate is more rapid on the Nazca plate side.

Seismic experiment. During the course of the MELT experiment, 51 ocean-bottom seismometers were deployed from November 1995 to May 1996 across the East Pacific Rise in two linear arrays (Fig. 1). Measurements of the velocities of seismic waves [compressional (*P*) waves, shear (*S*) waves, and surface (Rayleigh) waves] provide information on the presence of hot temperatures and melt, which decrease the seismic velocities. The delays of *P* and *S* waves from distant earthquakes and the pattern of Rayleigh-wave phase velocities are asymmetric (Fig. 1). In both surface-wave and body-wave tomography results, the velocities are low to the west of the axis and increase rapidly immediately to the east of the axis. The region of low velocities is several hundred kilometers across, and the magnitude of the observed changes are too large to be simply due to the effect of velocity increases caused by solid-state cooling of the mantle with increasing distance from the ridge axis. The results indicate that at least 1–2% melt is required over a region several hundred kilometers wide. Both modeling waveforms from regional earthquakes along the East Pacific Rise and the results of *P*- and *S*-wave tomography indicate that low velocities and melting extend to depths exceeding 100 km (60 mi). The observations suggest that the primary melt production in the upwelling mantle begins at about 100 km and that trace amounts of melting occur at depths of 150 km (90 mi) or more.

A limit on the depth extent of the low-velocity axial region is provided by *P*-to-*s* conversions at the primary discontinuities at 410 and 660 km (255 and 410 mi) depth. The 410-km discontinuity has been identified with the transition from the α to β phase of olivine, $(Mg,Fe)_2SiO_4$, and the 660-km discontinuity with the transition from the γ phase of olivine to perovskite plus magnesiowustite. The depths to the 410- and 660-km phase boundaries, respectively, increase and decrease with higher temperatures, so hot temperatures would thin the mantle transition zone between these two discontinuities. However, the time difference between conversions at these two discontinuities measured by the MELT experiment shows that the mantle transition zone is of normal thickness and that hot temperatures are not present at these depths. This observation as well as other MELT analyses suggest that ridge upwelling is not driven from deep (>300 km or 185 mi) within the mantle, but is shallow and passively driven by plate separation.

Because olivine, the most abundant mineral in the upper mantle, is anisotropic and develops lattice-preferred orientation in response to finite strain, observations of seismic anisotropy can be used to infer the pattern of flow-induced alignment of olivine grains associated with mid-ocean ridge spreading.

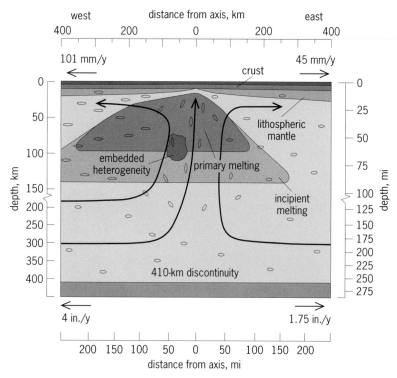

Fig. 2. Schematic cross section of the East Pacific Rise at the MELT region. The Pacific plate, moving at 101 mm/y (4 in./y) in the hot-spot coordinate frame, is to the left in the diagram and the Nazca plate is at the right. Darker tones indicate the region of melting; arrows indicate the direction of mantle flow; ovals denote the preferred orientation of olivine crystals. The broad asymmetric region of low seismic velocities is interpreted to be the primary melt production region. The embedded heterogeneity region represents additional melting created by anomalously enriched source material. Although the exact flow pattern is not known, the faster motion of the Pacific plate is expected to create greater shear in the upper mantle, inducing stronger anisotropy. The hot material and the faster motion of the Pacific plate may create the observed asymmetries in geophysical data. (*After MELT Seismic Team, Imaging the deep seismic structure beneath a mid-ocean ridge: The MELT experiment, Science, 280:1215–1218, 1998*)

The phases *SKS* and *SKKS* (types of *S* waves that travel through the Earth's core) show a phenomenon known as shear-wave splitting, which provides an unambiguous indication that the upper mantle across the MELT array is anisotropic. The direction of fast shear-wave polarization is aligned parallel to the spreading direction. The magnitude of anisotropy is asymmetric, with greater anisotropy off axis on the Pacific plate than on the Nazca plate. Rayleigh-wave phase velocities also indicate anisotropy parallel to spreading and greater anisotropy on the Pacific plate.

Model of East Pacific Rise. The MELT seismic results have been integrated into a schematic model of mantle flow and melting (**Fig. 2**). The region of melt production is distributed over a broad region consistent with passive flow models and is not confined to a narrow upwelling zone on axis as proposed by dynamic flow models. A surprising observation is that the lowest velocities are west of the ridge axis rather than immediately beneath it. One possible explanation is that there is anomalous melting of an embedded compositional heterogeneity that produces a local off-axis anomaly. The structure of the ridge system is strongly asymmetric: mantle densities and seismic velocities are lower and seismic anisotropy is stronger to the west of the rise axis. These features may be caused by both the more rapid absolute plate motion of the Pacific plate and high-temperature asthenosphere (mantle at depths greater than 100 km or 60 mi) return flow toward the ridge from the volcanic region of French Polynesia (also known as the Pacific Superswell) to the west.

For background information *see* ASTHENOSPHERE; EARTH CRUST; EARTH INTERIOR; FAULT AND FAULT STRUCTURES; HOT SPOTS (GEOLOGY); MID-OCEANIC RIDGE; OCEAN FLOOR; PLATE TECTONICS; SEISMOLOGY; TRANSFORM FAULT in the McGraw-Hill Encyclopedia of Science & Technology.   Cecily J. Wolfe

Bibliography. D. W. Forsyth, Geophysical constraints on mantle flow and melt generation beneath mid-ocean ridges, in J. Phipps Morgan et al. (eds.), *Mantle Flow and Melt Generation at Mid-Ocean Ridges*, American Geophysical Union, 1992; MELT Seismic Team, Imaging the deep seismic structure beneath a mid-ocean ridge: The MELT experiment, *Science*, 280:1215–1218, 1998.

Molecular machines

Macroscopic devices are used extensively in everyday life. A macroscopic device is an assemblage of components designed to achieve a specific function. Each component performs a simple act, while the entire device performs a more complex function, characteristic of the assembly. For example, the function performed by a hairdryer (blowing hot air) is the result of acts performed by a switch, a heater, and a fan, suitably connected by electric wires and assembled in an appropriate framework.

The concept of device can be extended to the molecular level. A molecular device is an assemblage

of a discrete number of molecular components (that is, a supramolecular structure) designed to achieve a specific function. Each molecular component performs a single act, while the entire supramolecular structure performs a more complex function, which results from the cooperation of the various molecular components. Molecular devices operate via electronic or nuclear rearrangements. Like macroscopic devices, they need energy to operate and signals to communicate with the operator. The extension of the concept of device to the molecular level is of interest not only for basic research but also for the growth of nanoscience and nanotechnology.

A molecular machine is a particular type of molecular device in which the component parts can display changes in their relative positions as a result of some external stimulus. Such molecular motions usually result in changes of some chemical or physical property of the supramolecular system, resulting in a "readout" signal that can be used to monitor the operation of the machine (**Fig. 1**). The reversibility of the movement, that is, the possibility to restore the initial situation by means of an opposite stimulus, is an essential feature of a molecular machine. Although there are a number of chemical compounds whose structure or shape can be modified by an external stimulus (for example, photoinduced *cis-trans* isomerization processes), the term "molecular machines" is used only for systems showing large-amplitude movements of molecular components.

Natural molecular machines. The concept of machines at the molecular level is not new. The human body can be viewed as a very complex ensemble of molecular-level machines that power motions, repair damage, and orchestrate an inner world of sense, emotion, and thought. Among the most studied natural molecular machines are those based on proteins such as myosin and kynesin, whose motions are driven by adenosine triphosphate (ATP) hydrolysis. One of the most interesting molecular machines of the human body is ATP synthase, a molecular-level rotatory motor. In this machine, a proton flow through

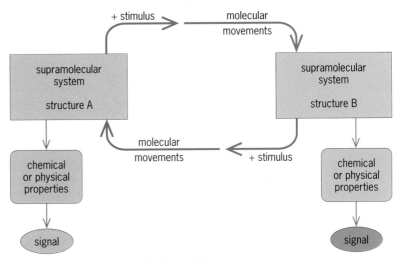

Fig. 1. Working scheme of a molecular machine.

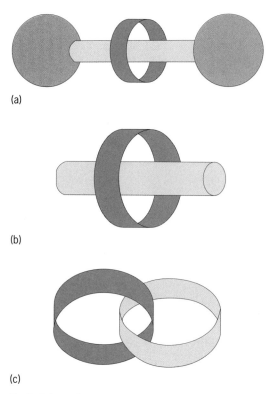

(a)

(b)

(c)

Fig. 2. Schematic representations of (a) rotaxanes, (b) pseudorotaxanes, and (c) catenanes.

a membrane spins a wheellike molecular structure and the attached rodlike species. This changes the structure of catalytic sites, allowing uptake of adenosine diphosphate (ADP) and inorganic phosphate, their reaction to give ATP, and then the release of the synthesized ATP.

Artificial molecular machines. An artificial molecular machine performs mechanical movements analogous to those observed in artificial macroscopic machines (for example, tweezers, piston/cylinder, and rotating rings). The problem of the construction of artificial molecular-level machines was first posed by Richard P. Feynman, Nobel Laureate in Physics, in his address "There is Plenty of Room at the Bottom" to the American Physical Society in 1959: "What are the possibilities of small but movable machines? ...An internal combustion engine of molecular size is impossible. Other chemical reactions, liberating energy when cold, can be used instead.... Lubrication might not be necessary; bearings could run dry; they would not run hot because heat escapes from such a small device very rapidly...." Although clever examples of a few artificial molecular machines (for example, a phototweezer for metal ions) were reported in the 1980s, substantial growth in this research field has occurred only recently, after the development of supramolecular chemistry. Supramolecular species such as pseudorotaxanes, rotaxanes, and catenanes are particularly suitable structures for the design of artificial molecular machines.

Analogously to what happens for macroscopic machines, the energy to make molecular machines work (that is, the stimulus causing the motion of the mole-

cular components of the supramolecular structure) can be supplied as light, electrical energy, or chemical energy. In most cases, the machinelike movement involves two different, well-defined and stable states, and is accompanied by on/off switching of some chemical or physical signal [absorption and emission spectra, nuclear magnetic resonance (NMR), redox potential, or hydronium ion (H_3O^+) concentration]. For this reason, molecular machines can also be regarded as bistable devices for information processing (Fig. 1).

Pseudorotaxanes, rotaxanes, and catenanes. Rotaxanes consist of a linear, dumbbell-shaped molecular component encircled by a macrocyclic component (**Fig. 2a**). Bulky groups (called stoppers) attached to both ends of the dumbbell-shaped molecule avoid the possibility of dethreading of the macrocyclic component. In pseudorotaxanes (Fig. 2b) there are no stoppers at the ends of the thread. Catenanes consist of mechanically interlocked macrocyclic rings (Fig. 2c). The rationale and efficient synthetic approaches for the preparation of complex supramolecular systems such as pseudorotaxanes, rotaxanes, and catenanes have been devised only recently. Such strategies usually rely on some kind of interaction (electron donor-acceptor, hydrogen bonding) which allows threading of the wirelike component into the macrocyclic one (pseudorotaxane structure), followed by blocking or cyclization reactions that lead to rotaxane and catenane structures, respectively.

In the late 1990s several molecular machines based on pseudorotaxanes, rotaxanes, and catenanes were designed and investigated. In all cases the mechanical movements performed by the systems are very simple, so that the term "machine" is perhaps not fully appropriate. However, the extension of the concept of machine to the molecular level is a very stimulating exercise that helps the development of chemistry and underlines new aspects of the bottom-up approach to nanotechnology.

The construction of supramolecular species performing as molecular machines is not an easy task since it requires careful selection of the component units, the design of an appropriate supramolecular structure, and complex synthetic work.

The stimulus causing the motion in molecular machines can be light, electrical energy, or chemical energy. Examples of molecular machines are a pseudorotaxane stimulated by light, a rotaxane stimulated by chemical energy, and a catenane stimulated by electrical energy. The structural changes described below are always accompanied by strong changes in the absorption spectrum, fluorescence, NMR signals, and electrochemical behavior.

Light-fueled "piston/cylinder" molecular machine. In the pseudorotaxane shown in **Fig. 3**, the driving force for self-threading is the interaction between the bipyridinium electron-acceptor units of the cyclic component and the dioxynaphthalene electron-donor unit of the wirelike component. In order to cause dethreading, the electron donor-acceptor interaction

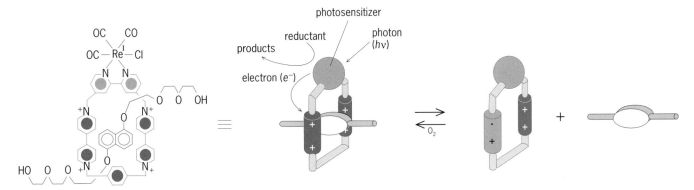

Fig. 3. Light-induced dethreading and oxygen-induced rethreading of a pseudorotaxane.

must be destroyed. This can be done by reducing the electron acceptor or oxidizing the electron donor. A rhenium (Re) bipyridine complex, which is an excellent electron-transfer photosensitizer, is incorporated into the cyclic component. Light excitation of the photosensitizer is followed by electron transfer from the excited photosensitizer to one of the bipyridinium units of the cycle. The back electron-transfer reaction is prevented by scavenging the oxidized rhenium complex with a suitable sacrificial electron donor. Since a bipyridinium unit is reduced, the donor-acceptor interaction responsible for the stability of the pseudorotaxane structure is weakened so that dethreading occurs. If oxygen is allowed to enter the irradiated solution, the reduced bipyridinium unit is back-oxidized and rethreading takes place. In conclusion, in this system a piston/cylinder-type movement can be caused by a light-fueled motor (namely, the electron-transfer photosensitizer).

Chemically driven molecular shuttle. Rotaxanes (**Fig. 4**) have been designed with the purpose of switching the ring component between two different positions (stations) along a wire-type component. The latter component contains two different recognition sites, namely, an ammonium center and an electron-acceptor bipyridinium unit. The ring component is a crown ether which incorporates two dioxybenzene units. Such a ring exhibits a strong affinity, based on hydrogen bonding, for the ammonium center and a lower affinity, based on an electron donor-acceptor interaction, for the bipyridinium unit. At the beginning, the rotaxane exists only as the translational isomer where the macrocyclic component resides on the ammonium recognition site. Upon addition of a base, the ammonium center is deprotonated and the hydrogen bonding interaction with the ring is destroyed, with consequent displacement of the ring on the bipyridinium station, where stabilization is obtained by donor-acceptor interaction. Addition of acid regenerates the ammonium center, and the cyclic component moves back to its original position.

Electrochemically driven rotation. The catenane shown in **Fig. 5** consists of a symmetric macrocycle containing two electron-acceptor bipyridinium units, and an asymmetric macrocycle containing two different electron-donor units (namely, a tetrathiafulvalene and a dioxynaphthalene unit). Initially, the catenane exists only as the translation isomer, with the better electron donor unit, namely, tetrathiafulvalene, inside the ring containing the two electron-acceptor units. Selective oxidation of the tetrathiafulvalene

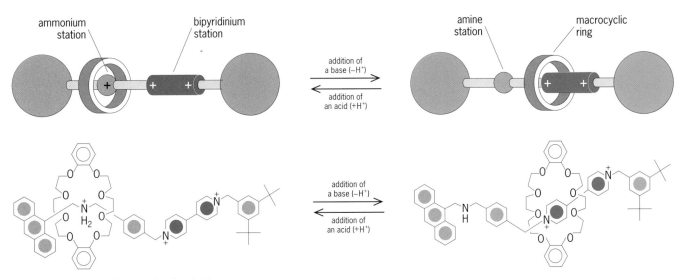

Fig. 4. Acid-base controllable molecular shuttle.

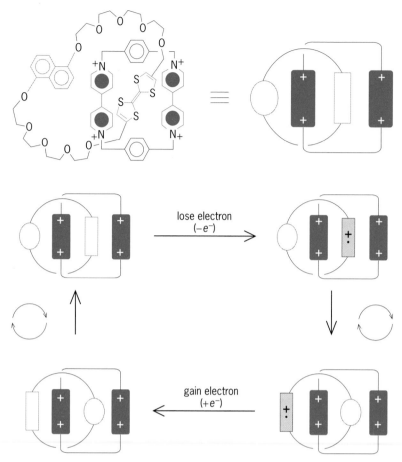

Fig. 5. Electrochemically driven rotation of a ring in a catenane.

MACROCYCLIC COMPOUND; NANOCHEMISTRY; NANO-TECHNOLOGY; PHOTOCHEMISTRY in the McGraw-Hill Encyclopedia of Science & Technology.   V. Balzani

Bibliography. V. Balzani, M. Gòmez-Lopez, and J. F. Stoddart, Molecular machines, *Acc. Chem. Res.*, 31: 405–414, 1998; A. Credi et al., Logic operations at the molecular level: An XOR gate based on a molecular machine, *J. Amer. Chem. Soc.*, 119:2679–2681, 1997; J.-P. Sauvage, Transition metal-containing rotaxanes and catenanes in motion: Toward molecular machines and motors, *Acc. Chem. Res.*, 31:611–619, 1998.

unit suppresses its electron-donor capacity and therefore destabilizes the original translational isomer with respect to the one in which the best electron donor, which is now the dioxynaphthalene unit, resides inside the electron-acceptor cyclophane. As a consequence, a circumrotation of the asymmetric ring with respect to the symmetric one takes place, with formation of the other translational isomer. Reduction of the oxidized tetrathiafulvalene unit regenerates the original structure.

Information processing. The interest in molecular machines arises not only from their mechanical movements but also from their switching aspects. Computers are based on sets of components constructed by the top-down approach. This approach, however, is now close to its intrinsic limitations. A necessary condition for further miniaturization to increase the power of information processing and computation is the bottom-up construction of molecular-level components capable of performing the functions needed (chemical computer). The molecular machines described above operate according to a binary logic and therefore can be used for switching processes at the molecular level. It has already been shown that suitable designed machinelike systems can be employed to perform complex functions such as multipole switching, plug/socket connection of molecular wires, and XOR logic operation.

For background information *see* ACID AND BASE; ADENOSINE TRIPHOSPHATE (ATP); CATENANES; LOGIC;

Moon

The Moon had long been thought to be dry and barren. Then, the discovery of ice on the planet Mercury raised the possibility that ice may exist in permanently shadowed regions of the Moon's poles. The *Clementine* spacecraft used radar observations of the lunar south pole in 1994 to detect water, but those data were not conclusive. The *Lunar Prospector* returned data in 1998 that proved more conclusively that water ice deposits exist on the Moon. Such deposits would be significant for any future human exploration and development of the Moon.

Theory. The Moon's axis is inclined by only $1.5°$, so many craters near the poles have regions that are permanently shaded from the Sun. These cold traps could permit water ice to exist at or near the surface, a theory first proposed in 1961. In 1979 this theory was expanded by James A. Arnold, who noted four sources of water ice on the Moon: reactions between the solar wind and the lunar regolith, meteorites, comet impacts, and outgassing from the lunar interior. Arnold concluded that there could be 10^{10}–10^{11} metric tons of ice on the Moon, enough to be spread into concentrations of 1–10% in the first 2 m (6.5 ft) of the regolith in the permanently shadowed regions.

A lack of data from the lunar poles hindered further work in this area. Interest in the field grew again after scientists detected evidence of ice at the poles of Mercury, a planet that has permanently shadowed regions at its poles. The discovery came from radar observations of the planet, when scientists noted an usually high reflectance from the poles as well as a ratio of circularly polarized radar waves characteristic of icy surfaces on the Earth and other bodies.

Clementine. In 1994 a radar experiment aboard the *Clementine* was used to look for ice on the Moon. This spacecraft was a joint mission of the Defense Department and the National Aeronautics and Space Administration (NASA) to study the Moon and test advanced sensor technologies. *Clementine* bounced radar signals off the south polar regions of the Moon, which were then received by antennas on the Earth. After more than 2 years of study, scientists reported in late 1996 that they had detected high reflectivity and polarization ratios that they had interpreted as deposits of water ice in shadowed regions of the lunar south pole. The amount of ice could not be inferred from the data, but scientists believed the ice consisted of small patches mixed with and covered by the lunar regolith.

The *Clementine* results were not conclusive. The radar data that supported the formation of ice came from a single orbit, leaving the possibility that the results were a fluke. Moreover, while separate radar observations from the ground-based Arecibo radio observatory turned up high polarization ratios at the lunar poles, such ratios were also seen elsewhere on the Moon, including locations in full view of the Sun. Additional data were necessary to prove that water ice existed on the Moon.

Lunar Prospector. *Lunar Prospector* was initially conceived in the mid-1980s as a small spacecraft dedicated to the search for water ice at the poles. The project was sponsored by NASA in 1995 as part of its Discovery program of inexpensive space science missions. *Lunar Prospector* was launched on January 6, 1998, and entered orbit around the Moon several days later.

Lunar Prospector was designed to look for water not through radar observations but by detecting neutrons with a spectrometer that separated neutrons based on their speed. The spectrometer consisted of two containers of helium-3, one wrapped in cadmium and the other in tin. When a neutron strikes a helium-3 atom, it creates a burst of energy recorded by the instrument. The cadmium-wrapped container detected only medium-speed neutrons, while the tin-wrapped container detected slow and medium-speed neutrons. A separate shield on the spacecraft's gamma-ray spectrometer detected fast neutrons.

The spectrometer detected neutrons generated by the collisions of cosmic rays with atoms in the crust of the Moon. Many of these high-energy, or fast, neutrons escape into space. Others bounce off other atoms in the crust before escaping into space. The latter neutrons do not slow down much unless they strike something similar in size to themselves: the nucleus of a hydrogen atom. The neutron spectrometer on *Prospector* would detect these slow, "cool"

neutrons, as well as medium-speed "epithermal" neutrons (that is, neutrons with energies in the range immediately above the thermal range) that bounce off atoms in the crust without striking hydrogen.

If the neutron spectrometer measured a drop in the level of epithermal neutrons and an increase in the number of cool neutrons, it would indicate hydrogen was in the regolith. That hydrogen could be locked up in water molecules or exist as atoms deposited by the solar wind, but any sizable amount of water would dwarf the signature from solar-wind hydrogen.

Initial spectrometer data released less than 2 months after launch showed strong evidence for the existence of water ice. A plot of the flux of epithermal neutrons showed significant decreases in the vicinity of the north and south lunar poles. The data showed a 2.2% decrease in the south polar region and a 3.4% dip in the north polar region. The greater decrease in the north was unexpected, as the north lacks a larger polar basin like the South Pole–Aitken Basin, which provides a large permanently shadowed region.

After collecting 6 months' worth of data, *Lunar Prospector* scientists were able to better estimate the amount and locations of water ice. The data showed dips in the epithermal neutron rate of 3.0 and 2.6% within $5°$ of the north and south poles, respectively (see **illus.**). The additional data also showed concentrations localized in several regions, including the Peary, Hermite, Rozhdestvenskiy, and Plaskett craters in the north and the South Pole–Aitken Basin. All of these craters have regions in permanent shade.

The form of the ice can be determined from the fast neutron data. If ice existed on the surface, the production of fast neutrons would decrease over the poles, as cosmic rays colliding with ice would not create fast neutrons. However, the *Lunar Prospector* data show no decrease in fast neutrons, implying that the ice is either finely mixed with the regolith or in concentrated deposits below the surface. If the radar observations from *Clementine* are correct, the latter condition is more likely.

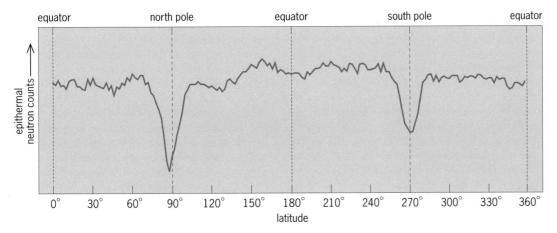

Epithermal neutron counts from *Lunar Prospector's* neutron spectrometer are plotted versus latitude. Dips at the north and south poles are interpreted as evidence of hydrogen, and hence water ice, at or below the surface. (*NASA Ames Research Center, Los Alamos National Laboratory*)

The amount of water on the Moon is dependent on a number of assumptions. Project scientists assumed that pure ice exists beneath a 40-cm (16-in.) layer of regolith down to a depth of 2 m (6.5 ft) in a region of 1850 km^2 (715 mi^2) around each pole. This gives about 3×10^9 metric tons at each pole, about an order of magnitude less than predicted by Arnold. However, these values could change considerably if any of the assumptions are adjusted, especially the form of the ice and the region of permanently shadowed regions where it is located.

Other possibilities. While it is widely believed that the data are evidence for water ice, the data are not entirely conclusive. The neutron spectrometer detects hydrogen, not water, so anything with hydrogen in it would be detected by the spectrometer. Some have proposed that the lunar poles are enriched with hydrogen from the solar wind rather than from water. This would require the regolith to contain many times more hydrogen than measured in *Apollo* samples. A lack of a hydrogen signature in the fast neutron data also argues against solar-wind hydrogen on the surface.

Other hydrogen-bearing ices, such as ammonia and methane, are also found in comets and could be transported to the Moon. However, water ice is the most stable of such ices and is the most likely to have survived on the Moon. The hydrogen may also exist as clays, or hydrated minerals, on the lunar surface. Such clays could have formed from comet collisions. Data from the *Galileo* spacecraft, collected during two lunar flybys en route to Jupiter, show evidence for such phyllosilicate minerals at the lunar poles.

Prospects. NASA currently has no missions to the Moon planned after *Lunar Prospector*. However, work is being done on the privately supported Ice Breaker mission proposal to send a rover to Peary Crater near the north pole. A rover would carry a ground-penetrating radar to look for ice, and a drill to extract samples for analysis. Pending private funding, the mission is scheduled for launch in July 2002.

A confirmation of lunar ice deposits would have a major impact on any future human exploration of the Moon. It has been estimated that from 3.3×10^7 metric tons of water deposits—the low end of early estimates from *Prospector* data—2.7×10^7 m^3 (7.2×10^9 gallons) of water could be extracted, enough to support 2000 people for nearly a century. Water can also be broken down into hydrogen and oxygen for life support and rocket propellant.

For background information *see* MERCURY (PLANET); MOON; NEUTRON; RADAR ASTRONOMY; SOLAR WIND; SPACE PROBE in the McGraw-Hill Encyclopedia of Science & Technology.   Jeffrey A. Foust

Bibliography. J. A. Arnold, Ice in the lunar polar regions, *J. Geophys. Res.*, 84:5659–5668, 1979; W. C. Feldman et al., Fluxes of fast and epithermal neutrons from *Lunar Prospector*: Evidence for ice at the lunar poles, *Science*, 281:496–500, 1998; S. Nozette et al., The *Clementine* bistatic radar experiment, *Science*, 274:1495–1498, 1996; N. J. S. Stacy, D. B. Campbell, and R. G. Ford, Arecibo radar mapping of the lunar poles: A search for ice deposits, *Science*, 276:1527–1530, 1997.

Multidimensional psychology

Objects (and events) in the environment vary in many different aspects. In psychology, aspects that vary continuously are called dimensions, and aspects that vary in all-or-none fashion are called features. For example, sounds vary continuously in loudness, so loudness is a dimension of sound. Animals either have or do not have wings, so wings is a feature of animals.

To predict how people will respond to different objects, it is important to describe the objects in terms of dimensions. There are many theories in psychology that use a multidimensional description of objects to account for how people respond to those objects in different situations. These situations include (1) categorization—how people assign the objects to categories, (2) identification—how individual objects are uniquely identified, (3) similarity judgment—which objects are perceived to be similar, (4) same-different judgment—whether two objects are judged to be the same or different, and (5) recognition memory—whether or not an object has been seen before.

In constructing such theories, it is important to distinguish between physical and psychological dimensions. Physical dimensions are the ways that objects can be varied during their construction or manipulation. Psychological dimensions are the ways that people perceive (or notice) the objects to vary. For example, the sounds emitted by tuning forks can be described by the physical dimensions of amplitude and frequency. However, the psychological dimensions of these sounds are loudness and pitch. The physical and psychological dimensions are roughly associated in the sense that when amplitude increases, loudness tends to increase, and when frequency increases, pitch tends to increase. However, amplitude and loudness (or frequency and pitch) are not the same. As evidence of this, if amplitude is held constant and frequency is varied, loudness changes as well as pitch. Multidimensional theories of how people respond in different situations are much better at predicting behavior when the objects are described in terms of their psychological dimensions rather than their physical dimensions.

In some cases, the problem of identifying the psychological dimensions is straightforward. For example, in considering a set of lines that vary continuously in length and orientation, it is simple to determine that the psychological dimensions are perceived length and perceived orientation. However, perceived length and orientation are not necessarily equal to physical length and orientation (for example, people are most accurate at judging orientations close to vertical and horizontal). In many cases, identifying the psychological dimensions is the most difficult part of constructing a dimensional theory of a specific task. For example, no one knows which dimensions of faces people attend to when categorizing faces by ethnicity or gender. To make matters worse, people almost certainly focus on different dimensions of the same objects when engaged in different tasks or judgments. For example, when identifying objects, people focus on dimensions along which the objects differ, but when judging similarity, people focus on ways in which the objects are alike. For example, the twins Manny and Moe are identical in every way except that Manny has a mole on his face. When identifying them, people will naturally attend to the mole and so never make an error, but when asked to judge the similarity of faces, people will discount the mole and judge the similarity of Manny and Moe to be high.

The most widely used multidimensional theories are based on a methodology called multidimensional scaling (MDS). The idea is that each object can be assigned a value on each of its psychological dimensions. The resulting values can be used to plot a point in a multidimensional psychological space in which the axes correspond to the various psychological dimensions. For example, a (pure) tone can be plotted as a point in a two-dimensional space (that is, a graph) in which the abscissa denotes loudness and the ordinate denotes pitch. According to MDS, the judged similarity between a pair of objects decreases with an increase in the distance between the points that represent the two objects. Objects that have point representations close together will be judged as very similar, whereas objects with point representations far apart will be judged as highly distinct.

Most of the popular statistical computer packages have routines that construct MDS spaces when given ratings of the judged similarity of pairs of objects. For example, a fragrance company might ask consumers to rate the similarity of various pairs of odors. These ratings would then be input into an MDS computer package, which would output a map of the psychological "odor space" (see **illus.**). In practice, such similarity ratings are well described by a two-dimensional space, in which one dimension corresponds to pleasantness and the other to arousal. MDS is widely used in industry for product design and testing.

Although MDS is still used in psychology, it is now generally recognized that perceived similarity is more complicated than assumed by MDS. One problem is that, according to MDS, similarities should satisfy all the properties of distances, but many demonstrations have shown this assumption to be false. For example, if point A is near point B, and B is near C, it must be true that point A is near C. This property of distances is called the triangle inequality. Well over 100 years ago, William James showed that similarities violate the triangle inequality. Specifically, he argued that a ball is similar to the Moon because both are round, and the Moon is similar to a flame because both are luminous, but in contradiction to the triangle inequality, a ball is not similar to a flame. Another problem is that MDS assumes that

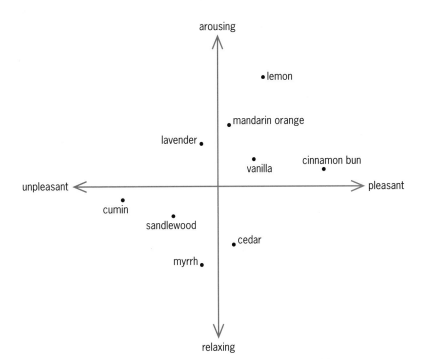

Possible two-dimensional scaling solution of similarity ratings of pairs of odors.

every time a person experiences an object, the perceived dimensional values are exactly the same. In reality, for a variety of reasons, there is some variability in perceived value. Sometimes the diet flavor in the same brand of diet cola tastes a little more intense than at other times. A number of studies have shown that explicitly modeling this variability can significantly improve the accuracy of multidimensional models (such as models of categorization and identification).

For background information *see* PERCEPTION; PSYCHOLOGY, PHYSIOLOGICAL AND EXPERIMENTAL in the McGraw-Hill Encyclopedia of Science & Technology.

F. Gregory Ashby

Bibliography. F. G. Ashby (ed.), *Multidimensional Models of Perception and Cognition*, Lawrence Erlbaum Associates, 1992; J. D. Carroll and P. Arabie, Multidimensional scaling, in M. H. Birnbaum (ed.), *Measurement, Judgment, and Decision Making: Handbook of Perception and Cognition*, Academic Press, pp. 179–250, 1998; P. Suppes et al., *Foundations of Measurement*, vol. 2: *Geometrical, Threshold, and Probabilistic Representations*, Academic Press, 1989; F. W. Young and R. M. Hamer (eds.), *Multidimensional Scaling: History, Theory, and Applications*, Lawrence Erlbaum Associates, 1987.

Multiservice broadband network technology

Multiservice broadband networks are on the verge of becoming a reality. The key drivers are rapid advances in technologies; privatization and competition; the convergence of the telecommunication, data communication, entertainment, and publishing industries; and changes in consumers' life-styles.

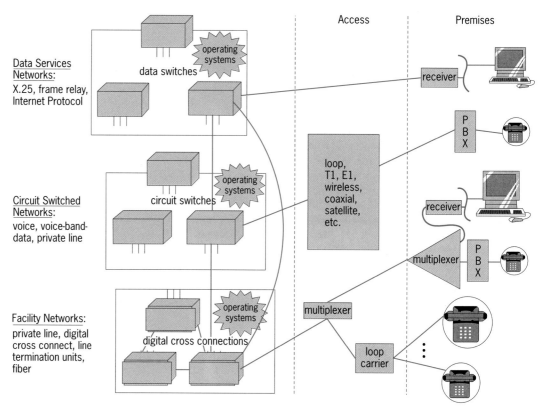

Fig. 1. Current network architecture.

Wide-area networks today consist of separate networks for voice, private lines, and data services, supported by a common facility infrastructure. Similar separation between voice and data networks exists on enterprise premises (**Fig. 1**). Residential users typically use copper loop to access voice networks as well as data networks (using voice-band modems).

Public voice and private line networks have been designed for very low latency and with an unrelenting attention to reliability and quality of service. Data networks (especially the public Internet) introduce longer and less predictable delays and are not suitable for highly interactive communication. For the most part, the Internet has not yet been shown to be reliable enough for mission-critical functions.

The time is ripe for these networks to change in fundamental ways. There is a demand for services involving multiple media, increasing intelligence, diverse speeds, and varied quality-of-service requirements. Increasing dependence on network services requires all forms of networking to be as reliable as today's voice network. The revolution taking place in electronics and photonics will provide ample opportunities to satisfy these requirements.

Technological advances. The storage capacity of a single dynamic read-only-memory (DRAM) chip increased from 64 kilobits in 1970 to 256 megabits in 1998, and 4-gigabit chips are possible in research laboratories (**Fig. 2a**). Development of multistate transistors and new lithographic techniques (enabling the fabrication of atomic-scale transistors) promise

the continuation of this trend. The computing power in a single microprocessor chip increased from 1 million instructions per second (MIPS) in 1981 to about 400 MIPS in 1998, and no limit is in sight (**Fig. 2b**). These advances are being translated into explosive growth in switching and routing capacities, information processing capacities, database sizes, and data retrieval speeds. Atomic-scale transistors also promise "system on a chip," resulting in inexpensive wearable computers, smart appliances, and wireless devices. They also promise less power consumption and longer battery life.

The advances in photonics are even more remarkable. The capacity of a single optical fiber increased from 45 megabits per second in 1981 to 1.7 gigabits per second in 1990 (**Fig. 3**). A major change occurred with the advent of dense wavelength-division multiplexing (DWDM) and optical amplifiers. Dense wavelength-division multiplexing allows the transport of many colors (wavelengths) of light in one fiber, while optical amplifiers amplify all wavelengths in a fiber simultaneously. These two innovations have made it possible to carry 400 gigabits per second (40 wavelengths each carrying 10 gigabits per second) on a single fiber and 1600 Gbps systems are on the horizon. Experimental work in research laboratories is pushing this capacity to over 3 terabits per second. Recent innovations have made it possible to put 432 fibers in a single cable. One such cable will be able to transport the daily volume of current worldwide traffic in 60 seconds. Many new and

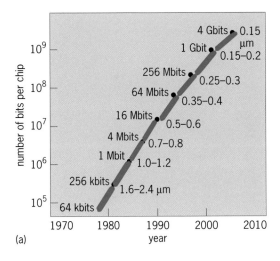

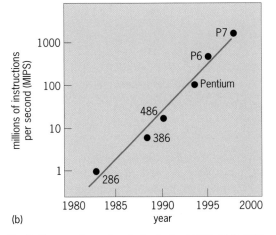

Fig. 2. Progress in semiconductor technologies. (a) Density of dynamic random-access memories (DRAMs). Lengths (in micrometers) are minimum internal spacings of chip components. Advances in manufacturing techniques allow narrower spacing, making possible higher density and hence larger storage capacity. (b) Microprocessor speeds.

up to gigabits per second to single homes by bringing fiber closer to the end users. Passive optical networks (PONs) will allow a single fiber from the central hub to serve several hundred homes by providing one or more wavelengths to each home using wavelength splitting techniques.

Advances in network infrastructure. Innovations in software technologies, network architectures, and protocols will harness this increase in networking capacities to realize true multiservice networks.

Narrowband voice [analog, ISDN (Integrated Services Digital Network), and cellular] and Internet Protocol (IP)–based data end systems are likely to continue operating well into the future, as are the Ethernet-based local-area networks. Most of the present growth is in IP-based end systems and cellular telephones, although many parts of the world are still experiencing enormous growth in analog and ISDN telephones. New services are being developed for the IP-based end systems (such as multimedia collaboration, multimedia call centers, distance learning, networked home appliances, directory assisted networking, on-line language translation, and telemedicine). These services will coexist with the traditional voice and data services. *See* DISTANCE LEARNING.

One challenge is to find the right network architecture for transporting efficiently the traffic generated by current and new services. An approach involving selective layered bandwidth management (SLBM) provides efficiency, robustness, and evolvability. SLBM uses all or some of the IP, Asynchronous Transfer Mode (ATM), SONET/SDH (synchronized optical network/synchronized digital hierarchy), and optical networking layers, depending on the traffic mix and volume (**Fig. 4**).

For the near future, ATM provides the best technology for a true multiservice backbone due to its

existing operators of wide-area networks have already started capitalizing on this growth in transport capacity. Transport capacity in wide-area networks around the world is expected to increase by a factor of up to 1000 by 2005.

Advances in electronics and photonics will also permit tremendous increases in access speeds, releasing the residential and small business users from the current speed restriction of the "last mile" on the copper loop that links the user to the network. New digital subscriber loop (DSL) technologies use advanced coding and modulation techniques to carry from several hundred kilobits per second to 50 megabits per second on the copper loop. Hybrid fiber-coaxial (HFC) technologies allow the use of cable television channels to provide several megabits per second upstream and up to 40 megabits per second downstream for a group of about 500 homes. Similar access speeds may also be provided by hybrid fiber-wireless technologies. Fiber-to-the-curb (FTTC) and fiber-to-the-home (FTTH) technologies provide

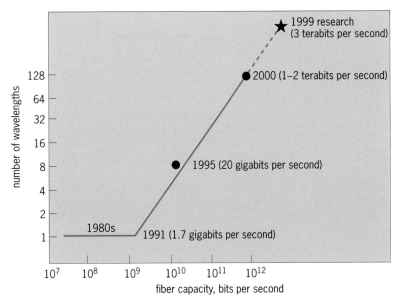

Fig. 3. Capacity growth in long-haul optical fibers. Evolution is plotted in both the total capacity and the number of wavelengths employed.

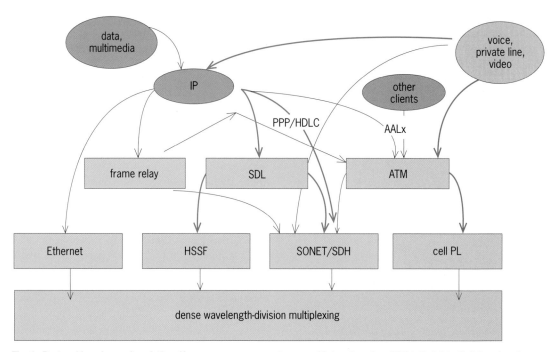

Fig. 4. Protocol layering and evolution. Heavy arrows represent new and future layering. PPP is Point-to-Point protocol; HDLC, High-speed Data-Link Control (used here only for the purpose of delineating the packets); SDL, Simplified Data-Link protocol; HSSF, High-Speed Synchronous Framing protocol; Cell PL, (ATM) Cell-based Physical Layer.

extensive quality-of-service and traffic management features such as admission controls, traffic policing, service scheduling, and buffer management. An ATM backbone network can support the traditional voice traffic, new packet voice traffic, various types of video traffic, Eithernet traffic, and IP traffic. The ATM Adaptation Layer protocols AAL1, AAL2, and AAL5 facilitate this integration. ATM also enables creation of link-layer (layer 2) virtual private networks (VPNs) over a shared public network infrastructure while providing secure communication and performance guarantees.

Traditional voice, IP, and ATM traffic will be transported over the circuits provided by the SONET/SDH network layer over the optical (wavelength) layer. SONET/SDH networks allow partitioning of the wavelength capacity into lower-granularity circuits which can be used by this traffic for efficient networking. The SONET/SDH layer also provides extensive performance monitoring. Finally, SONET/SDH networks, formed using ring topology, allow simple and fast (several tens of milliseconds) rerouting around a failure in a link or a node. This fast restoration of service makes such failures transparent to the service layers. The optical layer will consist of point-to-point fiber links over which many wavelengths will be multiplexed to allow very efficient use of the fiber capacity. Many public carriers and enterprise networks will use this multilayered networking. However, all these situations will change over time.

Rapid growth of IP end systems and high access speeds to the wide-area networks will create high point-to-point IP demands in the core backbone. At this level of traffic, the finer bandwidth partitioning provided by the ATM layer may not compensate for the additional protocol and equipment overhead. Thus, an increasingly higher portion of IP traffic will be carried directly over SONET/SDH circuits. At even higher traffic levels, even the partitioning provided by the SONET/SDH layer will be unnecessary. Meanwhile, optical cross connects and add-drop multiplexers will allow true optical-layer networking using ring or mesh topology. New techniques and algorithms being developed for very fast restoration at the optical layer will further reduce the need for the SONET/SDH layer. IP-over-wavelength networking is then expected to develop.

Of course, this simplification in the network infrastructure requires IP- and Ethernet-based networks to multiplex voice, data, video, and multimedia services directly over the SONET/SDH or dense wavelength-division multiplexing layer. Ethernet and IP protocols are being enhanced to allow such multiplexing.

New high-capacity local-area network switches (layer 2/3 switches) provide fast (10–1000 megabits per second) interfaces and large switching capacities. They also eliminate the need for contention-based access over shared-media local-area networks. Protocol standards (802.1p and 802.1q) are being defined to provide multiple classes of service over such switches by providing different delay and loss controls. Similarly, router technology and Internet protocols are beginning to eliminate the bottlenecks caused by software-based forwarding, rigid routing, and undifferentiated packet processing. New IP switches (layer 3 switches) use hardware-based packet forwarding in each input port, permitting very high capacities (60–1000 gigabits per second, or 50–1000 million packets per second). Protocols are being defined to use the Differential Service (DS) field in the header of each IP packet to signal the quality-of-service requirement of the application being supported by that packet. Intelligent buffering and packet scheduling algorithms in IP switches can use this field to provide differential quality of service as measured by delay, jitter, and losses. At the interface between layer 2 and layer 3 switches, conversion between 802.1p- and DS-based signaling can provide seamless quality-of-service management within enterprise and carrier networks. Resource reservation protocols are being defined to allow the application of signal resource requirements to the network. Traffic-policing algorithms will guarantee that the traffic entering the network is consistent with the resources reserved by the reservation protocols. Resource reservation protocols and traffic policing will further help quality-of-service management. Two other factors will add to the quality-of-service capability of IP networks. In particular, hierarchical classification of IP traffic using additional fields in the header will allow bandwidth guarantees and delay control for individual or aggregated flows based on the users as well as applications. Also, ongoing work on Multi Protocol Label Switching (MPLS) will allow flexible routing and traffic engineering in IP networks.

With these capabilities, public carriers can offer IP-layer (layer 3) virtual private networks to many enterprise customers over a shared infrastructure. Carrier-based policy servers, service and resource provisioning servers, and directories will interact with enterprise policy servers to map the enterprise requirements into carrier actions consistent with the overall virtual private network contract.

While backbone networks are likely to become more uniform, many different access technologies (such as copper loop, wireless, fiber, coaxial cable, fiber-cable hybrid, and satellite) will continue to play major roles in future networks. Standards are being defined for each of the access technologies to support diverse quality-of-service requirements over one interface. Thus, a true end-to-end solution is expected for multiservice broadband networking.

For background information *see* DATA COMMUNICATIONS; INTEGRATED CIRCUITS; INTEGRATED SERVICES DIGITAL NETWORK (ISDN); LOCAL-AREA NETWORKS; MICROPROCESSOR; OPTICAL COMMUNICATIONS; PACKET SWITCHING; SEMICONDUCTOR MEMORIES; TELEPHONE SERVICE; WIDE-AREA NETWORKS in the McGraw-Hill Encyclopedia of Science & Technology.   Bharat T. Doshi

Bibliography. F. Fenton and J. Sipes, Architectural and technological trends in access, *Bell Labs Tech. J.*, 1(1):3–10, Summer 1996; A. Netravali, The impact of solid state electronics on computing and communication, *Bell Labs Tech. J.*, 2(4):126–154, Autumn 1997; C. Netwon, Selective layered bandwidth management and the future of transport, *National Fiber Optics Engineers Conference*, 1998.

Mushroom (medicine)

Mushrooms have a long history of promoting health and vitality. In recent times, they have been proven to help the body strengthen itself and fight off illness, by restoring hormonal balance and natural resistance to disease (physiological homeostasis). Many of the medicinal mushrooms have demonstrated antitumor activity as well as antibiotic, antiviral, antiallergic, cytostatic (inhibiting cell development), and immunostimulating properties. Other benefits include lowering blood sugar, blood lipids (such as cholesterol), and blood pressure. Many of the compounds found in shiitake, reishi, maitake, cordyceps, and other mushrooms are classified as host defense potentiators. They include hemicellulose, polysaccharides, polysaccharide-peptides, nucleosides, triterpenoids, complex starches, and other metabolites. Various combinations of these compounds target the human immune system as well as aid in neuron transmission, metabolism, hormonal balance, and the transport of nutrients and oxygen.

Ergot. *Claviceps purpurea* (ergot) is a fungal parasite found in grain fields worldwide. It contains several alkaloids (nitrogenous bases), including ergotamine, which are used in the treatment of migraine headaches, precipitated by pertubation of intracranial blood vessels. These alkaloids cause smooth muscles in peripheral blood vessels to constrict, limiting blood flow. Although ergot can provide a major health benefit, it can also reduce blood flow to the point of death if not used properly. Most production (90%) is now on wheat.

Caterpillar fungus. *Cordyceps sinensis* (caterpillar fungus) is found on the cold mountain tops and snowy grass marshlands of China. The fungus infects insect larvae with spores that germinate before the cocoons are formed. It then fruits from the anterior end of the dead host. It is also grown in submerged fermentation culture. *Cordyceps sinensis* has been successfully used in clinical trials to treat liver diseases, high cholesterol, and loss of sexual drive. It contains various polysaccharides and amino acids that cause significant elevation of the number of helper T cells in mice and stimulation of phagocytotic functions. *Cordyceps ophioglossoides* (clubhead fungus) is a parasite on the fruiting bodies of the truffle found in the soil of bamboo, oak, and pine woods. It contains two protein-bound polysaccharides: CO—N is active against several tumors, and SN—C defends against a larger range of tumor systems and is an immune booster.

Reishi. *Ganoderma lucidum* (reishi or ling-zhi) grows throughout the United States, Europe, South America, and Asia. The mushroom is now cultivated but the wild type is superior in its medicinal qualities due to its natural environment. When cultivated, the mushroom grows prolifically on inoculated hardwood stumps or logs which are placed directly on the ground or shallowly buried in a shady, naturally moist location. A more elaborate method of reishi production on specially treated old plum tree sawdust was

Fig. 1. Production of maitake in bag culture.

developed in Japan. This process takes about 2 years and was designed specifically for red reishi. The rarest and most highly prized reishi, antlered reishi, is produced in a controlled environment high in carbon dioxide. Its potency is usually based on the level of triterpenoids (compounds whose molecular skeleton has 30 carbon atoms), which are bitter to the taste. Numerous clinical studies on humans indicate that the antlered reishi may have antiallergic, antiinflammatory, antibacterial, and antioxidant effects. A number of its polysaccharides, especially beta-D-glucan, appear to be responsible for antitumor and immunostimulating activity. These polysaccharides activate production of helper T cells, which attack infected cells. In addition, reishi contains 112 known triterpenes, including several ganoderic acids that have been shown to reduce blood pressure and cholesterol.

Maitake. *Grifola frondosa* (maitake) can be found in parts of the eastern United States, Europe, and Asia growing in masses at the base of stumps and on roots. It is also cultivated commercially in sterilized, supplemented sawdust (**Fig. 1**). A maitake D-fraction-containing extract was reported to reduce chemotherapy side effects and some symptoms of AIDS in human studies. Reports from clinical studies in China concluded that *Grifola* extract demonstrated an anticancer effect in patients with lung and stomach cancers or leukemia. The recovery rate of hepatitis B patients improved when they were given a concentrated polysaccharide extract of *Grifola*. Polysaccharides are also responsible for the immune response and anticancer activity of *G. umbellata* (zhu ling). It is thought that the D-fraction from maitake stimulates a cytokine, interleukin-1. In a pilot study evaluating weight loss, participants lost from 6 to 26 lb (2 to 10 kg) in 8 weeks with only the addition of the equivalent of 200 g (7 oz) of fresh maitake per day and no changes in diet.

Shiitake. *Lentinula edodes* (shiitake) is the second most widely cultivated mushroom in the world. It is native to the temperate climates of Asia. It is grown on hardwood logs (**Fig. 2**) and sterilized, supplemented hardwood sawdust substrate. It not only is touted for its medicinal properties but is a flavorful addition to many foods. In several animal studies, shiitake was shown to inhibit, by up to 60%, symptoms of type A influenza. Viruslike particles

and polysaccharides were thought to produce this effect, but no conclusive evidence was provided. Clinical studies that incorporated shiitake into the diet showed a reduction in cholesterol. After one meal of shiitake, toxic blood serum of chronic heart disease patients exhited a 30–41% reduction in the ability to cause cholesterol accumulation. Reduction of cholesterol and blood pressure has been attributed to the high fiber content of and lentinan in shiitake. Cholesterol is removed from the blood by freeing it from the plasma as well as temporarily accumulating lipids in the liver. Shiitake accelerates the conversion of low-density lipoprotein (LDL) to high-density lipoprotein (HDL) in the liver, prior to entry into the bloodstream. Lentinan and *L. edodes* mycelium extract (LEM) are the two most important components of shiitake. Both have demonstrated strong antitumor activity and work through stimulation of T cells, which activate macrophages and natural killer cells. The natural killer cells play a critical role in the destruction of tumors and viruses through stimulation of interferon (an antiviral glycoprotein produced by cells following exposure to a virus). Other polysaccharides, such as KS-2, have also shown antitumor and immunostimulating activities.

Turkey tail. *Trametes (Coriolus) versicolor* (turkey tail) is a common inhabitant of the woods and is found worldwide. This species may be readily identified by its bright colors. It is often found as a competitor in shiitake logs. It contains a protein-bound polysaccharide, PSK, which may be attributed with pharmacological activities including increased interferon production. Interferon can reduce the rate of harmful cell replication. It has also been shown in

Fig. 2. Shiitake mushrooms growing on a white oak log.

human and animal (mice) studies to possess antitumor activity, which is related to the ability of T cells to mediate the immune responses of the tumor-bearing host.

Snow fungus. *Tremella fuciformis* (snow fungus) prefers to grow on deciduous trees in the southern United States and in warm climates worldwide. It is produced commercially using the bag culture method and log method. While it was once primarily grown for its medicinal properties, it is now used mostly for food. It contains two polysaccharides, A and B, that boost immunological function and stimulate leukocyte activity.

Mode of action. The antiviral effects of mushrooms are believed to be ascribed to the induction of interferon in the host. The activity is due to the double-stranded ribonucleic acid from the mycophages attached to the spore and the fruiting body of the mushroom. Antitumor activity has been associated with various polysaccharides, one of which is KS-2 in shiitake. Polysaccharides found in fungi can stimulate immune function, inhibit tumor growth, reduce cholesterol and blood pressure, and demonstrate antiviral and antibiotic properties.

These and other medicinal properties of mushrooms often enhance the body's ability to fight the invader, rather than actually attacking it. As science and technology advance, more components of mushrooms will be discovered and evaluated. Their use in modern medicine is inevitable either as a dietary supplement or in drug therapy.

For background information *see* BASIDIOMYCOTINA; CELLULAR IMMUNOLOGY; ERGOT AND ERGOTISM; FUNGI; MUSHROOM in the McGraw-Hill Encyclopedia of Science & Technology.   Cathy Sabota

Bibliography. G. Z. Chen et al., Effects of *Cordyceps sinensis* on murine T lymphocyte subsets, *Chin. Med. J.* (English ed.), 104(1):4–8, 1991; C. Hobbs, *Medicinal Mushrooms*, Interweave Press, 1996; K. Jones, *Shiitake the Healing Mushroom*, Healing Arts Press, 1995; Y. Kusaka, H. Kondou, and K. Morimoto, Healthy lifestyles are associated with higher natural killer cell activity, *Prevent. Med.*, 21(5):602–615, 1992; J. E. F. Reynolds, *Martindale: The Extra Pharmacopoeia*, 30th ed., Pharmaceutical Press, London, 1993; Q. Y. Yang et al., Antitumor and immuno-modulating activities of the polysaccharide peptide PSP of *Coriolus-versicolor, Eos Rivista di Immunologia Ed Immunofarmacologia*, 12(1):29–34, 1992.

Nanochemistry

Molecular electronics is a relatively young field which explores the possibilities of using individual molecules as components in electrical circuits. After many years of speculation, this field has recently experienced its first experimental success. Such experiments have proven difficult, mainly due to the generally poor conductivity of molecules as well as the complications involved in attaching single molecules to macroscopic leads. These issues can now be addressed because of recently developed tools for nanometer-length scales, such as the atomic force microscope. Breakthroughs in the synthesis of suitable molecules, for example carbon nanotubes, have also been crucial. In the case of the carbon nanotubes, these developments have led to first experiments showing single-molecule transistors and wires, rendering the concept of molecular electronics more perceptible. Clearly, one of the long-term motivations for this type of work is the size reduction that might be obtained in integrated circuits. The smallest present-day transistors are typically over a hundred times larger than molecules, which have dimensions of roughly 1 nm. At this time, however, the research motivation is primarily scientific. It is expected that this cross-disciplinary research combining physics and chemistry will lead to new understanding of the molecular world.

Carbon nanotubes. Since the discovery in 1985 of the buckyball, research in fullerenes has generated much interest. Fullerenes are strikingly different from other carbon-based molecules. Their structure is a three-dimensional cage made entirely of carbon atoms. Their synthesis also stands apart: they are formed from the condensation of gaseous carbon atoms, instead of from solution. Single-walled carbon nanotubes (SWNT; **Fig. 1**), discovered in 1992, are cylindrical, like a sheet of rolled-up graphite, and capped with buckyball-type hemispheres. Their diameter is of the order of 1 nm, and they can be up to a few micrometers in length. The first nanotubes that were found are so-called multiwalled tubes, which consist of many tubes concentrically arranged like onion shells. For electron transport research, interest has focused on the single-wall variety, because of their better-defined structure and properties.

For physics experiments, nanotubes present important advantages when compared to other organic molecules considered for molecular electronics. Their electronic properties are exceptional. Depending on the angle with which the graphite sheet is rolled into a cylinder, the electrical properties can be either metallic or semiconducting; this has led to speculations of molecular wires as well as transistor-type functions. Their length, chemical inertness, and rigid structure make them relatively easy to handle. Because of the rigidity and symmetry of the molecules, their electronic structure seems robust and is affected little by disorder and vibrations.

Transistor manufacture of nanotubes. The first step in preparing molecules for use in electrical circuits is to attach macroscopic electrodes to an individual molecule. In the case of carbon nanotubes, many different routes have been explored. One strategy begins with the fabrication of small (for example, 100 nm wide and 25 nm thick) metal strips on a silicon/silicon dioxide (Si/SiO_2) substrate using electron beam lithography. The nanotube is put in a suspension using ultrasonic agitation. A drop of the suspension is then placed on the prepared substrate. After evaporation of the liquid, the resulting

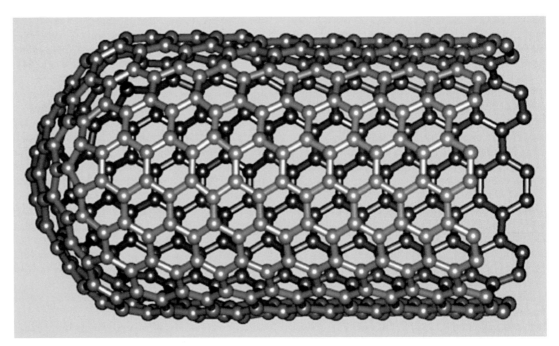

Fig. 1. Atomic structure of a single-walled carbon nanotube, basically consisting of a cylinder made from a single sheet of graphite. The diameter of this molecule is generally around 1 nm, and length ranges up to micrometers.

deposition of nanotube material can be investigated by atomic force microscopy. In these images, one observes individual nanotubes of about 1 nm high, bundles of nanotubes up to 100 nm in diameter, a low concentration of metal particles which are used

in the synthesis as catalysts, and amorphous carbon soot. In order to obtain a usable sample, where a single tube contacts only two or more electrodes, many electrode sets are made on a single substrate. After deposition, suitable samples are selected by first measuring the resistance between the electrodes followed by atomic force microscopy inspection. In another strategy, tubes are first deposited on the substrate, and their position is identified by atomic force microscopy. In a second route, the electrodes are deposited on top of the tubes. This technique has the advantage of higher control over the sample layout. However, the metal-tube interface is less well defined because more fabrication steps are performed in the presence of the tubes.

In a nanotube transistor fabricated using the first strategy (**Fig. 2**), the action can be modeled as follows. The sample layout is much like a traditional field-effect transistor (FET). The conduction channel is now composed of an individual semiconducting nanotube, which contacts the platinum (Pt) source and drain electrodes. The size of the gap in semiconducting nanotubes, which in this case should be around 0.6 electronvolt, depends on their diameter. Using the work functions for graphite and platinum, this leads to a situation where the nanotube valence-band edge is pinned to the Fermi energy of the electrodes. The bands bend downward in energy away from the electrode, creating a barrier for holes which are the majority carriers. Such pinning of the valence-band edge has also been observed in scanning tunneling microscope spectroscopy experiments of semiconducting nanotubes on a gold (Au {111}) surface. The bottom part of the device substrate is made of doped silicon, and can act as a gate. As in a regular FET, the application of a voltage

Fig. 2. Atomic force microscopy image of a nanotube transistor in a three-dimensional rendering. An individual semiconducting tube, with a measured diameter of about 1 nm, contacts two metallic source and drain electrodes. The platinum electrodes are fabricated with e-beam lithography on the silicon/silicon dioxide substrate. A voltage can be applied on the silicon layer, which acts as a gate.

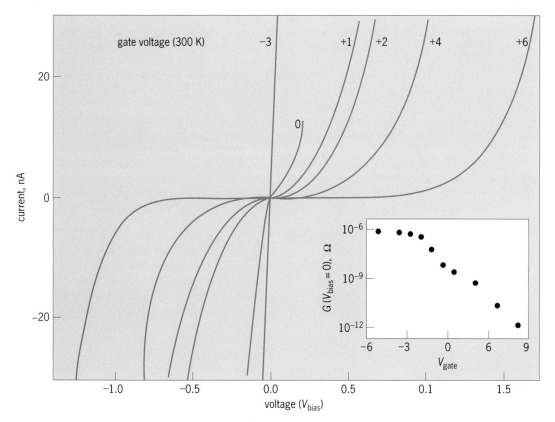

Fig. 3. Room-temperature current-voltage characteristics of a nanotube transistor at different gate voltages. For negative gate voltages the nanotube shows ohmic conductivity, whereas for positive gate voltages a clear voltage gap occurs, blocking the current at lower source-drain voltages. The inset indicates a six-order-of-magnitude variation of the conductance. The results demonstrate switching behavior at the level of a single molecule.

on the gate will modulate the charge carriers in the nanotube channel, and so affect the conductivity. Since the channel is p-type, a negative gate voltage should flatten the bands and lead to higher conductivity, whereas a positive gate voltage should result in a stronger barrier and thus decrease conductivity.

Electrical characteristics of nanotube transistor. Studies of current-voltage characteristics of a nanotube transistor at room temperature have been taken for different gate voltages (**Fig. 3**). For positive gate voltages, a voltage gap of various orders appears. For low voltages, current is blocked. For negative gate voltage, the characteristic becomes metallic. When the conductance of the device is plotted as a function of the gate voltage, the conductance is varied by six orders of magnitude. These results can qualitatively be understood with the standard FET-type model, explained above. This shows that it is indeed feasible to obtain transistor-switching behavior in individual nanotubes.

Considering the operation characteristics of these devices, it is observed that the voltage gain is still below 1. However, there is still room for improvement, by increasing the gate capacitance. From multipoint measurements, it is known that the resistance between carbon nanotubes and the metal electrodes is fairly high (approximately 500 kilohms). The result is a large device resistance ($\sim$1 megohm), even at negative gate voltages. This prohibits fast opera-

tion of the device. At present, the tube-metal interface is very poorly understood, so a clear route to controlling the contact resistance is not yet known. However, lower-resistance contacts have been achieved with different fabrication techniques.

Outlook for nanotubes. The recent results with devices based on carbon nanotubes have made a clear contribution to the field of molecular electronics. Experiments show that it is possible to make a molecular switch based on an individual nanotube. Carbon nanotubes roughly fall into two categories: those with semiconducting properties and those with metallic properties. Research on individual metallic tubes in a similar sample geometry has shown that they indeed act as molecular wires, and can carry current over long distances. These experiments have mainly focused on electron transport at very low temperatures. It has appeared that at these low temperatures the behavior is dominated by well-pronounced quantum-mechanical effects. It is hoped that future experiments will lead to a better understanding of the quantum mechanics of electrons in molecules.

Efforts in obtaining molecular circuits using other molecules have had some success. However, due to their generally bad conductivity and more difficult handling of the molecules used, results have been more difficult to obtain. Experiments on nanotubes are also still challenging, and the results are of use to a large extent as "proofs of concept," since practical applications of these strategies present major obstacles. In the synthesis of the tubes, the ability to make only metallic or semiconducting tubes with a desired length cannot be controlled. The question of integration of more devices has also not yet been addressed. Results have recently been obtained in functionalizing the end of the tubes with specific chemical compounds. This is a crucial first step for autoassembly, one of the routes that may be investigated to achieve integration.

Many fundamental questions are still not answered and present opportunities for scientific research. Some are related to the strongly one-dimensional nature of nanotubes. With their extremely small dimensions, nanotubes are not bulk systems. This contrasts with most current solid-state models and concepts, which assume bulk materials. These concepts, such as screening, need to be reconsidered for these small objects.

For background information *see* CARBON; FULLERENE; GRAPHITE; INTEGRATED CIRCUITS; NANOCHEMISTRY; NANOSTRUCTURE; PLATINUM; SCANNING TUNNELING MICROSCOPE; SEMICONDUCTOR; SILICON; TRANSISTOR in the McGraw-Hill Encyclopedia of Science & Technology.   Sander J. Tans

Bibliography. M. S. Dresselhaus, G. Dresselhaus, and P. C. Eklund, *Science of Fullerenes and Carbon Nanotubes*, Academic Press, 1996.

Nanoelectronics

Developments in integrated circuits and storage devices used in computers have proceeded at an exponential rate: at present it takes 2–3 years for each successive halving of the component size. Information storage has followed a similar trend in miniaturization of the size of the bits of magnetized material used in hard disks. However, these technologies have fundamental limits, below which the devices no longer function in a predictable manner. For instance, the oxide layers used in complementary metal oxide semiconductor (CMOS) devices are becoming so thin that they conduct electricity in a quantum-mechanical manner by electron tunneling. In 1998 it was estimated that microelectronics and magnetic storage technologies would reach their ultimate limits within 10–30 years. Projections for very large scale integration (VLSI) predict that a single chip will accommodate 90 million transistors with a feature size of 70 nanometers and a clock speed of 900 MHz by the year 2010. Currently, many critical dimensions in semiconductor devices are in the 100-nm range, with some insulating layers being tens of nanometers thick.

There is intensive effort to drive miniaturization even further. Miniaturization all the way down to the level of individual atoms and molecules would enable the fabrication of highly dense, fast, and energy-efficient devices. Energy dissipation in current devices can approach 100 W. As devices become more dense, the electronic elements must be designed to be more efficient. Another potential advantage of using individual molecules is that electronic circuitry could be prefabricated using chemical synthesis. This approach would permit techniques such as self-assembly to be used in fabrication, whereby molecules diffuse and dock onto specific connections.

The term nanoelectronics refers to electronic devices in which dimensions are in the range of atoms up to 100 nm. Nanoelectronics is regarded as the successor to microelectronics because it is capable of extending miniaturization further toward the ultimate limit of individual atoms and molecules. The first applications will probably be in the military sector.

The implementation schemes and device architectures for nanoelectronics may involve conceptually different approaches to future computational devices, such as DNA (deoxyribonucleic acid) computing and quantum computing, where atoms could act as quantum logic gates. Both of these concepts rely on controlling individual atoms or molecules, and may also be regarded as being in the realm of nanoelectronics.

Currently there are no established mass-production techniques for commercial nanoelectronic devices, nor has a well-defined fabrication strategy that can be scaled for massive integration of components been established. Whereas for microelectronic devices the transistor forms a basic building block, it is not clear what the basic (three-terminal) element of a nanoelectronic device will be. Likewise, a well-defined architecture, required to process data, has yet to be established.

Influence of quantum mechanics. Nanoelectronics research is currently looking not only for the successor to CMOS processing but also for a replacement for the transistor itself. On the scale of 10-nm dimensions, components have a wavelength comparable to that of an electron at the Fermi energy. The confinement and coherence of the electron gives rise to gross deviations from the classical charge transport found in conventional devices. Quantum-mechanical laws become increasingly dominant on the nanoscale, and it is probable that nanoelectronics will operate on quantum principles.

Single-electron devices. One popular approach has been to use small conducting islands in which electrons are confined and quantized in a definite state. These islands are typically connected to electrodes by thin tunneling barriers. Quantum dots, resonant tunnel devices, and single-electron transistors are examples of devices that use this basic concept, albeit in different ways. Single-electron transistors are an example of a three-terminal device in which the charge of a single electron is sufficient to switch the source-to-drain current. The tiny energy required to drive single-electron transistor devices makes this approach very appealing. Nevertheless, a variety of drawbacks and obstacles limit the application of such devices in solid-state nanoelectronics, for example, their sensitivity to small fluctuations in voltage and background charges, which tend to accumulate in semiconductors. Such problems suggest that single-electron devices may be used for storage rather than logic functions. Currently, such devices operate at cryogenic temperatures, although there are a few examples of room temperature operation. *See* QUANTUM-DOT LASERS.

Fabrication schemes. Owing to the dimensions in which nanoelectronics operates, there are two possible approaches to the fabrication of nanoelectronic devices: top down and bottom up (**Fig. 1**).

Top-down approaches. Top-down is the extension of established methods of engineering and microelectronics processing that relies on a patterning process often described in terms of depositing, patterning, and etching layers of material to define the circuitry and active elements. Top-down approaches rely on control of damage, and as the structures become smaller the defects make device operation increasingly problematic. Moreover, as these technologies are pushed to smaller sizes, the cost increases drastically, tolerances become more difficult to maintain, and the engineering laws of scaling become inapplicable. To scale such technologies to the atomic and molecular level, the use of softer methods with atomic tolerances based on bottom-up approaches will eventually become mandatory.

Present top-down fabrication is limited by the wavelength of the light used in the photolithography process, which illuminates proximity masks containing the pattern to be transferred to the substrate. To overcome this limit, shorter wavelengths are required, such as in extreme ultraviolet lithography. X-ray lithography is currently the leading technology in the drive to replace photolithography as a large-scale production tool because it uses masks, which are suited to high-volume production. Major challenges are mask fabrication and placement accuracy. Electron-beam lithography uses a finely focused beam of high-energy electrons to define a pattern. However, because the beam must be scanned to define the pattern, the technique in its present form is considered too slow for anything but the production of masks. Proposals to employ arrays of miniaturized electron guns are under investigation to alleviate this problem.

Bottom-up approach. This term describes device fabrication on an atom-by-atom basis. Molecules, which are prefabricated arrangements of atoms in a functional form, are also appealing for bottom-up fabrication. The advantage of the bottom-up approach lies in the design and chemical synthesis of functional molecules by the billions, which can then be assembled into nanoelectronic devices. The assembly of these perfect molecular units is more problematic, but two principal methods for engineering atoms and molecules currently exist.

One approach involves self-assembly, whereby specific intermolecular forces allow molecules to arrange themselves into more complex structures. Supramolecular chemistry is the art of designing molecules that arrange themselves into functional en-

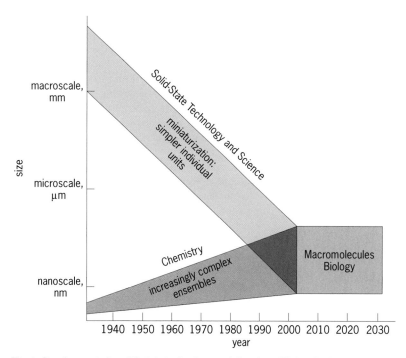

Fig. 1. Developments in solid-state technology and chemistry. Miniaturization in technology is based on ever smaller and simpler components. Macromolecular chemistry involves a bottom-up approach, in which units of atoms are connected to form structures of ever-increasing complexity. Understanding and control of biology is also approaching the scale of individual atoms and molecules (for example, in drug design), and so converges with chemistry and microelectronics. Concepts in all three fields may be used in the design, fabrication, and operational principles of nanoelectronic devices.

tities held together by intermolecular forces. Success has also been achieved in patterning surfaces using elastomeric stamps containing self-assembling molecules that transfer themselves to a surface upon contact.

The development of the scanning tunneling microscope and related scanning probe microscopy tools such as the scanning force microscope provide another approach to bottom-up fabrication. In scanning tunneling microscopy, a sharp tip, a needle of conical shape typically terminating in a single atom, is used to reposition, chemically change, or deposit individual atoms. It is an extension of mechanical assembly techniques on the atomic scale. Scanning tunneling microscopy is the highest-resolution member of the family of scanning probe microscopy techniques, which all use sharp tips in different modes of operation. A major drawback of scanning probe microscopy is that it is a serial process. To overcome this, a system employing 1000 scanning tips has been realized using micromechanics. It is anticipated that several thousand tips working in unison can be fabricated. These small devices also allow patterning to be conducted at higher speeds because of the higher mechanical frequencies of the components.

Clear evidence that scanning probe microscopy techniques are an ultimate engineering tool has been provided by demonstrations of the ability to finely control and assemble individual atoms at low temperatures, around $-270°C$ ($-454°F$). The low temperature is necessary to keep the atoms from becoming thermally agitated and diffusing away. This capability has also been extended to room temperature in experiments in which molecules were repositioned on surfaces.

Atomic and molecular devices. An abacus with individual molecules as beads having a diameter of less than 1 nm has been constructed. The "finger" to move these beads is the ultrafine tungsten tip of a scanning tunneling microscope, which also renders the result of such a "calculation" visible when operated in imaging mode. Stable rows of ten C_{60} (buckminsterfullerene) molecules were formed along steps just one atom high on a copper surface. These steps act as "rails," similar to the earliest form of the abacus, which had grooves instead of rods to keep the beads in line. Individual molecules were then approached by the scanning tunneling microscope tip and pushed back and forth in a precisely controlled way to count from 0 to 10 (**Fig. 2**). This work is a further step in bottom-up fabrication. It points the way to the assembly of more complex structures molecule by molecule, as nature does, and thus breaks ground for entirely new fabrication technologies in nanoelectronics.

Molecular amplifiers and transistors. In an extension of electrical transport studies of C_{60} molecules that involved electrically contacting individual molecules, an experimental electronic device with a tiny active part was demonstrated. It consists of a single C_{60} molecule, also known as a bucky ball, actuated by the ultrasharp tip of a scanning tunneling microscope, and represents a fully functional electromechanical amplifier. Applying a mechanical force to a single bucky ball using a scanning tunneling microscope tip can change its electrical conductance continuously and reversibly. By gently squeezing the bucky ball by lowering the tip only one-tenth of a nanometer, the molecular structure is deformed slightly, which in turn changes its electrical properties. The resistance of the squeezed bucky ball is 100 times lower than that of the uncompressed molecule, which allows electrons to tunnel more

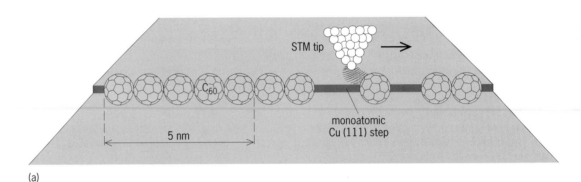

(a)

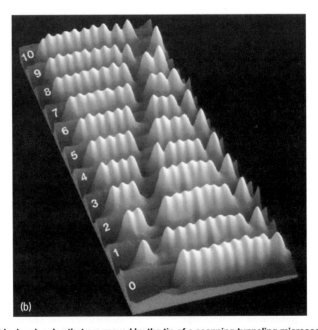

(b)

Fig. 2. Abacus with individual molecules that are moved by the tip of a scanning tunneling microscope (STM). (a) Repositioning of a C_{60} molecule adsorbed on Cu(111) step sites along the step direction. (b) Sequence of STM images of 10 C_{60} molecules adsorbed on the lower step, plotted in a pseudo-three-dimensional illuminated representation. Using the tip as a nanoactuator, the numbers 0 to 10 have been sequentially computed and read using molecules as counters.

easily through the molecule. A small voltage (10 mV) applied to a piezoelectric element of the tip to squeeze the molecule results in a fivefold voltage gain in the device (**Fig. 3**).

By externally deforming the molecule, the highest occupied and lowest unoccupied molecular orbital (HOMO and LUMO) levels (which are the equivalent of the valence and conduction bands in a conventional semiconductor) are broadened and shifted (Fig. 3*b*). This effect increases the weight of the tails of those levels at the electrode's Fermi level and therefore accounts for the increased conductance (transparency) of the molecule. (Electron transport at small bias voltage is still linear with voltage in this region.) One interesting consequence of the molecule's compression is that normally degenerate molecular levels (such as the fivefold HOMO and the threefold LUMO of a C_{60} molecule) are split in energy, which explains the five and three peaks of these orbitals that appear on the conductance spectrum under compression.

The next step in further decreasing the size of the device is to replace the macroscopic piezo-actuated electromechanical gate (the scanning tunneling microscope tip) with a smaller component. A micromechanical electrode actuated by electrostatic, thermomechanical, or other means might be used to vary the pressure on a molecule, thereby reducing the overall dimensions of the complete amplifier.

In another approach, researchers fabricated a transistor using nanotubes. Carbon nanotubes are molecularly perfect tubes with nanometer-scale diameters formed from carbon atoms. They are electrical conductors and have attracted much attention as candidates for molecular computers. Currently, carbon nanotubes are limited in their application to nanoelectronics by their variety of forms and types. They lack the uniformity of molecules synthesized by conventional chemical techniques, which can be purified to trillions of exactly identical copies. However, their good electrical characteristics do make them attractive candidates for nanoelectronics. Much research is being conducted to further develop methods to fabricate better tubes in larger quantities.

Atomic and molecular switches. The first atomic switch consisted of a xenon atom that was switched back and forth between a scanning tunneling microscope tip and a surface at low temperature. Associated with the motion of the atom was a change in the electric current flowing in the junction. This approach to the ultimate miniaturization of electronic components encouraged researchers to propose a number of devices.

In a simulated atomic relay based on this principle, a single atom in an atomic wire was switched in and out of registry by a third electrode, also made of an atomic wire. An atomic wire consists of single atoms lined up in a row. Such structures have been created in the laboratory using, for example, a surface with atomic grooves. A cheap and reliable method to fabri-

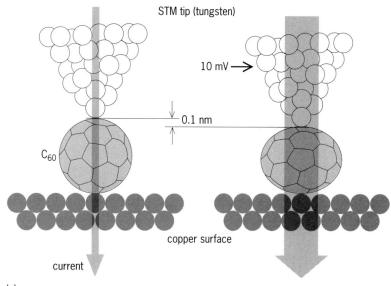

STM tip (tungsten)

10 mV →

0.1 nm

C_{60}

copper surface

current

(a)

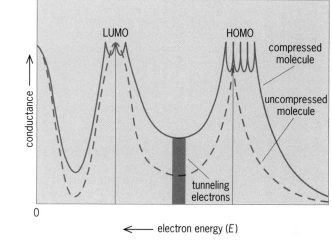

LUMO · HOMO

compressed molecule

uncompressed molecule

conductance

tunneling electrons

0

← electron energy (*E*)

(b)

Fig. 3. Effects of compressing the cage of a C_{60} molecule with the tip apex of a scanning tunneling microscope (STM). (*a*) Compression of 0.1 nm, producing a two-order-of-magnitude increase in the current intensity through the molecule. (*b*) Comparison of conductance spectra of compressed and uncompressed molecules, showing how external deformation causes broadening and shifting of the highest occupied and lowest unoccupied molecular orbital (HOMO and LUMO) levels. The vertical band indicates the path of the tunneling electrons from the Fermi level of the tip.

cate massive arrays of such wires just where they are needed is the focus of much research. A molecular relay based on similar concepts but with a differently shaped molecule has been proposed.

Molecular switches and other types of devices have been chemically synthesized from molecules that are mechanically interlinked. Although such devices are appealing in terms of their size, the fabrication of a reliable switch represents a major challenge in the quest for molecular-scale computers.

For background information *see* COMPUTER STORAGE TECHNOLOGY; FULLERENE; INTEGRATED CIRCUITS; MOLECULAR ORBITAL THEORY; NANOCHEMISTRY; NANOTECHNOLOGY; QUANTIZED ELECTRONIC STRUCTURE (QUEST); QUANTUM MECHANICS;

SCANNING TUNNELING MICROSCOPE; TRANSISTOR in the McGraw-Hill Encyclopedia of Science & Technology.   J. K. Gimzewski

Bibliography. S. Asai and Y. Wada, Technology challenges for integration near and below 0.1 μm, *Proc. IEEE*, 85:505-520, 1997; J. K. Gimzewski et al., Scanning tunneling microscopy of individual molecules: Beyond imaging, *Surf. Sci.* 386: 101-114, 1997; D. Goldhaber-Gordon et al., Overview of nanoelectronic devices, *Proc. IEEE*, 85:521-540, 1997; C. Joachim and J. K. Gimzewski, A nanoscale single-molecule amplifier and its consequences, *Proc. IEEE*, 86:184-190, 1998; E. S. Snow, P. M. Campbell, and F. K. Perkins, Nanofabrication with proximal probes, *Proc. IEEE*, 85:601-611, 1997; G.-L. Timp and R. E. Howard, Quantum mechanical aspects of transport in nanoelectronics, *Proc. IEEE*, 79:1188-1206, 1997.

Neuroethology

The study of the neural basis of animal behavior. Any type of behavior may be studied, but many researchers concentrate on complex behavior such as escape as opposed to simple movements such as scratching.

The field of neuroethology emerged in the 1970s. It combined neurobiology, the study of the nervous system, and ethology, the study of animal behavior. Neurobiology seeks to understand how the nervous system and its constituent cells work. Ethology seeks to understand behavior in the context of an animal's natural environment. Thus, neuroethologists study the neural basis of behavior in its natural context. Their research concentrates on biologically important behavior such as communication, reproduction, prey capture, and escape, and on explaining the means by which the nervous system is able to generate such behavior. Because the overt actions of an animal are the ultimate expression of the interactions of neurons in the nervous system, some investigators believe that careful study of a behavior and its natural context is a necessary step leading to a full understanding of the neural mechanisms that are responsible for it. Neuroethologists generally conduct studies in which the behavior of the whole animal is explicitly considered in their experiments; in which the animal retains at least limited freedom of movement; and in which sensory stimuli (if used) are chosen so as to be similar to the stimuli received by the animal in nature.

Escape. Behavior maneuvers that allow an animal to escape a predator have been a favorite subject for neuroethological research. An early study concerned how certain night-flying moths called noctuids evade predation by bats. Kenneth D. Roeder, considered by many the father of neuroethology, initiated this work in the 1960s. By combining laboratory studies of moth hearing with field studies of how these moths are able to avoid bats in nature, he set the standards for a reliable neuroethological study.

Bats hunt by emitting short ultrasonic cries as they fly. The bat hears the echoes of its cries bouncing off trees, moths, or other objects in its environment. When the bat hears an echo from a flying moth, it attempts to track and capture the moth, but is not always successful. *See* ECHOLOCATION.

Roeder showed in laboratory and field experiments that moths can hear bat cries at a distance of over 100 ft (30 m). A free-flying moth hearing the cry of a bat at a distance of about 20-100 ft (6-30 m) immediately turns and flies away from the sound. Since the echoes of objects as small as these moths are too faint for the bat to hear if the objects are more than about 20 ft (6 m) away, and since the bat rarely flies in a straight line for long, this strategy takes the moth away from the vicinity of the bat and minimizes the chance of being caught.

However, when a bat in its twisting flight encounters a moth within about 15-20 ft (4-6 m), the moth, upon hearing the bat, immediately engages in evasive maneuvers, twisting and turning sharply and unpredictably in its flight. The moth has no hope of outflying the bat in a straight race, but being smaller, it is much more maneuverable than the bat. Hence, evasion is its best strategy for escape, and is successful some of the time (**Fig. 1**).

Roeder's laboratory investigations revealed that the simple ears of moths contain only two sensory neurons each. One neuron has a low threshold of activation and begins to respond to sound at an intensity level that would be produced by a bat at a

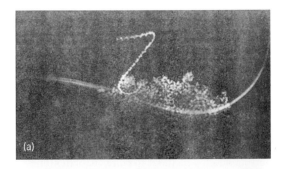

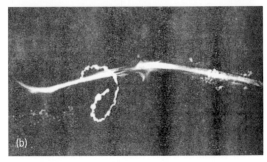

Fig. 1. Captured moths were tossed into the air in the vicinity of cruising bats. In these photographs the track of the moth appears as a fluttery path and that of the bat as a blurry streak. (*a*) The moth dives as it hears the bat entering from the left, and at the last instant makes a sharp turn to escape capture. (*b*) The moth attempts an evasive maneuver to avoid the bat flying in from the right. Since the track of the moth's flight terminates on the track of the bat, this moth did not escape.

distance of about 100 ft (30 m). The other neuron has a much higher threshold and does not begin to respond to a bat's cry until the bat is within about 15–20 ft (4–6 m) of the moth, when the cry is much louder. Roeder hypothesized that the difference in the moth's behavior when it heard a distant bat or a nearby bat was due to which sensory neuron was activated. Activation of only the low-threshold neuron induces a turn and steady flight away from the stimulus, whereas activation of the high-threshold neuron together with the low-threshold one overrides this response and causes erratic and unpredictable flight. Hence, an important and adaptive behavior could be explained, at least in part, as the result of the actions of certain neurons in the moth.

Sensory processing. Other neuroethological studies have revealed that land-based animals may also direct their escape away from the source of a potential threat. A cockroach, for example, can detect the movement of a toad's tongue or the swipe of a person's newspaper with delicate hairs on two appendages, the cerci, at the end of its abdomen. These hairs are exquisitely sensitive to the air currents generated by an object moving rapidly through the air. The air movement bends the hairs, which are connected to sensory neurons. Activation of these neurons causes excitation of fast-conducting neurons in the central nervous system, which in turn stimulate the motor control centers in the thorax that organize an escape run.

The ability of a cockroach to turn away from a potentially dangerous stimulus depends on the ability of the insect to detect the stimulus, determine the direction from which it comes, and react appropriately. Thirty years ago, it was thought that a cockroach that detected any air disturbance would simply start running straight ahead. However, behavioral and physiological experiments revealed a more accurate picture of events. First, the research showed that the hairs on the cerci are not only exquisitely sensitive but also polarized. That is, the hairs can respond to air moving slower that 12 mm/s, but do so only to air from a certain direction. Each hair sits in an oblong socket, which allows the hair to move mainly in one plane. Air moving along the length of the socket displaces the hair in its socket and excites a sensory neuron attached at the base. Air moving across the socket bends the end of the hair, but does not displace it in its socket sufficiently to excite the sensory neuron because the socket wall keeps it from moving very much. Different hairs have different orientations; hence, any air movement from a particular direction will activate only those hairs whose sockets are aligned more or less with the direction in which the air is moving.

Behavioral experiments in which cockroaches and hungry toads were placed together, revealed that the first movement of a cockroach under attack is not a straight-ahead run but a turn away from the toad. This initial turn is crucial to moving the insect away from the strike of the toad's tongue and therefore to survival (**Fig. 2**). The discrimination of wind direction

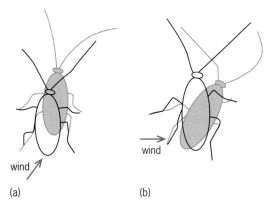

Fig. 2. A cockroach reacts to a wind stimulus (front legs are not shown). The outline shows the insect when the wind stimulus is first detected. The solid figure shows the insect's position a few milliseconds later. (*a*) When the air disturbance comes mainly from behind, both rear legs extend to move the insect forward. (*b*) When the disturbance comes from the left, the left rear leg extends, but the right rear leg flexes in order to turn the insect to the right.

becomes a factor. Hairs in similarly oriented sockets are sensitive to wind from the same direction; they activate sensory neurons whose axons extend to one or more specialized neurons in the insect's nerve cord. These neurons are called giant fibers because their axons are unusually thick, allowing conduction of nerve signals at high velocity. Some giant fibers receive input from hairs oriented in one direction, others from hairs oriented in a different direction. Air moving from a particular direction therefore preferentially excites only a few of the giant fibers, and it can be said that the pattern of neural activity of the giant fibers encodes wind direction.

The giant fibers in turn connect with other neurons in the insect's thorax. These neurons receive input from sense organs in and around the legs and integrate this input with the information from the giant fibers. The result is an appropriate response in leg muscles that will turn the insect away from the stimulus and initiate an escape run.

Behavioral choice. Neuroethological studies have illuminated other aspects of behavior, such as the basis of an animal's choice between two different actions. For example, toads and frogs strike at small moving stimuli but ignore or avoid large moving stimuli. In investigating the neural basis of this choice, researchers inserted electrodes (minute slivers of flexible wire connected to amplifiers) into selected parts of the brains of frogs and allowed the animals to look at moving squares of different sizes. The moving visual stimuli evoked strong activity in neurons in two brain regions. A small moving stimulus evoked a strong response in the brain region called the tectum. In contrast, large moving stimuli evoked a strong response in the pretectum, which is the posterior region of the thalamus, just anterior to the tectum.

Experiments in which small local lesions were made in the brain showed that the choice to approach or avoid a stimulus was made by interaction

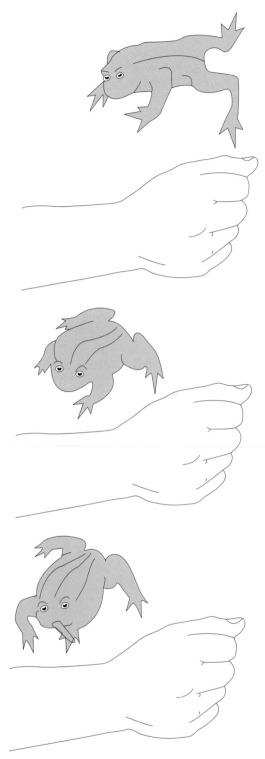

Fig. 3. A toad with a brain lesion in the pretectum will strike at virtually any large object, even the experimenter's hand.

of neurons in these two regions. Control of orienting and avoidance behavior seems to reside in the tectal and pretectal areas of the frog brain. All moving stimuli, not just small ones, excite neurons in the tectum. If the stimulus is sufficiently large, neurons in the pretectum are excited as well. These neurons inhibit the tectal cells, suppressing an approach toward the

stimulus and eliciting avoidance. If the regions of the pretectum that contain the neurons excited by large moving stimuli are destroyed, the frog loses all avoidance responses and if hungry will strike at anything that moves, even the hand of the researcher (**Fig. 3**).

Conclusions. Many neurobiologists are interested in the neural basis of behavior. The special contribution of neuroethologists has been to combine behavioral and physiological studies. By studying the behavior first, neuroethologists are able to identify its critical components, then to use this information to guide subsequent physiological experiments. Investigating the neural basis of a behavior without first fully characterizing the behavior can result in the misinterpretation of physiological results or in missing crucial elements of the neural mechanism underlying the behavior because subtleties of the behavior are not recognized.

Perhaps the most consistent finding of neuroethological studies is that behavior critical to an animal's survival is possible because of an integration of many elements. First, the animals must have a sensory system that is specialized to detect critical stimuli, whether it be an ear capable of detecting ultrasonic bat cries or hairs capable of detecting weak air movements. The specific role of the sensory system is to allow detection and discrimination of the appropriate sensory inputs. Second, the animals must have neural circuits capable of processing (interpreting) the sensory signals. In vertebrates such as toads, this processing takes place in the brain. In invertebrates such as cockroaches, it takes place elsewhere in the nervous system. Finally, the animal must have a motor system that can be adjusted and controlled to make rapid movements that are directed toward or away from an appropriate stimulus. The special value of an integrated neuroethological study is that it gives researchers a picture of how an entire sequence of actions is orchestrated, a picture that study of the individual components of the behavior in isolation would likely not reveal.

For background information *see* ECHOLOCATION; ETHOLOGY; NEUROBIOLOGY in the McGraw-Hill Encyclopedia of Science & Technology.   Fred Delcomyn

Bibliography. J. M. Camhi, *Neuroethology: Nerve Cells and the Natural Behavior of Animals*, 1984; J.-P. Ewert, *Neuroethology: An Introduction to the Neurophysiological Fundamentals of Behavior*, 1980; F. Huber and H. Markl, *Neuroethology and Behavioral Physiology: Roots and Growing Points*, 1983; K. D. Roeder, *Nerve Cells and Insect Behavior*, 1963; D. Young, *Nerve Cells and Animal Behaviour*, 1989.

Neurotechnology

Neurotechnology encompasses the application of microfabricated devices to achieve direct contact with the electrically active cells of the nervous system (neurons). Already used for some basic research in neuroscience, the technology is also being

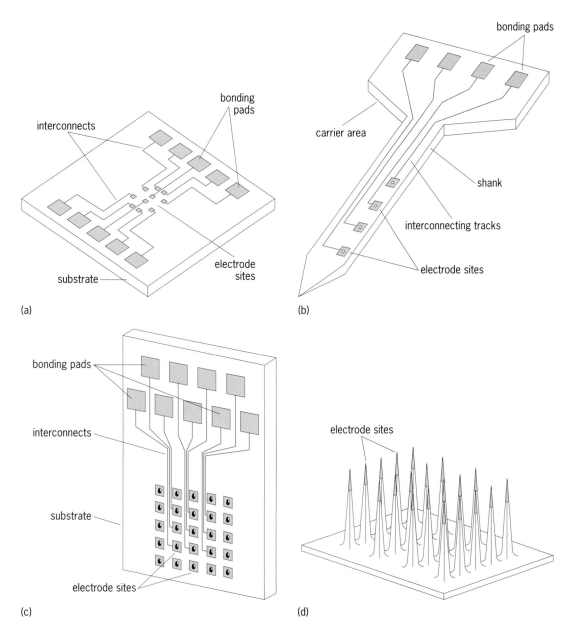

Fig. 1. Micromachined microelectrodes, not to scale. (*a*) Array-type microelectrode (*after IEE Colloquium on Medical Applications of Microengineering, 1996*). (*b*) Probe-type microelectrode (*after G. Ensell et al., Silicon-based microelectrodes for neurophysiology fabricated using a gold metallization/nitride passivation system, J. Microelectromech. Sys., 5(2): 117–121, 1996*). (*c*) Regeneration electrode; electrode sites have holes drilled through the device (*after IEE Colloquium on Medical Applications of Microengineering, 1996*). (*d*) Pin-cushion device, with electrode site at tip of each silicon needle.

developed for applications in neuroprosthetics.

Conventional approaches. Traditionally either metal wire microelectrodes or glass micropipettes have been used to study the electrical activity of individual neurons or small groups of neurons, either in intact tissue or in dissociated cell culture. The microelectrodes are fine metal wires that are insulated except for the electrode site at the tip. The micropipettes are glass capillary tubes that have been drawn to a narrow tip and filled with a conducting liquid. Tip dimensions are typically of the order of micrometers. While both approaches have been very successful, in some situations they are limited. It is difficult to accurately position large numbers of electrode sites (tips) in a small volume of tissue. Tip dimensions,

and hence electrical properties, are not easily reproducible. Bulk, fragility, and the difficulties involved in interconnecting many such microelectrodes hinder their implantation. This factor can be important when considering systems to replace or restore (with a neural prosthesis or orthosis, respectively) lost or impaired functions of the nervous system.

Neural prostheses and orthoses commonly utilize larger electrodes, which are placed either on the surface of the skin near the nerve trunk they are to electrically stimulate, or on the part of the brain or nerve trunk they are to stimulate or record from. In some cases, such as bladder voiding and exercise of paralyzed limbs, this approach has been successful. However, in others, such as visual prosthetics and

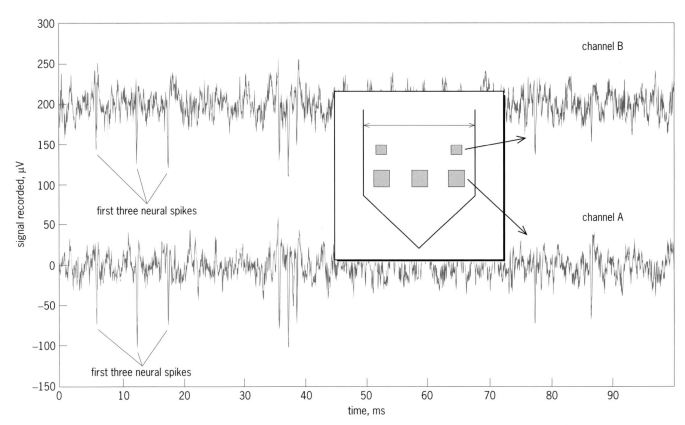

Fig. 2. Neural signals recorded through two adjacent electrode sites on the tip of a five-site probe (inset) from a peripheral nerve trunk of a locust. Channel A was recorded from a site 8 μm by 8 μm, and B from a site 4 μm by 4 μm. The two sites were spaced 20 μm apart. A 200-μV offset was added to channel B so that the two traces would not overlap in the plot. The first three neural spikes (signals) that are clearly above the noise level in each channel are indicated. This probe design was part of an investigation into the effects of site size and placement on the recorded signal. (*Data supplied by David Ewins, University of Surrey*)

paraplegic walking, more specific and intimate stimulation or recording of electrical activity is desirable.

Microfabricated devices. In the field of neurotechnology, microfabrication techniques that were originally developed in the semiconductor industry are applied to producing a variety of new microelectrode devices. Many of the devices fall into three general forms: array-type microelectrodes, probe-type microelectrodes, or regeneration electrodes (**Fig. 1***a–c*). Photolithography and microfabrication techniques allow many precisely defined electrode sites to be placed on these devices at known positions relative to one another.

Array-type microelectrodes. These consist of a flat insulating substrate upon which metal has been deposited and patterned to form electrode sites, bonding pads for connection to external circuitry, and tracks interconnecting the two (Fig. 1*a*). An insulating layer is deposited over the metal, and holes through this layer define the electrode sites and provide electrical access to the bonding pads. In some instances, transistors are fabricated at recording sites to provide immediate amplification of the signals. These devices were initially developed to monitor electrically active cells in culture. Cells would be grown on or placed over the array, which would form the bottom of a tissue culture chamber. The electrode sites would then be able to record electri-

cal activity from, or electrically stimulate, cells close to them (**Fig. 2**).

Such devices can potentially be used to explore electrical activity of neural networks in culture during development and learning. To some extent, they can also be used to study processing in semi-intact tissue. For instance, it is possible to dissect the retina from the eye of an amphibian, lay it on an array, and record simultaneous activity from up to about 100 individual neurons while a video screen is used to stimulate the photoreceptors. The data thus collected can be correlated with the images to explore how the retina processes information simultaneously in all dimensions (time, position, frequency, and intensity).

In one application, this technology is being developed as a retinal prosthesis. In some diseases, such as retinitis pigmentosa or age-related macular degeneration, photoreceptors in the eye cease to function. However, enough of the associated nerve cells remain responsive to electrical stimulation to make it reasonable to consider restoring some level of vision. It has been proposed that an array could be placed on the retina. A camera mounted on a pair of glasses would capture the image, and a laser could possibly be used to relay information and power to the array, which would stimulate neurons adjacent to the appropriate electrode sites. Although this is a very ambitious application, several research groups have started to work on it, and its potential has been demonstrated through brief experiments on human volunteers. However, it will probably be many years before a prototype system is developed.

Probe-type microelectrodes. These typically consist of one or more long, thin shanks that project from a larger carrier area (Fig. 1*b*). The shank, typically 1–2 mm long, 100 micrometers wide, and 15 micrometers thick, holds the electrode sites and is the part of the device that is inserted into the tissue. The carrier area holds the bonding pads and, in some cases, signal processing and conditioning circuitry. Probe-type microelectrodes will facilitate the investigation of processing in three-dimensional neural networks on the level of individual neurons. The development of active probes, that is, those with on-chip signal conditioning circuitry, will improve performance and reduce the problems of interconnecting many electrode sites. Such probes have been demonstrated, although further refinement is necessary.

While probe-type microelectrodes overcome many of the problems mentioned, their use in neurophysiology has been limited until recently by restricted access to the expensive facilities required for their fabrication. This contrasts with conventional microelectrodes, which are often fabricated by the researcher. The most readily available devices are currently passive, without on-chip signal conditioning electronics. It seems likely that their use will become more widespread as the technology is further developed. Areas being addressed include three-dimensional assemblies of multishank probes, active devices, and improved interconnection methods. Probes with integrated fluid channels that would allow responses to highly localized delivery of drugs or neurotransmitters to be monitored are also under development.

Regeneration electrodes. These are perforated devices (Fig. 1*c*). Peripheral nerve fibers, unlike those of the central nervous system, have some ability to regenerate after being cut. The regeneration electrode is designed to be placed between the surgically severed stumps of a peripheral nerve trunk. Nerve fibers then grow through the device, and electrode sites on the substrate provide an electrical interface. This type of device will have limited use for many neuroprosthetic or neurophysiological applications because a healthy nerve must be severed for its implantation. The main proposed application is for amputees. It is proposed that such devices can be utilized to provide control to prosthetic limbs. The control signals will be recorded from the nerve in which the device is implanted. Eventually, it may be possible to provide some kind of sensation from the limb. In this case, sensors on the limb will be used to direct stimulation of the nerve fibers.

The advantage of the regeneration approach lies in the fact that nerve fibers will probably remain in a fixed position with respect to the electrode sites over the life of the device, providing a relatively stable interface. Recording and stimulation have been demonstrated with prototype devices in animal models. However, the devices themselves provide a major impediment to nerve-fiber regeneration, with only a small fraction of fibers being in position to regenerate through a hole. Finding a solution to this problem that also fulfills the need for placement of more electrodes and possibly electronic circuitry on the device is likely to be difficult.

Visual prosthesis. Many novel devices cannot be classified as one of the above types. Among them, the pin-cushion device appears to be promising for visual prostheses (Fig. 1*d*). Arrays of relatively large electrodes have been implanted on the surface of the visual cortex of blind volunteers. These experiments demonstrated that such arrays could be accepted on the cortex for long periods of time, and also that stimulation with different electrodes elicited the impression of spots of light (phosphenes) in the visual field, although there was no obvious mapping between the stimulating electrode position and the apparent position of the phosphene. Problems faced by this approach included the need for high stimulation currents and the unpredictability of the results of simultaneous stimulation with closely spaced electrodes. It has been proposed that an array of fine needle-like electrodes inserted into the visual cortex could provide more specific stimulation and require less power. Initial experiments with human volunteers using simpler microelectrodes have suggested that it may be possible to partially realize some of the suggested advantages of a penetrating microelectrode array. However, much more experimental work is required to demonstrate that the approach is capable of providing any functionally useful kind of vision.

Status and challenges. Neurotechnology is beginning to make its impact in neurophysiological applications, and data acquisition and analysis systems are becoming more readily available, particularly array-type devices. There are, however, a number of significant problems that are likely to impede the long-term implantation of these devices and their use as neural prosthetics. These include the difficulty of passivating these devices (that is, depositing a thin film on the surface to isolate them from the environment) and encapsulating them so that they are biocompatible and the devices themselves are protected from the harsh environment in the body for long periods of time (10 or more years). For example, the long-term biocompatibility and reliability of continuously functioning devices have yet to be demonstrated. Problems regarding interconnecting, powering, and communicating with the devices also need to be addressed. The retinal-prosthesis projects have highlighted the need to match the mechanical properties of the device to those of the tissue in which it is implanted to avoid damage. The mechanical properties of the retina have been likened to wet tissue paper. These problems are now being addressed and progress is being made, particularly with regeneration electrodes and visual prosthetics.

For background information *see* BIOMEDICAL ENGINEERING; BIOPOTENTIALS AND IONIC CURRENTS; EYE

(VERTEBRATE); MEDICAL CONTROL SYSTEMS; NERVOUS SYSTEM (VERTEBRATE); NEURON; PROSTHESIS in the McGraw-Hill Encyclopedia of Science & Technology.

Danny Banks

Bibliography. D. Banks, Neurotechnology, *IEE Eng. Sci. Educ. J.*, 7:135–144, 1998; P. Heiduschka and S. Thanos, Implantable bioelectronic interfaces for lost nerve functions, *Prog. Neurobiol.*, 55:433–461, 1998; A. Manz and H. Becker (eds.), *Microsystem Technology in Chemistry and Life Sciences*, Springer-Verlag, 1998; Special issue, "Toward an Artificial eye," *IEEE Spectrum Mag.*, May 1996.

Neutrino

The recent discovery of neutrino oscillations, found in studies of atmospheric neutrinos, shows that neutrinos have a nonzero mass. The long-standing question of whether or not neutrinos have finite masses has finally been answered. Previously, there had been no firm evidence that neutrinos possess mass.

Since the present standard model of elementary particle physics, which describes the fundamental constituents of matter and their interactions, implies that neutrinos are massless, the new experimental evidence suggests that a new theory that goes beyond the standard model is needed. The finite neutrino mass may also change the present understanding of the universe and necessitate a reexamination of the evolution of the universe and of stellar objects. It especially affects views about dark matter, which is nonluminous and invisible but believed to constitute most of the matter in the universe.

Neutrino mass and oscillation. The neutrino was hypothesized in 1930 by W. Pauli to carry the missing energy required by the law of the conservation of energy in nuclear beta decays (the neutrino hypothesis). The existence of neutrinos was confirmed in 1956 when F. Reines and his colleagues detected neutrinos from a nuclear reactor.

Neutrinos are electrically neutral, subatomic particles. There are three kinds (flavors) of neutrinos, each associated with its charged partner: the electron, the muon, and the tau. The muons and taus are heavy electrons with masses 207 times and 3477 times larger than electron, respectively, and otherwise are identical to electrons.

Even in the 1930s the neutrino mass was known to be small through measurements of the maximum energy of electrons emitted in nuclear beta decay. The maximum energy is lowered by an amount corresponding to the mass of the neutrino. Today's much more sophisticated experiments place an upper limit on the mass of the electron neutrino that is about 10^5 times smaller than the electron mass.

If neutrinos have masses, it is possible that each neutrino state (flavor) is not a proper mass state but a mixture of those mass states. The electron neutrinos, muon neutrinos, and tau neutrinos may each have their own specific mixture. In a quantum-mechanical view, each mass state has its own wavelength determined by its specific mass and energy. If an electron neutrino, for example, is produced and moves through space, each of the mass states that compose the electon neutrino propagates with its own wavelength (phase). They become out of phase with each other while traveling. The mixture of the mass states which corresponded to an electron neutrino at the time of production becomes a different mixture which may correspond to a muon neutrino or a tau neutrino. This mixture changes periodically while neutrinos are traveling in the space; that is, it oscillates from one neutrino flavor to another, changing back and forth. This phenomenon is called neutrino oscillation in vacuum.

When the oscillation length (which is a function of the mass difference of the two proper mass states and the energy) is shorter than the distance from the source to the detector (**Fig. 1**, position *A*), only an averaged effect can be observed. The oscillation phenomenon is optimum when the oscillation length is similar to this distance (position *B*). If the oscillation length is much longer than the source-detector distance (position *C*), the effect can hardly be seen.

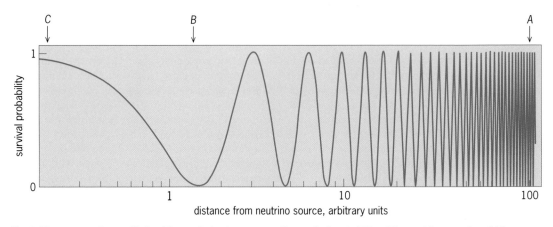

Fig. 1. Vacuum neutrino oscillation. The vertical axis measures the survival probability of the neutrinos produced. The horizontal axis measures the distance from the source of neutrinos to the detector. At position *A* the oscillation length is shorter than this distance, at *B* it is similar to this distance, and at *C* it is longer than this distance.

The neutrino oscillation can be identified either by the detection of an unexpected type of neutrino or by the deficit of the expected neutrinos at a reasonable distance. The neutrino oscillation gives information only about the mass difference between the two flavors and the mixing amplitude, and does not directly determine the mass itself.

Atmospheric neutrinos. In 1998, definite evidence of neutrino oscillations was found in studies of atmospheric neutrinos. Cosmic rays, which are high-energy particles from space, produce cascades of secondary particles when they strike the Earth's upper atmosphere. Those secondary particles subsequently decay and produce neutrinos and muons. The muon further decays and produces an electron, an electron neutrino, and a muon neutrino. Nearly all of those neutrinos pass through the entire Earth without any interactions. From counting the neutrinos produced in this decay chain, the ratio of muon-type neutrinos to electron-type neutrinos is expected to be 2:1 for the case in which the muons decay before they reach the surface of the Earth.

This ratio has been measured by several experiments, since the identification of electron neutrinos or muon neutrinos is relatively easy. The first measurement of such a flux ratio was made in 1988, and the result was 1.2:1. This deficit—suggestive evidence of neutrino oscillations—is called the atmospheric neutrino anomaly. Recently, this deficit has been confirmed with a high-statistics experiment.

In addition to this anomaly, the definitive evidence was obtained by measuring the flux as a function of distance from the production points. Cosmic rays enter the Earth very uniformly, and therefore the zenith-angle distribution of the incoming direction of atmospheric neutrinos is symmetric. This property does not depend upon how the neutrino flux is calculated. The zenith angle corresponds approximately to the distance from the neutrino source to the detector (**Fig. 2**).

Indeed, the measured zenith-angle distribution of electron neutrinos is very symmetric. However, the muon-neutrino distribution is very asymmetric

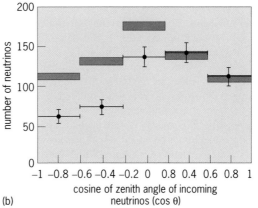

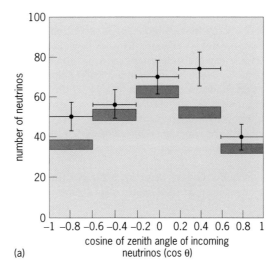

(a)

(b)

Fig. 3. Zenith-angle (θ) distribution of neutrinos observed at the Super-Kamiokande detector over a period of 535 days. Cross-bars show number of events observed in each range of values of cos θ, with the horizontal bars indicating the extent of the range and the vertical bars indicating the statistical uncertainty in the frequency of the neutrinos. Rectangles show the expected number of events predicted by a theoretical calculation; the thickness of the rectangles shows the range of uncertainty of the prediction. (a) Electron neutrinos. (b) Muon neutrinos.

(**Fig. 3**). The muon neutrinos coming up through the Earth, which have traveled about 13,000 km (8000 mi), are strongly suppressed, and those coming from above, with a flight distance around 10 km (6 mi), are not suppressed. This distance-dependence effect is seen only in the muon data, and cannot be explained by phenomena other than neutrino oscillations. The Earth is transparent to the atmospheric neutrinos, and therefore they are not reduced by absorption in the Earth.

The electron data suggest that the electron neutrinos are not oscillating. A recent experiment at a reactor, which can explore a similar neutrino mass range, also did not see the effect of electron neutrino oscillations. Therefore, the atmospheric neutrino oscillation is not an oscillation between muon neutrinos and electron neutrinos, but between muon neutrinos and tau neutrinos or another type of neutrino.

Solar neutrino problem. There is another indication that neutrinos oscillate: the long-standing solar

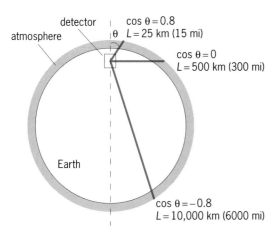

Fig. 2. Relation between zenith angle, θ, and the distance, L, between the point of production of atmospheric neutrinos and an underground detector.

neutrino problem. The problem is that the observed flux of neutrinos from the Sun is significantly lower than predicted. This problem has been known for longer than the atmospheric neutrino anomaly. The first solar neutrino experiment, using chlorine as a target, showed this deficit of solar neutrinos during the 1970s. The solar neutrinos are produced by a number of different nuclear reactions in the Sun and have a wide range of energies; the maximum solar neutrino energy is a little less than 20 MeV. Experiments based on the three different techniques which are used for detecting solar neutrinos cover almost the entire solar-neutrino spectrum, and show an energy-dependent suppression (about 30–50%) of the solar neutrino flux. This behavior is difficult to explain by modifying the standard solar models and strongly suggests the presence of neutrino oscillations.

It has been proposed that these oscillations of solar neutrinos originate in so-called matter-enhanced neutrino oscillations, or briefly, the MSW effect (after S. P. Mikheyeve, A. Yu. Smirnov, and L. Wolfenstein). When neutrinos are traveling through matter, they acquire an index of refraction, like light traversing matter. The index of refraction is related to the phase of the wave function and therefore, to neutrino oscillations. The solar neutrinos (which are electron neutrinos) have a greater ability to acquire this index of refraction than other neutrinos do. This difference enhances neutrino oscillations in matter and may resonantly convert electron neutrinos to another type of neutrino.

The current experimental results do not single out a particular solution to the problem of solar neutrino oscillations. There are several possible solutions, which include a vacuum-oscillation solution and MSW solutions.

The evidence for oscillations of solar neutrinos does not conflict with that for atmospheric neutrino oscillations. The two types of oscillations are complementary. One scenario is that the atmospheric neutrino oscillations are due to the oscillation between muon neutrinos and tau neutrinos, and the solar neutrino oscillations are due to the oscillation between electron neutrinos and muon neutrinos. This hypothesis will be tested by future experiments.

Consequences of neutrino mass. The neutrino masses are very small, but the discovery of this finite mass will have a major impact on both particle physics and astrophysics.

The immediate impact on particle physics results from the fact that the current standard model of elementary particle physics does not require massive neutrinos, and furthermore requires only left-handed neutrinos, spinning against their direction of motion, to explain all the interactions among elementary particles. The nonzero mass of neutrinos implies the existence of a reference frame moving faster than neutrinos, where neutrinos change their direction and thereby become right-handed. Therefore, the current model must be extended to include these right-handed neutrinos.

But the most significant impact on particle physics is related to the origin of the mass of neutrinos. The origin of a small neutrino mass is usually attributed to a so-called seesaw mechanism, in which the small mass is related to the inverse of a large mass scale, many orders of magnitude larger than the scale of the current particle physics model. Therefore, the finite neutrino mass naturally leads to the existence of a new huge energy scale that the standard model could never reach. A new theory beyond the standard model must be constructed to include physics at these very high energies.

Neutrino mass may also be related to the gravitational fate of the universe. A huge number of neutrinos, about 300 in every cubic centimeter, are believed to be left over from the early universe following the big bang. Even a small value of the neutrino mass would give rise to a huge amount of mass in the current universe.

The dark matter, which has not been identified, is believed to constitute more than 90% of the matter in the universe and to contribute significantly to the fate of the universe. The massive neutrinos may be a part of this dark matter. If the neutrino masses were about 1/50,000 of the electron mass (even though the mass differences could be much smaller), their contribution would be significant. *See* DEUTERIUM.

Prospects. The evidence for atmospheric neutrino oscillations is very strong, but the range of possible values of the oscillation parameters (mass difference and mixing) is broad. It is important to determine those parameters more precisely as well as to definitely determine the oscillation mode.

Experiments are being prepared in which beams of muon neutrinos from particle accelerators will pass through a few hundred kilometers of the Earth. These experiments will test the results from studies of atmospheric neutrinos and should pin down a smaller parameter region. Meanwhile, the ongoing and future solar neutrino experiments should identify a single solution for the solar neutrino oscillation problem by simultaneously observing the whole solar neutrino spectrum.

For background information *see* COSMIC RAYS; COSMOLOGY; ELEMENTARY PARTICLE; NEUTRINO; SOLAR NEUTRINOS; STANDARD MODEL; UNIVERSE in the McGraw-Hill Encyclopedia of Science & Technology.
Yoichiro Suzuki

Bibliography. Y. Fukuda et al., Evidence for oscillation of atmospheric neutrinos, *Phys. Rev. Lett.*, 81: 1562–1567, 1998; Z. Maki, M. Nakagawa, and S. Sakata, Remarks on the unified model of elementary particles, *Prog. Theor. Phys.*, 28:870–880, 1962; S. P. Mikheyev and A. Y. Smirnov, Resonance enhancement of oscillations in matter and solar neutrino spectroscopy, *Sov. J. Nucl. Phys.*, 42:913–917, 1985; C. Sutton, *Spaceship Neutrino*, Cambridge University Press, 1992; L. Wolfenstein, Neutrino oscillations in matter, *Phys. Rev.*, D17:2369–2374, 1978.

New carbon materials

Recent research with carbon has amplified knowledge of fullerenes, especially the C_{36} molecule and nanotubes. Carbon is an important element occurring in a staggering number of compounds, many with extensive industrial uses. Carbon is found in nature in two pure crystalline forms: as diamond, an insulator that is the hardest material known; and as graphite, a soft, electrically conducting material. Graphite is used in pencil lead and as a lubricant, among other things.

The difference between diamond and graphite lies in the bonding configuration of the carbon atoms. In diamond the carbon-carbon bonds lock the atoms into a strong three-dimensional (3D) covalently bonded network, while in graphite the carbon atoms are covalently bonded only into two-dimensional (2D) sheets. The sheets then stack one on top of the other to form a 3D solid. The binding between the sheets in graphite is due to relatively weak van der Waals bonds, and it is the easy sliding of one sheet over the next that gives graphite its slippery feel and lubricating qualities. Interestingly, the carbon-carbon bonds within a single sheet of graphite (sometimes called a graphene sheet) are even stronger than the 3D bonds in diamond, and so an ideal defect-free graphite sheet is theoretically the strongest (stiffest) material known. This property of graphite is exploited in graphite fibers, which can be made in different forms but often approximate sheets of (unfortunately highly defected) graphite scrolled up into long fibers with diameters of the order of micrometers.

Pieces of graphite sheets can be arranged to form small and perhaps more perfect carbon structures. In a graphite sheet, the carbon atoms are arranged in a honeycomblike lattice (**Fig. 1**). The bonds between the atoms form hexagons. It turns out to be geometrically impossible to roll or fold such a sheet, containing just hexagons, into a perfectly closed hol-

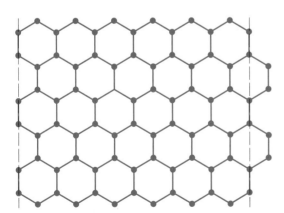

Fig. 1. Hexagonal "honeycomb" network, representative of arrangement of carbon atoms in a graphene sheet. The carbon atoms (dots) are at the vertices of the bonds. The broken lines delineate a strip that could be rolled into a tube; the particular tube segment thus formed would have six hexagons around its waist.

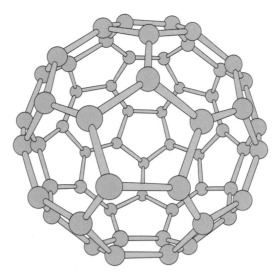

Fig. 2. Model of the buckyball, C_{60}. In this highly symmetrical carbon molecule, all 12 pentagons are isolated from each other.

low form. To accomplish closure, one must introduce pentagons. According to a theorem by Euler, one must introduce 12 pentagons to achieve a closed object. Suprisingly, this is independent of the number of hexagons in the final structure. The pentagons introduce the curvature needed for closure.

It might seem, then, that nature should abound with a huge number of different shell-like pure-carbon molecules, formed from various hexagon and pentagon arrangements of carbon atoms. This is certainly not the case. The reason is that the introduction of pentagons into a hexagonal carbon network is energetically very costly. Particularly costly is placing pentagons adjacent to one another. Therefore, the structures most likely to form would be those obeying an "isolated pentagon rule." Synthesizing even such isolated pentagon pure-carbon structures would necessarily require nonequilibrium methods, as graphite is the lowest-energy form of carbon and would be the natural product of any equilibrium growth process ("nonequilibrium" methods are of course also needed to produce diamond, which has a higher energy than graphite and is again, strictly speaking, metastable).

The field of pure-carbon cage molecules was born a decade ago when scientists synthesized and characterized small numbers of such structures in molecular beam experiments. An intense laser was used to vaporize a graphite target and create a nonequilibrium plasma or "soup" of hot isolated carbon atoms and carbon dimers, which upon rapid cooling coalesced into pure-carbon cage molecules. The most abundant and most stable of these molecules contained 60 carbon atoms in a highly spherical arrangement (**Fig. 2**). This molecule, C_{60}, was named buckminsterfullerene, or buckyball for short, by its discoverers (in honor of R. Buckminster Fuller, a practitioner of geodesic dome architecture). This discovery initiated the field of fullerenes, the name given to this entire class of pure-carbon cage molecules.

Higher- and lower-order fullerenes. The laser-vaporization synthesis method described above for fullerenes yields not only C_{60} but structures containing an even number N of carbon atoms less than or greater than 60. The structures with $N > 60$ are termed higher-order fullerenes, while those with $N < 60$ are termed lower-order fullerenes. C_{60} holds a special place in the fullerene family: it is the smallest fullerene which obeys the isolated pentagon rule. In fact, it has been suggested that C_{60} might be the smallest stable fullerene possible.

The laser-produced molecular beam experiments yield a rich spectrum of gas-phase fullerene molecules, including those with N greater than and less than 60, but that alone does not ensure that such molecules are stable enough to be produced in bulk, that is, in sufficient quantity that the molecules can be packed together into a solid. Achieving a solid (usually crystalline) arrangement of the molecules is of great importance for many characterization experiments, and having bulk quantities available is necessary for most applications as well.

Bulk quantities of fullerenes may be produced using a carbon-arc-generated plasma. This method easily produces significant amounts of C_{60}, and its application has been instrumental in the production and study of bulk samples of C_{60}. The solid form of pure C_{60}, which can be viewed as a molecular solid, is precisely that expected from close-packing hard spheres into a minimum volume. The diameter of a single C_{60} molecule is about 0.71 nanometer, and in the face-centered-cubic packing arrangement the C_{60} crystal lattice constant is 1.417 nm. The C_{60} crystal is held together by weak van der Waals bonds, and as such the overall solid is soft (although the buckyballs themselves are quite robust and hard). Interestingly, at room temperature the C_{60} molecules in the solid appear to be freely spinning about their equilibrium positions. Although it was first believed that these weak bonding properties might lead to superior lubrication properties, it turns out the the C_{60} molecular units are simply too small to act as efficient roller bearings between commonly machined surfaces.

The above method is currently the most efficient one not only for producing C_{60} but for producing bulk amounts of higher-order fullerenes as well, such as C_{70}, C_{76}, C_{82}, and C_{84}. It was believed that lower-order fullerenes were simply too unstable to be produced in bulk.

C_{36} fullerene. Lower fullerenes necessarily violate the isolated pentagon rule. However, a theoretical study determined that a spherical shell arrangement of 36 carbon atoms should nevertheless be stable. Of the many arrangements possible for placing 36 atoms in a shell using hexagons and pentagons, two have a particularly low energy and are thus the most likely to form. They are designated d_{6h} and d_{2d} (**Fig. 3**) to identify the particular symmetry of the molecule. The d_{6h} structure is simple: it consists of six hexagons wrapped around the belly of the molecule, followed

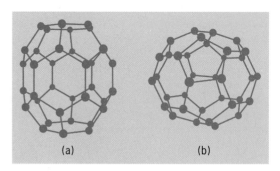

Fig. 3. Two isomers of C_{36}. (*a*) d_{6h}. (*b*) d_{2d}. Both versions contain adjacent pentagons. (*Courtesy of J. Grossman*)

by six adjacent pentagons on each cap. The very top and bottom of the caps are again a hexagon. The adjacent pentagons close the structure quickly, but at the expense of introducing much strain energy into the structure. The highly strained pentagon bonds are expected to be chemically very reactive. Thus, if C_{36} were formed, it should be particularly susceptible to bonding (with itself and with other atoms and molecules). This interesting (and possibly useful) feature has earned C_{36} the nickname "stickyball." In fact, because of its aggressive bonding nature, a solid formed from pure C_{36} is predicted to be very different from the van der Waals–bonded C_{60} solid. The C_{36} solid would be more covalently bonded, resulting in much greater mechanical strength. From theoretical predictions, the C_{36} molecule is 0.52 nm tall and 0.49 nm across at the waist. These dimensions make C_{36} significantly smaller than the C_{60} molecule (**Fig. 4**).

Bulk amounts of C_{36} can be synthesized using the nonequilibrium arc-plasma method. Scientists using such an approach, by "detuning" the synthesis conditions far away from those for optimal C_{60} production, were able to generate bulk quantities of C_{36}. One problem in producing the C_{36} is its propensity toward forming chemical bonds. If not handled properly, the material reacts with impurities in the environment, or even itself spontaneously polymerizes into a fused network of C_{36} molecules. This is of course exactly what is expected, given the high curvature and reactivity of the molecule.

So far, on an absolute scale only modest amounts of C_{36} have been produced, and many more experiments are needed to fully understand the properties of the isolated molecules and the C_{36}-based covalently bonded polymers and 3D solids. Interestingly, when thin films of C_{36} were produced, they were found to be extremely hard and scratch-resistant, quite unlike typical fullerene films or those based on other carbon types, such as amorphous carbon. This may have important industrial applications.

Additional theoretical studies have shown that C_{36} holds more surprises. It is predicted that for a particular crystal structure (arrangement of the molecules) of solid C_{36}, the material in its pure form will be electrically conducting, yielding an all-carbon metal. Furthermore, because of a favorable combination of suitable electronic states and interactions between the electrons and atomic vibrations within the solid, C_{36} might display superconductivity (the complete loss of electrical resistance) at moderately high temperatures. These predictions have not yet been confirmed. *See* SUPERCONDUCTIVITY.

Carbon nanotubes. If one of the caps of the d_{6h} C_{36} molecule is removed and an extra belt of hexagons is inserted, the molecule will be extended. Successive additions of such belts of hexagons will lead to a tubular structure, a carbon nanotube.

Geometrically, nanotubes can be constructed by cutting a strip from a graphite sheet and rolling this into a perfect seamless cylinder. The nanotube derived from the C_{36} molecule will be formed by using the strip outlined by the broken lines of Fig. 1. Obviously, this nanotube is not the only possibility. The geometry of a graphite sheet allows strips to be cut out not only with different (quantized) widths but also at different angles, leading to nanotubes with different diameters and chirality. An example of a chiral system is a barber pole or candy cane design; there is a certain helicity as one advances along the axis of the tube. As it turns out, the electrical properties of a carbon nanotube are extremely sensitive to the diameter and chirality of the tube. Some nanotubes are semiconductors, while others are metals.

A seamless defect-free nanotube would be a physical realization of the idealized graphite fiber discussed earlier. Such a system would have, in addition to the unusual electrical properties, outstanding mechanical and thermal properties.

Carbon nanotubes were first experimentally observed in high-resolution transmission electron microscopy studies of fullerene by-products. The nonequilibrium synthesis methods previously described for fullerenes are also quite efficient at nanotube production. Depending on the synthesis conditions (such as type of catalyst used), nanotubes with different geometries are produced (**Fig. 5**). Some are multiwalled, where many tubes are arranged coaxially and fit perfectly one inside the other (much like the collapsed segments of an old-fashioned mariner's telescope); others are single-walled. Often the single-walled tubes arrange themselves in an ordered bundle, resembling the closely packed strands of a steel cable. Nanotubes have also been produced from elements other than carbon, such as combinations of boron and nitrogen.

Typically, carbon nanotubes have an extremely high aspect ratio. They might be about 1–10 nm in diameter but many hundreds of micrometers in length. For comparison, if a typical garden hose of diameter 2 cm (0.8 in.) had the same aspect ratio, it would be several kilometers long. Despite their long relative length, nanotubes as currently produced are still too short for many mechanical applications, such as structural reinforcements. However, the unusual electronic properties have been exploited to form nanoscale electronic devices, such as rectifiers and transistors. Nanotubes might also be useful for energy storage systems. *See* NANOCHEMISTRY.

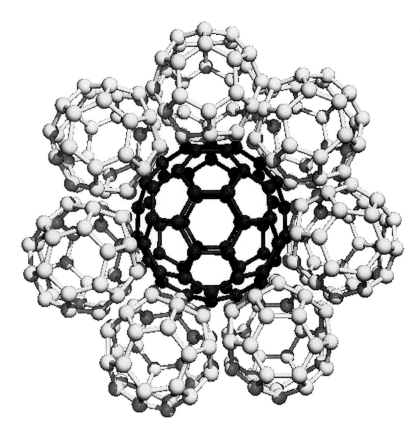

Fig. 4. Schematic representation of a C_{60} molecule surrounded by seven C_{36} molecules. C_{36} is significantly smaller than its buckyball cousin. (*Courtesy of C. Piskoti*)

Cut tubes. Part of the problem in exploiting nanotubes for many useful applications is that they often form a highly entangled mat. Untangling the mat is difficult, if not impossible, by known mechanical means. It is as if one had a hopelessly tangled ball of ultra-high-strength nylon fishing line. To extract useful pieces of the fishing line, it would be tempting to snip away at the ball with scissors and thus free up (shorter) pieces of the line.

The goal of systematically shortening tubes was recently achieved. Ultrasonic and chemical means were used to "cut" long pieces of nanotubes into shorter, manageable pieces. The cut tubes begin to resemble fullerene molecules. This molecular form for the tubes has tremendous advantages. For example, the short tubes dissolve in (or at least appear to be suspended in) various solvents, making available

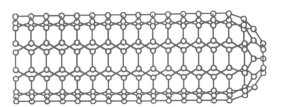

Fig. 5. Model of a carbon nanotube. The cylindrical body contains only hexagons, while the end cap contains some pentagons. Each end cap in a perfect tube would include six pentagons. (*After M. S. Dresselhaus, G. Dresselhaus, and P. C. Eklund, Science of Fullerenes and Carbon Nanotubes, Academic Press, 1996*)

many of the powerful techniques of wet chemistry. This allows, for example, different chemical species to be attached to the walls or ends of the tubes (such modifications are called functionalization), or it allows tubes to be easily mixed with other materials to form composites. It may even be possible, using appropriate templates or functionalization, to coax nanotubes in solution to self-assemble into useful patterns. This might be one way to make use of their unusual electronic and mechanical properties.

For background information *see* CARBON; FULLERENE; GRAPHITE; NANOCHEMISTRY; NANOSTRUCTURE; SCANNING TUNNELING MICROSCOPE; SEMICONDUCTOR; SUPERCONDUCTIVITY in the McGraw-Hill Encyclopedia of Science & Technology.   A. Zettl

Bibliography. M. S. Dresselhaus, G. Dresselhaus, and P. C. Eklund, *Science of Fullerenes and Carbon Nanotubes*, Academic Press, 1996; H. W. Kroto et al., *Nature*, 318:162, 1985; W. Krätschmer et al., *Nature*, 347:354, 1990; A. R. Kortan et al., *Nature*, 355:529, 1992; M. Cote et al., *Phys. Rev. Lett.*, 81:697, 1998; J. C. Grossmann et al., *Chem. Phys. Lett.*, 284:344, 1998; P. G. Collins et al., *Phys. Rev. Lett.*, 82:165, 1999; C. Piskoti, J. Yarger, and A. Zettl, *Nature*, 393:771, 1998; N. Hamada, S.-i. Sawada, and A. Oshiyama, *Phys. Rev. Lett.*, 68:1579, 1992; R. Saito et al., *Phys. Rev.*, B46:1804, 1992; S. Iijima, *Nature*, 354:56, 1991; S. Iijima and T. Ichihashi, *Nature*, 363:603, 1993; D. S. Bethune et al., *Nature*, 363:605, 1993; A. Thess et al., *Science*, 273:483, 1996; N. C. Chopra et al., *Science*, 269:966, 1995; P. G. Collins et al., *Science*, 278:100, 1997; S. J. Tans, A. Verschueren, and C. Dekker, *Nature*, 393:49, 1998; M. Bockrath et al., *Science*, 275:1922, 1997; J. Liu et al., *Science*, 280:1253, 1998.

Nobel prizes

The Nobel prizes for 1998 included the following awards for scientific disciplines.

Chemistry. Walter Kohn received one-half of the prize for his development of the density-functional theory. John A. Pople received the other half for his development of computational methods in quantum chemistry. Kohn is professor of physics at the University of California, Santa Barbara; and Pople is professor of chemistry at Northwestern University.

The theoretical work developed by Kohn helped to simplify the mathematics used to describe the bonding of atoms. Ignoring the motion of each individual electron, Kohn was concerned with the average number of electrons located at any one point in space. Called the the density-functional theory, it is a computationally simpler method, allowing for the study of very large molecules.

In 1964 Kohn was able to demonstrate a correlation between the total energy for a system as related to the known electrons' spatial distribution (electron density). He was able to theoretically calculate the energy depending on the density. It took several decades and modifications to this theory before the

equation for determining the energy could be used to study molecular systems of increased size. The density-functional theory is used in various chemical applications, from calculating the geometrical structure of molecules (providing bonding distance and angles) to outlining chemical reactions (such as enzymatic reactions).

Pople developed quantum-chemical methodology currently used in various branches of chemistry. His computational methods allowed for the theoretical study of molecules, their properties, and their behavior in chemical reactions. A computer is fed input data regarding specific details of a molecule or a chemical reaction. The output describes molecular properties or how a chemical reaction might take place. Results are generally used to explain or illustrate the outcome of various experiments.

As theoretical methodology was significantly improved at the end of the 1960s, Pople designed a computer program which was superior to others in a number of significant points. He made his computational methods easily accessible to researchers by designing the GAUSSIAN-70 computer program, first published in 1970. He refined the methodology during the 1970s and 1980s while producing advanced chemistry models. Pople included Kohn's density-functional theory in the GAUSSIAN program in the 1990s. The GAUSSIAN program is now used by chemists in universities and commercial companies the world over. These improvements are enabling chemists to analyze increasingly complicated molecules.

Physics. The physics prize was awarded to Robert B. Laughlin of Stanford University, Horst L. Störmer of Columbia University and Bell Laboratories (Murray Hill, New Jersey), and Daniel C. Tsui of Princeton University for their discovery of a new form of quantum fluid with fractionally charged excitations. This fluid is manifested in the fractional quantum Hall effect, which was discovered by Störmer and Tsui in 1982 and explained by Laughlin the following year. (All three were then at Bell Laboratories.)

If a strip of conducting material is placed in a magnetic field perpendicular to its surface, an electric current along the strip produces a voltage across it. This phenomenon, the Hall effect, results from the force of the magnetic field on the moving charge carriers (electrons or holes), and the Hall voltage is normally proportional to the magnetic field strength. However, in 1980 Klaus von Klitzing discovered that at low temperatures and high magnetic fields the Hall voltage of samples whose electrons are confined to motion in a plane varies with magnetic field strength in a series of steps. The values of the Hall resistance (the ratio of the Hall voltage to the current) are extremely close to a combination of fundamental physical constants divided by integers. This integer quantum Hall effect can be explained in terms of the quantum behavior of individual electrons, and reflects the filling of an integral number of quantum levels.

Störmer and Tsui studied the Hall effect using samples of very high purity, and at even lower temperatures and higher magnetic fields. They discovered many new resistance steps with values equal to the same constant divided by various fractions. Since fractionally charged particles have never been observed, this fractional quantum Hall effect can be explained only in terms of a strong correlation between the motions of the electrons. In Laughlin's theory, the electrons can be said to combine with the flux quanta of the magnetic field to condense into an incompressible quantum fluid, related to the quantum fluids of superconductors and liquid helium. Remarkably, coordinated motions of many electrons give rise to quasiparticles—entities that behave as if they were particles—whose charges are precise fractions of the electron charge. They account perfectly for the experimental results.

Recently, fractionally charged quasiparticles have been observed directly by measuring shot noise in tunneling currents. Fractional quantum Hall systems continue to be actively studied, revealing a rich array of phenomena, and provide a model for describing the effects of strong interactions in many different phases of condensed matter.

Physiology or medicine. Three pharmacologists shared the prize for their research on nitric oxide, demonstrating its function as a signaling molecule in the cardiovascular system. They are Robert F. Furchgott of the State University of New York Health Science Center in Brooklyn, New York; Louis J. Ignarro of the University of California, Los Angeles School of Medicine; and Ferid Murad of the University of Texas Medical School in Houston.

Nitric oxide is well known as a common air pollutant which is formed from the burning of nitrogen. In the human body, this gas is formed from the amino acid L-arginine by the enzyme nitric oxide synthase. Within the human body, nitric oxide has many different activities. In the circulatory system, nitric oxide functions to induce vasodilation, inhibit platelet aggregation, inhibit smooth muscle cell proliferation and migration, and maintain endothelial cell barrier function. Nitric oxide's many other physiological functions include neurotransmission and antimicrobial activity.

In 1977 Murad demonstrated that nitroglycerin (an unstable explosive chemical tamed by Alfred Nobel with his invention of dynamite) and related compounds relaxed smooth muscle through the production of nitric oxide. Simultaneously and independently, Furchgott found that endothelial cells, which line the interior of blood vessels, produced an unknown signal molecule that induced relaxation of vascular smooth muscle cells. This signaling molecule was termed endothelium-derived relaxing factor (EDRF). In 1980 Furchgott consulted Murad about these findings. Murad began to work on the endothelium and discovered that EDRF activates the same enzyme (guanylate cyclase) as nitric oxide. Further research revealed that both EDRF and nitric oxide are unstable and respond to the same inhibitors and activators. In 1986 Furchgott and Ignarro independently proposed that EDRF and nitric oxide are the

same molecule. Based on analysis of the two substances, these findings were later confirmed.

The work by Furchgott, Ignarro, and Murad has led to a great increase in nitric oxide research. It has been shown that the gas has many more functions than first believed, and plays a role in memory formation, tumor suppression, and immunity. Its therapeutic applications include bronchodilation (the result of bronchial smooth muscle relaxation) and the amplification of its vasodilation property to increase blood flow (for example, as targeted by new drugs to treat impotence).

Nuclear fusion

Since about 1951, physicists and engineers in many countries have searched for a practical way to use nuclear fusion reactions to produce electrical power. A fusion reaction occurs when two isotopes of hydrogen, deuterium (one proton and one neutron) and tritium (one proton and two neutrons), join together to produce helium plus a free neutron. Due to the difference in nuclear binding energies, about 17.6 megaelectronvolts (MeV) of energy is released in the form of the kinetic energy of the alpha particle (3.5 MeV) and the neutron (14.1 MeV). As an energy source, fusion has several important advantages. Enough tritium fuel could be produced from lithium to power the world for many centuries, and more advanced fusion reactors operating with the more difficult deuterium-deuterium reaction could power the world for billions of years since deuterium is a component of all water. Because no fission products or actinides are produced, the waste can be much safer to store than that from a fission plant. The fusion reactor's structure would be activated by the neutrons, but proper choices of the construction materials could result in fusion reactors whose activity would rapidly decay after decommissioning. Moreover, because fusion does not occur as a chain reaction like fission does, and because the total energy stored in even a very powerful fusion reactor is small, malfunctions cannot lead to uncontrolled increases of the power output.

Making fusion a practical power source is probably the most difficult technological challenge ever attempted. The basic difficulty arises from the fact that, like all atomic nuclei, the deuterium and tritium nuclei both carry positive electrical charge and thus repel each other. In order to have a useful probability of fusing, the nuclei must approach each other with energies of at least several kiloelectronvolts (1 keV corresponds to a temperature of about $1.06 \times 10^7\,°C$). The Sun, like all stars, uses different (less energetically favorable) fusion reactions to produce its power. The Sun's core is believed to operate at a temperature of 1–1.5 keV. However, the Sun is very large by human standards. In order to make a reactor of practical size for a power plant, conditions more extreme than those in the Sun's core must be produced.

Magnetic confinement. Most controlled fusion research has been directed toward using a magnetically confined plasma as the reacting medium. Plasma is the fourth state of matter, in which the electrically negative electrons have been detached from positively charged ions. The plasmas of interest for fusion have approximately equal amounts of positive and negative charge within them, so that the plasma as a whole is nearly electrically neutral. Because all of the ions and electrons are charged, they will spiral around the lines of force of an applied magnetic field. Thus, a suitably shaped magnetic field can be used to confine a very hot plasma and prevent it from touching the material walls of the vacuum vessel. If the plasma were to touch the wall itself, the impurities it ingested would extinguish the plasma within a minute fraction of a second. In order to produce a practical magnetically confined plasma reactor, the plasma must be heated to temperatures of 10–20 keV (several times the temperature of the Sun's core), and the energy leakage across the confining magnetic field must be sufficiently small to allow the plasma to rise to a temperature where its fusion reactions release much more energy than is being applied to it.

Energy leakage. Accordingly, magnetic confinement research has been dominated by a quest to understand and control the energy leakage across the confining magnetic field, and by the development of methods to heat the confined plasma. Most of this research since about 1970 has concentrated upon toroidal geometries in which the magnetic field lines responsible for confinement close upon themselves. The many ions and electrons gyrating along these field lines scatter off each other, giving rise to transport of energy and particles across the confining magnetic field. The energy leakage associated with these two-body scattering events can be made quite low in toroidal confinement systems such as tokamaks or stellarators. Both these devices produce a toroidally spiraling magnetic field in which the pitch of the spiral changes from one flux surface to the next as one goes from the outside to the center of the plasma. In a tokamak, these field lines are produced partly by a toroidal array of external coils and partly by an electrical current flowing within the confined plasma. In a stellarator, the detailed structure of the confining fields is governed to a larger extent by carefully shaped external coils.

If the energy leakage were due only to the cross-field transport arising from two-body collisions, which is referred to as neoclassical transport, then the problem of attaining sufficiently good energy confinement for a reactor would have been challenging but conceptually straightforward. Instead, the first decades of magnetic confinement research encountered energy leakage across the magnetic field which was many times greater than predicted by neoclassical theory. This leakage was referred to as anomalous transport because its origin was not well understood, but was suspected to arise largely from various types of turbulence. Turbulence, which could be driven by various wave instabilities, entrains large numbers of particles over significant distances in the plasma, which means that it can produce a large energy leak.

Improved confinement regimes. Within recent years a great deal of progress has been made both experimentally, in discovering new operating regimes with improved energy confinement, and theoretically, in developing an understanding of how these new regimes reduce energy leakage across the magnetic field. These enhanced confinement regimes allowed tokamaks for the first time to release several megawatts of fusion power when a deuterium-tritium fuel mix was used in the TFTR tokamak at Princeton, New Jersey, and in the JET tokamak at Abingdon, United Kingdom. The largest Japanese tokamak, JT-60U, located in Naka, has used one of these advanced confinement operating regimes to achieve a fusion energy output in a deuterium plasma which for the first time reached equivalent scientific breakeven. Scientific breakeven is defined as the condition during which the output of fusion energy equals the external heating energy being supplied to the plasma. "Equivalent" indicates that the JT-60U tokamak is not equipped to use tritium, so the actual energy released by the fusion of deuterium nuclei with each other was multiplied by a factor to account for the large increase in fusion reactivity which would have occurred if a mix of deuterium and tritium had been used.

Most of these improved confinement operating regimes are characterized by localized transport barriers occurring either near the outer edge of the plasma or deeper within the plasma. The energy leakage across these transport barriers is less, in some cases much less, than that which occurs in a standard confinement discharge. Diagnostic measurements reveal that these barriers exist where there is rapid variation, as a function of the plasma minor radius, in parameters such as the radial electric field strength, the plasma rotation velocity around the minor radius, the density, and the temperature. In some cases, experimental measurements reveal a drop in the magnitude of plasma fluctuations within these regions, suggesting that the turbulence is suppressed. Based in part upon these experimental results and in part upon results from theory and computational modeling, an understanding has emerged of the mechanism by which these transport barriers reduce energy leakage. A strong radial electric field builds up along some of the magnetic flux surfaces. This growth can be driven by several mechanisms, including differences in loss from the plasma of positive charges versus negative charges, plasma flows, and pressure gradients. If the parameters driving the electric field production vary strongly from one flux surface to the next, this electric field will have shear and vary in strength. This electric field interacts with the magnetic field to produce sheared plasma rotation, in which the rotational velocity varies from one flux surface to another. The sheared rotational velocity rips apart turbulence structures connecting the inner plasma with the outer plasma, thereby suppressing the heat leakage arising from turbulent transport.

The most successful field configuration of this sort was observed on the TFTR tokamak to result in the total or near-total suppression of anomalous energy loss among the confined ions, meaning that essentially the only leakage through the ions was the minimum possible conduction arising from two-body collisions. Under these conditions, the particle confinement also appears to become nearly perfect within the transport barrier, although the energy leakage through the electrons remains higher than that which would arise solely from two-body collisions. This configuration is being studied on such tokamaks as the DIII-D tokamak in La Jolla, California, and the JT-60U tokamak, which has succeeded in maintaining the condition for several seconds. If it should prove possible to fully exploit the excellent confinement of this operating mode, it might be possible to design fusion power reactors which would be smaller and cheaper than would be the case with designs using conventional scaling of confinement time.

Plasma heating. The science and technology of heating the plasma and driving current have also pro-gressed. Most of the externally supplied heating power in tokamaks comes either from radio-frequency waves or from beams of energetic neutral atoms, which are ionized in the plasma and confined by the magnetic field while they give up their energy through collisions to the bulk plasma. Until recently, all of the neutral beams were produced from positive-ion beams which picked up an electron in a gas cell after electrostatic acceleration to become neutral. However, this neutralization process becomes very inefficient at the beam energies which will be needed for large tokamak reactors. If, instead, negative ions are accelerated and the extra electron is then stripped away to produce energetic neutral atoms, the efficiency remains high over a very large range of beam energies and well beyond any requirement for a fusion reactor. The hindrance has been that negative ions of hydrogen isotopes are much more difficult to produce in large quantities than are positive ions, and they are also harder to accelerate without heavy losses.

In 1996, the first negative-ion-based neutral-beam system went into operation on the JT-60U tokamak. As the important physical processes governing the performance of this system are becoming

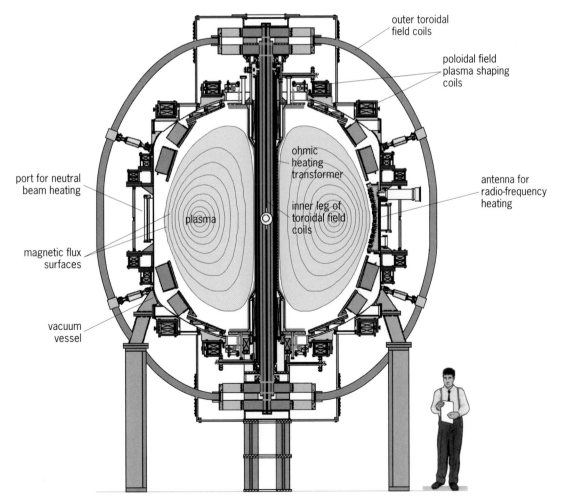

National Spherical Torus Experiment under construction at the Princeton University Plasma Physics Laboratory in New Jersey. (*Drawing by J. Robinson*)

understood, its capabilities are increasing. It has injected several megawatts of beams at energies up to 360 keV into the plasma. Another negative-ion-based neutral-beam system also began to come on-line in 1998 at the Large Helical Device, a type of stellarator located near Nagoya, Japan.

Alternative configurations. Although much progress has been made in understanding, controlling, and improving the energy confinement properties of tokamaks, there remain a number of concerns, such as whether the minimum unit size for a commercial reactor will be large. This concern is spawning experimental and theoretical efforts in alternative magnetic configurations such as compact stellarators and spherical toruses. A spherical torus lies at the low-aspect-ratio end of the tokamak family of designs. Aspect ratio is the ratio of the major radius to the minor radius of the plasma toroid, where the major radius is measured from the center of the machine to the center of the magnetic flux surfaces, and the minor radius is the average of the distance from the center of the flux surfaces to the plasma boundary. Spherical toruses may lend themselves to smaller unit sizes for commercially attractive fusion reactors, and they can operate with much lower externally applied magnetic fields than conventional tokamaks of larger aspect ratios, a capability which should reduce capital costs. Two spherical torus experimental devices are under construction at the Plasma Physics Laboratory of Princeton University and at the Culham Laboratory in England. The **illustration** shows the design of the National Spherical Torus Experiment, which is scheduled to begin operation at Princeton in 1999. The plasma can be heated inductively by the ohmic heating transformer, by neutral-beam injection, or by radio-frequency heating. The toroidal field coils produce the magnetic field component which runs the long way around the torus. The poloidal field coils produce part of the magnetic field component which runs the short way around the plasma toroid. (The rest of this field component is produced by the electric current flowing in the plasma.) The poloidal field coils also control the shape of the plasma.

Inertial confinement. Aside from magnetic fields, the other long-researched branch of confinement employs the inertia of a strongly compressed capsule of fusion fuel to confine the dense plasma long enough for fusion reactions to occur. This is similar to the process which works so well in hydrogen bombs, but it has been much more challenging to reproduce in a controlled and energy-efficient fashion for a reactor. Present experiments employ arrays of high-power lasers to drive implosions of the fuel capsules. The National Ignition Facility, which is under construction at the Lawrence Livermore National Laboratory in California, will be by far the largest and most powerful such laser array ever built.

For background information *see* MAGNETOHYDRODYNAMICS; NUCLEAR FUSION; PLASMA PHYSICS; TURBULENT FLOW in the McGraw-Hill Encyclopedia of Science & Technology.   Larry R. Grisham

Bibliography. B. A. Carreras, Progress in anomalous transport research in toroidal magnetic confinement devices, *IEEE Trans. Plasma Sci.*, 25:1281–1321, 1997; L. R. Grisham et al., The scaling of confinement with major radius in TFTR, *Phys. Rev. Lett.*, 67:66–69, 1991; L. R. Grisham et al., Scaling of ohmic energy confinement, *Phys. Plasmas*, 1:3996–4001, 1994; R. J. Hawryluk, Results from deuterium-tritium tokamak confinement experiments, *Rev. Mod. Phys.*, 70:537–585, 1998.

Nuclear structure

Nuclear structure studies with electrons have entered a new phase of development. In addition to an ongoing program of selected experiments at very high energy at the Standard Linear Accelerator Center (SLAC), an 800-MeV continuous-beam microtron is in operation at Mainz, Germany, with a complement of spectrometers for detecting the scattered electrons and a tagged-photon facility. The Bates Laboratory at the Massachusetts Institute of Technology in Cambridge, where pioneering studies in this field have been carried out since the early 1970s, now has a storage ring where large circulating currents can be directed on a variety of thin gas targets. This configuration has led to an entirely new class of experiments, as first demonstrated in Novosibirsk, Russia, and at the NIKHEF Laboratory in Amsterdam.

The leader of this field is the Thomas Jefferson National Accelerator Facility (TJNAF, formerly known as the Continuous Electron Beam Accelerator Facility, or CEBAF) in Newport News, Virginia. Since the mid-1970s such a facility has been a top priority for the field of nuclear physics in the United States and has formed an integral part of the Long-Range Plan for the field. The facility is now fully operational and carrying out its physics program. TJNAF produces a 4-GeV beam of electrons using superconducting niobium cavities cooled by superfluid helium. It delivers 200 microamperes of electrons to three independent experimental areas which are equipped with a pair of high-resolution spectrometers, a high-momentum spectrometer and various additional detectors, and a large-acceptance spectrometer, respectively (**Fig. 1**).

Nuclear physics background. The nucleus is a unique form of matter, consisting of many nucleons (protons and neutrons) in proximity. All of the forces of nature are present in the nucleus—strong, electromagnetic, and weak. Even gravity is important in the case of neutron stars, which are enormous nuclei held together by gravitational attraction. The nucleus provides a unique microscopic laboratory to test the structure of the fundamental interactions. The nucleus manifests remarkable properties as a strongly interacting quantum-mechanical many-body system. Furthermore, most of the mass and energy in the visible universe comes from nuclei and nuclear reactions. In addition, there are underlying degrees of freedom in the nucleus—quarks and gluons—interacting through the remarkable forces described by quantum chromodynamics (QCD). Thus, the

Fig. 1. Aerial view of Thomas Jefferson National Accelerator Facility (TJNAF) site, Newport News, Virginia. (*Courtesy of Mary Beth Stewart*)

nucleon itself is a complicated nuclear many-body system. Nuclear physics is also crucial for understanding various aspects of the universe, for example, the early universe, the formation of the elements, supernovas, and neutron stars. In sum, nuclear physics is really the study of the structure of matter.

Advantages of electron scattering. There are several reasons for using the scattering of energetic electrons to study the nucleus. First, the interaction is known. It is given by quantum electrodynamics (QED), the most accurate physical theory ever developed. A known entity is being measured; the interaction is with the local charge and current density in the target. Second, the interaction is relatively weak, of order $\alpha \approx 1/137$, the fine-structure constant, so that measurements can be made on the target without greatly disturbing its structure. If only the electron is observed, there are three variables in electron scattering: the three-momentum transfer squared, the energy transfer, and the scattering angle. It is possible to map out the Fourier transform of the transition charge and current densities in the nucleus by varying the three-momentum transfer at fixed energy transfer. Upon inversion of the Fourier transforms, the microscopic spatial distribution of these quantities can be identified.

In effect, an electron accelerator and its detectors form a huge microscope for looking at the nucleus. The situation is directly analogous to the diffractive scattering of light from a small aperture. The macroscopic diffraction pattern seen on a screen and the knowledge of the wavelength of the light make it possible to deduce the properties of the tiny aperture itself. In electron scattering, the three-momentum transfer and wavelength with which the spatial structure is probed bear an inverse relation to each other. The scale of interest for nuclei is tens of fermis (1 fermi = 10^{-15} m), down to tenths of fermis, where quarks and gluons play an explicit role. To achieve these conditions, momentum transfers ranging from hundreds of MeV/c to many GeV/c are needed (where c is the speed of light).

By varying the scattering angle at fixed values of the three-momentum transfer squared and the energy transfer, the polarization of the virtual quantum of radiation (a virtual photon) exchanged between the electron and nucleus is varied. Through this procedure, it is possible to separate the electron's interaction with the charge and current in the target. Furthermore, in addition to the electron's electromagnetic interaction with the target through the exchange of a virtual photon, there is a weak interaction through the exchange of a heavy neutral Z boson. This interaction is negligible in the nuclear domain unless searches are made for effects that are completely absent in quantum electrodynamics; parity violation is one such effect. The difference is cross section for right- and left-handed

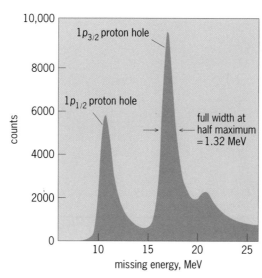

Fig. 2. Missing energy spectrum in coincidence reaction $^{16}O(e,e'p)^{15}N$ taken at an incident electron energy of 2.445 GeV and four-momentum transfer squared of $Q^2 = q^2 - (\omega/c)^2 = 0.81$ (GeV/c)^2, where q^2 is the three-momentum transfer squared, ω is the energy transfer, and c is the speed of light. Overall energy resolution in ^{15}N is $\Delta E/E = 1.32/2445 = 5.4 \times 10^{-4}$. (*TJNAF, Experiment E89-033, C. Perdrisat*)

polarized electrons, a difference present only if parity is violated, measures the interference between the electromagnetic amplitude and that arising from the weak neutral current. From such measurements, the Fourier transform, and hence the detailed microscopic spatial distribution, of the weak neutral current in the nucleus can be deduced. Such knowledge effectively doubles the power of electron scattering.

Coincident electron scattering. The new generation of continuous-beam electron accelerators allows coincidence experiments, where other reaction products can be detected in coincidence with the scattered electron. Coincident electron scattering can provide much new information. First, measurement of the energy and momentum of an ejected proton makes it possible to determine the Fourier transform of the initial proton wave function in the nucleus, and its energy. In this way, the nucleus can be examined orbit by orbit. Second, polarization measurements, which involve a second scattering, can now be made. Furthermore, interference terms involving interesting small amplitudes can be determined. New reactions, such as the production of a K^+ meson, can be examined; this particular reaction leaves a hypernucleus with negative strangeness in a precisely defined configuration. In addition, although the basic electromagnetic interaction in a nucleus occurs with quarks, it is the hadrons (mesons and nucleus) which are actually emitted; in free space, quarks are confined to the interior of hadrons. Coincident electron scattering on a nucleus provides an ideal laboratory to study quark hadronization.

Applications. Of the numerous current applications, a few examples will be briefly discussed. The deuteron consists of a neutron and a proton; it is the only bound state of the two-nucleon system. Elas-

tic scattering measurements out to very high values of the three-momentum transfer squared, together with measurements of the tensor polarization of the recoiling deuteron, provide an unprecedented picture of this fundamental nuclear system. Although the neutron has no net charge, the positive charge and negative charge are slightly separated inside of it. The charge distribution inside the neutron is being measured with polarization transfer and polarized target experiments on the deuteron. (There is no free neutron target.) Such measurements, as do those on the proton, form benchmarks for quark models of the nucleon and, more fundamentally, for the solution of quantum chromodynamics in the strongly coupled nuclear domain.

The internal dynamics of the nucleon are being studied through coincidence experiments on the decays of its resonant excited states. Such studies provide additional benchmarks for quantum chromodynamics.

The reaction whereby a polarized electron incident on a nucleus produces a scattered electron and an ejected polarized proton, makes it possible to study how the nucleon spin propagates out from the nuclear interior. In this way, direct test is available of relativistic models of nuclear structure which describe the spin dependence of the nuclear shell model. A missing-energy spectrum (that is, the energy of proton removal) taken at TJNAF for this reaction on an oxygen-16 (^{16}O) target producing nitrogen-15 (^{15}N) in the unpolarized case is shown in **Fig. 2**. The two peaks are proton holes in the shell structure of ^{16}O.

A parity-violation experiment to determine the distribution of weak neutral charge in the proton has

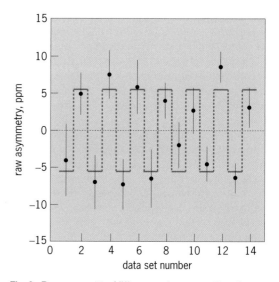

Fig. 3. Raw asymmetry (difference of cross sections for right- and left-handed electrons divided by the sum of these cross sections) observed with a longitudinally polarized electron beam on a proton target. Average value of raw asymmetry is shown for each data set. Alternating data sets have a half-wave plate inserted in the laser beam (at the injector) and are expected to have the opposite asymmetry. Horizontal lines are the best fit to the data. Raw asymmetry determined from this experiment is $A_{raw} = -5.64 \pm 0.75$ parts per million. (*TJNAF, HAPPEX Experiment, M. Finn*)

been carried out at TJNAF. The experiment used a longitudinally polarized electron beam on a proton target. The asymmetry in the cross sections for right- and left-handed electrons was measured while the incident photon polarization was reversed at the injector on a macroscopic time scale, using a half-wave plate, and nothing else was varied (**Fig. 3**). The asymmetry depends on the weak neutral form factor, which measures the spatial distribution of this quantity. The current understanding is that in the nuclear domain, with only the light u and d quarks and their antiquarks, the weak neutral charge distribution should be identical to that of the electromagnetic charge. Any difference must arise from s (heavy) quarks. No difference was found, a result that has profound implications for the understanding of the structure of matter.

For background information *see* NUCLEAR STRUCTURE; PARITY (QUANTUM MECHANICS); PARTICLE ACCELERATOR; QUANTUM CHROMODYNAMICS; QUANTUM ELECTRODYNAMICS; QUARKS; SCATTERING EXPERIMENTS (NUCLEI); WEAK NUCLEAR INTERACTIONS in the McGraw-Hill Encyclopedia of Science & Technology.   John Dirk Walecka

Bibliography. F. Gross and R. Holt (eds.), 1992 CEBAF Summer Workshop, *AIP Conf. Proc.*, no. 269, 1992; J. D. Walecka, *Theoretical Nuclear and Subnuclear Physics*, 1995.

Nuclear testing

A global ban on the explosive testing of nuclear devices is the motivation behind the Comprehensive Test Ban Treaty. The treaty is viewed by its advocates as a means to avoid nuclear warfare and to lessen international tensions generally. Practically speaking, it is assumed that any nation with access to plutonium can fabricate, without explosive testing, a nuclear weapon like that dropped on Nagasaki by the United States in 1945. A ban on explosive testing deters an emerging nuclear state from developing more sophisticated weapons that employ plutonium fission or thermonuclear fusion or that are sufficiently compact to deliver on a missile. For countries with nuclear arms, a ban on testing does not render their weapon stockpiles ineffective. Rather, the ban is intended to deter them from modernizing weapon design. The test ban treaty is only one tool available to the international community to deter the proliferation of nuclear weapons. Export controls on fissile materials and nuclear technology, safeguards for weapon stockpiles, and regulations on plutonium disposal from nuclear power plants are also necessary.

For the treaty to be effective, compliance by the signatory nations must be verifiable. An International Monitoring System (IMS) has been established to detect and locate nuclear explosions worldwide (see **illus.**). Buried explosions send shock waves through the rock of the Earth's interior; the IMS detects these with a seismometer network that overlaps existing networks for monitoring earthquake activity. Explosions at sea generate sound waves that can travel the global ocean; the IMS detects these with hydroacoustic sensors. The IMS monitors atmospheric vibrations associated with nuclear explosions with an infrasound sensor network. It also employs radionuclide sensors to detect minute amounts of radioactive fallout from explosions that are not completely contained underground.

Status of test ban treaty. The treaty has been signed by over 150 nations. Over 30 nations have also ratified the treaty, including 17 of the 44 nations deemed nuclear-capable. The treaty takes effect once ratified by all nuclear-capable nations. September 24, 1999, was set as a deadline after which a conference would be convened to discuss ways to facilitate ratification. Hearings on ratification in the United States have been delayed by political linkages to other foreign policy issues, such as revisiting the antiballistic missile defense treaty and implementing the 1997 Kyoto Protocols on global warming. India and Pakistan have indicated a willingness to sign the treaty after some conditions have been met.

International Monitoring System. The Comprehensive Test Ban Treaty Organization (CTBTO) maintains an ongoing monitoring effort in Vienna, Austria, funded through the United Nations. The central clearing house for IMS data is the International Data Center (IDC), also in Vienna. Data analysis procedures and preliminary monitoring efforts have been performed since January 1995 at the prototype IDC in Arlington, Virginia. Signatory nations may also maintain National Data Centers (NDCs) to collect IMS data for submittal to the IDC, and to serve as an auxiliary outlet for the IDC data products. The United States' NDC is sited with the Air Force Technical Applications Center at Patrick Air Force Base, Florida. The IMS, when complete, is projected to collect data from sites in nearly 90 countries.

Seismic monitoring. Both earthquakes and underground nuclear explosions generate vibrational waves that propagate worldwide. Their locations can be found by timing the wave arrivals at a network of seismometers. Measurements of ground motion in three dimensions can determine the arrival direction of the wave. Fast computers can process data from an array of closely spaced sensors to improve arrival-direction estimates and to enhance weak signals in ambient ground noise. Underground nuclear tests can rival moderate earthquakes in size; the May 21, 1992, thermonuclear test by China rated magnitude $M = 6.6$ on the Richter scale. A nuclear device similar to the enriched uranium bomb dropped on Hiroshima (10–20 kilotons or 4.18–8.36×10^{13} joules yield) corresponds to a magnitude on the Richter scale between 4.7 and 5.4, depending on the geologic region. Seismic events of this size in continental regions leave clear signals in dozens, perhaps hundreds, of seismological observatories worldwide. A 1 kiloton (4.18×10^{12} J) explosion corresponds to a magnitude between 3.8 and 4.3 and is detectable readily at regional distances, within 1000–1500 km

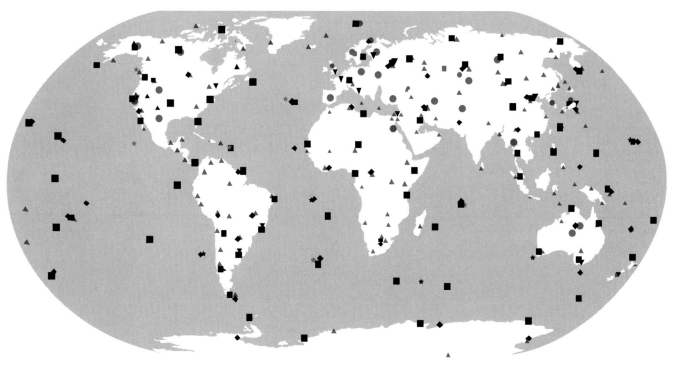

Key:
● seismic primary array
▲ seismic primary three-component
• seismic auxiliary array
▲ seismic auxiliary three-component

★ hydroacoustic (hydrophone)
⋆ hydroacoustic (T phase)
◆ infrasound
■ radionuclide stations
▼ radionuclide laboratories

International monitory system network, consisting of 321 monitoring stations and 16 radionuclide laboratories. (*Comprehensive Test Ban Treaty Organization*)

(620–930 mi), and at many more-distant monitoring stations. A clandestine tester might try to evade detection by exploding a nuclear device in a large subterranean cavity. Decoupling the explosion from the surrounding rock can lower its apparent size by a factor of 10 for high-frequency waves detectable locally, and up to a factor of 70 for longer-period waves detected at greater distance. Decoupling evasion is most feasible for yields of 1 kiloton or less, bringing the desired detection threshold to $M = 3.0$ or less.

Earthquakes and explosions can be discriminated either by location or by features in their seismic signatures. Explosions occur near the earth surface. Earthquakes occur deeper and are often associated with known geologic faults. Explosions are near-instantaneous and compressional in nature; their seismic signals are often dominated by high-frequency P (pressure) waves. Earthquakes rupture rock that slides along a planar fault for a finite time interval; their seismic signals are often dominated by relatively low-frequency S (shear) waves. The P wave travels faster than the S wave (which has a slower propagation speed) and arrives first at the monitoring station. From this information, the ratio of P-wave amplitude to S-wave amplitude is used to differentiate between an explosion and an earthquake. Events at or near the detection threshold are difficult to discriminate.

The IMS monitors seismic activity with a mix of sensor arrays and single three-component seismometers. Of these, 50 primary stations report data in a continuous stream to the IDC for real-time detection and discrimination of potential explosions. Supplemental data from 120 auxiliary seismic observatories can be retrieved by the IDC whenever a suspicious event is detected by the network of primary stations. Many stations in the IDC network perform double duty in global or regional earthquake-monitoring networks.

Hydroacoustic monitoring. Explosions in water generate sound waves similar to the P waves in solid rock. At depths of 1.5 km (0.9 mi) or less, the SOFAR (sound fixing and ranging) channel in the ocean captures and transmits sound globally with very little loss of energy (although the sound waves can be weakened in cold or shallow water). The IMS projects a sparse global network of 11 observatories to monitor clandestine undersea nuclear tests. Six of these observatories will collect data from underwater hydrophones. Five will be seismometers sited on islands, intended to record seismic T phases, solid-rock waves generated when an underwater acoustic signal strikes the submerged portion of an island.

Acoustic sensors can detect a 1 kg (2.2 lb) explosion of TNT in the SOFAR channel at distances of several thousand kilometers. A 1 kiloton (4.18×10^{12} J) nuclear blast near the surface of the open ocean would be detected worldwide. Blind spots for detection occur along many continental coastlines, where landmasses screen and distort acoustic waves. Source locations may be accurate to only a few hundred kilometers, and may require complementary data from satellite surveillance. Underwater explosions must also be distinguished from other sound sources in the SOFAR channel, such as airgun blasts and drilling noise from offshore oil exploration, ship noise, and whale calls. Large explosions generate large bubbles of gas that have characteristic transient long-period vibrations.

Infrasound monitoring. Under- and aboveground explosions generate atmospheric pressure waves with periods of 0.2–100 s. Pulses from distant 1 kiloton (4.18×10^{12} J) blasts are weak (one-millionth of average atmospheric pressure), but detectable at 2000–5000 km (1240–3110 mi) range. Distant signals can be distinguished by microbarographs from local disturbances by summing pressure from a network of air ducts connected by hoses. An array of four such systems, with 1–3 km (0.6–1.9 mi) station spacing, can be used to determine the arrival direction of infrasound signals. The IMS is projected to maintain 60 infrasound miniarrays.

The location accuracy of infrasound sources is degraded by Doppler effects, as the pressure waves typically travel within atmospheric wind systems. Their propagation speeds up or slows down, depending on wind direction. A suspected nuclear blast must be distinguished from a variety of other infrasound sources, such as volcanoes, severe weather, and large mining blasts. Meteors from space are also common, most of which explode in the upper atmosphere. Observational studies suggest that meteor blasts equivalent to a 15-kiloton (6.27×10^{13} J) nuclear bomb may occur once per year on average.

Radionuclide monitoring. Nuclear blasts generate a wide variety of short-lived radioactive isotopes. If released to the air, radioactive by-products are transported worldwide by winds in the form of particulate aerosols and gases. Particulate fallout reaches the surface via precipitation or gravitational settling. Radioactive isotopes of the noble gas xenon are primary monitoring tools, as they do not react chemically with other gas molecules. The IMS is projected to maintain 80 radionuclide monitoring stations, 40 of which would be capable of detecting xenon radionuclides. A network of 16 radionuclide laboratories are stipulated for sample analysis.

Current technology for radionuclide monitoring is difficult to maintain at remote sites, as the gamma-ray detectors for particulate fallout often require liquid-nitrogen cooling. The short half-lives of many radionuclides are also a challenge to detection. Standalone atmospheric sampling and analysis systems are being developed, however. Once a radionuclide signal is detected, tracing it back to its source depends on several factors. The transport of radionuclides, especially those subject to rainout effects, has been modeled only crudely thus far in computer models of atmospheric circulation.

On-site inspection. If any signatory nation of the test ban treaty charges another signatory nation with a nuclear test, an on-site inspection is authorized if 30 of the 51 nations of the CTBTO Executive Commission vote for it. The on-site inspection may cover a 1000 km^2 (386 mi^2) area. Evidence of radionuclide by-products may be sought via airborne or ground measurements. Underground explosions may leak radioactive noble gases over time along cracks in the basement rock. Close-in seismic monitoring of the suspected blast area may reveal aftershocks associated with an explosion, which relieve residual stress in the blasted rock. Seismic prospecting methods and ground-penetrating radar may be used to find an underground blast cavity.

False alarms and undetected tests. On August 16, 1997, a small seismic event ($M = 3.5$) was detected in the vicinity of Novaya Zemlya, Russia, an island in the Arctic Ocean with one of the nuclear testing facilities of the former Soviet Union. A satellite observation of coincidental activity at the test facility caused United States intelligence officials to conclude that a nuclear test had occurred in violation of Russia's announced moritorium on test blasts. The United States made a diplomatic protest to the Russian government, which denied the charge. News of this exchange reached the press in September, after which independent seismologists obtained relevant seismic data for analysis. Both they and the prototype IDC located the event in the Kara Sea, at least 80 km (50 mi) from the Novaya Zemlya nuclear facility. Comparison with older seismograms from known Soviet test blasts and small local earthquakes revealed the event to have the characteristics of an earthquake. The P (pressure) wave was too small, relative to the S (shear) wave, to identify an explosion as its source. This affair demonstrated the power of seismic monitoring data to resolve disputes over the nature of small suspicious seismic events. However, the tightly constrained location of the event required the use of data from seismic observatories outside the official, but then incomplete, IMS seismic network.

On May 11, 1998, India announced the detonation of three simultaneous nuclear explosions, with yields of 0.2, 12, and 43 kilotons (0.8, 50, and 180×10^{12} J). A seismogram recorded in Pakistan 740 km (460 mi) away showed a P-wave to S-wave ratio that was representative of an explosion. India announced two more nuclear tests on May 13 with yields of 0.2 and 0.6 kiloton (0.8 and 2.5×10^{12} J). Pakistan responded with its own nuclear tests on May 28 and May 30. These events illustrate the diplomatic use of nuclear tests to threaten potential adversaries, echoing the Cold War arms race. No attempt was made to conceal these tests; all but the May 13 blasts were detected on seismometers worldwide. However, seismic data suggest either that the Indian tests were unsuccessful or that the devices detonated were not those claimed. In particular, the maximum published

Richter magnitude for the May 11 event was 5.2, corresponding to a yield of only 12 kilotons (5×10^{13} J). This is considered too small for the thermonuclear device that India claimed to have detonated. Likewise, seismic data collected at Nilore, Pakistan, 750 km (466 mi) from the Indian test site, has no signal corresponding to the claimed Indian explosions on May 13. The ground-noise level at Nilore suggests that any test explosion that day would have had a yield of less than 0.1 kiloton (4.2×10^{11} J), and probably much smaller. The 1998 India-Pakistan nuclear tests demonstrate that the monitoring tools arising out of the Comprehensive Test Ban Treaty can be used to check claims of weapons capability, and therefore can restrain high-stakes diplomatic bluffs.

For background information *see* ATMOSPHERIC ACOUSTICS; ATOMIC BOMB; INFRASOUND; NUCLEAR EXPLOSION; NUCLEAR FISSION; RADIOACTIVE FALLOUT; SEISMOLOGY; SHOCK WAVE; SOFAR; SONAR; WAVE MOTION IN FLUIDS in the McGraw-Hill Encyclopedia of Science & Technology.   Jeffrey Park

Bibliography. National Research Council, *Research Required to Support Comprehensive Test-Ban Treaty Monitoring*, National Academy Press, 1997; P. G. Richards and W. Y. Kim, Testing the Nuclear Test Ban Treaty, *Nature*, 389:781–782, 1997; G. van der Vink et al., False accusations, undetected tests and implications for the CTB Treaty, *Arms Control Today*, pp. 7–13, May 1998.

Obesity

Obesity is an excessive accumulation of body fat which confers health risks (diabetes, cardiovascular diseases, arthritis, some types of tumors, among others). The health risk is determined by the amount of fat, the distribution of body fat, and the presence of other risk factors.

As an example of food consumption, a man of average weight consumes and utilizes approximately 55 million kilocalories (230 million kilojoules) in his first 60 years. Even a minor increase in daily energy intake or a minor decrease in daily energy expenditure over several years results in obesity. Indeed, obesity is one of the most frequent and serious metabolic diseases. Furthermore, strategies for long-term reduction of body weight are largely ineffective. More than 90% of people who lose weight eventually regain it. However, recent research suggests that an effective treatment for obesity may be found.

Health implications. The amount of fat stored in adipose tissue can be estimated by the body mass index (BMI), calculated by dividing body weight in kilograms by the square of the height in meters (BMI = kg/m^2). Individuals with a BMI of 20 to less than 25 have the lowest health risk (normal BMI). Individuals with a BMI of 25 to less than 30 are classified as overweight and have a moderate health risk. Individuals with a BMI of 30 or more are considered obese and have a high health risk. Approximately

32% of the United States population is overweight, and of these 22% is obese.

Another determinant of health risk, body fat distribution can have a peripheral or central pattern. The peripheral pattern is associated with females, and is also called gynoid, lower-body obesity, or pear shape. The central pattern is typically male and is called android, upper-body obesity, or apple shape. In women and men, upper-body obesity confers a greater health risk because of its association with high cholesterol, hypertension, and insulin resistance. Body fat distribution can be calculated by dividing the waist circumference by the hip circumference. A ratio of more than 1 for men or more than 0.9 for women is characteristic of upper-body obesity.

The association of obesity with other health risk factors is the ultimate determinant of morbidity and mortality. For example, an overweight individual (BMI of 25 to less than 30) with apple shape and a family history of diabetes may be at greater risk than an obese individual (BMI of 30 or more) with pear shape without any personal or family health risk factors.

Etiology. Of unknown etiology, human obesity belongs to a large group called complex diseases. Included in this group are type II diabetes, hypertension, and ischemic heart diseases (characterized by deficient supply of blood to cardiac muscle). The inheritance of these diseases is different from the classic mendelian mode of transmission of monogenic diseases (related to or controlled by one gene). The phenotype of complex diseases reflects the multifactorial effect of all contributing genes (polygenic) and all environmental factors. For instance, the prevalence of obesity in the United States population has doubled in the last hundred years. Yet it is clear that the gene pool of the population has not changed in just a century. What has changed is the environment and life expectancy. Humans have moved from a lean environment into an obese environment (effortless availability of plenty of high-fat food). Furthermore, twin studies fully support the influence of genetics in human obesity. The intrapair correlation coefficient of the values for BMI of identical twins reared together is identical to that of identical twins reared apart.

The characteristics of complex diseases render the search for contributing genes a major challenge. Currently, none of the genes contributing to the common form of human obesity are known. However, the genes contributing to many monogenic forms of obesity in rodents have been identified in the last 5 years. This information has provided a strong biochemical foundation for the pathophysiology of obesity.

Pathophysiology. The closed loop that regulates body weight has at least three systems: (1) the messenger system, in the periphery, informs the brain of the amount of body fat; (2) the translation system, in the brain, receives the information from the periphery and compares it with an internal standard of body fat (the set point of body weight or lipostat); and

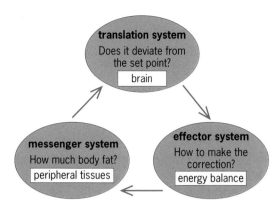

Fig. 1. Defense of the set point of body weight in lean and obese individuals. The messenger, translation, and effector systems interact. The treatment of obesity may need to modify one or more of these systems.

(3) the effector system regulates energy balance. The equation of energy balance includes energy intake by regulating satiety and appetite mechanisms; and energy expenditure, that is, physical activity, resting metabolic rate, and the thermic effect of food (energy required to absorb and store the ingested calories) [**Fig. 1**]. When the amount of body fat deviates from the set point of body weight, compensating efforts come into play to return body weight to its status quo.

Less than 5 years ago, the discovery of the hormone leptin, the key player of the messenger system, catalyzed understanding of the loop system (Fig. 1). The experimental evidence for the existence of leptin goes back almost 50 years. G. R. Hervey first demonstrated the presence of a hormone that regulated body weight through an interaction with the hypothalamus. The production of obesity by the destruction of the ventromedial hypothalamus in one member of a parabiotic (physiologically and anatomically associated) rat pair led to death by starvation in the unlesioned animal. Hervey proposed that a circulating satiety factor was produced in excess by the lesioned parabiont as body fat accumulated. This animal was rendered insensitive to the factor by destruction of the ventromedial hypothalamus; thus, the unlesioned parabiont became hypophagic (undernourished) in response to the high level of the satiety signal transmitted across the parabiotic union.

In 1994, researchers identified the gene responsible for obesity in one of the most intensively studied rodent models of obesity, the ob/ob mouse, homozygous for a mutant form of the obese (*ob*) gene. The gene product is a 146-amino-acid protein secreted into the blood by the adipose tissue. This hormone, leptin, is the circulating satiety factor identified by Hervey. Subsequent cloning of the leptin receptor revealed it to be a single-membrane-spanning protein which shares homology with the cytokine class I receptor family. It is abundantly expressed in the hypothalamus, which is the major site of leptin action.

It is clear that obesity in the ob/ob mouse is due to leptin deficiency. Leptin replacement in ob/ob mice results in normalization of body weight and other disturbances such as hyperglycemia (high blood glucose level), low metabolic rate, and low fertility. In contrast, in the db/db mouse and the fa/fa rat, which possess mutations in the leptin receptor, leptin levels are elevated, and administration of leptin does not improve obesity. Except for the ob/ob mouse and two rare patients who have mutations in the leptin gene, obesity in animals and humans is associated with hyperleptinemia (high leptin level). There is a strong positive correlation between serum leptin concentration and body fat. Thus, the adipocyte (fat cell) appears to correctly send the message to the brain on the amount of body fat. Since the action of leptin is to decrease appetite and increase energy expenditure, the finding of increased serum leptin concentrations in obese subjects indicates leptin resistance (**Fig. 2**).

The great majority of obese subjects do not have a defect in the production of leptin. There are no intravascular defects, such as leptin antibodies, leptin antagonists, or increased production of leptin-binding proteins, to limit the concentration of free leptin that reaches the brain.

In obesity, an excess of leptin may not easily reach the hypothalamic leptin receptor because of the blood-brain barrier. Serum leptin is about 300% higher in obese individuals than in lean individuals. However, the cerebrospinal-fluid leptin concentration in obese individuals is only 30% higher than in lean subjects. Consequently, the leptin cerebrospinal fluid/serum ratio in lean individuals is much higher than in obese individuals. The reduced efficacy of brain leptin transport in obese individuals may provide a mechanism for leptin resistance. Obesity

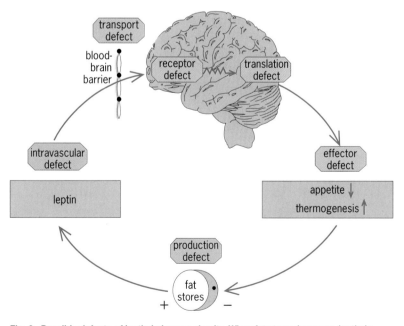

Fig. 2. Possible defects of leptin in human obesity. When fat stores increase, leptin is produced and secreted by the adipose tissue. It circulates in the blood and crosses the blood-brain barrier. It interacts with its receptor in the hypothalamus, where it regulates several brain hormones. The appetite decreases and thermogenesis increases, thereby reducing fat stores.

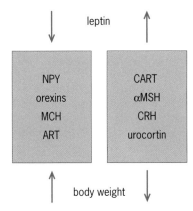

leptin

NPY	CART
orexins	αMSH
MCH	CRH
ART	urocortin

body weight

Fig. 3. Effect of leptin on brain neuropeptides. The neuropeptides in the left box increase body weight, and those in the right box decrease body weight. Leptin decreases body weight by both decreasing neuropeptides on the left and increasing those on the right. NPY, neuropeptide Y; MCH, melanocyte concentrating hormone; ART, agouti-related transcript; CART, cocaine-amphetamine-related transcript; αMSH, melanocyte stimulating hormone; CRH, corticotrophin releasing hormone.

induced in rodents by a high-fat diet became resistant to maximal concentrations of leptin administered peripherally. In sharp contrast, intracerebroventricular administration of maximal concentrations of leptin results in a strong response of decreased food intake and weight loss. However, suboptimal leptin concentrations result in a diminished response compared with normal rodents. Thus, diet-induced obesity is associated with both peripheral leptin resistance and central leptin insensitivity.

Since the great majority of obese subjects do not possess leptin receptor mutations, human obesity appears to be characterized by a post–leptin receptor defect. The leptin system defect might include the fast-growing list of neuropeptides that are influenced by leptin and are believed to be involved in body weight regulation (**Fig. 3**). The defect in leptin action in human obesity might be far from leptin itself or even far from the transduction system; it might reside in the effector system. The distal biochemical and behavioral mechanisms that control appetite, physical activity, and thermogenesis may turn out to be the critical defects.

Treatment possibilities. The discovery of the role of leptin in obesity (Fig. 1) provides the opportunity to develop an effective therapy. Patients with type II diabetes are insulin-resistant, yet insulin replacement is an effective therapy. Clinical trials are under way to determine if leptin injections would be an effective therapy in obese subjects who are leptin-resistant. Ideally, an oral treatment similar to the oral hypoglycemic agents used in type II diabetes should be developed to enhance leptin sensitivity. This effort might require a better understanding of the possible defects of leptin action (Fig. 2).

Alternatively, the plethora of brain hormones (Fig. 3) known to increase or decrease body weight could be targets for rational drug design by developing antagonists or agonists, respectively. The effector system itself could become a target for drug development. Agonists to the β3 adrenergic receptor could stimulate the release of fat from the adipose tissue (lipolysis) and uncouple metabolism so that the intrinsic energy of the fat is not stored but dissipated as heat (thermogenesis). Regardless of the strategy, it is imperative that the treatment of obesity be directed not only to decrease body weight but also to improve metabolism and preserve health and quality of life.

For background information *see* DIABETES; LIPID METABOLISM; METABOLIC DISORDERS; OBESITY; TWINS (HUMAN) in the McGraw-Hill Encyclopedia of Science & Technology.   José F. Caro

Bibliography. J. F. Caro et al., Leptin: The tale of an obesity gene, *Diabetes*, 45:1455–1462, 1996; I. Wickelgren, Obesity: How big a problem?, *Science*, 280:1364–1367, 1998; Y. Zhang et al., Positional cloning of the mouse obese gene and its human homologue, *Nature*, 372:425–432, 1994.

Ocean currents

The first section of this article discusses the shallow-water or surf-zone coastal currents, specifically alongshore and rip currents. The second section discusses the findings of the World Ocean Circulation Experiment, a long-term effort, as part of the World Climate Research Programme, to measure the flows of ocean currents. Research on ocean currents is providing critical information in a broad range of environmental topics, from localized coastal processes to changes in global climate.

Coastal Currents

As waves approach the shoreline, they preferentially break in areas of shallow water, particularly over sandbars and shoals. In breaking, the waves transfer some of their momentum to the water column and drive coastal currents and circulation. Alongshore currents and rip currents are members of the class of coastal currents which predominantly occur in shallow-water areas known as the surf zone or nearshore region (**Fig. 1**).

Alongshore currents. Alongshore (or simply, longshore) currents travel parallel to the shoreline. Their direction of travel and strength are dictated by the height and arrival direction of nearshore waves. The larger the waves and the higher the angle of their approach to shore, the stronger the longshore currents. Strong winds can play a significant role in modifying both the strength and direction of longshore currents, depending on whether the winds are aligned with the waves or opposed to them.

The strength of longshore currents varies with distance from the shoreline, which is confined within the active surf zone, and can vary in width from 100 to 1000 m (328 to 3280 ft), the larger widths occurring during storm conditions. Maximum longshore current speeds can exceed 1 m s^{-1} (3.3 ft s^{-1}) but are typically 25–50 cm s^{-1} (0.8–1.6 ft s^{-1}). The maximum longshore current usually occurs in the middle portion of the surf zone and decreases with strength both shoreward (toward the beach) and seaward (toward deeper water). In this sense, longshore currents are very similar to a river flowing parallel to shore, with the weakest currents occurring at the lateral boundaries and the maximum in between.

Recent studies indicate that longshore currents can have quite complex behavior. Continuing the analogy to rivers, longshore currents meander as they progress down the coast. That is, the maximum longshore current can be closer to shore at one beach location and farther seaward at another beach location. These meanders migrate with the longshore current, so for a fixed observer or swimmer the longshore current will be seen to pulsate, as the stronger and weaker portions of the current migrate past. This phenomenon of longshore currents is termed shear waves. Shear waves were first observed in field experiments on the North Carolina coast at the Army Corps of Engineers Field Research Facility. Computer modeling of these meandering longshore currents has shown that they might be a source of the strong cross-shore flows, commonly known as rip currents.

Rip currents. Rip currents travel seaward, cross-shore, roughly perpendicular to the shoreline and the time-averaged longshore current. Rip currents are one form of nearshore circulation, the "plumbing" that returns water from the surf zone–nearshore region to deeper water, seaward of where the waves start to break. As such, rip currents have close ties to longshore currents and nearshore waves, and are commonly called riptides. Riptide is an incorrect term because the tides, driven by the Sun and Moon, can change the overall water depths in the surf zone–nearshore region and modify nearshore currents, but they do not directly force cross-shore currents by themselves.

Rip currents are known for their strong, offshore-directed flows with speeds in excess of 1 m s^{-1} (3.3 ft s^{-1}). They are fed by longshore current flows and typically are quite narrow, jetlike currents with widths of 10–50 m (33–164 ft). Rip currents are broad near the shoreline, narrower and stronger through the mid and outer surf zone, only to become broad and diffuse seaward of the surf zone. Rip currents are particularly hazardous to swimmers as the speed of their flows exceed those of even the best swimmers.

Rip currents show a variety of behaviors. In the lee of coastal structures, such as jetties or piers, or in gaps in a sandbar, they can be fixed in longshore location and be quite steady in their flow. Surfers typically ride them out seaward of the breakpoint in order to catch waves. On open coasts, however, rip currents have a reputation for being somewhat ephemeral, changing alongshore location, strength, and duration on time periods as short as 10–15 min. This is an additional reason why they can be such hazards to swimmers.

There are two main hypotheses about their formation, which are both likely to be correct in different

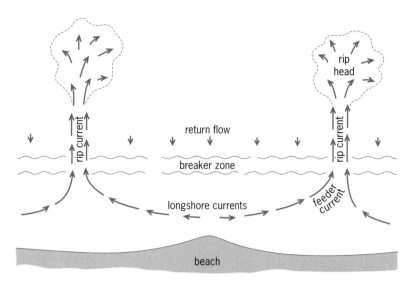

Fig. 1. Nearshore water circulation is driven by breaking waves which transfer momentum to currents. These currents comprise longshore currents which flow parallel to the shoreline, and rip currents which are cross-shore-directed flows that return water to the region seaward of the breaking waves. (*After P. D. Komar, Beach Processes and Sedimentation, 2d ed., Prentice Hall, 1997*)

situations. As waves approach shore, break, and traverse the surf zone, they tend to push water shoreward, leading to a slightly raised mean sea level, constituting a wave setup or seaward tilt, near the shore. If there are variations in the height of waves along the shore, this wave setup has a shore-parallel tilt as well, which causes currents to flow parallel to the shore (another source of longshore currents). Where two oppositely directed longshore flows meet, the flow turns seaward and generates a rip current. As the pattern of offshore waves varies over time (height and direction), the location and strength of rip currents also vary.

The second hypothesis on rip current formation stems from the recent work on longshore currents and shear waves mentioned earlier. As the longshore current meanders parallel to shore, it can become unstable. That is, because of variations in the wave forcing, the patterns of the sandbars, and the roughness of the seabed, there will be times and locations when the flow will squirt seaward. These seaward squirts are typically quite strong, narrow, and ephemeral, the same characteristics used to describe rip currents. It is the objective of ongoing research to ultimately resolve the underlying dynamics of both longshore currents and rip currents.

Reginald A. Beach

Global Ocean Currents

Although the general paths of major ocean currents have long been known, measuring their flow rates is not easy. A major component of the World Ocean Circulation Experiment (WOCE) was studying such flows, which allows calculations on how much heat, fresh water, and dissolved components are transported within and between ocean basins. WOCE was established in the 1980s as part of the World Climate Research Programme to improve knowledge of global ocean circulation and hence to improve models to forecast climate change. The WOCE field program took place from 1990 to 1998; assembly and interpretation of the data will continue through about 2005.

Results from WOCE. In the mid-1980s, it was suggested that the ocean behaves as a global conveyor belt, having an essentially steady, laminar, long-term flow, so that data from different times can be combined relatively easily into a global circulation pattern. In this view, cold deep water originating in the northern Atlantic flows southward and is transported via the northern edge of the Antarctic Circumpolar Current to the Indian and Pacific oceans. Here it spreads northward, rises (upwells) to the surface, warms, and flows back to the Atlantic through the Indonesian throughflow between Asia and Australia and around the southern tip of Africa. WOCE and other data have shown this view to be overly simplistic. A recent overview of the general circulation (**Fig. 2**) makes it clear that much interaction occurs between water at different depths and that upwelling and downwelling (sinking) occur in all three oceans. Also, most flow within individual ocean basins comes from water recirculating within that basin, rather than being transported through it.

The main flow of ocean currents occurs along the western boundary of each basin because of fluid dynamics and the Earth's rotation. Away from these regions, current flows are generally low. New WOCE measurements from neutrally buoyant floats drifting at about 1000 m (3280 ft) depth have for the first time provided data on midwater velocities across whole ocean basins; these are generally zonal (east-west) and slow (<3 cm s^{-1} or 1.2 in. s^{-1}). At the bottom, however, flows can increase, with velocities up to 20–30 cm s^{-1} (7.8–11.8 in. s^{-1}) found along the western edges of deep ocean basins. These bottom flows, together with the passage of water across rough topography along the midocean ridges, produce considerable mixing within the lower 500–1000 m (1640–3280 ft) of the water column.

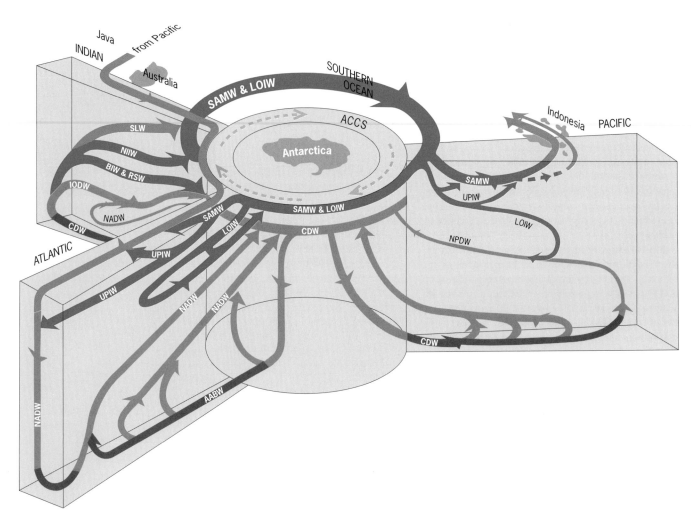

Fig. 2. Three-dimensional circulation within the Atlantic, Indian, and Pacific oceans showing the vertical recirculation within each basin and the interchange between oceans within the Antarctic Circumpolar Current and via the Indonesian throughflow. AABW, Antarctic Bottom Water; ACCS, Antarctic Circumpolar Current System; BIW, Banda Sea Intermediate Water; CDW, Circumpolar Deep Water; IODW, Indian Ocean Deep Water; LOIW, Lower Intermediate Water; NADW, North Alantic Deep Water; NIIW, North Indian Intermediate Water; NPDW, North Pacific Deep Water; RSW, Red Sea Water; SAMW, Subantarctic Mode Water; SLW, Surface Layer Water; UPIW, Upper Intermediate Water. (*From W. J. Schmitz, On the World Ocean Circulation, vol. II, Woods Hole Oceanographic Institution, 1996*)

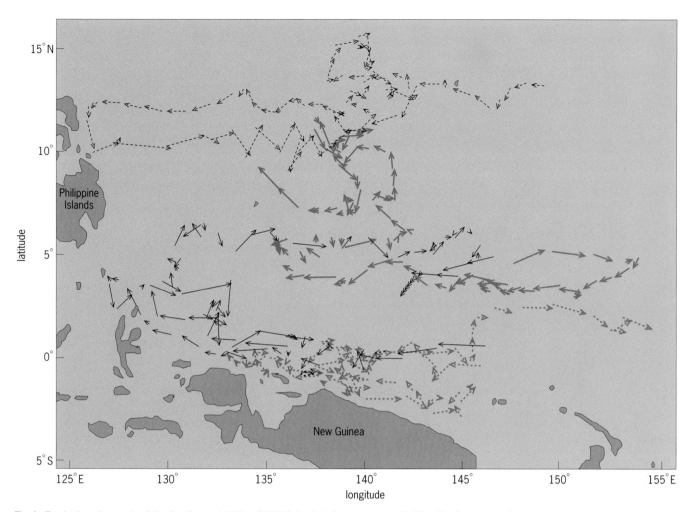

Fig. 3. Tracks from four water-following floats at 1000 m (3280 ft) depth in the western tropical Pacific. Each arrow shows the displacement of a particular float over 25 days. Most flow is zonal (east-west), and directions reverse on a scale of several months. (*From Russ Davis, Scripps Institution of Oceanography, La Jolla, CA*)

WOCE data have shown that the idea of steady currents also is incorrect. Measurements, using several different techniques, have shown high variability at all scales examined. For example, near the Equator in the Pacific, currents measured directly at about 1000 m (3280 ft) depth reverse direction over several months in a presently unpredictable manner (**Fig. 3**). Similarly, a major feature of the "conveyor belt" is the transport of warm surface water from the Pacific to the Indian Ocean via the Indonesian throughflow. WOCE data suggest that about 7 sverdrups (1 Sv = 10^6 m^3 s^{-1}) of water are involved, but with variability on both seasonal and interannual scales of about the same magnitude. Westward flow is reduced during the winter (northwest) monsoon; similarly, during periods of El Niño activity, the pressure head of warm water in the western Pacific is shifted eastward and flow is again reduced. Other possible ways to close the circulation loop during such times include cold surface water flowing through Drake Passage from the southeast Pacific into the southwest Atlantic, or east-west transport from the Pacific to the Indian Ocean between Australia and the Antarctic Circumpolar Current (the vol-

ume involved here is likely small). All three routes probably contribute to global circulation, which route is most important at any given time being controlled by large-scale climate variability. *See* EL NIÑO.

Variability at longer time scales is also important. The northern basins of the Atlantic supply cold, dense water that feeds middepth and bottom circulation; for example, in the Labrador Sea, winter temperature differences between the air and sea result in localized deep convection. Cold, dense surface water, formed as salt is excluded from sea ice during the freezing process, is quickly transported to 1500–2000 m (4920–6560 ft) depth. Measurements of water properties such as temperature, salinity, and dissolved chemical tracers have shown that the formation rates of such water masses change over decadal periods and can apparently cease for several years at a time. WOCE included many measurements to define water properties at particular sites; analysis has shown that consistent changes in these properties have occurred over several large regions of the ocean during the past few decades. For example, water at about 1500–2500 m (4920–8200 ft) depth in the subtropical North Atlantic has warmed

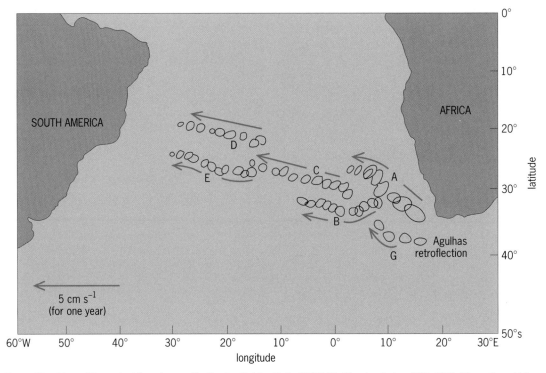

Fig. 4. Monthly positions of eddies observed in the South Atlantic by GEOSAT altimetry during 1986–1987. All are thought to have originated at the Agulhas retroflection south of Africa and denote a transfer of water from the Indian to the South Atlantic Ocean. Note the arrow showing how far an eddy traveling at 5 cm s⁻¹ would go. (*After A. L. Gordon and W. F. Haxby, Agulhas eddies invade the South Atlantic, J. Geophys. Res., 95:3117–3125, 1990; used with permission of the American Geophysical Union*)

by about 0.3°C (0.5°F) since 1920. Conversely, large areas of the southern Indian and Pacific oceans have cooled and freshened in the last 20–30 years. Reasons for such large-scale changes are unknown at present.

An exciting development in WOCE is the confirmation that eddies, rings of water moving either clockwise or counterclockwise depending on their temperature structure, are responsible for much of the variation in interbasin transport. These eddies, which may be up to 300 km (186 mi) across and extend down to about 1200 m (3960 ft) depth, are particularly obvious in regions adjacent to major surface currents such as the Gulf Stream, Kuroshio, East Australian, and Agulhas currents. They may be observed directly in satellite-derived sea surface height images (eddies that are "warm" relative to their surroundings show as elevations, while "cold" eddies show as depressions) and can be followed across ocean basins over several months as they decay. Good examples are found south of Africa in the Agulhas retroflection region, where they pinch off from the southward flowing Agulhas Current and transport large volumes of water from the Indian Ocean into the southeast Atlantic (**Fig. 4**). Some of these Agulhas eddies have been followed across the Atlantic to the Brazil Basin.

Model developments. Recent models developed as part of WOCE also are helping to elucidate ocean current flow and are becoming important for forecasting purposes (for example, the onset and magnitude of El Niño). Latest-generation computers allow scientists to run models at scales fine enough (20–50 km or 12.4–31 mi) to include eddies. At these scales, model results are converging with observations, showing relatively narrow (50–100 km or 31–62 mi) main current flows, eddies being important for transporting mass and dissolved components (**Fig. 5**), and varying positions and transports of

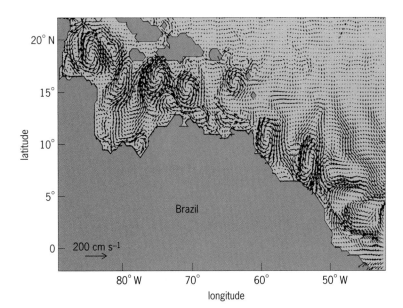

Fig. 5. Eddies in the surface layer of the North Brazil Current as predicted by a model of the Atlantic Ocean with a resolution of about 18 km (11.2 mi). The arrows show surface current velocities. (*From Yi Chao, Jet Propulsion Laboratory, Pasadena, California*)

major current streams. The ability both to model the global ocean and to constrain model outputs by directly incorporating hydrographic and other data permits estimations of model accuracy. While errors remain large at present, continued increases in computing power suggest future reduction in errors and thus improved forecasts. However, it is clear that the simple "conveyor belt" view of ocean circulation is seriously wrong; while the large-scale (1000–2000 km or 620–1240 mi) current patterns are generally stable, the ocean is essentially turbulent and variable, with the circulation dominated by relatively small-scale phenomena. To fully understand long-term changes in ocean current variability will require very long time series of measurements (50–100 years) at selected key sites.

For background information *see* ANTARCTIC OCEAN; ARCTIC OCEAN; ATLANTIC OCEAN; FLOW MEASUREMENT; INDIAN OCEAN; INSTRUMENTED BUOYS; LAMINAR FLOW; OCEAN CIRCULATION; PACIFIC OCEAN; TIDE; UPWELLING in the McGraw-Hill Encyclopedia of Science & Technology.   Piers Chapman

Bibliography. P. D. Komar, *Beach Processes and Sedimentation*, 2d ed., Prentice Hall, 1997; Open University Course Team, *Ocean Circulation*, Pergamon Press, 1989; S. Pond and G. L. Pickard, *Introductory Dynamical Oceanography*, Pergamon Press, 1983; W. J. Schmitz, *On the World Ocean Circulation*, vols. I and II, WHOI 96-03 and 96-08M, Woods Hole Oceanographic Institution, 1996; M. Tomczak and J. S. Godfrey, *Regional Oceanography: An Introduction*, Pergamon Press, 1994.

Ocean optics

The first section of this article discusses use of satellite pictures to detect subtle changes in color reflected by the ocean, due to phytoplankton concentration. Studies based on these color changes are used to determine ocean productivity. The second section discusses underwater laser line scan systems and obtaining of high-quality, panoramic images of submerged objects at extended viewing ranges.

Ocean Color Remote Sensing

Space-borne instruments known as ocean color sensors are able to map phytoplankton distributions by detecting subtle shifts in the spectrum of sunlight reflected from the upper ocean. The color of the ocean is affected by light-absorbing pigments in the phytoplankton. As the concentration of chlorophyll and other photosynthetic pigments increases, there is a spectral shift in the reflectance of the ocean from blue to green. The feasibility of detecting this color shift from space was first demonstrated by the Coastal Zone Color Scanner (CZCS), launched in 1978. A new era began in 1996 with the launch of Japan's Ocean Color and Temperature Sensor (OCTS), and continues with the United States' Sea-viewing Wide Field Sensor (SeaWiFS). These ocean color remote sensing missions were designed to obtain a multidecadal time series of ocean chlorophyll.

With such satellite records, scientists are seeking to understand processes that affect the ocean's capacity for biological productivity. They seek to predict how marine ecosystems will respond to climate change and how changes in marine ecosystems will, in turn, affect climate.

Phytoplankton and climate. About half of all photosynthesis occurs in the ocean and is carried out almost entirely by phytoplankton within 100 m (330 ft) of the surface. Given sufficient light and nutrients, phytoplankton reproduce rapidly, converting inorganic carbon into organic carbon and (in some species) calcium carbonate. The phytoplankton are then consumed by zooplankton, fish, and other animals in the food chain. Some of the organic carbon and calcium carbonate produced near the surface sinks into the deep ocean. This biologically mediated carbon flux is known as the biological pump. Carbon-rich particles sinking below the upper wind-mixed layer of the ocean enter a very stable environment where the carbon is stored for many decades or centuries. About 80% of all mobile carbon on Earth is stored in this deep reservoir.

Rising levels of atmospheric carbon dioxide are not expected to affect the biological pump directly because carbon does not limit biological productivity in the ocean. However, there may be indirect climate-induced changes resulting from changes in ocean and atmospheric circulation. These changes could occur rapidly because the phytoplankton carbon pool has an extremely rapid turnover time (in the order of days rather than years). Hence, perturbations to the ocean's biological pump could cause rapid changes to climate through the resultant feedback on the global carbon cycle. Climate models are now beginning to incorporate parametrizations of ocean biological processes. These models show that biological processes tend to ameliorate the effects of physical climate forcings. Whatever the response of marine ecosystems to climate change, it is clear that a significant change in the biological pump would have a large influence on climate. Scientists do not know what maintains the past and present level of oceanic primary production, and, indeed, whether the level remains in steady state, how it changes in response to climate change, and how it influences climate. A time series of global ocean color measurements spanning decades would shed light on these issues. *See* GLOBAL WARMING.

Mapping phytoplankton from space. The CZCS was an experimental sensor onboard the *Nimbus* 7 satellite. Operating intermittently (only 10% of the time) between 1978 and 1986, it provided clear evidence that phytoplankton could be mapped from space. The present era of global measurements began in 1996 with the launch of Japan's *ADEOS* satellite carrying the OCTS. The OCTS provided global coverage at 700 m (2300 ft) resolution for 8 months beginning in November 1996, but then *ADEOS* experienced a catastrophic power failure on June 30, 1997. On August 1, 1997, ORBIMAGE launched SeaWiFS onboard the *Orbview 2* satellite, and on September 18 SeaWiFS began producing systematic global chlorophyll maps. SeaWiFS is the first commercially owned and operated satellite sensor supplying data to a major science mission of the National Aeronautics and Space Administration (NASA). NASA purchased the rights to distribute SeaWiFS data to the research community, while ORBIMAGE retains the exclusive right to market the data for commercial and operational purposes.

SeaWiFS is not just an ocean color sensor. An innovative bilinear gain gives SeaWiFS the capability to image land and cloud patterns, thus making it the first truly global imager. This was achieved by having four detectors per band. Three have a sensitive gain required to detect low radiances typical of the ocean, and their signals are averaged to improve signal-to-noise. The fourth detector has a less sensitive gain but wider dynamic range for viewing bright targets such as land and clouds. For example, subtle variations in the color of nearshore waters can be seen alongside adjacent land features.

A rigorous program of calibration and validation is required by all sensors contributing data to a long time series. The CZCS had no built-in mechanism to characterize instrument changes over time. Thus, although CZCS provided the first climatology showing seasonal patterns in chlorophyll, there was no way to separate interannual variability in pigments from trends in the instrument's performance. SeaWiFS employs a lunar calibration technique to monitor detector trends. Once a month, the *Orbview 2* spacecraft is rotated to allow SeaWiFS to scan the full face of the Moon. This procedure produces a monthly time series used to adjust calibrations for temporal trends in the detectors. For absolute calibration, SeaWiFS relies on precise in-water optical measurements made from moorings and ships. These "vicarious calibrations" are compared with water-leaving radiances derived from SeaWiFS. The principal mooring used for this purpose is the MOBY (Marine Optical Buoy), located in very clear waters off the Hawaiian island of Lanai. The principal ship-based program is the Atlantic Meridional Transect conducted onboard a supply ship that runs twice a year between England and the Falkland Islands.

The SeaWiFS chlorophyll algorithm, the mathematical formula for deriving chlorophyll concentration from water-leaving radiance, was based on coincident measurements of chlorophyll and radiance. Unlike the CZCS algorithm based on fewer than 50 "sea truth" measurements, the SeaWiFS algorithm is based on data from over 1200 stations around the world. The currently used algorithm enables the chlorophyll concentration to be derived from the ratio of reflectances (Rrs) in two SeaWiFS bands (**Fig. 1**). This algorithm applies to open-ocean areas where phytoplankton pigments are the only substance affecting the color of the water. In coastal areas, suspended sediments and dissolved organic matter also affect the color of the water. Thus, the development of algorithms for complex nearshore waters is a major focus of ongoing research in this field.

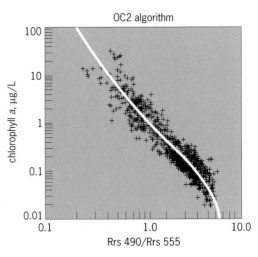

Fig. 1. SeaWiFS chlorophyll algorithm derives chlorophyll concentration from a ratio of reflectance in a blue channel (Rrs 490) to reflectance in a green channel (Rrs 555). The crosses are measurements made at the sea surface throughout the world's oceans. The solid line is the SeaWiFS chlorophyll algorithm.

Future sensors. A number of global imagers with ocean color capability are scheduled to begin operating in 1999 and 2000. India's OCM is scheduled for launch in early 1999 on the *IRS P4* satellite. NASA will launch the Moderate Resolution Imaging Spectroradiometer (MODIS) on the *EOS-AM1* satellite in the summer of 1999, and the European Space Agency will launch the Medium Resolution Imaging Spectometer (MERIS) onboard *ENVISAT*. A second MODIS will fly on the *EOS-PM1* satellite, and Japan's Global Imager (GLI) will be launched on *ADEOS 2* in 2000. Many of these will have bands with higher spatial resolution than the 1 km (3280 ft) resolution bands of SeaWiFS. Beyond 2005, a bridging mission has been proposed to transition from MODIS to operational instruments deployed by the National Polar Orbiting Environmental Satellite System (NPOESS).

Janet W. Campbell

Underwater Laser Systems

The idea of using laser-illuminated sensors for underwater search and survey has been an intriguing possibility for over 30 years. Possible architectures for laser-illuminated underwater imaging systems, for example, were first discussed publicly at the Ocean Optics Conference sponsored by the Society of Photographic Instrumentation Engineers (SPIE) in 1961. Initial interest in these ideas was high, and several programs to develop underwater laser systems were initiated and actively funded during the late 1960s and early 1970s. These programs, unfortunately, failed to meet their technical objectives or the expectations of the sponsors and were discontinued. Supporting technologies such as commercial lasers and digital electronics continued to mature, however, and by the mid-1980s it became appropriate to reevaluate the operational potential of underwater laser systems. Several development programs have been initiated since 1986 with the result that reliable,

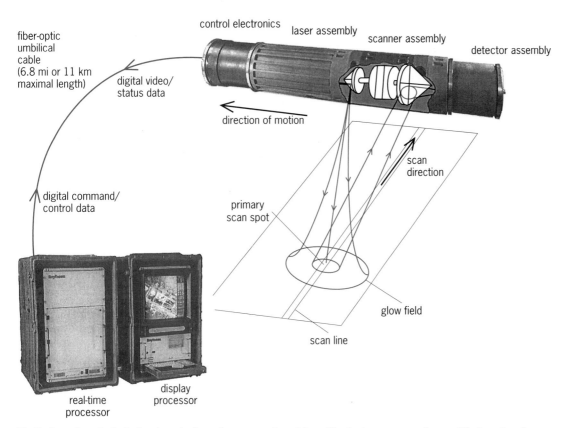

Fig. 2. Operation of a typical underwater laser line scan system. A laser illuminates a very small area of the target, and an optical detector, such as a photomultiplier tube, detects a portion of the light that is reflected by the target and turns that optical signal into an electrical signal. A combination of spinning mirrors and sensor motion scans the target in two dimensions. Sampling the detector output in synchronicity with the scanning process allows collection of data that can be used to generate a two-dimensional reflectance map on a computer monitor. This reflectance map looks much like a conventional image.

operational systems are now commercially available.

System description. In its basic form, an underwater laser line scan system (**Fig. 2**) consists of the underwater optical sensor and the topside control console. Several other components, such as an umbilical cable and power supplies, must be added to make the system fully operational. These supporting components are generally application-specific.

Optical sensor. The underwater optical sensor consists of the sensor control electronics, laser, scanner, and detector subassemblies. The integrated assembly is installed inside a watertight cylindrical pressure vessel. Two rotating mirrors, which are part of the scanner subassembly, are rigidly attached to a common rotating shaft, and a laser is oriented such that its output beam is incident on one of the mirrors. The laser beam is reflected toward the target and illuminates a very small portion which is shown as the primary scan spot. The laser provides an intense, highly collimated beam that makes it possible to minimize the size of the primary scan spot. This is important since the size of the scan spot ultimately determines the maximum spatial resolution of the system.

As the laser beam travels toward the target, it is scattered and absorbed by the intervening seawa-

ter. Absorption and scattering are the processes that have historically made it very difficult to generate high-quality images of submerged objects at acceptable operating ranges.

The unscattered, unabsorbed laser light that manages to reach the target illuminates the relatively small, localized area of the primary scan spot. This is the area of greatest interest because from it the optical signal is generated.

As the mirrors rotate, the scan spot traces a line across the target (thus the name "line scanners"). When there is relative motion between the scanner and the target perpendicular to the scan direction, sequential scan lines will be displaced slightly and the target will be effectively scanned in two dimensions. By coordinating the scan rate with the relative velocity that exists between the sensor and the target, it is possible to control the spacing between sequential scan lines and thereby ensure that the desired rectangular sample pattern is maintained.

The target reflects a portion of the incident energy back toward the sensor. When the scan spot is incident on a bright portion of the target, a relatively large fraction of the incident energy is reflected. Conversely, when the scan spot is incident on a dark

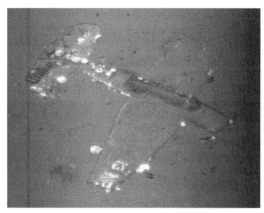

Fig. 3. Laser line scan systems can provide panoramic images of submerged objects under challenging conditions. This image was generated by an early prototype laser line scan system. This system used analog data acquisition circuitry and a 2-watt argon ion laser as light source. These images were generated during cruises of the USS *Dolphin*, a Navy diesel-electric research submarine.

portion of the target, a relatively small fraction of the incident energy is reflected. The reflected energy is scattered and absorbed as it travels back toward the sensor, and a small amount of the remaining energy enters the sensor's receiver aperture through the cylindrical input window. If the power in the laser beam is held constant and certain geometric factors are ignored, it can be assumed that the amount of energy reflected by the target, and therefore the amount of energy received by the sensor, is proportional to the local reflectance of the target. Local reflectance means that the reflectance of the target is

averaged over the area illuminated by the primary scan spot.

The light that passes through the receiver aperture is incident on the photocathode of a photomultiplier tube which converts the optical energy to an electrical signal. By sampling the output of the photomultiplier tube at times that are synchronized with the mirror rotation, it is possible to build up a reflectance map of the target. This is accomplished by digitizing the sampled photomultiplier tube signal and storing the resulting digital data in a video random access memory (VRAM) module. This stored data can then be read and displayed by a cathode-ray tube controller. With the whole process of data sampling, storage, and display properly integrated and synchronized, there is a direct relationship between each physical location on the screen of the cathode-ray tube and a specific location on the target. As the sensor passes over a stationary target, the display can be controlled in such a way that the data from each new scan line is always displayed at the top of the screen. This requires that all previous lines are sequentially moved down the screen to make room for the new line, and results in a "waterfall" display that gives the operator a very realistic view of the passing target.

The image displayed on the cathode-ray tube is not really an image in the conventional sense. Rather, it is the data resulting from many individual reflectance measurements displayed in a visual format. These measurements are made sequentially and are therefore independent of each other, which is an important point. In conventional imaging systems, all

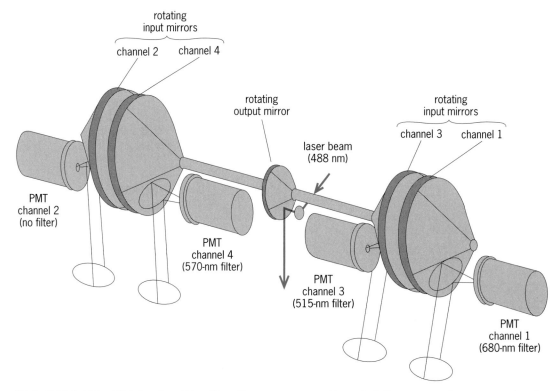

Fig. 4. Optical configuration for a four-channel laser line scan sensor used to evaluate fluorescence phenomena produced by underwater plants and animals stimulated with blue laser light.

picture elements, or pixels, are processed simultaneously and the entire image appears instantaneously in the focal plane. Since the pixels are processed in parallel, each pixel affects every other pixel in the image. The various scattering processes that are ubiquitous in the underwater environment cause the point spread function of the imaging system to broaden significantly, and the interaction among neighboring pixels therefore increases dramatically. Since scanning systems develop an image one pixel at a time, the potential for pixel-to-pixel interaction is reduced and the systems become much less sensitive to image degradation caused by environmental scattering processes.

Laser-illuminated underwater imaging systems have become an operational reality, with laser line scan imaging systems commercially available. Optical systems are being integrated with appropriate acoustic sensors and the necessary support sensors to realize well-balanced, fully integrated systems for underwater search and survey applications (**Fig. 3**).

Most laser line scan systems measure the optical energy that is incident on the receiver aperture. Significant information can be extracted from these measurements, but that intensity is only one aspect of the available light field. Other aspects of the light field, such as its polarization characteristics, can be measured and potentially contain additional valuable information. To evaluate this possibility, a special laser line scan sensor was built with four independent receivers (**Fig. 4**). Each receiver can be fitted with various optical elements such as polarization analyzers or narrowband optical filters to measure the polarization or polychromatic characteristics of the underwater light field that is reflected from laser-illuminated targets. The outputs from red, green, and yellow channels of this sensor can be combined to produce a pseudo-color image that is proving to be useful in the investigation and evaluation of underwater communities and habitats.

For background information *see* APPLICATIONS SATELLITES; IMAGE PROCESSING; LASER; OPTICAL DETECTORS; OPTICS; PHOTOMULTIPLIER; PHYTOPLANKTON; SCIENTIFIC SATELLITES in the McGraw-Hill Encyclopedia of Science & Technology.   Bryan W. Coles

Bibliography. T. H. Dixon et al., A range gated laser system for ocean floor mapping, *MTS J.*, 17:3, 1984; P. G. Falkowski et al., Biogeochemical controls and feedbacks on ocean primary production, *Science*, 281:200–206, July 10, 1998; P. J. Heckman and P. D. McCardell, A real-time optical mapping system, *SPIE Proc.*, 160:189–196, 1978; J. Leatham and B. W. Coles, Use of laser sensors for search and survey, *Underwater Intervention '93 Conference*, Association of Diving Contractors, New Orleans, January 18–21, 1993; C. R. McClain et al., Science quality SeaWiFS data for global biosphere research, *Sea Technol.*, September 1998; J. E. O'Reilly et al., Ocean color chlorophyll algorithms for SeaWiFS, *J. Geophys. Res.*, 103(C11):24,937–24,953, 1998; Underwater photo-optics, *SPIE Sem. Proc.*, Santa Barbara, October 10–11, 1996.

Oil production

These have been recent improvements in two technologies employed in oil production: coiled-tubing drilling and horizontal drilling. Coiled-tubing drilling has been used in petroleum operations since the late 1960s, but only in the 1990s has it been employed to drill wells. Horizontal wells emerged in the late 1980s as the preeminent advance in hydrocarbon recovery technology.

Coiled-Tubing Drilling

The earliest wells drilled with coiled tubing led to the expectation of direct cost benefits. That expectation was not realized, and except for some special cases (such as offshore and remote environments) there is still no direct economic advantage over conventional drilling. However, coiled tubing has unique capabilities, performing some operations very well that conventional equipment cannot.

Advantages. Coiled tubing can make fast trips into and out of wells, circulate continuously while in motion, operate in flowing wells and small-diameter wellbores, and form a compact portable unit. Thus, coiled tubing is the choice for underbalanced drilling and lateral branches.

Underbalanced drilling. Underbalanced drilling allows the well to flow during drilling. Conventional drilling requires an overbalance, in which drilling fluid exerts a hydrostatic pressure on the porous rock formations sufficient to prevent formation fluids from flowing into the wellbore. Some sensitive formations are damaged in overbalanced drilling due to clay swelling or plugging of the formation pore spaces. Coiled tubing routinely operates in flowing wells and can drill with a lower-density fluid, thus allowing the well to flow during drilling and preventing invasion and damage by higher-density drilling fluids.

Lateral branches. Coiled-tubing drilling is quite successful in the reentry of existing wells. It has the ability to drill new lateral branches into nearby reservoirs that cannot be accessed economically using conventional means. In these situations, coiled tubing can drill a new branch from an existing well into another reservoir, and complete the well without having to remove any of the tubing or equipment already in place. This technology is new, but as offshore fields continue to mature and decline in production this type of operation is expected to become much more common.

Equipment. A drilling unit consists of a coiled-tubing unit, drilling fluid system, pressure containment equipment, and downhole equipment (**Fig. 1**). The continuous tubing is up to about 16,000 ft (4900 m) in length. Common diameters for drilling are 2.375 in. (60 mm) and 2.875 in. (73 mm), but 2.0 in. (51 mm) and 3.5 in. (89 mm) are also used. The coiled-tubing unit provides the power and controls for running the tubing into and out of the borehole and the controls and monitoring devices for the drilling operation. It consists of a reel for storing the tubing, a hydraulically driven gripper/injector that

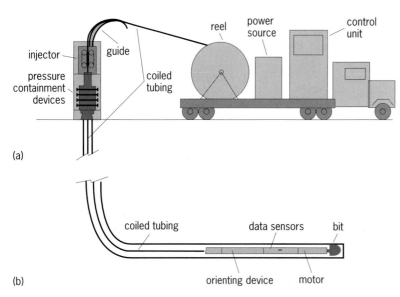

(a)

(b)

Fig. 1. A typical coiled-tubing drilling unit consists of (a) surface equipment and (b) downhole equipment.

lowers the tubing into the wellbore and pulls it back out, a power supply, and a control unit. A drilling fluid system circulates fluid through the coiled tubing to remove the drilled rock particles from the borehole. Pressure containment equipment on top of the well allows the coiled tubing to be run into and out of the wellbore under pressure. The downhole equipment drills the borehole, controls the borehole trajectory, and provides various data on the borehole and the rock formations being drilled. A drilling bit, a motor operated by the drilling fluid to rotate the bit, a sensor to detect the orientation of the borehole, and a device to change the directional orientation of the bit make up the basics. There are optional sensors to evaluate rock and fluid characteristics, temperature, pressure, vibrations, torque, and motor speed. Most coiled-tubing drilling operations use a continuous electric conductor inside the tube to provide real-time two-way signals between the surface and downhole equipment.

Limitations. In the process of going from the reel into the wellbore, out of the wellbore, and back onto the reel, the tube is subjected to three elastic-plastic bending cycles. The cyclic strain is an order of magnitude beyond the strain at yield, so the metal is cycled well into the plastic range every time it is run into and out of the wellbore. Consequently, the service life of coiled tubing is quite short, only one to six wells depending on the conditions. Directional control with coiled tubing requires a special approach because the tube cannot be rotated. A downhole device must be used to rotate the motor/bit assembly into a specific position in order to drill in the desired direction, and this device is often problem-prone. Also, the inability to rotate makes the removal of rock cuttings in highly deviated wellbores difficult because rotation tends to keep the cuttings propelled into the flow stream of the drilling fluid. However, none of these difficulties is insurmountable, and coiled tub-

ing will continue to have a niche in drilling operations.
           Ted G. Byrom

Horizontal Drilling

The primary horizontal drilling technologies include measurement-while-drilling (MWD; a technique to communicate measurements downhole to the surface typically by mud pulse telemetry), extended-reach medium-radius directional drilling equipment, and reliable mud motors. Horizontal wells are designed to essentially parallel rock strata, encountering much greater intervals of oil- and gas-bearing rocks than conventional vertical wells. They have been most successful in reservoirs with natural fractures, with coning problems, or with thin pay zones.

Natural fractures. Natural fractures in reservoirs are often nearly vertical, with the preponderance of the open fractures running in one direction. Such fractures often provide the majority of the flow capacity of a reservoir, while most of the storage is contained in the rock matrix. Vertical wells may miss many of these fractures; horizontal wells may intersect hundreds of fractures and substantially increase oil and gas recoveries.

Coning. Coning is the mechanism by which underlying water enters an oil well, producing gas cap. Coning is inherently rate-sensitive and occurs when the vertical component of viscous forces exceeds the net gravity forces. Horizontal wells offer the advantage of a much larger cresting volume (**Fig. 2**) and lower pressure drops for a given rate. The most critical parameter in assessing coning improvement by horizontal wells is the vertical permeability. This value is difficult to estimate and is a function not only of the near-wellbore variability in permeability (in the vertical direction) but of the spatial variability in shale lenses and other extreme values of permeability. Premature entry of water into vertical wells may be due to coning, or by behind-pipe communications, or by influx of aquifer water without coning.

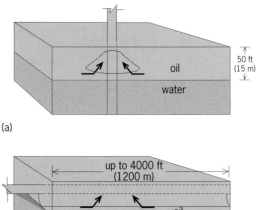

(a)

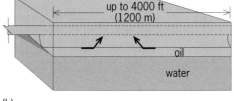

(b)

Fig. 2. Concepts of (a) coning and (b) cresting for vertical and horizontal wells.

As a general rule, if vertical wells actually cone water or gas, the vertical permeability is high enough to consider horizontal wells.

Thin pay zones. Horizontal wells in reservoirs with thin pays substantially increase the well's flow capacity compared to vertical wells by dramatically increasing the total area available to flow into the well. Other successes include thermal recovery applications, reservoirs with large horizontal permeability anisotropies, sand control applications, Productivity Index (PI) improvement, and marginal field development.

Drilling technology. Drilling technology for horizontal wells evolved from directional drilling technology developed for offshore and other centralized drill-site applications. Most horizontal wells are characterized by their turning radius, reflecting how quickly the well angle goes from vertical ($0°$) to horizontal ($90°$). In general, the horizontal wellbore length possible increases with increased radius.

Ultrashort radius systems. These are deep, large-diameter perforating systems that may be coupled with a sand control system. They go from vertical to horizontal in a foot or less. Some systems use high-pressure fluids to jet horizontal holes that are typically 10 ft (3 m) in length. Other systems ream out large open-hole intervals and use drills to place multiple 5–20-ft (1.5–6-m) horizontal wells into the reservoir. These wells may actually be partially cased with slotted or perforated liners or prepacked screens. A typical application of this technique is utilized in steam injectors for tar sands.

Short-radius systems. These can accomplish the turn from vertical to horizontal within a 20–150-ft (6–46-m) radius. The drilling equipment is combined with small-diameter tubulars and is typically used to reenter vertical wells. Short-radius horizontal wells typically drill only 500–1500 ft (150–460-m) of horizontal hole. Directional control is not as good as in medium- and long-radius systems. It is generally impossible to log, selectively complete or stimulate, or successfully work over the resulting small-diameter holes. This technique is often attractive when a vertical wellbore is available or when a more expensive horizontal well does not appear economic.

Medium-radius systems. These systems use conventional tubular goods and a variety of directional drilling motors. The turning radius can vary from about 200 to 500 ft (60 to 150 m). Extended-reach medium-radius wells can achieve horizontal lengths in excess of 8000 ft (2440 m). Medium-radius systems have greatly improved directional accuracy compared to short-radius methods, and may be as accurate as the long-radius techniques. Hole diameters and turning radius are sufficient to permit the use of measurement-while-drilling and steering tools for directional control and either logging-while-drilling (LWD) or drill-pipe-conveyed logs for formation evaluation. Small-diameter (say, less than 7-in. or 18-cm) holes will not be able to run the full suite of LWD tools; however, smaller-diameter LWD tools are being developed.

Long-radius systems. These systems use extensions of conventional deviated well technology to drill 1.5–8°/100 ft (30 m) build rate wells. Displacement length potential is very long, with true horizontal sections of more than 9000 ft (2740 m). However, a long build section (800–4000 ft or 245–1220 m) coupled with long "hold" sections is used in long-radius systems to drill extended-reach wells in excess of 20000 ft (6100 m) vertical offset. Long-radius drilling assemblies require only conventional rotary tools and motors. Almost all logging, measurement-while-drilling, and logging-while-drilling tools can be run in these wells with drill-pipe-conveyed logging or logging-while-drilling required for hole deviations greater than about $60°$.

Multilateral horizontal wells. These wells employ a variety of techniques to incorporate additional horizontal laterals. They are used to access multiple reservoirs from a single vertical takepoint, to access stacked reservoirs, to evaluate multiple exploration targets from a single well, and in enhanced recovery projects. Completing and producing independent horizontal laterals remains a challenge; however, very complex wellbore configurations are being implemented in areas with high reserves. Multilateral wells are most attractive for deep reservoirs where the vertical well portion of the drilling is high, where the need for downhole intervention during the life of the well is low, and when conventional laterals generate inadequate flow rates and recovery efficiency.

Steering horizontal wells. A drill bit can be rotated by either surface drives that transmit torque using the entire drilling string or by downhole motors. Surface drives can be either on the rig floor (the rotary table) or at the top of the drill string (top-drive). Top-drive rigs have several advantages in directional drilling because they allow longer stands of drill pipe, greater power, and circulation and rotation while pulling out of the hole.

Downhole motors are used to generate power and rotate the drill bit without requiring rotation of the drill pipe. This mode is referred to as slide drilling. A positive displacement motor consisting of an eccentric rotor (a helical shaft) turning in a rubber stator is operated by mud flow through the channel in the rotor/stator interface.

For surface-rotated drilling, the bottom-hole assembly (BHA) typically consists of stabilizers to control wellbore trajectory and prevent the BHA from getting stuck to the borehole; and drill collars, stiffer than conventional drill pipe and often treated to eliminate magnetic interference with the measurement-while-drilling/logging-while-drilling equipment. Elaborate combinations of stabilizers are used to increase, decrease, or maintain hole angle. Adjustable stabilizers provide real-time control of downhole bit inclination. This type of bottom-hole assembly provides good control of the inclination but less control of wellbore azimuth.

Modern horizontal wells frequently utilize a combination of rotary and slide drilling. Rotary drilling is generally faster than slide drilling and results in fewer doglegs (variability in hole trajectory) and kinks. The amount of torque and drag are related directly to dogleg severity, the length of the build section, drilling fluids, and the length of horizontal displacement.

Drill bits for horizontal wells include standard tricone bits and polycrystalline diamond compact (PDC) bits. The aggressive PDC bits outlast conventional bits in shale and clean limestone intervals but are less attractive in more abrasive formations. Roller-cone bits have more moving parts than PDC bits; this increases the possibility of fishing for lost cones in a horizontal interval. Roller-cone bits have a greater tendency to walk than do PDC bits.

Geological control. Horizontal well planning requires all the planning, safety, and environmental controls used for vertical wells and a variety of new challenges. Many of these challenges are related to the difficulty of directional control. Vertical and conventional deviated wells require a well to be drilled deep enough to penetrate the horizons of interest at a target location. A small error in estimating the actual depth of a target horizon is not usually significant. Even large errors of a few hundred feet can usually be corrected for dynamically, based on sample analysis, or by a log run.

Horizontal wells approach the target formation nearly parallel to the bed direction. Medium- and long-radius wells have excellent geometrical depth-control capabilities. However, a small error in estimating the depths and dip of a target horizon may cause significant problems in horizontal wells. For example, if a horizontal well is initially in the center of a 40-ft-thick (12-m) target with trajectory parallel to the estimated dip, a $2°$ dip error will cause the well to be out of zone in less than 600 ft (180 m).

Here, two problems are evident. First, the spatial geometry of the reservoir and identification of the actual target zone must be very accurately defined. Second, true vertical depth (TVD) errors induced by inaccurate estimates or small faults require significant additional drilling lengths (usually nonproductive) to get the well back in zone. Both problems require the operator to be able to rapidly and accurately assess both well path and geologic variations. The use of geological controls (including mud logs, logging-while-drilling, drill-pipe-conveyed logs, and pumpdown logs) for the well course is called geosteering.

Geosteering is the planned interactive navigation of a wellbore using geological criteria. It implies feedback, with the model of the wellpath and reservoir being continuously informed by all available data. Potential gains in production must be balanced with additional drilling and formation evaluation costs. Proper characterization requires knowledge of where the wellpath is located, where the current trajectory will take the wellpath, and where the wellbore should go. Uncertainty in geological modeling and the need to maximize profitability require an interdisciplinary team approach.

Successful geosteering is more likely when lateral continuity of the target interval is good. Detection of lateral discontinuities is improved by incorporation of three-dimensional (3-D) seismic technologies. When the goal is to maintain an optimal separation from a fluid contact, successful geosteering requires a detectable variation in the formation evaluation response across the contact. The decision to use geosteering must consider the type and quality of data available and the ability to accurately steer the wellbore.

Geosteering can be categorized into five types as follows.

I. Landing in zone. Logs obtained while performing the build or hold sections of the well are converted to true vertical depth format and correlated to predict the depths and dips of the target interval. This information is used to determine the subsequent directional drilling required. This method is used exclusively for many wells, and in conjunction with virtually every other geosteering method.

II. Boundary detection. This method steers a horizontal well trajectory based on log indications of exiting or reentering specific formations. The principle is somewhat analogous to driving a car on a highway in which each lane is outlined with small bumps. The driver can detect leaving a zone by the characteristic sound and feel of the bumps. Similarly, in type II geosteering, one must be able to detect bed boundaries that help identify the appropriate zones. Some methods use deep reading devices to make it possible to detect boundaries from a distance, while others do not detect the boundary until it has been crossed.

III. Magnitude maintenance. This method uses the absolute magnitude of the log response to identify when a wellbore remains in zone. Thus, reverting to the driving analogy, as long as the road is smooth, the driver is assured that he has not driven off into the dirt. Type III methods offer no suggestions as to how to get back into zone, and are used primarily to maintain the status quo and to alert as to when an updated geologic model may be required.

IV. Model methods. Model methods require characterizing information from an offset location that is transformed into a model of the response that should be obtained from the horizontal well. Measured log information from the horizontal well can then be compared directly to the calculated model results. This comparison is typically in measured depth or vertical section along the horizontal interval. A series of reservoir dips is used to calculate a series of modeled log responses. The actual dips are determined by selecting the model of the horizontal well response that most closely resembles that from the actual wellbore.

V. True stratigraphic depth modeling. This method involves comparing characterizing information from a horizontal well being drilled in a stratigraphic interval with information obtained from an offset vertical well. The true stratigraphic depth (TSD) format is inherent in this comparison, and apparent bed dips (as well as faults) are interpreted by the comparison process used to develop the TSD display. The horizontal well log is transformed to TSD format and compared directly to the vertical log.

For background information *see* BORING AND DRILLING, GEOTECHNICAL; OIL AND GAS, OFFSHORE; OIL AND GAS FIELD EXPLORATION; OIL AND GAS WELL COMPLETION; OIL AND GAS WELL DRILLING; PETROLEUM ENGINEERING; PETROLEUM GEOLOGY; TURBODRILL; WELL LOGGING in the McGraw-Hill Encyclopedia of Science & Technology.   Nathan Meehan

Bibliography. P. Buset and J. O'Neil, Coiled tubing sidetrack taps bypassed reserves, *World Oil*, pp. 87–93, April 1998; T. G. Byrom, Coiled tubing drilling—in perspective, *J. Petrol. Technol.*, 1999; C. Clavier, The challenge of logging horizontal wells, *Log Analyst*, pp. 63–84, March-April 1991; R. Cox, *Horizontal Underbalanced in a Sour Gas Carbonate Using Coiled Tubing*, Society of Petroleum Engineers, SPE 37075, 1996; L. R. B. Hammons et al., Stratigraphic control and formation evaluation of horizontal wells using MWD, *1991 SPE Annual Technical Conference*, Dallas, SPE 22538, October 6-9, 1991; T. L. Koonsman and A. J. Purpich, Ness horizontal well case study, *1991 SPE Offshore Europe Conference*, Aberdeen, SPE 23096, September 3-6, 1991; D. G. Kyte, D. N. Meehan, and T. R. Svor, Method of Maintaining a Borehole in a Stratigraphic Interval During Drilling, U.S. Patent No. 5,311,951, May 17, 1994; D. N. Meehan, Advances in horizontal well technology, *JNOC-TRC 2d International Symposium*, Chiba, Japan, 1993, Spec. Publ. TRC, JNOC, no. 4, 1994; D. N. Meehan, Rock mechanical aspects of petroleum engineering, *Proceedings of the 1st North American Rock Mechanics Symposium*, Balkeema, 1994; D. N. Meehan and T. R. Svor, Quantifying horizontal well logs in naturally fractured reservoirs, Part II, *1991 Fall SPE Meeting*, Dallas, SPE 22792; D. N. Meehan and S. K. Verma, Integration of horizontal well log information in fractured reservoir characterization, *1992 Fall SPE Meeting*, Washington, D.C., SPE 24697; A. B. Ramos, Jr., et al., Horizontal slimhole drilling with coiled tubing: An operator's experience, *J. Petrol. Technol.*, pp. 1119-1125, October 1992; S. T. Solomon et al., A multidisciplined approach to designing targets for horizontal wells, *J. Petrol. Technol.*, February 1994; T. R. Svor and D. N. Meehan, Quantifying horizontal well logs in naturally fractured reservoirs, Part I, *1991 Fall SPE Meeting*, Dallas, SPE 22634.

Oil reserves

Most oil is produced from giant oil fields found before the 1970s and is collectively termed conventional oil. This oil will continue to dominate supply until well past peak production. Some nonconventional oil is already in production and will be increasingly important after peak. At that time a discontinuity will mark the onset of a decline in the supply of this fuel that has powered the world economy for most of the twentieth century. To obtain an accurate estimate of the oil reserves, it is necessary to determine how much oil has been found and when it was found. The amount discovered at any reference date comprises how much had been produced to that date (cumulative production) and how much remains to be produced from known fields (reserves).

Production. The measurement of oil production involves simply reading a meter. However, the public database on oil production is not a reliable source due to inconsistent reporting resulting from the inclusion or exclusion of natural gas liquids and synthetic fuels; the inclusion or exclusion of nonconventional oils, variously defined; the metering of two or more fields together; war loss, for example 2 billion barrels (Gb) in Kuwait; and national boundary changes. Approximately 23 Gb of conventional oil, as herein defined, were produced in 1998, and cumulative production to the end of that year was about 817 Gb (see **table**).

Reserves. The reporting of reserves is estimated using geological or engineering knowledge and criteria. The size of the potential reserves, in a prospect prior to drilling, is routinely estimated by mapping the volume of the geological trap on the basis of seismic surveys and regional information to obtain a value for oil-in-place, of which not more than 20–50% is normally recoverable. Since most oil reserves occur far underground and cannot be measured directly, reserve estimates are subject to various degrees of probability. High case estimates are generally applied, having a probability ranking of not more than 30–40% (termed P_{30}-P_{40}), especially where the cost may be offset against tax with high marginal rates.

An entirely different approach goes into action if the first well on the prospect, termed a wildcat, is successful. Engineers have to wrestle with many complex issues such as the designing of the facilities, setting the investment levels, and maximizing the profit (financial investments are at stake and often in substantial amounts). It makes good sense to use the lowest reserve estimate giving an acceptable economic return. In the case of a large field with a low economic threshold, the reserves announced at the onset of commercial development may be as little as half those proposed by the explorers. Such conservative assumptions can be said to have a probability of 90–95% (P_{90}-P_{95}). It is obvious that they will grow over time to a P_{50} value, which by definition is the value established on final abandonment.

The conservative (P_{90}-P_{95}) value is commonly termed proved or proven and is the number reported by companies in their financial statements. A further constraint is imposed by the U.S. Securities and Exchange Commission, which accepts as proved only such oil as is in the catchment area of a producing well, normally 40 acres (16 hectares). Consequently, the reserves of a field appear to "grow" as it is drilled up. The P_{50} value, which is the best estimate of what a field will actually deliver, is commonly termed proved and probable. A realistic estimate for P_{50} conventional reserves as of the end of 1998 is about

Conventional oil endowment, 1998*

Rank	COUNTRY	Production		Reserves, median probable (P$_{50}$)	Discovered, Gb	Yet-to-find, Gb	Yet-to-produce, Gb	Ultimate, Gb	Year of peak production
		Thousand barrels per day	Cumulative, Gb						
1	Saudi Arabia	8058	83.23	207.85	291.08	18.92	226.77	310	2008
2	Former Soviet Union, except Caspian	7021	133.13	77.31	210.44	19.56	96.87	230	1988
3	United States (48)	4544	163.76	16.86	180.62	9.38	26.24	190	1971
4	Iraq	2114	23.96	89.38	113.34	16.66	106.04	130	2008
5	Iran	3597	49.14	62.79	111.93	8.07	70.86	120	1974
6	Venezuela	2331	48.77	28.60	77.37	7.63	36.23	85	1970
7	Kuwait	1796	28.37	53.86	82.23	2.77	56.63	85	2008
8	Abu Dhabi	1893	15.40	60.71	76.12	3.88	64.60	80	2008
9	Mexico	3048	25.52	22.82	48.34	6.66	29.48	55	2000
10	China	3196	23.77	24.44	48.21	2.79	27.23	51	1999
11	Libya	1395	20.90	23.77	44.66	3.34	27.10	48	1970
12	Nigeria	2080	19.43	18.28	37.71	2.29	20.57	40	1974
13	Caspian	200	1.82	5.00	6.82	23.18	28.17	30	2030
14	United Kingdom	2660	16.06	12.98	29.03	0.97	13.94	30	1998
15	Norway	3049	11.71	14.19	25.90	3.10	17.29	29	2000
16	Canada	618	16.26	6.90	23.16	4.84	11.74	28	1972
17	Algeria	818	10.96	16.04	24.17	2.83	16.04	27	1978
18	Indonesia	1289	18.04	8.02	26.06	0.94	8.96	27	1977
19	Neutral Zone	547	5.78	7.66	13.44	0.56	8.22	14.0	1999
20	Egypt	842	7.52	4.38	11.90	1.10	5.48	13.0	1995
21	Oman	895	5.66	6.34	12.52	0.39	7.34	13.0	1999
22	India	659	4.63	5.96	10.59	0.41	6.37	11.0	1999
23	Colombia	743	4.76	5.15	9.91	1.09	6.24	11.0	2000
24	Argentina	848	7.15	2.88	10.03	0.47	3.35	10.5	1998
25	Qatar	664	5.79	3.83	9.62	0.38	4.21	10.0	1998
26	Malaysia	731	4.25	4.00	8.25	0.75	4.75	9.0	1999
27	Australia	592	4.88	3.47	8.36	0.64	4.12	9.0	1998
28	Brazil	419	4.22	2.11	6.33	0.67	2.78	7.0	1990
29	Angola	726	3.46	2.51	5.97	1.03	3.54	7.0	1998
30	Romania	135	5.57	1.00	6.57	0.18	1.18	6.75	1976
31	Syria	555	3.03	2.49	5.52	0.48	2.97	6.0	1995
32	Ecuador	383	2.71	2.71	5.42	0.58	3.29	6.0	1997
33	Dubai	315	3.33	1.08	4.41	0.34	1.42	4.75	1991
34	Gabon	360	2.40	1.64	4.05	0.45	2.10	4.50	1996
35	Brunei	136	2.73	1.44	4.17	0.08	1.52	4.25	1978
36	Trinidad	121	3.02	0.59	3.61	0.14	0.73	3.75	1981
37	Peru	110	2.18	0.77	2.96	0.54	1.32	3.50	1983
38	Yemen	382	1.08	1.53	2.61	0.39	1.92	3.00	2001
39	Congo	238	1.14	1.37	2.51	0.24	1.61	2.75	2000
40	Germany	58	1.84	0.51	2.34	0.16	0.66	2.50	1967
41	Denmark	235	0.83	1.03	1.86	0.24	1.27	2.10	2000
42	Vietnam	226	0.46	1.02	1.47	0.33	1.34	1.80	2003
43	Tunisia	79	1.09	0.44	1.53	0.22	0.66	1.75	1985
44	Italy	109	0.75	0.63	1.38	0.27	0.90	1.65	2000
45	Cameroon	125	0.91	0.31	1.23	0.27	0.59	1.50	1986
46	Bahrain	102	1.08	0.19	1.27	0.23	0.42	1.50	1993
47	Turkey	64	0.75	0.29	1.03	0.17	0.45	1.20	1991
48	Netherlands	57	0.77	0.15	0.92	0.08	0.23	1.00	1987
49	Sharjah	70	0.40	0.40	0.81	0.19	0.60	1.00	2001
50	Papua	79	0.24	0.43	0.68	0.27	0.71	0.95	1993
51	France	34	0.68	0.16	0.84	0.06	0.22	0.90	1988
52	Austria	21	0.75	0.09	0.84	0.06	0.15	0.90	1955
53	Hungary	26	0.67	0.07	0.74	0.16	0.23	0.90	1984
54	Thailand	84	0.27	0.44	0.71	0.19	0.63	0.90	2003
55	Albania	5	0.53	0.14	0.67	0.13	0.27	0.80	1983
56	Pakistan	55	0.38	0.20	0.59	0.11	0.32	0.70	1992
57	Bolivia	28	0.38	0.17	0.54	0.16	0.32	0.70	1974
58	Phillippines	1	0.05	0.34	0.39	0.21	0.55	0.60	2010
59	Chile	9	0.41	0.13	0.54	0.15	0.19	0.60	1982
	REGIONS[†]								
1	Middle East Gulf	18,005	205.88	482.26	688.13	50.87	533.12	739.00	2008
2	Eurasia	10,583	165.49	107.96	273.45	46.00	153.96	319.45	2001
3	North America	5162	180.02	23.76	203.78	14.22	37.98	218.00	1972
4	Latin America	8040	88.09	65.93	154.02	29.03	94.96	183.05	1999
5	Africa	6664	67.81	65.90	133.71	11.79	77.69	145.50	1997
6	Western Europe	6222	33.39	29.72	63.11	4.94	34.66	68.05	1998
7	East	3852	35.93	25.33	61.26	3.94	29.27	65.20	1997
8	Middle East. Other	3046	21.12	16.68	37.80	2.65	19.33	40.45	1998
9	Other	380	3.36	3.54	6.90	1.10	4.64	8.00	
10	Unforeseen	0	0.00	0.00	0.00	13.30	13.30	13.30	
	WORLD	61,953	817	821	1638	162	983	1800	

*Conventional oil excludes heavy, tar, deepwater (≥500 m), polar, enhanced, natural gas liquids. Values are shown as computed but should be generously rounded.
[†]Middle East Gulf includes Abu Dhabi, Iran, Iraq, Kuwait, Saudi Arabi. Eurasia includes former Soviet Union, Eastern Europe, China. North America includes Canada and United States (48). Western Europe includes Norway to Italy.

821 Gb. Higher estimates have been published, but questionable reporting practices, imprecise definitions, and the lack of auditing cast doubt on their validity (see table).

Published reserves are generally described as proved, although in reality they are normally far from that. Some are indeed P_{90-95}; most are closer to P_{50}; and some have a lower probability ranking, as in the case of the former Soviet Union, several members of the Organization of Petroleum Exporting Countries (OPEC; where quota was based on reserves), and Mexico (where reserves were used as debt collateral).

Another important issue is the dating of the reserve revisions. Most upward revisions are attributed to advances in technology, cost cutting, and good management, when in fact much of the revision is just a correction of earlier underreporting. These revisions should be backdated to the discovery of the field concerned, because nothing was dynamically added at the date of the reported revision. Failure to backdate gives a very misleading impression of the discovery trend (**Fig. 1**).

Adding to the difficulty in estimating reserves is the widely held assumption that advances in technology permit more oil to be extracted from known fields. Recovery factors are relevant only during the exploration and development planning stage. Later, recovery is determined from the performance of the wells themselves, and what percentage that might be of an assumed amount of the oil-in-place is largely irrelevant. The amount cannot in any case be measured accurately because much of it is held forever in the ground by capillary pressures. A depletion plot of the giant Prudhoe Bay field in Alaska demonstrates the negligible impact of technology on the decline trend that has been constant for many years (**Fig. 2**).

Discovered oil. The sum of the cumulative production and P_{50} reserves gives total discovered oil of 1638 Gb through 1998. World discovery peaked in the 1960s (Fig. 1), and wildcat drilling is delivering ever less (despite all the technological achievements). Discovery rate has been falling for decades, and in 1998 was down to about 6 Gb per year. Consequently, four barrels of oil are consumed for each new barrel found.

Extrapolation of this discovery trend points to an ultimate recovery of about 1800 Gb, which is confirmed by other statistical techniques. Of this, approximately 162 Gb are yet-to-find. Production is likely to peak when about half (900 Gb) of the ultimate recovery has been produced, which based on current trends will be in the early years of the twenty-first century. During this time, the peak of production will likely be a discontinuity with great political and economic consequences. It will be necessary to tap other sources of energy and to conserve supplies of conventional oil. Including gas liquids and nonconventional oil, an overall peak of hydrocarbon production occurs around the year 2008. At the present rate of consumption, there would

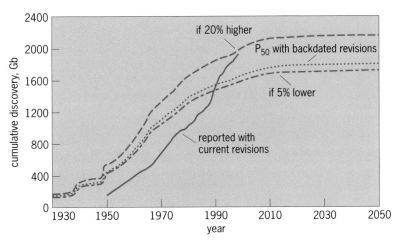

Fig. 1. Impact of backdating reserve revisions on discovery trends. Peak discovery occurred in the mid-1960s.

be an appreciable loss of all hydrocarbons by 2050 (**Fig. 3**).

Nonconventional oils. The production of nonconventional oils, which are less well known and at an earlier stage of depletion, is likely to be stepped up after global peak. The resource base of some categories is large, although producible only at slow rates.

Oil shales. These are actually neither oil nor shale but immature source rocks that have to be retorted at temperatures above $600°C$ ($1112°F$) to give up their hydrocarbons. Oil shales are more akin to coal than conventional oil; have a poor energy equation, consuming almost as much energy as they yield; and carry high environmental costs.

Tarsands and extra-heavy oils. These are viscous and dense oils which represent biodegraded conventional oils forming some huge deposits on the margins of prolific basins; the largest are in western Canada and eastern Venezuela. The extraction process, which partly involves open-cast mining, is slow and costly. The deposits, though very large, are far from homogeneous, with only the most favorable locations being currently viable. Even if production were increased three- or fourfold (with an enormous effort), it would still be insufficient to impact global peak. Under some classifications, heavy oils (10–20° API) are also treated as nonconventional, but the

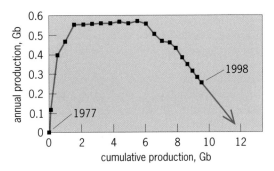

Fig. 2. Depletion plot of oil reserves in the Prudhoe Bay field showing that depletion is unaffected by advances in technology. The ultimate recovery was estimated at 12.5 Gb, which is still an accurate approximation.

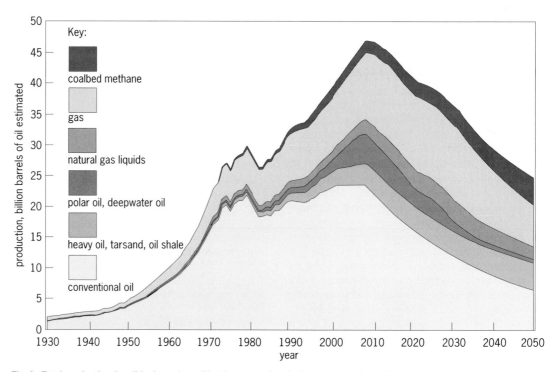

Fig. 3. Total production for all hydrocarbons (that is, conventional oil, nonconventional oil, and gas liquids).

public database does not allow a clear differentiation.

Deepwater oil. This is found in greater than 500 m (1640 ft) of water and is technologically within reach. Some 400 deepwater wildcats have been drilled yielding about 25 Gb of reserves. Most of the deepwater areas are nonprospective for well-understood geological reasons, but the margins of the South Atlantic, the Gulf of Mexico (including the Mexican sector), and a few other areas offer an ultimate potential of perhaps about 100 Gb. Only fields with high flow rates per well are remotely viable, and the disposal of gas is a further constraint.

Polar oil. This is treated as nonconventional because it lies in an inhospitable environment. Antarctica is closed by agreement, and it is in any case not very prospective. The Arctic is more promising. Alaska is already in production, and vast tracts in the Siberian offshore have been barely explored. However, evidence suggests that much of the province is gas-prone.

Enhanced recovery. This is classified as nonconventional oil, as it means revisiting a difficult field that was developed with primitive technology long ago. Enhanced recovery is mainly practiced in the onshore United States, where about one-tenth of the fields are susceptible and yield up to 10% more oil with the help of steam injection and other well-known techniques. The category is not to be confused with improved recovery, which covers the proceeds of various technological advances, including highly deviated wells (termed horizontal). Improved recovery is routinely applied in all modern fields, and its contribution is to be included in the P_{50} estimates of conventional oil.

For background information *see* PETROLEUM ENGINEERING; PETROLEUM ENHANCED RECOVERY; PETROLEUM GEOLOGY; PETROLEUM RESERVES; PETROLEUM RESERVOIR MODELS; OIL AND GAS FIELD EXPLORATION; OIL SHALE in the McGraw-Hill Encyclopedia of Science & Technology.   Colin J. Campbell

Bibliography. C. Bond Hatfield, Oil back on the global agenda, *Nature*, 387:121, 1997; C. J. Campbell, *The Coming Oil Crisis*, Multi-Science Publishing Co. & Petroconsultants, 1997; C. J. Campbell and J. H. LaHerrère, The end of cheap oil, *Sci. Amer.*, 278(3):78–83, 1998; L. F. Ivanhoe, Updated Hubbert curves analyze world oil supply, *World Oil*, 217(11): 91–94, 1996; W. Youngquist, *Geodestinies: The Inevitable Control of Earth Resources over Nations and Individuals*, National Book Co., 1997.

Open pit mining

Global Positioning System (GPS) technology is rapidly becoming an integral part of open pit mining. Mines around the world use GPS for surveying, tracking mobile equipment, and providing real-time precision guidance for excavators, drills, and bulldozers (**Fig. 1**). GPS provides a level of control never before available, allowing mine operators to effectively compete even when commodity prices drop.

Global Positioning System. GPS is a navigational system developed by the U.S. Department of Defense and operated by the Air Force. It allows users to determine position, velocity, and time, using GPS receivers and radio signals broadcast from satellites orbiting Earth at an altitude of 20,200 km (12,552 mi). The GPS constellation comprises 24 satellites, over

Fig. 1. Equipment monitoring using GPS.

six orbital planes consisting of four satellites each. Each satellite orbits Earth roughly every 12 h. Together, the satellites cover the entire planet continuously. In order to determine its position (latitude, longitude, and elevation), a GPS receiver requires an unobstructed view of at least four GPS satellites. *See* GLOBAL POSITIONING SYSTEM.

Two navigational services are provided by GPS: Standard Positioning Service (SPS) for civilian use, and Precise Positioning Service (PPS) for military applications. The civilian receiver can be accurate to roughly 10–20 m (33–66 ft) horizontally, but the U.S. military degrades the accuracy to about 100 m (330 ft).

Achieving better accuracies, like those necessary for mining applications, requires the use of a relative positioning technique called differential Global Positioning System (DGPS). In DGPS, a GPS receiver is installed at a fixed location and its position is surveyed. Next, the position that a receiver computes is compared to its known (surveyed) position to determine the measurement inaccuracies. Correction factors, known as differential corrections, are then broadcast by radio link to the receivers on the "rovers" (surveyors, trucks, shovels, drills, and so on) to fix their calculations and bring accuracy to within meters.

Obtaining even higher precision requires more complex techniques. Then, accuracies of around 1–2 cm (0.4–0.8 in.) are routinely obtained with good satellite geometries. Real-Time Kinematic (RTK) systems achieve these accuracies on-the-fly.

Other factors can affect GPS accuracy in mining: mine topography, overhead satellite geometry (measured as Position Dilution of Precision, or PDOP), multipath (satellite signals bouncing off surrounding objects), and atmospheric effects.

In deep pit mines, because of high walls and other obstructions, the GPS constellation alone might not support tracking and positioning systems on a continuous basis. Such mines can dramatically increase the number of hours of available use when GPS is augmented with the Russian Global Navigation Satellite System (GLONASS; **Fig. 2**). GLONASS provides up to 24 additional satellites when fully deployed, thereby increasing the number of visible satellites. Units that receive both GPS and GLONASS transmissions are now available.

Tracking mobile equipment. Many mines now use GPS integrated with computerized dispatching systems to track haul-truck, shovel, and loader locations. Because mining equipment is expensive to operate, idle time needs to be minimized; optimal

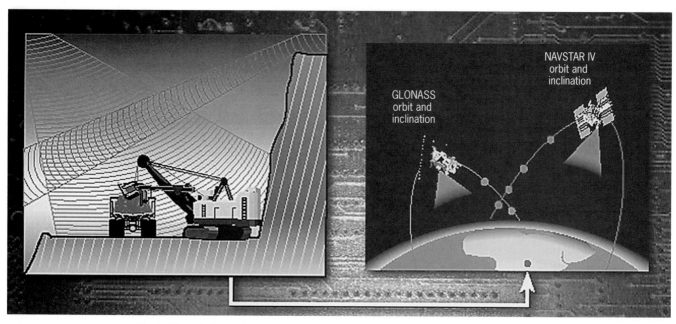

Fig. 2. GPS with GLONASS augmentation.

haul-truck assignments are possible only when using exact truck positions. With GPS, truck locations are continuously monitored (**Fig. 3**), and if shovel dig rates change, trucks can be reassigned to minimize truck and shovel idle time.

GPS information is collected from equipment at regular intervals (every 30 s) when the mine is operating, and is stored in a database. Mines can use this historical data to generate "dot traces" of equipment movement for selected periods of time. These plots have numerous uses, including monitoring shovel advances, truck dumping patterns, and road use. Additionally, GPS readings for trucks and shovels collected throughout the shift can be replayed (animated) later to help identify bottlenecks.

Normally, an accuracy of up to 2 m (6.6 ft) is

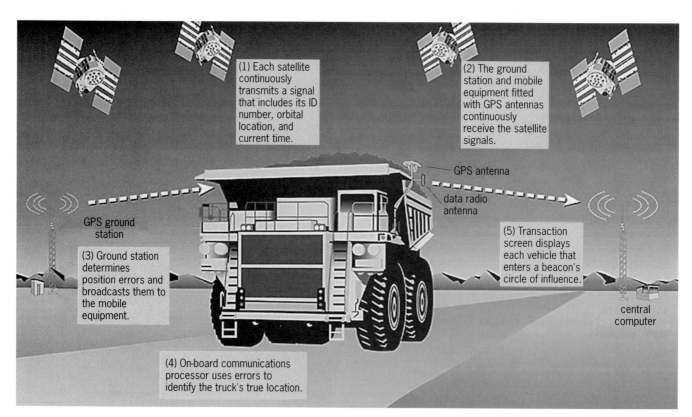

Fig. 3. Tracking mine haul trucks using GPS.

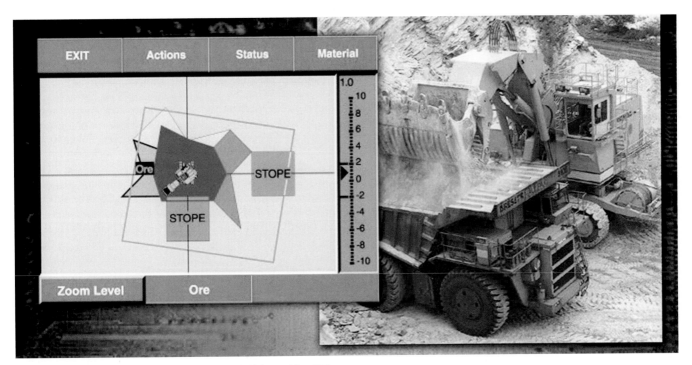

Fig. 4. Excavator display for monitoring a shovel with high-precision GPS.

sufficient for most mobile equipment applications. This is achieved using low-precision, low-cost receivers and DGPS. This level of accuracy allows mines to determine, for example, the crusher into which a truck is dumping.

In addition to tracking primary load and haul equipment using GPS, mines are now installing receivers on support equipment such as graders, bulldozers, and water trucks to track and optimize work assignments.

Precise positioning. Mine surveyors traditionally have had laborious jobs. They survey areas that the shovels work, and stake the limits of each mine block to help operators make sure that they are digging the right material. They continuously "spot survey" shovel elevation to make sure the shovels are on target grade. All holes on drill patterns need to be staked, as do cut and fill ramps developed by bulldozers. Precision GPS is helping to provide better control and minimize field survey work.

Shovel positioning. High-precision GPS receivers are used on mining shovels to provide real-time pit floor grade control and to make sure that material from each mine block goes to the correct destination. Pit or bench floor grade control is important to mines. If a shovel is digging below plan grade, it will be excavating unblasted material. This results in increased wear and maintenance cost for the shovel. If the shovel is digging above plan grade, then when next drilling in the area the productivity will suffer because drilling through broken material causes problems. Additionally, benches need to be kept as flat as possible in order to minimize wear on haul-truck frames; GPS helps to achieve this, therefore minimizing rework by bulldozers to flatten benches.

To help control pit or bench floor grade, GPS systems show an elevation bar on a graphical operator display (**Fig. 4**) that indicates whether a shovel's tracks are high or low in relation to the target elevation for the bench. The elevation bar is also useful for maintaining level benches or for creating sloping benches for improved drainage or to follow the dip of the orebody.

For material tracking, an outline of mine blocks (each block containing similar material) is sent over a radio telemetry link from the planning system in use at the mine, and is displayed to the shovel operator. As the shovel digs material from nearby mine blocks, the block being excavated is highlighted, and a dig line is updated on the display to show material dug; a shovel icon displayed on the console swings as the shovel swings. Additionally, GPS position updates are sent back to a workstation in the mine office, where a display similar to the operator's display is shown. This allows planners to monitor shovel progress in real time, and allows operators to verify they are digging the correct mine blocks at all times, even at night and in inclement weather, without the requirement for survey stakes to show block delineation. This tracking is particularly important in gold mines or in mines that blend their material.

Horizontal bucket position accuracy of under 2 m (6.6 ft) is common using these systems; this amounts to much less than one bucket width. Vertical position accuracy is normally within 30 cm (1 ft).

Because shovels operate against the wall of the open pit, a large portion of the sky is often blocked from view. Further, shovels have superstructures that exacerbate multipath problems, and constant swinging that causes satellites to drop in and out of view behind these superstructures. However, even if Position Dilution of Precision is poor, Real-Time Kinematic systems provide the necessary accuracy for shovel operation. In particularly deep pit mines, however, GPS may need to be augmented to increase availability.

Drill positioning. Similar to shovel positioning systems, blast hole drills require Real-Time Kinematic–level GPS and on-board consoles for operator guidance. Normally surveyors must stake entire drill patterns or the grid of holes to be drilled and blasted. Using GPS on drills, the drill pattern is sent over a radio telemetry link from the mine planning system to on-board consoles. The console displays "virtual drill hole stakes" that help guide operators to each hole with no physical staking required. When a drill is in position to start, the plan hole depth, current drill pit depth, penetration rate, and depth from bottom of hole are displayed. Operators use this information to drill the blast hole to exact depth.

Accurate drill positioning is important; a deviation from the drill pattern might cause large blocks or boulders of unbroken material after blasting if hole spacing is too great. It is also important to drill down to planned elevation to ensure a flat bench below after blasting and excavating.

Open pit drills are track-mounted and include leveling jacks to make sure that the drill steel is vertical before starting to drill. To help with GPS positioning over virtual stakes, inclinometers are used to monitor drill roll, pitch, and yaw so that leveling effects can be anticipated.

Bulldozer positioning. Like shovel and drill positioning systems, bulldozers require Real-Time Kinematic–level GPS and operator consoles. Because bulldozers can work a relatively large area, large construction plans are sent over the data radio link. The operator console shows areas to be cut or filled, and plan versus actual track elevation. Because machine attitude also needs to be monitored, two GPS receivers or inclinometers are sometimes used to track roll, pitch, and yaw. Because of the real-time control, mine operators claim near-perfect road construction the first time when using GPS as a guide.

Common practice with GPS. Many mines are using GPS as a labor-saving productivity tool to make them more competitive. The most common application is for heavy-equipment fleet control; however, precise shovel-positioning systems are gaining popularity. Mines in the Canadian west and American southwest are routinely using GPS-based fleet control and shovel, drill, and bulldozer systems. Because of the real-time control, planners are being creative. For example, benches that slope to drain water are being planned and implemented using shovel systems. This further enhances productivity in the area, and was previously an achievable with traditional methods.

GPS is fast becoming an indispensable part of the open pit mining environment. It is an ideal solution for open pit mines seeking to gain greater insight into and control over operations. Systems under development include unmanned dump trucks that use a combination of precise GPS and inertial guidance for navigation; and unmanned bulldozers, excavators, and drills using GPS as a key sensor for navigation. These systems are highly productive, and keep humans out of potentially unsafe areas.

For background information *see* COORDINATE SYSTEMS; MINING; OPEN PIT MINING; REMOTE SENSING; SATELLITE NAVIGATION SYSTEMS; SURFACE MINING; TERRESTRIAL COORDINATE SYSTEM in the McGraw-Hill Encyclopedia of Science & Technology.

Les T. Zoschke

Bibliography. J. Chadwick, Satellite positioning, *Min. Mag.*, pp. 312–321, May 1998; R. G. Graber, Challenges for GPS mine integration, presented at the SME Annual Meeting, Orlando, Florida, March 1998; R. Sheremeta, A year of GPS at Fording Coal Limited, Greenhills Operations, *CIM Bull.*, 89(1002):66–70, July–August 1996.

Optical coherence tomography

Optical coherence tomography (OCT) is a recently developed, noninvasive technique for imaging subsurface tissue structure with micrometer-scale resolution. The principles of time gating, optical sectioning, and optical heterodyning are combined to allow cross-sectional imaging. Depths of 1–2 mm (0.04–0.08 in.) can be imaged in turbid tissues such as skin or arteries; greater depths are possible in transparent tissues such as the eye.

Optical coherence tomography complements other imaging modalities commonly used to image subsurface tissue structure, including ultrasound and confocal microscopy. It has a resolution about an order of magnitude better than ultrasound, although the depth of imaging is less. Unlike ultrasound, it does not require a coupling medium between the instrument probe and tissue, facilitating endoscopic applications and imaging of sensitive structures such as the eye. Optical coherence tomography is a type of confocal system, although device parameters are usually set for a lower resolution than is available in commercial confocal microscopes. However, it has a depth of imaging several times that of confocal microscopy.

Principles of operation. In a typical optical coherence tomography system (**Fig. 1**), light from a broadband, near-infrared source and a visible aiming beam is combined and coupled into one branch of a fiber-optic Michelson interferometer. Broadband sources include superluminescent diodes, fiber amplifiers, and femtosecond pulse lasers in the wavelength range of 800–1550 nanometers. The light is split into two fibers using a 2×2 coupler, one leading to a reference mirror and the second focused into the tissue. Light reflects off the reference mirror and is recoupled into the fiber leading to the mirror. Concurrently, light is reflected from index-of-refraction mismatches in the tissue and recoupled into the fiber leading to the tissue. Reflections result from changes in the index of refraction within the structure of the tissue, for instance between intercellular fluid and collagen fibers. Light that has been back-reflected from the tissue and light from the reference arm recombine within the 2×2 coupler.

Because the broadband source has a short coherence length, only light which has traveled very close to the same time (or optical path length) in the reference and tissue arms will interfere constructively and destructively. By changing the length of the reference arm, reflection sites at various depths in the tissue can be sampled. The depth resolution of the optical coherence tomography system is determined by the effectiveness of this time gating and hence is inversely proportional to the bandwidth of the source. An optical detector in the final arm of the Michelson interferometer detects the interference between the reference and tissue signals. During optical coherence tomography imaging, the reference-arm mirror is scanned at a constant velocity, allowing depth scans (analogous to ultrasound A-scans) to be made. Either the tissue or the interferometer optics is mounted on a stage so that the beam can be scanned laterally across the tissue to build up two- and three-dimensional images, pixel by pixel.

The optical sectioning capability of optical coherence tomography (its ability to image thin slices of tissue) is determined by the tissue-arm optics. Light from the single-mode optical fiber is essentially a point source, which is focused to a small spot in the tissue. The highly turbid nature of many biological tissues tends to scatter light away from this focus, but the scattered light is not efficiently coupled back into the fiber. Therefore the lateral resolution of the optical coherence tomography system is approximately equal to the focused spot size. Often, the focus-spot diameter and lateral resolution are made approximately 10–20 micrometers to match the depth resolution of the system (using superluminescent diodes) and to provide a relatively long depth of focus. This long working distance is desirable for in vivo systems, since it is difficult to control the exact distance between the instrument optics and the tissue.

Optical coherence tomography has extremely high sensitivity; reflections of less than 10^{-10} of the incident optical power can be detected by using the optical heterodyne technique. Movement of the reference arm mirror induces a modulation of the interferometric signal at the Doppler frequency. In the optical heterodyne detection technique, the envelope of the detector current is recorded by demodulating at this frequency, thus rejecting noise outside the signal bandwidth. The magnitude of the detected signal at the modulation frequency is proportional to the reflectivity of the tissue.

Color Doppler OCT. Several instruments have been built based on variations of the basic optical coherence tomography system. For instance, polarization-sensitive optical coherence tomography uses polarization-altering optics in the arms of the interferometer to determine the sample birefringence from the magnitude of the back-reflected light. Op-

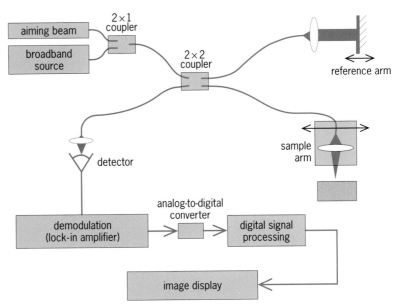

Fig. 1. Typical optical coherence tomography system, based on a broadband source and fiber-optic Michelson interferometer.

tical coherence microscopy uses a system of high numerical aperture to achieve resolutions comparable to confocal microscopy but with increased depth of penetration.

Color Doppler optical coherence tomography (CDOCT) is an augmentation capable of simultaneous blood flow mapping and spatially resolved imaging. This technique (also called optical Doppler tomography) makes use of the interferometric phase information ignored in conventional optical coherence tomography. Doppler shifts in light backscattered from moving objects (such as red blood cells) add or subtract from the modulation frequency. If the returned signal is coherently demodulated, a short-time Fourier transform can be performed to find the signal strength at various Doppler-shift frequencies. The mean velocity of the scatterers, $\overline{V}_s$, within each Fourier window can be estimated from the centroid, $\overline{f}_s$, of the localized Doppler frequency spectrum, according to the equation below. Here, c is the speed

$$\overline{V}_s = \frac{\overline{f}_s c}{2 v_o n_t \cos \theta}$$

of light, v_o is the center frequency of the broadband light source, n_t is the local index of refraction, and $\cos \theta$ is the angle between the instrument probe and the flow direction. The lateral resolution of a Doppler image remains the same as that of the magnitude image obtained in conventional optical coherence tomography; however, the depth resolution is often limited by the Fourier transform window size. A mean velocity can be calculated for each three-dimensional pixel in the image and combined to create a Doppler image separate and complementary to the magnitude image obtained from the amplitude of the signal alone. Alternatively, a threshold can be placed on the signals contributing to the Doppler image in order to eliminate noise, and the

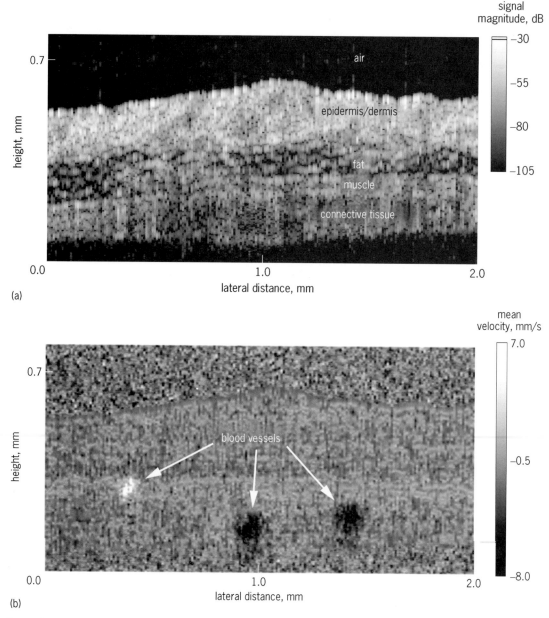

Fig. 2. Optical coherence tomography images of healthy hamster skin in vivo. (*a*) Conventional magnitude image. Various layers of skin can be seen in cross section. (*b*) Result after Doppler processing of the same signal used to create the magnitude image. Three blood vessels are visible because of the Doppler shift caused by their moving red blood cells.

Doppler image can be overlaid on the magnitude image.

Applications. Optical coherence tomography can be used to probe the structure of any accessible tissue. Noninvasive studies of the eye and skin are being performed, and a commercial device has been developed for retinal imaging. Using optical coherence tomography, parameters such as eye length can be accurately measured, and the cross-section images of the retina give a clear and quantifiable assessment of retinal separation and macular degeneration, among other pathologies. In skin, the morphology of normal skin layers and components, and disorders such as psoriasis, can be imaged. For example, the expected layers of skin, including epidermis and dermis, fat,

muscle, and connective tissue, are seen in a structure image of healthy hamster skin in vivo (**Fig. 2***a*). Blood vessels located in the connective tissue and fat layers are not seen in this image, because blood and other skin tissues have similar optical properties in the near infrared. When the same signal is processed to distinguish Doppler shifts, however, the moving blood is readily apparent (Fig. 2*b*).

Endoscopes and catheters are used to facilitate imaging of the cardiovascular system and gastrointestinal, respiratory, urinary, and reproductive tracts. Being fiber-based, optical coherence tomography is readily adapted to these applications. Miniature electromechanical and piezoelectric scanning devices have been developed to perform lateral scanning of the interferometer tissue arm optics. In blood vessels, the layered structures of the vessel wall are visible, and distinction can be made between fibrous and calcified plaques. Various mucous membranes of the body have also been successfully imaged. Differences in backscattering properties allow mucosal tissue layers to be diferentiated. Cancerous regions are often marked by a disordering of the layered structure and a more homogeneous appearance of the optical coherence tomography image. Because of this distinction, optical coherence tomography has the potential to guide excisional biopsy and improve the random technique frequently used today. Samples of tissue would be taken from areas which appeared suspect on optical coherence tomography images, for standard histological analysis.

Potential. Research groups are working on improving the performance of this novel imaging modality. Sources with broader bandwidths and higher powers will increase system resolution and depth of imaging. Signal processing techniques that efficiently extract parameters of interest and display data in an easily comprehended fashion are being developed. Efforts are under way to better understand optical coherence tomography images in terms of the morphology of the tissue under study.

For background information *see* COHERENCE; COMPUTERIZED TOMOGRAPHY; CONFOCAL MICROSCOPY; DOPPLER EFFECT; FIBER-OPTICS IMAGING; HETERODYNE PRINCIPLE; INTERFEROMETRY; MEDICAL ULTRASONIC TOMOGRAPHY; ULTRASONICS in the McGraw-Hill Encyclopedia of Science & Technology.

Jennifer Kehlet Barton

Bibliography. D. Huang et al., Optical coherence tomography, *Science*, 254:1178–1181, 1991; J. A. Izatt et al., Optical coherence tomography and microscopy in gastrointestinal tissues, *IEEE J. Selected Top. Quantum Electr.*, 2:1017–1028, 1997; X. J. Wang, T. E. Milner, and J. S. Nelson, Fluid flow velocity characterization by optical Doppler tomography, *Opt. Lett.*, 20:1337–1339, 1995.

Organic electrical conductors

Conducting polymers have received much attention since the early 1980s. They have inherent electrical conductivity and are different from other polymeric materials, which are made conducting by blending polymers with carbon or metal powders. In particular, the demonstration that films of conducting polymers on electrodes can be switched electrochemically between the insulating and conductive state with a corresponding change in the electrical, optical, and chemical properties attracted the attention of investigators from various disciplines. The brief overview below presents selected examples to describe the structural characteristics and properties of conducting polymers.

Polymer structure. Conducting polymers are normally composed of aromatic rings linked together to preserve the aromatic nature of the monomer.

Thus, they are known as polyaromatics. They are primarily linear polymers and have minimal branching. The bonding of the aromatic units in the polymer is usually para, or 1,4, for the phenylene (structure **1**) and aniline (**2**) polymers and 2,5 for the heterocyclic pyrrole (**3**) and thiophene (**4**) polymers (**Table 1**).

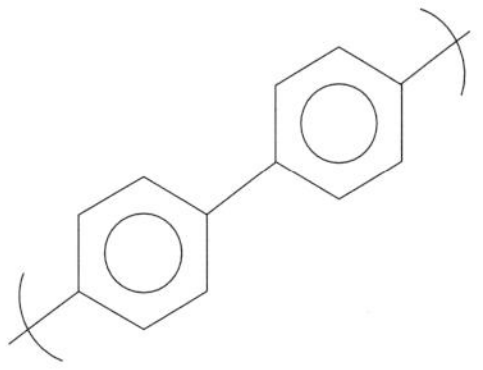

(1)

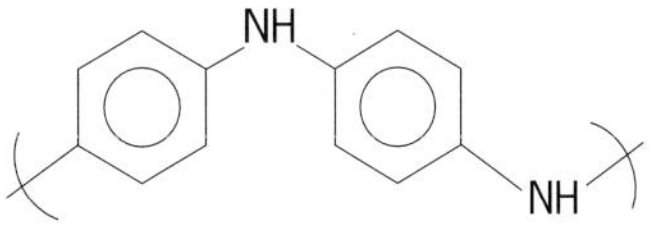

(2)

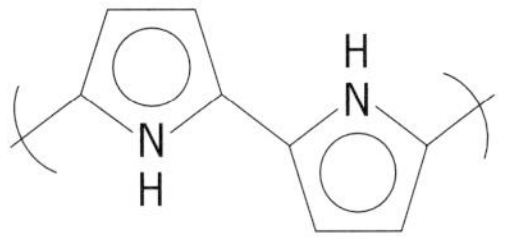

(3)

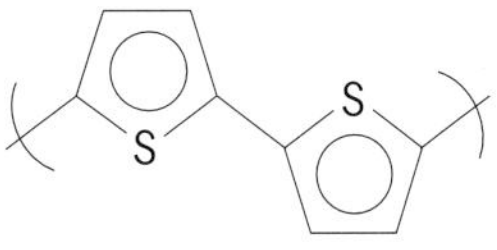

(4)

Polymers containing substituent side groups have been prepared to alter the properties. Some of the more common substituents are the long-chain alkyl and aryl groups.

Conductivity. The conductivity of these polymers is normally in the range of 0.1–1000 siemens/cm. There are conductivity differences between the polymers, and the conductivity of any one polymer often depends on the preparation and handling procedures. These polymers are significantly less conductive than metals (around 1,000,000 S/cm) but are several orders of magnitude more conductive than semiconductors. These polymers conduct electric current for very long periods of time when they are protected from environmental factors. The conductivity usually increases with temperature. The

Table 1. Examples of conducting polymers with no substituent side groups			
Polymer	Structure	Charge/ monomer	Conductivity, siemens/cm
Polyphenylene	1	0.2	100–500
Polyaniline	2	0.025–0.5	0.1–10
Polypyrrole	3	0.25–0.33	100–400
Polythiophene	4	0.05–0.15	1–20
Polyacetylene	5	0.05–0.2	100–1000

polymers function in a limited temperature range because most of them begin to decompose above 200°C (392°F). Some of these polymers have limited stability even at ambient conditions. For example, dry free-standing films of polythiophene and of polyphenylene and polythiophene lose conductivity upon standing in ambient conditions. Polypyrrole is less sensitive, and polyaniline appears not to be affected. The presence of substituents on the polymer often depresses the conductivity.

Charged polymer. Conducting polymers also differ from conventional polymers in that the polymer chains are charged and more commonly charged positively (cationically) due to the loss of electrons from the π structure. The degree of oxidation charge varies with the polymer type and ranges from one charge for every 2–20 aromatic rings. For polyacetylene (structure **5**) the charge level can be much

$$\left(\!\!=\!\!\diagdown\!\!=\!\!\diagup\!\!\right) \qquad \textbf{(5)}$$

higher. In most cases, the conductivity of the polymer increases with the level of oxidation. The polymers have relatively low oxidation potentials and are easily oxidized with mild oxidizing agents to generate the cationically charged form. The charge is delocalized along the π structure, and it is this characteristic which gives the structure a fair amount of stability. The π-delocalized charge in these polymers should be distinguished from the charge on polyelectrolytes or ionomers where the charge centers along the polymer are provided by a "salt" center, for example, a sulfonate group, and therefore, are localized on the "salt."

Polymer composition. The conducting polymers have a π-delocalized positive charge; therefore, the actual material is a combination of the polymer plus anions to maintain overall charge neutrality. The anion content is significant and can vary from 10 to 50% by weight. For example, the anion content in percent by weight is in the range of 10–35% for polypyrrole, 7–25% for polythiophene, and 22–40% for polyaniline. Since these materials are often prepared in electrolyte solution, either electrochemically or with a chemical oxidant, the recovered material has an anion content in excess of the stoichiometry due to inclusion of excess salt from the electrolyte solution. These materials can be considered polymer salts with a very large cation. The accompanying anion affects both the composition and properties of the conducting polymer. Finally, the saltlike nature of these materials makes them hydroscopic, and they normally contain varying amounts of moisture depending on the polymer and the anion.

Electrochemical properties. While the conductive polymers are actually salts where the polymer is in the oxidized form (cationically charged), the polymers are identified by the name of the neutral polymer. The respective neutral polymers are electrically insulating and are not to be confused with the oxidized, conductive polymer form. The neutral polymers have relatively low oxidation potentials and can be oxidized with mild oxidizing agents or by elec-

trochemical means to generate the oxidized form which has conductivity. Most of the polymers, for example, polypyrrole, polythiophene, polyacetylene, and polyphenylene, are electrochemically oxidized in nonaqueous solvents. Polyaniline and other aromatic amine polymers are oxidized in aqueous solutions. These amine-based polymers can also be switched by protonation and deprotonation reactions of the amine groups. With the proper experimental conditions (moisture and oxygen are often concerns), these polymers can be repeatedly driven ("switched") between the cationic and the neutral form. The cationic form is reduced to the neutral form, and the neutral form is oxidized to the cationic form as in the reaction below. Switching the

polymer is accompanied by a big change in its properties (**Table 2**).

Solubility, thermal stability, and mechanical properties. The parent polymers are effectively insoluble, severely limiting the handling and processing of these materials. The solubility is improved by incorporating substituents onto the polymer chain or by using certain anions. For example, a polyaniline with a surfactant anion which can be suspended in aqueous solutions is commercially available. Derivatives of polythiophenes, polypyrrole, and polyphenylene with alkyl-chain substituents dissolve in various organic solvents and, in some cases, water. The solubility limitation has also been circumvented with the use of a variety of alkyl-substituted oligomers which have the electrochemical, optical, and electrical properties of the polymers. The oligomers are soluble in common solvents and can be processed. The conducting polymers are not stable at temperatures much above 100–200°C (212–392°F). Polypyrrole, for example, loses conductivity with heating to around 200°C (392°F), then decomposes without significant softening or melting. Several of these polymers have a similar decomposition pattern. The mechanical properties of these parent polymers are not good, being hard and brittle. For example,

Table 2. Characteristics of a polyrrole in various forms

Neutral	Oxidized form (cationically charged)
Electrical insulator	Electrical conductor
Transparent	Opaque
Light yellow	Dark, blue black
Hydrophobic	Hydroscopic
Strong	Hard, brittle

polypyrrole stretches only 5% at break and has a tensile strength of 1200–8600 lbf · in.$^{-2}$ (81.7–585.2 atm). Therefore, these polymers are not easy to handle as free-standing materials and are often used as films on a substrate.

Polymer processing. Many of the substituted polymers are solvent-processible and can be spin-coated or cast from solutions to produce coatings or free-standing films. This is not the case with the parent polymers, which are insoluble. These solvent-blended polymers are often phase-separated composites rather than polymer blends. Again, the conducting polymers are quite incompatible with conventional polymers, as they are with conventional solvents. The blends with polyaniline are phase-separated, and contain a connective path network of polyaniline through the mixture at a very low polyaniline content. Thus, conductivity can be achieved with little change in the mechanical properties of the host polymer. Alternatively, th neutrale polymers are often more soluble than the conducting form, and can be processed and then oxidized to produce the conductive component. Significant progress has been made in developing process methods for these materials. The literature contains a variety of patents describing formulations, for example, suspensions, or processes for producing conductive polymer coatings. Finally, conventional melt processing is a less attractive option for these materials because of their instability at higher temperatures.

Applications. The strong interest in these materials arises from the curiosity about the transport mechanism plus the commercial possibilities of using these lightweight conductors whose properties could be potentially adjusted by modifying the polymer structure. In this regard, many reports describe prototypes or experiments demonstrating the feasibility of using these polymers in certain technologies. These feasibility studies often lack the full parametrization of the performance to demonstrate the success of these materials in the commercial application. These materials have been proposed for use in light-emitting diodes, electrochromic display devices, rechargeable batteries, chemical sensors, enzyme biosensors, antistatic applications, radiation-blocking shields, artificial muscles of actuators, chemical-mechanical micromachines, and so on. In many of these applications the conductive polymer is utilized in a thin-cell electrode configuration. Many other applications have been considered, for example, microlithography patterning, electromagnetic shields, conductive adhesives, electrostatic dissipaters, and electronic devices such as diodes and field-effect transistors.

Another measure of the commercial value of these polymers is the number of patent applications describing their use. Since 1993 there have been many patents describing formulations and processes for the application of conducting polymers onto solid substrate and fabrics for specific end uses. Other patents describe the use of conducting polymers in lubricants, adhesives, energy storage devices, flex-

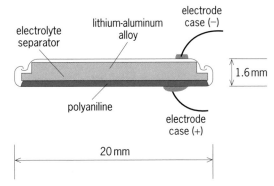

Cross section of the electrode configuration of a button battery.

ible plastic electrodes, printed circuit boards, electrochromic devices, and ion-conducting membranes.

Finally, there are many commercial products incorporating conducting polymers since the commercialization of the button-type, 3-V battery with polyaniline and polypyrrole developed by Bridgestone and BASF (see **illus**). This "sandwich" or thin-cell electrode configuration is fairly common in many of the applications which utilize the active properties of conducting polymers. The number of products which incorporate a conducting polymer continues to increase. For example, the polymers are used in capacitors (Nippon Electric), antistatic layers in film (Bayer), and odor sensors (Neotronics and AromaScan). Research and development efforts continue.

In summary, the area of conducting polymers has come a long way in two decades with coherent films of these materials readily available on the laboratory scale by the electrochemical polymerization of the pyrrole, thiophene, or aniline. This advance provided material for the research on the electrochemical, electrical, and optical properties of conducting polymers. The participation of the innovative synthetic chemist was needed to modify these materials and improve their solubility to provide processing latitude. This aspect is now well on its way, and the availability of conducting polymers in processible formulations permits testing in product applications to determine the performance issues, such as response, reliabilty, and limitations.

For background and information *see* BATTERY; CONJUGATION AND HYPERCONJUGATION; ELECTROCHEMISTRY; ELECTRON-TRANSFER REACTION; FILM (CHEMISTRY); ORGANIC CONDUCTOR; OXIDATION-REDUCTION; POLYMER; SEMICONDUCTOR; SOLID-STATE CHEMISTRY in the McGraw-Hill Encyclopedia of Science & Technology.   Arthur F. Diaz

Bibliography. A. F. Diaz, M. T. Nguyen, and M. Leclerc, *Physical Electrochemistry*, Chap. 12, Marcel Dekker, 1995; J. E. Frommer and R. R. Chance, *Encyclopedia of Polymer Science and Engineering*, 2d ed., vol. 5, p. 462; M. E. Galvin, *JOM*, p. 52, March 1997; T. A. Skotheim (ed.), *Handbook of Conducting Polymers*, Marcel Dekker, 1986.

Paleoecology

Paleoecology is a discipline that deals with interactions of fossil organisms in their ancient environments. However, current research indicates that paleoecological data (the fossil record) are inherently different from neoecological data (the present-day world). Due to postmortem processes, such as transport by water currents or reworking by burrowing organisms, skeletal remains of various ages are often mixed together during their fossilization. Thus,

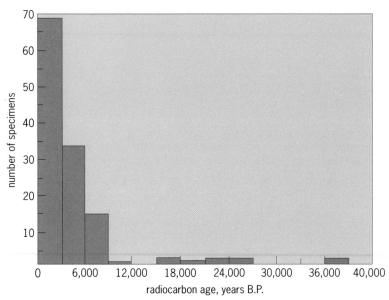

(a)

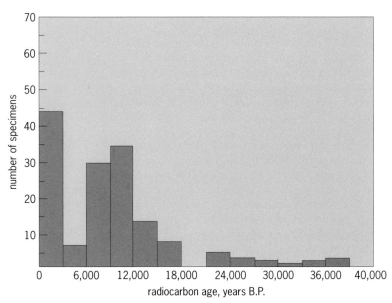

(b)

Time averaging estimated by age distributions of radiocarbon-dated shells collected from the coastal and shelf areas of the modern ocean floor. (a) An average empty shell found on the coastal sea bottoms is about 2500 years old (median age = 2465 years). (b) In the deeper, shelf environments, shells are even older (median age = 8810 years). (Data compiled from K. W. Flessa and M. Kowalewski, Shell survival and time-averaging in nearshore and shelf environments: Estimates from the radiocarbon literature, Lethaia, 27:153–165, 1994)

in contrast to the living organisms studied by ecologists, individual fossils found preserved together in geological strata are not necessarily contemporaneous. As in human cemeteries, fossiliferous horizons mix together remains of organisms that lived at different times and never interacted with one another. The fossil record tends to be time-averaged.

Time averaging. Time averaging (temporal mixing) is the process by which events that happened at different times appear to be synchronous in the geological record. This critical parameter determines what type of information can or cannot be retrieved from the fossil record. A fossil horizon that contains remains from an interval of 10 years may potentially provide some information about a single ancient ecosystem. However, a horizon time-averaged over a period of 10,000 years that mixes organisms that may have lived in various environments, perhaps even under different climatic conditions, certainly cannot inform about a single ecosystem. Using such a horizon for ecological analyses may potentially be as erroneous as digging up a graveyard to determine the demography of a population alive in a specific year.

The level of time averaging in the fossil record and the variation in age of individual fossils contained within an average sample are important questions of modern paleoecology. Unfortunately, in order to evaluate time averaging, individual fossils must be dated with a high level of precision (about 100 years or better). Such resolution cannot usually be achieved in the fossil record. Despite this high age uncertainty inherent in paleontological dating, time averaging is worth investigating: regardless of the fact that the age estimate of a given site may be off by millions of years, it is still important to know whether or not the fossils from that site were all contemporaneous. Consequently, researchers have turned to the youngest fossil record (the late Pleistocene and Holocene), where individual shells or bones can be dated using high-resolution techniques such as the radiocarbon method. Age differences among the dated specimens from the same sites can be used to measure temporal mixing in the youngest and currently forming shell or bone accumulations. Those estimates can then be used (by analogy) to infer potential levels of time averaging that may be expected for the fossil record in general.

In the late 1980s, researchers from the University of Arizona used radiocarbon to date individual mollusk shells that were randomly collected from the present-day tidal flats of Cholla Bay in the Gulf of California. The outcome was surprising. Out of 30 "modern" shells that were dated, 5 shells were more than 1000 years old; one of them, in reasonably good shape, was more than 3000 years old.

Although paleoecologists had anticipated time averaging for many years prior to the Cholla Bay study, the discovery of so many old shells, scattered among living and recently dead mollusks, was unexpected. Since the development of the radiocarbon method in the early 1950s, hundreds of shells collected from coastal and shelf areas of the ocean floor had already been dated. Thus, it was relatively simple to compile the radiocarbon literature and evaluate the Cholla Bay study on the global scale. The compilation showed that extensive temporal mixing is normal on modern sea bottoms (see **illus.**). In addition, several case studies undertaken in recent years have documented extensive time averaging in mollusk accumulations from various settings, including terrestrial, lacustrine, and marine environments. One study, conducted on the Colorado River delta in the northern Gulf of California, showed that even at very high sampling resolution, when laterally adjacent specimens are collected from within single sedimentary layers, shells may still vary in age by several hundred years (careful collecting cannot improve the temporal resolution of paleoecological data).

The importance of the dating projects was not so much in proving the existence of time averaging: accumulations with millions of years of temporal mixing are well known to paleontologists because they can be easily recognized by incongruent composition (for example, Cretaceous dinosaur teeth preserved with Tertiary mammal bones). The projects demonstrated that all accumulations, even those that consist of ecologically congruent species, known to have been extant at the same time, may be time-averaged at some scale.

Extensive temporal mixing is not restricted to mollusks. Time averaging among pollen grains can reach thousands of years. Even minute tests of benthic foraminifers, found on the modern sea floor, can be a few thousand years old. Notably, given its potential implications for archeology, substantial time averaging has been documented in terrestrial vertebrate accumulations. Whether found in river deposits, on land surfaces, or in caves, bone assemblages often contain skeletal remains that differ in age by hundreds to thousands of years. Dating projects have thus unmasked time averaging as a pervasive phenomenon that affects a great variety of skeletal remains in a wide spectrum of depositional environments.

For various reasons, some types of fossils are much less prone to time-averaging processes than shells or bones. For example, leaves are fragile, decay rapidly, and thus require rapid burial to be preserved in the fossil record. Consequently, they are much less susceptible to postmortem mixing, and their fossil accumulations are almost unaffected by time averaging. Also, structures left by encrusting epibionts (organisms that colonize shells of other organisms) often provide records which can be resolved at very fine temporal scales.

Problems. One negative aspect is that due to time averaging, the fossil record offers chronologically disordered, unresolved, or even spurious patterns, with noncontemporaneous organisms, populations, and ecosystems tumbled together into incomprehensible mixtures. Time averaging makes it unacceptable to use the models, methods, or paradigms that were derived from or developed for neoecological data. Apart from paleoecology, time averaging may also cause problems in other paleobiological and geobiological analyses. For example, intraspecific morphological variability may be increased owing to temporal mixing of noncontemporaneous populations. Similarly, environmental signals recorded by stable isotopes retrieved from a particular specimen may not be typical for a period of time represented by a given fossil horizon.

Advantages. However, there is also a positive view on time averaging. Temporal mixing may be advantageous because it can remove short-term noise. The record of long-term ecological systems or long-term environmental conditions can therefore be freed of the short-term fluctuations. For example, the Colorado delta case study showed that time-averaged accumulations can have a uniform and continuous age structure. This means that individuals from all time intervals recorded in the accumulation are equally represented. Consequently, shell accumulations can provide optimal records for a long-term history of ancient ecological systems.

The positive view is methodologically sound because ecological processes may potentially operate at varying time scales and thus may be visible at various levels of resolution. After all, except for anthropocentric bias, the scale and resolution of observation are arbitrary parameters. The positive view is attractive because it implies that paleoecological data offer different (and not poorer) resolution than neoecological data. The long-term averages are potentially as valid and as informative as the short-term data, and the fossil record can provide insights at resolutions and time scales beyond the reach of ecology. Indeed, paleoecologists now put more emphasis on long-term patterns such as evolutionary trends in prey-predator systems or recurrent fossil communities.

For background information *see* ANIMAL EVOLUTION; CARBON-14; ECOLOGY; FOSSIL; GEOCHRONOLOGY; ORGANIC EVOLUTION; PALEOECOLOGY; TAPHONOMY in the McGraw-Hill Encyclopedia of Science & Technology.   Michał Kowalewski

Bibliography. P. A. Allison and D. E. G. Briggs (eds.), *Taphonomy: Releasing Data Locked in the Fossil Record*, Plenum Press, 1991; K. W. Flessa, A. H. Cutler, and K. H. Meldahl, Time and taphonomy: Quantitative estimates of time-averaging and stratigraphic disorder in a shallow marine habitat, *Paleobiology*, 19:266–286, 1993; S. M. Kidwell and A. K. Behrensmeyer (eds.), *Temporal Resolution in the Fossil Record*, Paleontological Society, 1993; M. Kowalewski, G. A. Goodfriend, and K. W. Flessa, High-resolution estimates of temporal mixing within shell beds: The evils and virtues of time-averaging, *Paleobiology*, 24: 287–304, 1998.

Permo-Triassic mass extinction

The history of life on Earth has been punctuated by explosive radiations, catastrophic mass extinctions, and postextinction recoveries. There have been at least six major mass extinctions in the past 540 million years. Although the Cretaceous-Tertiary mass extinction, which took place 65 million years ago (Ma) and led to the disappearance of the dinosaurs, is perhaps the most well known, the Permo-Triassic extinction is by far the most profound. Paleontologists have long divided the history of life, as preserved in fossil-bearing rocks, into three major eras: Paleozoic, Mesozoic, and Cenozoic. The boundaries between the eras are defined by two major extinctions, the Permo-Triassic (P-T) and the Cretaceous-Tertiary (K-T), respectively. Some 251 Ma, the Paleozoic Era ended with the most profound extinction of life since the origin of animals and brought life closer to complete extinction than ever before or since. This extinction has long been recognized as bringing about a fundamental change in the history of life in the fossil record. *See* GEOLOGICAL TIME SCALE.

Definition of mass extinction. Extinctions have occurred throughout the history of life. However, during mass extinctions, massive reorganizations of entire ecosystems and the demises of dominant groups typically occur. A mass extinction is a relatively rapid event that causes widespread extinctions of many groups. Mass extinctions and subsequent recoveries have played fundamental roles in animal evolution, perhaps comparable to natural selection. Understanding mass extinction and recovery is crucial for improving models of evolutionary processes. In particular, it is important to know whether all mass extinctions were caused by similar mechanisms. For example, given the strong evidence for involvement of a bolide impact in the extinction at the end of the Cretaceous, some have proposed such an event as the cause of all mass extinctions.

The event and its causes. The Permo-Triassic extinction triggered the most widespread reorganization of ecosystems and animal diversity in the past 540 million years and is largely responsible for much of the structural biodiversity in the oceans today. From about 480 Ma to the end of the Permian, most marine animals that lived in shallow oceans were largely immobile and attached to the sea floor. Following the extinction, minor groups of mobile predators that are relatives of modern-day fish, squids, snails, and crabs were able to expand into the newly emptied habitats. Put simply, the major groups living in shallow oceans of the Late Triassic were more similar to those of the present day than to those of the preceding Paleozoic Era. Following the extinction, dinosaurs and mammals also arose. Life would probably be much different if the extinction at the end of the Permian had not occurred.

As with all major events in Earth history, geologists, paleontologists, and biologists have speculated about the enigmatic causes of this great extinction. The findings that may have to be explained in any model are (1) extinction of 85% or more of all species in the oceans, approximately 70% of land vertebrates, and significant extinctions of plants and insects; (2) widespread evidence for low oxygen levels (anoxia) in both deep and shallow levels of the oceans before, during, and after the extinction interval; (3) dramatic changes in the carbon cycle of the atmosphere and oceans, indicating sequestration of organic carbon in the deep ocean before the extinction and a rapid change in the isotopic composition of carbon in carbonate rocks during the extinction; (4) a rise in sea level coincident with the extinction; (5) eruption of a large-volume (3–5×10^6 km^3; 7.2–12×10^5 mi^3) flood basalt province in Siberia contemporaneous with the extinction; (6) significant global warming following the extinction; and (7) a drop in the percentage of atmospheric oxygen, which favors groups with active metabolisms, consistent with the pattern of extinction of species in the oceans. A better understanding of the end-Permian extinction and its recovery should allow for new insights into the role of mass extinctions in evolution. Possibly, some of these findings are irrelevant to the extinction.

Mass extinctions are recognized in the fossil record by the abrupt disappearance of taxa, sometimes associated with a discrete boundary bed. For example, the end-Cretaceous extinction is marked by a layer rich in ejecta from the impact of a bolide with Earth. A critical question about any mass extinction is its rapidity, and determination thereof must involve statistical analysis of the stratigraphic and fossil record. Differences in sediment accumulation rate and preservation potential of organisms can lead to an artificially abrupt or drawn-out extinction signal, especially if the extinction is of short duration (less than 1 million years). Time is not linearly distributed in the rock record, and thus stratigraphic thickness cannot be converted to time. To understand the tempo of an extinction requires a combination of high-precision geochronology and paleontological studies. *See* FOSSIL; STRATIGRAPHY.

Dating the event. High-precision geochronology has revolutionized understanding of the rates of geological and biological processes. The uranium-lead (U-Pb) method applied to zircon minerals is a powerful method for dating because it exploits two independent decay schemes (^{238}U to ^{206}Pb and ^{235}U to ^{207}Pb). Volcanic ash beds often contain zircons and are commonly interstratified with fossil-bearing rocks. Thus, they provide a way of directly constraining the amount of time represented by a given thickness of rock. *See* RADIOISOTOPE DATING.

In south China, several stratigraphic sections preserve the Permo-Triassic boundary, including a spectacular occurrence at Meishan that serves as an international standard for comparison. Interlayered with the fossil-bearing rocks are a series of thin volcanic ash beds that have been dated using the U-Pb zircon technique. Ages of individual ash beds were determined with uncertainties of 200,000–500,000 years.

The age and duration of the Permo-Triassic boundary has been the subject of considerable debate. The geochronological results from south China (see **illus.**) allow three major conclusions: (1) The age of the boundary is within the same error margins for two sections 1500 km (932 mi) apart. (2) The

Permian-Triassic boundary exposed at Meishan, China, showing the U-Pb dates (in millions of years) for ash beds interlayered with fossil-bearing rocks. The extinction is recorded within the last meter directly below the lowest ash bed (251.4 ± 0.4 Ma). The upper two ash beds are in Lower Triassic rocks. (*From S. A. Bowring et al., Geochronology of the end-Permian mass extinction, Science, 280:1039–1045, 1998*)

extinction occurred in less than 1 million years. At Meishan, the extinction occurred over about a 4-cm (1.6-in.) thickness of limestone that could have been deposited in less than 200,000 years based on average accumulation rates for the section. The only other extinction for which an estimate of the duration has been made is the end-Cretaceous, which is estimated to be 200,000–300,000 years. The short duration is a powerful constraint on potential mechanisms and seemingly rules out gradual environmental change, related to plate tectonics, for example. (3) The age of the boundary is within the same error margins as the eruption of the Siberian flood basalts.

Ocean chemistry and environmental change. An important aspect of understanding extinctions is the relationship between mass extinction and environmental change. Variations in the isotopic composition of carbon, specifically, the ratio of carbon-13 to carbon-12 ($^{13}C/^{12}C$), can be used to infer changes in the amount of organic carbon burial in the oceans. In general, the greater the amount of organic carbon burial, the higher the ratio $^{13}C/^{12}C$ in the oceans. The reason is that the carbon isotopic composition of limestone precipitated from seawater is a good monitor of inorganic carbon because marine organisms are depleted in ^{13}C relative to inorganic carbon. Thus, the variation in the isotopes of carbon with stratigraphic position can be used to evaluate temporal changes in seawater chemistry and to correlate between stratigraphic sections.

The Permo-Triassic extinction is characterized by an abrupt disappearance of Permian fossils followed by the appearance and rapid diversification of Triassic assemblages. Studies of carbon isotopes from many different sections, both marine and terrestrial, indicate an abrupt shift in carbon isotopic values at the boundary. However, whether the isotopic shift exactly coincides with the extinction and whether the extinction is globally synchronous are still open questions, resolvable only with high-precision geochronology.

The shift in carbon isotopes implies a global pulse of isotopically light carbon (low $^{13}C/^{12}C$). A variety of causes for the isotopic shift have been suggested, including (1) marine anoxia associated with flooding of shelf areas with carbon dioxide (CO_2) enriched bottomwaters during rapid overturn; (2) onshore migration of a dysaerobic sea layer during transgression; (3) extinction-related decrease in primary productivity resulting in export of less organic carbon to the deep oceans and a shift of whole ocean values to lower values; (4) oxidation and erosion of late Permian peat and coal deposits which have very low $^{13}C/^{12}C$ values; and (5) catastrophic release of methane deposits from shelf areas due to warming of oceans. Methane has very low $^{13}C/^{12}C$ values.

It seems clear that any one of these mechanisms cannot on its own explain the existing observations. On the basis of high-precision geochronology, three speculative scenarios can explain the events at the Permian-Triassic boundary. In the first, eruption of the Siberian flood basalts in the latest Permian released large amounts of carbon dioxide (and possibly sulfates, producing acid rain) and initiated a period of global warming. Warming of shallow seas was sufficient to melt methane hydrates. A short volcanic winter may have first been triggered by volcanic aerosols and followed by greenhouse conditions and warming. This cooling-warming cycle may have triggered convective overturn of the oceans, dumping deep carbon dioxide–rich bottomwater onto the shelf regions leading to hypocapnia (death by carbon dioxide poisoning) and a rise of atmospheric carbon dioxide levels. *See* GAS HYDRATES; GLOBAL WARMING.

In the second scenario, extinction, perhaps related to the Siberian traps, caused collapse of primary productivity and cessation of export of light carbon (low $^{13}C/^{12}C$) to the deep ocean, which produced a transient isotopic shift before the oceans returned to more normal values during recovery. It is possible that the export of sequestered light carbon and carbon dioxide–charged water by upwelling, combined with volcano-induced extinction, could explain the observations. No single mechanism, however, is sufficient to explain all the geologic and paleontologic data, but the massive eruption of the Siberian traps may well have been the proximal cause for a cascade of events leading to the apparent synchroneity of marine and terrestrial extinctions.

The third possibility is that the latest Permian biota were declining as a result of the above scenarios and that the collision of Earth and a bolide was the final straw that pushed the planet to the brink of total extinction. All of the mechanisms discussed above have been operative at other times in Earth history and may have even caused other extinctions. However, none before or since have been as profound as the truly singular events of the Permo-Triassic extinction.

Future work. The terrestrial record is also marked by profound extinction at the end of the Permian. A critical aspect of understanding the end-Permian extinction is the comparison of the terrestrial and marine records. For example, if the terrestrial extinction corresponds exactly in time with the marine event, this would require a mechanism that could simultaneously affect both marine and terrestrial ecosystems. If the appropriate ash beds are present in terrestrial sections, this possibility can be tested with a resolution of approximately 200,000–300,000 years. Negative spikes in the carbon isotopic composition of carbonates and organic matter have already been noted at or near the boundary in terrestrial rocks from Australia. Lastly, more effort will be necessary to search for evidence of a catastrophic impact.

For background information *see* CRETACEOUS; DATING METHODS; EXTINCTION (BIOLOGY); FOSSIL; GEOLOGICAL TIME SCALE; MESOZOIC; PALEONTOLOGY; PALEOZOIC; PERMIAN; STRATIGRAPHY; TRIASSIC in the McGraw-Hill Encyclopedia of Science & Technology.   Samuel A. Bowring

Bibliography. S. A. Bowring et al., Geochronology of the end-Permian mass extinction, *Science*, 280:1039–1045, 1998; D. H. Erwin, *The Great Paleozoic Crisis: Life and Death in the Permian*, Columbia University Press, 1993; D. H. Erwin, The Permo Triassic extinction, *Nature*, 367, 231–236, 1999; A. Hallam and P. B. Wignall, *Mass Extinctions and Their Aftermath*, Oxford University Press, 1997.

Peroxisome

Peroxisomes, along with the nucleus, mitochondria, endoplasmic reticulum, lysosomes, Golgi apparatus, and chloroplasts, are among the basic structures in animal and plant cells (**Fig. 1**). Of all these organelles, which carry out many vital biochemical processes, peroxisomes are the most recently discovered. Originally called microbodies, they were first noted in mouse kidney proximal tubule cells by J. Rhodin, a Swedish graduate student, in 1954. Much of the pioneering work in peroxisome biology was done in the laboratories of the cell biologist Christian de Duve. He and his colleagues demonstrated that these organelles contain several hydrogen peroxide (H_2O_2)-producing enzymes (for example, urate oxidase and D-amino acid oxidase) and catalase, the enzyme responsible for degrading toxic H_2O_2 to water and oxygen. In 1965 de Duve suggested the name peroxisomes for Rhodin's microbodies. Some lower organisms and plants contain microbodies that carry out specialized functions, and so the microbodies were named accordingly. Examples are glyoxysomes (fungi and plants), which contain the five enzymes of the glyoxylate cycle, and glycosomes (trypanosomes), which house the enzymes of glycolysis.

Peroxisomes have been found in all animal and plant cells, with the exception of mature erythrocytes. They are spherical or ovoid and range in size from about 0.1 to 1.0 micrometer. They are bounded by a single membrane bilayer and contain a finely granular matrix and, in some cells, a microcrystalline core. Some believe that peroxisomes, like mitochondria, originated as endosymbionts early in evolution. However, unlike mitochondria and chloroplasts, peroxisomes contain no deoxyribonucleic acid (DNA).

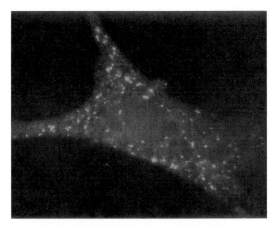

Fig. 1. Peroxisomes in human skin cells (fibroblasts) visualized by immunofluorescence of the matrix protein catalase. Magnification 1000×.

Biogenesis. New peroxisomes arise from the import of matrix and membrane proteins into preexisting peroxisomes, which then divide by fission. It is estimated that mammalian cell peroxisomes contain over a hundred different proteins. Proteins destined for the peroxisome are encoded by nuclear genes and are synthesized on cytoplasmic polyribosomes. Two amino acid sequences that function as peroxisome targeting signals (PTS1 and PTS2) have been identified in several peroxisomal matrix proteins. The mechanisms that target other matrix proteins and all membrane proteins to peroxisomes are as yet undefined.

Yeasts have proven to be excellent model organisms for investigating peroxisome biogenesis. By identifying mutants lacking peroxisomes or mutants with defective import of matrix proteins, at least 20 distinct proteins required for normal assembly of the organelle have been identified. These proteins, called peroxins, have diverse functions. Some are cytosolic proteins, such as the receptors for PTS1- and PTS2-containing proteins, while others such as docking factors and pore-forming proteins are found in the peroxisome membrane. Higher animals contain peroxins that are homologous to their yeast counterparts.

Proliferation and carcinogenesis. The number of peroxisomes in a cell can, in some cases, be regulated by external stimuli. For example, there is a dramatic increase in peroxisome number in yeast if fatty acids are used as a primary fuel source. A similar phenomenon is seen in livers of rats treated with a number of drugs and chemicals collectively called peroxisome proliferators. Several drugs known to decrease blood lipids in humans (hypolipidemic drugs) and some industrial plasticizers belong to this group of compounds. These chemicals work by interacting with intracellular peroxisome proliferator activated receptors (PPARs). PPARs are closely related to steroid hormone receptors and control the transcription of specific nuclear genes. Interestingly, peroxisome proliferation in rats has been associated with increased incidence of liver tumors. While human tissues contain PPARs, there is little evidence for either peroxisome proliferation or hypolipidemic drug-related carcinogenesis in humans.

Biochemical functions. Peroxisomes are vital to the normal functioning of cellular metabolism. Both catabolic and anabolic processes take place in these organelles. Catabolic peroxisomal processes include the β-oxidation of fatty acids and fatty acyl-like compounds, the α-oxidation of 3-methyl branched-chain fatty acids, the degradation of L-pipecolic acid, and the detoxification of hydrogen peroxide. Anabolic processes include the synthesis of plasmalogens (ether phospholipids), cholesterol, dolichol, and other isoprenoids. Peroxisomes also contain enzymes involved in amino acid, purine, and pyrimidine metabolism.

Mitochondria degrade dietary and stored fatty acids by β-oxidation, a well-known pathway that is tightly coupled to energy production. Peroxisomes contain two complete β-oxidation pathways, which are not involved in energy production but function either to degrade potentially toxic compounds or to remodel certain fatty acyl groups (**Fig. 2**). In both mitochondrial and peroxisomal pathways, fatty acids are first activated by forming a thioester with coenzyme A (CoA; the active form of the B-vitamin pantothenic acid.) Once fatty acyl-CoA enters the organelle, a series of four oxidative enzymatic reactions takes place, producing acetyl-CoA or propionyl-CoA and a fatty acyl-CoA shortened by two or three carbons. In many cases, the shortened fatty acyl-CoA undergoes additional cycles of β-oxidation, resulting in further degradation. The substrate specificity of the peroxisomal β-oxidation pathways is unique and distinct from that of the mitochondrial pathway. Very long chain fatty acids (containing 24 or more carbons) are degraded in peroxisomes. These fatty acids are quite toxic, producing severe human disease when they accumulate. 2-Methyl branched-chain fatty acids (the product of α-oxidation of 3-methyl branched-chain fatty acids) are also degraded in peroxisomes. Catabolism of prostanoids (prostaglandins, leukotrienes, and thromboxanes), dicarboxylic fatty acids, and some polyunsaturated fatty acids such as arachidonic acid require peroxisomal β-oxidation. One cycle of peroxisomal β-oxidation is required for the biosynthesis of docosahexaenoic acid, a polyunsaturated fatty acid that is important for normal development of the brain and retina. Many xenobiotic compounds (chemicals not normally found in an organism, such as drugs) contain fatty acyl side chains that undergo metabolism by peroxisomal β-oxidation; again, this may represent a detoxification reaction. Cholesterol contains a methyl-branched acyl side chain that must undergo one cycle of peroxisomal β-oxidation in its conversion to bile acids in the liver. Acetyl-CoA and propionyl-CoA produced by peroxisomal β-oxidation ultimately leave the organelle and may be converted to energy by mitochondria. Thus, while not directly coupled to energy metabolism, peroxisomal β-oxidation performs many vital functions.

Fatty acids with a branch on the third carbon cannot be degraded by β-oxidation without removing the first carbon (as formyl-CoA) by a process known as α-oxidation. Dietary phytanic acid, derived from the phytol in chlorophyll and also toxic if not degraded, is a 3-methyl branched-chain fatty acid. α-Oxidation degrades phytanic acid to a 2-methyl branched-chain compound, pristanic acid, which can then undergo β-oxidation. Certain amino acids are also degraded in peroxisomes by the action of D-amino acid oxidase (which acts on basic and neutral amino acids) and D-aspartate oxidase (which acts on acidic amino acids). L-Pipecolic acid, an imino acid product of lysine catabolism, is degraded by a specific oxidase in peroxisomes of higher primates. Most of the peroxisomal oxidase enzymes consume molecular oxygen and produce H_2O_2, which is toxic to cells. The abundant peroxisomal enzyme catalase detoxifies H_2O_2 either by decomposition to water

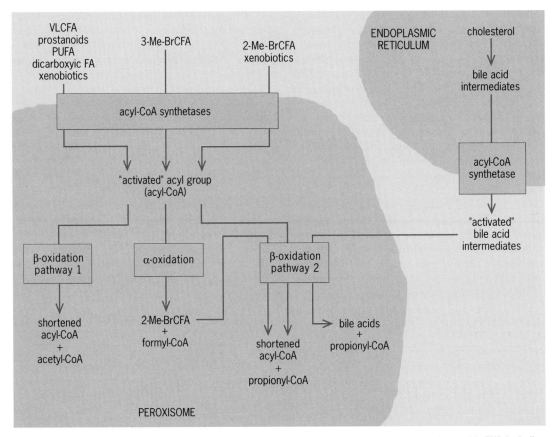

Fig. 2. Peroxisomal fatty acid metabolism. Peroxisomes are important for the degradation of many fatty acids (FA), including very long chain FA (VLCFA), polyunsaturated FA (PUFA), 2- and 3-methyl branched-chain FA (2-Me-BrCFA; 3-Me-BrCFA), and prostanoids (prostaglandins, leukotrienes, thromboxanes). Peroxisomal fatty acid oxidation pathways are also involved in the degradation of many xenobiotic compounds and in the synthesis of bile acids from cholesterol.

and oxygen (catalic activity) or by oxidation of compounds such as ethanol to acetaldehyde or formic acid to CO_2 (peroxidatic activity).

Another major function of peroxisomes is the biosynthesis of alkyl glycerophospholipids, which differ from the more common acyl glycerophospholipids by the substitution of an ether-linked alkyl chain for the ester-linked acyl group on the first carbon of glycerol. Two enzymes unique to peroxisomes, dihydroxyacetonephosphate acyltransferase and alkyldihydroxyacetonephosphate synthase, play a critical role in alkyl glycerophospholipid synthesis. Introduction of a double bond between carbon 2 and 3 of the alkyl chain of alkyl glycerophospholipids yields alk-1-enyl glycerophosopholipids, or plasmalogens, which account for about 18% of cellular phospholipids and are mainly structural components of biologic membranes. In addition, plasmalogens are thought to act as free-radical scavengers. Another important alkyl glycerophospholipid is platelet activating factor (alkylacetyl glycerophosphocholine), a hormonelike signaling molecule.

Peroxisomes are known to contain several enzymes for the synthesis of isoprenoids such as cholesterol and dolichol. Like phospholipids, cholesterol is a necessary component of cellular membranes and is also the precursor of steroid hormones

and bile acids. Dolichol is a membrane constituent that also plays an important role in the glycosylation of proteins. The isoprenoid building blocks of both cholesterol and dolichol are the same, and several enzymes catalyzing their biosynthesis have been found in peroxisomes. While many of these enzymes are also found elsewhere in the cell, all of the enzymes that are required for the conversion of the two-carbon precurosor acetate to farnesyl pyrophosphate, a 15-carbon isoprenoid, are thought to be present in peroxisomes. Sterol carrier protein 2, thought to be important for the synthesis of steroid hormones and probably involved in intracellular cholesterol transport, is also found in peroxisomes.

Another peroxisomal oxidase (L-hydroxyacid oxidase) converts glycolic acid, a product of hydroxyproline and ethylene glycol metabolism, to glyoxylate. If glyoxylate is not converted to glycine by peroxisomal alanine:glyoxylate aminotransferase, L-hydroxyacid oxidase degrades it to oxalic acid. Oxalate excess is a major cause of kidney stones and can lead to renal failure. A role for peroxisomes in gluconeogenesis has also been proposed because they contain both alanine:glyoxylate aminotransferase and serine:pyruvate aminotransferase, enzymes that convert amino acids derived from tissue protein breakdown during starvation to ketoacids which can ultimately be converted to glucose.

Defective structure or function. Much of the knowledge about peroxisomal biochemistry and metabolism has come from the study of human diseases in which either the organelle fails to form normally or a protein involved in a peroxisomal metabolic pathway is deficient. All of these diseases are genetic disorders, most with autosomal recessive mode of inheritance. Their most commonly observed feature is mental retardation, and many victims die in infancy or early childhood. Diseases in which peroxisomes fail to form normally result from a genetic defect in one of the peroxins; failure to properly import matrix proteins results in defects in multiple biochemical processes. The most common (1 in 50,000 individuals) peroxisomal disease is X-linked adrenoleukodystrophy, in which a defective peroxisomal membrane protein results in impaired β-oxidation of very long chain fatty acids. For almost all peroxisomal metabolic pathways, patients with a defective enzyme or protein have been described.

For background information *see* CELL (BIOLOGY); CELL METABOLISM; LIPID METABOLISM in the McGraw-Hill Encyclopedia of Science & Technology.

Paul A. Watkins

Bibliography. P. B. Lazarow and H. W. Moser, Disorders of peroxisome biogenesis, in C. R. Scriver et al. (eds.), *The Metabolic and Molecular Bases of Inherited Disease*, 7th ed., McGraw-Hill, 1995; C. Masters and D. Crane, *The Peroxisome: A Vital Organelle*, Cambridge University Press, 1995; J. K. Reddy et al., Peroxisomes: Biology and role in toxicology and disease, *Ann. N. Y. Acad. Sci.*, vol. 804, 1996; H. van den Bosch et al., Biochemistry of peroxisomes, *Annu. Rev. Biochem.*, 61:157–197, 1992.

Pest management

Nematodes are one of the most abundant groups of multicellular organisms in the biosphere, occurring in all habitats and ecosystems. They are nonsegmented roundworm invertebrates that range in size from microscopic inhabitants of soil to 27-ft-long (8-m) parasites that live in the placenta of sperm whales. The 15,000 species described to date include marine inhabitants and parasites of plants, insects and other animals, and humans. In addition to the 1500 species parasitizing plants, many other species living in the soil feed on bacteria, fungi, algae, and even other nematodes. All plant-parasitic nematodes have a stylet or spear which they use to feed on plant cells (**Fig. 1**). Widely known animal parasites include heartworms, pinworms, and hookworms. While thousands of nematode species are parasites, many others are beneficial and contribute to soil-nutrient cycling or serve as biological control agents of plant pests.

Economic losses. Of the total economic losses to crops, turf, and ornamentals, the losses due to plant-parasitic nematodes vary considerably but often range 5–10% annually. Estimates of crop loss due to nematodes on a wide range of food and fiber crops average about 12% annually, with total estimated losses for 21 crops amounting to more than $77 billion per year worldwide. Under severe infestations, crop losses to nematodes such as root-knot species (which induce galls or knots to form on roots) may be in the range of 50%. Plant-parasitic nematodes are often the hidden enemy, usually in the soil: they cause stunting and patchy crop growth similar to that associated with other soil problems, and so their negative impact on crop productivity is frequently overlooked (**Fig. 2**). Precise estimates of crop-yield loss to nematodes are difficult to obtain, but experiments involving comparisons of effective nematode management treatments such as resistant cultivars, effective nematicides, or crop rotations have proven very useful in this regard (**Fig. 3**). Such experiments provide a means of developing reasonably reliable damage functions, damage thresholds, and estimates of yield losses due to nematode infestations in given environments.

Sustainable agriculture. Intensive modern agriculture involves the application of new technology and exploitation of biotic and abiotic natural resources. In this process, plant genotypes for many agroecosystems often have greater uniformity and consequently a narrower genetic base; this may favor the buildup of associated pests and pathogens. In contrast, sustainable agriculture is designed to prevent the depletion or loss of the Earth's resources while rebuilding the productive capacity of the soils and limiting pest and pathogen damage.

Integrated pest management is a component of sustainable agriculture. It deals with three processes: (1) determining how the life system of the nematode (or the pest) must be modified to reduce its density to tolerable levels, (2) effecting this modification through biological knowledge combined with current technology, and (3) with present technology, developing new methods of management that are increasingly compatible with economic and environmental requirements. Ideally, integrated pest management as well as sustainable agriculture should focus on integrated crop production systems as well as management of all associated pests or pathogens.

Strategies and tactics. Early nematologists provided a strong, conceptual framework for the development of integrated pest management systems for plant-parasitic nematodes. As early as 1889, the range of tactics for nematode management included sterilization of soil by starvation, the use of nonhost plants, and the potential of trap crops, compost, nematicides, and soil amendments. These recommendations were expanded in the early 1900s to include the control of fungi-induced diseases that are often enhanced by root-knot and other nematodes. The discovery of effective nematicides (especially soil fumigants), characterization of associated disease complexes, development of highly nematode-resistant cultivars, and effective crop rotation systems in the 1950s and 1960s provided profitable options for growers to manage plant-parasitic nematodes.

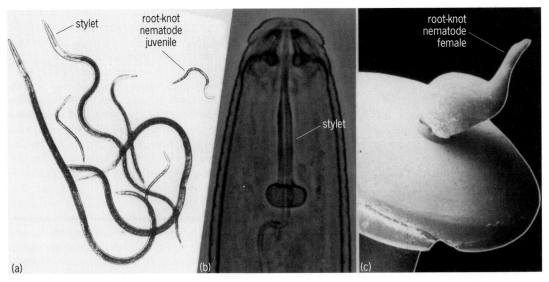

Fig. 1. Root-knot nematodes. (*a*) Vermiform plant-parasitic nematodes from a soil sample. The stylet, or feeding spear, characterizes "dagger" nematodes. The juvenile is the infective stage. (*b*) Highly magnified head. (*c*) Adult, pear-shaped female on a pinhead. (*Courtesy of J. D. Eisenback*)

The ultimate objective of nematode management is the prevention of yield loss or damage of current and future crops. For nematode management, the primary strategies include (1) exclusion-avoidance, (2) reduction of initial population densities of target nematodes, (3) suppression of nematode reproduction, and (4) restriction of current and future crop damage. The tactics used under these four strategies can be extensive. Government-regulated quarantine and the simple practice of using clean, pest-free seeds and plants are tactics falling under exclusion-avoidance. A wide range of cultural practices include rotation or timing of planting or harvesting, vertical host resistance (usually one gene involved), chemical nematicides, physical means of control such as root destruction at the end of the season, and use of biological control measures. These may be used singly or in combination to reduce the initial population densities of nematodes below a damaging level. Host resistance, postplanting supplemental nematicides, application of organic amendments, and practices that favor the buildup of nematode antago-

nists are used to suppress nematode reproduction on crops. Nematode-tolerant cultivars may also prove useful. Tolerant crop cultivars, in contrast to resistant plants, may allow high rates of nematode production without sustaining significant yield losses. This tactic can be useful in the short term but could be detrimental to subsequent crops because of the associated increased population densities of the nematodes.

Rotation. Cropping systems, based on crop rotation with nonhosts, have been important in subsistence and traditional agriculture for centuries. The primary principle in crop rotation for nematode management is the reduction of population densities of target species to allow the subsequent production of a susceptible crop. Crop rotation, often the preferred means of nematode management, also provides diversity in time and space. This tactic, however, is often of limited value in fields infested with a number of damaging nematode species or with species that have wide host ranges. Grass fallows and various green manure cover crops have proven useful in managing several nematode species. Other plants, such as *Crotalaria* spp., serve as effective trap crops resulting in the death of nematodes, and African marigolds (*Tagetes erecta*) produce nematicidal compounds. In addition to rotation, destruction of residual host-plant roots immediately after harvest can be a very effective means of nematode management. This practice prevents the continued reproduction of nematodes on plants with perennial growth and greatly enhances population declines. Other cultural practices include adjustment of timing of planting and harvest, modified tillage practices, fallow, and flooding.

Host resistance. The use of resistant crop cultivars is the most economical means of nematode management. However, most currently available resistant cultivars are limited to nematodes that induce their

Fig. 2. Severe damage to soybean, involving stunted plants and patchy growth pattern, caused by a combination of the Columbia lance and sting nematodes. (*Courtesy of S. R. Koenning*)

Fig. 3. Comparison of the growth of soybean-cyst-resistant soybean (four rows at upper left and four rows at lower right) to poor growth of susceptible soybeans (left foreground and upper right).

hosts to develop specialized feeding cells. Included are root-knot, cyst, reniform, and citrus nematodes. The few exceptions include the stem nematode on alfalfa, the dagger nematode on grape, and the burrowing nematode on banana. To ensure the durability of resistance genes, appropriate integration of resistant cultivars in cropping systems must be followed.

Nematicides. From the late 1950s through the 1970s, nematicides became the first line of defense against plant-parasitic nematodes. These low-cost and highly effective compounds were often used as an insurance measure by risk-averse farmers. Because of environmental and human health concerns, the more effective and less expensive materials have been removed from the market. Continued usage of other materials such as fenamiphos results in enhanced microbial decomposition of the product, negating the expected benefits. Even with these difficulties, nematicides likely will remain an indispensable tool for nematode management on high-value crops and perennials.

Biological control. The use of nematode pathogens and antagonists is an attractive means of nematode management. Numerous fungi and bacteria have potential for biological nematode management, but success to date has been limited to small test systems. Most research has focused on the introduction of the nematode antagonist into soil or on crop seeds. A more promising approach for biological nematode management involves the use of certain legume cover crops such as velvet bean (*Stizolobium deeringianum*) that favor the buildup of growth-promoting, nematode-antagonistic rhizobacteria. In fact, certain of these bacteria may induce acquired disease resistance in plants as well as enhancing plant growth.

Integrated nematode management. Many success stories of single- and multiple-pest management have been recorded. In Europe, the potato cyst nematodes have been controlled by an integrated management program that includes rotation with nonhosts, crop

resistance, and nematicides. A similar system has been very effective in managing root-knot on tobacco in North Carolina. With the addition of shoot and root destruction, periodic discing (a type of soil tillage involving a series of disc-shaped plows), and a rye cover crop, most associated weeds, insects, viruses, fungi, and bacterial pathogens are controlled on tobacco. Cyst and root-knot nematodes on soybean are routinely controlled in the United States by integrating rotation and nematode-resistant soybean cultivars. In these rotations, it is critical that the sources of nematode resistance not be overused, for resistance-breaking genotypes of the target nematodes will develop.

Information emerging from research in soil biology offers the potential for facilitating the development of new strategies and tactics for nematode, general pest, and crop management on a sustainable basis. In contrast to the traditional focus on plant-parasitic nematodes only, the roles of microbivorous nematodes in soil nutrient cycling and soil health are now being addressed. The interactions of associated microflora and microfauna with beneficial nematodes as well as plant-parasitic nematodes, and the effects of growth-promoting rhizobacteria on pests and crop performance in a range of cropping systems come into play. Thus, the agroecosystem framework for soil and plant health, including nematode management, is evolving to encompass different trophic groups of nematodes (plant parasites, bacterivores, fungivores, omnivores, and predators) as well as other soil organisms.

Related interactions. Sustainable nematode management also must address the associated nematode-pathogen-parasite-pest complexes often encountered in crop production. Early research on nematode-microflora interactions typically involved the predisposition of crop plants to attack by nematodes and associated fungi and bacteria, and the role of nematodes as vectors of plant viruses. Weed management can be equally important in this regard. A wide range of weeds associated with crops are frequently found to be hosts for plant-parasitic nematodes. Furthermore, stunting of crop plants by plant-parasitic nematodes may accentuate associated weed and insect problems. A number of weeds also may serve as efficient reservoirs for viruses that are often vectored by plant-parasitic nematodes. Green manure cover crops must be selected with care as some plants may support the buildup of nematodes.

Nematode interactions with beneficial microflora warrant attention. For example, the soybean cyst nematode often interferes with nodulation and fixation of atmospheric nitrogen by nitrogen-fixing bacteria on soybean. Vascicular-arbuscular mycorrhizal fungi enhance the ability of plants to absorb phosphorus and other nutrients as well as water. These fungi may also limit crop damage and yield losses induced by nematodes and other pathogens.

Emerging technologies. Emerging technologies have great potential for enhancing the effectiveness and sustainability of integrated pest management systems, including plant-parasitic and plant-beneficial nematodes.

Nematode diagnostics. Reliable and rapid identification of nematode species is critical for nematode management. Routine identification of nematode species and often host races of these pathogens already have benefitted greatly from modern biotechnology. For example, differential isozyme patterns of esterases and malate dehydrogenase are widely used to aid in identifying root-knot nematode species. Deoxyribonucleic acid (DNA) probes are available for diagnosis of certain plant-parasitic nematode species and many beneficial microbivorous species.

Adaptation of soil-sampling machines, as being developed for use in precision agriculture, should reduce nematode population assessment costs and increase reliability. The Global Positioning System and geostatistical analyses are opening new avenues for the development of more precise nematode population–crop response models. The Global Positioning System, combined with computerized harvesters, also facilitates accurate record keeping for spatial crop yields as related to infestations of nematodes, other crop pests, and soil factors. This combination of new and improved diagnostic and assessment technologies will enhance the use of previously neglected beneficial nematodes, soil bacteria, and fungi. *See* PRECISION AGRICULTURE.

Molecular crop breeding. New developments for host resistance in crop-pest management allow for advanced products, as evidenced by the availability of transgenic crops resistant to certain insects or corps tolerant to herbicides. Although this on-the-farm use is still to come for nematode management, significant progress has been made in developing model systems for transgenic host resistance to nematodes, as well as using marker-assisted selection for host resistance.

One approach for engineering nematode-resistant plants encompasses the identification and cloning of natural plant-resistance genes and their transfer into susceptible crops. Scientists were able to clone a sugarbeet cyst nematode–resistant gene from a wild beet (*Beta*) species and transfer the gene to a susceptible sugarbeet line which then became resistant to this pathogen. Similar success was achieved with the *Mi* gene for root-knot nematode resistance in tomato. During this research, the *Mi* gene also was found to confer resistance to aphids in tomato. Although the cloned *Mi* gene has great potential utility for many crops, problems related to the transformation process of given crop species must be overcome.

A second promising approach involves the transformation of plants with one or more transgenes encoding a product that is detrimental to the nematode or that suppresses the action of key genes necessary for the host-nematode interactions. For example, constructs of a root-specific gene in tobacco have been used to develop promising root-knot nematode–resistant genotypes of tobacco and cotton. Tobacco transformed with an antisense construct gave about 70% less root-gall development than did susceptible plants. Another model system involving the transformation of *Arabidopsis thaliana* with a modified rice cystatin resulted in resistance to root-knot and sugarbeet cyst nematodes.

The application of molecular markers in traditional plant breeding is proving to be invaluable in facilitating the selection process for crop resistance to nematodes. Molecular markers are now available for crop resistance genes for many host-nematode combinations, including cereal and cyst, potato and cyst, peanut and root-knot, soybean and cyst, soybean and root-knot, sugarbeet and cyst, tomato and root-knot, and tobacco and root-knot. This technology for tracking resistance genes also facilitates the pyramiding of resistance genes, as is being done in soybean and other crops. In addition to nematode- and other pest-resistant genes, beneficial genes (for high yield or quality), as well as detrimental genes (producing poor yield associated with some sources of nematode resistance) also can be tracked via molecular markers and be added or deleted from given crop genotypes.

For background information *see* AGRICULTURE; FOREST PEST CONTROL; PESTICIDE; PLANT PATHOLOGY in the McGraw-Hill Encyclopedia of Science & Technology.   Kenneth R. Barker; Richard S. Hussey

Bibliography. K. R. Barker and S. R. Koenning, Developing sustainable systems for nematode management, *Annu. Rev. Phytopathol.*, 36:165–205, 1998; K. R. Barker, G. A. Pederson, and G. L. Windham (eds.), Plant and Nematode Interactions., *Agron. Monogr.*, no. 36, American Society of Agronomy, 1998; W. R. Nickle (ed.), *Plant and Insect Nematodes*, Marcel Dekker, 1984; V. M. Williamson, Root-knot nematode resistance genes in tomato and their potential for future use, *Annu. Rev. Phytopathol.*, 36:277–293, 1998.

Phospholipases

Phospholipases are enzymes that hydrolyze membrane phospholipids to generate products that are involved in the regulation of many cellular processes. Their activity is controlled by a wide variety of agonists, including hormones, neurotransmitters, growth factors, and cytokines. The principal phospholipases are designated PLA_1, PLA_2, PLC, and PLD, and they cleave phospholipids at different sites (**Fig. 1**). PLA_1 releases fatty acids from the first position (F_1) of phospholipids, whereas PLA_2 cleaves them from the second position (F_2). PLC hydrolyzes phospholipids at the phosphodiester bond to produce diacylglycerol (DAG) and a phosphorylated head group, as found in phosphocholine or phosphoinositol. PLD hydrolyzes the phosphodiester bond at a different site to yield phosphatidic acid and a free head group, as found in choline or ethanolamine. Phospholipases are widely distributed in nature and are found in almost all mammalian cells. In mammals, the principal substrates of PLA_1, PLA_2, and PLD

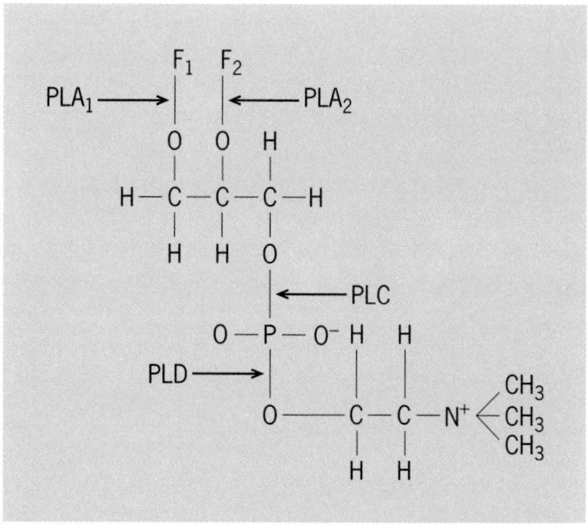

Fig. 1. Sites as which different phospholipases cleave the typical phospholipid phosphatidylcholine. F_1 and F_2 are long-chain fatty acids esterified at positions 1 and 2 of the glycerol backbone, respectively.

are phosphatidylcholine and phosphatidylethanolamine, whereas PLC acts on phosphatidylinositol (PI) and its mono- and bisphosphates (PIP and PIP_2). In plants and bacteria, the phospholipases are less selective toward their substrates.

Phospholipase A_1. PLA_1 has not been extensively studied. It is involved in the remodeling of phospholipids in cells. It may play a role in the release of arachidonic acid through the consecutive actions of PLA_1 and PLA_2 or a lysophospholipase on phosphatidylcholine or -ethanolamine.

Phospholipase A_2. PLA_2 is physiologically important because it releases fatty acids (principally arachidonic acid) from phosphatidylcholine and -ethanolamine. Arachidonic acid is then metabolized to various eicosanoids (prostaglandins, thromboxanes, leukotrienes, lipoxins). The type of eicosanoid produced depends on the cell type. Eicosanoids have major effects on platelets, blood vessels, the stomach, and the female reproductive system. They are also involved in inflammation. PLA_2 occurs in different isoforms that can be grouped into small (14 kilodaltons) secreted forms and large (29–85 kDa) cytosolic forms. The secreted forms of PLA_2 are found in snake and bee venoms, pancreatic juice, platelets, and synovial fluid. They require millimolar concentrations of Ca^{2+} and are rich in disulfide bonds. They have a rigid three-dimensional structure which is similar in all forms of secreted PLA_2. One type of secreted PLA_2 is involved in the sustained release of arachidonic acid from the surface of macrophages and other cells. It is also elevated in serum and tissue exudates in several human inflammatory diseases, where it participates in the production of lipid mediators of inflammation.

The cytosolic forms of PLA_2 are widely distributed in mammalian cells and occur in Ca^{2+}-dependent and -independent forms. Ca^{2+}-independent cytosolic PLA_2 is mainly involved in the remodeling of cellular phospholipids or in the degradation of platelet-activating factor. Ca^{2+}-dependent cytosolic

PLA_2 plays a major role in arachidonic acid release and eicosanoid formation in cells stimulated by a variety of cytokines, growth factors, and other agonists. These agonists cause either long-term or short-term effects on arachidonic acid release. The sustained effects are due to an action on cytosolic PLA_2 synthesis that is exerted at the transcriptional level and is suppressed by glucocorticoids. In addition, the enzyme is regulated more acutely by agonists that elevate intracellular Ca^{2+} and stimulate mitogen-activated protein kinase (MAPK). MAPK is an important regulatory enzyme in the cell that not only directs cytosolic PLA_2 but is involved in the regulation of cell growth at the transcriptional level. MAPK is regulated by many agonists through a cascade of protein kinases and other signaling proteins initiated by the occupancy of cell surface receptors. The increase in Ca^{2+} induced by the agonists causes translocation of cytosolic PLA_2 from the cytosol to the cell membrane, where it acts on phospholipid substrates. Membrane association of the enzyme occurs through a Ca^{2+}-dependent phospholipid binding domain. MAPK phosphorylates cytosolic PLA_2 on a specific serine residue, thereby enhancing enzymatic activity. In most cells, activation of cytosolic PLA_2 requires both Ca^{2+} mobilization and activation of MAPK. Cytosolic PLA_2 is the rate-limiting reaction for eicosanoid production and is therefore the major point at which agonists exert their control on this process.

Phospholipase C. PLC is a widely distributed enzyme of considerable physiological significance. In mammals, its primary substrate is phosphatidylinositol bisphosphate, a phospholipid in the cell membrane. PLC hydrolyzes this phospholipid to diacylglycerol and inositol trisphosphate (IP_3). Both of these substances are important intracellular signals. The function of diacylglycerol is to activate most isozymes of protein kinase C (PKC), which is an important regulator of cell function. (A kinase is an enzyme that transfers a phosphate group to another protein.) Activation of protein kinase C leads to the phosphorylation of cellular proteins, resulting in the generation of physiological responses, including altered cell growth, ion channel activity, and secretion, and potentiation of synaptic activity in the brain. The role of inositol trisphosphate is to release Ca^{2+} from intracellular stores in the endoplasmic reticulum (**Fig. 2**). The resulting elevation in cellular Ca^{2+} concentration increases the activity of calmodulin (Cam) and other Ca^{2+}-binding proteins. The phosphorylation of cell proteins by Ca^{2+}-calmodulin-dependent protein kinases and the actions of Ca^{2+} on other proteins lead to other physiological responses. Such responses include the contraction of smooth muscle, the breakdown of glycogen, and the secretion of enzymes and neurotransmitters.

There are several isozymes of PLC, which are grouped into β, γ, and δ types. There are four different β isozymes ($PLC\beta_{1-4}$), which are regulated by agonists acting through heterotrimeric G proteins (proteins that bind guanine nucleotides). These G proteins consist of α, β, and γ subunits and act

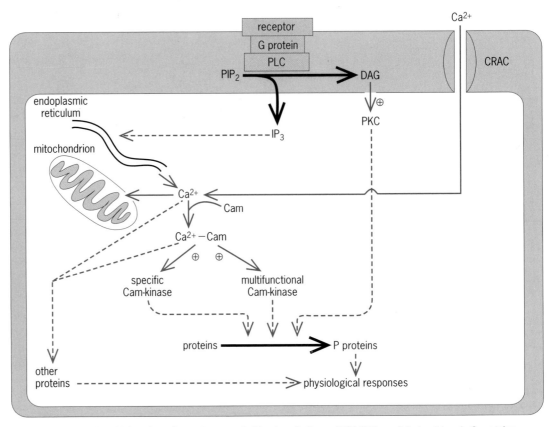

Fig. 2. Mechanisms by which activated receptors coupled to phospholipase C (PLC) through heterotrimeric G proteins produce their physiological responses. PIP_2 is phosphatidylinositol bisphosphate; DAG, diacylglycerol; PKC, protein kinase C; IP_3, inositol trisphosphate; Cam, calmodulin; Cam-kinase, Ca^{2+}-calmodulin-dependent protein kinase; P proteins, phosphorylated proteins; CRAC, Ca^{2+} release-activated channel.

as signal-transducing proteins for a large number of hormones, neurotransmitters, and other intercellular agonists. The agonists that function through G proteins interact with membrane receptors having seven transmembrane segments. Agonist binding stimulates the receptors to undergo a conformational change which, in turn, activates specific types of G proteins. The G-protein superfamily consists of several families (G_s, $G_{i/o}$, G_q, G_{12}), but only the $G_{i/o}$ and G_q families regulate PLCβ. The G_q family regulates the β_1 and β_3 isozymes of PLC through their α subunits, which bind to a specific domain on the isozymes. However, the $G_{i/o}$ family regulates the β_{2-4} isozymes through their $\beta\gamma$ subunits, which bind to a separate domain. The activation of PLCβ isozymes by the G-protein subunits accounts for the contractile effects of vasoactive agents such as epinephrine, angiotensin, and vasopressin on vascular smooth muscles. It also explains the effects of many agonists on contraction or secretion in the gastrointestinal and other systems.

PLCγ exists in two isozymic forms, with PLCγ2 being confined to hemopoietic (blood-forming) cells. These isozymes are not regulated by G proteins but are activated by growth factors or cytokines whose receptors subsequently become tyrosine-phosphorylated. Receptor phosphorylation can result from activation of intrinsic receptor tyrosine activity or

the action of soluble tyrosine kinases. Phosphorylation results in the generation of phosphotyrosine residues at certain sites in the cytoplasmic domain of the receptors. Specific cytosolic proteins and enzymes then become associated with select phosphotyrosines in the receptors through special binding sites termed SH2 domains, prompting activation. The PLCγ isozymes are some of the cytosolic proteins that are activated by this mechanism, and their activation accounts for some of the cellular responses to growth factors and cytokines.

PLCδ is smaller (85 kDa) than the other PLC isozymes (140–150 kDa) and exists in four isoforms. It is not regulated by G proteins or tyrosine phosphorylation. It is strongly stimulated by Ca^{2+}, and this response is believed to be the major factor in its regulation.

Phospholipase D. PLD is very widely distributed in nature and is found in almost all mammalian cells. It has been proposed to function in the regulation of cellular growth and secretion and in processes involving the actin cytoskeleton. PLD is regulated by a large number of hormones, neurotransmitters, growth factors, and cytokines. Two mammalian isozymes (PLD1 and PLD2) have been cloned, but other isoforms may exist. They hydrolyze phosphatidylcholine to produce the signaling molecule phosphatidic acid, which acts on many regulatory enzymes and other proteins in the cell. It can also be converted to diacylglycerol by a phosphatidic acid phosphatase and to lysophosphatidic acid by PLA_2. The formation of diacylglycerol from phosphatidic acid results in a second wave of diacylglycerol production and protein kinase C activation that is seen when many cell types are stimulated by agonists. Thus, PLD activation can result in long-term cellular changes due to prolonged protein kinase C stimulation. Lysophosphatidic acid formed from phosphatic acid is an important intercellular messenger. Many cell types have receptors for lysophosphatidic acid and show rapid morphological and functional changes in response to this lipid.

The regulation of PLD is complex. It requires phosphatidylinositol bisphosphate for activity and can be regulated positively by protein kinase C and by low-molecular-weight (20–25 kDa) G proteins of the ADP-ribosylation factor and Rho families. These small G proteins act at different sites on the enzyme. Fatty acids can either stimulate or inhibit PLD, depending on the isoform. The effects of ADP-ribosylation factor on PLD are related to the function of this small G protein in the trafficking of proteins from the endoplasmic reticulum to and through the Golgi apparatus. This is an essential component of the process by which cells distribute newly synthesized proteins to the cell surface or organelles. The effects of protein kinase C and Rho on PLD are involved in the regulation of the enzyme by hormones, neurotransmitters, and growth factors. Further investigation is required to define the roles and participating molecules involved in PLD activation.

For background information see CELL MEMBRANES; EICOSANOIDS in the McGraw-Hill Encyclopedia of Science & Technology.   J. H. Exton

Bibliography. J. H. Exton, Phospholipase D: Enzymology, mechanisms of regulation and function, *Physiol. Rev.*, 77:303–320, 1997; H. E. Hamm, The many faces of G protein signaling, *J. Biol. Chem.*, 273:669–672, 1998; C. C. Leslie, Properties and regulation of cytosolic phospholipase A_2, *J. Biol. Chem.*, 272:16709–16712, 1997; T. Pawson, Protein modules and signalling networks, *Nature*, 373:573–580, 1995; S. G. Rhee and H. S. Bae, Regulation of phosphoinositide-specific phospholipase C isozymes, *J. Biol. Chem.*, 272:15045–15048, 1997.

Physical anthropology

The first Neanderthal (also spelled Neandertal) skeleton was scientifically described just 2 years before Darwin's *The Origin of Species* was published. This revelation not only marked the beginning of the field of paleoanthropology but ignited an intense debate about the relationship of Neanderthals to living people, especially those of European descent. Some early workers believed that the peculiar aspects of Neanderthal morphology were due to degenerate skeletal pathologies. However, additional finds from Belgium, France, and Gibraltar made it increasingly clear that such remains represented a population of archaic humans who inhabited Europe during the earlier part of the last glacial stage. Subsequent finds demonstrated that Neanderthals also extended into western Asia, as far as Israel and Uzbekistan, although they are not known from Africa or from the Far East. Now, it is known that the Neanderthals originated in Europe over 200,000 years ago and disappeared about 30,000 years ago, but experts remain fiercely divided over the question of whether Neanderthals might be included among modern human ancestors.

Characteristics. This distinctive population of archaic humans was a persistent presence in Eurasia for some 200,000 years. In agreement with what is known of their environments and lifestyle, their physique was muscular and thickset, with body proportions apparently reflecting adaptation to cold, and a skeleton built for strength and endurance. Although the morphological diversity of Neanderthal populations may well have been as great as that of modern humans, they were characterized by a number of distinctive anatomical features. Their brain was as large as that of modern humans and was enclosed within a longer, lower skull, with a strong browridge over the eye sockets. The face was large and high and was dominated by an enormous nose, perhaps an adaptation for breathing in relatively cold, dry air. The front teeth were relatively large and often heavily worn, and the lower jaw lacked a prominent chin. Recent research using three-dimensional x-rays has revealed that the inner ear of Neanderthals was shaped slightly differently from that of modern humans.

Even older types of archaic humans have been found that are distinct from both Neanderthals and recent humans. From the fossil evidence it appears that about 500,000 years ago there was a common ancestral population in Africa and Europe, which subsequently divided. North of the Mediterranean, its descendants evolved into the Neanderthals. South of the Mediterranean, the evolution of the species *Homo sapiens* occurred in Africa. About 100,000 years ago, early *Homo sapiens* began to emerge from Africa and spread across the Old World, first expanding eastward. By 40,000 years ago, they had also spread westward to enter Europe, where they encountered the last of the Neanderthals. It is generally assumed that *Homo sapiens* played a role in the eventual extinction of Neanderthals.

Molecular data. In 1997 a team of researchers published a molecular analysis which contributed novel data about the Neanderthals by sequencing part of the deoxyribonucleic acid (DNA) of the original Neander Valley individual. A great deal of inferential information about recent human evolution had already been reconstructed from the DNA of living humans. Such studies are able to use combinations of data from many different gene systems, or they can look at variation in a particular segment of DNA in great detail. There are three kinds of DNA which can be studied. The first is nuclear DNA, which makes up the chromosomes contained within the nucleus of human body cells. This DNA includes the blueprints for most human body structure, and a combination of it is inherited from both parents. For example, analysis of global variations in a nuclear DNA strand called the CD4 locus on chromosome 12 shows that African populations display many different patterns of variation, while those from the rest of the world have basically only one pattern. The results suggest that non-African populations are descended from ancestors who emerged from North or East Africa about 90,000 years ago.

The second type is mitochondrial DNA, which is found outside the nucleus of cells and is passed on through females. Studies of this DNA in living humans made a tremendous impact in 1987 with the suggestion that a female ancestor (popularly dubbed Eve) for all present-day mitochondrial variation lived in Africa about 200,000 years ago.

The third type is Y-chromosome DNA, which is part of the chromosome that determines the male sex in humans. This DNA can be used to study evolutionary lines in males only. Such studies are still in their infancy, but instructive results have already been obtained. Recent research suggests that variation in Y-chromosome DNA is very low, and that a hypothetical Adam for present-day males may also have lived in Africa about 200,000 years ago. However, some data also suggest an important Asian contribution to present-day Y-chromosome variation.

Population history. The degree of DNA variation within modern populations can also be used to reconstruct something of ancient population histories. For example, using estimates for the age of common ancestors for particular DNA types, it has been argued from the relatively high nuclear DNA variation that the breeding size of the human population through most of its evolution was about 100,000 individuals. This may not seem a large number compared with the enormous world population now, but it was a significant size for a large-bodied mammalian species. However, from the much lower variation inherent in human mitochondrial DNA, this number must have dropped to about 10,000 breeding individuals in the recent evolutionary past, producing a bottleneck which filtered out some previous genetic variability. This contraction in numbers may reflect divisions in the formerly large and widespread common ancestral population of Neanderthals and modern humans. Vast distances, repeated climatic extremes, and the intrusion of geographical barriers such as ice caps and deserts during the last 300,000 years may have progressively isolated human populations from each other, leading to increasing differentiation and eventual separation as species. Perhaps only one of these isolated groups, about 10,000 in breeding size and restricted to Africa, gave rise to all living *Homo sapiens*. It seems likely that the most important physical barrier was the Sahara Desert, increasing in its extent and impact through each cold period over the last 500,000 years, and that human origins were influenced by its growth in regions to the south.

The 1997 recovery of meaningful DNA from a Neanderthal fossil provided the first glimpse of the genetic makeup of an extinct human form. Remarkably, the mitochondrial DNA sequence was recovered from the Neander Valley skeleton itself. Analyses of the DNA and comparisons with that obtained from humans and chimpanzees imply that the evolutionary divergence between modern humans and Neanderthals might have begun about a half million years ago, far earlier than generally believed. The new mitochondrial DNA evidence also failed to support any particularly close relationship between modern Europeans and Neanderthals, since recent European DNA is as distinct from the Neanderthal sequence as is that of Africans, Australians, or Asians.

Genetic exchange. With respect to the contentious issue of whether Neanderthals and early modern humans might have exchanged genes when they co-existed in western Eurasia for at least several millennia, these new results diminish, but do not rule out, the possibility of some interbreeding. It seems clear that Neanderthal populations represented by the type specimen did not contribute discernible mitochondrial DNA to human populations sampled today. However, other Neanderthal or early modern samples might show evidence of admixture, and this result based on mitochondrial DNA does not exclude the possibility of exchange via nuclear genes. However, if such exchanges were the exception rather than the rule, hybrids may have contributed little or nothing to the DNA of succeeding generations. The separation time of the Neanderthal and modern human lineages inferred from this first successful analysis certainly fits well with one view of the Neanderthals already clearly articulated from the fossil record—that they were a distinct race, or more likely species, of human which evolved separately from *Homo sapiens* over several hundred thousand years, and which was not part of modern humans' ancestry.

However, the degree of mitochondrial difference between the Neanderthal and recent human sequences cannot, on its own, demonstrate that the distinction is at the level of species, since such a level of variation can be found both within and between living mammalian species. The inferred separation time must also be viewed with some caution as a phylogenetic marker, since the estimated date of about a half million years would apply to the divergence of what is essentially one complex gene. Mitochondrial DNA has been diverging in modern humans for perhaps 150,000 years without speciation, so a figure of around 500,000 years really provides a kind of upper limit for Neanderthal separation. Gene divergence precedes population divergence, which precedes any specific separation. Thus, this mitochondrial divergence date is not inconsistent with a separation time of the *Homo neanderthalensis* and *Homo sapiens* lineages, based on fossil data, of about 300,000 years.

For background information *see* DEOXYRIBONUCLEIC ACID; FOSSIL HUMAN; NEANDERTALS; ORGANIC

EVOLUTION; PHYSICAL ANTHROPOLOGY; PRIMATES in the McGraw-Hill Encyclopedia of Science & Technology.   Chris Stringer

Bibliography. R. Cann, M. Stoneking, and A. Wilson, Mitochondrial DNA and human evolution, *Nature*, 325:31–36, 1987; M. Krings et al., Neandertal DNA sequences and the origin of modern humans, *Cell*, 90:19–30, 1997; M. Nei, Genetic support for the out-of-Africa theory of human evolution, *Proc. Nat. Acad. Sci. USA*, 92:6720–6722, 1995; C. Stringer and R. McKie, *African Exodus*, Henry Holt, 1998; S. Tishkoff et al., Global patterns of linkage disequilibrium at the CD4 locus and modern human origins, *Science*, 271:1380–1387, 1996.

Planetary atmospheres

In the inner solar system, Mercury, Venus, and Mars are similar in size to the Earth, so the four are collectively called terrestrial planets. Their small masses and sizes are in marked contrast to those of the giant planets—Jupiter, Saturn, Uranus, and Neptune—at far greater distances from the Sun. The four terrestrial planets and Earth's Moon offer a rich variety in physical makeup and environment. Traditionally, Mercury and the Moon were described as airless bodies, classic cases of objects too small in mass to hold on to gases to form a permanent atmosphere. With advances in modern technology for ground-based remote sensing and space-flight opportunities for direct samplings, both Mercury and the Moon are now known to have transient atmospheres, continuously being formed and lost.

Apollo lunar experiments. A new perspective on the rocky, dusty worlds of Mercury and the Moon began with the Apollo program of lunar exploration in the 1970s. Sensors brought to the surface of the Moon by astronauts detected a very thin atmosphere, one still comparable to a vacuum in comparison to the dense atmosphere at the surface of the Earth (10^{19} particles/cm^3). Apollo instruments detected a concentration of gases of approximately 10^7 particles/cm^3 during the day and about 10^5 particles/cm^3 at night. Initial concerns that the instruments were merely detecting gases that evaporated off the lunar modules were ultimately dismissed in favor of accepting a weak atmosphere produced by capture of solar-wind particles (such as helium) or the radioactive decay of elements in the lunar soil (such as argon from potassium). As interests turned to other areas in the solar system, the Moon's minor atmosphere received little attention for many years.

Remote sensing of Mercury. In 1985 the ground-based detection of sodium and potassium gas above the limb of the planet Mercury occurred. Long considered to share common characteristics, the Moon and Mercury were examples of how small, primitive bodies can be formed and suffer little change for billions of years. At Mercury, the *Mariner 10* spacecraft fly-bys in 1974 and 1975 had found evidence for a weak atmosphere, whose characterization and monitoring were rather limited. The use of ground-based telescopes with spectrographs to record the signals of specific chemical species signaled a new era of ground-based remote sensing of primitive bodies. Much like comets, these rocky surfaces somehow emitted gases that escaped into space, providing a transient atmosphere to study.

Lunar sodium and potassium gases. The spectroscopic technique applied to Mercury was then turned to the Moon by two different groups in 1988, and sodium and potassium gases were detected there as well. The elements sodium and potassium are not particularly abundant in the solar system but are relatively easy to detect because they scatter sunlight very efficiently. They are not the major constituents of the atmospheres of Mercury and the Moon but serve as excellent tracers of other gases presumably there but more difficult to detect. For example, the total number of sodium plus potassium atoms detected just above the Moon's limb is barely 100 atoms/cm^3, far below the concentrations suggested by the Apollo data. At Mercury, sodium and potassium are far more abundant, approaching 10^4–10^5 atoms/cm^3 in an atmosphere estimated by *Mariner* data to be near 10^6 particles/cm^3.

Sputtering sources. The source of the sodium gases on both of these bodies is still debated actively. To liberate gases from the surface material (regolith) requires the impact of micrometeors or solar-wind ions and electrons, or sunlight. These are called sputtering agents, and laboratory experiments show that they indeed can free atoms and molecules from surfaces with sufficient energy to move away from the surface. Both hot and cold gaseous populations are possible from these processes, and the degree to which the thermal and superthermal components dominate close and distant regions is still under study. Sputtered gases are pulled back to the regolith by gravity, or pushed away by solar radiation pressure, or lost by photoionization and removal by the magnetic field in the solar wind (**Fig. 1**). The term surface-boundary exosphere is applied to an atmosphere produced by vaporization of surface material under conditions where collisions aloft are so rare that the liberated gases can have long parabolic trajectories back to the surface, or can directly escape from it.

Imaging results. The next step in understanding the sodium and potassium atmospheres of the Moon and Mercury came from new low-light-level imaging techniques capable of taking a picture of the full extent of these atmospheres. In the lunar case, images of the sodium brightness in two dimensions (**Fig. 2**) show that the atmosphere extends to several times the radius of the Moon. This suggests that sodium atoms have relatively high speeds, close to the escape speed (2.3 km/s or 1.4 mi/s); yet, the pattern of brightness decreasing with distance also shows that there are slower-speed sodium atoms as well. To test the mechanisms responsible for the sputtering, several research groups are actively making lunar observations during meteor showers and at times when

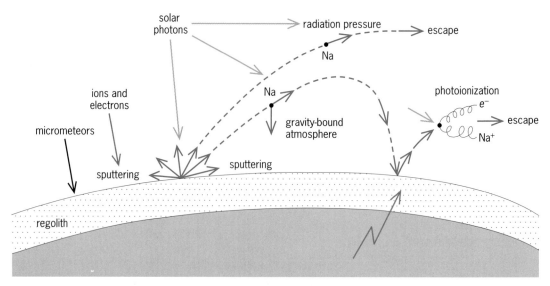

Fig. 1. Processes that release and govern the subsequent motion of atoms sputtered from the regoliths of the Moon and Mercury. Sodium atoms are used to illustrate the possibilities.

the solar wind is shielded from the surface by the Earth's magnetic field.

At Mercury, attempts to photograph its two-dimensional atmosphere involve far more complex imaging systems. Because it is the planet closest to the Sun, when Mercury is viewed from the Earth's position it can never be seen at a large angular separation from the Sun (Mercury's so-called maximum elongation angle is about 28°). If viewed prior to sunrise as a "morning star" in the east or just after sunset as an "evening star" in the west, the light from Mercury must pass through a long, slanted path in the

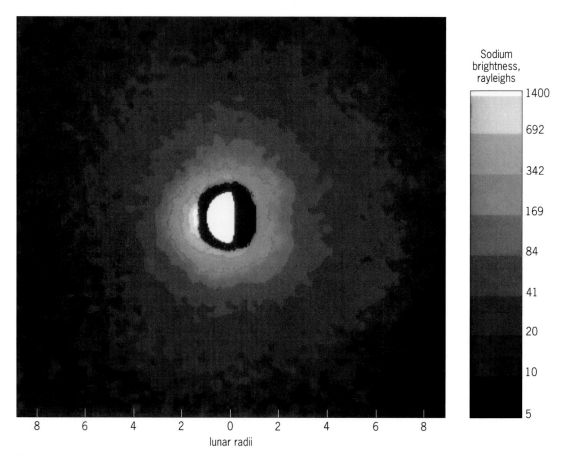

Fig. 2. Image of the sodium atmosphere surrounding the Moon. The sodium brightness units shown are far below those capable of being seen by the unaided human eye. 1 rayleigh $= 10^{10}/4\pi$ quanta per square meter per second per steradian.

Earth's dense atmosphere. Such low-elevation-angle observations suffer from scattering, absorption, and turbulence. To view Mercury when it is high in the sky (near zenith) so that the path of light through Earth's atmosphere is shorter necessitates that observations be conducted near local noon. Pointing a large telescope with sensitive equipment so close to the Sun is highly undesirable. For example, the Hubble Space Telescope is never allowed to make observations of Mercury or of Venus. The task thus falls to ground-based systems that can handle daytime viewing, and these are few. Yet, some new high-resolution imaging spectrographs actually make this possible. These advanced daytime systems are starting to yield two-dimensional images of Mercury's sodium atmosphere, showing it to extend to distances that are surprisingly large (several times the radius of the planet). The challenge for future observers will be to improve these capabilities to accommodate the very low light levels required to map the full extent of Mercury's extended atmosphere.

Other lunar gases. The search for other gases on the Moon has been frustrating. Given the relative ease of detecting sodium and potassium by their bright emission lines, it was expected that other gases could be found using similar ground-based spectrographs operating in the visible-light portion of the spectrum. Yet, unambiguous detections for silicon, aluminum, calcium, iron, titanium, barium, or lithium have not been reported. Several of the more interesting candidates to search for, such as water (inferred by detecting the hydroxyl molecule, OH), are not possible using ground-based telescopes because they do not have emission lines in the visible. They do, however, have ultraviolet spectral lines. Ultraviolet light is absorbed by the ozone in the Earth's stratosphere, and so space-based observing methods must be used. In a recent observation conducted from the Hubble Space Telescope, the hydroxyl molecule, aluminum, and magnesium were not detected. A more promising set of observations using a small satellite (*ORFEUS-SPAS*) deployed during a space shuttle mission succeeded in the first ultraviolet remote detection of argon above the dayside hemisphere on the Moon. The number densities (approximately 10^5 atoms/cm^3) were higher than anticipated; if confirmed by future observations, the long-sought "missing mass" to account for the Apollo total gas-pressure measurements would be identified.

Observations of the Moon's atmosphere, in both minor and major species, are thus on the threshold of becoming a valuable tool for studies of the origin and evolution of the Earth's closest surface-boundary exosphere. The results and insights obtained from such laboratory-in-space investigations could then be applied to other supposedly atmosphereless bodies in the solar system.

For background information *see* MERCURY (PLANET); MOON; SOLAR WIND; SPUTTERING; ULTRAVIOLET ASTRONOMY in the McGraw-Hill Encyclopedia of Science & Technology.   Michael Mendillo

Bibliography. B. Flynn and M. Mendillo, A picture of the Moon's atmosphere, *Science*, 261:184–186, 1993; A. E. Potter and T. H. Morgan, Discovery of sodium and potassium vapor in the atmosphere of the Moon, *Science*, 241:675–680, 1988; A. E. Potter and T. H. Morgan, Discovery of sodium in the atmosphere of Mercury, *Science*, 229:651–653, 1985; P. R. Weissman, L. McFadden, and T. V. Johnson (eds.), *Encyclopedia of the Solar System*, Academic Press, 1999.

Plant evolution

The origin and early evolution of plants on land was an interval of unparalleled innovation in plant life. From simple beginnings as aquatic green algae, land plants evolved an elaborate life cycle and an extraordinary array of complex organs and tissue systems. By the end of the Devonian Period [354 million years ago (Ma)] they had developed specialized sexual organs (gametangia), stems with an intricate fluid transport mechanism (vascular tissue), structural tissues (wood), modifications to the epidermis for respiratory gas exchange (stomates), leaves and roots of various kinds, diverse spore-bearing organs (sporangia), seeds, and the tree habit. The evolution of the land flora marked the beginning of the development of modern terrestrial ecosystems, and these events had far-reaching consequences for the environment. Knowledge of this early phase of plant evolution is based on evidence from the fossil record, as well as on data on the relationships (phylogeny) of living species. Improvements in understanding the relationship among plants have been important in clarifying the early phases of plant evolution and the interpretation of evidence from the fossil record. *See* TREE.

Origin. Phylogenetic studies provide compelling evidence that land plants originated within a small but highly distinctive group of aquatic green algae of the family Charophyceae (**Fig. 1**). Unlike most green algae, the Charophyceae are predominantly fresh-water plants. It is therefore believed that the ancestors of land plants evolved in similar environments (terrestrial rivers or lakes, rather than marine locales). It is also clear from improved understanding of phylogeny that the evolution of plants on land was a unique event. In contrast to the transition to the land by animals, which happened independently in numerous different groups, all land plants share a common land-dwelling ancestor. This common ancestry is reflected in remarkable similarities in reproduction, life cycle, and molecular genetics among otherwise highly distinctive groups.

A consistent and increasingly well-documented picture of the morphology of these early plants is emerging from the study of living and fossil species. The earliest plants were small, simple organisms that are most closely related to modern bryophytes such as liverworts or hornworts (**Fig. 2**). Furthermore, the morphology of charophycean algae indicates that the land plant ancestor was even simpler, probably a unicellular or filamentous plant, comprising only a handful of cells. These observations show that in

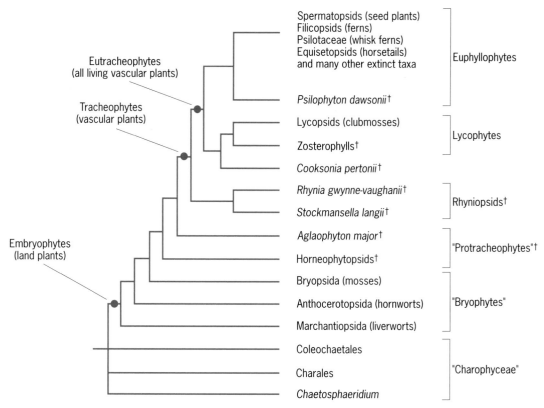

Fig. 1. Relationships among land plants and closely related green algae (Charophyceae). Phylogenetic relationships depicted are consistent with a broad range of data, including gene sequences and comparative morphology. Relationships are indicated for some important early fossil groups or species by the symbol †.

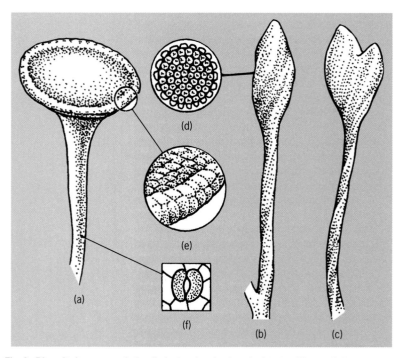

Fig. 2. Diversity in some early fossil plants, showing terminal parts of incomplete specimens. (a–c) Stem with enlarged end region containing spores (sporangium), magnified ×15, ×40, and ×30, respectively. (d) Transverse section through sporangium showing spore mass, ×70. (e) Details of epidermis at rim of sporangium, ×45. (f) Stomate comprising two guard cells, ×120.

contrast to land-dwelling vertebrates and arthropods, the transition to the land by plants did not involve modifications to an already present, well-developed, and highly differentiated plant body in the aquatic ancestor. Despite many similarities in cell morphology and biochemistry, land plants inherited little above the cellular level from the green algae. The diverse array of tissues and organ systems in living land plants is entirely a product of evolution in the terrestrial environment.

Early fossil evidence. The earliest fossil evidence of land plants comes from microscopic spores found in sediments of mid-Ordovician age (470 Ma). In land plants, spores are unicellular reproductive propagules that possess a distinctive cell wall, which is impregnated with a complex, highly resistant polymer (sporopollenin). Spores are the products of meiotic cell division and are produced in groups of four (tetrads). There is clear evidence of this tetrahedral configuration in early fossils, which are either dispersed as tetrads or exhibit characteristic marks in their cell walls.

Evidence from the fossil record documents an increasing diversity of spore types through the Late Ordovician and Silurian (458–412 Ma). These data are consistent with an increase in diversity of land plants at the bryophyte level of organization. Fossil evidence of the spore-producing plants (megafossils) first appears in the Late Silurian (428 Ma). This hiatus between the earliest spore evidence and confirmation from megafossils has been widely discussed, and it probably reflects a combination of limited sampling, collector bias, and preservational bias. One complicating factor is that spores are produced in vast quantities, and so they greatly outnumber the plants that produce them. Also, because they are wind-dispersed, spores are widely distributed and incorporated into a broad range of facies types. These two properties ensure that spores are much more likely to be represented in the fossil record than are megafossils.

The earliest megafossils document small, leafless plants that strongly resemble the sporophytes (spore-producing phase) of modern mosses or liverworts (Fig. 2). Spores were produced in expanded terminal regions called sporangia, but unlike in bryophytes the stems branched dichotomously. Mineralized fossils at some Devonian localities preserve remarkable details down to the cellular level, enabling these early plants to be reconstructed with great accuracy. At the Early Devonian (about 412 Ma) Rhynie Chert locality in Scotland, early plants are preserved in their growth position, and this has enabled life cycles to be reconstructed in some detail (**Fig. 3**). Unlike living land plants, life cycles in these early extinct groups are strongly isomorphic (gamete-producing and spore-producing phases are very similar).

The size of land plants had increased significantly by the Early Devonian, although all species were still herbaceous. The complexity of branching, sporangial form and distribution, and anatomical differentiation had also increased by this time. Vascular plants dominate the fossil record, and preservation of bryophytes is extremely rare. Herbaceous clubmosses, with small, scalelike leaves, were the earliest vascular plants with recognizable modern counterparts. The ancestors of important living groups such as ferns, horsetails, and seed plants (including conifers and flowering plants) had evolved by the Late Devonian. Evidence from the fossil record shows that many of the organ systems of living vascular plants (for example, stem, root, many types of leaf, cones) are derived ultimately from modified branching systems. The leaves of ferns and seed plants have clearly evolved from simpler dichotomous branching systems similar to those in many Devonian fossils. Other types of leaf appear to have completely different origins, perhaps developing through sterilization of sporangia (for example, the small scalelike leaves of clubmosses) or gametangia (for example, mosses).

Adaptations. Terrestrial plants interact principally with the gaseous medium of the atmosphere. This transition from an aqueous to a predominantly

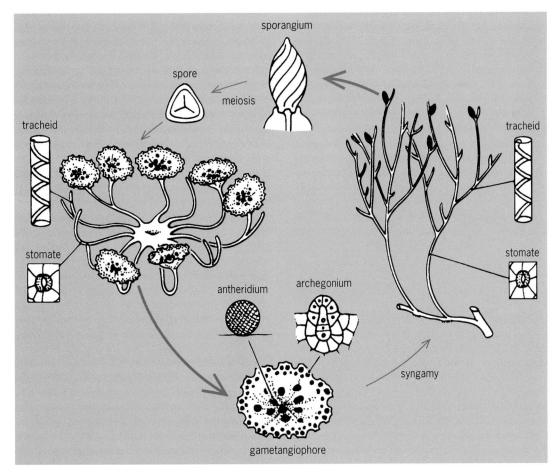

Fig. 3. Probable life cycle of some extinct Early Devonian plants.

gaseous environment had profound consequences for the morphology and physiology of plants. For organisms with aquatic origins, life on land requires the ability to conserve water. All land plants have a waxy cuticle that significantly reduces water loss through evaporation. Land plants also have reduced ratios of surface area to volume, brought about through the development of multicellular parenchymatous bodies. Most land plants also possess variable aperture pores in the epidermis (stomates) that facilitate and regulate the exchange of water and gases with the atmosphere. Physiological data show that plants over about 10 cm (4 in.) in height require a vascular system for the transport of water and metabolites. In most plants, this comprises an interconnected system of elongate cells (xylem, phloem). Cuticle, stomates, and a vascular system were key innovations enabling plants to manage their water economy and thus to thrive on land.

Land plants have inherited a form of sexual reproduction from green algae that, in primitive groups, requires the presence of free water. Male gametes have two or more flagella that propel the sperm toward the egg cell through a superficial layer of water. While this transfer of sperm in the process of fertilization is a legacy of an aquatic ancestry, land plants have additionally evolved gametangia, which are specialized structures that protect egg and sperm until optimal conditions for fertilization. The dispersal phase of the life cycle in land plants has undergone dramatic changes. Instead of the swimming biflagellate zoospores of green algae, flagella have been lost from the unicellular dispersal phase, the cell wall has become impregnated with the highly resistant polymer sporopollenin, and the cell (spore, pollen) is dispersed through the air.

With the development of larger plants, the addition of structural support to the stems was increasingly important to avoid buckling and collapse. Aquatic plants are more or less neutrally buoyant and require little skeletal support. Land plants are much heavier than air, and in addition to supporting their own weight they must be able to withstand the shearing effects of wind. Small plants can maintain posture through a combination of simple turgor pressure and robust cells with thick cellulose cell walls. Plants greater than a few tens of centimeters in height require additional structural support. Fossil evidence shows that this was achieved in early plants by stiffening the cellulose cell walls in the cortical regions of the stem with a resistant biopolymer called lignin. Support from this internal skeleton is supplemented in many groups by the massive development of additional lignified secondary cortex or secondary wood. The development of roots was also critical to the evolution of large plants. For larger plants, roots are essential for anchorage and for obtaining water and minerals from the soil. These innovations led ultimately to the evolution of trees and the widespread development of forests by the Late Devonian (370 Ma).

For background information *see* PALEOBOTANY; PLANT ANATOMY; PLANT EVOLUTION; PLANT PHYSIOLOGY; POLLEN; SEED in the McGraw-Hill Encyclopedia of Science & Technology.   Paul Kenrick

Bibliography. W. A. DiMichele et al., Paleozoic terrestrial ecosystems, in A. K. Behrensmeyer et al. (eds.), *Terrestrial Ecosystems Through Time: Evolutionary Paleoecology of Terrestrial Plants and Animals*, pp. 205–325, University of Chicago Press, 1992; P. Kenrick and P. R. Crane, The origin and early evolution of plants on land, *Nature*, 389:33–39, 1997; K. J. Niklas, *The Evolutionary Biology of Plants*, University of Chicago Press, 1997; T. N. Taylor and E. L. Taylor, *The Biology and Evolution of Fossil Plants*, Prentice Hall, 1993.

Plant–fungi interactions (paleobotany)

Fungi are organisms that live entirely by the absorption of nutrients from either living or dead organisms. They represent an important component of the ecosystem, where they function as decomposers and in the recycling of carbon and minerals. Thus, fungi played a pivotal role in the establishment of early terrestrial ecosystems. The oldest evidence of life on Earth is septate filaments, like those of certain cyanobacteria, dated at approximately 3.6 billion years. Evidence of the oldest terrestrial fungi includes hyphae approximately 430 million years old, but it is impossible to determine how these fungi interacted with other organisms in the ancient ecosystem. The oldest fossil fungi in which interactions can be accurately documented are from the Lower Devonian Rhynie Chert, dated at approximately 400 million years.

Rhynie Chert fungi. The Rhynie Chert consists of a series of siliceous beds that represent an ancient ecosystem composed of periodically desiccated fresh-water ponds. The Rhynie Chert ecosystem is unique in that the cells of the organisms are preserved in silica so it is possible to document and examine interaction among organisms, such as fungi and their hosts (**Fig. 1**). Small plants inhabited the sandy substrate along the edge of the ponds, along with algae, cyanobacteria, and eubacteria. Interacting with these organisms was a diverse assemblage of fungi that functioned as decomposers (saprobes), parasites (organisms that live at the expense of another, often invading it), and pathogens (a parasite able to cause disease). Other fungi entered into mutualistic associations in which two organisms lived in close juxtaposition and were beneficial to each other.

Saprophytes. Today all major groups of fungi contain saprobes, with the basidiomycetes and ascomycetes the most common. During Rhynie Chert time, major decomposers were chytrids, which include the only fungi with flagellated cells. Saprophytic interactions were the most widespread in the Rhynie Chert ecosystem. However, because of the nature of this nutritional interaction in which cellulose and lignin are broken down, saprophytism is

Fig. 1. Fungal hyphae and spores in the stem; ×120.

difficult to document. In the chert blocks there are numerous examples of partially degraded stems and other plant material and associated fungi. Some of these fungi not only are morphologically identical to their modern counterparts but also possess the same type of life history.

Parasites. Parasites were also a common component among the Rhynie Chert fungi. In order to demonstrate the presence of biotrophic parasites, it is necessary to document a response by the host. One host response present in the Rhynie Chert ecosystem occurs in some cells of the aquatic plant *Palaeonitella*. Normal cells that make up the main axis of the plant are approximately 75 micrometers in diameter. Other cells that have been infected by a small, top-shaped chytrid are swollen to more than four times the size of the normal cells. This abnormal growth is termed hypertrophy, and it is common in modern relatives of *Palaeonitella* that are also parasitized by chytrids. Another host response in the Rhynie Chert ecosystem involves fungi that parasitize other fungi, a condition termed mycoparasitism. Some chytrids are attached to the outer surface of their fungal host, while others occupy the interior of the host cells. Still others occupy the space between wall layers of spores. Each of these fossil parasites is quite different, suggesting that a variety of nutritional modes had evolved by the Lower Devonian. In these mycoparasites, it is difficult to determine whether the host was alive or dead when the interaction first occurred. Other mycoparasites were associated with large

spores thought to be the resting stage of mutualistic fungi. Extending in from the inner surface of the wall in these spores are conical projections termed papillae (**Fig. 2**). Such structures represent a response by the host in which new wall material is synthesized by the spore protoplast as a defense against the invading parasite. The papillae indicate that the host fungus was alive when the infection took place. Another host–parasite interaction involves abnormal cellular proliferation in the host. This response is termed hyperplasia; it also occurs in some of the Rhynie Chert macroplants. Finally, certain modern chytrids disrupt the organelles of the cells they infect. One example involves abundant starch grains being produced in the chloroplasts of the host in the presence of certain chytrids. Dark granular bodies occur in some fossil cells in which a particular chytrid thallus is also present. These cell inclusions may also represent another example of a host response in the Rhynie Chert ecosystem.

Mutualists. Mycorrhizae form mutualistic associations with the roots and absorbing organs of more than 80% of all land plant species. Arbuscular mycorrhizae represent the most common type of association today and include approximately 50 species of zygomycetous fungi. In this form of mutualism, the fungus gains a source of carbon, while the plant benefits from increased absorption of certain nutrients, especially phosphorus. Other benefits may include disease resistance and drought tolerance. Endomycorrhizae have been hypothesized as an integral stage in the colonization of the land surface because these symbioses afforded plants the opportunity to move into new niches that were otherwise unavailable. Information from molecular biology suggests that endomycorrhizae evolved up to 460 million years ago, which is consistent with the Rhynie Chert fossil evidence. In this symbiosis, specialized hyphae, termed arbuscules, occur in certain cells of the absorbing organs. Arbuscule branches greatly increase

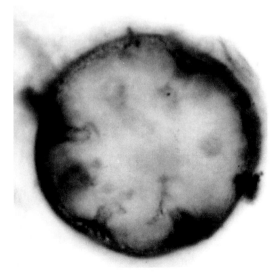

Fig. 2. Fungal spore with host response in the form of papillae; ×700.

the surface area between the fungus and the host, and through these structures there is a bidirectional transfer of metabolites and nutrients by the mycorrhizal partners. Arbuscules have recently been documented in the Rhynie Chert macroplant *Aglaophyton*, and there is now abundant evidence that all of the land plants in the Rhynie Chert ecosystem were mycorrhizal, including the free-living gametophytes. Arbuscules occur in a narrow band in both the above and below portions of the fossil, and are identical to those found in modern endomycorrhizae. Other plant cells contain coils and loops of hyphae that may represent the arbuscules of different endomycorrhizae.

Other fungi termed endophytes or mycophylla invade the leaves and stems of various living plants and function as mutualists. Many of these fungi cause no observable symptoms to the host but produce various toxins that protect the plant from certain herbivores and pathogenic fungi. While many of the Rhynie Chert plants possess fungi in the aerial parts, it is not possible at the present time to determine where these fungi represent endophytes and, therefore, early stages in terrestrial plant defenses.

Lichen. Lichens are mutualistic associations that unite a fungus with a cyanobacterium or green alga, or both. This association involving a photosynthetic partner results in a body form that is unlike that of either participating organism. Because cyanobacteria are ancient, it might be concluded that lichens are extremely old and would be present in early terrestrial ecosystems. However, to date, the only convincing fossil lichen comes from the Rhynie Chert. *Winfrenatia* consists of mats of fungal hyphae that form numerous shallow depressions on the surface of the thallus. Within each depression is a loosely arranged net of hyphae, with each space occupied by a cyanobacterial cell (**Fig. 3**). It is suggested that the fungus parasitized some of the cyanobacterial cells to maintain a continuous source of carbon, while in other depressions cyanobacterial cells continued to divide to maintain the photosynthesizing partner of the mutualism. As a result of this symbiosis, the cyanobacterial colonies can invade new habitats and, perhaps through the production of lichen acids, discourage invertebrate herbivory.

Fossil evidence indicates that at least 400 million years ago several fungal groups interacted with other organisms in terrestrial ecosystems much as they do today, including a variety of symbiotic associations with photosynthesizing organisms. As a result of these interrelationships, plants were able to exploit new niches on the land surface. Several theories have been advanced to explain how fungal–plant symbioses may have originated. One suggests that saprophytic fungi increasingly became nutritionally specialized with their hosts, while another argues that fungi were initially parasitic but became partners as land plants evolved effective defense mechanisms to hold parasitism in check. It is doubtful that one scenario fits all of the fungal–plant interactions documented in the Rhynie Chert. Rather, the biochemical and genetic factors that are required in these interactions have a long and complex evolutionary history that predates the invasion of the land by photosynthesizing organisms and fungi. Modern molecular approaches directed at deciphering fungal–plant interactions, in addition to more ancient, well-preserved fossil biotas, represent the best opportunities for understanding the many evolutionary steps involved.

For background information *see* FUNGAL ECOLOGY; FUNGI; MYCOLOGY; MYCORRHIZAE; PALEOBOTANY; PALEOECOLOGY in the McGraw-Hill Encyclopedia of Science & Technology.   Thomas N. Taylor

Bibliography. C. J. Alexopoulos, C. W. Mims, and M. Blackwell, *Introductory Mycology*, 1996; S. P. Stubblefield and T. N. Taylor, Recent advances in palaeomycology, *N. Phytol.*, 108:3–25, 1988; T. N. Taylor et al., Fossil arbuscular mycorrhizae from the Early Devonian, *Mycologia*, 87:560–573, 1995; T. N. Taylor, H. Hass, and H. Kerp, A cyanolichen from the Lower Devonian Rhynie Chert, *Amer. J. Bot.*, 84:992–104, 1997; T. N. Taylor and E. L. Taylor, *The Biology and Evolution of Fossil Plants*, Prentice Hall, 1993.

Plasma displays

The development of plasma display panels has progressed greatly in recent years, and this technology now appears to be foremost in the effort to develop a television receiver that hangs on a wall. A plasma display panel offers intrinsic advantages: (1) a flat screen, permitting a television receiver as little as 5 cm (2 in.) thick, even with a 140-cm-diagonal (55-in.) screen; (2) an emissive display with very wide viewing angles, well suited for a family audience; (3) high-quality images, with vivid color rendition; and (4) relatively simple fabrication technologies to produce large screens (107–140 cm or

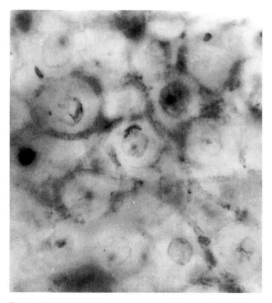

Fig. 3. Lichen hyphal net and cyanobacterial cells; ×800.

42–55 in.). While extensive resources have been committed to the mass production of such screens, considerable research is continuing in order to improve brightness, contrast, and image quality, to reduce power consumption, and to lower manufacturing costs for both panel technology and driving electronics.

A plasma display panel is an emissive display that uses gas discharges to convert electrical energy into light for image presentation. Ionized gas (plasma) is used to produce electromagnetic radiation. A discharge in a gas can emit radiation in either the visible spectrum (in monochrome displays) or at near-ultraviolet wavelengths (in color displays). In the color displays, fluorescent materials (red, green, and blue) are used to convert invisible ultraviolet light into the three primary visible colors. Rare gases are used in order to avoid chemical reactions during discharge; neon in monochrome displays gives a red-orange light, and a mixture of neon and xenon supports ultraviolet photon production in color panels.

Structure. Many display-cell designs have been invented and used since the appearance of plasma displays in the 1940s, from very simple numeric displays (Nixie tubes) to monochrome matrix displays (either direct-current or alternating-current types) and, recently, color video displays. At present, there are three contending color video display structures: the alternating-current matrix sustain structure, the alternating-current coplanar sustain structure, and the direct-current structure with pulse memory drive. The principles of these devices are similar, although the devices operate in slightly different ways. The alternating-current matrix sustain structure, the simplest of the three, will be used to illustrate the physical principles of a plasma display panel (**Fig. 1**).

The panel is fabricated from two glass substrates. Generally each panel is 3 mm (0.12 in.) thick, giving a 6-mm (0.25-in.) thickness for the device whatever its size. On the front substrate an array of row metallic electrodes is formed and then covered with a layer of dielectric (transparent enamel) and a thin layer of magnesium oxide (MgO). On the rear substrate an array of column electrodes is formed and covered with a dielectric (white enamel) and magnesium oxide. Barrier ribs are deposited between each pair of adjacent column electrodes in order to insulate discharge sites laterally. Then phosphors are deposited in the channel formed by the ribs. The two substrates are assembled and sealed after the front substrate has received the spacers that will maintain a uniform gas space between the two plates. The gas space is then evacuated and filled with a mixture of neon and xenon at a pressure of approximately 400 torr (53 kilopascals) in order to maintain the two plates below normal atmospheric pressure (760 torr or 101 kPa).

The display is a matrix display comprising a great number of elementary picture elements (pixels). For example, a standard-definition television display with 480 rows of 640 white pixels (red, green, and

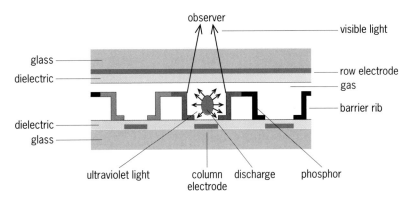

Fig. 1. Section through three pixels of an alternating-current color plasma display panel structure. Each pixel is located where a column electrode and a row electrode cross. (*Thomson Plasma*)

blue) accounts for $480 \times 640 \times 3 = 921,600$ elementary pixels.

Gas discharge process. The gas discharge process relies on the breakdown of gas which occurs when a voltage applied to the gas causes a transition from the usual insulating stage (neutral state) to a conducting stage. The electrons in the gas excite and ionize gas molecules, which in turn extract secondary electrons from the negative electrode. Under certain conditions, enough electrons and ions are produced to create a quasineutral plasma between the electrodes. Collisions between the electrons and ions can then excite the gas particles to higher energy states, and these excited particles release their excess energy as intense light emission (the invisible or ultraviolet region of the spectrum). This phenomenon is called the glow discharge. The voltage at which the discharge is initiated is called the breakdown voltage and is a function of various parameters such as the nature and pressure of gas, the length of the gas gap, and the secondary emission coefficient of the surface material (magnesium oxide) in contact with the gas. In normal structures the breakdown voltage is around 150 V.

Two properties of gas discharges are important to plasma displays:

1. The gas discharge is an avalanche process. This means that, without a limiting device such as a resistor or a capacitor in the loop, the discharge current can drastically increase until the cell is destroyed.

2. Due to the electric field distortion in the gas gap that results from the presence of charged species (ions and electrons), the discharge can be maintained at a voltage below the breakdown voltage. This feature is a hysteresis effect.

Memory effect. The memory effect was discovered in 1964. The basic idea is to introduce a capacitive device consisting of dielectric layers (enamel and magnesium oxide layers) between the electrodes and the gas (Fig. 1).

One function of the enamel is to limit the discharge current increase. The discharge current grows up to the glow discharge regime, as explained above, but the ions on the cathode side and the electrons at the anode side are stopped by the insulating magnesium

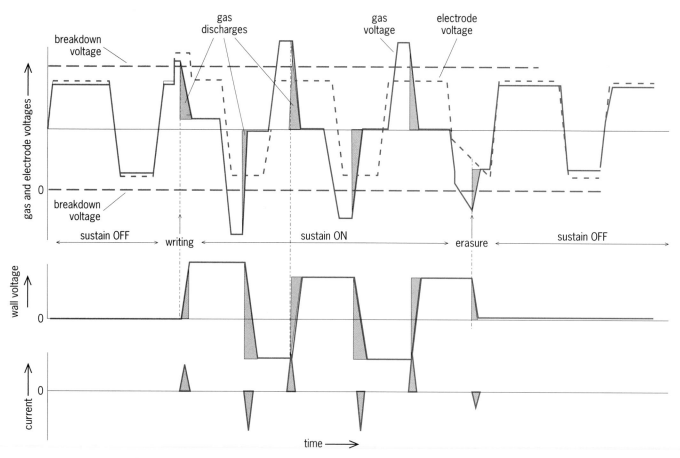

Fig. 2. Voltage and current waveforms of alternating-current plasma display panel driving signals. The elapsed times for the waveforms shown are of the order of microseconds.

oxide and enamel layers and accumulate on these surfaces. The field produced by these surface charges tends to diminish the field inside the gas gap to such an extent that the discharge is quenched. Typically the duration of the discharge current pulse is a few tens of nanoseconds. Once the discharge has been quenched, it is necessary to reverse the voltage applied to the electrodes to initiate a new discharge in the other direction, and so on. The plasma display panel is called an alternating-current structure because it can work only with alternating waveforms. The cathode of the device is, in turn, the front-substrate magnesium oxide surface and the rear-substrate magnesium oxide surface.

A second function of the insulating layers is to introduce the memory effect. The charges stopped by the dielectric layers during a particular discharge remain stored for the entire time during which the voltage is maintained at the level that has produced the discharge. When the voltage is inverted, the applied electric field points in the same direction as the field induced by the stored charges. Hence the following discharge can be produced by applying a voltage, V_s, lower than the voltage, V_w, which was required to initiate the first discharge. The next discharge will also need only the voltage V_s to occur.

However, if a voltage V_s is applied to an uncharged cell, it will not initiate any discharge. Hence, for the same applied alternating voltage, $\pm V_s$, the cell can be in either an ON state or an OFF state, depending on the previous occurrence of a "write pulse" of voltage V_w higher than V_s. This behavior is called the memory effect. V_s is called the sustain voltage because it is able to sustain or maintain the cell in the state (ON or OFF) imposed by previously applied "addressing pulses": a write pulse of higher amplitude in order to initiate a first discharge, or an erase pulse of lower voltage in order to create a "half discharge" that will draw charges to the dielectric by an amount just sufficient to neutralize the charges stored in the previous sustain discharge (**Fig. 2**).

Properties. Based on the operating principles described above, the alternating-current plasma display panel is characterized by several features. It is a memory device which requires special driving sequences and signal generation, which are provided by electronic circuitry. It is also a bistable device; that is, cells have only two states. In the ON state, light is emitted twice at each sustain period until an erase voltage is applied, while in the OFF state there is no light at all. This last property imposes severe constraints on the generation of half-tone levels of light intensity.

Generation of gray levels. Indeed, there is no way to change the amount of light that is emitted at each discharge. Hence, the only way to generate gray levels is to modulate the duty cycle of the ON state phase. For example, if the image is refreshed at 50 Hz, that is, every 20 ms, and a brightness of 10% of the peak value is desired, then the ON state will last 2 ms and the OFF state will be 18 ms long. The human eye will integrate the brightness and perceive an average brightness of 10% of the peak value. With appropriate addressing schemes, it is possible to get $2^8 = 256$ levels of brightness for each cell and $256^3 = 16$ million colors by combining the three primary colors. Such information treatment requires fast addressing of the screen and relatively complex driving electronics to provide the different signals (write, sustain, and erase) with the appropriate timing and amplitude to the $480 + (3 \times 640) = 2400$ electrodes of the screen.

For background information *see* BREAKDOWN POTENTIAL; ELECTRONIC DISPLAY; GAS DISCHARGE; GLOW DISCHARGE; PLASMA (PHYSICS) in the McGraw-Hill Encyclopedia of Science & Technology.

Jacques Deschamps; Henri Doyeux

Bibliography. J. Deschamps, Colour display panels, *Proceedings of the 16th International Display Research Conference* (Birmingham, UK), Society of Information Display, 1996; A. Sobel, Television's bright new technology, *Sci. Amer.*, 278(5):48–55, May 1998; L. F. Weber, Plasma display, in L. E. Tannas, Jr. (ed.), *Flat-Displays and CRT's*, 1985.

Podded propulsion

Podded propulsion is a breakthrough in ship propulsion systems. Naval architects worldwide are implementing designs using pods for a variety of ships. From icebreakers to drill ships, product tankers to cruise vessels, these designs are producing new hydrodynamic hull forms allowing shipbuilders to construct more efficient, quiet, and environmentally friendly vessels. Electric propulsion has already provided for cleaner and more environmentally friendly ships because of lower emissions and more efficient energy use. Podded propulsion moves the process a stage further in increasing hydrodynamic efficiency. This article gives a short history of the evolution of electric propulsion leading to the introduction of the pod, describes basic constructional details, and explores some of the major advantages achieved by this new form of marine propulsion.

Electric propulsion. In the early 1980s, cruise ships were predominantly the first type of vessels to switch from diesel mechanical drives to diesel electric ones. This development led to the more general usage of diesel electric systems in the marine industry in the late 1980s. Modern electrically driven ships utilize a number of diesel engines, each with an electrical generator controlled by an associated power management system that allows a precise matching of the power generation capability to the sum of the electrical loads: ship service load, propulsion load, and so forth. *See* INTEGRATED ELECTRIC PROPULSION.

With recent advances in power switching devices, such as thyristors and power transistors, the pro-

peller shaft has come to be driven by a variable-speed electric synchronous or induction motor using the power devices to switch varying amounts of electrical power to the motor. The desired ship speed is thereby achieved by altering the rotational speed with a shaft to which a fixed-pitch propeller is attached.

The popularity of diesel electric cruise ships has grown, and the electric propulsion concept is now applied to many other commercial ships, such as product tankers, drill ships, and ferries. Even some of the world's navies are building "electric" ships. Furthermore, gas turbine prime movers are sometimes used in place of, or in combination with, diesels. However, since 1995 a further evolution in electric ships has begun to take place, namely, the "podding" of the electric motor outside the hull form of the vessel (**Fig. 1**).

In a traditional electric motor–driven ship, the motor is mounted on the shaft line, within the hull. This configuration has led to shorter shaft line requirements, reducing the size of the machinery space and allowing for the design of either a smaller ship or a ship with additional cargo space (in a cruise ship, additional cabin space). These reduced space requirements, along with the more efficient use of the diesel generators, have offset the additional initial cost of the electric propulsion system, and with the advent of a podded propulsion system, this has been improved further since even less space is required.

Configuration. Podded propulsion consists of the same components as in a traditional electric ship, but with the electric motor mounted in a pod under the hull of a ship, attached to the ship by a strut (**Fig. 2**). The podded propulsor can be attached to the ship hull in such a manner that the entire assembly can be rotated 360°. This rotation is known as azimuthing, a technique that had been previously used in the mounting and control of small ship thrusters, typically in the bow or stern of a ship.

However, thruster technology is limited to about

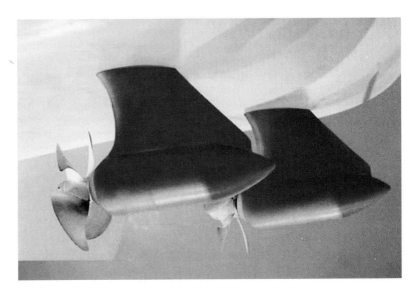

Fig. 1. Podded propulsors mounted on ship hull. (Kamewa Group)

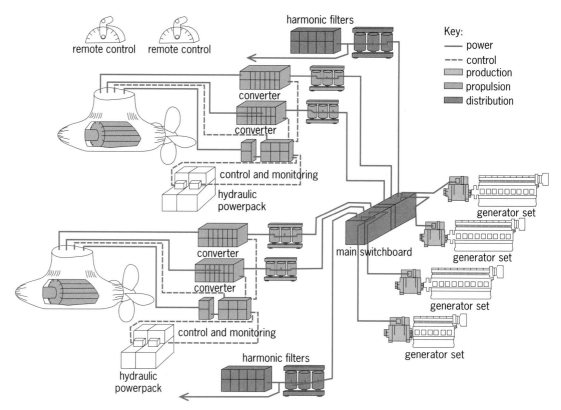

Fig. 2. Diesel-electric podded propulsion system. (*Kamewa Group*)

5 megawatts, whereas podded propulsors can be as large as 19 MW and may even increase to 25 MW in the future. This allows podded propulsion to provide the main propulsion source for even the largest ships.

Propulsor construction. The typical pod design consists of four major components: the pod housing, the electric motor, the fixed-pitch propeller, and the strut (**Fig. 3**).

The pod housing, normally constructed of cast metal, surrounds the installed equipment. The electric motor is mounted within the pod, and the latest designs "shrink-wrap" the motor stator to the pod shell, using seawater to enhance the cooling of the motor. Synchronous alternating-current motors are typically used. Such motors, being maintenance-free, have been widely used in conventional electric configurations with either a cycloconverter or synchroconverter (load commutated inverter) electronic converter. As the technology is developed, both permanent-magnet machines and advanced alternating-current induction machines are expected to be used. These machines have performance, space, and noise advantages over the synchronous motor, particularly important in ships which require a low noise profile and where space is at a premium.

A fixed-pitch propeller is connected directly to the motor shaft through water- and oil-tight seals and thrust bearings in the pod housing. For environmental reasons, double seals are often used to ensure that no lubricant can escape into the seawater and, conversely, no water enters the pod.

The strut connects the pod to the hull, and the upper mount contains the azimuthing equipment along with an electrical slip-ring arrangement. The azimuthing equipment uses hydraulic motors to rotate the pod, while the slip rings allow the electric power to be passed from the variable-speed electronic controllers mounted in the ship to the electric motor mounted in the pod. The strut also allows additional cooling of the motor by circulating air between the ship and the pod, as well as personnel access to the inside of the pod for inspection purposes. A complete pod can weigh as much as 180 metric tons (200 short tons).

Advantages. Podded propulsion has advantages over conventional systems in the areas of maneuverability, efficiency, vibration, cost, design, space requirements, and construction time.

Maneuverability. The use of azimuthing propulsors eliminates the necessity for steering equipment and associated rudders, since the ship can be driven in any direction. This arrangement allows for enhanced dynamic performance, since the propulsion forces can be directed precisely as needed for the navigational requirements; it also allows for high maneuverability during docking, crash stops, and so forth.

Increased efficiency. Any propeller system is most efficient when operating in clean, undisturbed water flows. A conventional ship arrangement, with the propeller mounted on the shaft line at the stern, requires a compromise. The shaft line and the hull form have to be designed to reduce the amount of disturbance so that the cleanest flow of water over the propeller is achieved, thereby keeping cavitation around the propeller to a minimum. With a podded propulsor, the propeller can be mounted in front of the pod, thereby eliminating any inflow disturbances from the shaft-line extremities. In this configuration, the hull form can be redesigned to ensure that the cleanest possible inflow of water enters the propeller and, as a result, increased propulsion efficiency is achieved. This efficiency gain can be of the order of 8–14%, depending on the hull design, and equates to lower fuel requirements.

Reduced vibration. One source of vibration is the transmission of propeller pressure pulses to the adjacent hull. Due to the design configuration of a podded propulsion system, the clearance between propeller tip and the hull is significantly larger than in a conventional shaft arrangement. As the clearance is larger, the pressure pulse transmitted to the hull is greatly reduced, minimizing the vibration to almost zero. This reduction is a major advantage in the case of passenger-carrying vessels such as cruise ships.

Equipment cost. The cost of a podded electric propulsion system is almost the same as that of a conventional shaft-line arrangement. However, for the same price, a far more efficient vessel design is achieved.

Simplified design. The podded propulsion system simplifies the design of the aft of the ship. The ship lines are simpler, eliminating the rudder and steering gear, the shaft line and associated bearings, stern-tube shaft brackets, stern thrusters, and, in some ship designs, reduction gearboxes.

Reduced space requirement. The "all-electric ship" previously pioneered a reduction in the space requirement on board a ship for the propulsion system. The pod takes this process a step further since the motor is now outboard, thereby freeing up even more space. This development, coupled with the elimination of the steering equipment, makes possible a very compact design, which reduces the cost of construction.

Shorter construction time. In a traditional design, with the electric motor mounted on the propeller shaft, the motor and shaft are among the first items to be installed. However, these items are not supplied on short notice, and construction time must be lengthened to allow for their procurement. A podded propulsion system has the advantage of reduced construction time, since the pod can be mounted at the end of the construction period rather than at the beginning.

In many ways, podded propulsion allows the ultimate flexibility in ship construction since the major propulsion components, including the pod, the prime movers, the azimuthing equipment, and the electronic controller, can be assembled in module form and need only be installed at the end of the construction period (the rest of the equipment is installed during the construction period), and most of the necessary connections are made by electrical conductors.

Disadvantages. There are trade-offs associated with all propulsors, and pods are no exception. As

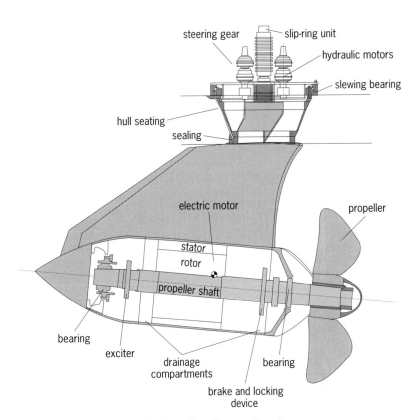

Fig. 3. Cutaway view of a podded propulsor. (*Kamewa Group*)

the vessel speed is increased to greater than 25–28 knots (46–52 km/h), the gains in hydrodynamic efficiency decrease, thereby reducing some of the economic advantages in operating cost. Moreover, for major repair work on a podded propulsor, the vessel generally requires a dry dock. Also, the use of pod propulsion requires an electrical propulsion system, and there are still some vessel types for which such a system may not provide the optimum solution for its propulsion plant.

Prospects. The technology of the pod is continually being developed. The latest designs incorporate the ability to remove and replace the pod underwater in the event of damage to the pod or major equipment failure within the pod.

The ability to attach the propeller to either end of the pod, along with the possibility of two-propeller designs, allows the pod configuration to be optimized for different ship applications, such as semi-submersible ships, high-speed cruise ships, and medium-speed tankers (**Fig. 4**). Already, pods using permanent-magnet machines are being introduced, and smaller and more powerful pods are expected using advanced alternating-current induction motors.

Podded propulsion is now a viable option for medium to large commercial vessels (with greater than 5 MW of propulsion power). However, for naval combatants more development is needed to meet the shock requirements and the requirements for low acoustic noise signatures and reduced radiated magnetic fields. The future may show electric podded propulsion to be the choice for most large vessels.

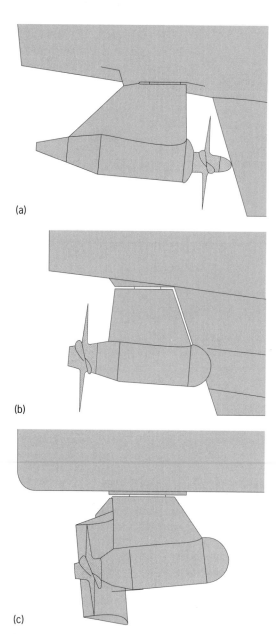

Fig. 4. Optimization of pod configuration for (*a*) high speed, (*b*) medium speed, and (*c*) low speed and maximum thrust. (*Kamewa Group*)

For background information *see* ALTERNATING-CURRENT MOTOR; MARINE ENGINE; MARINE MACHINERY; SHIP DESIGN in the McGraw-Hill Encyclopedia of Science & Technology.   Brian Pope

Bibliography. Institute of Marine Engineers, *Proceedings of All Electric Ship '98: Developing Benefits for Maritime Applications*, 1998.

Population dynamics

A metapopulation is a set of geographically distinct local populations occupying discrete habitat patches. In some patches, the local population can become extinct—for example, if there are not enough reproductive females, or if there is a drought or other environmental hazard that depletes the resources used by the local population. Alternatively, individuals migrating from other patches can colonize an empty patch. Thus, the dynamics of a metapopulation depends on the relation between extinction and colonization events.

Humans are modifying the landscape at high rates. Large areas of previously continuous habitat, such as some tropical forests, are now a mosaic of patches surrounded by human-modified habitat. In order to conserve biological diversity, it is important to understand how the discontinuous distribution of available habitat influences population dynamics. Metapopulation theory provides a framework for investigation.

Richard Levins introduced the concept of a metapopulation during the late 1960s in connection with crop pest control. He proposed a model describing how the fraction of patches occupied depends on the relation between the extinction of previously occupied patches and the colonization of empty patches. There are two ways to eradicate a pest: increase local extinctions, or reduce colonization from one patch to another. The Levins model is at the core of metapopulation theory and its application to conservation biology. In conservation biology the goal is the preservation of the target species, rather than its eradication as in biological control.

Work on metapopulation theory has emphasized the conditions necessary for a metapopulation to persist. In particular, colonization must exceed extinction. Another consideration is the permanent destruction of a certain fraction of patches. Here, too, are well-defined thresholds below which the metapopulation will become extinct despite the fact that some patches are still available.

There are different levels of detail in metapopulation models, and a trade-off between realism and simplicity. For example, the Levins model is a spatially implicit model in which an average variable is described, that is, the fraction of patches occupied by a local population, but there is no information on their spatial distribution. Such models are analytically tractable but are only a simple representation of reality since they assume homogeneous space and global colonization. In other words, they assume that colonization is distance-independent.

Spatially explicit models are those in which each patch has its specific spatial coordinates and colonization can be described as a local process. Such models are more realistic, but they usually cannot be solved analytically. Computer simulations must be used. The spatial structure of the landscape in which the species is found can be incorporated only into spatially explicit models. These details can be important in understanding population dynamics and species diversity.

Dispersal among local populations, involving the migration of individuals from their patch into another, can increase metapopulation persistence if there is asynchrony in the fluctuations of the local populations, that is, if these populations fluctuate in an independent way. Also, competitive and prey-predator interactions can exclude species at local spatial scales, but species can still coexist at larger spatial scales. This coexistence is associated with self-organized spatial patterns in the species' spatial distributions. An example of such spatial patterns is a demographic explosion of rodents spreading as a wave through a vast region of habitat containing fewer individuals. Thus, the spatial component can be very important in understanding population regulation.

Spatially explicit models of population dynamics can generate an unexpected range of large-scale spatial patterns in population density as traveling waves, despite the short-range scale of dispersal and the initial homogeneous setting. The main lesson from such models is that biological interactions alone can explain the high levels of spatial heterogeneity observed in nature. In a similar way, under certain circumstances dispersal among patches can modify the dynamics observed in each patch when it is isolated.

Dispersal among small patches of habitat can also have important effects at a genetic level. First, dispersal increases the gene flow, as individuals can arrive from different patches under different selective regimes. Second, frequent extinctions in a metapopulation context increase random drift, that is, random changes in allelic composition. An additional factor related to the small size of patches of available habitat is the inbreeding of local and isolated populations. The reproduction of genetically related individuals in small and isolated groups results in a lack of genetic diversity. Colonization of individuals from one patch to another can be important in introducing genetic diversity, thus avoiding fitness reduction (and subsequent population extinction).

Considerable empirical work on metapopulations has taken place in the last few years. Several situations in nature are well described by the metapopulation concept. An example studied by Ilkka Hanski and colleagues is the Glanville fritillary butterfly (*Melitaea cinxia*) in the Åland Islands of Finland. This species occupies small dry meadows and faces a high risk of local extinction. Its persistence can be understood only within the metapopulation context. The butterfly suffers from two competing parasitoid enemies. Both parasitoids are present within patches, although one species is competitively superior to the other. The coexistence of the interaction also can be understood only at a metapopulation scale. The less competitive parasitoid is a superior disperser. Different dispersal abilities, together with the patchy distribution of available habitat, are important factors in understanding species diversity and persistence.

For background information *see* ECOLOGICAL COMMUNITIES; ECOLOGY; POPULATION DISPERSAL; POPULATION ECOLOGY in the McGraw-Hill Encyclopedia of Science & Technology.   Jordi Bascompte

Bibliography. J. Bascompte and R. V. Solé (eds.), *Modeling Spatiotemporal Dynamics in Ecology*, 1998; I. Hanski, *Metapopulation Ecology*, 1999; I. Hanski and M. Gilpin (eds.), *Metapopulation Biology: Ecology, Genetics and Evolution*, 1997; D. Tilman and P. Kareiva (eds.), *Spatial Ecology: The Role of Space in Population Dynamics and Interspecific Interactions*, 1998.

Precision agriculture

Precision farming involves adjusting soil and crop management according to small-scale variability of soil and other production factors within fields. Such site-specific management requires the collection and interpretation of large quantities of data collected at different spatial scales in order to predict the impact of variable management practices on yield, occurrence of disease, nutrient losses, or any other parameter of interest. This task is done through the use of sensors placed on farm equipment, satellites, or aircraft scanners. Numerous computer programs are available to handle large data sets. Information is converted into a digital format that allows accurate representation of spatial patterns as they occur in individual fields, across fields or regions, or over even larger areas. The greatest challenge is not the collection and transfer of data into maps that reflect spatial patterns or features of interest, but the interpretation of the data and prediction of the impact of variable management practices.

Spatial variation and scale. Farmers know that certain areas of their fields show higher productivity than others. Differences in yield among areas can vary over time, and may be caused by temporal and spatial variability and interactions of all crop yield factors: microclimate, nutrient and water availability, soil moisture and organic matter content, competition with weeds, and incidence of disease and insect attacks. Precision farming uses information about the variability of soil properties and topography, the spatial patterns of diseases, cropping practices, and weather conditions to obtain and adapt management practices that are best suited for each specific area within a field. Spatial and temporal variability of several key factors that control a cropping system's performance becomes a central part of the site-specific management process. Because of the diminishing-returns response of yield and profits to most factors, a precise matching of input rates (of fertilizers and pesticides, for example) with crop needs should result in optimal economic and biological efficiency, as well as in reduced release of chemicals into the environment. Until recently, high labor and low capital and fertilizer costs made uniform field-level management the best choice for large-scale farming in developed countries. Usually, fields ranged from a few tens to hundreds of acres (10–100 hectares), and rates of application were uniform and based on average characteristics within each field. Technological advances in remote sensing, the Global Positioning System, automated yield monitoring, and information management have triggered the potential for optimizing the application of fertilizers, pesticides, and other inputs at resolutions much finer than fields, at the level of a few meters. The actual benefits of site-specific management depend on the intensity and pattern of spatial variability of factors that are most limiting to each cropping system's performance. Fields with a high degree of variability at scales that are too fine, as well as those with large-scale patterns of little intensity, do not offer the potential to improve efficiency by more precise management.

Applications and benefits. The first application of precision farming was largely geared toward varying the rate of fertilization across the field. However, precision farming includes all inputs that can be varied across the field, such as seeds, pesticides, water, tillage operations, microorganisms that promote biological nitrogen fixation, and control of soil-borne pathogens that affect the rate of seed emergence. For example, pests often occur in the field in localized areas determined by their dispersal, reproductive, and survival abilities. After determining the spatial heterogeneity of weeds or diseases by using aerial photos or ground scouting, pesticides are used only in areas where the pest surpasses a threshold value. The rate of application can be dependent on the actual level of infestation in each location.

Environmental benefits of site-specific management are expected for two reasons. First, a better spatial matching of applications with needs will result in reduced release of chemicals into nontarget parts of the environment, even if the total amount applied remains constant. For example, by shifting fertilization from areas where the soil supplies enough nitrogen for the crop to those areas where an average application rate would be deficient, a greater proportion of the total nitrogen will be absorbed and retained by the crop. Second, it is possible that under site-specific management the total amount of each input applied will be smaller than under traditional field-level uniform management. Site-specific tillage and residue management may contribute significantly to erosion control and reduction of nonpoint-source pollution. Although positive environmental consequences are expected from site-specific management, empirical proof is sparse and will require considerable research efforts.

Economic benefits from site-specific management are also expected for reasons similar to those behind environmental benefits. Site-specific management should result in a saving of inputs, but will also imply higher hardware, software, information-gathering, and management costs. In spite of not including the costs of training and adoption of the new technology, a majority of the empirical evidence is not conclusive but appears to indicate that site-specific management is not always more profitable than traditional uniform management. This statement is qualified by the fact that the technology is at an early stage of development; reductions in costs are expected as the technology evolves, and increases in benefits are expected as agricultural decision-making theory and practice shift from uniform to precise management. For example, current fertilization recommendations and the underlying data may not be directly applicable to site-specific management and must be thoroughly reviewed.

Component technologies. The success and further development and adoption of site-specific management depend on a variety of technological and conceptual advancements, such as the Global Positioning System, geographic information systems, remote sensing, yield monitors, computers, and soil sensors. These technologies are evolving rapidly, but many still face significant challenges to realize the full potential of site-specific management.

Because of its focus on spatial variability, site-specific management is based on the link between specific values for multiple variables with precise locations within fields. For example, the exact location of soil samples has to be known in order to describe the spatial pattern of soil variability. The Global Positioning System makes this possible. A device determines the exact position of the equipment, sensors, or soil samples by measuring its distance to multiple geostationary satellites that are part of the Global Positioning System. Satellites transmit a signal at a certain wavelength and frequency that is picked up by the receiver on Earth. In simplified terms, the distance from the receiver to each satellite is determined by measuring the time the signal takes to travel between satellite and receiver. Extremely accurate atomic clocks are involved. The orbits of the satellites are known, and thus, by combining the distances to four satellites, the exact geographical position and elevation of the Global Positioning System receiver can be calculated. For security reasons, the U.S. Department of Defense introduces an error called selective availability in the Global Positioning System signal. Selective availability produces position errors of up to 100 m (328 feet), a value that makes uncorrected readings useless for site-specific management. The error vector can be determined by having a second Global Positioning System station in a known location. This vector can be applied to correct the location of interest either after the data are gathered (postprocessing differential correction) or as the positions are gathered (real-time differential correction). These services are widely available. For example, the Global Positioning System is used when the crop yield is determined by using a yield monitor installed on the harvester. By constructing yield maps for several years, the variability in crop yield across space and time can be analyzed to assess the possible underlying causes of differences in yield.

A geographic information system assists the farmer in transforming spatial data into organized information on spatial and temporal patterns of soil resources and crop yield. The software and hardware involved in geographic information systems provide tools to identify relationships between the observed soil variability and crop performance, so that precise applications can be carried out. Data on spatial pattern are obtained from existing soil maps, direct field sampling, airborne video, aerial photographs, and satellite remote sensing. For example, level of soil moisture can be derived from video and satellite information. Typically, farmers contract geographic information system services from consulting companies that provide customized analyses.

Technology to carry out real-time sensing of soil

properties in the field by using specially developed sensors is undergoing explosive development, but still has a long way to go before affordable and reliable real-time sensing equipment is available. Sensors that can measure soil properties almost continuously as the machinery moves at 2–4 km/h (1–2.5 mi/h) are being developed to measure organic mater, moisture, texture, pH, and nutrient concentrations. Most of these sensors determine the variables of interest indirectly. For example, organic matter measurements are based on soil color, and texture is correlated with acoustical and mechanical properties of the soil. In the most developed version, a sensor is installed on the tractor, and soil samples are analyzed almost instantaneously for available nitrogen. The information is processed by an on-board computer, which then recommends how much nutrient should be applied to optimize profits, and sends the finding to the fertilizer applicator at the back of the machine, where rates are adjusted continuously.

Challenges. A major challenge posed by site-specific management is the interaction between spatial and temporal variability. Intensity, scale, and pattern of spatial variation must be determined in a timely manner for each field. These descriptions have to be interpreted to optimize management unit size and combination of inputs to be applied in each management unit.

The underlying causes of the variability in yield across a farmer's field can be numerous and are often linked to differences in soil characteristics such as the availability of nutrients, which causes areas with high and low fertility. Variability in fertility across a field can be induced by management practices such as tillage operations and leveling for irrigation purposes.

Successful site-specific management systems should be able to identify the key parameters that limit optimum return across the field. Information and knowledge must be integrated into a decision-making package that optimizes enterprise economic return and environmental impact as a function of the spatial patterns of both soil characteristics and application of inputs. This approach to management is in stark contrast to traditional field management. Yet, it has been increasingly accepted by researchers and farmers alike that fields should not be treated as homogeneous entities that lead to inefficient use of input resources and have negative impact on the environment because of excessive use of chemicals.

Before precision farming practices can be fully successful, farmers, extension agents, and researchers have to be able to predict site-specific management's effects on economic yield, pest control, nutrient losses, and other aspects. This cannot be achieved without a basic and thorough understanding of the underlying mechanisms that cause variability of key processes within fields. Agricultural research will need an explicit focus on variability and spatial scale. For example, fertilization rate recommendations prescribed for uniform management will likely not be appropriate for site-specific management. Thus, although site-specific management may have been

driven by technological change, it provides a unique opportunity for renovation and development of agricultural sciences from very practical to highly theoretical fields.

For background information *see* AGRICULTURAL SCIENCE (PLANT); AGRICULTURAL SOIL AND CROP PRACTICES; AGRICULTURE; AGRONOMY; APPLICATIONS SATELLITE; GEOGRAPHIC INFORMATION SYSTEMS; SATELLITE NAVIGATION SYSTEMS in the McGraw-Hill Encyclopedia of Science & Technology.

Chris van Kessel; Emilio A. Laca

Bibliography. F. J. Pierce and E. J. Sadler, *The State of Site Specific Management for Agriculture*, American Society of Agronomy, 1997; F. C. Stevenson and C. van Kessel, A landscape-scale assessment of the nitrogen and non-nitrogen rotation benefits of pea, *J. Soil Sci. Soc. Amer.*, 60:1797–1805, 1996.

Production lead time

Production (or manufacturing) lead time is commonly defined as the total time required to convert raw materials into a finished product. More than the simple sum of machine operation and assembly times, it also includes time spent waiting for machines to be available, moving between machines, and performing quality inspections and other nonoperation functions. Short production lead times are ideal, as long times reduce the responsiveness of a company and result in high inventories.

While customer lead time is the interval of time between placing the order and receiving the order, production lead time typically focuses on the time required to produce the order once all the necessary materials are present. It does not take into consideration the time required to obtain the necessary materials or the time the materials wait in the plant before production begins.

Production lead time is very important in make-to-order industries (such as the aerospace industry). There, lead time can be a deciding factor in obtaining orders. In addition to low cost and superior quality, companies win orders by promising to deliver them quickly. Lead time is not as much of an issue in make-to-stock industries (such as small appliances and other retail products) in that these industries are building inventories against future demand, thus providing a buffer against longer than expected production lead times.

Components. The primary components of production lead time are setup, processing, handling, and nonoperation. Setup time (t_s) is required to prepare the machine for the particular order of parts. Processing time (t_p) includes the time spent loading and unloading the part from the machine as well as actual machining time. Handling time (t_b) is the time an order spends being transported between machines. Nonoperation time (t_n) includes wait time (for both machines and transportation) and time spent on certain other activities such as inspection. Assuming that an order consists of Q identical parts and that there are N operations in the processing sequence

Data for lead time calculation				
Operation (i)	Setup time (t_s), h	Processing time (t_p), h/part	Handling time (t_h), h	Nonoperation time (t_n), h
1	2	0.10	0.25	3
2	3	0.05	0.25	4
3	1	0.07	0.25	2
4	2	0.15	0.25	6
5	4	0.20	0.25	5

(with i indicating the operation in the processing sequence), production lead time (LT) is expressed by Eq. (1).

$$\text{LT} = \sum_{i=1}^{N}[t_{si} + Qt_{pi} + t_{hi} + t_{ni}] \tag{1}$$

For example, consider an order that consists of 20 identical parts. Five operations must be performed to produce this part. Setup, processing, handling, and nonoperation times are provided in the **table**. The production lead time for this order as computed from Eq. (1) is 44.7 h. Assuming a 40-h work week, the production lead time for the order of 20 parts is 1.1 weeks.

Setting due dates. Knowledge of the lead time for a product is important for setting due dates for orders. When a customer places an order, either the customer states when the order is needed or asks for an estimated time of completion. In either case, a method of estimating lead time is required. If the customer requires a specific delivery date, the company needs to be able to determine if the customer due date is achievable given current production orders. If the customer requested date is not feasible, an estimate of when the order will be done is necessary for negotiating a new due date.

One method for setting due dates requires knowledge of the average production lead time and its standard deviation (σ_{LT}). Given this information, an estimated lead time is calculated such that there is a good probability of completing the order on time. The percentage of orders completed on time is known as the service level (s), where z_s denotes the value on the measurement axis (for which s of the area under the z curve lies to the right of z_s). The z curve is the standard normal curve. The standard normal table lists the values of z_s for values of s ranging from 0 to 100%. To be more precise, the table actually works in reverse in that for values of z_s ranging from -3.49 to $+3.39$, the value of s is given. Assuming that production lead time is normally distributed, the quoted lead time for an order can be computed by Eq. (2).

$$\text{Estimated lead time} = \text{LT} + z_s\sigma_{\text{LT}} \tag{2}$$

For example, suppose the average production lead time is 2 weeks with a standard deviation of 0.75 week. If a 95% service level is desired, an estimated lead time for the order is computed to be 3.23 weeks

in Eq. (3). The value of 1.645 for z_s was obtained by

$$\text{Estimated lead time} = 2 \text{ weeks} + (1.645)(0.75)$$
$$= 3.23 \text{ weeks} \tag{3}$$

referencing a standard normal table. First, the value 0.95 is found in the body of the table (corresponds to $s = 95\%$). The corresponding row and column label values are added together to obtain the value of z_s. The additional 1.23 weeks (1.645×0.75) is often called the safety lead time. Caution should be exercised when using this method to estimate lead times when the standard deviation is very high. A large standard deviation will lead to a large safety lead time. Since one-half of all orders are expected to complete in less than the average lead time, a large safety lead time can lead to a large finished goods inventory if orders cannot be delivered early.

Managing lead time. Production lead time can make the difference between receiving customer orders and going bankrupt. In all industries, a short production lead time is preferred to a long time. One way to reduce production lead time is to reduce the individual components, such as setup, handling, and nonoperation times. In some industries (such as metalworking factories), these three components constitute 95% of the total production lead time. Setup time can be reduced through the use of common fixtures and by scheduling similar orders to run back to back. Handling time can be reduced by improving the layout of the facility. Machines that follow one another in the processing sequence of many products should be located near one another. Finally, nonoperation time can be reduced by efficient scheduling practices and the elimination of unnecessary inspections and other nonoperation activities.

Production lead time is also related to the amount of work in process (orders currently being worked on in the plant) on the production floor and the production rates of the machines. This relationship, shown in Eq. (4), is known as Little's law. For

$$\text{Production lead time} = \frac{\text{Work in process}}{\text{Production rate}} \tag{4}$$

industrial plants operating at less than 100% capacity, production lead time is largely unaffected by the amount of work in process in the plant. However, as the production rate reaches its maximum value (capacity), increasing the work in process merely serves to increase the production lead time. Therefore, controlling the release rate of materials to the plant is an important part of controlling production lead time. Little's law also shows that production lead time can be reduced by increasing the production rate of the plant. This approach typically requires an increase in the plant's capacity. Capacity can be increased by purchasing additional equipment or by working more hours (for example, authorizing overtime or adding another shift).

Production lead time is essential to the success of a company. In the global market place, customers demand low-cost, high-quality products in a short period of time. As a result, effective management of production lead time is as important as striving for low cost and high quality. Short production lead times allow a company to be responsive to its customers' needs.

For background information *see* INDUSTRIAL ENGINEERING; INVENTORY CONTROL; MANUFACTURING ENGINEERING; MATERIAL RESOURCE PLANNING; MATERIALS HANDLING; PRODUCTION PLANNING; QUALITY CONTROL in the McGraw-Hill Encyclopedia of Science & Technology.   Elin M. Wicks

Bibliography. M. P. Groover, *Automation, Production Systems, and Computer Integrated Manufacturing*, Prentice Hall, 1987; W. J. Hopp and M. L. Spearman, *Factory Physics*, Richard D. Irwin, 1996.

Protein

Proteins are cellular machines. These molecules perform many functions, such as catalyzing biochemical reactions for metabolism, providing structural support, regulating biological processes, and transporting other molecules. The function of a protein is closely tied to its folded structure, that is, the three-dimensional arrangement of its atoms. Each protein structure is formed by a polypeptide chain for which the sequential arrangement of amino acids is encoded by the information of the genetic material, deoxyribonucleic acid (DNA) [**Fig. 1**].

Folding process. Protein structure is achieved through a complex folding process (**Fig. 2**). Proteins are synthesized as a long, continuous strand of amino acids covalently linked through a peptide-bond backbone. The amino acid side-chain moieties extending from this backbone interact to form local secondary and long-range tertiary and quaternary structures. The final structure is often called the native folded state; it is maintained through noncovalent interactions such as hydrophobic interactions, electrostatic interactions, and hydrogen bonds. Since a single protein chain can have thousands of amino acids, and 20 different amino acids are commonly used as assembly units, the number of possible sequences and configurations is immense. Even so, a single amino acid change can significantly alter the folding and the structure of a protein. Understanding the folding process and the forces that maintain the native structure is one of the oldest and most challenging problems in biochemistry.

The native state of a protein is generally 5–10 kcal/mol more stable than the denatured state, which is actually an ensemble of loose structures. Levinthal's paradox theorizes that a random search of all possible configurations to find the one with lowest energy (most stable) would take a minimum of 10^{77} years. In reality, helical formations can occur in 10–100 nanoseconds, loop formations can fold in 500 ns, and the minimum time for the folding of a small protein is 1 millisecond. Therefore, protein folding must be a directed process. Folding intermediates have been proposed to limit the conformational search by defining one or more pathways to the native state. In addition, intermediates are essential for certain processes such as translocation across or insertion into membranes, release of nonpolar ligands, and association with other cellular components.

Discrete intermediates in the folding process have been isolated for several proteins. Due to the extremely fast time scales of protein folding, partially folded states are often studied at equilibrium where the native state has been perturbed by high or low temperatures, chemical denaturants, pH extremes, or high pressures. Often these intermediates have well-formed local structures but fluctuating or absent long-range structures. The average radius of folding intermediates (the extent to which the structure is well constrained) lies between those of the native and unfolded states (Fig. 2). Moreover, local structures may not be evenly dispersed throughout the folding intermediate; some regions may remain unfolded. Equilibrium intermediates are thought to be relevant to the kinetic folding process for several reasons. Conditions that favor the formation of the intermediates also speed up overall folding rates. Also, kinetic intermediates (structural forms that occur transiently during protein folding) often have structures similar to equilibrium intermediates for those few proteins for which kinetic structural data are available.

Models. Several theoretical models have been proposed to describe the sequence of intermediates assumed by a protein in the course of its folding reaction. In the hydrophobic collapse model, hydrophobic residues are excluded from the aqueous solvent, driving an initial collapse of the polypeptide chain.

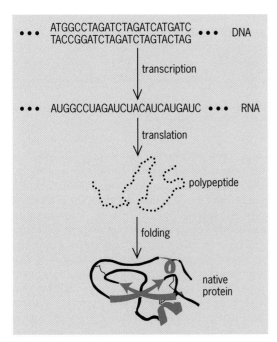

ATGGCCTAGATCTAGATCATGATC
TACCGGATCTAGATCTAGTACTAG ••• DNA

transcription

••• AUGGCCUAGAUCUACAUCAUGAUC ••• RNA

translation

polypeptide

folding

native protein

Fig. 1. Protein folding, the terminal step in protein production.

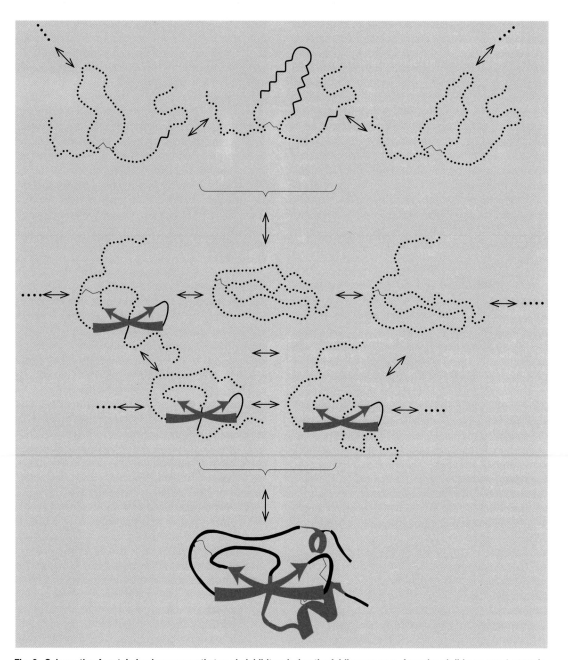

Fig. 2. Schematic of protein bovine pancreatic trypsin inhibitor during the folding process. Arrowhead ribbons are β-strands, and coiled ribbons are α-helices. (*Structures provided by Elisar Barbar and Clare Woodward*)

The native structure is then achieved by reorganizing and forming noncovalent interactions within the collapsed state. In the nucleation-condensation model, local structures form and propagate, followed by fusion of these various folded regions to form the native protein. Finally, in the diffusion-collision model, elements of local structure form and encounter each other by the process of diffusion. Experimental evidence supports each of these models for different proteins. Thus, the model that best describes the folding process for any single protein may vary significantly.

A generalized description of protein folding has been developed which statistically characterizes the energy landscape of folding as a funnel in which the wide top represents the ensemble of unfolded configurations and the deepest valley represents the native state (**Fig. 3***b*). Intermediates along the pathway are described as local minima along the sides of the funnel and are separated from the native state by energy barrier peaks. A funnel with smooth sides describes a single, cooperative phase transition (the transition proceeds from starting structure to final structure) with no intermediates (Fig. 3*a*). The opposite extreme is a funnel with very rough sides and a large number of low-energy structures separated by tall barriers. Such a funnel characterizes the folding of a heteropolymer (a compound comprising two or more molecules that are different from one another) that does not adopt a discrete native state (Fig. 3*c*).

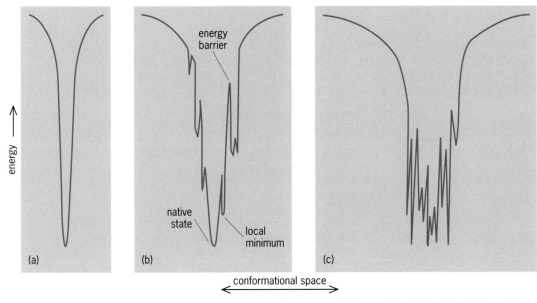

Fig. 3. Two-dimensional cross section through representative folding funnels. (*a*) A protein that folds without intermediates. (*b*) A protein that folds through discrete intermediates. (*c*) A random heteropolymer with no unique structure.

In vivo, a protein can be trapped in a structure corresponding to a local minimum in its energy funnel. Therefore, cells have evolved "chaperone" systems that lower the energy barriers and thereby speed folding and prevent aggregation. A commonly studied bacterial chaperone system is the GroEL/GroES complex. This complex binds unfolded protein in its hydrophobic center, which is cylindrical in shape, and facilitates folding into the native structure.

Much progress has been made toward understanding folding phenomena in individual proteins. The computer revolution has enabled modeling of the process, but simulating the folding of even a small protein still requires a great deal of time, despite using the largest supercomputers. Modeling is limited to a large degree by incomplete descriptions of the force fields describing the noncovalent interactions that drive folding. With current tools and understanding, structure prediction is possible for peptides and very small proteins, and simple proteins have been designed and created. The current challenge is to extend this knowledge to the large, multidomain proteins that perform the processes crucial to living organisms.

The comprehension and manipulation of protein structure and folding have significant potential impacts in many areas of modern life. Industrially, existing proteins have been modified to enhance function or to create new functions, although current production systems can yield misfolded or aggregated proteins. Medically, defects in protein structure and folding are implicated in many diseases, among them cystic fibrosis and Alzheimer's disease. A comprehensive understanding of protein folding and structure will contribute toward the logical design of therapies for these diseases. Finally, the ability to predict folding patterns is critical for interpreting data from the Human Genome Project, which will generate the amino acid sequences, but not the structural information, of all proteins in the human body.

For background information *see* NUCLEAR MAGNETIC RESONANCE (NMR); PEPTIDE; PROTEIN; X-RAY CRYSTALLOGRAPHY in the McGraw-Hill Encyclopedia of Science & Technology.

Sarah Bondos; Liskin Swint-Kruse; Kathleen Matthews

Bibliography. E. Barbar et al., Dynamics of the conformational ensemble of partially folded bovine pancreatic trypsin inhibitor, *Biochemistry*, 37:7822–7833, 1998; J. D. Bryngelson et al., Funnels, pathways, and the energy landscape of protein folding: A synthesis, *Proteins Struc. Func. Genet.*, 21:167–195, 1995; V. E. Bychkova and O. B. Ptitsyn, The molten globule in vitro and in vivo, *Chemtracts Biochem. Mol. Biol.*, 4:133–163, 1993; T. E. Creighton, *Proteins: Structures and Molecular Properties*, 2d ed., W. H. Freeman, 1993.

Pterosaur

The first remains of pterosaurs were found in Germany in the 1780s, and it was not clear what sort of animals they were. Regarded as aberrant birds by some and aberrant bats by others, they were established as reptiles by the French comparative anatomist Georges Cuvier in the early 1800s. He noted they were not at all like typical reptiles: they stood erect like birds, they had hollow bones and probably an insulatory covering, and they flew.

Noteworthy finds. Since that time, hundreds of pterosaur skeletons and isolated bones have been discovered on most continents. The best records originally came from Germany, where the Late Jurassic Solnhofen Limestones, the lithified remains of ancient lagoons, preserved many complete and beautiful specimens of *Rhamphorhynchus*, *Pterodactylus*,

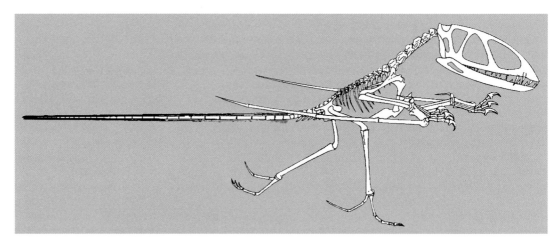

Skeleton of *Dimorphodon*, an Early Jurassic pterosaur, restored as a biped. Wingspan is about 1.6 m (5.3 ft).

and other forms. Equally striking were pterosaurs such as *Dorygnathus* and *Campylognathoides* from the Early Jurassic of the Holzmaden region of southern Bavaria (see **illus.**). In the late 1800s, many worn and broken pterosaur bones were recovered from the newly explored badlands of Kansas and Nebraska; they included *Pteranodon* and *Nyctosaurus*. In recent decades, some of the most interesting finds have come from China and from a series of limestone nodules in the Santana region of Brazil, where fossil remains have included three-dimensional skeletons, some preserved soft parts, and bizarre skull crests. Specimens discovered in the lake district of Kazakhstan and Kirgizia preserve a kind of epidermal covering that appears to include both skin and a furlike covering, not comparable to mammalian fur.

Classification. Pterosaurs were traditionally divided into two subgroups. The Rhamphorhynchoidea comprised the early pterosaurs with long tails, a long fifth toe, and short metacarpal (palm) bones in the hand; the Pterodactyloidea reduced the tail and fifth toe to stumps and lengthened the metacarpals. The first category has been abandoned because some rhamphorhynchoids are more closely related to pterodactyloids than to other rhamphorhynchoids. (The term pterodactyl, first used to describe these animals, now informally denotes any pterodactyloid; the term pterosaur describes the whole group.) Several conflicting schemes of the relationships of pterosaurs have been proposed; none seems clearly preferable.

Several analyses of the relationships of pterosaurs have shown that they have close common ancestors with dinosaurs. They share long legs, a thigh bone with an offset head, a reduced fibula (the splint bone on the side of the shin in humans), a hinge-like ankle, long foot bones and toes, and other features that no other animals share. But it has also been observed that all these features are related to locomotion, which could be independently derived (though there is no evidence for this). Unfortunately,

pterosaurs are so modified through the rest of their skeletons that they leave little trace of features that might link them to other reptiles. In order to advance this question, further analyses must include more kinds of pterosaurs and other reptiles, and more features held in common.

The hindlimb features just mentioned are also shared by birds and other dinosaurs. Accordingly, it has been suggested that pterosaurs also walked upright, on their toes, with a narrow gait, using the hindlimbs exclusively or nearly so. An alternative view suggests that the hindlimbs were too fragile, that the foot was connected to the wing and so impeded terrestrial progression, or that the joints of the feet did not allow walking on the toes. Indistinct trackways ascribed to pterosaurs in the 1950s were later shown to be very similar to crocodile trackways, and the forefoot tracks were too close to the body to have been made by the long forelimbs. More recent discoveries in France of tracks with forefoot impressions far outside the body midline appear to have been made by small pterosaurs slowly foraging for seaside invertebrates. Other indistinct tracks continue to be attributed to pterosaurs, and functional explanations given of how the forelimbs could twist to make tracks near the body. But these tracks do not have clear features of pterosaur anatomy, and the forefoot prints seem too close to the body, so the issue remains unresolved.

Flight. Experts agree that pterosaurs were true flying reptiles, powered by wings that flapped vigorously. The flight muscles had extensive attachments to the wing bones and the expanded breastbone, and the shoulder girdle was modified to brace the force of the flight stroke against the skeleton. The wing itself was made of thin skin, attached to the bones of the forearm and wing-finger. It was reinforced by thin, perhaps keratinous fibers only a few micrometers in diameter but several centimeters long. These radiated over the undersurface of the wing in a pattern similar to the feather shafts of birds, and they provided the stiffness and flexibility that made the wing membrane airworthy. All but the largest pterosaurs, like the largest birds, were active flappers; the largest relied on soaring.

Pterosaurs have other modifications for flight. The bones are extremely thin-walled and hollow, and many of them have openings on their surfaces that are remnants of channels through which the respiratory tissues ran. These tissues opened inside the bones, apparently providing further respiratory surface and helping to cool the blood. Birds have similar structures and functions, and their dinosaurian ancestors have at least the beginnings of these structures, which may reflect relatively high metabolic levels. The shoulder girdles of pterosaurs are braced against their large, expanded breastbones that anchored the flight muscles. Many large pterodactyloids have a series of fused vertebrae in the upper back, often showing evidence of sockets in which the ends of the scapulae inserted as a further brace. Specialized structures such as internal struts, plywoodlike layers, and torsion-resistant tubular design helped the skeleton resist the forces of flight.

Anatomical studies demonstrate that different pterosaurs produced various flight strokes. The most basic is roughly the shape of a lazy figure-8, at the shoulder joint, and an up-and-down motion was also possible. The elbow is a hinge, the wrist is blocky but slightly flexible, and most of the length of the wing is formed by a single elongated finger. The wing-finger dwarfs the first three, which were free and could still manipulate objects. The base of the wing-finger has a pronounced pulley joint, modified so as to flex the finger to the side, rather than up and down like the other fingers. The rotation of this joint is very important in flight.

The shape of the wings of pterosaurs is an open question. Some specimens of *Rhamphorhynchus* suggest a narrow, gull-like wing that tapered toward the waist. Soft tissues preserved in some specimens of *Pterodactylus* suggest an attachment of the wing to the upper leg. But without evidence of the supporting fibers, it is difficult to tell which soft-part remains represent the true wing and which are merely body remains squashed into two dimensions. Claims have been made that the Kirgiz specimens of *Sordes*, which preserve both wing fibers and other epidermal structures, demonstrate that the wings were attached to the ankles. This would have given the wing a much broader, more batlike planform than suggested by the *Rhamphorhynchus* specimens. In aerodynamic terms, the aspect of the wing (roughly the relationship of the length to the mean breadth) would be lower, and less like those of many vigorously flying birds. This might suggest different styles of flight and different ecological roles. But the *Sordes* specimens have the same problems of flattening and ambiguous interpretation as the *Pterodactylus* specimens do. At least one study has suggested that the specimens are systematically distorted, and merely create an illusion of wing membranes that are nothing but skin.

Much more needs to be learned about the soft tissues of pterosaurs. Some filamentous epidermal structures appear not only on and near the wing areas but all over the body. These are apparently different from the structural fibers of the wing membranes, but it is not clear whether they grade into each other or are entirely different. The best-preserved specimens of *Rhamphorhynchus* indicate that the wing's structural fibers occur only on the surface of the underside of the wing, and they are embedded superficially in the epidermis, not free. The ability to differentiate the various kinds of fibers would be important in discerning whether a given preserved tissue in a pterosaur fossil is from the wing or another part of the body. Such information could have broad implications for flight style, terrestial locomotion, and physiology.

For background information *see* DINOSAUR; FLIGHT; REPTILIA in the McGraw-Hill Encyclopedia of Science & Technology.   Kevin Padian

Bibliography. S. C. Bennett, Terrestrial locomotion of pterosaurs: A reconstruction based on *Pteraichnus* trackways, *J. Vert. Paleontol.*, 17:104–113, 1997; K. Padian and J. M. V. Rayner, The wings of pterosaurs, *Amer. J. Sci.*, 293A:91–166, 1993; P. Wellnhofer, *The Illustrated Encyclopedia of Pterosaurs*, 1991.

Quantum-dot lasers

Bulk semiconductor materials such as silicon, gallium arsenide (GaAs), and indium phosphide (InP) have a continuum of electronic states within certain energy ranges, and there is a distinct gap in energy between filled valence-band states and empty conduction-band states. An electron in the conduction band can propagate from one end of the crystal to the other when an electric field is applied. It behaves like a free electron, which can occupy states with continuous values of momentum and energy. Making the semiconductor smaller in all directions such that the electrons are completely confined within a region about 10 nanometers in diameter changes the energy spectrum drastically. The charge carriers occupy discrete energy levels similar to electrons in an atom. Quantum dots are small islands of semiconductor material with only a few thousand atoms. *See* NANOELECTRONICS.

In an ideal quantum dot the small region of low-band-gap material should be completely embedded in large-band-gap material so that electrons and holes are localized. The separation of the discrete energy levels depends inversely on the size of the quantum dot. It turns out that the size must be as small as about 10 nm in order to achieve a level splitting of more than the thermal energy at room temperature. Quantum dots can be regarded in many respects as artificially made atoms; therefore they are interesting for fundamental studies and for special devices. The most interesting device

Fig. 1. Atomic force microscope surface image (1×1 μm) of islands formed after deposition of 2.5 atomic layers of indium phosphide (InP) on gallium indium phosphide (GaInP).

applications are single-electron transistors and quantum-dot lasers.

Discrete energy spectrum. Semiconductor lasers consist of a *pn* junction which is forward biased such that electron-hole recombination takes place in the depletion region. Polished surfaces (called facets) perpendicular to the junction are used to form the resonant cavity needed for laser operation. Decreasing the dimensionality of the space available to the charge carriers, from bulk material to quantum films, quantum wires, and quantum dots, increases the optical gain because of the lower and lower spread of occupied energy states. In quantum dots the electronic states are completely squeezed into discrete transition energies.

The discrete energy spectrum promises high-efficiency lasing at a distinct energy. Bulk semiconductors have a continuum of electronic states within certain energy ranges. In laser diodes, only the electrons in the conduction-band minimum recombining with the holes in the valence-band maximum contribute to the stimulated light emission. Since the electronic states in quantum dots are squeezed into discrete transition energies, population inversion is expected to be achieved with fewer carriers, and thus the threshold current should be lower than for currently used laser diodes. A further improvement is expected with regard to the temperature stability. The spread of energetic states usually causes a significant shift of the threshold current with changing temperature. For a quantum-dot laser the temperature dependence is almost eliminated, again as a consequence of the discrete energy spectrum and the large separation between the energy levels.

Preparation. The experimental challenge is to prepare the semiconductor structures, typically 10 nm in size, in an efficient and reproducible way. Since the mid-1990s, there has been considerable work on the direct synthesis of semiconductor nanostructures by applying the phenomenon of island formation during strained-layer heteroepitaxy, a process called the Stranski-Krastanov growth mode. Cluster formation is observed during epitaxial growth of a semiconductor material on top of another one that has a lattice constant several percent smaller. For the first few atomic layers the atoms arrange themselves in a planar layer called the wetting layer. As the epitaxial overgrowth proceeds, the atoms tend to bunch up and form clusters. The cluster formation is energetically favorable, because the lattice can elastically relax the compressive strain and thus reduce strain energy within the islands. Since the quantum dots appear spontaneously during growth, they are said to be self-assembling. Stranski-Krastanov growth is observed for several material combinations. The most intensively investigated examples are indium arsenide (InAs) on gallium arsenide, indium phos-

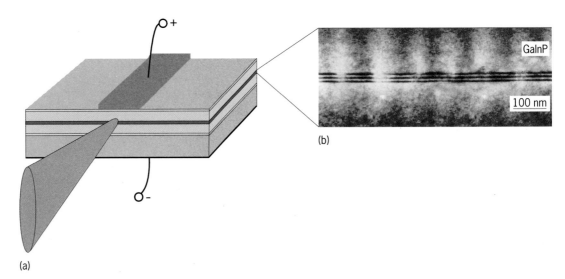

Fig. 2. Indium phosphide (InP) quantum-dot laser diode. (*a*) Structure. (*b*) Transmission electron micrograph of gallium indium phosphide (GaInP) waveguide layer with a stack of three layers of InP quantum dots.

phide on gallium indium phosphide (GaInP), and germanium on silicon.

For example, the lattice constant of indium phosphide is about 4% larger than that of gallium indium phosphide. Cluster formation is observed (**Fig. 1**) after the deposition of more than 1.5 monolayers of indium phosphide on gallium indium phosphide (where a top layer that is half populated is counted as half a layer). After the growth of three atomic layers, clusters are seen on the surface, sitting on top of the wetting layer. The clusters nucleate preferentially at atomic step edges, which are always present on the surface. The size of these dots is about 15 nm in diameter and 3 nm in height with a truncated pyramidlike shape. The dot size varies by less than 10%; this uniformity is essential for the realization of a quantum-dot laser. After the growth of the indium phosphide clusters, the large-band-gap material gallium indium phosphide is deposited on top.

The size of the islands can be adjusted within a certain range by changing the amount of deposited dot material. The change in size corresponds to an energy shift of the emitted light, which is due to the altered carrier confinement and strain in the dots. Smaller dots emit photons of shorter wavelengths, corresponding to higher energies. Overall, Stranski-Krastanov growth makes it possible to prepare extremely small quantum dots in a maskless process without lithography and etching. This capability makes it a promising technique to realize quantum-dot lasers.

Structure. A quantum-dot laser should have a dense array of equal-size dots within the active region where the emitted light travels between a pair of mirrors. The energy of the laser light depends on the band gap of the dot material, on the strain, and inversely on the size of the dots. A waveguide structure is used to keep the photons close to the active material. For example, an indium phosphide quantum-dot laser may have three layers of indium phoside quantum dots positioned in the middle of a gallium indium phosphide waveguide layer (**Fig. 2**). Aluminum indium phosphide (AlInP) layers with a smaller refractive index below and above the waveguide layer confine the light beam and keep it close to the layer of dots to stimulate further light emission. In other words, a photon initiates the emission of a further coherent photon from a quantum dot by electron-hole recombination if the energy separation of the discrete states agrees with the photon energy. Electrons and holes can be introduced in the dot by optical excitation or by an electric current if the waveguide layer is incorporated in a *pn* junction. For indium phosphide quantum dots, lasing has been observed in the red visible range (**Fig. 3**).

Dense stacking of individual dot layers helps to increase the total amount of active material in the laser structure. Significant linewidth reduction has been observed in photoluminescence from stacked indium arsenide or indium phosphide dot layers.

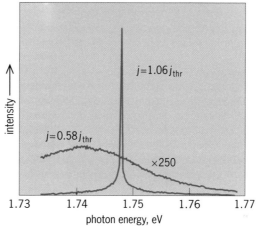

Fig. 3. Light emission from indium phosphide (InP) quantum-dot laser diode shown in Fig. 2. (*a*) Red laser emission emerging from the mirrors of the laser structure. (*b*) Spectrum of laser output below and above threshold current, j_{thr}. Spectrum below threshold current ($j = 0.58\,j_{thr}$) is magnified 250 times relative to spectrum above threshold current ($j = 1.06\,j_{thr}$).

This is explained by improved size homogeneity due to correlated nucleation of the dots. The atoms arriving during growth of the upper layer tend to accumulate and form new dots just above those in the underlying layer because of the strain field around the embedded dots. This correlated nucleation is observed if the separation between dot layers is not too large.

Status. Self-assembling quantum-dot lasers are attracting increasing attention. Laser diodes have been fabricated with stacks of ten layers of dots. However, device performance superior to that of the highly developed quantum-film lasers in current use remains to be demonstrated. In particular, better size distribution is the subject of continuing investigations. Quantum-dot laser diodes may find application in the field of fast optical data transfer.

For background information *see* BAND THEORY OF SOLIDS; CRYSTAL GROWTH; ELECTRON-HOLE RECOMBINATION; LASER; QUANTIZED ELECTRONIC STRUCTURE (QUEST); SEMICONDUCTOR; SEMICONDUCTOR HETEROSTRUCTURES in the McGraw-Hill Encyclopedia of Science & Technology.   Karl Eberl; Markus K. Zundel

Bibliography. K. Eberl, Quantum-dot lasers, *Phys. World*, pp. 47–50, September 1997.

Quark-gluon plasma

A quark-gluon plasma is a phase of matter that is predicted to exist when nucleons made up of quarks and gluons are compressed and heated to high pressures or temperatures. The protons and neutrons that make up conventional matter may be thought of as bags containing three quarks held together by numerous gluons, which carry the strong force. Virtual quark-antiquark pairs are also present, momentarily created by gluon interactions. When nuclear matter is heated or compressed, the individual nuclei may disappear, forming a plasma composed of free quarks and gluons (**Fig. 1**). At normal nuclear density, a temperature of 150–200 megaelectronvolts (2×10^{12} °C) is thought to be required. A cooler quark-gluon plasma may form at high densities; at about ten times normal nuclear density (normal nuclear matter is 0.15×10^9 eV/fermi^3 or 2.7×10^{17} kg/m^3), a quark-gluon plasma may form without heating. In this condition, the individual nucleons overlap enough to lose their identity. This is foreshadowed in normal nuclei by the EMC (European Muon Collaboration) effect: The properties of individual nucleons change when they are bound together in a nucleus.

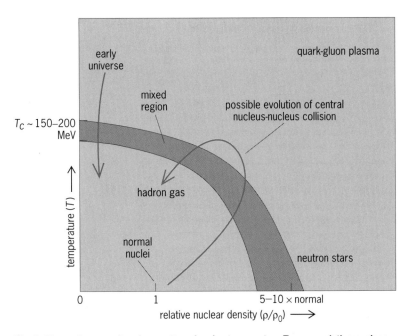

Fig. 1. Phase diagram of nuclear matter, showing temperature *T* versus relative nuclear density ρ/ρ_0, where the normal nuclear density $\rho_0 = 0.16$ nucleon/fm^3. Various regions of matter are shown, with normal nuclei being at *T* = 0 and $\rho/\rho_0 = 1$. A possible path of the early universe and a possible trajectory followed by matter produced in a central nucleus-nucleus collision are shown.

A quark-gluon plasma is believed to have formed shortly after the big bang. When the universe cooled to a temperature of about 150–200 MeV, about 10 microseconds after the big bang, the quark-gluon plasma froze into normal matter. A static, relatively low-temperature quark-gluon plasma may also be present in the core of neutron stars, where gravity produces huge pressures.

Laboratory studies. A quark-gluon plasma might be created in a laboratory by colliding beams of heavy nuclei. Accelerators have been used to study relativistic heavy-ion collisions, where the nuclei move with a large fraction of the speed of light (see **table**). The earliest, relatively low-energy experiments focused on how the colliding nuclei broke up; relatively few new particles were created in the collision. At these energies, the data could be explained quite well by hydrodynamics, with the nuclear matter described as a fluid.

At the higher energies present at the Alternating Gradient Synchrotron (AGS), the number of produced particles is comparable to the number of incident protons and neutrons (collectively known as nucleons). The particle production can be explained by models based on individual proton-proton collisions, except that the initial-state nucleons and produced particles can further interact with each other. This interacting mix is known as a hadron gas. At the Alternating Gradient Synchrotron, a mixture of hadron-gas and hydrodynamic variables are used to describe the collisions. At this energy, head-on collisions bring the nuclei to an almost complete stop, producing a very high nucleon density.

At the Super Proton Synchrotron (SPS) at the Center for European Nuclear Research (CERN), particle production is even larger, with produced particles outnumbering incident nucleons six to one (**Fig. 2**). However, hydrodynamic behavior is still clearly observable. At this energy, the colliding nuclei are partially transparent, and most of the nucleons emerge from the collisions and continue downstream, retaining a fraction of their energy. The collision will leave a central region with a high energy density but low nucleon density (**Fig. 3**). However, the density quickly rises as the particles present interact and create more particles.

Future experiments will reach much higher energies by colliding circulating beams of heavy ions. At collider energies, the nucleon-free region should be quite free of initial particles. There, the quark-gluon plasma may be treated as an excited state of the vacuum. The higher energies also allow hard interactions between constituent quarks and gluons to be observed. These newly observable reactions allow for many new ways to search for a quark-gluon plasma.

Experimental signatures. The fireball formed when two relativistic nuclei collide, quickly expands and cools. If a quark-gluon plasma is formed, it lives for only a few times 10^{-23} second, until the collision products expand and cool sufficiently to revert to a hadron gas. As the hadron gas expands, the density

Accelerators used to study heavy-ion collisions

Laboratory	Accelerator	Year of first operation	Mode	Center-of-mass energy (per nucleon), GeV[†]	Nuclei accelerated
Lawrence Berkeley Laboratory (http://www.lbl.gov/)	Bevalac (http://www-rnc.lbl.gov/EOS/)	1974	Fixed target	2.4	Up to uranium
Brookhaven National Laboratory (http://www.bnl.gov/)	Alternating Gradient Synchrotron* (http://www.rhichome.bnl.gov/AGS/ags.html#exp)	1986	Fixed target	4.8	Silicon, gold
CERN (http://www.cern.ch/)	Super Proton Synchrotron*	1986	Fixed target	17	Oxygen, sulfur, lead
Brookhaven National Laboratory	Relativistic Heavy-Ion Collider (http://www.rhic.bnl.gov/STAR/) (http://www.rhic.bnl.gov/phenix/)	1999[‡]	Collider	200	Up to gold
CERN	Large Hadron Collider (http://www.cern.ch/ALICE/)	2005[‡]	Collider	3400	Up to lead

*Synchrotrons began with lighter ion beams and were later upgraded to accelerate heavy ions.
[†]The center-of-mass energies listed are for heavy ions; lighter ions can be accelerated to slightly higher energies. 1 GeV = 10^9 eV = 10^{-10} joule.
[‡]Scheduled.

drops and the interactions stop at a time known as freezeout. Particles present at freezeout continue expanding, and can be detected by experimental apparatus. The thermodynamic parameters of the system can be found from these particles. The temperature T is given by the particle momentum spectrum. The energy flow transverse to the beam direction is a measure of the energy density. The entropy, or number of degrees of freedom (ways the system can behave), as measured by the produced particle multiplicity, should rise sharply if a transition occurs. However, these thermodynamic parameters are muddied by the transition back to the hadron gas, which precedes freezeout; the temperature measured is that at freezeout.

Conventional thermodynamics considers only systems in equilibrium. Because of the brief lifetime of the quark-gluon plasma, this assumption may be problematic in heavy-ion collisions. Calculations and data indicate that in heavy-ion collisions the system interacts rapidly enough that thermal equilibrium is reached, with all parts of the system at the same temperature. However, chemical equilibrium, where the relative number of up, down, and strange quarks becomes constant, may not be reached. The cross sections for quark interactions are larger than those for hadrons, so equilibrium should be reached quickly in a quark-gluon plasma. In a hadron gas, there are many different types of particles present instead of just the three quark flavors. Between the additional particle types and the slower reaction rate, calculations are not definitive, and equilibrium may not be reached by freezeout. Some data from the Super Proton Synchrotron indicate that chemical equilibrium is probably present at freezeout. However, because of the nonequilibrium mixture early in the collision, nonequilibrium thermodynamics may be necessary to explain some aspects of heavy-ion collisions.

One sign of a phase transition is long-range ordering, where the system can act collectively over a large distance, such as the entire collision fireball. This collective action may produce large fluctuations in some experimentally observable quantities, such as the overall produced particle multiplicity, or the number of produced strange particles, from one collision to the next. It could also introduce observable correlations between pairs (or higher multiples) of produced particles.

Because the properties of a quark-gluon plasma and a hadron gas are poorly known, sophisticated

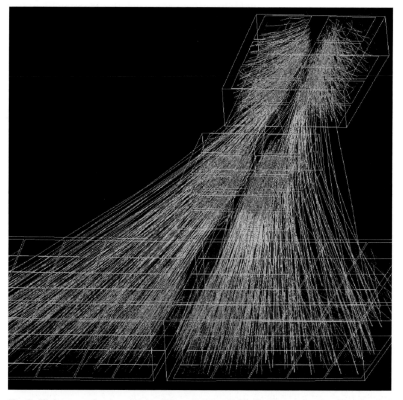

Fig. 2. High-energy lead-on-lead collision, as recorded in the time projection chambers of the NA-49 experiment at the Center for European Nuclear Research.

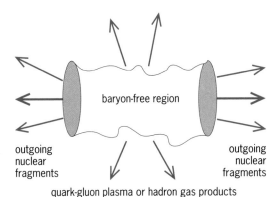

Fig. 3. High-energy nuclear collision. The colliding nucleons retain some of their energy and continue downstream after the collision, leaving a baryon-free region in the middle where high-energy densities are present.

analyses are required to determine if a quark-gluon plasma is present early in a collision. Two general strategies have been used. The first studies the global parameters discussed above. The second uses probes that are sensitive to the conditions early in the collision. Conceptually, the simplest are high-energy photons. These may be produced electromagnetically during the collision, and escape from the fireball without further interaction. Although a clean measurement of the early dynamics, these photons are rare, and may be overshadowed by background.

Another example is the J/psi, a meson composed of a charmed quark and an anti–charmed quark. The J/psi production rate can be extrapolated from measurements in proton-proton collisions. The likelihood of a J/psi interacting and being destroyed is much higher in a quark-gluon plasma than a hadron gas, so the absence of J/psi mesons may signal the quark-gluon plasma. The NA-50 experiment at the Super Proton Synchrotron observed an apparent suppression of J/psi mesons in central (head-on) lead-on-lead collisions, compared to expectations from proton-proton collisions, proton-lead collisions, lighter-ion collisions, and less central lead-on-lead collisions.

High-energy quarks are another probe: They lose energy faster in a quark-gluon plasma than in a hadron gas. Energy loss can be measured in events where a quark is produced opposite a high-energy photon. The photon should escape from the collision intact, so the photon-quark momentum imbalance measures the quark energy loss.

Although current Super Proton Synchrotron data provide some intriguing hints, all of the signals suffer from significant uncertainties. While more data might help clarify the situation, it is likely that higher-energy collisions are required to cleanly observe the quark-gluon plasma. Any quark-gluon plasma produced at these higher energies will be larger and longer-lived. Between the increased quark-gluon plasma production and the wider range of accessible signatures, data interpretation should become significantly clearer.

Theoretical predictions. Because quantum chromodynamics involves strong couplings, traditional perturbation-theory approaches are inapplicable, so theoretical studies of the quark-gluon plasma are difficult. Currently, the most promising approach is to solve the equations of quantum chromodynamics numerically on a four-dimensional space-time lattice, using relaxation techniques. Tremendous computing power is necessary to get meaningful results, and significant compromises are required with current computers. Still, some interesting results have emerged. Most groups agree that at normal densities temperatures of about 150 MeV are required to form a quark-gluon plasma. The transition to a quark-gluon plasma is probably a second-order phase transition, with a weak first-order transition not excluded. This means that there is no latent heat at the transition, but the specific heat changes because of the increased number of degrees of freedom.

Theorists have also explored analogies between the quark-gluon plasma phase transition and other transitions. In normal matter, the vacuum may be thought of as a insulator, while in a quark-gluon plasma, it becomes a conductor. An analogy to the transition to superconductivity may be usefully drawn.

Cold quark-gluon plasma has been studied mostly in the context of neutron stars. The pressure gradient there may allow for the existence of a number of different phases and mixtures of phases of quark-gluon plasma and hadron-gas matter at different radii. It has also been suggested that neutron-star "earthquakes," whereby the rotational speed is seen to change rapidly, may be caused by phase changes which manifest themselves as changes in the moment of inertia of the neutron star.

For background information *see* BIG BANG THEORY; GLUONS; J/PSI PARTICLE; NEUTRON STAR; PARTICLE ACCELERATOR; PHASE TRANSITIONS; QUANTUM CHROMODYNAMICS; QUARK-GLUON PLASMA; QUARKS; RELATIVISTIC HEAVY-ION COLLISIONS; SUPER CONDUCTIVITY; THERMODYNAMIC PRINCIPLES in the McGraw-Hill Encyclopedia of Science & Technology.

Spencer Klein

Bibliography. K. K. Geiger, Space-time description of ultra-relativistic nuclear collisions in the QCD parton picture, *Phys. Rep.*, 258:237–376, 1995; J. Harris and B. Müller, Search for the quark-gluon plasma, *Annu. Rev. Nucl. Part. Phys.*, 46:71–107, 1996; Proceedings of Quark Matter '97, *Nucl. Phys. A*, A638 (1 and 2):1c–610c, August 1998.

Radioisotope dating

Knowledge of the age and evolution of the solid Earth depends in large part on the ability to measure isotopes formed by the radioactive decay that occurs in rocks and minerals. Radioactive isotopes decay naturally by one of several mechanisms. For example, alpha decay involves the loss of a helium nucleus (two protons and two neutrons), and beta decay involves the loss of an electron. The field of isotope geochemistry developed rapidly in the 1960s and 1970s with the widespread use of solid-source mass spectrometers to measure rubidium-strontium (Rb-Sr), samarium-neodymium (Sm-Nd), uranium-lead (U-Pb), and lutetium-hafnium (Lu-Hf) isotopes. However, it was not until the 1990s that the isotopes of rhenium (Re) and osmium (Os) could be measured easily and precisely enough to allow widespread use of a new chronometer based on the beta decay of ^{187}Re "parent" to ^{187}Os "daughter." Due to analytical developments and improvements, data for this isotope pair have been accumulating at an impressive rate and have been used to date meteorites, define the age and nature of different reservoirs in the mantle, define the role of crustal material in magma genesis, and place constraints on the amount of oceanic crust that may be recycled at subduction zones.

Mathematically, the increase of ^{187}Os in a rhenium-bearing system can be expressed as ^{187}Os $= (^{187}$Os$)_i$ $+ {}^{187}$Re $(e^{\lambda t}-1)$, where i refers to the amount of ^{187}Os present initially, λ is the decay constant for ^{187}Re, and t is the time elapsed since the system became closed to diffusion of rhenium and osmium. Because mass spectrometry allows precise measurements of isotope ratios, the growth of ^{187}Os is more commonly presented in a form in which ^{187}Os is normalized to a stable isotope such as ^{188}Os:

$$^{187}\text{Os}/^{188}\text{Os} = (^{187}\text{Os}/^{188}\text{Os})_i + {}^{187}\text{Re}/^{188}\text{Os}(e^{\lambda t} - 1)$$

For isotope system pairs such as Rb-Sr, Sm-Nd, U-Pb, and Lu-Hf, both elements are incompatible (mantle-silicate partition coefficient $D \ll 1$) in most mantle minerals, and are thus enriched in the melt during a melting event (see **table**). Rhenium and osmium behave differently than other isotope systems and thus offer unique perspectives on long-standing geochemical problems. Both rhenium and osmium are highly siderophile ("iron-loving") elements, and so are compatible ($D > 10,000$) in metal relative to silicates. Also, in mantle and crustal melting environments, rhenium is thought to be incompatible and osmium strongly compatible (**Fig. 1**). Rhenium is incompatible in most silicate phases (such as clinopyroxene and olivine) and compatible in only a few phases (such as garnet, sulfide liquid, and magnetite). There is evidence for the compatibility of osmium in a variety of common igneous minerals (such as olivine and chromite), in addition to sulfide liquid. Because of the differences in partitioning

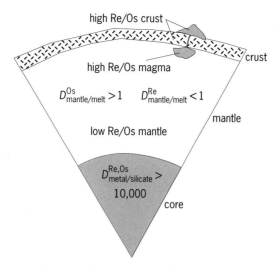

Fig. 1. Behavior of rhenium and osmium in the Earth. Rhenium and osmium are partitioned strongly into the metallic core of the Earth due to their high metal-silicate partition coefficients ($D > 10,000$). Rhenium is concentrated into the crust because it is incompatible ($D < 1$) during melting, while osmium is compatible ($D > 1$).

behavior, Re/Os ratios will become fractionated and develop over time into distinct ^{187}Os/^{188}Os ratios for the lithosphere, asthenosphere, and crust (**Fig. 2**). This unique characteristic of the parent (^{187}Re) being much more incompatible than the daughter (^{187}Os), combined with their highly siderophilic behavior, makes this isotope system applicable to many problems in solid-Earth geochemistry—for example, the origin of continental flood basalts and the source of hot spot magmatism.

Application of the Re-Os system to the problem of continental flood basalt genesis has led to determining whether crustal or mantle lithospheric contamination is responsible for producing the isotopic and compositional characteristics of these magmas. For

Comparison of the Re-Os system to other common geological isotope pairs

Isotopic pair	Decay type	Decay constant (λ)	Bulk $D^\dagger$ parent	Bulk $D^\dagger$ daughter
^{87}Rb-^{87}Sr	beta	1.42×10^{-11}	0.0012	0.016
^{147}Sm-^{143}Nd	alpha	6.54×10^{-12}	0.06	0.02
^{176}Lu-^{176}Hf	beta	1.94×10^{-11}	0.80	0.09
^{187}Re-^{187}Os	beta	1.666×10^{-11}	0.22	8.4

† Bulk D is defined as the sum of the partition coefficients for all solid phases present: D (bulk) $= D$ (phase 1) $* X$ (phase 1) $+ D$ (phase 2) $* X$ (phase 2) $+ D$ (phase 3) $* X$ (phase 3), and so on, where X is the fraction of the phase present. Phase is just a select word for mineral. Therefore, phases 1, 2, and 3 may respectively be olivine, pyroxene, and garnet for the Earth's mantle.

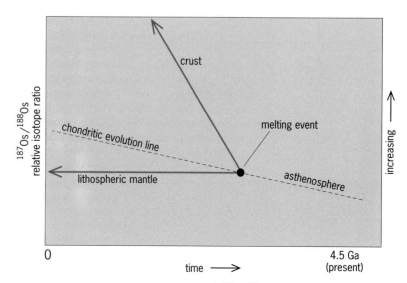

Fig. 2. Evolution of mantle and crustal values of ^{187}Os/^{188}Os with time (Ga = billion years). If the mantle remained unchanged (did not melt) during the history of the Earth, osmium isotopic values would fall along the line labeled asthenosphere or chondritic evolution line. A melting event will fractionate rhenium and osmium such that the crustal (or magmatic) osmium values will evolve to high values and the residual mantle or lithospheric values will remain lower than the asthenosphere.

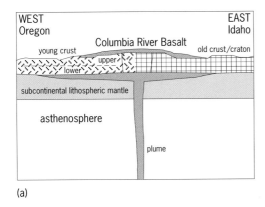

(a)

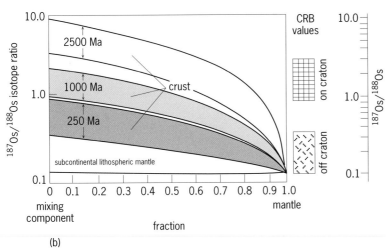

(b)

Fig. 3. Cross section showing Columbia River Basalt (CRB). (*a*) Continental flood basalt genesis, involving young (off craton) and old (on craton) crust, upper and lower crust, subcontinental lithospheric mantle, and asthenospheric mantle. (*b*) Elevated values of $^{187}Os/^{188}Os$ from the continental flood basalt (range of values is defined by the boxes on the right), compared to calculated values of $^{187}Os/^{188}Os$ for mixing between mantle-derived magma and crustal material (Ma = million years). Note that mixing of subcontinental lithospheric mantle with the mantle-derived magma would produce very low $^{187}Os/^{188}Os$; because these low values are not observed, this lithospheric mantle can be ruled out as an important source in continental flood basalt magmatism. (*After J. T. Chesley and J. Ruiz, Crust-mantle interaction in large igneous provinces: Implications from the Re-Os isotope systematics of the Columbia River flood basalts, Earth Planet. Sci. Lett., 154:1–11, 1998*)

Radiogenic osmium measured in oceanic island basalt (OIB) has focused attention on specific models for the structure of Earth's mantle, while enabling researchers to determine sources for hot spot magmatism. In the past 30 years, five different mantle reservoirs have been proposed based on major- and trace-element studies (**Fig. 4**), and isotopic strontium, neodymium, lead, hafnium measurements. These are (1) depleted modern mantle (DMM); (2) enriched mantle 1 (EM1); (3) enriched mantle 2 (EM2); (4) old recycled oceanic crust (HIMU); and (5) a lower mantle source called focal zone (FOZO). The radiogenic osmium measured in oceanic island basalt seems attributable to the HIMU component, but the specific material providing this signature is debated. Four materials that have been proposed are (1) recycled oceanic crust (1 to 2 billion years in age); (2) lower mantle (with a higher Re/Os ratio than the upper mantle); (3) ancient mantle heterogeneities (such as garnetite or trapped residual silicate melt from a magma ocean); and (4) small amounts of core metal.

Recycled oceanic crust is appealing for several reasons. First, subducted basaltic crust is introduced into the upper mantle with an initially high Re/Os ratio. Within the stability field of garnet at a range greater than 90–100 km (56–62 mi), rhenium could be retained during dehydration or melting of the slab, and thus garnet may play a role in preserving the high Re/Os ratio (and, with time, high osmium isotope ratio) of slab material during subduction zone processing. Thus, elevated osmium isotopes should be a sensitive indicator of recycled basaltic material. Second, it is supported by correlations of osmium isotopes with isotopes of strontium, neodymium, lead, and oxygen (O) in oceanic island basalt, as well as numerous other trace element and isotopic data which are all consistent with recycling of basaltic crust. Third, mass balance calculations indicate that the crustal abundance of rhenium is too low (by a factor of 10) to balance the depletion of rhenium in the depleted modern mantle, suggesting that there is an unidentified reservoir of rhenium. Recycled oceanic crust could be such a reservoir because it could easily account for the missing mass of rhenium over 4 billion years of oceanic plate subduction. The high osmium isotopic compositions of oceanic island basalts, together with the evidence that rhenium is compatible in garnet, implies that the sources of mantle plumes contain most of the rhenium, which has been apparently removed from the upper mantle, and that this rhenium resides in recycled oceanic crust.

There are, however, other ways to interpret the data. One possibility is that part of the oceanic island basalt source comes from a lower mantle with higher Re/Os ratios, and thus more radiogenic osmium, than the upper mantle. Not enough is known about the high-pressure partitioning behavior of rhenium or osmium to determine if the lower mantle has a higher Re/Os ratio than the upper mantle. Furthermore, molten mantle trapped in the earliest stages of terrestrial differentiation likely had high Re/Os ratios. This scenario depends on whether or not the

previously discussed reasons, crustal material will have a higher Re/Os ratio than the lithosphere and thus will become rich in ^{187}Os in a short period of time. Such material can then be easily traced in magma genesis or crustal assimilation. The large continental flood basalt province in the northwest United States, the Columbia River Basalt, has been shown to have formed in part by melting substantial amounts of crustal material, because the $^{187}Os/^{188}Os$ measured for the basalt is much higher than that available in the lithospheric mantle or convecting mantle (**Fig. 3**). Two aspects of the crustal material involved have been determined. First, because these basalts have relatively high osmium and Re/Os contents, the crustal material involved was most likely from the mafic lower crust, rather than evolved upper crust. Assimilation of low-osmium upper crust would yield basaltic magmas with lower bulk osmium, and this has not been observed. Second, basalt which has erupted through the crust of old cratons has more elevated $^{187}Os/^{188}Os$ values than those erupted through young noncratonic crust.

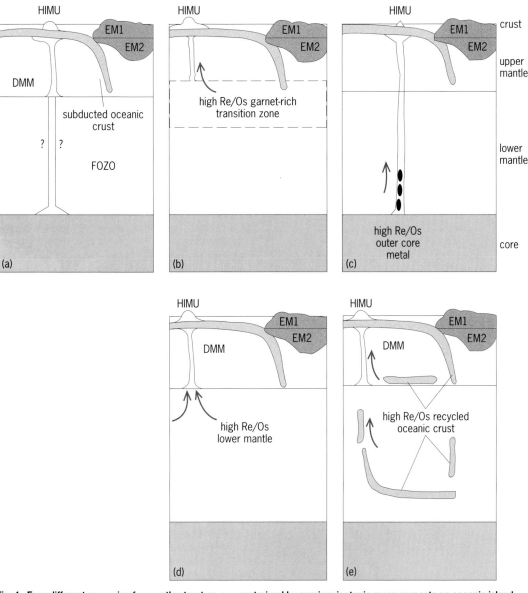

Fig. 4. Four different scenarios for mantle structure as constrained by osmium isotopic measurements on oceanic island basalts; (a) Standard mantle reservoirs are defined by the neodymium, strontium, lead, and oxygen isotopic systems—depleted modern mantle (DMM), enriched mantle 1 (EM1), enriched mantle 2 (EM2), old recycled oceanic crust (HIMU), and focal zone (FOZO). (b) Garnet-rich transition zone or ancient trapped magma from an early magma ocean is present in the deep upper mantle. (c) High Re/Os metallic liquid from Earth's outer core is added. (d) Lower mantle has high Re/Os. (e) Recycled oceanic crust has high Re/Os.

Earth went through a magma ocean stage in its early history. In addition, it has been proposed that the radiogenic osmium signature in oceanic island basalt is derived from small amounts of outer-core metallic liquid added to the Earth's mantle. This intriguing hypothesis works for a very specific, but perhaps unlikely, set of conditions for the differentiation of the Earth. It also requires further testing after acquisition of liquid metal/solid metal, liquid metal/liquid silicate, and solid silicate/liquid silicate partitioning experiments at high temperatures and high pressures.

The Re-Os isotopic system offers unique new explanations for many problems of geochemistry, and especially the evolution of the solid mantle. This system should help provide new perspectives on long-term problems such as the genesis of island or continental arc magmas, and the evolution of cratonic belts.

For background information *see* ALPHA PARTICLES; ARCHEOLOGICAL CHRONOLOGY; BETA PARTICLES; CHEMOSTRATIGRAPHY; DATING METHODS; GEOCHRONOMETRY; ISOTOPE; RADIOISOTOPE; RADIOISOTOPE (GEOCHEMISTRY); ROCK AGE DETERMINATION in the McGraw-Hill Encyclopedia of Science & Technology.

Kevin Righter

Bibliography. C. J. Allegre and J.-M. Luck, Osmium isotopes as petrogenetic and geological tracers, *Earth Planet. Sci. Lett.*, 48:148–154, 1980; A. D. Brandon et al., Coupled [186]Os and [187]Os evidence for core-mantle interaction, *Science*, 280:1570–1573, 1998; J. T. Chesley and J. Ruiz, Crust-mantle interaction in large igneous provinces: Implications from the Re-Os isotope systematics of the Columbia River flood basalts, *Earth Planet. Sci. Lett.*, 154:1–11, 1998; S. B. Shirey and R. J. Walker, The Re-Os isotope system in cosmochemistry and high-temperature geochemistry, *Annu. Rev. Earth Planet. Sci.*, 26:423–500, 1998; K. Righter and E. H. Hauri, Compatibility of rhenium in garnet during mantle melting and magma genesis, *Science*, 280:1737–1741, 1998.

Regeneration (ear)

In vertebrates, the senses of hearing and balance are mediated primarily through the perception of pressure waves within the inner ear. While many structures in the ear are required for the perception of these signals, the actual conversion of pressure waves into electrical impulses that can be understood by the central nervous system occurs in small groups of microscopic sensory cells in the inner ear. Each of these sensory cells contains a bundle of 30–200 highly specialized microvilli (called stereocilia) that project into a fluid-filled space (**Fig. 1**). These stereocilia bundles resemble tufts of hair when observed at the microscopic level, and so are referred to as hair cells. However, each stereocilium is approximately 1000 times thinner than a human hair and is so sensitive that even extremely low energy pressure waves can cause it to vibrate. In fact, it is the movement of these stereocilia that

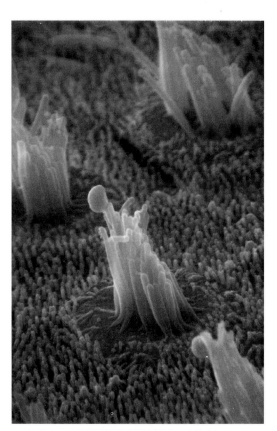

Fig. 1. Scanning electron micrograph of the stereociliary bundle on a hair cell from the sacculus of a bullfrog.

leads to the generation of an electrical signal that is carried along nerve fibers to the brain. *See* VESTIBULAR SYSTEM.

Hair cells. In humans and most other mammals, auditory signals are detected by approximately 17,000 hair cells located within the spiral-shaped cochlear region of the inner ear. Similarly, the sensations of balance and acceleration are mediated through five small patches of hair cells located within the vestibular portion of the inner ear. The development of the inner ear of many mammals, including the production of the full complement of hair cells, is complete at the time of birth. However, as an individual ages the number of hair cells within the cochlea slowly declines. The factors that lead to a decrease in the number of hair cells are not completely understood, but some drugs, short-term exposure to very intense sounds, and long-term exposure to loud noises have been identified as causes. Lost hair cells are not replaced and, as a result, hearing acuity decreases with age. In this way, cochlear hair cells are similar to many types of neurons within the central nervous system. The full complement of cells develops during embryogenesis, and subsequent losses lead to permanent deficits. Since estimates suggest that as much as 10% of the human population suffer from some degree of hair cell loss, the development of techniques and treatments to enhance hair cell regeneration has great potential benefit.

The first clues toward understanding hair cell regeneration were found in the ears of nonmammalian vertebrates. The basic structure of the inner ear, including the presence of hair cells, is similar throughout all vertebrates. However, in nonmammalian vertebrates the total number of hair cells does not decrease with age. In fact, studies initiated in the early 1980s demonstrated that in fishes and amphibians the production of new hair cells continues throughout life. Since these animals also continue to grow in size throughout their lives, newly generated hair cells are continuously added. New hair cells are also continuously produced in the vestibular epithelia of birds. In birds the production of new hair cells is matched by the spontaneous loss of existing hair cells, so there is no net change in hair cell number. The inner ear of the bird contains a basilar papilla, a structure that is very similar in function to the mammalian cochlea. Interestingly, there is no ongoing production of new hair cells in the basilar papilla under normal circumstances. However, if hair cells are lost as a result of injury or trauma, the epithelium responds by regenerating hair cells. Recent results have demonstrated that there is the potential for some hair cell regeneration in vestibular epithelia in mammals as well. Although the recovery time is slow (on the order of months), research indicates that the loss of hair cells in vestibular epithelia induces a limited amount of hair cell recovery. These results have demonstrated that the production of mammalian hair cells may not be limited to the embryonic period, and they have set the stage for an examination of the factors that regulate hair cell regeneration.

Mitosis. An important issue related to the regeneration of hair cells is the identification of their source. In nonmammalian vertebrates, most regenerated hair cells are produced through the mitotic division (cell division) of nonsensory cells referred to as supporting cells. Supporting cells are found in all hair cell epithelia and are evenly distributed between the hair cells (**Fig. 2**). The nuclei of these cells are located in a layer between the hair cells and the basement membrane, but each supporting cell contains a cytoplasmic projection that extends upward to interdigitate between the hair cells. Each hair cell is surrounded by the projections from four to eight supporting cells, and as a result individual hair cells do not contact one another. The exact functions of the supporting cells have not been completely determined, but it is believed that these cells play a role in the electrical and physical isolation of individual hair cells.

Two markers that have been used to study cellular mitosis are ^{3}H-thymidine (tritiated thymidine) and bromo-deoxyuridine (BrDU). ^{3}H-thymidine is an analog of thymine that contains a radioactive tracer. The thymidine is incorporated into cells in the S-phase of mitosis, during which time deoxyribonucleic acid (DNA) is synthesized. Cells that are proliferating can be identified based on the presence of the radioactive signal. Alternatively, using bromo-deoxyuridine allows uridine to be incorporated, instead of thymine, during the S-phase. An antibody that is specific for the bromo group can then be used to detect incorporation into cells that are proliferating.

Studies using markers for cellular mitosis have demonstrated that in response to hair cell loss, nearby supporting cells undergo mitotic divisions to generate new cells. Some of these new cells will develop as regenerated hair cells, while others will develop as supporting cells. In epithelia that produce hair cells continuously, the situation is similar; new hair cells develop from cells that have been generated through mitotic division. Therefore, cellular mitosis of supporting cells represents the primary mechanism for the production of regenerated hair cells.

Recent studies have attempted to identify biochemical or environmental factors that might play a role in triggering cellular mitosis following hair cell loss in either birds or mammals. Among the best candidates are compounds, such as growth factors, that have been shown to induce mitosis in other cell types. Results indicate that in birds the rate of cellular mitosis of supporting cells is increased if the cells are exposed to increased concentrations of insulin or of insulinlike growth factor. Similar studies in mammals have demonstrated that supporting-cell mitoses are increased by insulinlike growth factor, as well as by transforming growth factor alpha (TGFα) and fibroblast growth factor-2 (FGF-2). In addition, treatments that lead to increased concentrations of cyclic adenosine monophosphate (cAMP) within supporting cells induce increased levels of mitosis in birds. Finally, some studies have demonstrated that many of the cell types that are recruited to sites of cellular in-

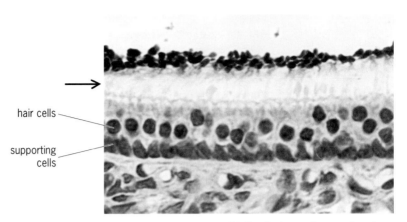

Fig. 2. Cross section through a hair cell sensory epithelium in the vestibular region of an adult mouse ear. Hair cells and supporting cells are located throughout the epithelium but are segregated into two distinct layers. Stereocilia bundles are present above the apical surfaces of the hair cells (arrow).

juries, such as macrophages, are also recruited to the sites of injured hair cells. Since these cells have been shown to play multiple roles in cellular recovery, including the release of factors that promote cellular mitosis, it is possible that the appearance of these types of cells at sites of hair cell injuries could play a role in triggering regenerative responses.

Despite the encouraging results that have been obtained in both the avian auditory and vestibular systems and in the mammalian vestibular system, there is still little evidence that cellular mitosis or regeneration of hair cells occurs in the mature mammalian cochlea. The factors that prevent hair cell regeneration in the mature cochlea remain unclear. However, one factor may be the relative complexity of the cellular structure in the cochlea. All hair cell epithelia contain both hair cells and supporting cells, and in all hair cell epithelia (except the cochlea) the arrangement of hair cells and supporting cells within each sensory structure is relatively disorganized. By contrast, in the cochlea both hair cells and supporting cells are arranged in clearly defined rows, and the number of cells that develop as either cell type appears to be strictly regulated. Therefore, it is possible that changes related to this greater level of complexity have caused this structure to lose its ability to regenerate.

Although most regenerated hair cells develop through cellular mitosis, a second possible mechanism has recently been proposed. Results indicate that in some cases new hair cells can develop in the absence of mitotic cell division. These findings suggest that some new hair cells may arise through the direct conversion of existing supporting cells into new hair cells. In particular, studies of mammal vestibular epithelia have demonstrated that the number of new hair cells is much greater than the number of cells that have been generated through mitosis. While this type of evidence has provided growing support for the conversion theory in both mammals and other vertebrates, one caveat must be considered. Several recent studies have demonstrated that under some circumstances hair cells can sustain sublethal injuries. The injured cell loses its specialized stereocilia bundle as well as many of the other structures that are characteristic of hair cells. Over time these cells recover and begin to reexpress all of the structures required to function as hair cells. Injured hair cells look very similar to supporting cells; therefore it is possible that under some circumstances, hair cells that are classified as converted supporting cells are simply hair cells that have sustained an injury and have subsequently recovered.

Regardless of the source of regenerated hair cells, researchers are beginning to understand the cellular and biochemical factors that initiate and mediate the formation of these cells. This increased knowledge should lead to the development of possible regenerative therapies that could help restore hearing in individuals with either congenital or acquired hearing loss.

For background information *see* EAR; HEARING (HUMAN); HEARING IMPAIRMENT; MITOSIS; PHYSIOLOGICAL ACOUSTICS in the McGraw-Hill Encyclopedia of Science & Technology.   Matthew W. Kelley

Bibliography. J. T. Corwin and J. C. Oberholtzer, Fish n′ chicks: Model recipes for hair-cell regeneration?, *Neuron*, 19:951–954, 1997; D. A. Cotanche, Hair cell regeneration in the avian cochlea, *Annu. Otol. Rhinol. Laryngol. Suppl.*, 168:9–15, 1997; A. Forge et al., Hair cell recovery in the vestibular sensory epithelia of mature guinea pigs, *J. Comp. Neurol.*, 397:69–88, 1998; B. M. Ryals and E. W. Rubel, Hair cell regeneration after acoustic trauma in adult *Coturnix* quail, *Science*, 240:1774–1776, 1988; H. M. Sobkowicz et al., Cellular interactions as a response to injury in the organ of Corti in culture, *Int. J. Dev. Neurosci.*, 15:463–485, 1997; P. S. Steyger et al., Calbindin and parvalbumin are early markers of non-mitotically regenerating hair cells in the bullfrog vestibular otolith organs, *Int. J. Dev. Neurosci.*, 15:417–432, 1997; J. S. Stone et al., Recent insights into regeneration of auditory and vestibular hair cells, *Curr. Opin. Neurol.*, 11:17–24, 1998; M. E. Warchol, Macrophage activity in organ cultures of the avian cochlea: Demonstration of a resident population and recruitment to sites of hair cell lesions, *J. Neurobiol.*, 33:724–34, 1997; M. E. Warchol et al., Regenerative proliferation in inner ear sensory epithelia from adult guinea pigs and humans, *Science*, 259:1619–1622, 1993; J. L. Zheng et al., Induction of cell proliferation by fibroblast and insulin-like growth factors in pure rat inner ear epithelial cell cultures, *J. Neurosci.*, 17:216–226, 1997; J. L. Zheng and W. Q. Gao, Analysis of rat vestibular hair cell development and regeneration using calretinin as an early marker, *J. Neurosci.*, 17:8270–8282, 1997.

Restoration ecology

Restoration ecology is the investigation of ecological principles and field methodologies that increase the ability to successfully alter ecosystems for restoration. This process requires transforming an area into a precisely defined, indigenous historic ecosystem that replicates the structure, function, diversity, and dynamics of the ecosystem. It is practiced across a wide variety of natural communities (such as grasslands, forests, wetlands, lakes, and rivers), and is done for a number of purposes (such as detoxifying polluted lands, creating habitat for endangered species, and restoring natural beauty to parks for human enjoyment). Restoration is a challenging task, and to date, most restoration projects have failed.

History and scope. Through the 1970s, restoration ecology was largely a science of revegetating and rehabilitating ecosystems degraded and contaminated by human use. However, even in these early years, restoration ecology was distinct from simple rehabilitation because it was explicitly directed toward recreating natural biological and geochemical conditions of a site, including the structure, species composition, function, and dynamics of natural ecosystems. The first restoration project to restore prairies was begun in the 1930s, at the University of Wisconsin-Madison arboretum. Since then, prairies have become one of the best-studied ecosystems in restoration ecology. In tallgrass prairies, fire is necessary to maintain diversity and dynamics. High-nutrient soils, such as abandoned agricultural fields, often support high densities of a few plant species and are much less diverse than native prairies. Restoring grasslands often requires lowering soil fertility by mixing sawdust with soil or adding carbohydrates to increase microbial consumption of nitrogen.

Other pioneering work was done to restore plant communities at abandoned minesites in Great Britain. Mine sites can be negatively affected in several ways, including compaction of soil, absence of necessary nutrients (particularly nitrogen—the opposite problem from abandoned agricultural fields), and the presence of toxic levels of heavy metals and salts. These problems may be overcome by a combination of direct manipulation of the environment and introduction of plants and animals that will increase the rate of rehabilitation. For example, compacted soil can be loosened mechanically by tilling and biologically by introducing earthworms to tunnel in soil. Soil nitrogen can be increased directly through fertilization and by planting nitrogen-fixing plants, such as legumes. Soil toxicity is often best dealt with by finding native plants able to evolve tolerance to toxic compounds. The discovery of rapid plant evolution in response to soil toxicity is one of the greatest contributions of restoration ecology to basic science.

In aquatic systems, early research was directed toward restoring contaminated water for human health and safety, typically by removing toxic substances or lowering nutrient loading. The most common ways of dealing with contaminated waters are restricting inputs or diluting inputs. For example, in the 1960s the city of Seattle, Washington, successfully diluted nutrient loading by diverting sewage from Lake Washington, a relatively small (87 km^2) lake, to the

Pacific Ocean. Recent projects, such as controlled flooding of the Grand Canyon in 1996 and ongoing restoration of the Kissimmee River in central Florida, highlight the importance of restoring hydrological patterns to channeled and dammed rivers, as well as restoring water quality.

Restoration ecology is becoming important in conservation biology, as well as in treating extremely damaged systems. There appear to be several reasons for this shift. First, habitat loss and degradation are the primary causes of species' endangerment and extinction. Because there is increasingly little pristine land, there is a growing need for conservation organizations to consider restoring degraded lands. Second, many governments now require ecological restoration as mitigation for development. In the United States, wetlands lost to development must be replaced by either creating new wetlands or improving the condition of existing wetlands. Finally, it is increasingly apparent that some kind of restoration is necessary even in protected natural areas. For example, many protected grasslands in North America have been invaded by plants that are not native. These plants may not be weeds in the traditional sense; many are flowers, naturalized from gardens, and others are European and Asian grasses, planted for pasture improvement. Controlling invasive plant and animal species is now a major component of restoration ecology.

Research areas. Because of the focus on restoring native, indigenous species, restoration ecologists must define criteria for success. In terrestrial systems, researchers often compare restoration projects to reference communities, undisturbed (or less disturbed) areas that are otherwise similar to the restoration site. An advance in this area is the recognition of temporal changes in plant communities. For example, fire is an integral component of many forest and grassland ecosystems; natural landscapes are a mosaic of sites that have burned at different times in the past, and plant and animal communities change with time after fire. In aquatic systems, indicator species have been identified by comparing biota in heavily disturbed and less disturbed systems. Restoration success requires return to the species composition of reference communities, with the caveat that in natural or restored systems species composition will change over time. For single-species projects, an alternative approach is to define aspects of the community to restore around the needs of the target species.

One central paradigm of restoration ecology is a bottom-up approach—the idea that if appropriate conditions for species at lower trophic levels (plants and other primary producers) are created, species at upper trophic levels (herbivores, predators, and decomposers) will follow. Nearly all restoration projects focus primarily on plants. However, in some cases plants cannot persist in isolation. Many grasses and trees require mycorrhizae, underground fungi that supply soil nutrients to plants. Herbivorous mammals can be important in shifting the balance of

Single-species versus ecosystem-level approaches in terrestrial restoration. (*a*) Restoration of prairie habitat for populations of the Fender's blue, an endangered butterfly in Oregon, is based upon the species' needs: native lupines, which are the required larval host plants; wildflowers, which provide nectar sources for adult butterflies; and controlling weeds, which outcompete the native plants. (*b*) Community and ecosystem factors (nutrient- and sediment-holding capacity, as well as overall biodiversity) are the focus of restoration at the Elkhorn Slough in California; experimental test plots (on the hillslope) show contrasting results of restoration with native grasses versus revegetation with nonnative annual plants. (*Courtesy of C. Schultz and F. Reil*)

competition between plant species. In lakes, presence or absence of key predatory fish can noticeably change algal abundance, species composition, and nutrient dynamics. At the present time, these appear to be exceptions to a general bottom-up rule.

Restoration ecology is a young science, and far more restoration projects are abandoned or fail than succeed. Few projects compare possible restoration methods, and few restoration treatments are experimentally replicated across sites. Therefore, little information exists about how consistently different methods work. Restoration attempts are seldom investigated sufficiently to determine the cause of failure. Current advances in statistical methodology may allow comparison across independently conducted projects to determine patterns of success and failure. These comparisons, combined with increased emphasis on quantitative restoration research and recordkeeping, should lead to increased success.

For background information *see* ECOLOGY in the McGraw-Hill Encyclopedia of Science & Technology.

Elizabeth Crone

Bibliography. A. D. Bradshaw, The reconstruction of ecosystems, *J. Appl. Ecol.*, 10:1–17, 1983; W. T. Edmonson, *The Uses of Ecology: Lake Washington and Beyond*, University of Washington Press, 1991; S. Packard and C. F. Mutel, *The Tallgrass Restoration Handbook*, Island Press, Washington, D.C., 1997; J. B. Zedler, Ecological issues in wetland mitigation: An introduction to the forum, *Ecol. Appl.*, 6:33–37, 1996.

Sensors

A new class of sensors is based on environmentally responsive hydrogels. These sensors are made by spatially modulating the chemical nature of gels or by depositing a different material onto certain gel surface areas selected with a mask. These novel gels can change their shape or surface pattern in response to environmental stimuli such as temperature and pH. The gels' controlled sensitivity to a variety of environmental stimuli offers new opportunities in the development of sensor technology.

Bi-gels. A bi-gel consists of two gels that are interlinked. Since each gel has different properties, depending on the temperature, pH, or solution, one gel can act upon the other. By changing conditions, the bi-gel can take on different forms and perform various functions.

The bi-gel strip (**Fig. 1***a*) is synthesized by making an ionic *N*-isopropylacrylamide (NIPA) gel slab, with the addition of polyacrylamide gel (PAAM) to form NIPA/PAAM interpenetrating networks. The end product is 2.0 to 3.0 mm (0.08 to 0.12 in.) thick with layered network structure, that is, a 0.8 to 1.8 mm (0.03 to 0.07 in.) thick PAAM network, and a 1.2 mm (0.05 in.) PAAM network interpenetrated by a NIPA network. The volume of the ionic NIPA gel shrinks drastically at temperatures higher than 37°C (98.6°F), whereas the volume of the PAAM gel does not. Similarly, the volume of the PAAM gel shrinks much more in acetone-water mixtures (>34 wt %) than does the NIPA gel. The fully swollen bi-gel strip at room temperature (22°C or 71.6°F) bends slightly toward the PAAM gel side. Upon increasing the temperature above 37°C, the NIPA gel shrinks drastically, whereas the PAAM gel is insensitive to the temperature change. As a result, when the bi-gel strip is heated uniformly, it gradually bends into an arc (Fig. 1*b* and *c*). Because the PAAM gel shrinks drastically in acetone-water mixtures and the NIPA gel does not, the bending in the opposite direction occurs as the bi-gel strip is immersed in an acetone-water solvent (Fig. 1*d* and *e*). The transition between the straight and the bending states is reversible.

The mechanism of this bi-gel bending is different from that produced by an electric field or by infrared laser heating. In those previous cases, the gradient of the electric potential field or the gradient of the temperature field has to be applied to trigger the bending of the gels, which are macroscopically homogeneous. In the case of bi-gels, the external environment is uniform, while the samples are internally modulated. Therefore, the sample bends directly in response to the environmental changes such as temperature and solvent composition without the need of having a gradient environment.

The difference of the thermal expansion coefficients between two gels can be much larger than that between two metals. For example, the difference of expansion coefficients for a typical bimetallic strip of brass and steel is about $7 \times 10^{-6}\ °C^{-1}$. However, for the PAAM-NIPA bi-gel, such difference can be as high as $0.6\ °C^{-1}$ [the ionic NIPA gel with 8 millimoles sodium acrylate shrinks to a half of its linear size per degree near the transition point at 36°C (96.8°C), while the PAAM gel has no volume change in this temperature range], which is five orders of magnitude greater. Therefore, bi-gels can be far more sensitive to their environment than bimetals.

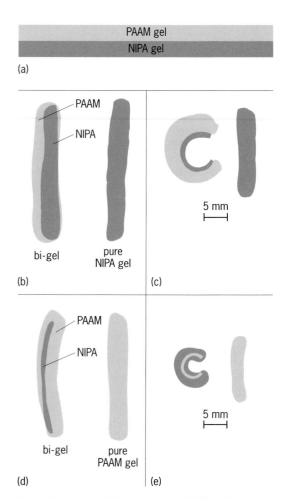

Fig. 1. Bi-gel strip. (*a*) Cross section; a PAAM gel is modulated by interpenetrating one side with NIPA gel. (*b*) Bending of the bi-gel in water at a temperature of 30.0°C [86.0°F] and (*c*) 37.8°C [100°F]. The pure NIPA gel is shown for reference. (*d*) Bending of the bi-gel in water-acetone mixture at an acetone concentration of 20 wt % and (*e*) 45 wt %. The pure PAAM gel is shown for reference.

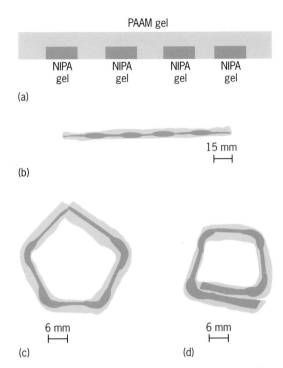

Fig. 2. Shape memory gel. (*a*) Cross section; a PAAM gel is modulated with NIPA gels at four locations. It was made by polymerizing the PAAM gel in the presence of four NIPA gel pieces on a glass plate. The separation between NIPA gel pieces is about 12 mm. (*b*) Gel shape at 22°C [71.6°F], (*c*) 39°C [102.2°F], and (*d*) 41°C [105.8°F].

Shape memory gels. When conditions such as temperature, pH, or solution change, a shape memory gel transforms into a predestined form. Similarly to the bi-gel, by changing forms, the shape memory gel can perform various functions.

A gel with a more complex modulated structure (**Fig. 2a**) can be prepared. The PAAM gel is modulated with the NIPA gel at four locations so that the gel can bend only at these modulated sites upon increasing the temperature. At room temperature, the modulated gel is straight (Fig. 2b). Upon an increase of temperature, the gel parts at modulated sites start bending, forming a pentagon (39°C or 102.2°F; Fig. 2c) and a quadrangle (41°C or 105.8°F; Fig. 2d). The change from pentagon to quadrangle is related to the higher shrinking ratio of NIPA at 41°C than that at 39°C. As a result, the bi-gel at 41°C bends more (about 90°, that is, forming a quadrangle) than one at 39°C (about 70°, that is, forming a pentagon). The transition between different shapes is reversible. A variety of shapes, including sinusoidal, cylindrical, and spiral, of the gels at various temperatures can be obtained by designing the modulation pattern of the system.

The novel gel functions obtained from the modulation method are based on the fact that the volumes of different gels are sensitive to different environmental aspects. The applications of the modulation method could be extensive. For example, a modulated bi-gel of a pH-sensitive gel (polyacrylic acid) with a temperature-sensitive gel (NIPA) should bend in a pH environment as well. Because a large amount of strain can be produced, the modulated gels may be used to make gel sensors that will change their shape in response to the change of environments.

Gels with shape memory properties could be administered through small incisions in vivo, subsequently regain their original shape, and then remain to perform a desired function. Compared with well-established shape memory materials such as shape memory alloys, shape memory gels have some advantageous properties such as very large deformation and availability to many external stimuli. The major controlling environment for shape memory alloys is limited to temperature, while shape memory gels are responsive not only to temperature but also to pH, salt solution, acetone concentration, electric field, light, and so on.

Thin gold film array on gel surface. A thin gold film array has been made on a gel surface by sputtering-deposition technique. NIPA gel slabs are made by free-radical polymerization. The NIPA gels are then dried slowly in a partially sealed glass container so that they shrink uniformly. A dry NIPA gel is covered by a grid and placed on the anode in the chamber of a sputter coater. During the coating operation, the chamber is pumped to a low-level vacuum with an argon gas pressure of 0.04 mbar (0.03 torr). A sputtering voltage of 2.5 kV is applied to ionize the argon. This process is continued for approximately 3.5 min, giving a gold thickness of approximately 53 nanometers. After coating, the mask is removed and the gel is immersed in water. The affinity between gold and the polymer gel may be due to the interaction between the lone pair electron of the amine group in the NIPA gel and the empty orbitals of gold, and to entrapment of gold in the gel.

For example, NIPA gel film in water at 30°C (86°F) is swollen. The surface array of the gel contains a dark area which is the gold-covered surface and a light area which is the bare gel surface. The spaces between the small gold squares in the gel surface can serve as slits for light diffraction. An increase from room temperature to 33.6°C (92.5°F) causes the NIPA gel to shrink due to the volume phase transition, and causes the array periodic constant to decrease. As a result, the periodic surface array and its diffraction pattern for the gel at 33.6°C are much different. The gel is thermally cycled at least seven times, and the temperature-induced change in the diffraction pattern is observed to be reversible. The periodic array produced on the gel surface may serve as a grating with a temperature-tunable grating constant.

The gel with periodic surface patterns can be used as a sensor. For example, the deformation of a gel under external force can be easily monitored. In one experiment, a gel with a periodic surface array is suspended in water and subjected to a uniaxial force at its bottom end along the y direction. Diffraction patterns of the gel are recorded for various loads. For zero force, the separation (Δl_x) between neighboring diffraction spots along the x direction is equal to that (Δl_y) along the y direction. The ratio $\Delta l_y / \Delta l_x$ decreases, however, as the external force increases. For a load of 5.85 g (0.21 oz), the value of $\Delta l_y / \Delta l_x$ is about 0.72.

This method incorporating a metal array to a hydrogel surface opens a new avenue to make microelectrode array devices which can greatly enhance sensor performance. The combination of microelectrode arrays and smart gels has potential applications for ion chromatography, for detecting enzyme activity, and for monitoring cell electrical activity.

Gels with engineered surface pattern. Synthesis of an engineered surface pattern is also accomplished by depositing one polymer gel (NIPA) on the surface of another polymer gel (PAAM). An interpenetrating polymer network of NIPA and PAAM is formed with the estimated depth of about 0.5 mm (0.02 in.).

Since both PAAM substrate and NIPA gel in the surface of the PAAM gel are transparent at room temperature, no surface structure can be observed. When this surface-engineered sample is warmed to 37°C (98.6°F), the NIPA-deposited areas become cloudy while the PAAM gel remains in the transparent state, so that a designed solid image appears. This imaging can be turned on or off by simply switching the temperature below or above the low critical solution temperature ($T_c = 34°C$ or 93.2°F) of the NIPA gel. The switching time is approximately 10 s.

Such a gel sensor display is responsive to the temperature. By changing the chemical composition of the gels, the displays can also be responsive to other external stimuli such as pH and ion concentration. The image of the gel display can be clearly seen at any angle in contrast to a liquid crystal display that requires head-on viewing.

For background information *see* BIOSENSOR; GEL; pH; POLYMER; POLYMERIC COMPOSITE; SHAPE MEMORY ALLOYS; SOL-GEL PROCESS in the McGraw-Hill Encyclopedia of Science & Technology.

Zhibing Hu

Bibliography. Z. Hu et al., Polymer gels with engineered environmentally-responsive surface patterns, *Nature*, 393:149–152, 1998; Z. Hu et al., Synthesis and application of modulated polymer gels, *Science*, 269:525–527, 1995; Y. Osada and S. B. Ross-Murphy, Intelligent gels, *Sci. Amer.*, 268:82, 1993; T. Tanaka, Gels, *Sci. Amer.*, 244:124–138, 1981.

Ship design

Recent advances in high-speed computing have benefited the field of computational fluid dynamics. Ship designers can now analyze by computer complex water- and air-flow behavior, previously poorly understood through expensive and time-consuming testing with scaled ship models. These improved computational fluid dynamics techniques are revolutionizing the ship design process and enabling better designs to be produced more quickly. This article provides an overview of the use of computational fluid dynamics in ship design; and then its use, along with theory and experiment, in design of sailing vessels for racing.

Computational Fluid Dynamics

Ship design is an evolutionary process where the designer largely relies on past experience. As in all design work, the final product is a compromise between conflicting demands, and in the hydrodynamics area the main trade-off for merchant ships is between large loading capacity and small power consumption. Important constraints are the length of the ship, which strongly affects the production cost, and the speed required to meet the intended schedule. Further, the ship has to be seaworthy and have adequate maneuvering capabilities.

The traditional tool available to the designer, apart from empirical data, is water basin experiments, where a scaled model of the ship is tested under self-propelled or towed conditions. This is an expensive and time-consuming process which needs to be carried out at the initial design stage. Since many other design problems cannot be solved without knowledge about the shape of the hull, time is especially critical at this stage. Therefore, there is great interest in another approach, taking advantage of the developments in computational fluid dynamics.

RANS methods and panel methods. All flows of interest in ship hydrodynamics are governed by the Navier-Stokes equations, which express the relation between the three components of velocity and the pressure. The computer power required for their solution is prohibitive, but a modification known as the Reynolds-averaged Navier-Stokes (RANS) equations is often quite useful. These equations must be solved with a suitable model for the turbulence.

To further reduce the computational effort, it may be assumed that the flow is frictionless, which is a good approximation if the immediate neighborhood of the hull is avoided. Under this assumption, the velocity may be represented by a potential governed by the Laplace equation. In practice the problem is often solved by taking advantage of the linearity of the Laplace equation and superimposing elementary solutions to the problem on panels on the hull and (in most methods) on the free surface. These panel methods are considerably faster than the more exact RANS methods and are much more used at present.

In the following, the use of the two methods is presented for each of the four main areas of ship hydrodynamics, namely resistance, propulsion, seakeeping, and maneuvering.

Resistance. Steady-state local flow and hull-generated waves are normally included in the first area. Here there has been a breakthrough, and many shipyards and consultants are using computational fluid dynamics to optimize their designs. By far the most common approach is to compute the wave pattern for different hull alternatives using a panel method and to systematically change the shape so as to minimize the generated waves (**Fig. 1**). By analyzing the wave pattern and other local flow quantities such as the pressure and velocities, the designer may spot weak points in the design. There is also a growing interest in the environmental effects of ship waves, particularly from high-speed ferries which often operate in restricted waters. To minimize the wash, automatic shape optimization has been applied (**Fig. 2**).

Due to limitations in computer capacity, most wave predictions in practical ship design are made using potential flow methods. Viscous effects are thus neglected, which may result in exaggerated stern waves. To compute the boundary-layer flow near the hull and the trailing wake, RANS methods are required. The local flow in these regions is important for the design of the propeller and for the positioning of appendages, such as antirolling devices, shaft brackets, and fins (**Fig. 3**).

Propulsion. Potential flow methods have been used extensively in propeller design for several decades. Complex blade shapes, unsteady flows, and certain types of cavitation on the blades may be handled in many of the methods available today. Although panel methods are increasingly popular, a simpler approach known as the vortex lattice method is still the most common one. The basic difference between the two methods is that the boundary condition is applied on the mean camber surface (midway between the pressure and suction sides) in the vortex lattice method but on the real surface in the panel method, which is thus more exact.

Several viscous effects are not accounted for in the potential flow methods. Apart from the obvious need to include friction in the thrust and torque calculation of the propeller, the boundary layer developing on the blades reduces the lift force that produces the thrust. The cavitation is also influenced, most importantly by changing the point of detachment of the cavity from the blade surface. To account for these effects, simple viscous methods have been applied on the blade surface with some success.

However, during the past decade several RANS methods have been applied to propulsor flows as well. The simplest approach is to model the propeller by a distribution of forces in the propeller disk. In this way, the interaction between the propeller and the wake flow behind the hull can be investigated. More advanced methods, where the blade flow is computed, have also been attempted. Most of this work is for the open water case, that is, where the propeller is operating in an undisturbed flow, but in 1998 the first predictions of the unsteady blade flow of a propeller operating behind a ship hull were presented. The practical use of RANS methods for propeller flows, as well as for other propulsors, is likely to increase substantially. At present, the methods may be regarded as research tools.

Seakeeping. The term seakeeping includes a number of important hull-wave interaction effects, including ship motions, hydrodynamic loads, added resistance, water impact, water on deck, capsizing, and safety. Numerical methods for computing motions and loads have been available for several decades,

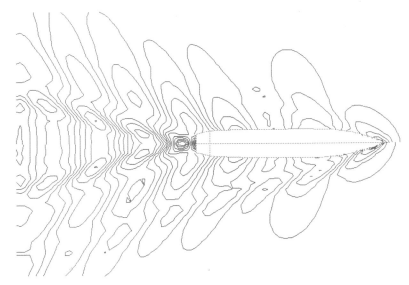

Fig. 1. Comparison of wave patterns from two different ship designs. Wave heights are represented by contour lines. The port side of one hull with a large bulb is shown in the lower half of the diagram, while the starboard side of the other hull with a small bulb is shown in the upper half. Detailed study of wave contours indicates that the lower hull produces the smaller waves.

and it was in these areas that practically useful results from numerical hydrodynamics were first obtained. One reason is that viscosity plays a relatively small role for most phenomena of interest. Methods based on the potential flow approximation thus have good prospects for success.

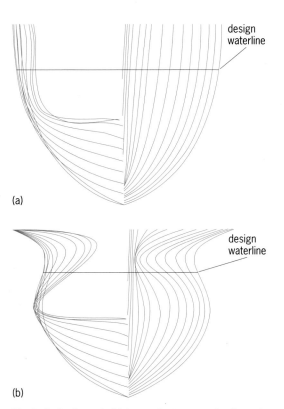

Fig. 2. Body plans of a high-speed catamaran. Sections of the forebody are shown to the right, and of the afterbody to the left. (*a*) Original hull. (*b*) Hull that has been optimized for a certain speed in shallow water. Essentially, hull volume has been moved downward beneath the waterline.

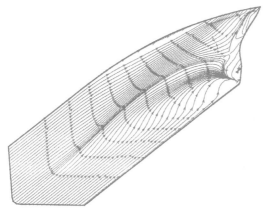

Fig. 3. Predicted flow close to the surface of a tanker hull (limiting streamlines). Afterbody, port side is shown. The pattern is quite complex, with the flow converging toward two lines, one pointing downward toward the end of the keel and one upward toward the extreme end of the hull. Behind the lines of convergence the flow goes essentially forward. This flow pattern should be avoided since it increases the resistance.

All early seakeeping methods were based on the assumption of slender hulls, and the flow problem was solved in a quasi-two-dimensional mode. More recently, three-dimensional panel methods similar to those described above have been developed, relaxing the requirement on slender hull shapes. In principle, the panel methods may be divided into two categories: frequency-domain and time-domain. The former may be obtained from a relatively straightforward generalization of the steady-state methods, and the results are obtained as a function of the frequency of the waves encountered by the ship. Frequency-domain methods are accurate and efficient for smaller ambient waves, but for steeper waves the time-domain approach is required. Here, the solution is integrated in time, and the motions and forces are obtained as a function of time. More computer effort is required in this case. The more complex phenomena, such as water impact (slamming), are not yet computed in the fully three-dimensional methods but in simplified two-dimensional models.

Although viscosity may often be neglected, it plays an important part in the damping of certain motions, particularly rolling. This effect may be increased by external devices, such as bilge keels, along the side and bottom of the hull. To account for this and for the friction between the hull and the water, viscous flow methods must be applied. Viscosity is also important in the flooding of compartments in damaged conditions and during capsizing. Mostly these effects are estimated using simpler methods, but recently Navier-Stokes methods have been applied also to the seakeeping problem. The first complete RANS solution for a ship moving in head seas was presented in 1998.

Maneuvering. The hydrodynamic coefficients required to simulate ship maneuvering are normally obtained from either semiempirical models or test data, while the use of computational fluid dynamics is less frequent. Potential flow methods of the same kind as those in seakeeping have been used to some

extent, but a major problem is that viscous effects are more important when the hull moves at an angle of attack, that is, slips sideways or rotates. Therefore the interest in RANS methods has increased, and several research studies have been presented since the mid-1990s. One remarkable achievement in the application of computational fluid dynamics to ship hydrodynamics is a recent project where the RANS equations are coupled with a six-degree-of-freedom rigid-body motion method and solved for a fully appended submarine with a rotating propeller.

Lars Larsson

Sailing Vessels

Design is a critical element in the creation of a sailing vessel that is intended for racing. Very small differences in the fluid forces on the vessel can have a profound effect. The most widely known sailboat competition is the America's Cup. A change in the drag of the water on an International America's Cup Class yacht of only 1% will alter the time for the vessel to sail the course by an amount ranging from 67 s in a 5-knot (9-kilometer-per-hour) breeze to 19 s in an 18-knot (33-kilometer-per-hour) breeze (**Fig. 4**). These times represent very substantial margins of victory or defeat. Therefore, all aids are brought to bear at the highest possible level of precision in the design of these vessels.

Computational fluid dynamics, in the complete sense, is the science of obtaining numerical (computer-based) solutions to the complete equations governing all aspects of the fluid flow, air, and water, around all the geometric elements of the yacht, including all the interactions between the elements. For example, the flow around the keel influences the flow around the hull. The term computational fluid dynamics usually connotes numerical, computer-based solutions of approximations to the fundamental Navier-Stokes equations for fluid flow, including the effects of fluid viscosity. As yet, there are no complete solutions including both the air and water flow. However, there has been work on solutions for the water flow that are based on the RANS procedure, discussed above, which uses an approximate average of the interrelated viscous and turbulence effects. In

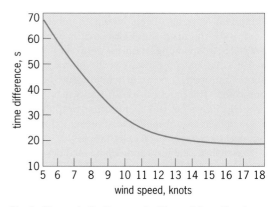

Fig. 4. Change in the time required for an International America's Cup Class yacht to sail the course when the hull resistance is changed by 1% at various wind speeds. 1 knot = 1.85 km/h.

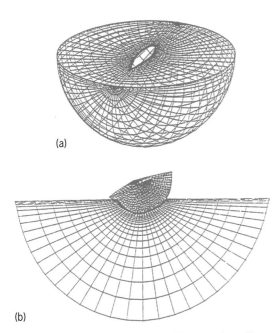

(a)

(b)

Fig. 5. Grid system for the hull of a sailing vessel used in a computational fluid dynamics calculation based on the Reynolds-averaged Navier-Stokes (RANS) procedure. (*a*) Three-dimensional view. (*b*) One cross section of the grid. (*From H. Miyata, Time-marching CFD simulation for moving boundary problems, 21st Symposium on Naval Hydrodynamics, Office of Naval Research, Trondheim, Norway, 1996*)

this method, the entire flow domain is divided into small grid elements, and the interaction between all the elements is computed (**Fig. 5**).

Balance equations. Although results using advanced computational fluid dynamics in a qualitative way often aid in the design process by showing trends among candidate designs, they do not permit solutions with force errors of less than a fraction of a percent, which are needed in the design of a racing vessel. The fundamental reason for this failure is that sufficiently accurate and tractable mathematical models have not been developed for the turbulent flow which invariably forms very close to the surface of the vessel and proceeds into the wake. Because of this difficulty, it is necessary to decompose the total

force into a sum of subsidiary forces and to treat each of these in the most accurate way that is applicable to that force.

A partial subdivision of the forces (and moments) and their names is conveniently given in a force diagram (**Fig. 6**). When a vessel is sailing at constant speed on a fixed course, corresponding hydrodynamic forces and aerodynamic forces are in equilibrium. For example, three balance equations are satisfied: (1) Aerodynamic forward force = hydrodynamic resistance. (2) Aerodynamic heel force = hydrodynamic heel force. (3) Aerodynamic heeling moment = hydromechanical righting moment.

There are three additional balance equations which have a smaller influence on the boat speed. These equations (balances of yaw moment, vertical force, and trim moment) are sometimes considered and sometimes not. For simplicity, only the three primary balances are considered here. For this approximation, when the boat is sailing, there are specific values of speed, heel angle, and leeway angle for which the balances occur and the three equations are true. An extremely important class of computer programs, called velocity prediction programs, which have been developed since the mid-1980s, contain mathematical models that approximate the six terms, three on each side of the equations, and computationally adjust the hypothetical speed, heel angle, and leeway angle such that the three balance conditions are met. Providing the speed, heel angle, and leeway angle to the user is essentially equivalent to providing the hypothetical performance of the vessel.

Since the forces and moments cannot be determined exactly by theoretical or numerical methods, the validity of a velocity prediction program is strongly dependent on the accuracy of the approximate mathematical models used for them. A brief description of the models will be given.

Aerodynamic forces. Numerical implementation of thin airfoil theory for sails has been in use since 1960. This theory predicts the lift, which is the force perpendicular to the relative airstream, and the induced drag, which is a portion of the force parallel to the

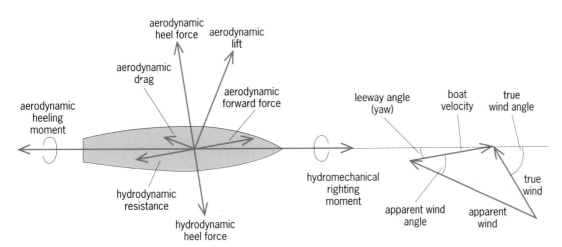

Fig. 6. Forces and moments on a sailing vessel.

airstream, as well as the moments associated with these forces. However, the theory neglects viscosity, so the surface friction drag and effects of the separation of flow from the sail surfaces are not included. To model complete sail forces, the computational results of the thin airfoil theory are augmented by an empirical formulation for the viscous effects. Very recently, thin airfoil theory has been coupled with viscous boundary-layer theory. The boundary layer is the thin region next to the sails where viscous forces are most important. The result is an improved numerical-theoretical prediction for sail forces. However, before these new methods can be used in a routine way, further development is needed to make them more robust when there is a substantial amount of flow separation on the sails. At present, they give good results when the amount of separated flow is small.

Hydrodynamic forces. The heel force, which is the side force measured perpendicular to the centerplane of the heeled vessel, can be reasonably well approximated by panel methods. As discussed above, these methods provide a computer-based implementation of ideal fluid theory including lifting surfaces. Even a symmetric hull produces a heel force when sailing in any direction other than downwind, because the hull moves with a leeway angle through the water. In other words, the direction of forward motion differs slightly from the direction that the vessel is pointing. The results of the theory that is usually employed are slightly in error because most of the theoretical implementations consider the water surface as flat, whereas in actuality the motion of the vessel causes waves on the surface. A practical finding is that the results of the usually employed lifting-surface theory can be used effectively for determining the difference in heel force and the associated dynamic heeling moment between two different designs, but are not reliable on an absolute basis.

Some computational implementations of the theory have included the dynamics of the air-water interface (the free surface). Although this approach provides a better estimate of the heel force, its most important contribution is the ability to compute the wave-making drag of the vessel. This is the component of the hydrodynamic drag force required to generate the ship waves made by any moving object at the free surface. The present methods are generally accurate enough to evaluate design trends. However, their inaccuracies prevent choosing with confidence one of two similar boats as having better performance.

The panel methods ignore the effects of viscosity, the most important of which are the skin friction drag and the influence of viscosity on the wake of the vessel and the wake of each appendage, such as the keel or rudder. Empirical methods have been developed to estimate the effect of viscosity on drag.

Physical modeling and testing. Due to uncertainties in completely numerical methods, experimental programs are conducted in which a scale model of the hull is towed in a tank over a wide range of speeds, yaw angles, and heel angles. The results of these experiments are then used to make a mathematical model of heel and forward forces, and of the heel moment, for use in the velocity prediction program.

The scaling process between a full-size boat and a smaller tank-test scale model includes scaling of the hydrodynamic forces. Scaling these forces requires specified ratios of full-scale speed to model-scale speed. The model test speed is chosen so that inertial and wave-making forces are properly scaled. This procedure is called Froude scaling. However, a different speed ratio would be required for proper scaling of the viscous forces, which is called Reynolds scaling. As a result, the model tests do not provide the correct viscous forces. To cope with this problem, empirical methods are used to subtract the viscous forces from total measured model-scale forces and to add full-scale viscous forces to the scaled-up inertial forces to get the best possible estimate of the full-scale total force. The error in this approximation is minimized by making the model as large as possible.

Integration of modeling processes. Numerical modeling is a useful tool in sailing vessel design and is essentially a requirement when designing racing vessels for which small variations in aerodynamic and hydrodynamic forces have a major impact on the success of the design process. The numerical modeling process is imperfect, largely because of the effects of turbulence and viscosity. Therefore, a certain amount of physical modeling and testing is required as well. Since the physical model is generally smaller than the full-scale vessel, there are uncertainties in viscous and turbulent forces in this process as well. From experience, it appears that the physical models are more robust and accurate than the numerical models, but careful integration of both processes is the best way to obtain results. Advances are expected in the areas of improved computational fluid dynamics and better inclusion of boundary-layer effects in numerical panel methods.

For background information *see* AERODYNAMIC FORCE; AIRFOIL; BOUNDARY-LAYER FLOW; CAVITATION; COMPUTATIONAL FLUID DYNAMICS; FLUID-FLOW PRINCIPLES; NAVIER-STOKES EQUATIONS; POTENTIAL FLOW; SHIP POWERING AND STEERING; TOWING TANK in the McGraw-Hill Encyclopedia of Science & Technology.   Jerome H. Milgram

Bibliography. L. Larsson, Scientific methods in yacht design, *Annu. Rev. Fluid Mech.*, 22:349–385, 1990; J. H. Milgram, Fluid mechanics for sailing vessel design, *Annu. Rev. Fluid Mech.*, 30:613–653, 1998; J. H. Milgram, Naval architecture technology used in winning the 1992 America's Cup match, *SNAME Trans.*, 101:399–436, 1993; M. Ohkusu (ed.), *Advances in Marine Hydrodynamics*, Computational Mechanics Publications, Southampton, 1996; P. van Oossanen, Predicting the speed of sailing yachts, *SNAME Trans.*, 101:337–397, 1993; *Proceedings of the 21st Symposium on Naval Hydrodynamics*, Trondheim, Norway, 1996, and *22d Symposium*, Washington, D.C., 1998; *Proceedings of the 22d International Towing Tank Conference*, Seoul, 1999.

Simulation

Simulation involves the use of the computer as an aid in modeling physical systems. One common method for thinking about a taxonomy of systems is organized around a dichotomy of continuous versus discrete. If a system variable is continuous, the interval between one value and another can be made arbitrarily small. Discrete variables denote a countable set of values with the intervals between the values often being irregular. Time, state, and event are three key variables of any simulation and can be either continuous or discrete. Time marks the temporal flow of a system, state reflects an attribute of an object embedded within the system, and event (as in physics) defines a state at a point in time. It is both convenient and practical to consider events as demarcating changes in state, although, strictly speaking, it is not essential that the state must change when an event exists.

The focus of this article is discrete-event simulation, where there are a countable number of events along a time axis, but where the value of state can be either continuous or discrete. An example of the continuous-state case is where a ball bounces on a surface. Events are defined as times when the ball strikes a boundary (that is, the ground or a wall). The state space (that is, position) is continuous. In contrast, a discrete-state case is found in digital logic where the state space is binary. There is also the situation where a system may have both continuous and discrete events; such simulations are often called hybrid or combined.

Phases. All forms of simulation, whether discrete-event or continuous, involve the phases of modeling, execution, and analysis. Modeling is the process of creating objects that describe the physical object or phenomenon under study. A simulation program must be generated from the model, and that program is then executed. In some cases, this generation is automatic, freeing the modeler from having to program. In other cases, a program is created using the model as reference. Analysis is an iterative process involving the use of the model and generated program output. An example is a study of an aircraft flight path to see if it follows an acceptable trajectory. Analytic questions are posed of the model, such as whether the simulated plane behaves correctly in response to simulated ground and pilot maneuvers, and the simulation is used to answer the questions.

Geometric and dynamic models. Modeling lies at the core of simulation. The sort of modeling involved with either discrete-event or continuous simulation is called dynamic modeling. This terminology is used because the objects from which the model is created define the behavior (or dynamics) of the system. Another type of model, involved in geometric modeling, defines the shape and structure of the system. Modeling defines the role that one object plays toward another in the same way that a scaled-down model plays the role of the full-size physical phenomenon to be studied by simulation. There-fore, modeling is a special relationship between two objects with one object literally modeling the other.

Modeling with events. The simulation of an aircraft may appear at first to involve a continuous movement through space and, therefore, to require a continuous model. At one level of abstraction, this is true; however, at a more abstract level, its behavior can be studied in regard to the plane passing key discrete points along its route and during its maneuvers while it is grounded at the airport. It is possible to regard the problem from a broader perspective and review all aspects of a flight. The flight involves not only actual plane behavior once airborne but also the plane's behavior while parked at the airport, while taxiing, and when landing. This set of behaviors can be broken up into discrete events: parking, taking-off, landing, and taxiing.

Events reflect distinct changes in state or phase. For the aircraft, the beginning or ending of activities such as taxiing and parking are events, whereas the activities themselves reflect states. A common method of modeling with events can be found in queueing models. An example of such a model is provided by the Orlando International Airport in Florida. A geometric model (that is, a two-dimensional map; **Fig. 1**) shows the structure of the airport, but it is also necessary to show the behavior of the planes. A queueing model emphasizes the discrete nature of planes waiting in line in the taxi area and then proceeding to wait in line in the parking area (**Fig. 2**). Individual airplanes are identified with tokens. The movement of a token into a queue of tokens represents a plane moving toward a queue of waiting planes. Each plane is waiting to use a resource (that is, a facility). Braces represent facilities, and rectangles inside each set of braces represent servers. In the above example, there are two facilities (for

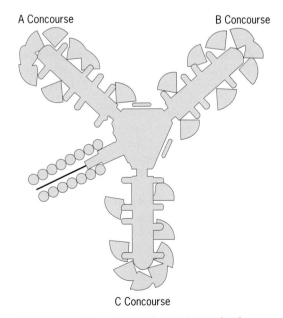

Fig. 1. Geometric model of the Orlando International Airport. (*Delta Airlines*)

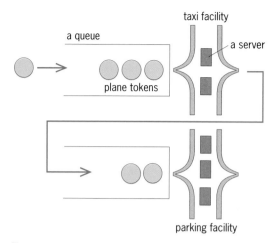

taxi facility

a queue

a server

plane tokens

parking facility

Fig. 2. Dynamic discrete-event model (queueing model) of the Orlando International Airport.

taxiing and parking), each of which represents a shared resource that each plane must use. There are two taxi lanes and three gates. As a token arrives in the queue, it checks to see if a server is free. If a server is available, the token obtains service by that facility and subsequently proceeds through the queueing network. The greater the number of servers per facility, the fewer planes must wait in a queue.

The queueing model involves time moving forward in discrete jumps since each server in a facility takes a certain amount of time for service. This time is frequently specified using a probability distribution peculiar to the application. After some study of service times for taxiing and parking, it might be found that taxiing involves a uniform probability distribution and that parking involves a normal distribution. Only a careful empirical study of the airport can determine which distributions are appropriate. As the discrete-event simulation proceeds forward in time, events denoting individual plane parking and taxiing activities are logged and provided to the simulation user, often in the form of a graph or graphical display. Even though the activity of taxiing involves continuous behavior, the system has been abstracted to be concerned with only the discrete events, and not with the lower-level positional data of each plane over time.

The queueing model abstracts away the actual geometry of the airport, airplanes, runways, and parking areas to focus specifically on the dynamics of taxiing and parking. This sort of abstraction is very common during the modeling process, whether the model reflects geometry or dynamics. Some parts of the system are removed or simplified so that the model delivers only those aspects of the system that are of primary interest to the simulation user.

Operation. The internal structures of all discrete-event simulation programs involve the key method of scheduling. Events are placed into a future event list sorted by event time, and then the next event is executed. The feedback loop, comprising event execution and scheduling, forms the crux of discrete-event simulation programs. For the plane example,

the arrival and departure of planes at resources form the set of possible events. An arrival event is scheduled, and then a routine associated specifically with this event is executed. The arrival routine involves a request for a resource such as a parking spot. The execution involves scheduling a departure event after some period of time has passed, indicating the time needed for accessing the requested resource.

Issues and choices. Discrete events correspond to discrete objects in a domain. Airplanes are discrete and come in distinct units. The presence of units is an indicator that discrete-event modeling and simulation is appropriate for the task at hand.

Other applications requiring discrete events include transportation systems, queues of manufactured parts, or waiting lines of people. If the questions to be asked of a system dictate a high-level answer involving the determination of statistics (the mean waiting time for planes, for example), then discrete-event simulation is appropriate. If, however, a question involves geometry (for example, whether a plane will fit within a space or collide with other planes), a continuum approach with a continuous set of events is more appropriate. In deciding on whether to implement a discrete-event approach, it is necessary to first determine what questions are to be asked of the system. Then the right level of abstraction needed to answer these questions must be chosen.

Software. Discrete-event simulation should thus be viewed as a technique used to model and execute the more abstract parts of systems that have lower-level continuous components. In this sense, the only difference between studying the discrete versus the continuous is based on level of granularity. Two types of software are available for modeling and simulating systems of events: generic and application-specific. A generic simulation package allows the user to simulate any type of system involving any number of resources, servers, and events. However, a user whose goal is only to model aircraft will want to visualize planes, runways, and terminals as icons or three-dimensional objects, with application-specific software. Some packages allow users to customize a generic package to suit a variety of application domains.

Hybrid systems. The future of discrete-event systems is in combined or hybrid systems. All physical systems have both discrete and continuous components. The challenge is to integrate these components into a whole so that the resulting multilevel model is a good representation of the physical system.

For background information *see* DISTRIBUTION (PROBABILITY); MODEL THEORY; QUEUEING THEORY; SIMULATION in the McGraw-Hill Encyclopedia of Science & Technology.   Paul A. Fishwick

Bibliography. J. Banks (ed.), *Handbook of Simulation: Principles, Methodology, Advances, Applications, and Practice,* John Wiley, 1998; P. A. Fishwick, *Simulation Model Design and Execution: Building Digital Worlds,* Prentice Hall, 1995; A. M. Law and

W. D. Kelton, *Simulation Modeling & Analysis*, McGraw-Hill, 1991; B. P. Zeigler, *Multifacetted Modeling and Discrete Event Simulation*, Academic Press, 1984.

Small computer system interface (SCSI)

The small computer system interface (SCSI, pronounced "scuzzy") provides a linkage between a computer central processor environment and peripheral devices such as disk drives and compact-disk–read-write (CD-RW) devices (**Fig. 1**). The SCSI bus connects the peripheral devices to the host bus through a host bus adapter. Expander circuits are used to break up the bus into segments that have better electrical transmission properties.

This article describes the evolution and capabilities of SCSI in terms of peak speed and configuration enhancements. During the 1990s, SCSI increased its peak speed by 32 times and extended its reach from a few meters to as far as 100 m (330 ft) with hublike architectures. This mainstay storage interconnect technology has continued to evolve without severing the interoperability with prior generations. Except for very long distances, SCSI can do nearly everything required for a storage interconnect across a very broad range of device types: disks (fixed and removable), tapes, compact-disk–read-only memories (CD-ROMs), digital video disks, scanners, printers, hubs, switches, and other specialized devices.

SCSI consists of up to 16 data lines, 2 parity lines, 9 control lines, and a few power and ground lines. It is a parallel multidrop bus, meaning that these lines connect to every attached device simultaneously and have a terminator on each end. During the time in a data transaction when the SCSI devices are determining who will talk to whom, the data lines are used not for transmitting data but as an expanded set of control lines. In this mode (or phase), each data line serves as a device identification line. The maximum number of devices on a bus is therefore 16 since there are only 16 independent data lines. After it is agreed which two devices on the bus will communicate, the data lines revert to carrying data and each data line carries a single bit stream. SCSI is said to be in a data phase when the data lines are being used to carry data.

Peak speed. The peak speed of SCSI is determined by the data-phase bandwidth and the efficiency of the implementation of the host bus adapter, such as the PCI (peripheral component interface), which joins the host bus to SCSI. As with any system, the workload must be able to take advantage of the peak speed if end-user benefit is to be seen from increases in peak data rates. When workloads demand that the disk or device spend milliseconds to retrieve the requested data from the storage media to the SCSI port (as when doing a seek operation), the peak speed does not influence the performance of the system. In other words, the system performance does not benefit from increases in peak data-phase speed unless the workload is bandwidth-limited. There are many real workloads that are bandwidth-limited, and the entire system benefits if the data can be collected into a relatively large chunk before transmitting the data across the SCSI bus.

The SCSI data-phase bandwidth is directly related to the number of parallel lines used for data transmission and the time between adjacent data bits on the same line. Slow synchronous SCSI delivers a maximum of 5 megabytes per second in a 9-bit-wide path (8 data bits and 1 parity bit) with 5 megabits per second per line. (A data byte equals 8 bits of data.) The most aggressive version presently specified is 160 megabytes per second in an 18-bit path (16 data bits and 2 parity bits) operating at 80 megabits per second per line. Thus, there has beeen a 32-fold increase in total bandwidth, which was achieved in five stepped twofold increases during the 1990s. The doubling took place roughly every 2 years, and there is no reason to believe that this pattern will not continue for several more generations. The methods to produce the doubling at each step may change, however.

The synchronous content derives from a defined timing relationship between the data bits transmitted on the data lines and the data latching signal transmitted on a separate control line. This latching signal is used to commit the data bits to the receiver. There is no synchronization between the transmit clock and the receive clock.

Three methods have been used to achieve the doubling required at each step: doubling the number of bits in the parallel data path, increasing both the frequency of the data and latching signals, and using both threshold crossings per cycle instead of one for the latching signals. The base frequency has doubled three times. These methods give rise to several

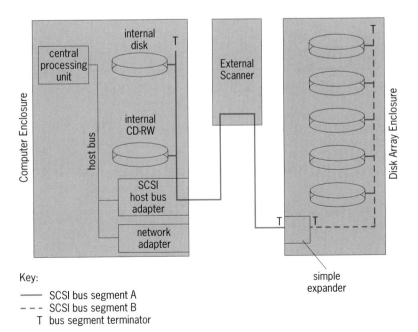

Key:
—— SCSI bus segment A
- - - SCSI bus segment B
T bus segment terminator

Fig. 1. Small computer system interface (SCSI) in a typical computer system. Both segment A and segment B are part of the same SCSI bus.

Relationships between data path width, signal frequencies, and latching methods of SCSI protocols

Peak delivered data rate, megabytes per second	Path width, bits	Latching signal frequency, MHz	Number of latching events per latching cycle	Peak data signal frequency, MHz	Designation
5	8	5	1	2.5	Slow narrow
10	16	5	1	2.5	Slow wide
10	8	10	1	5	Fast (10) narrow
20	16	10	1	5	Fast (10) wide
20	8	20	1	10	Fast 20 (Ultra SCSI) narrow
40	16	20	1	10	Fast 20 (Ultra SCSI) wide
40	8	40	1	20	Fast 40 (Ultra2 SCSI) narrow* only
80	16	40	1	20	Fast 40 (Ultra2 SCSI) wide*
160	16	40	2	40	Fast 80 (Ultra3 SCSI) wide*

*Low voltage differential (LVD) transmission only.

options for implementing these different rates (see **table**).

Although 36-bit-wide SCSI (32 data bits and 4 parity bits divided between two separate parallel cables) is defined in the standards, it is virtually not used in practice and is not expected to be described in future generations of SCSI.

Three types of transmission are used in SCSI: single ended (SE), high voltage differential (HVD), and low voltage differential (LVD) for the highest speeds.

Configuration enhancements. Every device attached to a multidrop bus disturbs the signals just by its presence. When the signals pass the attached device, they see an electrical path into the device and another path continuing down the bus path. This fork in the electrical road causes some of the energy in the signal to be temporarily diverted into the attached device, disrupting the signal that continues down the bus. Also, as the signal propagates down the bus, it experiences loss of high-frequency content, loss of amplitude, and loss of precision in the timing relationship with its partner data bits because the propagation speed is not exactly the same down every data line.

SCSI physical configuration management is the science of working with the signal disturbances and bus parameters so that both usable signals and usable configurations are possible. One key concept recently introduced into SCSI configuration management is the breaking up of a single bus into smaller pieces while still maintaining the logical bit transmission patterns between the pieces. If this can be done, a huge expansion of the extent and complexity of the possible configurations will result without the need for any software changes. A class of active circuits called expanders has evolved to provide this logically invisible coupling between different SCSI buses. When the total SCSI bus, or SCSI domain, is separated using expanders, the resulting bus fragments are called bus segments.

With separation of the SCSI domain into bus segments comes the ability to build complex domains that use segments with no devices attached except at the very ends (point-to-point), the ability to use different transmission types for different segments in the same domain, and the ability to create networklike hubs. These features are made possible by using expanders that are completely invisible to the SCSI software. While they are sometimes called smart wires, they are properly known as simple expanders.

A general configuration using simple SCSI expanders consists of a fully populated SCSI domain, where every expander has only one device attached.

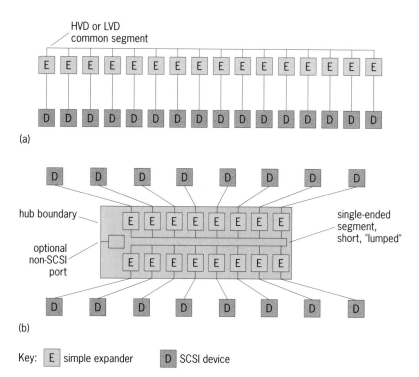

Fig. 2. Generalized SCSI configuration using simple expanders. (*a*) Distributed architecture using simple expanders (E) and devices (D). Common segment is up to 25 m (82 ft) long with high voltage differential (HVD). (*b*) Hub architecture using same elements.

Key: E simple expander   D SCSI device

In the simplest physically distributed architecture, the expanders are placed along a common segment (**Fig. 2***a*). By folding the common segment into a single enclosure, a centralized wiring scheme, or hub, becomes possible (Fig. 2*b*). All kinds of special configurations can be created by reducing the number of expanders and adding more than one device to an expander segment. As long as all segments in the domain operate well and the total number of devices does not exceed 16, any combination is possible. Each expander may use a different transmission type on the device side but must use the same type on the common segment.

There is a simple set of rules for configuring systems using simple expanders. Simple expanders are available for a variety of transmission types and speeds. They are exceptionally useful as isolation elements for disk arrays that attach to a host computer through an external cable (Fig. 1).

Another general class of expanders comprises bridging expanders, which have at least two SCSI ports (each with SCSI identifications). Bridging expanders are capable of expanding the configurations for parallel SCSI in two ways: using many more devices in a single SCSI domain, and multiplexing different parts of the domain so that performance can be distributed between different segments. Simple and bridging expanders can be cascaded to create very complex configurations.

For background information *see* COMPUTER PERIPHERAL DEVICES; COMPUTER STORAGE TECHNOLOGY; COMPUTER SYSTEMS ARCHITECTURE in the McGraw-Hill Encyclopedia of Science & Technology.

William E. Ham

Bibliography. W. E. Ham, Recent advances in basic physical technology for parallel SCSI: UltraSCSI, expanders, interconnect, and hot plugging, *Digital Tech. J.*, 9(3):6–31, 1998; P. M. Ridge and D. Deming, *The Book of SCSI: A Guide for Adventurers*, No Starch Press, 1995; F. Schmidt, *The SCSI Bus and IDE Interface: Protocols, Applications and Programming*, 2d ed., Addison-Wesley, 1997; *SCSI Enhanced Parallel Interface*, EPI Project T10/1143DT, Global Engineering (15 Inverness Way East, Englewood, CO 80112-5704), 1998; *SCSI Parallel Interface-2*, SPI-2 Project T10/1142D, Global Engineering, 1998.

Smart structures and materials

Fundamental research on and development of smart structures and materials have shown great potential for enhancing the functionality, serviceability, and life-span of civil, aerospace, and mechanical infrastructure systems. This could contribute significantly to the improvement of productivity and quality of life. Moreover, the recent explosive growth in computer power and connectivity provides unprecedented opportunities for rapid and efficient access to enormous amounts of knowledge and data, including real-time data for monitoring and control of smart structures, and for studying vastly more complex systems than was hitherto possible.

Smart materials and intelligent structures mimic biological systems in that they possess intelligence, having their own sensors (sense organs), processor (brain), and actuators (muscles). These sensors include optical fibers, corrosion sensors, environmental sensors, and sensing particles. Actuators include shape memory alloys, hydraulic systems, and piezoelectric ceramic polymer composites. The processor or control aspects are based on microchip, computer software, and hardware systems.

Researchers from diverse disciplines have developed smart or intelligent structures that can monitor their own condition, detect impending failure, control damage, and adapt to changing environments. The potential applications of such systems are abundant—for example, smart aircraft skin embedded with fiber-optic sensors to detect structural flaws; bridges with sensing and actuating elements to counter violent vibrations; tiny flying micro-electromechanical systems (MEMS) with remote control for surveying and rescue missions; and stealth submarine vehicles with swimming "muscles" made of special polymers. Such a multidisciplinary research front, represented by material scientists, physicists, chemists, biologists, and engineers, has created a new entity called smart structures/materials, generally by combining sensing, processing, and actuating elements integrated with conventional structural materials such as steel, concrete, or composites. Some of these structures/materials currently being researched or in use are piezoelectric composites, which convert electric current to (or from) mechanical forces; electro-rheological (ER) fluids, which can change from liquid to solid (or the reverse) in an electric field, altering basic material properties dramatically; and shape memory alloys (SMA), which can generate force through changing the temperature across a transition state. *See* MICRO-ELECTROMECHANICAL SYSTEMS (MEMS).

With advanced computing and new developments in material sciences, researchers can now characterize processes and design and manufacture materials with desirable performance and properties. One challenge is to model short-term microscale material behavior, through mesoscale and macroscale behavior, into long-term structural systems performance (see **table**). Accelerated tests to simulate various environmental forces are needed. Supercomputers or workstations used in parallel are useful tools to solve this scaling problem by taking into account the large number of variables and unknowns to project microbehavior into infrastructure systems performance, and to model realistically short-term test results into long-term life-cycle behavior.

Research. Another challenge is to achieve optimal performance of the total system rather than just in the individual components. Among the topics requiring study are energy-absorbing and variable dampening properties, as well as structures/materials having a stiffness that varies with changes in stress, temperature, or acceleration. Among the characteristics

Scales in materials and structures			
Smart materials		Intelligent structures	Infrastructure (smart systems)
nano/micro-level	*meso-level*	*macro-level*	*systems integration*
molecular scale	micrometers	meters	up to kilometer scale
micromechanics	mesomechanics	beams	bridge systems
nanotechnology	interfacial	columns	lifelines

sought in smart structures/materials are self-healing for cracks and in-situ repair of damage to structures such as bridges and water systems so that their useful life can be significantly extended. There is the associated problem of simply being able to detect or predict when repair is needed and when it has been satisfactorily accomplished. Through sensing, feedback, control, and use of the measured structural response, the structure (such as the Space Station) adapts its dynamic characteristics to meet the performance objectives at any instant. A futuristic smart bridge system incorporates some new concepts, including wireless sensors, optical fiber sensors, data acquisition and processing systems, advanced composite materials, structural controls, dampers, and geothermal energy for bridge deck deicing (**Fig. 1**).

Applications. The many new applications for smart materials/structures include the following.

Self-healing concrete. In research by grantees of the National Science Foundation, hollow fibers filled with crack-sealing material are placed into the concrete. If cracking occurs, the fibers break and release the sealant.

Smart paints. At a National Science Foundation Engineering Research Center, researchers from Lehigh University developed smart paints. These paints release red dye, contained in capsules, when the un-derlying structure cracks. The red dye highlights the cracks for easy detection.

Optical fibers. Optical fibers which change in level of light transmission during stress are useful sensors. They can be embedded in concrete or attached to existing structures. Researchers at Rutgers University studied optical fiber sensor systems for on-line and real-time monitoring of critical components of structural systems (such as bridges) for warning of imminent failure. Others, at Brown University and the University of Rhode Island, investigated the fundamentals and dynamics of embedded optical fibers in concrete. Japanese researchers developed glass and carbon fiber–reinforced concrete, which provides the stress data by measuring the changes in electrical resistance in the carbon fibers. Fiber-optic sensors in bridges are being developed at New Mexico State University. Sensing fiber-optic cables strung under the bridge with epoxy will communicate stresses by sending light beams down the cable at regular intervals, and the bending of the light beams will be measured. These gauges can also be used to monitor general traffic patterns. The sensors serve as data collectors as well as wireless transmitters.

Electro- and magneto-rheological fluids. Researchers at the University of California, Berkeley, recently completed a study of the application of electro-rheological

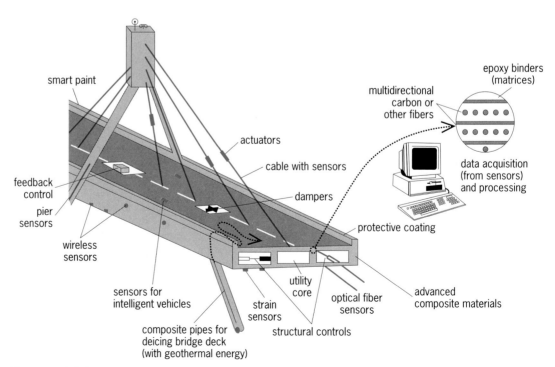

Fig. 1. Futuristic intelligent bridge system.

fluids for the vibration control of structures. These fluids stiffen very rapidly (changing elastic and damping properties) in an electric field. Similar controls were achieved under magnetic fields on magneto-rheological fluids at Notre Dame University.

Shape memory alloys. Shape memory alloys have many applications. For example, a French sculptor created a novelty statue of a skier using shape memory alloys. Every morning the statue stands up, then in the afternoon when the temperature is higher it bends into a downhill skiing position. Shape memory alloy frames for eyeglasses always spring back to original position, even if someone sits on them. In the medical area, shape memory alloy tubes triggered by body temperature have been used to open up the urinary tract blocked by an enlarged prostate. Researchers at the University of Texas, Virginia Tech, and the Massachusetts Institute of Technology are studying shape memory alloys which have a wide range of applications, including biosensors. *See* BIOSENSOR.

Superelastic microalloys. Surface superelastic microalloys are being used as sensors and microactuators (Michigan State), and elsewhere magnetostrictive active vibration control is being researched (Iowa State, Virginia Tech).

Semiactive vibration absorbers. Researchers at the University of Oklahoma have developed a smart microcontroller coupled with hydraulic systems that reduces large vibration amplitudes over 50% that are produced by heavy trucks passing over a highway bridge. The result is 15% more load capacity and extension of bridge life by 20 years.

Fig. 2. Tubular transport system for goods. (*Courtesy of William Vandersteel*)

Actuator systems. Researchers at the University of Toledo have developed actuator systems for low- and high-frequency macro motion. For example, motions of a few centimeters can be achieved by smart material actuator systems. Potential configurations made of piezoceramic and electrostrictive materials for providing larger motions at reasonable force levels are being evaluated.

Structural monitoring systems. A structural integrity monitoring system at a San Diego sports stadium

(a)

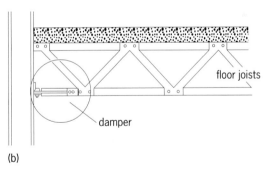

(b)

(c)

Fig. 3. Dampers, (*a*) New York's World Trade Center, fitted with dampers. (*b*) Placement. (*c*) Damper. (*3M Company*)

includes 17 sensors constantly measuring peak strains and one accelerometer measuring forces, installed at three locations in the stadium. It is wirelessly connected to the central communication module equipped with cellular telephones. The purpose is to detect and monitor any accumulated damages due to earthquakes and winds as well as to other causes such as dynamic live loads associated with large-scale rock concerts. At the University of South Carolina a system is being developed that includes distributed monitoring of structures, active buckling control of structures, and active damage control of composites. Micrometer-size magnetostrictive particles are mixed with the composite for monitoring. Low-velocity-impact-damage resistance of composites can be improved by hybridizing them with shape memory alloys.

Intelligent transport systems. With wireless communication and intelligent controls, automobile can travel at closer intervals in high speed. Researchers at the University of California, Berkeley, demonstrated that the traffic volume can be increased by a factor of 4 with this system, equivalent to building four times more highways. To reduce traffic congestion on interstate highways and in cities, tubular transport of goods has been proposed. The method utilizes linear induction propulsion, whereby an electromagnetic thrust is induced in each goods-carrying capsule as it passes over magnetic induction coils, freeing highways and streets of trucks (**Fig. 2**). *See* INTELLIGENT HIGHWAY TRANSPORTATION SYSTEM.

Viscoelectric dampers. To diminish swaying due to high wind, the World Trade Center twin towers in New York City have been fitted with 1000 viscoelastic dampers at the bottom of floor joists (**Fig. 3**). The dampers, functioning like shock absorbers in cars, lessen the vibration and swaying of the building during windy days, providing a better working environment.

Disclaimer. This paper reflects the personal views of the author, not necessarily those of the National Science Foundation. Input and additions by his NSF grantees and colleagues, including Drs. C. S. Hartley, S. Saigal, S. C. Liu, O. W. Dillon, C. A. Rogers, W. N. Patten, T. T. Soong, P. Grayson, and W. Vandersteel are gratefully acknowledged.

For background information *see* FIBER-OPTIC SENSOR; HYDRAULIC ACTUATOR; MICROCOMPUTER; MICROSENSOR; OPTICAL COMMUNICATIONS; SHAPE MEMORY ALLOYS; TRAFFIC-CONTROL SYSTEMS; VIBRATION in the McGraw-Hill Encyclopedia of Science & Technology.   Ken P. Chong

Bibliography. K. P. Chong et al., Engineering research in composite and smart structures, *Composites Eng.*, 4(8):829–852, 1994; K. P. Chong et al. (eds.), *Intelligent Structures*, Elsevier, 1990; C. A. Rogers and R. C. Rogers (eds.), *Recent Advances in Adaptive and Sensory Materials and Their Applications*, Technomic Publishing, Lancaster, Pennsylvania, 1992; H. S. Tzou and G. L. Anderson (eds.), *Intelligent Structural Systems*, Kluwer, 1992; E. Udd (ed.), *Fiber Optic Smart Structures*, John Wiley, 1995.

Soil chemistry

The natural cycling of carbon and nitrogen between organic and inorganic compounds provides the basis for healthy ecosystem function. The uptake of atmospheric carbon dioxide (CO_2) by vegetation begins the carbon cycle (as in food webs), and mineralization of this fixed carbon back to CO_2, primarily by soil microorganisms, completes the cycle. Locally the balance between fixation and mineralization is a major control on the quality and sustainability of an ecosystem, while globally this cycling impacts climate by affecting the concentration of CO_2, an important greenhouse gas, in the atmosphere. Nitrogen is the most common nutrient limiting the fixation of carbon by terrestrial vegetation, and as a result nitrogen compounds are widely used in fertilizers (for example, urea and ammonium nitrate). Nitrogen can be found in compounds considered atmospheric pollutants resulting from the combustion of fossil fuels, and the deposition of these nitrogen compounds (for example, nitrogen oxides, NO_x) through acid rain or snow is significant. Unlike carbon, which is returned or recycled to the atmosphere as CO_2, mineralized nitrogen often is lost along hydrologic flow paths as precipitation moves through soil to surface water. Hydrologic nitrogen export from soils with high nitrogen inputs (either as fertilizer or through atmospheric deposition) has overloaded the ability of many natural ecosystems to process or cycle this nitrogen, resulting in degraded streams, lakes, and forests.

Seasonally snow-covered soils in high-elevation catchments of western North America provide an opportunity to examine the effects of variable snow cover on the natural cycling of carbon and to identify how changes in climate may affect the soil-atmosphere CO_2 balance. These areas also are the primary source of water for both agricultural and urban uses at lower elevations, and are subject to degradation by increasing levels of atmospheric nitrogen deposition. Therefore, research on soil carbon and nitrogen cycling improves understanding of both the controls on regional water quality and the potential effects of global climate change.

Physical environment. High-elevation ecosystems in the Rocky Mountains lie between 2500 and 4400 m (8200 and 14,400 ft). Orographic lifting of regional air masses significantly increases precipitation in these environments relative to surrounding lower elevations. Soils are typically young and relatively shallow compared to lower-elevation sites at similar latitudes. The dominant climatic characteristic of these ecosystems is the long period of snow cover (6 to 9 months), with the redeposition of snow by wind characterizing the pattern of snow distribution across the landscape. The length of the growing season is strongly constrained by climate, primarily mediated by the timing of accumulation and melt of seasonal snow cover but also by occasional freeze events and snowfall during the growing season. Soils often freeze in late autumn before a consistent snow

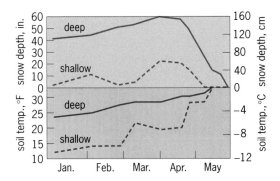

Fig. 1. Effect of snow depth on soil temperature under the snow. Deep snowpacks result in much warmer soil temperatures.

cover develops, and gradually warm under the snowpack during the winter due to a bottleneck in the transfer of geothermal energy under the insulating snow cover. Because solutes in soil depress the freezing point of water, soils thaw when temperatures are between 0 and $-6°C$ (32 and 21°F).

The depth and timing of snowpack accumulation is the primary control on soil temperature and the presence of thawed soil during the winter. Continuous snow cover insulates the soil surface from much colder air temperatures, while shallow, discontinuous snow cover allows the soil to continue to cool (**Fig. 1**). The depth necessary to insulate the soil varies based on snow density and air temperature but typically ranges between 30 and 40 cm (12 and 16 in.). The earlier this snowpack depth is reached, the warmer the minimum soil temperature at the site. The shallow snow site demonstrates this warming with an approximately 30-cm (12-in.) snowpack in January and February, followed by a cooling in March as the depth of snowpack decreased. While soil under the deeper snowpack was thawed by January, soil at the shallow site did not thaw until late April shortly before melt.

Soil cycling under snow. Thawed soil under snow provides a unique environment for soil biological activity to affect carbon and nitrogen cycling. The efflux of CO_2 through the seasonal snowpack is the most visible result of this activity and provides a method to quantitatively evaluate the effects of climate change on these systems. Because this loss is the direct result of heterotrophic activity respiring organic carbon in the snow-covered soil, the magnitude of the flux depends on soil temperatures being warm enough to allow unfrozen water to be available for use by soil organisms. For example, during a heavy snow year, fluxes began when soil thawed in midwinter, and continued throughout the remainder of the snow-covered season (**Fig. 2**). During the following year there was less snow, soil remained frozen throughout much of the winter, and CO_2 fluxes from the same area did not begin until very late in the winter. The result of this relatively small change in seasonal snow cover was that approximately 25 times more carbon was lost from the system during the winter with heavy snow than in the light snow year.

The timing of snowpack development and the presence of thawed soil also affect the cycling of nitrogen within the soil environment under snow. In contrast to the pattern observed with carbon, however, increasing snow depth decreases the loss of nitrogen from the system. The higher CO_2 fluxes observed under deep, continuous snow cover result from growing microbial communities in the soil. These growing communities require nitrogen for the synthesis of proteins, resulting in a strong, biological sink for nitrogen in these soils. This microbial nitrogen sink results in lower levels of nitrogen available for export. This inverse relationship can be observed in the field by using winter soil CO_2 flux as an index of microbial activity and nitrogen leachate (**Fig. 3**).

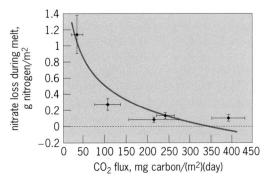

Fig. 3. Relationship between winter CO_2 efflux, an index of soil microbial activity, and nitrogen (N) loss during melt. Nitrogen loss from soils during snowmelt is much lower when soil microbial activity is high.

The nitrogen stored in microbial biomass subsequently is transferred to vegetation as sites become snow free. While the mechanism for this transfer remains unknown, a possible explanation is that freeze-and-thaw events under a waning snowpack trigger the release of nitrogen from the microbial biomass and this nitrogen is subsequently used by vegetation at the beginning of the growing season. The amount of nitrogen transferred at this time may be as much as 75% of the annual vegetation uptake.

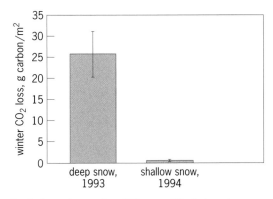

Fig. 2. Overwinter carbon (C) loss as CO_2 during a heavy snow year, and from the same sites during a light snow year. Winter CO_2 efflux was much higher when the soil was insulated under the deeper snowpack.

Effects of climate change. The magnitude of future increases in global temperature is a matter of debate, but any increase in energy retention by the atmosphere is likely to result in an increase in water vapor within the atmosphere. Because high-elevation systems scavenge water vapor as precipitation, any global warming scenario is likely to result in increased precipitation, primarily as snow. For example, the annual precipitation rate has increased 14 mm (0.6 in.) per year at a site in central Colorado over the last 30 years, possibly resulting from a warming of the plains east of the site.

If this trend in precipitation continues, the increased duration of snow cover may result in increased carbon mineralization and winter CO_2 efflux. If the increased duration of snow cover is sufficient to shorten the growing season and reduce carbon fixation by growing vegetation, the ecosystems could switch from net sinks to net sources of atmospheric CO_2. In contrast, the increase is snow cover may increase ecosystem nitrogen retention, which could increase vegetation growth and carbon fixation and have a beneficial effect on surface water quality by reducing the export of nitrogen. The inability to predict the effects of climate change on these relatively simple systems underscores the need for research designed to improve the understanding of the basic structure and function of ecosystems.

For background information *see* ACID RAIN; BIOGEOCHEMISTRY; CARBON; CARBON DIOXIDE; CARBON-NITROGEN-OXYGEN CYCLES; FERTILIZER; NITROGEN; NITROGEN FIXATION; SOIL; SOIL CHEMISTRY in the McGraw-Hill Encyclopedia of Science & Technology.

Paul O. Brooks

Bibliography. P. D. Brooks, S. K. Schmidt, and M. W. Williams, Winter production of CO_2 and N_2O from alpine tundra: Environmental controls and relationship to inter-system C and N fluxes, *Oecologia*, 110: 403–413, 1997; R. A. Sommerfeld, A. R. Mosier, and R. C. Musselman, CO_2, CH_4, and N_2O flux through a Wyoming snowpack, *Nature*, 361:140–143, 1993.

Soil remediation

The use of vegetation for environmental applications can be advantageous. For example, plants are used to stabilize soil and lessen erosion. Grass planted along waterways used in agriculture acts as a filter, collecting sediment and enhancing water quality. A relatively new use of vegetation is for the removal of harmful contaminants in the soil. This article discusses how plants are used to remediate soil (phytoremediation) containing organic contaminants and radioactive isotopes.

Organic Contaminants

The root zone of plants (rhizosphere) contains large populations of microorganisms. These microorganisms biodegrade organic contaminants to utilize the carbon and energy. Most of these microbial transformations occur in the soil and water external to the plant roots. Some organic contaminants are transformed to harmless by-products within plants. This plant process may be viewed as a solar-driven pump-and-treat system, which moves contaminants to the rhizosphere and helps to contain them on site. Thus, vegetation provides an environment for biodegradation of organic contaminants as well as a pumping system to move contaminated water through the rhizosphere and plant.

Evapotranspiration. Photosynthesis requires that plants use large quantities of water for growth. For example, a large tree can use 300 gallons (1200 liters) in one day. The usage rate depends upon the availability of water in the soil, humidity, temperature, wind speed, and solar flux. Alfalfa (*Medicago sativa*) and poplar trees (*Populus*) have deep roots allowing water to be extracted from depths of 20 ft (6 m) or more. Plants that use large quantities of water are important when vegetation is used to limit the extent of a plume of contamination by drawing water from the saturated zone below the water table into the rhizosphere. In the root zone, microorganisms remove some organic contaminants from the water before it enters the plant. After entering the plant, most of this water is discharged as water vapor into the atmosphere (evapotranspiration). Simulation studies based on prior temperature, humidity, and precipitation data for the site can be used to estimate the vegetated area required to prevent contaminated water from moving away from the site. If possible, the site should be designed so that surface water from significant precipitation events drains from the area.

Vegetation is used to contain contaminant plumes at many sites. An example is maintenance of sufficient vegetation to evapotranspire all leachate which flows into the area from a landfill. If trees lower the water table relative to the surrounding region, ground water will flow into the vegetated area and the leachate plume will be contained.

Biodegradation. The microorganisms in the rhizosphere feed on dead roots and root exudates as well as any organic contaminants which can be utilized as carbon and energy sources. These microbial processes account for the mineralization and transformation of many organic contaminants. The major end products are carbon dioxide and water, which are used by the plants. Decaying plant organic matter and root exudates sustain the microbial population and account for the diversity of genetic capability found in the rhizosphere. Some organic matter remains in the soil as humus. While it is known that microbial enzymes and plant enzymes act as catalysts in the transformation of organic compounds, the extent to which they work together in the rhizosphere is not clear.

Transport processes. Some volatile organic compounds (such as trichloroethane, trichloroethylene, carbon tetrachloride, and benzene) are not biodegraded in the rhizosphere or plant because necessary enzymes are not present. Plants release these volatile compounds to the atmosphere, where they may be degraded by other processes. Water is vaporized along with the volatile compounds, and the rate of release of the volatile compounds depends on the rate of evapotranspiration. There is very significant dilution because the water must be distributed as vapor in air. Consequently, volatile contaminants are not usually detected in the gas phase above the soil surface.

Applications. Since vegetation has been planted at many contaminated sites, there has been considerable success with easily degradable compounds such as nitrates and some hydrocarbons. Riparian buffers have been used to reduce concentrations of agricultural chemicals entering streams. For example, poplar trees and grasses have been very successful in reducing to very low values the concentration of nitrate near waterways.

Poplar trees and alfalfa are being used as treatment systems for landfill leachate. For example, in Oregon a 14-acre (6-hectare) field of commercially grown poplar trees receives irrigation water from a landfill leachate collection system. Poplar trees have also been planted on top of closed landfills to act as a sink for precipitation and to provide a more active landfill closure. Vegetation has also been planted around the edges of landfills to evapotranspire any contaminated water that is released from under the landfill. Phytoremediation field studies have demonstrated that many petroleum hydrocarbons are biodegraded in the rhizosphere.

Remediation costs. The search for inexpensive technologies for remediation of contaminated soil and ground water has led to plant-based remediation. Natural vegetation is often present at contaminated sites. Early studies on the fate of compounds in vegetated soil were motivated by health concerns and the desire to understand the effects on pesticides. Compared with alternatives, designing and using phytoremediation processes at contaminated land sites has the potential to be the most cost-effective way to achieve soil remediation. Larry E. Erickson

Radioactive Contaminants

Uptake of radionuclides into plants has been extensively studied. Most of the earlier work was done to estimate quantities of radionuclides that would be accumulated by plants, animals, and ultimately humans as a result of nuclear fallout. In recent times, observations were made that plants were taking up radionuclides at nuclear waste disposal sites, indicating that some contaminants were mobile. Two radionuclides are of particular concern: cesium-137 (^{137}Cs) and strontium-90 (^{90}Sr). These radionuclides are among the most abundant fission products formed and have half-lives of 30 and 28 years, respectively. Large areas of land have been contaminated with these radionuclides—from the Chernobyl nuclear power plant accident, at nuclear weapons production facilities, and various other nuclear reactor sites around the world.

The use of plants to decontaminate soil and water has become attainable. For radionuclides, work is ongoing to determine if isolating contaminants inside a plant for later disposal (phytoextraction) is a practical alternative to excavating contaminated soil and shipping it to a radioactive waste disposal facility. In phytoextraction, the plants containing the radionuclides would be harvested, dried, and volume-reduced by high-pressure compaction, and the plant waste (a much smaller volume than the soil) would be sent for disposal. Phytoextraction is less developed than other methods of metal extraction and still presents a challenge. However, it is possible to use this method for reducing the concentrations of these contaminants in the soil to acceptable levels.

A plant's ability to take up a contaminant is measured by the concentration ratio (CR), the concentration of the element of interest in the dried plant material divided by its concentration in the dried soil. The higher the CR, the greater the concentration of contaminant in the plant compared to that of the soil. An average, derived from 51 studies based on cesium-137, gives a CR for vegetables of 0.13, while the National Council on Radiation Protection default value is 0.12. Compared to these CR values, experiments have shown relatively little uptake of cesium-137 into plants. The reason is that cesium-137 tends to be strongly sorbed on many soil minerals and not to move into solution where it can be taken up by plants. For example, cesium becomes strongly bound to illite or mica minerals, which are common. Consequently, uptake of cesium into plants is generally low.

The chemical behavior of cesium is very much like that of potassium (K), a major component of soils and a nutritional requirement for plants. Some radionuclides present as contaminants in soils (such as cesium-137) are at concentrations typically in the order of 10^{-12} mole/kg. Because of the very low radionuclide concentrations in soils, there can be significant competition for uptake into plants from similar elements. In the case of cesium-137, there is ample evidence that high potassium concentrations in soil inhibit uptake of cesium by plants, making phytoextraction less efficient. An inverse correlation between potassium content of soil and the concentration of cesium-137 has been demonstrated in edible plants grown on soil from a contaminated cooling reservoir for a nuclear reactor. High potassium concentrations can also be beneficial. Potassium chloride (KCl) salt was applied at a ratio of 600–1000 kg/ha (535–892 lb/acre) onto coral soils at one of the Bikini islands, Atoll, which was contaminated by a nuclear weapons test. The application of potassium chloride salt reduced cesium-137 concentrations in coconut (*Cocus nucifera*) meat by a factor of 10 after only 20 months.

The most important factor controlling cesium-137 concentrations in plants is the initial concentration in the soil. In addition to the competition between cesium-137 and potassium, other correlations with soil parameters that can influence uptake of cesium-137 into plants have been observed. The ammonium ion, like potassium, can also replace cesium on some minerals. The presence of organic matter in the soil appears to enhance the uptake of cesium-137 into plants, while the use of fertilizer tends to inhibit uptake. This is probably due to the high potassium content of many fertilizers. Also, a direct correlation between uptake of cesium-137 into plants when concentrations of calcium and magnesium are increased in the soil has been observed (uptake of cesium-137 is reduced). A poor correlation exists between the total elemental cesium in soils and cesium-137 uptake. The small quantities of natural cesium that are found in soils are probably contained within some of the minerals and are not mobile.

In phytoextraction two factors balance each other: the concentration of the contaminant in the plant and the mass of the plants growing on the site. It is necessary to optimize each of these factors to provide the greatest mass of contaminant in the harvested plant matter. Thus, the use of fertilizer on contaminated sites is typically necessary to achieve adequate plant growth. Nonetheless, the potassium required by plants may inhibit cesium-137 uptake.

Inducing enhanced accumulation (hyperaccumulation) of metals in plants through the use of chelating (bonding with various metals) solutions to mobilize the contaminant from the soil and increase its availability for uptake is a significant recent advance. This concept has been applied successfully to lead (Pb) in soils. Also, plant uptake of uranium (U) has been enhanced by mobilization of uranium with citric acid solutions. This method has shown a concentration ratio as high as 6.7 (that is, uranium present in soils at concentrations of 750 mg/kg compared to concentrations of uranium in plants of 5000 mg/kg). Greenhouse studies using soil contaminated with cesium-137 have shown that solutions of ammonium nitrate (up to 80 millimoles per kilogram of soil) can increase the CR by factors up to 12 times that of control samples.

Many types of plants have been examined for their capacity to take up cesium-137. These experiments are typically done in pots in greenhouses or in small field cells. It is difficult to compare results of these studies because of the variety of soils and conditions. CR values of 0.1 are typical. In one study, tomato (*Lycopersicon*), chard (*Beta vulgaris* var. *cicla*), cucumber (*Cucumis sativis*), pea (*Pisum sativum*), and maize (*Zea mays*) were found to take up the greatest mass of cesium-137. In a field lysimeter study of nine plant types, CRs ranged from 0.005 to 0.3, with the leaves of sugarbeets (*Beta vulgaris*) having the highest values. In field studies on a contaminated lake bed, okra (*Hibiscus esculentus*) was observed to have high uptake, with CRs averaging 6.8. For Indian mustard (*Brassica juncea*), a commonly used plant in phytoremediation, typical CR values are between 0.5 and 0.7.

Field studies were performed at a contaminated waste management facility, where the soil was contaminated in the 1950s. It was found that cesium-137 uptake was greatest in cabbage (*Brassica oleracea* var. *capitata*) with CRs ranging from 2 to 3 in related pot studies, and in red root pigweed (*Amaranthus retroflexus*) with CRs ranging from 2.2 to 3.2. In neither plant did application of ammonium nitrate solu-

tions in the field have any influence on the uptake of cesium-137. Based on this field work, about 3% of the cesium-137 in the soil can be removed by each crop and often two crops can be harvested per year. For plants that are currently known to take up significant quantities of cesium-137, it would take at least 10–15 years to meet soil cleanup goals. This length of time needs to be substantially reduced before phytoremediation can become a tool for decontamination of sites containing cesium-137. Research continues into plants with higher CR values and greater mass, and on the application of reagents that may help induce greater uptake of cesium by plants.

For background information *see* BIODEGRADATION; CESIUM; DECONTAMINATION OF RADIOACTIVE MATERIALS; PHOTOSYNTHESIS; PLANT; RADIATION; SOIL in the McGraw-Hill Encyclopedia of Science & Technology.   Mark Fuhrmann

Bibliography. T. A. Anderson and J. R. Coats, *Bioremediation through Rhizosphere Technology*, ACS Symp. Ser. 563, 1994; E. L. Kruger, T. A. Anderson, and J. R. Coats, *Phytoremediation of Soil and Water Contaminants*, ACS Symp. Ser. 664, 1997; S. K. Sikdar and R. L. Irvine (eds.), *Bioremediation: Principles and Practice*, Technomics, 1998; S. Trapp and J. C. McFarlane, *Plant Contamination*, CRC Press, 1995.

Space flight

In 1998 a number of achievements highlighted both human space flight and automated space exploration and commercial utilization (**Table 1**). There were 77 successful launches, down from 85 in 1997 (**Table 2**). United States accomplishments included the successful conclusion of the joint shuttle/*Mir* program with Russia; the beginning of the assembly of the International Space Station (ISS) in Earth orbit; the second space voyage of 77-year-old senator John Glenn; the possible discovery of ice on the Moon; and surprising discoveries by the Hubble Space Telescope.

For the first time in history, not only were there more commercial satellites on orbit than military satellites, but also more commercial space launches than military ones. Increasingly, the commercial market, rather than the government, is driving the space industry, particularly in direct-to-consumer applications. The space segment continues to play the major role in global telecommunications, with mobile services advancing. All the proposed big-LEO (low-Earth-orbit) satellite network systems made major progress, particularly Iridium and Globalstar. *See* COMMUNICATIONS SATELLITE.

The United States space shuttle and Russia's Earth-orbiting *Mir* added two rendezvous-and-docking flights to the seven previous missions, bringing phase 1 of the space station development program to a successful conclusion.

A total of seven crewed flights (three less than in 1997) from the two major space-faring nations

TABLE 1. Significant space events in 1998

Mission designation	Date	Country	Event
Lunar Prospector	January 6	United States	NASA's return to the Moon after 25 years; probe went into lunar orbit on January 11 and began mapping mission on January 15; apparently discovered water ice at poles.
STS 89 (*Endeavour*)	January 22	United States	Eighth *Mir* link-up; delivered Andrew Thomas and supplies; returned David Wolf (126 days in space) and cargo; carried Spacehab double module in payload bay.
Soyuz TM-26/Mir 25	January 29	Russia	Launch of Mir 25 crew of three, including French researcher Leopold Eyharts, to relieve Mir 24 crew.
Trace	April 1	United States	*Transitional Regional and Coronal Explorer*, launched on a Pegasus XL, air-dropped from an L-1011, into a Sun-synchronous orbit at altitude of 400 mi (640 km).
STS 90 (*Columbia*)	April 17	United States	Research mission Neurolab with crew of seven; conducted microgravitation research on the human nervous system.
STS 91 (*Discovery*)	June 2	United States	Last (ninth) *Mir* link-up, picked up seventh U.S. occupant, Thomas (135 days in space); crew included Russian phase 1 director and cosmonaut V. V. Ryumin; ended phase 1 of International Space Station Program.
Planet B/Nozomi	July 3	Japan	First Japanese Mars probe; propulsion malfunctioned during second gravity-assist, and probe was replanned to reach Mars in December 2003 or January 2004.
Soyuz TM-28/Mir 26	August 13	Russia	Launch of Mir 26 crew of three to relieve Mir 25 cosmonauts; crew included Yuri Baturin, a former aide to President Yeltsin.
Ariane 5	October 3	Europe	Second successful launch of ESA's heavy-lifter Ariane 5 (*503*); carried *Maqsat 3* payload and Atmospheric Reentry Demonstrator; qualified Ariane 5 for operational phase.
DS-1	October 24	United States	First of new millennium high-technology test missions; targeted for asteroid 1992KD in July 2000; powered by xenon-fueled low-thrust ion engine; autonomous navigation.
STS 95 (*Discovery*)	October 29	United States	Ninety-second shuttle mission; science mission with highly diverse research disciplines; with 77-year-old John Glenn, first American in orbit in 1962.
ISS-1A/R (Zarya)	November 20	United States/ Russia	First element launch for International Space Station (ISS): Russian-built FGB/Zarya on Proton booster, from Baikonur in Kazakhstan.
ISS-2A (Unity)/STS 88	December 4	United States	Second ISS element, Node 1/Unity, launched on *Endeavour* with crew of six, including cosmonaut S. Krikalev; successful Unity/Zaryalink-up and activation in orbit.
Mars Climate Orbiter	December 11	United States	Mars orbiter; to enter Mars orbit on September 23, 1999, to begin study of Mars climate and atmospheric circulation, and act as radio relay for *Mars Polar Lander*.

carried 39 humans into space, including 6 women. The total number of people launched into space since 1958 (counting repeaters) then totaled 809, including 79 women (or 392 different individuals, including 34 women).

United States Activity

The four reusable shuttle vehicles of the U.S. Space Transportation System continued carrying people and payloads to and from Earth orbit for science, technology, and operational research. After overcoming a number of last-minute technical and schedule problems on the ground, the International Space Station program moved briskly into actual orbital assembly.

Space shuttle. During 1998, the National Aeronautics and Space Administration (NASA) completed five space shuttle missions, three less than in 1997.

STS 89. *Endeavour*, on its twelfth flight, January 22–31, conducted the eighth docking mission to *Mir* (January 24–29). The combined crew of 10 transferred a record cargo, including water, a new computer, and air-conditioning equipment. Andrew S. Thomas became the seventh U.S. astronaut to board *Mir* for a long-duration mission, replacing David A. Wolf, who had spent 126 days aboard *Mir*. Other payloads on STS 89 included the Spacehab double module with science experiments and new research equipment for Thomas.

STS 90. *Columbia*'s twenty-fifth flight, April 17–May 3, carried the research module Neurolab. The prime mission for the Spacelab module, on its sixteenth flight, was to conduct microgravity research on the human nervous system, including the brain, spinal cord, nerves, and sensory organs. Acting both as subjects and operators, the crew performed 26 experiments, including studies on a variety of species.

STS 91. The *Discovery* mission of June 2–12 was the last of nine docking missions with *Mir* (June 4–8), thus ending phase 1 of the International Space Station program. It also marked the first use of the superlightweight external tank developed to save about 7500 lb (3400 kg) of structural weight and improve the shuttle's payload capacity on flights to the space station. Veteran cosmonaut and Russian phase 1 director Valery V. Ryumin was among the crew members. The crew transferred batteries and other resupply items to the station, and science samples and equipment to the shuttle for return. Other

TABLE 2. Space launches and attempts in 1998

Country	Number of launches*	Number of attempts
United States (NASA, Department of Defense, commercial)	34	36
Russia	24	25
Europe (European Space Agency, Arianespace)	12	12
People's Republic of China	6	6
Japan	1	2
Israel	0	1
North Korea	0	1
Total	77	83

*Successful launches to Earth orbit and beyond.

payloads included a Spacehab single module, the Alpha Magnetic Spectrometer, to study high-energy particles from deep space, and several experiments in advanced technology, human life sciences, space-station risk mitigation, and microgravity physics. *Discovery* brought home Thomas, the last of seven long-duration United States astronauts, after his 135-day stay on *Mir*.

STS 95. On the twenty-fifth flight of *Discovery*, October 29–November 7, the crew included payload specialist Glenn, at 77 years the oldest human ever to fly into space and also the first United States astronaut to fly into Earth orbit in 1962. To take scientific advantage of his advanced age and unbroken medical history records with NASA over 40 years, about 10 of the flight's 83 experiments and payloads investigated questions pertaining to geriatrics in space, that is, the phenomena of natural aging versus weightlessness-induced effects closely resembling them. The other experiments made up a diverse set of commercial and scientific research investigations addressing biotechnology, life sciences, solar physics, and astronomy. The Spartan 201-04 free-flyer, which had suffered a failed deployment on mission STS 87, was successfully deployed on November 1 and retrieved 2 days later.

STS 88. The primary mission cargo on the December 4–15 flight of *Endeavour* was the first United States-built station element for the International Space Station, the connecting module Node 1/Unity. The crew of six included Russian cosmonaut Sergei K. Krikalev, on his second shuttle mission. Rendezvous with Zarya (Dawn), launched by Russia on November 20, occurred on December 6. The crew also deployed an Argentinean satellite (*SAC-A*) and a U.S. Air Force test satellite (*MightySat*).

International Space Station. The International Space Station program successfully concluded phase 1 and began the launch and assembly of station elements in orbit.

The program is organized into phase 1, the joint shuttle/*Mir* program; phase 2 for space station assembly up to initiation of orbital research capability with a permanent crew; and phase 3 for further expansion and completion. A total of 45 assembly flights are planned, with 33 to be launched by the United States shuttle and 12 on Russian boosters.

Phase 1. Phase 1 of the program, the joint United States–Russian effort on *Mir* to expand cooperation in human space flight, conduct science research, and gather risk-mitigating experience for the subsequent assembly and operations phases, came to a successful conclusion in 1998 after a total of nine shuttle-*Mir* missions. Seven United States astronauts accumulated over 966 days aboard *Mir*.

A range of activities provided the framework for United States/Russian cooperation in space and on the ground, and the mechanism for facilitating the integration of Russia into full partnership in the space station. Phase 1 also provided essential opportunities to test assumptions and validate models in the actual environment that the space station will experience. The knowledge and experience gained has demon-strated its value in space station design decisions as well as in the planning and operational procedures for crews.

The United States also gained experience in joint international ground operations in real time. This implied development, often from scratch, of common flight procedures and procedural terminologies as well as adequately scoped preflight science planning and definition for research activities on board a long-duration space laboratory. To conduct crewed missions on *Mir*, as is also anticipated on the space station, the United States got first-hand experience with realistic approaches to crew training for long-duration flights, as opposed to the short-duration shuttle missions, and learned how to set up an effective central management for crew training, operations, and multilateral integration.

FGB/Zarya. Funded and owned by the United States and built by Russia, the power and control module FGB (*Funktionalnyi-grusovoi blok*) named Zarya (Dawn) was launched uncrewed on a three-stage Proton rocket from Russia's cosmodrome in Kazakhstan on November 20, as the first component of the space station. The 44,000-lb (20-metric-ton) spacecraft is essentially an unpiloted tugboat that provides early propulsion, steering, power, and communications for the station's first months in orbit. Later, Zarya will become little more than a stationary passageway, stowage module, docking port, and fuel tank. After its successful launch, Zarya deployed its solar arrays and communications antennas and performed a series of thrust maneuvers to place it into its circular target orbit at an altitude of about 250 mi (400 km) and $51.6°$ inclination to the Equator.

Node 1/Unity. The first United States–built component of the space station, a six-sided connecting module and passageway, or node, named Unity, was the primary cargo of the space shuttle mission STS 88 launched on December 4. Unity was successfully mated to Zarya on December 6. On December 7, astronauts Jerry L. Ross and James H. Newman conducted a spacewalk to establish full functionality between the two elements by connecting 40 cables over a distance of 75 ft (23 m) and to install aids for later spacewalks. A second spacewalk was performed on December 9 to install two S-band antennas, remove covers from Unity's four side hatches, mount sunshades over two electronics boxes, and fully deploy two backup rendezvous antennas on Zarya which had not properly unfolded. On December 10, astronauts entered the node and Zarya for internal preparations. A third spacewalk by Ross and Newman brought the total time spent outside the space station to 21 h 22 m (**Fig. 1**). On December 13, the mated components were separated from the shuttle.

The first of three Multi-Purpose Logistics Modules (MPLMs) was delivered to Kennedy Space Center, in Florida, on August 1. The element, built in Italy, was the first hardware to arrive from Europe.

Advanced transportation systems activities. NASA is engaged in a cooperative effort with industry to develop a reusable space launch vehicle to eventually

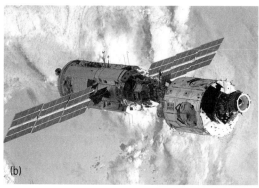

Fig. 1. First two components of the International Space Station, Node 1 (Unity) and FGB (Zarya). (*a*) Components being assembled in the cargo bay of the space shuttle *Endeavour* during a spacewalk by astronauts Jerry L. Ross and James H. Newman. (*b*) Mated components flying free after separation from the shuttle. (*NASA*)

take over launchings for a fraction of today's cost of space transportation with turnaround rates considerably lower than those of the space shuttle. Lockheed Martin is developing the X-33 as a technology demonstrator for a single-stage-to-orbit reusable launch vehicle. The smaller air-launched X-34 will test reusable launch vehicle technologies. During 1998, key rocket engine tests got under way in the development of the propulsion systems, the 60,000-lb-thrust (267-kilonewton) liquid oxygen (LOX)/kerosene Fastrac engine for the X-34, and the 500,000-lb (2.2-meganewton) XRS-2200 LOX/liquid hydrogen linear aerospike engine for the X-33.

NASA is also developing a Crew Return Vehicle to eventually take over the emergency lifeboat function for the space station from the currently chosen Russian Soyuz three-seater capsules. The vehicle will accommodate seven crew members. It must have a minimum lifetime of one year attached to the space station, and must be capable of supporting medical equipment, carrying out a 9-h orbital free-flight mission, and allowing access for a medical officer to an injured or ill crew member during such a mission. In 1998, NASA conducted the first full-scale flight tests of the X-38, the prototype for the Crew Return Vehicle, by dropping it from a B-52 aircraft. An unpiloted test vehicle is eventually planned to be deployed from a space shuttle for descent from orbit.

Space sciences and astronomy. Several automated and remotely controlled research and exploration

missions continued to provide a wide range of significant discoveries.

Hubble Space Telescope. The Hubble Space Telescope continued to probe far beyond the solar system. It discovered what may be the first planet observed outside the solar system, some 450 light-years from Earth. The object, known as TMR-1C, appears to lie at the end of a strange filament of light, suggesting it has been flung away from the vicinity of a newly forming pair of binary stars.

Hubble's infrared instruments penetrated a wall of dust girdling the Centaurus A galaxy (NGC 5128), revealing blue clusters of newborn stars and silhouettes of dust filaments interspersed with orange-glowing gas. Located 10 million light-years from Earth, the galaxy was revealed to contain the gravitational whirlpool of a black hole of 10^9 solar masses, which is sweeping up a twisted disk of hot gas. The gas disk comprises the remnants of a smaller galaxy that collided with NGC 5128, was devoured by it, and is now fueling the black hole in its center.

Using the Hubble, astronomers viewed the violent birth of extremely massive stars, each 300,000 times brighter than the Sun, in the Small Magellanic Cloud. The telescope pinpointed 50 individual stars packed densely into the nebula's core within a 10-light-year diameter. Over a 10-day observation period in October, the Hubble also peered down a 12-billion-light-year corridor of deep space in the constellation Tucana near the celestial south pole, viewing thousands of galaxies never before seen. Nearer to Earth, the Hubble helped to reveal that Neptune's largest moon, Triton, seems to have heated up significantly since the *Voyager 2* spacecraft visited it in 1989.

Trace. The *Transitional Regional and Coronal Explorer* (*Trace*) was launched by NASA on April 1 on a Pegasus XL launch vehicle air-dropped at Mach 0.8 from an L-1011 aircraft. *Trace* was placed in a Sun-synchronous orbit of 97.8° inclination at about 400 mi (640 km) mean altitude. Its purpose is to explore the connections between the Sun's magnetic fields and its plasma structures.

NEAR. Almost 2 years after its liftoff in February 1996, the *Near Earth Asteroid Rendezvous* (*NEAR*) spacecraft, NASA's second Discovery program mission, returned for a close pass of Earth on January 22–23. The 2-h swingby, 7 months after *NEAR* flew past and imaged the asteroid 253 Mathilde, provided the probe with a critical gravity assist to put it on a trajectory toward its main target, asteroid 433 Eros, which it was to reach and orbit in January 1999. But a failure of the first of several rendezvous maneuvers on December 20 changed plans. *NEAR* was quickly guided to a contingency flyby of the asteroid on December 23 within 2320 mi (3830 km), which resulted in 1100 images of the 18-by-8-mi (30-by-14-km) potato-shaped cratered rock showing details down to about 1600 ft (500 m) resolution. A firing of the spacecraft's engine on January 3, 1999, to gain a second chance to meet and orbit Eros in February 2000 was successful.

Galileo. *Galileo* in 1998 continued to return unprecedented data on Jupiter and its satellites. The spacecraft accomplished its basic mission in December 1997. In 1998 it operated in the extended Galilei Europa Mission, making six flybys of Jupiter's satellite Europa. However, during two of these flybys the spacecraft went into a "safe" mode, missing most of the science data. More encounters with Jupiter's satellites were planned for 1999: one with Europa, four with Callisto, and two with Io. *See* JUPITER.

Cassini. *Cassini*, NASA's 6-ton (5.6-metric-ton) spacecraft heading to the planet Saturn on a 6.7-year journey, flew by Venus on April 26 in the first of three gravitational assists. During the flyby, two science instruments returned good data, a radar system and a radio and plasma wave instrument which listened for lightning sferics from Venus's atmosphere.

A second Venus flyby was planned on June 24, 1999, followed by an Earth flyby on August 18. *Cassini* should pass Jupiter on December 30, 2000, finally reaching Saturn on July 1, 2004. *Cassini*, which consists of an orbiter and the Titan entry probe *Huygens*, will remain within the Saturn system for the exploration period of at least 4 years.

Ulysses. In April 1998, *Ulysses* completed its first full circle of the Sun and embarked on its second solar orbit, to study the Sun's polar regions under conditions of high solar activity. Polar passes in 2000 and 2001 will occur close to the maximum of the solar cycle. Before that, *Ulysses* will make coordinated observations of the Sun's corona and the solar wind.

DS-1. NASA's *Deep Space 1* (*DS-1*) technology test satellite was launched on October 24, beginning a 2-year mission into a stretched-out elliptical orbit around the Sun which will take it, in July 2000, to the asteroid 1992 KD then 120 million miles (193 million kilometers) from Earth. The spacecraft's mission is to test several advanced technologies for interplanetary science missions, particularly an ion engine which uses electric power to accelerate ionized xenon fuel to over 18 mi/s (30 km/s) for high efficiency and low thrust. Other advanced technologies on board test autonomous navigation, power generation with concentrator solar arrays, and K_a-band communications. After the rendezvous with 1992 KD, *DS-1* is tentatively scheduled for a flyby in January 2001 at the mysterious object Wilson-Harrington and, in September, at Comet Borelly. *See* AUTONOMOUS NAVIGATION.

Lunar Prospector. As the third flight in NASA's Discovery Program, the small (660 lb; 300 kg) spacecraft *Lunar Prospector* was launched on January 6 to provide the first global maps of the Moon's elemental surface composition and the lunar gravitational and magnetic fields. Its instrument groups are a magnetometer-electron reflectometer and spectrometers for gamma rays, neutrons, and alpha particles. Using the Doppler effect, mass concentrations (mascons) can be detected, helping to create an exact model of the gravitational field. On January 11, *Lunar Prospector* successfully entered into a lunar orbit via a 32-min engine firing. On January 15, the spacecraft reached its final mapping orbit at 62 mi (100 km) altitude, which will be adjusted and modified from time to time.

In March, NASA reported preliminary findings indicating, with high probability, that water ice exists at both the north and south poles of the Moon. The ice appears to be not concentrated in polar ice sheets but only in a 0.3–1% mixing ratio in combination with the Moon's rocky soil, or regolith. *See* MOON.

Mars exploration. Mars remained of great interest following the *Pathfinder/Sojourner* mission's phenomenal success. *Mars Global Surveyor* produced the highest-resolution images of Mars to date, while its gravitational and magnetic field experiments revealed unexpected local phenomena that have changed interpretations of the planet's subsurface structure. Continuing this characterization work on a global scale will be the tasks of the *Mars Climate Orbiter*, launched in 1998, and the *Mars Polar Lander*, to be launched in 1999. Additional robotic missions to Mars are in preparation.

In June 1998, NASA presented a selected set of results gleaned from data collected from *Mars Pathfinder* and *Sojourner*. The data show that Mars has undergone a dramatic swing from wet and warm in its infancy to an arid world unchanged over the last 2 billion years. Catastrophic floods had raced across *Pathfinder*'s landing region 1–3 billion years ago, and little has changed there since then.

Mars Global Surveyor was the first of a planned series of surveyor-type Mars explorers. After its arrival at Mars on September 11, 1997, it started a long series of aerobrake passes around the planet. At the end of 1998, it had successfully completed about 950 aerobraking orbits, settling into an orbit of 68 mi (109 km) low point and 2635 mi (4216 km) high point, and about 290 science phasing orbits. Although attainment of its final mapping orbit had been delayed until March 1999, *Mars Global Surveyor* had already transmitted remarkable surface pictures (**Fig. 2**). Among them were new images of the landing zone for the *Mars Polar Lander*, showing strange, layered terrain in the south polar region that represents a dramatic departure from the landscapes observed by the *Viking* landers and *Pathfinder*. After the arrival of *Mars Climate Orbiter* and *Mars Polar Lander*, images from *Mars Global Surveyor* will be used in concert with data from the other spacecraft to better characterize the geology of Mars, particularly its south pole.

Mars Global Surveyor captured the full evolution of a Martian dust storm. Data from its thermal emission spectrometer also supported the finding from *Pathfinder* that the planet once had abundant water and thermal activity. Using the orbiter's laser altimeter, a three-dimensional picture of Mars's north pole was assembled from 2.6 million laser-pulse measurements (**Fig. 3**), with a spatial resolution of 0.6 mi (1 km) and a vertical accuracy of 15–90 ft (5–30 m). It allowed a precise estimate of the volume of the water ice cap, as well as study of surface variations and cloud heights in the region.

NASA's *Mars Climate Orbiter* took off on December 11 on a Delta 2 rocket. The robotic explorer is scheduled to arrive at Mars on September 23, 1999, to study the planet's climate and atmospheric

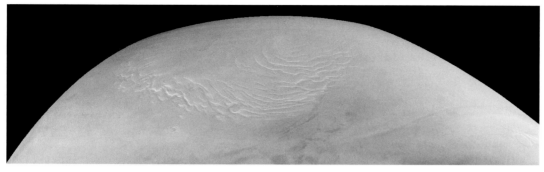

Fig. 2. Simulated view of the north polar region of Mars from an altitude of 740 mi (1200 km), obtained by *Mars Global Surveyor* in September 1998. The swirled pattern at top center is an area of polar layered deposits covered in part by the permanent north polar ice cap. (*NASA*)

circulation. For the first several months after its arrival, however, its primary job will be to serve as a communications relay for the *Mars Polar Lander* (launched on January 3, 1999), which has a 90-day planned lifetime.

Pioneer 10 and Voyager. Science instruments aboard *Pioneer 10*, the Earth's longest-lived interplanetary explorer, were switched off in 1997 due to the weakness of its nuclear batteries, but contact is maintained via the Deep Space Net, a worldwide network of 230-ft (70-m) parabolic antennas. On February 17, *Voyager 1*, launched in 1977, overtook *Pioneer 10*, becoming the most distant human-created object in space. At the end of 1998, *Voyager 1* was over 6.7 billion miles (10.8 billion kilometers) from the Sun, and its signal took over 10 h to reach the Earth at the speed of light.

Voyager 1 and *Voyager 2* are studying the space environment in the outer solar system. Data indicate that before 2004 they may travel through the termination shock of the heliopause, where the solar wind abruptly slows down from supersonic to subsonic speed, and on into interstellar space.

Earth science. In 1998, Earth-observing missions of the civilian space program coordinated under the auspices of NASA's Earth Science program continued to provide excellent information.

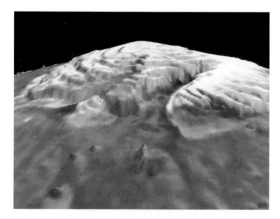

Fig. 3. Three-dimensional view of Mars's north pole during spring and summer of 1998, obtained from approximately 2.6 million measurements by a laser altimeter aboard *Mars Global Surveyor*. The ice cap is about 750 mi (1200 km) across, with a maximum thickness of 1.8 mi (3 km). (*NASA*)

TRMM. Data from the joint United States/Japan *Tropical Rainfall Measuring Mission (TRMM)* exceeded expectations for accuracy and resolution. The satellite studies tropical and subtropical rainfall by using microwave and visible-infrared sensors and airborne radar. The project also studies how El Niño–related rainfall anomalies correlate with other processes in the ocean and the atmosphere. In one instance, the three-dimensional *TRMM* radar showed better vertical resolution of storm structure (critical for determining overall intensity) than ground-based radar. Presently, only 2% of the area covered by the orbiting *TRMM* is covered by ground-based radar.

TOPEX/Poseidon. The joint United States/France *TOPEX/Poseidon* satellite, launched in 1992, detected a temporary change in average global sea level during the 1997–1998 El Niño event. By measuring the sea level height around the world at 10-day intervals with a precision of 0.16 in. (0.4 cm), the spacecraft was able to show the average global sea level rising about 0.8 in. (2 cm) before it returned to normal levels.

Department of Defense activities. United States military space organizations continued their efforts to make space a routine part of military operations. Military launches from Cape Canaveral, Florida, and Vandenberg Air Force Base, California, in 1998 totaled eight missions: two satellites for signal intelligence and reconnaissance on two Titan launchers, a signal intelligence satellite and a communications satellite on two Atlas 2A launchers, a communications satellite on an Atlas 2, a technology development satellite and a geophysics research satellite on two Taurus vehicles, and a remote-sensing satellite on an air-launched Pegasus XL. The "Mercury" signal intelligence payload on one of the Titans was lost on August 12 when its booster, a Titan 4A/Centaur, exploded on a self-destruct signal, approximately 42 s into powered flight. The failure was reportedly the most costly crewless accident in the history of Cape Canaveral launch operations.

Commercial space activities. Of the 36 launches conducted by the United States in 1998, 19 were commercial, with only one failure, including ten Delta 2 vehicles, four Pegasus XL launches from L-1011 aircraft, three Atlas 2A's, and one Titan 2. There were 81 United States commercial payloads, with 49 launched on United States carriers and 32 on foreign launchers (China, 11; Russia, 21). The inaugural launch of the powerful Delta 3 with nine strap-on boosters, carrying the *Galaxy X* communications satellite, failed catastrophically with an explosion of the vehicle shortly after liftoff on August 26.

A commercial communications satellite, *AsiaSat 3*, which had been placed in a highly elliptical Earth orbit on December 25, 1997, when the fourth stage of a Proton booster malfunctioned on its second burn, was salvaged by Hughes Global Services by redirecting it to the Moon to employ two gravity-assist flybys to help establish a new operational Earth orbit. The satellite, now called *HGS 1*, established final geosynchronous orbit on June 16.

The development of large communications satellite networks continued to advance rapidly. In May, Iridium completed deployment of its 66-satellite constellation. Its 13-month launch program included 15 launches using Delta, Proton, and Long March boosters. Globalstar LPP had 8 of its 56 spacecraft in low Earth orbit when 12 additional satellites were lost on a Ukrainian Zenit rocket on September 9, dealing the program a severe setback. A third project, ICO Global Communications, plans to launch the first of its 12 medium-Earth-orbit satellites in 1999 on Atlas 2A, Proton, and Delta 3 boosters.

Russian Activity

Despite a chaotic economic situation, Russia in 1998 showed little slack in its space operations from 1997. Its 24 successful launches (out of 25 attempts) fell short of the previous year's 27 (out of 29 attempts): 8 Soyuz-U (two crewed), 7 Protons, 3 Zenit 2's (one failed), 3 Molniyas, 1 Tsiklon, 2 Kosmos 3M's, and 1 submarine-launched Shtil.

Space station Mir. By the end of 1998, Russia's seventh space station, *Mir*, in operation since February 20, 1986, had circled the Earth approximately 73,513 times. It had been visited 37 times, including 9 times by a United States space shuttle. To resupply the occupants, the space station was visited in 1998 by three automated Progress cargo ships, bringing the total of Progress ships launched to *Mir* and the two preceding stations, *Salyut 7* and *Salyut 6*, to 83, with no failure (except for the collision of one of the drones with the station in 1997).

Mir continued to depend on the space shuttle for its logistics until phase 1 of the space station program concluded with the departure of STS 91/*Discovery* on June 8. During their long-duration stays, the United States crew members conducted the majority of scientific research on orbit, over a wide range of subjects.

Soyuz TM-27 and Mir 25. Soyuz TM-27 was launched on January 29, with the new Mir 25 crew, Talgat Musabayev and Nikolai Budarin, accompanied by a French researcher, Leopold Eyharts. After docking to *Mir* 2 days later, the space station supported six occupants until February 19, when the Mir 24 crew, along with Eyharts, departed in *Soyuz TM-26*.

Soyuz TM-28 and Mir 26. Soyuz TM-28 was launched on August 13 with the new Mir 26 crew, Gennady Padalka, Sergei Avdeev, and Yuri Baturin, a former aide to President Yeltsin. The docking to *Mir* on August 15 used the manual backup system due to prior failure of one of two automatic systems on *Mir*. Baturin returned to Earth with the Mir 25 crew on August 25.

Russian commercial activities. The Russian space program's efforts to enter the commercial market continued to progress in 1998. Between 1985 and 1998, 128 Proton and 346 Soyuz rockets were launched, with only seven failures of the Proton and nine of the Soyuz. Of the seven Protons launched in 1998, four were for commercial customers. After the successful launch of two Ukrainian Zenit 2 vehicles, the third launch on September 9 failed catastrophically, losing its commercial payload of 12 Globalstar communications satellites.

European Activity

With France's reliable Ariane 4 family of expendable launchers, the commercial operator Arianespace in 1998 carried out 10 launches from Kourou, French Guyana, one less than in 1997. Nine of the 1998 flights carried a total of 13 commercial satellites. One carried the French commercial remote-sensing satellite *SPOT 4*.

Ariane 5. The most significant event for the European Space Agency (ESA) in 1998 was the full success of the third launch of the Ariane 5 heavy booster, after a partial success in 1997 and the explosion of vehicle *501* in June 1996. The flight of *503* on October 21 accomplished all objectives and successfully finished the qualification program, carrying a 6019-lb (2730-kg) dummy satellite, *Maqsat 3*, and the Atmospheric Reentry Demonstrator (ARD). The latter, released 12 min after launch, performed Europe's first controlled reentry into the atmosphere and was retrieved intact from the Pacific Ocean.

Scientific spacecraft. The *Solar and Heliospheric Observatory* (SOHO), built by the European Space Agency, launched in 1995 by NASA, and operated from NASA's Goddard Space Flight Center, encountered severe problems. The spacecraft spun out of control on June 25 and lost power and radio contact because of erroneous commands from ground controllers. A delicate recovery operation began, and signals from the dormant *SOHO* were acquired in early August. After extensive work reactivating the satellite, attitude control was recovered in September, and *SOHO* was able to face fully to the Sun. The unexpected malfunction of two gyroscopes muted the effects of this repointing, but *SOHO* was subsequently returned to service, with 9 of the 12 instruments reactivated. High-quality images of the Sun were taken in October.

On May 16, the European Space Agency switched off its *Infrared Space Observatory* (ISO), thereby bringing its highly successful mission to a close. The observatory, launched in 1995, made more than 26,000 observations. Its supply of liquid helium lasted much longer than expected, but ran out on April 8, essentially terminating the instruments utility. Its "last light" observation, on May 10, of hydrogen emission lines from the hot supergiant star Eta Canis Majoris showed, surprisingly, that this star is probably surrounded by a disk of matter. *See* INFRARED ASTRONOMY.

Asian Activity

The Asian economic crisis in 1998 had deep impacts on Asia-Pacific commercial satellite communications and government space programs, but the emergence from the crisis could be the basis of more solid growth in these areas.

Japan. Japan's space activities in 1998 accomplished several milestones, but also suffered a severe blow when an H-2 booster with a Japanese research spacecraft malfunctioned on February 21, marking the worst launch vehicle failure in the history of the Japanese space program. Because of an aborted second burn of the LE 5 oxygen-hydrogen engine in the H-2's second stage, the Japanese Communications and Broadcast Engineering test satellite *Comets* was left in an essentially useless orbit. Until this incident, the LE-5 had maintained a perfect record spanning five earlier missions on the H-2 and nine flights on the H-1 booster. It was the third major space program setback for Japan in the last 5 years, following the loss of the *ETS 6* (*Engineering Test Satellite*) in 1994 and the failure in 1997 of the *Advanced Earth Observation Satellite* (*Adeos*).

On July 7, an autonomous rendezvous-and-docking test was conducted successfully on the *ETS 7* satellite. The first of five tests checked out the performance of the rendezvous-and-docking system after a 6-ft (2-m) separation between the main craft and the chaser. Results are relevant to future spacecraft operating in conjunction with the International Space Station, such as Japan's transfer vehicle, HTV. In October, loss of attitude and electrical power in Japan's earth resources satellite *JERS 1*, launched in 1992, forced the satellite's shutdown.

On July 4, Japan launched its first spacecraft to Mars, *Nozomi* (formerly *Planet B*), on an M 5 booster. The 1177-lb (535-kg) probe underwent the first of two gravity-assist Earth swingbys on September 24. Following the second flyby on Dec. 18, one of its thrusters stuck open, wasting fuel and producing insufficient acceleration for the Mars insertion. After a correctional burn on December 21, the mission was redesigned, calling now for three trips around the Sun in the coming 4 years and two swingbys of Earth, in December 2002 and June 2003. It will now arrive at Mars in December 2003 or January 2004 rather than October 11, 1999, as originally planned.

Work continued on the H-2A series of launchers in the Ariane 4 class, which are intended to be more cost-effective to operate for commercial use. Among the main features of the H-2A are strap-on boosters and an improved oxygen-hydrogen LE-7A first-stage engine. Meanwhile, Japan stayed on track in the de-velopment of its contribution to the space station, the Japanese Experiment Module (JEM), along with its ancillary remote manipulator system and porch-like exposed facility.

China. The space program of the People's Republic continued its recovery after its commercial launcher setbacks in 1996, when two Long March (Chang Zheng) 3 (CZ-3) rockets exploded after liftoff, with casualties. During 1998, China succeeded in launching six Long March rockets with no failures: two CZ-3B's with Chinese communications satellites, and four CZ-2C's, each carrying two satellites for the Iridium constellation. China also announced the planned development of an upgraded version of the Long March CZ-2E as well as of the first Chinese crewed spacecraft, Project 921. The CZ-2E(A) will be able to carry payloads of up to 26,500 lb (12,000 kg) to low Earth orbit. The crewed spacecraft, displayed in April, bears a strong resemblance to Russia's Soyuz and would be launched in the early 2000s on a human-rated version of the CZ-2E(A).

North Korea. In the first space mission attempt of its multistage Taepo Dong 1 missile, North Korea on August 29 tried to launch a satellite, *Kwangmyong-song 1*. The test failed, but the attempt attracted considerable attention.

For background information *see* ASTEROID; COMMUNICATIONS SATELLITE; INFRARED ASTRONOMY; JUPITER; MARS; MILITARY SATELLITES; MOON; REMOTE SENSING; SATELLITE ASTRONOMY; SPACE BIOLOGY; SPACE FLIGHT; SPACE PROBE; SPACE SHUTTLE; SPACE STATION; SPACE TECHNOLOGY; SUN in the McGraw-Hill Encyclopedia of Science & Technology.

Jesco von Puttkamer

Bibliography. *Jane's Space Directory, 1997-1998*; NASA Public Affairs Office, News releases, 1998; Space Publications LLC and A. T. Kearney, *State of the Space Industry—1998*.

Species stability

Darwin's *On the Origin of Species* linked processes observable in the present to patterns in the fossil record and formulated a coherent theory of evolution by natural selection. That link has been challenged by paleontologists on the grounds that the fossil record actually demonstrates long periods of little change in lineages interspersed with brief periods of relatively rapid change (punctuational equilibria). Stephen Jay Gould has proposed that evolutionary theory should include processes at three separable tiers of time: ecological moments, geological time (millions of years), and mass extinctions. Whatever progress is made at the first tier can be undone by the processes at the second tier (by punctuational equilibria) or third tier.

However, this hierarchy overlooks the significance of global climatic changes on the time scale of 20,000–100,000 years resulting from astronomical forcing of variations in the Earth's orbit, known as Milankovitch oscillations. These variations have profound impacts on the abundance and distribution of organisms, best seen in the fossil record of the Quaternary Period (the last 1.6 million years). Such climatic changes must have been present throughout Earth history, and thus phenomena seen in the Quaternary fossil record may illuminate processes in the history of life more generally on time scales intermediate between those of ecology and paleontology.

Milankovitch oscillations. The Earth is subject to perpetual quasi-cyclical changes over a wide range of frequencies because of its position and movement relative to other bodies in the solar system. Diurnal and annual cycles derive from the Earth's rotation about its own axis, the tilt of that axis, and an elliptical orbit around the Sun. Tidal cycles result from weak gravitational attractions between the Earth, Sun, and Moon. More generally, the Earth's orbit is influenced by gravitational attractions between the Earth and all the other bodies of the solar system, compounded by the Earth's own slightly irregular shape. The gravitational attractions are strong enough to affect the eccentricity of the orbit, the angle of tilt of the Earth's rotational axis, and the precession of the equinoxes (changing position in the elliptical orbit of the equinoxes, when the Sun crosses the Equator). These variables fluctuate with frequencies of 20,000–100,000 years. They are known as Milankovitch oscillations, after the mathematician who made the first modern calculations of these changes. They were probably established, with minor subsequent change, by the initial orbit of the Earth.

It has now been shown that Milankovitch oscillations control the pace of Quaternary ice ages. They produce variation in the amount of solar radiation received by the Earth and the latitudinal and seasonal variation of this radiation. This variation produces climatic changes, affecting temperatures, precipitation, and other aspects of the atmospheric environment. Computer models of these effects indicate that at the time of the last glacial maximum (around 20,000 years ago), continental surface temperatures were 2.5–40°C (4.5–72°F) colder than today, depending upon location and proximity to subcontinental scale glaciers. Rainfall was also affected, with important changes in the strength and location of subtropical monsoons.

Milankovitch oscillations are a consequence of gravitational attractions between celestial bodies and have, therefore, been a permanent feature of Earth history. Sedimentary rocks with regularly repeating sequences that have periodicities corresponding to orbital variations have been reported throughout the geological record, from the Precambrian to the present.

Biological time scales. The two relevant time scales here are the generation times of organisms and the duration of species, relative to the time scales of astronomical forcing of environmental change. Generation times may be from a few minutes for bacteria up to a hundred years for some trees. For most organisms, diurnal changes take place on a shorter time scale than their lives and can be considered as part of their environmental background. For those organisms that live longer than one year, the annual cycle of the seasons is a recurrent, predictable event. No organism, however, lives for periods of time exceeding the periodicity of Milankovitch oscillations, and so for no organism do these oscillations form part of its predictable environmental background. An ordinary winter kills off most of the populations of insects that have life-spans shorter than one year, and mass extinctions remove a high proportion of taxa (at any level) at approximately 26-million-year intervals. In the same manner, orbital variations have a profound influence on individuals and populations that persist within shorter time scales.

The other relevant time scale is the duration of species. Species typically persist for periods of 1–30 million years, depending on taxonomic group. They thus persist much longer than the period of Milankovitch oscillations. Insofar as the duration of species is concerned, these oscillations might just as well not be happening. This is true even for the current (Quaternary) series of oscillations with climatic changes enhanced to yield massive, subcontinental-scale glaciation across much of the Northern Hemisphere.

Effects of Quaternary oscillations. The Quaternary has long been recognized as a period of fluctuating climates. There have been several continental glaciations within the last 425,000 years and many during the last 2.3 million years. The distinctive Quaternary feature of major glaciation may be due to the present configuration of the continents or to the relatively recent uplift of the mountains of western North America and the Tibetan plateau. Apart from the direct effects of ice sheets, the major environmental aspect of the Quaternary ice ages was repeated sea-level changes on the order of 100 m (about 300 ft), as water from the oceans became locked up on land as ice.

There is an excellent fossil record of the response of terrestrial plants to Quaternary climatic oscillations, in the form of pollen and macrofossils in sequences of peat and lake sediments. In eastern North America and western Europe, the present natural deciduous forests did not exist 20,000 years ago but developed during the period from about 15,000 to 5000 years ago by the spread of the individual species concerned from full-glacial refugia. These species appear to have spread as much as 1000 km (about 600 mi) at rates of 100–2000 m (about 300–7000 ft) per year. Each species appears to have responded individually to climatic change. During this period of change, some communities persisted for only 1000–2000 years and then broke up. The forest communities of eastern North America and western Europe have no history longer than 10,000 years. Forest communities formed and broke up repeatedly during the intervals between successive glacial periods.

Eastern North America and western Europe were

Temporal hierarchy of dominant processes controlling evolutionary patterns seen in the geological record			
Tier	Periodicity, years	Cause	Evolutionary process
First	—	Natural selection	Microevolutionary change within species
Second	20,000–100,000	Orbital forcing	Disruption of communities, loss of accumulated change
Third	—	Isolation	Speciation
Fourth	~26 million	Mass extinctions	Sorting of species

subjected to extreme temperature changes during the Quaternary because of their proximity to continental ice sheets. However, substantial changes in flora and vegetation have been identified in nonglaciated regions that experienced much smaller temperature changes. These regions were probably more representative of the likely effects of pre-Quaternary Milankovitch oscillations on flora and vegetation. There is no doubt that tropical forests, even in lowlands, experienced substantial changes during the climatic shifts of the Quaternary.

The late Quaternary fossil record for animals is less complete than for plants, but it has led to the same conclusions. Communities of mammals and beetles, for example, formed and broke up as the climates changed, because each species responded individually. Similarly, marine organisms were also highly mobile, on spatial scales of whole oceans, in response to Quaternary climatic changes. Such change happened repeatedly and regularly on time scales of only a few thousand years, much less than typical species durations.

Evolutionary consequences. The environmental history of the Quaternary is undoubtedly of great interest for the history of modern communities. This history assumes a greater significance, however, if it can be read as a model for the way that species have responded to climatic changes of the frequency of Milankovitch oscillations.

Over periods of ecological time, climates may be constant enough for stable communities to develop. In such circumstances, adaptation and evolution by Darwinian natural selection takes place. As climate changes in response to Milankovitch oscillations, these communities break up and new communities develop under the new conditions of climate and other aspects of the environment. Many species will have shifted their distributions on a subcontinental or suboceanic or even larger scale. Adaptations accumulated under the previous relatively stable conditions are likely to be lost unless they prove workable. It is unlikely that adaptation could proceed in the same direction as before: the climate has changed, the species may be living in a different environment, and its competitors have changed. Thus, orbital forcing of climatic and environmental change on time scales of 20,000–100,000 years may undo a substantial fraction of any progress accumulated at a microevolutionary level in ecological time, leaving mass extinctions (on time scales of tens of millions of years) to undo any lineage trends resulting from speciation events in the 1–10-million-year time scale.

Gould's three-tier hierarchy should thus be expanded to four tiers for a more complete understanding of the processes controlling evolutionary patterns seen in the geological record and for a proper integration of paleontological and ecological evidence (see **table**).

Most paleontological research takes place at the time scales of the third tier, which is too coarse to delineate what is happening at lower tiers. All ecological research takes place at the first tier, and the problem has been to integrate the two time scales. Until recently, the Quaternary ice ages seemed to form a barrier to this integration, because the one slice of time available for the intermediate time scale seemed to be atypical of Earth history as a whole. Now that it has been accepted that the ice ages are merely accentuated expressions of perpetual variations in the Earth's orbit, it is clear that the data of Quaternary research can be used to illuminate processes in the history of life on time scales between those of ecology and paleontology: the type of change that has taken place in the Quaternary has been operating throughout Earth history. The result reinforces the concept of punctuational equilibria and makes it more difficult to maintain the original thesis of Darwin that processes visible in ecological time build up into the macroevolutionary trends seen in the paleontological record.   K. D. Bennett

Bibliography. K. D. Bennett, *Evolution and Ecology: The Pace of Life*, Cambridge University Press, 1997; C. Darwin, *On the Origin of Species by Means of Natural Selection, or the Preservation of Favoured Races in the Struggle for Life*, John Murray, 1859; M. B. Davis, Pleistocene biogeography of temperate deciduous forests, *Geosci. Man*, 13:13–26, 1976; FAUNMAP Working Group, Spatial response of mammals to late Quaternary environmental fluctuations, *Science*, 272:1601–1606, 1996; S. J. Gould, The paradox of the first tier: An agenda for paleobiology, *Paleobiology*, 11:2–12, 1985; J. D. Hays, J. Imbrie, and N. J. Shackleton, Variations in the earth's orbit: Pacemaker of the ice ages, *Science*, 194:1121–1132, 1976; H. E. Wright, Jr., et al. (eds.), *Global Climates Since the Last Glacial Maximum*, University of Minnesota Press, 1993.

Stratigraphy

Stratigraphy is the science of layered rocks, including the study of their succession, relations, physical and chemical properties, and origins. Most layered rocks are sedimentary in origin but may include interbedded volcanic strata (such as flows and ash beds) in some settings.

Origins. Modern work on the origins, composition, and geometry of strata has determined that the concept of accommodation space is fundamental. This is the space made available for potential sediment accumulation below base level (above this level, erosion will occur). In the marine environment, sea level corresponds closely to base level, although some processes, such as the accumulation of beaches, deltas, and reefs, can generate limited thicknesses of deposit slightly above base level. In nonmarine settings the profile of a graded river is the surface that controls sedimentary accommodation. A sedimentary basin is an area of the Earth's surface where accommodation space has been continuously added over long periods of time by sinking of the Earth's crust (subsidence) and has been modified by sea-level change. The accumulation and removal of sediment in a basin also depend on external forcing processes, including global sea-level change (eustacy), earth movements (tectonism), and climate change. The internal composition and organization of the sediments in a basin depend on the nature of the sediment supply, the rate at which sediment enters the basin, and how this rate compares to the rate at which new sedimentary accommodation space is generated. For example, in a basin undergoing slow subsidence but with a large sediment supply, coastal depositional systems (such as shorelines, deltas, and carbonate platforms) may build seaward (this process is termed regression), and water depths will decrease with time. Large depositional systems, such as major deltas, may build laterally into the sea (this process is termed progradation). Conversely, where the basin undergoes rapid subsidence but with a limited sediment supply, water depths will increase with time, the sea will rise across the basin margins (transgression), and coastal sedimentary environments will gradually shift landward (retrogradation).

All these processes and forcing functions have been intensively studied by geologists and geophysicists, and the science of stratigraphy has been revolutionized in recent years, with the development of seismic stratigraphy in the 1970s and of computer modeling of sedimentary basins in the 1980s. Developments in the field can be grouped under four headings: geodynamic models, process-response models and facies analysis, chronostratigraphy, and sequence stratigraphy.

Geodynamic models. The development of plate tectonics assisted in the development of geodynamic models. Modern plate-tectonic interpretations of the geologic past help to explain the history of basins within the context of moving plates. Geophysical modeling of the Earth's crust has determined four principal reasons why the crust subsides to form sedimentary basins: (1) The crust thins as a result of stretching (as at a continental margin), erosion, or the removal of magma, for example, by a major series of volcanic eruptions. (2) The mass of the crust is increased by

loading with sediments or volcanic rocks or by the piling up of rocks in a new mountain belt. (3) The mass of lower levels of the crust is increased by the intrusion of molten magma or by rocks being pushed downward at the edges of oceans (the process is termed subduction). (4) Subsidence occurs over places in the middle of large continental areas where convection currents in the mantle cause "cold spots." Cool mantle is denser than its surroundings and tends to sink.

Process-response models and facies analysis. Sediments contain internal evidence of their origins, in the form of dynamic structures (such as animal borings and ripple marks) and textures (such as graded bedding, indicating changing current energies). Facies analysis is the study of these features, and the accumulated evidence from interbedded sedimentary units leads to the building of process-response models (also termed facies models), in which the various interpreted processes reveal the nature of the local depositional geography, such as a reef, delta, lake, and submarine fan.

Chronostratigraphy. It is rarely possible to trace individual beds very far. A knowledge of regional stratigraphic variation, which is important for the investigation of regional geologic history and for tracing economic deposits through the subsurface, depends on the geologist's ability to correlate strata from place to place. Comparison of the age of individual layers, by use of fossils (biostratigraphy), magnetic remanence (magnetostratigraphy), isotopic composition (chemostratigraphy), and radiometric dating is an essential stratigraphic tool and has led to the development of the geological time scale of absolute ages. *See* GEOLOGICAL TIME SCALE; RADIOISOTOPE DATING.

Sequence stratigraphy. Refinements in processing techniques and conceptual breakthroughs in the interpretation of reflection seismic data led to the development of seismic stratigraphy, a tool now widely used in petroleum exploration and development. Very large scale stratigraphic features are revealed by this technique, which has shown that most stratigraphic successions may be subdivided into successions of sequences. A sequence is a package of strata bounded by unconformities (and their stratigraphic equivalent) that indicate periods of erosion (negative sedimentary accommodation) and between which the internal arrangements of sedimentary facies reveal a systematic cycle of base-level rise and fall (**Fig. 1**). This systematic facies arrangement gives sequences considerable predictability, which is useful in areas of limited data, such as coastal offshore frontier basins.

Quantitative dynamic stratigraphy. Much success has been achieved in the integration of the various stratigraphic data sets and the concepts and methodologies noted above, leading to the evolution of a new form of science recently termed quantitative dynamic stratigraphy. Much can now be quantified through a better understanding of physical and chemical principles and through the acquisition of new kinds of data, including geodynamic processes, the stresses and resulting strains of mountain

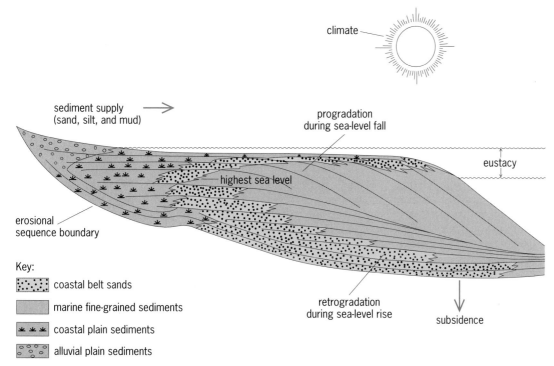

Fig. 1. Generalized model of sedimentation on an extensional continental margin, where the rate of subsidence increases in an offshore direction (to the right) and sediment supply is dominated by sand, silt, and mud fed into the basin from the coast (from the left). A complete cycle of sea-level rise and fall is shown. Note the architecture of prograding and retrograding coastal sands formed as sea level rises and falls. (*After P. R. Vail, Seismic stratigraphy interpretation using sequence stratigraphy, Part 1: Seismic stratigraphy interpretation procedure, in A. W. Bally, ed., Atlas of Seismic Stratigraphy, Amer. Ass. Petrol. Geologists Stud. Geol., 27, 1:1–10, 1987*)

building (tectonics), and some of the broader features of the sedimentary response to climate change and surface distribution processes (such as currents, waves, and wind; **Fig. 2**). More is known about rates of subsidence, rates of sedimentation, and the shape of prograding and retrograding continental margins, due to numerical documentation and simulation with the use of computers. There are two main types of model: forward models, which simulate sets of processes and responses, given predetermined input variables (**Fig. 3**), and inverse models, which use the structure of a forward model to simulate a specific result, such as an observed basin architecture. A particular value of such models is that they provide the opportunity to test different interpretations by variation in the input data. The values for unknown variables can be estimated and used as input into repeated model runs in order to simulate stratigraphic architecture. The results are illustrated graphically with ranges of values for the different variables used as input in order to develop families of possible solutions. These graphs are then compared with the actual basin architectures. Numerical modeling of this type permits the application of sensitivity tests to gauge the relative importance of the variables.

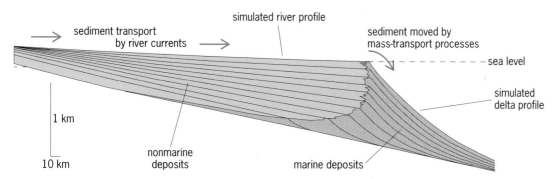

Fig. 2. Computer-simulated model of sedimentation on a coastal plain that undergoes differential subsidence, with the rate of subsidence increasing to the right (offshore). Vertical exaggeration is ×100. Sediment is fed into the basin from the left and distributed according to a diffusion equation that models the transport down a river slope. The steeper slope of the delta reflects the more efficient sediment transport in the marine environment. Bedding traces mark increments of 1 million years. (*After T. E. Jordan and P. B. Flemings, Large-scale stratigraphic architecture, eustatic variation, and unsteady tectonism: A theoretical evaluation, J. Geophys. Res., 96B:6681–6699, 1991*)

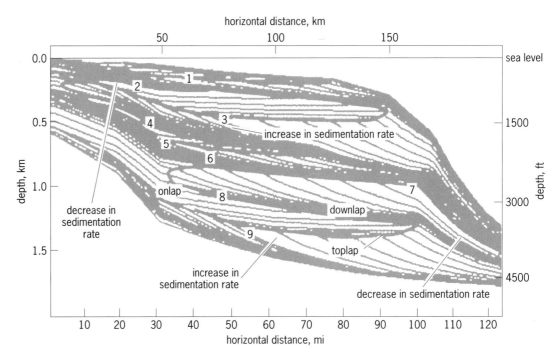

Fig. 3. Model of continental-margin sedimentation with subsidence rate increasing offshore (to the right), variable sediment supply, and three complete cycles of sea-level change. The heavy line tracks the position of the shoreline. Onlap, toplap, and downlap refer to the architecture of bedding terminations that develop in response to changes in accommodation. (*After D. T. Lawrence, Evaluation of eustasy, subsidence, and sediment input as controls on depositional sequence geometries and the synchroneity of sequence boundaries, in P. Weimer and H. W. Posamentier, eds., Siliciclastic Sequence Stratigraphy, Amer. Ass. Petrol. Geologists Mem. 58, pp. 337–367, 1993*)

Figures 2 and 3 illustrate the development of models for clastic sedimentation on an extensional continental margin, such as those bordering the Atlantic and Indian oceans. In Fig. 2, each million-year increment of sediment represents the output of an equation that simulates the transport of a set volume of sediment down a river profile and then across a prograding delta. The addition of a cycle of sea-level change and varying the sediment input rate lead to the model shown in Fig. 3. The shoreline moves seaward during times of falling sea level, when deltaic sediments actively prograde and the beds step out across earlier deposits, an architectural configuration termed downlap. The beds may be shaved off by erosion from waves or rivers during falling sea level as accommodation space is removed, leading to toplap. Onlap occurs during rising sea level, as accommodation increases, sedimentation rises up the continental margin, and the shoreline shifts landward.

One use of these models has been to test the sensitivity of depositional systems under various conditions to changes in the major controls: subsidence rate, sea-level change, and sediment supply. It has been demonstrated that the effects of each of these controls may mask the others, making it difficult to extract clear, simple signals of global change. For example, early work had held out the hope of being able to build a curve of global sea-level change from basin architectures, but it is now realized that this cannot be done. Current work is focusing on two main areas: (1) the effects of climate change on sediment supply and the resulting variations in basin-

fill geometry and (2) detailed examination of the effects of regional tectonic events, such as faulting, on subsidence patterns, sediment supply, and basin architectures. More powerful computers with much larger memories are permitting the development of models that simulate sedimentary processes in three dimensions.

For background information *see* FAULT AND FAULT STRUCTURES; GEOCHRONOMETRY; PALEONTOLOGY; SEDIMENTOLOGY; SEISMIC STRATIGRAPHY; STRATIGRAPHIC NOMENCLATURE; TECTONOPHYSICS in the McGraw-Hill Encyclopedia of Science & Technology.
                                                                    Andrew D. Miall

Bibliography. T. A. Cross (ed.), *Quantitative Dynamic Stratigraphy*, Prentice Hall, 1990; A. D. Miall, *The Geology of Stratigraphic Sequences*, Springer-Verlag, 1997; A. D. Miall, Whither stratigraphy?, *Sediment. Geol.*, 100:5–20, 1995.

Superconducting devices

Superconducting fault-current limiters use the transition of superconductors from zero to finite resistance to limit the fault currents that result from short circuits in electric power systems. Such short circuits can be caused by aged or accidentally damaged insulation, by lightning striking an overhead line, or by other unforeseen faults. If not deliberately checked, the subsequent fault current is limited only by the impedance of the system between the location of the fault and the power sources. This current can reach as much as 100 times the nominal current of the system (the largest current in normal operation), and leads to high mechanical and thermal stresses, both of which are proportional to the square of the current's value.

Power-system components must be designed to withstand these stresses for a certain time period. This time interval is usually determined by the time needed for circuit breakers to interrupt the short circuit, and can range from 20 to 300 ms. The higher the fault current anticipated, the higher are the costs of all types of equipment in the system, especially the circuit breakers. Besides additional initial costs, high fault currents increase maintenance costs because they accelerate the aging of system components.

Conventional measures to limit fault currents include the use of artificially increased impedances in the system, which can be realized via air-coil reactors or high transformer impedances. Another approach is to limit the number of power sources that will feed the fault current, by artificially splitting the grid. However, these measures conflict with the increasing demand for higher power quality, which requires strongly interconnected grids with low impedances under normal operation.

Fault-current limiters. Fault-current limiters fulfill this requirement. They are devices which under normal operation have negligible influence on a power system but, in case of a short circuit, will limit the fault current to a value that is not much higher than the nominal current. Approaches to realizing fault-current limiters have been based on fast (less than 1 ms) current interruption (for example, explosive fuses or power electronics), on the detuning of LC-resonance circuits (only for alternating-current applications), or on components with strongly nonlinear current-voltage characteristics (such as semiconductors, iron-core reactors, or superconductors).

Among the nonlinear materials, superconductors stand out because of their unique transition from zero to finite resistance. This article discusses superconducting fault-current limiters, which utilize this transition. Fault-current limiters that are based on other concepts but might still use superconductors, for example, in order to realize low-loss coils, are not discussed. There are essentially two types of superconducting fault-current limiters: the resistive type and the shielded-core type.

Superconductors. Superconductors lose their electrical resistance below certain critical values of temperature (T), magnetic field (B), and current density (j). The phase diagram of a superconductor (**Fig. 1**) can be characterized by three regions. Inside a surface S_1, the material is in its true zero-resistance state. Between the inner surface S_1 and an outer surface S_2, the material is in a transition region, where its resistivity increases very rapidly with T, B, and j. While this transition extends over a large range along the B-j plane, it is rather narrow on the T axis. Outside S_2, the material is a normal conductor, whose resistivity is essentially independent of B and j.

Although superconductivity was discovered in

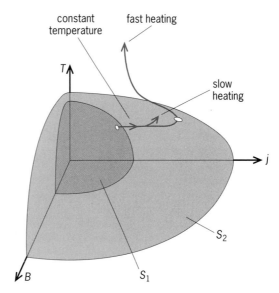

Fig. 1. Simplified phase diagram of a superconductor in the space of temperature (T), magnetic field (B), and current density (j). Inside surface S_1, the superconductor has zero-resistance; between S_1 and S_2, it is in a transition state; outside S_2, it is a normal conductor. The curved arrows represent the transition from superconducting to normal conducting state, with the positions of the arrows corresponding to the j value after about 5 ms for the constant-temperature, slow-heating, and fast-heating designs. The transition region on the B-j plane can be more than 10 times wider than indicated.

1911, the highest critical temperature known, until 1986, was 23 K ($-418°$F). The high cooling cost of these so-called low-temperature superconductors, which are mostly metals and alloys, has prohibited many commercial applications. With the advent of the high-temperature superconductors, which have a critical temperature of the order of 100 K ($-280°$F), the prospects for commercialization of superconductivity have improved. However, whereas products of low-temperature superconducting materials have reached a very mature state, high-temperature superconducting products are still under development. Since high-temperature superconducting materials are complicated ceramic systems, the fabrication of long-length, flexible conductors is difficult.

Resistive units. The resistive type of superconducting fault-current limiter has a straightforward design in which the superconductor is directly connected in series to the line to be protected (**Fig. 2**). The superconductor is immersed in a coolant (such as liquid helium for a low-temperature superconductor or liquid nitrogen for a high-temperature superconductor), which is chilled by a refrigerator. In order to limit cooling costs, the connection from the line at room temperature to the superconductor is provided by special current leads, which are designed to minimize the heat transfer to the coolant.

Normal operation. The cross section of the superconductor is chosen so that at nominal current the superconductor is operated inside the surface S_1 of the phase diagram (Fig. 1), that is, at zero resistance. The superconductor's interference with the network is thus neglegible. Both the current and the magnetic

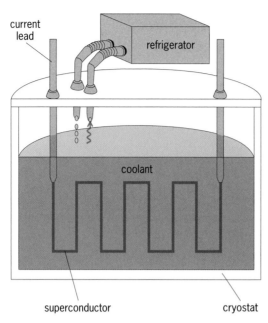

Fig. 2. Resistive superconducting fault-current limiter.

field (which, in this case, is generated entirely by the current) vary with the load in normal operation, whereas the temperature is constant.

The superconductor has true zero impedance only for direct currents. For the more common alternating-current (ac) applications, the finite length of the superconductor leads to a certain reactance, and the magnetic ac field generated by the current produces so-called ac losses. They both depend very strongly on the geometry of the superconductor, and can be reduced by optimized conductor architecture. The contribution of the ac losses to the total impedance of the fault-current limiter is negligible. However, in contrast to the reactance, they dissipate energy in the superconductor and thus increase cooling costs.

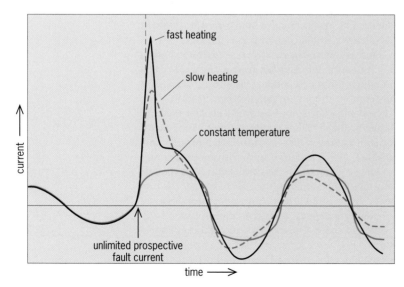

Fig. 3. Current-limiting behaviors of constant-temperature, slow-heating, and fast-heating designs of fault-current limiters when a fault occurs.

Fault operation. In case of a fault, the increase of current and magnetic field will move the superconductor out of the superconducting region into the transition region between surfaces S_1 and S_2 (Fig. 1). The rapidly increasing resistance will limit the current to a value below the unlimited prospective fault current. After some time, such as 100 ms, a breaker will interrupt the circuit. The actual current-limiting behavior strongly depends on the length of the superconductor and on the type of superconducting material. The three main types of current-limiting behavior can be realized simply by varying the length of the same superconducting material (**Fig. 3**).

Designs. By using a very long conductor, the electric field to which the superconductor is exposed during the fault can be made very small. Thus, the corresponding current density exceeds the nominal value by only a small amount. The power density (electric field times current density) dissipated in the superconductor is negligible so that the superconductor will essentially not warm up (the constant-temperature curves in Figs. 1 and 3) until the breaker interrupts the current. The design leads to very good current limiting behavior. However, the long length of the conductor is very expensive and causes rather high ac losses.

The other extreme design uses a very short conductor, leading to a relatively high voltage drop during the fault. As the fault occurs, the current increases. Essentially, it is limited not by the small resistivity of the intermediate state (between S_1 and S_2) but only by the inductance of the system. However, because of the very high energy density dissipated in the superconductor, the material heats up rapidly and, after a few hundred microseconds, passes through S_2 (Fig. 1). Finally in the normal-conducting state, the superconductor limits the fault current close to or even below the nominal value (the fast-heating curves in Figs. 1 and 3). This design needs a rather small amount of superconducting material. However, the height of the first peak depends on the unlimited prospective fault current, and instantaneous recovery to normal operation is not possible since the superconductor first has to cool down. Furthermore, severe overvoltages may arise from the abrupt current reduction, caused by the rapid transition into the normal-conducting state. To reduce these overvoltages, the employment of a normal resistor or reactor in parallel to the superconducting fault-current limiter has been suggested.

Using a conductor of intermediate length leads to an intermediate level of energy dissipation in the superconductor. For the first 10 ms, the maximum of the fault current is limited to several times (for example, 5–10 times) that of the nominal current. The superconductor keeps warming up and after a few tens of milliseconds enters the normal conducting region (the slow-heating curves in Figs. 1 and 3).

In the above discussion it is assumed that during the limitation process the voltage drop is uniform over the whole length of the conductor. In reality, however, superconductors tend to develop thermal instabilities called hot spots, which are due to the strong current and temperature dependence of their resistivity in the state between S_1 and S_2. If, caused by any inhomogeneity, a part of the superconductor has a higher resistivity, this part will heat up faster, leading to an even higher resistivity at that point and further accelerated heating, which may finally lead to a burn-through. The common measure to reduce this problem is to attach a normal conducting bypass in close electrical contact to the superconductor, which allows the current to bypass the hot spot. Of course, the bypass reduces the total normal resistance of the conductor, and this reduction may have to be compensated by increasing the conductor's length. The problem of excessive heating also limits the maximum possible electric field to which the conductor can be exposed during the fault; that is, it also sets a lower limit on the length of the superconductor.

Shielded-core units. In the shielded-core or shorted-transformer type of superconducting fault-current limiter, the current-limiting superconductor is not connected galvanically in the line but is magnetically coupled into the power network. The device is essentially a transformer (**Fig. 4**), with its primary normal conducting coil connected in series to the line to be protected, while the secondary side is a superconducting tube (that is, a one-turn coil). In normal operation the iron core sees no magnetic field because it is completely shielded by the superconductor. Because of the inductive coupling between the line and the superconductor, the device is sometimes also referred to as an inductive superconducting fault-current limiter. The voltage on the secondary is reduced by the number of turns of the primary coil, while the current is increased by the same factor. The superconductor on the secondary has to be designed for these values. Assuming an ideal transformer, the device will behave exactly like a resistive superconducting fault-current limiter. For ceramic high-temperature superconductors, this geometry is usually more easily realized than long-length conductors on which the resistive superconducting fault-current limiter relies.

Comparison of limiter concepts. In order to achieve a certain current limiting behavior, both types of superconducting fault-current limiter need the same amount of any particular superconducting material. Compared to the resistive superconducting fault-current limiter, the shielded-core superconducting fault-current limiter shows the major disadvantages of being applicable only to alternating currents and having much larger size and weight. However, the device needs no current leads. This feature is especially attractive for the protection of high-current systems, since the losses of the leads are proportional to their current-carrying capability. Both limiter concepts have been actively pursued for their respective advantages.

Research activities. The development of superconducting fault-current limiters based on low-temperature superconducting materials has followed the resistive concept, utilizing the fast-heating design. A 6.6-MVA (6.6 kV nominal voltage, 1000 A nominal current) single-phase prototype has been built, with a 2–4-GVA device under development. Another group has developed and tested a 7.6-MVA (36 kV, 210A) single-phase device.

At the beginning of 1999, there remained only one large superconducting fault-current limiter project that utilized low-temperature superconductors, but there were worldwide more than 10 major projects involving high-temperature superconducting devices. The reason, of course, is the lower cooling costs required for high-temperature superconductors. However, high-temperature superconducting materials are much less advanced than low-temperature superconductors. Essentially, only three types of high-temperature superconducting materials are available: silver sheathed wire (with a cross section of less than 1 mm²), thin films (about 1 μm thick), and bulk material (with a cross section of several square millimeters). They are suited differently for applications in resistive and shielded core superconducting fault-current limiters.

The largest prototype superconducting fault-current limiters with high-temperature superconductors utilizes the so-called Bi-2212 (bismuth) bulk material. A three-phase 1.2-MVA prototype based on tubes of this material is of the shielded-core and slow-heating type. It was successfully operated for one year under actual conditions in a Swiss hydropower plant.

Because of their high critical current density (1 MA/cm²), YBCO (yttrium-barium-copper-oxygen) films are especially suited for the fast-heating design. In a resistive 100-kVA model based on this material, the films are deposited on planar ceramic substrates, covered with a gold bypass, and patterned into a meander (a labyrinthine pattern) in order to create a long conducting path.

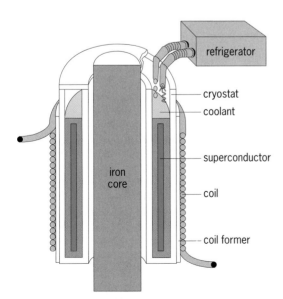

Fig. 4. Shielded-core (shorted-transformer or inductive) superconducting fault-current limiter. Only the superconducting tube is cooled by the refrigerator.

Labels in figure: refrigerator, cryostat, coolant, superconductor, coil, coil former, iron core

The development of high-temperature superconducting materials has focused on wire technology. Long lengths of Bi-2223 wire have been produced. Even though this wire might be suitable for cable, motor, and transformer application, it is at present not particularly suited for superconducting fault-current limiters because of its low normal resistance, caused by the high portion of silver in the material. However, if the resistivity of the silver matrix could be increased, the wire would be quite suitable for resistive superconducting fault-current limiters and could even enable the incorporation of current limiting functionality into other devices. Following this approach, a project has been launched to develop a current-limiting transformer.

For background information *see* CIRCUIT BREAKER; ELECTRIC POWER SYSTEMS; ELECTRIC PROTECTIVE DEVICES; LIGHTNING AND SURGE PROTECTION; SUPERCONDUCTING DEVICES; SUPERCONDUCTIVITY; TRANSFORMER in the McGraw-Hill Encyclopedia of Science & Technology.   Willi Paul

Bibliography. B. Gromoll et al., Resistive current limiter with YBCO films, *Applied Superconductivity 1997* (EUCAS '97), Inst. Phys. Conf. Ser. Nr. 158, pp. 1243–1246, 1997; W. Paul et al., Test of 1.2 MVA high-Tc superconducting fault current limiter, *Supercond. Sci. Technol.*, 10:914–918, 1997; T. Verhaege et al., Experiments with a high voltage (40 kV) superconducting fault current limiter, *Cryogenics*, 36: 521–526, 1996; T. Yazawa et al., Experiments with a 6.6 kVA/1 kA single-phase superconducting fault current limiter, *Applied Superconductivity 1997* (EUCAS '97), Inst. Phys. Conf. Ser. Nr. 158, pp. 1183–1186, 1997.

Superdeformed nuclei

The nucleus is a unique many-body quantal system consisting of a finite number of strongly interacting fermions (neutron and protons). While the interactions between nucleons are sufficiently weak to allow both neutrons and protons to behave as independent particles, the number of nucleons is large enough to allow for collective behaviors such as rotations and vibrations. Consequently, nuclei exhibit a large variety of shapes, and the equilibrium shape is determined by the interplay between macroscopic (liquid-drop) and microscopic (quantal shell-correction) contributions to the total energy of the nucleus. The microscopic shell corrections arise from the occupation of nonuniformly distributed energy levels. Rotation also plays a role in the shape of a nucleus by modifying both the liquid-drop moment of inertia and the relative energies of the single-particle levels. Currently, a major thrust of nuclear structure research is to study how the nuclear shape evolves as a function of both excitation energy and angular momentum in different areas of the nucleonic chart.

Superdeformation refers to one of the more exotic shapes the nucleus can acquire. A nucleus becomes superdeformed when it is trapped in a metastable potential minimum corresponding to an elongated football-like shape with a major-to-minor axis ratio of roughly 2:1. The nuclear structure associated with this minimum provides a unique laboratory for testing nuclear models. Superdeformation was first proposed in the 1960s to explain the fission isomers observed in heavy-actinide nuclei with atomic mass (A) of around 240. In the late 1980s, three more regions of superdeformation were discovered near $A \sim 130$, 150, and 190, and more recently, superdeformed bands have been discovered around $A \sim 60$ and 80 (**Fig. 1**). These five latter regions have been characterized by identifying and studying superdeformed rotational bands.

Superdeformed bands. Superdeformed bands are one of the best examples of quantum rotors. Their gamma-ray energy spectrum is characterized by a long sequence of gamma rays (10–20), regularly spaced in energy. These gamma rays correspond to transitions between consecutive levels in the rotational band (**Fig. 2**). In all superdeformed regions except $A \sim 240$, superdeformed bands have been identified using heavy-ion-induced fusion evaporation reactions. In these types of measurements, an accelerated projectile nucleus collides and fuses with a target nucleus to form a highly excited compound nucleus. This system first cools by evaporating particles, namely, neutrons, protons, or alpha particles. When particle evaporation is no longer energetically favored, the nucleus releases the rest of its energy by emitting gamma rays. When gamma emission begins, the nucleus possesses 50–70 h of angular momentum, where h is Planck's constant divided by 2π. Since gamma-ray transitions typically connect nuclear states differing in angular momentum by 1 or 2 h, a nucleus excited in this manner can emit 30 or more gamma rays as it cools down.

Gamma-ray spectrometers. In a nucleus with a superdeformed minimum, the sequence of gamma-ray transitions will pass through any one superdeformed band at most only 2% of the time. In order to distinguish this rare decay mode from the multitude of pathways that the nucleus takes to the ground state during gamma deexcitation, it is important to measure as many of the emitted gamma rays as possible. This is accomplished by using advanced gamma-ray spectrometers which consist of high-purity germanium detectors. Each germanium detector has an energy resolution of about 0.2% for detecting 1-MeV gamma rays. The two largest gamma-ray spectrometers in the world are Gammasphere, in the United States, and Euroball, operated by the European Community. Each detector array has a 10% efficiency for detecting 1-MeV gamma rays and, on average, detects about 20% of the gamma rays (4–6) emitted in a heavy-ion-induced fusion reaction. Both of these spectrometers have been in use since 1992, and they have provided a gain in sensitivity of more than a factor of 100 over previous spectrometers. By using these arrays, the $A \sim 130$, 150, and 190 superdeformed regions have been extended, and two new

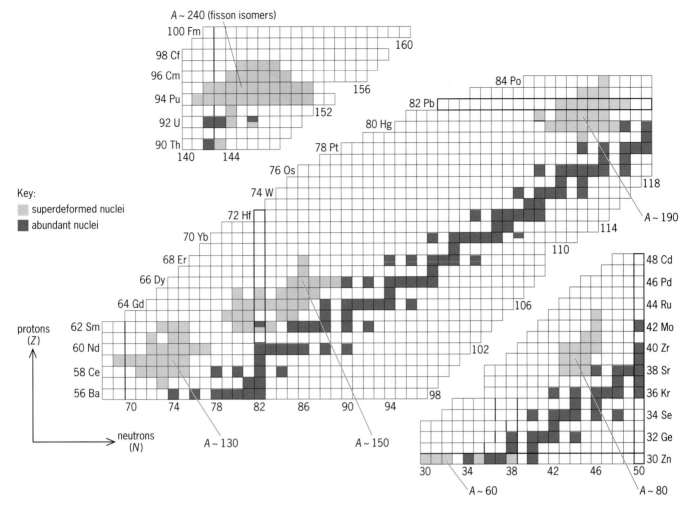

Fig. 1. Regions of the nuclear chart for neutron number (*N*) and proton number (*Z*) where superdeformed bands have been observed.

regions have been discovered near $A \sim 60$ and 80 (Fig. 1). In many instances, multiple superdeformed bands are observed in a single nucleus.

Structure and evolution. Superdeformed bands are populated at both high excitation energy (10–15 MeV) and high angular momentum (40–60 $\hbar$), when the excitation energy of the superdeformed minimum is roughly equal to the excitation energy of the first minimum (**Fig. 3**). Once a superdeformed band is populated, regularly spaced gamma rays are emitted as the superdeformed band spins down. As this occurs, the excitation energy of the superdeformed band becomes much greater than that of the lowest-lying states of the same spin in the first well (the yrast envelope). When the barrier between the first and superdeformed well becomes small enough, the superdeformed band can mix with states in the first well. Even though this interaction is small, it is sufficient to cause the superdeformed band to depopulate by tunneling through the potential barrier and decay toward the ground state.

Topics of interest in the study of superdeformation include attempts to measure and describe (1) how the gamma decay becomes trapped in the second minimum, (2) the microscopic structure of individual superdeformed bands, (3) the response of the nucleus as the superdeformed band spins down, and (4) the mechanism responsible for the rapid decay out of the superdeformed band to the ground state. For most superdeformed bands, basic quantities such as exact excitation energies, spins, and parities are not known because direct transitions between the superdeformed band and known levels in the nucleus are not observed. Nevertheless, the underlying structure of individual superdeformed bands is well understood since the level spacing of a rotational band is partially determined by its underlying microscopic structure. By comparing the measured level sequence of a superdeformed band to what is calculated for a particular microscopic configuration, the underlying single-particle nature of the band can be inferred.

Identical bands. One fascinating aspect of superdeformation involves the observation of identical bands, that is, superdeformed bands in neighboring nuclei sharing identical level spacings. This phenomenon is surprising because mass differences between neighboring nuclei are expected to result in

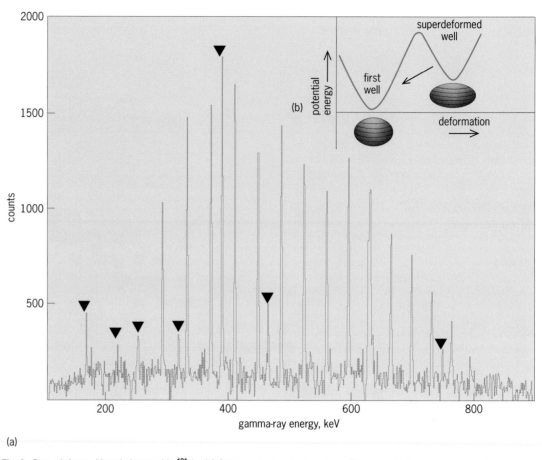

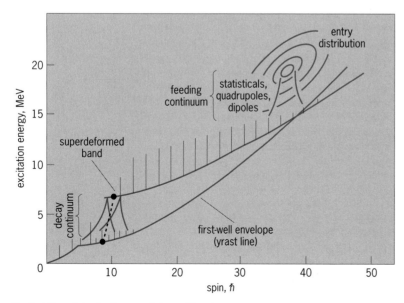

Fig. 2. Superdeformed band observed in ^{191}Hg. (*a*) Gamma-ray energy spectrum. The unmarked, regularly spaced gamma-ray lines link transitions between consecutive members of the superdeformed band. The lines marked with a triangle correspond to transitions between levels associated with states in the first well. Gamma-ray transitions connecting states in the first well with superdeformed levels are not observed for this superdeformed band. (*b*) Schematic plot of the potential energy of the nucleus as a function of deformation.

Fig. 3. Life and death of a superdeformed band.

differences between transition energies that are ten times larger than those observed between identical bands. Identical bands are found in all the superdeformed regions (except $A \sim 240$), and at present there

is no universally accepted explanation of the phenomenon. Some theories suggest that these bands are a manifestation of symmetry restoration in nuclei, while other explanations are more mundane, such as relating the phenomenon to cancellation effects involving small changes in deformation and nucleonic interactions.

Experimentally, this phenomenon has been best characterized in the $A \sim 150$ superdeformed region, where many pairs of identical bands have been identified. In general, the deformations of $A \sim 150$ superdeformed bands are found to vary significantly; however, identical bands are observed to have equal deformations. Theoretical calculations can reproduce these observations by considering only contributions from the addition or subtraction of nucleons and ignoring the residual interactions between them. For identical bands, differences in the microscopic structure are related to changes in single-particle occupancies that do not affect the measured properties, such as the deformation and the level structure. This extreme single-particle description which has emerged for superdeformed bands is in contrast to other types of collective phenomena in nuclei which are strongly affected by the residual interactions among nucleons.

Deviations from axial symmetry. While superdeformed bands are associated with a prolate shape, there is evidence that some bands deviate from axial symmetry. For example, the level spacing for a number of superdeformed bands in the $A\sim 80$ region can be understood in microscopic calculations only by assuming that the nuclear shape is triaxial (all three principal axes having different lengths). In the mass 190 region, several superdeformed bands are observed to decay directly to superdeformed levels in other bands. This phenomenon can be explained by assuming the superdeformed band lying higher in energy is coupled to an octupole (pear-shaped) surface vibration. Finally, in the $A\sim 150$ region small deviations are observed in the level structure of several superdeformed bands. These deviations occur when alternate levels in the superdeformed band are perturbed in opposite directions by approximately 100 electronvolts. It has been suggested that a hexadecupole deformation, which flattens the ellipsoidal surface of the nucleus, might be responsible for this phenomenon.

Decay to first well. When superdeformed bands decay into the first well, their excitation energies are found to lie between 2 and 6 MeV. At these excitation energies, the superdeformed levels are in proximity to many less-deformed levels lying in the first minimum. When the barrier between the two minima is small enough, a superdeformed level is able to interact with many states in the first minimum. The superdeformed band then decays to the first well by tunneling through the potential barrier. The gamma-ray spectrum associated with this decay is smooth and nearly featureless, reflecting the fact that there are no favored pathways for the nucleus to take as it makes its way to the ground state. As a result, the probability to observe a direct decay from the superdeformed band to known levels near the ground state is quite low. In the mass 190 region, direct gamma-ray transitions have been observed for only three superdeformed bands. No such links have been established in the $A\sim 150$ or 80 regions, and one example has been found in the $A\sim 60$ region. A number of cases have been reported in the mass 130 region, where the deformations and the excitation energies are the smallest for superdeformed shapes.

Prospects. Outstanding problems associated with identical bands, new regions of superdeformation, and decay out of the superdeformed minimum continue to occupy both experimental and theoretical studies. There are predictions of more elongated nuclear shapes with an axis ratio of 3:1. Many attempts have been made to produce and observe states with such a deformation, but thus far none have been successful. If such nuclear shapes do exist, it may take an even more sensitive gamma-ray spectrometer to observe the signal associated with such an exotic phenomenon.

For background information *see* GAMMA-RAY DETECTORS; GAMMA RAYS; NUCLEAR FISSION; NUCLEAR REACTION; NUCLEAR STRUCTURE in the McGraw-Hill Encyclopedia of Science & Technology.

Michael P. Carpenter

Bibliography. C. Baktash, B. Haas, and W. Nazarewicz, Identical bands in deformed and superdeformed nuclei, *Annu. Rev. Nucl. Part. Sci.*, 45:485–541, 1995; R. V. F. Janssens and T. L. Khoo, Superdeformed nuclei, *Annu. Rev. Nucl. Part. Sci.*, 41:321–355, 1991; P. J. Nolan, F. A. Beck, and D. B. Fossan, Large arrays of escape-suppressed gamma-ray detectors, *Annu. Rev. Nucl. Part. Sci.*, 44:561–606, 1994; P. J. Nolan and P. J. Twin, Superdeformed shapes at high angular momentum, *Annu. Rev. Nucl. Part. Sci.*, 38:533–562, 1988.

Tea (health)

Health benefits of green and black teas in relation to cardiovascular disease and cancer are being studied. Catechins of green tea as well as the oxidized catechin conjugates of black tea (thearubigens and theaflavins) are strong antioxidants that scavenge most oxygen-centered free radicals, including the highly reactive hydroxide radical ($OH^{\bullet}$). Epicatechin, epicatechin-3-gallate, epigallocatechin, and epigallocatechin-3-gallate are the most potent antioxidants in tea (**Fig. 1**). During the production of black tea, enzyme-catalyzed oxidation of catechins leads to the formation of catechin quinones that form structurally complex and darkly pigmented theaflavins and thearubigens. Caffeine, another component of tea, has been reported to have cancer-preventing action in animal studies, but it is relatively ineffective in inhibiting the growth of cancer cells. Uncomplexed catechins are more concentrated in green tea than in black tea. Epidemiological studies indicate health benefits from consumption of either kind of tea. In addition, catechins and polyphenols have specific effects on the growth of cancer cells and

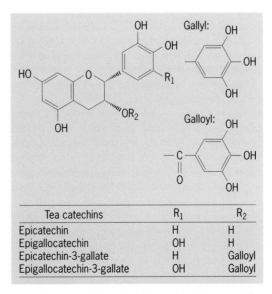

Tea catechins	R_1	R_2
Epicatechin	H	H
Epigallocatechin	OH	H
Epicatechin-3-gallate	H	Galloyl
Epigallocatechin-3-gallate	OH	Galloyl

Fig. 1. Structures of the principal catechin of green tea. Tea polyphenols are made up largely of the epicatechin group of flavanols, which differ from other catechins by the orientation of the hydroxyl group on the pyran ring. Gallocatechins have three hydroxyl groups on the b ring. Catechin gallates are gallic acid esters of the hydroxyl group on the pyran ring.

growth-associated proteins that extend beyond their antioxidant properties to help explain their anticancer activity.

Cardiovascular disease reduction. Evidence that drinking tea offers protection against cardiovascular disease comes both from epidemiological studies with tea drinkers in Japan and Norway and from animal studies. The Japanese and Norwegian studies showed a significant inverse relationship between tea drinking and plasma cholesterol levels without any effect on the levels of high-density lipoprotein (HDL, the so-called good cholesterol). The Japanese study involved predominantly green tea drinkers, and the Norwegian study mainly black tea drinkers. Lowered blood pressure and reduced coronary heart disease were noted. Trends toward reduced cardiovascular risk were also noted in studies in Israel and Holland. In these studies, tea consumption was only one of the possible lifestyle variables, and a clear separation of the effects of tea from other factors was not always possible. Similar criticisms emerged in human studies relating to tea consumption and cancer.

Rats fed excessive dietary fat to induce hypercholesterolemia exhibited reduced blood pressure and lower blood cholesterol levels in response to green tea polyphenols. Some of this effect may be attributed to the reduced solubility of cholesterol in mixed bile salt micelles (submicroscopic structures made of polymeric molecules or ions) when tea catechins are present, resulting in reduced cholesterol uptake from the intestine.

Cancer protection. The health benefits of tea in cancer have been broadly attributed to three areas: antioxidant properties, effects on intestinal microflora and nutrient absorption, and effects on the cell's biochemical machinery specific to the growth of cancer cells. The anticancer effects of tea are indicated both from human epidemiological observations and from animal in vivo studies. In general, these effects have been attributed to epigallocatechin gallate, the major tea catechin. Other antioxidant catechins and polyphenols present in tea may have less marked anticancer properties.

Human studies. Epidemiological studies showed that cancer onset of Japanese patients who had consumed ten cups of green tea per day was 8.7 years later among females and 3 years later among males, compared with patients who had consumed under three cups per day. A possible relationship between high consumption of green tea and the low incidence of prostate and breast cancer in some Asian countries has been postulated. Little is known about the absorption and clearance of tea catechins from the circulation. They are highly water soluble and effectively absorbed in the intestine.

Animal and in vitro studies. Animal studies involving known chemical carcinogens as well as transplanted tumor cells indicate reduced cancer risk from the consumption of green or black tea or tea catechins. These studies have suggested cancer-protective responses that span the entire process of carcinogenesis (including the formation and activation of carcinogens and cancer initiation, promotion, and progression) or that diminish tumor growth and metastatic spread. The former are most often attributed to antioxidant effects, whereas the latter may be related to more specific mechanisms of cancer cell growth inhibition.

To establish a mild protective or therapeutic effect of tea or its constituents on cancer may prove difficult from retrospective human studies or even from prospective intervention studies. The preferred approach has been to carry out controlled studies in animals and in vitro to establish the candidate mechanisms underlying the proposed anticancer effect of tea. Once appropriate biomarkers for anticancer effects of tea become established, they may be used effectively in controlled human intervention trials.

tNOX. A cancer- (tumor-) specific growth protein, tNOX, has been identified as a target for epigallocatechin gallate to help explain the anticancer benefits of tea. NOX proteins are located at the cell surface. They are responsible for the increase in cell size following cell division. Once cells divide, they must reach some minimum size in order to divide again. Cells in which NOX activity is blocked are unable to enlarge. Therefore, they cease to divide and, after several days, undergo programmed cell death (apoptosis).

NOX activity in noncancer cells is constitutive but regulated and growth factor–responsive. Conversely, a NOX activity found in cancer cells (tNOX) is unregulated and constitutively activated. Therefore, enlargement of cancer cells is also unregulated and constitutively activated. Because cancer cells express tNOX, an unregulated NOX form, cell enlargement no longer limits multiplication of cancer cells.

The reactions carried out by NOX proteins are hydroquinone and nicotinamide adenine dinucleotide (NADH) oxidation and protein disulfide-thiol interchange (**Fig. 2**). These functions, in concert with other membrane proteins, allow cells to enlarge.

The cancer-specific NOX protein, which is the potential tea target, differs from the noncancer NOX protein in several respects. Unlike the noncancer NOX protein, tNOX is inhibited by agents that induce cells to differentiate and by antitumor drugs that act at cell surface targets. The antitumor drugs targeted to tNOX have no effect on the activity of the noncancer CNOX protein at therapeutic doses. At therapeutic doses, they slow the growth of cancer cells but not that of noncancer cells.

Not only do green and black tea catechins slow the growth of cancer cells but inhibition of tNOX may be the underlying mechanism. NOX activities of human mammary cancer cells and human cervical carcinoma (HeLa) cells grown in culture as well as partially purified tNOX from HeLa and mammary cancer cells were inhibited by tea dilutions of 1:100 to 1:1000. Green teas were about 10 times more effective than black teas.

Epigallocatechin gallate. Equivalent results were achieved using only the principal catechin of green tea, epigallocatechin gallate (EGCg). EGCg inhibited tNOX activity by about 50% at extremely low

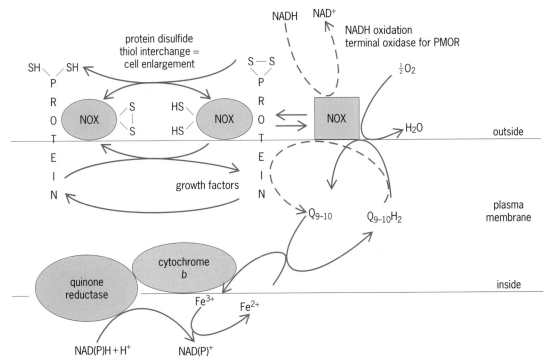

Fig. 2. Physiological and multifunctional properties of the drug- and hormone-responsive plasma membrane NADH oxidase. The right side of the diagram illustrates the spatial relationships of the inside NAD(P)H:quinone reductase, the membrane pool of coenzyme Q (Q_{9-10}), and the external NADH oxidase (NOX) protein across the plasma membrane (PMOR indicates the plasma membrane oxido-reductase electron transport chain). The NOX protein can function as a terminal oxidase of plasma membrane electron transport, donating electrons from cytosolic NADH either to molecular oxygen in a two-electron transfer or to reduction of protein disulfides. The left side of the diagram summarizes the two reactions catalyzed by the protein disulfide-thiol interchange and thiol-disulfide oxido-reductase activities. The interchange activity, which occurs in the absence of external reductants, is postulated to play a role in growth. Growth factors appear to serve as switches to emphasize this part of the mechanism.

concentrations of 1–10 nanomoles per liter. Green tea contains more EGCg than does black tea, which explains the greater effectiveness of green teas compared to black teas in inhibiting tNOX.

Tea preparations as well as EGCg were without effect on the noncancer NOX protein activity. Cancer cells express both noncancer NOX and cancer-specific NOX proteins and are inhibited by EGCg. Noncancer cells express only noncancer NOX protein and are not inhibited by EGCg.

Not only does EGCg inhibit the NADH oxidase of plasma membrane vesicles from cancer cells and not that of normal cells, but the substance exerts a parallel inhibitory response on growth. The growth of HeLa cells and mammary cancer (BT-20) cells was almost completely inhibited by EGCg, whereas the growth of Chinese hamster ovary cells and of noncancerous human mammary (MCF-10A) cells was much less affected by EGCg. The growth of the MCF-10A mammary epithelial cells was unaffected by EGCg except at very high doses of 100 micromoles per liter concentration, whereas growth of the tumorigenically transformed BT-20 mammary cancer and HeLa cells was 50% inhibited at about 5 micromoles per liter concentration. The MCF-10A cells quickly recovered from any adverse responses to EGCg and grew normally. In contrast, HeLa and BT-20 cancer cells did not recover and died. With treated HeLa cells, nuclei exhibited patterns of fluorescence characteristic of programmed cell death (apoptosis).

The anticancer activity of EGCg is related to growth inhibition, regulation of cell cycle progression, and induction of apoptosis rather than to nonspecific antioxidant function. Preferential inhibition of growth and induction of apoptosis by EGCg in cancer cell lines compared to noncancer cell lines has been reported previously. In general, tNOX inhibitors lead to both interrupted cell cycle progression and induction of apoptosis. The activity is related in some measure to the regulation of cell cycle progression and induction of p53-dependent apoptosis.

Anticancer benefit. Controlled clinical trials are being planned to determine the amount of tea that is required to have anticancer benefit. Studies in China suggest 4–10 cups of tea per day as beneficial. Regular spacing of tea consumption throughout the day may be the best strategy. If caffeine is a problem, a decaffeinated product may be used or regular tea may be extracted for 10 s in hot water and the first extract discarded to remove much of the caffeine. Brewing may then be continued to extract the beneficial catechins and tea flavor.

Until controlled clinical trials are completed, the therapeutic benefits of tea in cancer management remain undetermined. Tea consumption is generally regarded as extremely safe. However, tea consumption should be in moderation and never as a substitute for proven therapy or in combination with a therapy where tea constituents such as vitamin K are counterindicated.

For background information *see* ANTIOXIDANT; CANCER (MEDICINE); CELL (BIOLOGY); TEA in the McGraw-Hill Encyclopedia of Science & Technology.

D. James Morré; Dorothy M. Morré

Bibliography. Z. P. Chen et al., Green tea epigallocatechin gallate shows a pronounced growth inhibitory effect on cancerous cells but not on their normal counterparts, *Can. Lett.*, 129:173–179, 1998; H. Fujiki et al., Cancer inhibition by green tea, *Mutation Res.*, 402:307–310, 1998; I. E. Dreosti, Bioactive ingredients: Antioxidants and polyphenols in tea, *Nutr. Rev.*, 54:551–558, 1996; D. J. Morré, P. -J. Chueh, and D. M. Morré, Capsaicin inhibits preferentially the NADH oxidase and growth of transformed cells in culture, *Proc. Nat. Acad. Sci.*, 92:1831–1835, 1995.

Tree

The appearance of trees by the Middle to Late Devonian profoundly altered the ecology and complexity of terrestrial ecosystems. The tree habit is believed to have resulted from selective pressures to enhance light capture and propagule (a structure that has the ability to give rise to a new plant) dispersal. However, large size, tall stature, and the development of a canopy necessitated a number of evolutionary novelties which would permit the development, survival, and ecological competitiveness of large, mechanically stable structures. From the Mid-Devonian to Mid-Carboniferous, treelike growth forms independently evolved in five groups of plants: lignophytes (progymnosperms and seed plants), calamites (Carboniferous sphenopsids), lycopsids (Devonian and Carboniferous representatives of the clubmosses), cladoxylaleans (an enigmatic group of extinct Paleozoic pteridophytes), and tree ferns. Today, trees are represented by only two of these groups among seed plants and tree ferns. While all five produced effective trees, differing in mechanical attributes and probably representing a range of habitat preferences, each was constructed mechanically from different developmental processes. Interestingly, the developmental pattern characterizing lignophytes involving a bifacial vascular cambium was probably the simplest in terms of developmental complexity and ultimately the most successful as seen in today's range of habitats dominated by gymnosperm and angiosperm trees.

Mechanical constraints. Trees must withstand mechanical forces from two main sources: static loading caused by the force of gravity on the plant body, and dynamic loads caused by wind. Wind forces are believed to particularly influence tree growth, and wind pressures become more prevalent with increasing height from the ground. Whereas early herbaceous land plants would have been protected by the ground–air interface, treelike forms would have needed to adapt to resist bending or torsional forces caused by wind. Failure by excessive wind force can occur in two ways: mechanical failure of the trunk or branches, or uprooting. Early land plant ecosystems were probably further inhospitable to tree growth and root anchorage in unstable or thin soils supporting herbaceous plants. Increased height, size, and transpiration surface would necessitate modifications in hydraulic conductance within the plant body. Under these mechanical and hydraulic constraints, simple scalar enlargement of body plans with existing tissue repertoires of early land plants would not form the basis for effective trees. Mechanical and hydraulic constraints would require specialized tissues for these functions: tissues imparting stiffness and strength correctly placed for mechanical support, and modified tissues specialized for transporting water and nutirents.

Modern trees. Most modern trees have evolved a number of mechanisms to resist static and dynamic loads. A widespread strategy is controlled flexibility, where the lower parts of the branches and trunk are stiffly built and capable of resisting bending while the tips of the branches are more flexible and so deflect in the wind providing less sail area. Among modern trees, whether the crown comprises woody branches (conifers, angiosperms) or large fronds (palms, tree ferns), the principle is similar in that the apical part of the crown deflects wind forces. Interpreting the question of how and why trees evolved can therefore be suitably addressed from a mechanical standpoint.

Lignophyte trees. A watershed in the evolution of plant architecture was reached with the appearance of the bifacial cambium among basal members of the progymnosperms, a paraphyletic group (a group that consists only of members that share a single recent common ancestor) of tracheophytes believed to be ancestral to all seed plants. The bifacial cambium is a meristematic band of tissue producing secondary xylem (wood) toward the inside and secondary phloem toward the outside. This innovation permitted a plant stem to increase in girth (width) and develop different mechanical properties along the stem for controlled flexibility.

Because of the fragmentary state of the fossil record, it is unknown whether earliest representatives of lignophytes with a secondary vascular cambium produced trees. The probability is that they did not, partly because such stems are confined to 0.5–5 cm (0.2–2 in.) in diameter and because of uncertainty of the efficiency of the outer periderm or bark of these early plants. Significant increases in stem diameter require a second meristematic tissue outside the secondary phloem which continuously compensates for the increasing stem girth and protects living tissues within. By the Mid to Late Devonian, such a system is observed in *Archaeopteris*, in which genuine treelike proportions with secondary wood formed trunks up to 2 m (7 ft) in diameter. Woody axes in the fossil record are more common, and evidence from the branching patterns indicates large crowned trees, possibly over 30 m (98 ft) in height.

The simple combination of a secondary vascular cambium and an efficient periderm or bark has not been bettered mechanically; it formed a template for a wide variety of developmental variations and allowed further adaptive refinements to the type of tree observed in *Archaeopteris*. Although many subsequent groups show this pattern of development, a precise understanding of how it became established in consequent seed plant lineages is unclear. Many seed plants in Late Devonian and Carboniferous ecosystems possessed secondary vascular cambia but relied on an armor of strong fibers in the outer part of the stem for support. Significantly, these forms produced large frondlike leaves, but they did not reach treelike proportions. Exceptions include the Mid-Carboniferous Medullosales, in which tree-sized plants produced crowns of large fronds above stems strengthened by outer bands of fibers. While these plants might have been self-supporting, the secondary wood was probably not important for support of the stem but provided hydraulic supply to the leaves. A similar organization is seen among some cycads and the extinct group of Mesozoic plants known as Bennettitales.

The only other mode of tree development which is ecologically important and widespread now is that shown by palms. These treelike growth forms are, nevertheless, descendants of wood-forming lignophytes, but members of the monocotyledons which have since lost their ability to produce secondary xylem. In these plants, mechanically stable trunks are produced by a compound structure of leaf traces and fibers surrounding the leaf bases around the outside of the stem. Wide stems up to a meter in diameter are produced by primary lateral expansion of the growing apex before the plant reaches a significant height.

Several other monocotyledons have evolved a treelike habit such as that of bamboo (*Dendrocalamus*), which may reach up to 40 m (130 ft) high, in addition to the aloes, dracaenas, and yuccas. One drawback of these "alternative" tree designs is that both hollow stems and palmlike stems face mechanical constraints in the degree of branching possible. Branches are normally scarce and at low angles to the vertical stem so as to reduce local buckling in hollow stems and splitting in nonwoody stems of palms and similar forms.

Lycopsid trees. From the Mid-Devonian to the Late Carboniferous, treelike lycopsids were important components, especially of humid habitats. Compared with typical lignophytes, their development and mechanical functioning was bizarre. The Carboniferous *Diaphorodendron vasculare*, like many other arborescent lycopsids, produced a unifacial cambium. Biomechanical analyses indicate that the outer secondary periderm, and not the wood, was the mechanically significant tissue contributing over 98% to the flexural stiffness of the lower stem. Indeed, the development of the trunk during ontogeny was highly complex and involved the eventual replacement of a photosynthetic "leaf cushion surface" by mechanically important fiberlike bark. Many arborescent lycopsids were branched, and mechanical models have demonstrated that the apical parts of the tree were proportionately less stiff than the main stem, showing evidence of controlled flexibility. Treelike lycopsids suffered widespread extinction at the end of the Carboniferous, after which they never reestablished such ecological importance. Compared with the functional design of lignophytes, the arborescent lycopsid organization was relatively inefficient in producing two principal tissues (fibrous periderm, wood) for support and hydraulic conduction, whereas the lignophytes produce only one (for example, wood).

Calamite trees. Large-bodied sphenopsids of tree-size proportions were probably in existence by the Late Devonian and well established in Mid-Carboniferous humid environments, where remains of trunks at least 10 m (33 ft) long have been reported. Unlike extant sphenopsids known from *Equisetum*, *Calamites* produced trunks up to 1 m (3.3 ft) in diameter with secondary xylem and an outer periderm. Like modern horsetails, leaf and branch architecture was built around nodal whorls. A striking feature of the *Calamites* stems is that main trunks and branches were hollow. These trees were, therefore, "lightweight" structures affording mechanically stable stems for relatively little energetic cost. Mechanical analyses of *Calamites* indicate that even though hollow, the organization was resistant to local buckling partly because of the ring of secondary xylem. The maximum height, size, and shape of these plants was dictated by a range of factors: a series of "what-if" calculations demonstrate that stems with a basal diameter of 0.5 m (1.6 ft) could have achieved heights of 20–30 m (66–98 ft). Arborescent sphenopsids, like their lycopsid counterparts, largely died out by the end of the Permian.

Ferns. Besides lycopsids, sphenopsids, and lignophytes, other plant groups that "experimented" with a significantly increased body size include the Devonian and early Carboniferous Cladoxylales and later groups of true ferns, including Marattiales, Cyatheaceae, and Dicksoniaceae. Paleozoic Cladoxylales such as *Pseudosporochnus* might possibly have achieved nearly treelike proportions via stems with extremely complex lobes of xylem tissue surrounded by an outer zone of fibers. Even if approaching relatively treelike proportions, the overall size of such forms would have been determinate.

By the Mid-Carboniferous, treelike ferns are best known from the marattialian genus *Psaronius*, which probably reached heights of up to 10 m (33 ft). Superficially resembling a palm, the base of the obconical (oblong and cone-shaped) stem comprised a "false trunk" supported by a tightly formed mantle of adventitious rootlets. Elsewhere, the stem was bounded by a zone of fibers surrounding a ground tissue of leaf traces and parenchyma.

Modern tree ferns show a determinant growth

structure and are largely constrained in the size of trunk they can develop. Mechanical stiffness of the trunk is afforded by a combination of leaf or frond traces and the fibers surrounding leaf bases, which are still effective after the leaves have died. Some tree ferns, similar to the fossil *Psaronius*, also produce a mantle of random rootlets near the base.

Mechanics. By standards, trees are complex evolutionary structures and need to fulfill a number of mechanical and hydraulic requirements to be viable. The fossil record indicates that many "designs" of tree have evolved. Today, only one model of development, that of lignophytes, dominates most terrestrial ecosystems. There is still much to infer from the evolutionary biology of plant growth forms and their related life histories and why one form of tree should rise to dominance. This brief mechanical synthesis suggests that the bifacial vascular cambium provided the means of developing a hydraulically and mechanically "smart" biomaterial with a high potential for modulating stem stiffness and water conduction. This provided an escape from tree forms with uneconomical or highly determinate growth forms which are prone to extinction.

For background information *see* PALEOBOTANY; PLANT EVOLUTION; PLANT TRANSPORT OF SOLUTES; PRIMARY VASCULAR SYSTEM (PLANT); STEM; TREE; TREE FERNS; TREE GROWTH in the McGraw-Hill Encyclopedia of Science & Technology.   N. P. Rowe; T. Speck

Bibliography. K. J. Niklas, *Plant Biomechanics*, 1992; N. P. Rowe, T. Speck, and J. Galtier, Biomechanical analysis of a Palaeozoic gymnosperm stem, *Proc. Roy. Soc. London Ser. B*, 252:19–28, 1993; T. Speck, A biomechanical method to distinguish between self-supporting and non self-supporting plants, *Rev. Paleobot. Palynol.*, 81:65–82, 1994; T. Speck and N. P. Rowe (eds.), Modelling form and function in fossil plants, *Rev. Palaeobot. Palynol.*, 102(spec. issue):1–114, 1998.

Trilobita

Trilobites, a group of extinct marine arthropods, are among the most familiar fossils. Their name is derived from the threefold, lengthwise division of their shieldlike external skeleton into a central axial lobe (which covered the body) and two flanking pleural lobes (which protected the walking legs and associated gill branches). Trilobites typically crawled over or burrowed through the sediment on the sea floor. With a few exceptions, they were likely to have been sluggish swimmers. Most trilobites probably fed by ingesting sediment or filtering organic particles out of the water column, although some were generalized predators capable of feeding on soft-bodied organisms such as worms. Trilobites occupied the oceans for more than a quarter billion years. Their record is a rich store of information that can be used to develop and test hypotheses about the processes that have shaped the history of life. These data are helping to throw light on the nature of large-scale changes in the composition of marine communities

and on factors that may influence the survival of groups during mass extinction events.

Historical record. Appearing more than half a billion years ago, during the Cambrian explosion of multicellular marine animals, trilobites were among the early success stories in the oceans. During the Cambrian Period [544–490 million years ago (Ma)], an interval often popularized as the Age of Trilobites, they were the dominant skeletonized invertebrates in the shallow continental-shelf seas that covered much of North America and other continents. The Cambrian history of the group was a "boom-and-bust" affair, punctuated by four major extinction events that are well documented in the rocks of North America (a tropical continent during the Cambrian) but may also be recorded in at least some other regions, including China and Australia. The causes of these extinctions remain uncertain, although habitat destruction in the outer part of the continental shelf during intervals of sea-level rise may have played a role. It has been argued that changes in water temperature or the levels of dissolved oxygen may have contributed to the extinctions. Whatever the cause, trilobites recovered rapidly from each event and continued to dominate marine communities into the early part of the Ordovician Period (490–439 Ma).

In the Middle Ordovician (an interval termed the Whiterockian), a new wave of origination changed the fabric of marine ecosystems. Several groups, including articulate brachiopods, corals, and crinoids (the Paleozoic Fauna), diversified rapidly and became

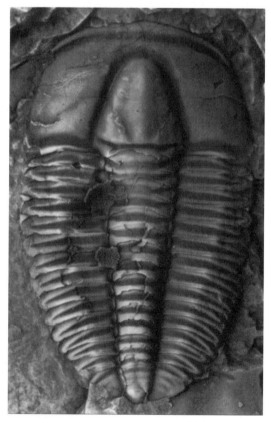

Complete skeleton of a typical trilobite (*Bailiella*) from mid-Cambrian rocks of eastern Newfoundland, Canada.

dominant animals in a spectrum of marine community types that were to persist until the great extinction at the end of the Permian Period (250 Ma). As community membership expanded, trilobites became reduced in importance, although the way in which this reversal of fortune was accomplished remains controversial. When viewed on a global scale, the number of different groups of trilobites (in evolutionary parlance, clade diversity) declined from the Mid-Ordovician, reaching minimum diversity during a major extinction at the end of the period. This decline might be viewed as reflecting some kind of negative interaction (for example, competition) with the rapidly expanding Paleozoic Fauna, but the diversity history of trilobites at a different scale suggests a different story. Interactions such as competition occur between populations of different species in local habitats. However, recent data show that the number of trilobite species occupying the marine shelf habitats as recorded by different types of sedimentary rocks was essentially constant between the Late Cambrian and the Silurian. This constancy does not fit a scenario characterized by ecologic displacement. In local habitats, it appears that trilobites were in effect diluted by the appearance of newly diversifying groups such as the articulate brachiopods. Trilobites no longer stood out among the growing crowd of new neighbors in Ordovician communities.

The global Ordovician decline in trilobite clade diversity must reflect other factors, one of which may be a reduction in the number of biogeographic regions. The world's flora and fauna are not evenly distributed but are divided into major biogeographic regions, referred to as realms or provinces, that are related to factors such as climate and the geographic dispersal of the continents. As exemplified by the contrast between Australia's marsupial-dominated mammal fauna and the placental-dominated faunas of Eurasia or North America, biogeographic regions contribute to global diversity because different groups fill the same ecologic role in each region. The number of biogeographic regions has varied over time as global geography has changed in response to plate tectonics and continental drift. During the Ordovician Period, the number of trilobite biogeographic regions declined as at least two ancient oceans narrowed and three distinct continental blocks approached each other. Exchange of species between continents across the narrowing oceans progressively broke down biogeographic differences. Evidence suggests that, in many cases, immigrants first occupied deeper marine environments around the margins of the continents and invaded the shallower waters of the continental shelves later during periods of sea-level rise. Toward the end of the Ordovician, reduced biogeographic differentiation was also related to the spread of cold-water faunas during an interval of climatic cooling.

When looked at in more detail, the Ordovician decline in clade diversity takes on a different appearance. Trilobite clades that diversified at the beginning of the Ordovician (termed the Ibex Fauna) declined dramatically in the later stages of the period

and were almost completely extinguished at a major Late Ordovician mass extinction (one of the five largest in the history of life) which seems to have been precipitated by a global decline in temperature. In contrast, several groups of trilobites that constitute the Whiterock Fauna began radiating alongside the Paleozoic Fauna, beginning in the Middle Ordovician. Most of the Whiterock Fauna survived the Late Ordovician extinction event and went on to become the dominant trilobites in the Silurian and Devonian periods, an interval of approximately 110 million years. The reasons for the disparate pattern of Ordovician diversification between the Ibex and Whiterock faunas are not understood but, whatever the cause, it did stack the deck in favor of the Whiterock Fauna during the Late Ordovician extinction. Rapid expansion of the Whiterock Fauna during the Ordovician would have resulted in an increase in the number of component species. This factor may have buffered the Whiterock Fauna against the effects of climatic change during the extinction event: the more representative species a group has, the greater are the odds of it surviving during a time of major environmental change, especially if they are spread over a wide geographic region. The remnants of the Ibex Fauna did not fare so well.

Mass extinction. The next major crisis in trilobite history came at a major mass extinction near the end of the Devonian Period (the Late Frasnian extinction, about 360 Ma), which also influenced several other invertebrate groups. Only one major clade of trilobites (the proetids) survived, and among the casualties were clades that had been important members of marine communities for an interval of almost 150 million years. Trilobites never fully recovered from this blow. The Devonian also saw profound changes in marine ecosystems as new jawed vertebrate predators, including sharks and a variety of bony fishes, appeared on the scene. This was the first act in a story of escalating predation that was to be increasingly important later in the history of the marine biosphere. Some workers have claimed that this initial faunal revolution was reflected in a variety of invertebrate groups with the appearance of skeletal features (such as spines) designed to ward off predators. Healed injuries show that even the earliest trilobites of the Cambrian Period suffered predation, but it is likely that the appearance of large, sophisticated predators in the Devonian had a particularly negative effect on the group. Certainly, it is a viable explanation for the failure of the group to bounce back from the Late Frasnian extinction event.

The Carboniferous and Permian periods (354–251 Ma) represent the twilight of trilobite history. The proetids occupied a range of marine shelf environments but reached only a fraction of the diversity that trilobites had attained earlier in their history. Whereas trilobites in older strata encompassed a wide spectrum of morphologies, Carboniferous and Permian faunas were rather monotonous. The group declined steadily during the Carboniferous and Permian, and Late Permian sequences in places such as the southwestern United States indicate that diversity had dwindled to only a handful of species that were confined largely to outer-shelf environments. The Late Permian was also the time of the greatest mass extinction of the last half billion years—more than double the magnitude of the Late Cretaceous event. Given the steady downward track of diversity during the preceding 100 million years, it was perhaps inevitable that trilobites would be among the casualties. By the dawn of the Triassic Period, the group was extinguished completely.

For background information *see* ARTHROPODA; CAMBRIAN; DEVONIAN; ORDOVICIAN; PERMIAN; TRILOBITA in the McGraw-Hill Encyclopedia of Science & Technology.   Stephen R. Westrop

Bibliography. J. M. Adrain, R. A. Fortey, and S. R. Westrop, Post-Cambrian trilobite diversity and evolutionary faunas, *Science*, 280:1922–1925, 1998; S. R. Westrop and J. M. Adrain, Trilobite alpha diversity and the reorganization of Ordovician benthic marine communities, *Paleobiology*, 24:1–16, 1998; H. B. Whittington, *Trilobites*, Boydell Press, 1992.

Turbidite

Turbidites are sediments that were initially deposited on the ocean floor in deep marine water below the effects of storm wave action (**Fig. 1**). Turbidites are transported to the deep ocean floor, under the influence of gravity, by downslope flow of turbulent sediment-water mixtures called turbidity currents. In the rock record, turbidite strata often occur in the deep (often more than 9000 ft or 3000 m) parts of sedimentary basins and sometimes also at great

water depths (1500–9000 ft or 500–3000 m).

In the late-1980s, turbidite reservoirs were envisioned as the last great frontier in petroleum exploration; that prediction has turned out to be true. At present, there is active exploration and development in many parts of the world, including the Gulf of Mexico, California, offshore West Africa, the North Sea, and Brazil. Turbidite reservoirs can be very large. There are 43 giant turbidite fields worldwide, and more than 500 million barrels of oil equivalent (MMBOE) are expected to be ultimately recovered from each. In the current Gulf of Mexico "Deepwater Play," over 50 confirmed discoveries have been made, totaling 5 billion barrels of oil equivalent (BBOE) in estimated reserves.

Technology integration. With the increased economic viability of turbidite reservoirs has come increased research effort to better understand where these reservoirs occur in the stratigraphic record (for exploration), their stratigraphic architecture (for reservoir development), and their origin. Knowledge of the complexity of turbidite deposits has advanced through the integration of a number of disciplines and data types, including two- and three-dimensional (2-D and 3-D) seismic reflection records, wireline logs, biostratigraphy, cores, petrophysics and rock properties, petroleum reservoir simulation, and outcrop studies.

Through this effort, understanding has evolved from the simple 1970s conceptual model of a fan-shaped body of turbidite strata in which grain size and bed thickness progressively decrease in a downcurrent direction, to more complex—but realistic—models illustrating the variety and complexity of

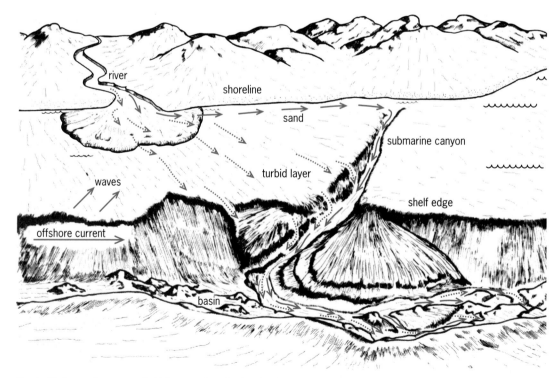

Fig. 1. Marine environments. Turbidites are deposited in the deep-water slope and basin environments. (*After D. G. Moore, Reflection profiling studies of the California continental borderland: Structure and Quaternary turbidite basins, Geol. Soc. Amer. Spec. Publ., no. 107, p. 142, 1969*)

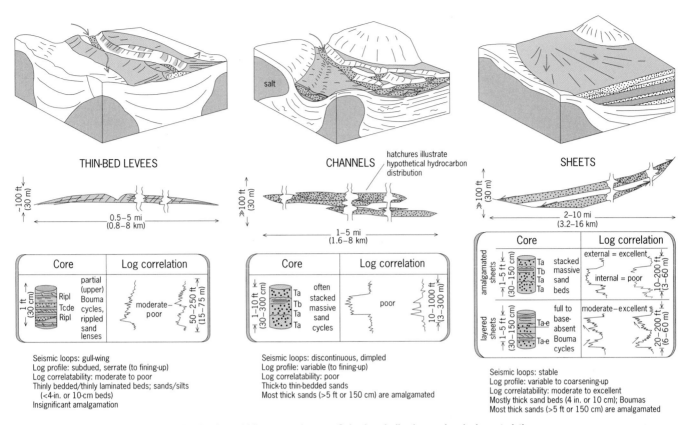

Fig. 2. Generalized end-member classification for turbidite reservoir types. Seismic, wireline log, and rock characteristics are listed beneath each turbidite reservoir type. Seismic loops refer to characteristics of seismic reflection records. Log profile refers to the characteristic shape of wireline logs, as illustrated under Log correlation. Log correlatability levels of moderate, poor, and excellent refer to the ability to correlate turbidite intervals in different wells based upon their wireline log characteristics. Boumas refer to specific types of turbidite beds (Ta, Tb, Tc, Td, Te), each with unique distinguishing features. (*After M. A. Chapin et al., Reservoir architecture of turbidite sheet sandstones in laterally extensive outcrops, Ross Formation, Western Ireland, in P. Weimer et al., eds., Submarine Fans and Turbidite Systems, Gulf Coast Section—Society of Economic Paleontology and Mineralogy, 15th Annual Research Conference Proceedings, pp. 53–68, 1994*)

architectural elements that are commonly encountered in outcrop and in the subsurface.

Regional petroleum exploration scale. At the regional scale, there has been considerable research regarding the relative roles of cyclic sea-level fluctuations and tectonic movements on the distribution of turbidite strata. These strata generally exhibit a distinct occurrence within a depositional stratigraphic sequence. The sandier (more petroliferous) parts of turbidite strata usually rest upon an erosional surface which forms during a lowering of sea level, when the supply of sediment to deep water is relatively voluminous. Overlying strata become progressively muddier as a result of subsequent sea-level rise, when more sediment is trapped in the shoreline zone and less is supplied to deep water. Tectonic movements also influence sedimentation by producing significant sea-floor topography at the time of deposition; examples include intraslope salt withdrawal basins in the northern Gulf of Mexico and West Africa, shale tectonics and faulting of offshore Nigeria, and wrench- and rift-related basement faulting in the North Sea and California. In addition, it is recognized that complex depositional patterns can occur without major fluctuations in sea level.

Petroleum reservoir scale. At the smaller scale, much research has alluded to the architectural make-up of turbidite deposits. For example, turbidite elements have been classified into channels, overbank, lobes, channel-lobe transition, and erosional features. A similar classification has been proposed comprising channel complexes, levees, and sheets; also provided were some basic core, log, and seismic criteria for their recognition in the subsurface (**Fig. 2**). Another proposed classification is based on two primary controls: (1) amount of gravel, sand, and mud in gravity flows; and (2) the nature of the submarine conduit through which the gravity flows traveled on their way to deep water (through single channels, through multiple channels, or as nonchannelized spillover). The various combinations result in a set of end-member building blocks: wedges, channels, levees, lobes, and sheets (**Fig. 3**).

The particular type of architectural element which is prevalent in a turbidite petroleum reservoir greatly affects how that reservoir will produce oil or gas. At the subseismic scale, such factors as lateral bed continuity and vertical connectivity—as well as the degree to which porosity and permeability have been occluded during burial—ultimately will control reservoir performance. For example, there have been several cases in which very thin, porous, permeable sands can be more productive than thick, large sands. These thin sands are more difficult to recognize and

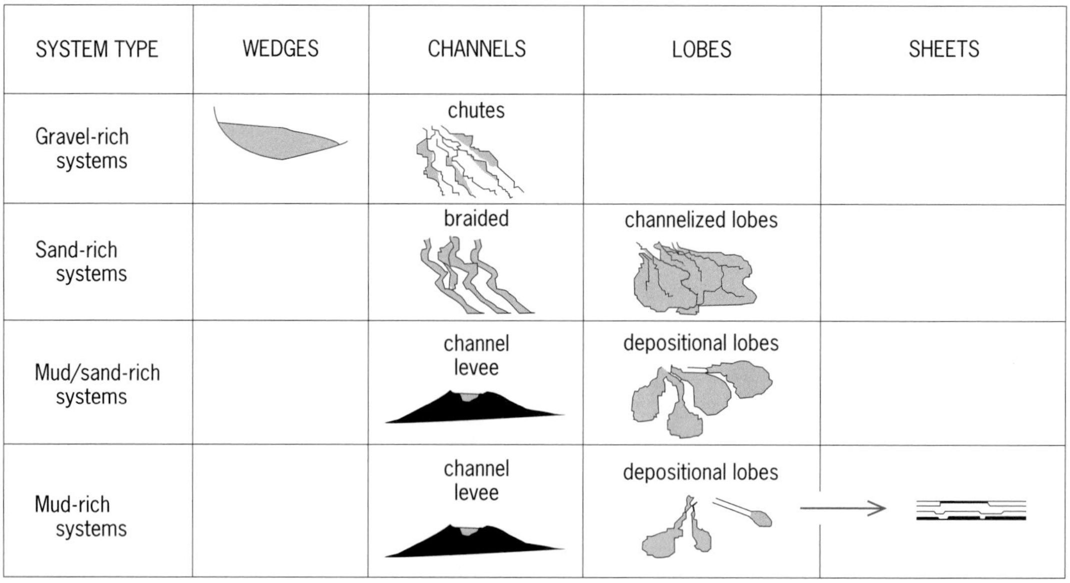

SYSTEM TYPE	WEDGES	CHANNELS	LOBES	SHEETS
Gravel-rich systems		chutes		
Sand-rich systems		braided	channelized lobes	
Mud/sand-rich systems		channel levee	depositional lobes	
Mud-rich systems		channel levee	depositional lobes	

Fig. 3. Reservoir architectural elements of deep-water depositional systems. (*After H. G. Reading and M. T. Richards, Turbidite systems in deep-water basin margins classified by grain size and feeder system, Amer. Ass. Petrol. Geol. Bull., 78:792–822, 1994*)

correlate on well logs and are also beneath seismic resolving power. Work is being conducted to quantify these important architectural variables, mainly from long, continuous outcrops where good measurements can be made.

Research areas. Concerning the origin of turbidites, there has been a resurgence in research to understand the processes by which sediments are transported to deep water. The long-held view that the principal mechanism of sediment transport to deep water is periodic turbidity currents has recently been challenged. It is maintained that the presence of small amounts of clay in a gravity flow increases the viscosity of the flow, giving rise to a cohesive, laminar (rather than turbulent) sandy debris-flow deposit which can be mistaken for a turbidity current deposit. Arguments against this concept have several grounds, including the implication that such flows may not be very laterally extensive, contrary to observation.

This controversy has spurred new interest and research into turbidite hydrodynamics. Experimental work has begun in a number of laboratories that address the mechanics of sediment transport in different types of gravity flows. For example, it has been shown that turbulent gravity flows exhibit differing properties of flow steadiness and uniformity, thus complicating the nature of the resultant deposit.

Another area of research is the nature of sinuous (and probably meandering) channels in deep-water depositional systems. Sea-floor images from both modern fans and subsurface horizon images derived from 3-D seismic have revealed sinuous channels that, at first glance, exhibit many similarities to continental, meandering river systems. However, differences exist between the two kinds of channels. Deep-water sinuous channels do not appear to have many oxbow-shaped cutoffs, or as much meander-

ing within one meander belt. Another enigma is that within some 3-D seismic horizon slices, both straight and sinuous channels exist within the same general stratigraphic interval. The origin of sinuosity may be related to more continuous, dense undercurrent flows originating from major continental river systems, as opposed to the long-held view that turbidity current flow is episodically induced by sediment failure in the shallow-water marine shelf and upper-slope environments.

Turbidite systems are major petroleum reservoirs in many sedimentary basins in the world and will remain so well into the twenty-first century. The favorable economics associated with this type of deposit will ensure a sustained research effort, leading to quantification of the occurrence and variability of turbidite systems and a more thorough understanding of their origin.

For background information *see* FACIES (GEOLOGY); MARINE GEOLOGY; MARINE SEDIMENTS; PETROLEUM RESERVOIR ENGINEERING; SUBMARINE CANYON; SUBMARINE FAN; TURBIDITE in the McGraw-Hill Encyclopedia of Science & Technology.

Roger M. Slatt; Paul Weimer

Bibliography. M. A. Chapin et al., Reservoir architecture of turbidite sheet sandstones in laterally extensive outcrops, Ross Formation, Western Ireland, in P. Weimer et al. (eds.), Submarine Fans and Turbidite Systems, *Gulf Coast Section—Society of Economic Paleontology and Mineralogy, 15th Annual Research Conference Proceedings*, pp. 53–68, 1994; B. C. Kneller, Beyond the turbidite paradigm: Physical models for deposition of turbidites and their implications for reservoir prediction, in A. J. Hartley and D. J. Prosser (eds.), Characterization of Deep Marine Clastic Systems, *Geol. Soc. Spec. Publ.*, no. 94, pp. 31–49, 1995; E. Mutti and W. R. Normark, An integrated approach to the study of turbidite systems, in P. Weimer and M. H. Link (eds.), *Seismic Facies and Sedimentary Processes of Submarine Fans and Turbidite Systems*, Springer-Verlag, 1991; H. G. Reading and M. T. Richards, Turbidite systems in deep-water basin margins classified by grain size and feeder system, *Amer. Ass. Petrol. Geol. Bull.*, 78:792–822, 1994; G. Shanmugam, The Bouma Sequence and the turbidite mind set, *Earth Sci. Rev.*, 42:201–229, 1997; P. R. Vail, Seismic stratigraphy interpretation using sequence stratigraphy, Part 1, in A. W. Bally (ed.), Atlas of Seismic Stratigraphy, *Amer. Ass. Petrol. Geol. Stud. Geol.*, no. 27, pp. 1–10, 1987.

Two-photon absorption

Chemists have developed tremendous expertise in designing and synthesizing dye molecules which efficiently absorb light of a specified wavelength. Such strongly absorbing molecules have been used for many applications, including dyeing, photography, initiation of chemical reactions, and optical filtering. The probability of a molecule absorbing one photon is proportional to the intensity of the input beam. With sufficiently intense light, such as a laser beam, it is possible for a molecule to absorb simultaneously two photons, each of approximately half the energy (twice the wavelength, λ) normally required to reach an excited state. This two-photon absorption process is intrinsically weak relative to one-photon excitation. The probability of a molecule absorbing two photons simultaneously is proportional to the square of the intensity of the input beam. Advances in recent years have provided a rational basis for the design of efficient two-photon-absorbing molecules and have demonstrated their application in photonics, materials science, and biology.

Excitation processes. Two-photon absorption allows for the excitation of molecules with precise three-dimensional (3D) spatial confinement. This 3D control of excitation arises from the fact that the intensity of a focus laser beam decreases quadratically with distance from the focal plane. Since the probability of two-photon absorption is proportional to intensity squared, two-photon absorption falls off as the fourth power of distance from the focus. As a result, two-photon absorption at distances appreciably above and below the focal plane is negligible (**Fig. 1***a*). In addition, with two-photon absorption it is possible to excite molecules at a greater depth in a normally absorbing medium since the target molecules can be selectively excited with photon energies well below that at which the medium exhibits single-photon absorption (Fig. 1*b, c*). Furthermore, improved depth penetration is possible in scattering media since two-photon absorption involves use of light at about twice the wavelength used for single photon excitation, and light scattering scales as λ^{-4}; light at 2λ will suffer only one-sixteenth the degree of scattering that would have been incurred by use of light at λ. Because of these advantages, various schemes to exploit two-photon absorption have been proposed. However, many of these schemes employ chromophores developed for one-photon excitation and have rather low efficiency for two-photon absorption (called the two-photon absorption cross section, δ). Accordingly, these schemes would be greatly aided by the development of molecules with large δ: this would decrease the need for very high power and rather costly lasers, thus making two-photon absorption applications more economically feasible. In particular, molecules with large δ are currently in great demand for a variety of applications, including two-photon excited fluorescence microscopy, optical power limiting and switching, 3D microfabrication, and optical data storage. The key to the design of molecules with high sensitivity in two-photon absorption is the knowledge of how δ depends on molecular structure.

Molecule design. Two-photon absorption and fluorescence have been studied for decades. However, until 1998 there had been few systematic studies

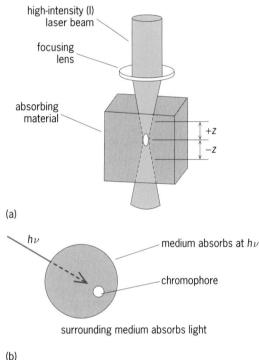

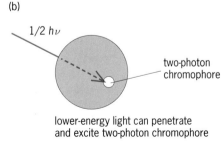

Fig. 1. Laser beam with various intensities demonstrating (*a*) localized excitation in 3D by two-photon excitation; (*b*) limited penetration depth of one-photon excitation in an absorbing medium; and (*c*) improved penetration depth through the use of lower-photon-energy (*hν*) light and two-photon excitation.

of structure-property relationships for the optimization of two-photon absorption in conjugated organic molecules. In most cases, the measured δ values for such molecules were relatively small ($<100 \times 10^{-50}$ cm^4 s/photon), although laser dye Rhodamine B was found to have a δ value of roughly 300×10^{-50} cm^4 s/photon, which presumably was the largest value reported up to that time for any dye. In 1998 it was reported that 4,4'-bis(di-n-butyl)amino-E-stilbene (structure **1**) exhibited a value of $210 \times$

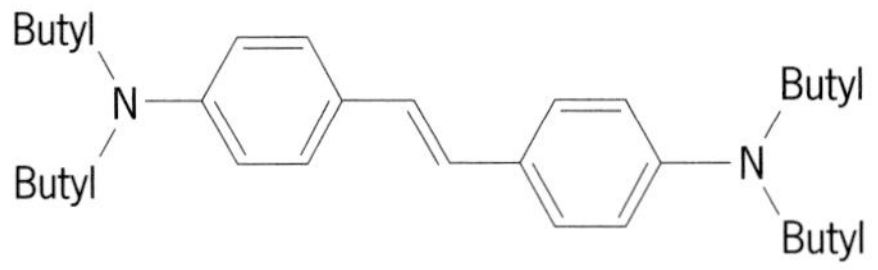

(1)

10^{-50} cm^4 s/photon at 605 nanometers, which is almost 20 times larger than δ of the unsubstituted parent molecule *trans*-stilbene (**2**). From this initial

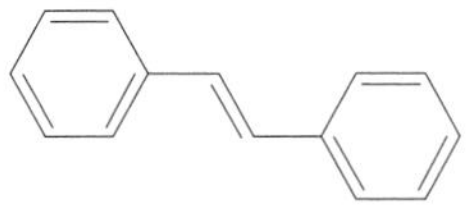

(2)

observation, it was conjectured that the large increase in two-photon absorption for (**1**) relative to (**2**) was related to a symmetrical charge transfer from the amino nitrogens to the conjugated bridge of the molecule. These results suggested a strategy to enhance δ and tune the wavelength of the two-photon absorption peak for π-conjugated organic chromophores: design symmetric donor/acceptor substituted molecules in which the distance and degree of intramolecular charge transfer upon excitation are facilitated. This can be achieved by extending the conjugation length with the insertion of additional double bonds or of phenylene-vinylene or phenylene-butadienylene groups (**3–5**) to increase

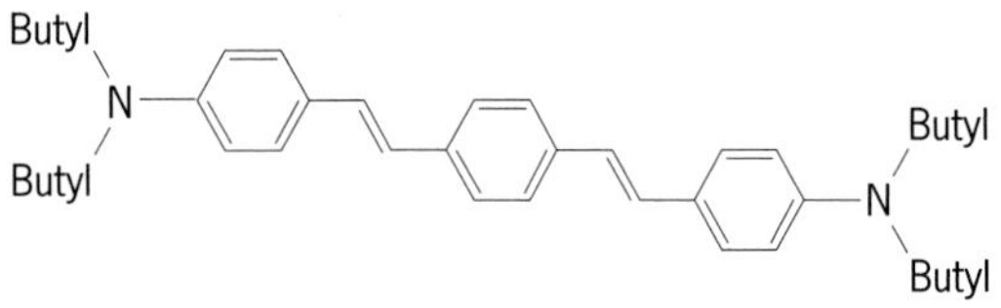

(3)

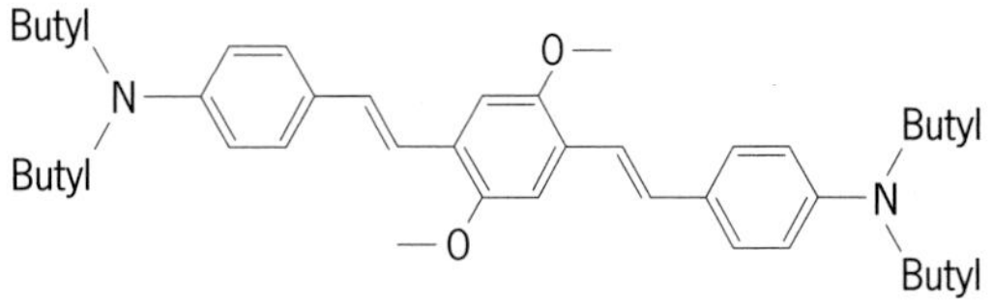

(4)

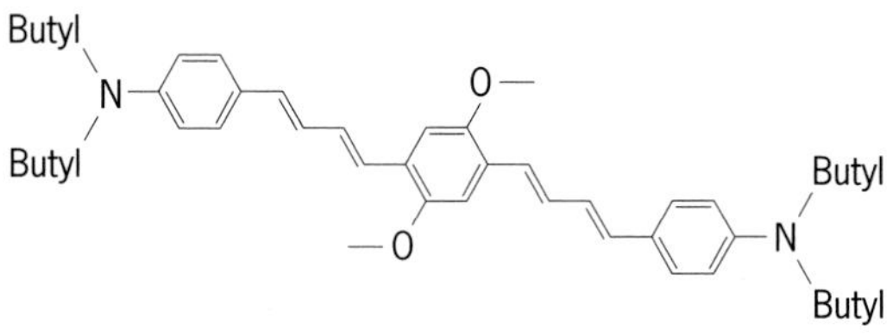

(5)

the distance over which charge can be transferred. Additionally, the extent of charge transfer from the ends of the molecule to the center can be increased by creating a donor-acceptor-donor (D-A-D) motif wherein electron accepting groups are attached to the central ring of a bis-styrylbenzene backbone (**6**).

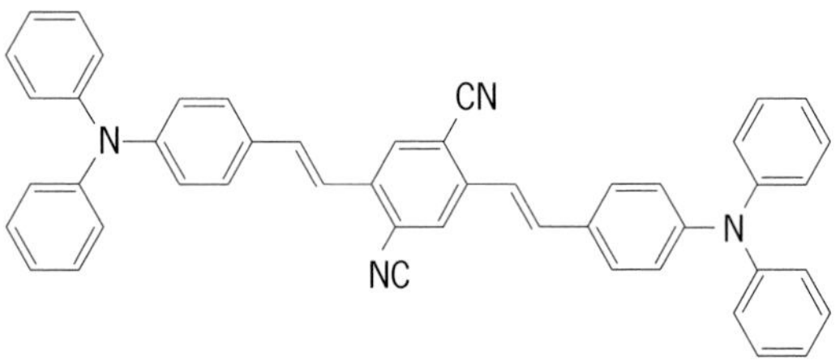

(6)

The direction of the symmetric charge transfer can be reversed by substituting donor groups along the conjugated bridge and acceptors on the ends (**7–9**),

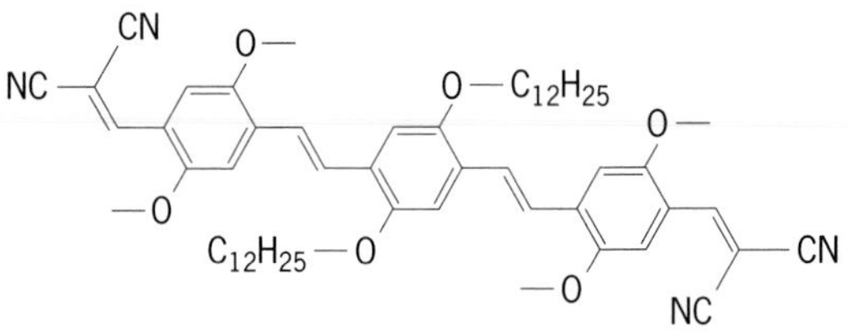

(7)

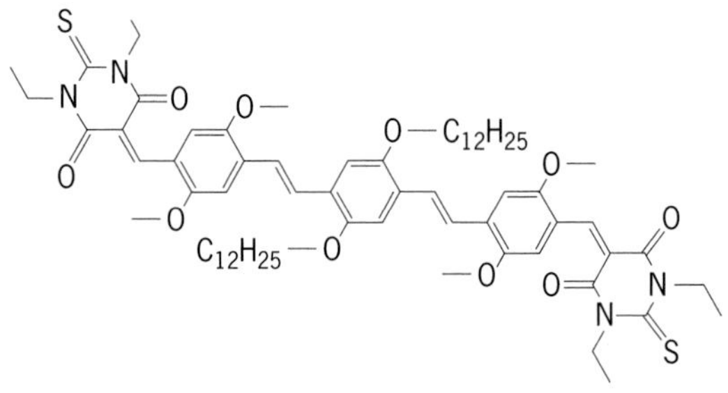

(8)

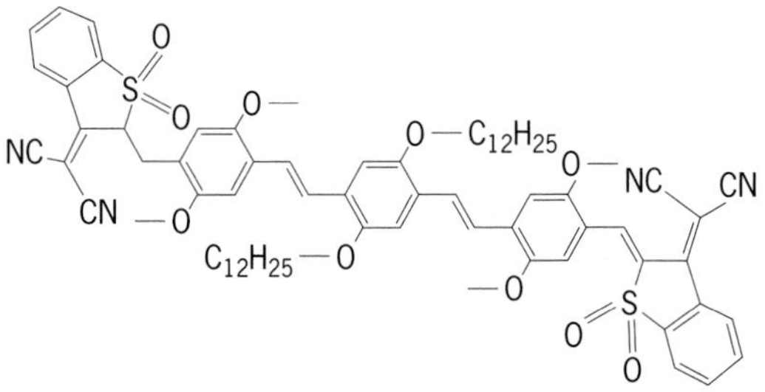

(9)

creating an acceptor-donor-acceptor (A-D-A) motif. The data show that both D-A-D and A-D-A designs

Experimental two-photon absorption data for structures (1) through (9)		
Structure	Wavelength ($\lambda^{(2)}_{max}$), nm	Cross section (δ), 10^{-50} cm^4 s/photon
1	605	210
2	514	12
3	730	995
4	730	900
5	775	1250
6	835	1940
7	825	480
	940	620
8	970	1750
9	975	4400

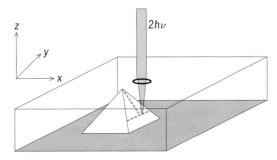

Fig. 2. Scanning laser exposure of a two-photon photopolymer resin 3D lithographic microfabrication.

are effective in enhancing δ and in red-shifting the position of the two-photon absorption (see **table**). They also show that δ increases with chain length. The value of δ for structure (9) [4400×10^{-50} cm^4 s/photon] is the largest pure two-photon absorption response for organic materials reported to date.

Two-photon-initiated chemistry. Because two-photon absorption provides a means of activating chemical or physical processes with high spatial resolution in three dimensions, it has enabled the development of 3D optical data storage and 3D lithographic microfabrication. Under tight-focusing conditions, the absorption is confined at the focus to a volume of order λ^3, where λ is the laser wavelength, resulting in a volume element which is typically about 1 μm^3. Under appropriate conditions, subsequent chemical reactions, such as photoinitiated polymerization, can also be localized in this small volume. Three-dimensional data storage and microfabrication have been demonstrated using two-photon-initiated polymerization (TPIP) of resins incorporating commercial ultraviolet-absorbing initiators. Such photopolymer resins exhibit low photosensitivity since the initiators have small δ. Two-photon excitable resins incorporating high δ molecules such as structures (4) and (5) as initiators have been used to demonstrate two-photon-initiated polymerization with an order of magnitude increase

in photosensitivity relative to previous resins. The high-sensitivity two-photon-initiated polymerization resins have been employed for 3D optical data storage schemes and for 3D lithographic microfabrication (3DLM). *See* MOLECULAR MACHINES.

In the 3D lithographic microfabrication scheme, photoexcitation initiates a cross-linking reaction in the photopolymer resin, thereby reducing the solubility of the exposed material. In this process, an arbitrary 3D pattern is impressed into a photopolymer by scanning the focus of an intense laser beam within the material (**Fig. 2**). The exposed 3D pattern can then be developed to give a free-standing 3D microstructure by dissolving away the unexposed material in a single exposure/development cycle. In contrast, other current schemes for defining 3D microstructures typically involve building up an object through sequential layer-by-layer processes.

The 3D lithographic microfabrication approach can be used to produce a variety of interesting 3D microstructures such as integrated optical elements and micromechanical structures. Three-dimensionally periodic structures are of interest as potential photonic bandgap (PBG) materials, which have unique optical reflection and filtering properties. For example, a "stack-of-logs" structure can have a cubic symmetry 3D periodic structure with a period spacing of 5 μm, as appropriate for diffraction of infared (IR) radiation (**Fig. 3**).

Researchers are now beginning to use 3D

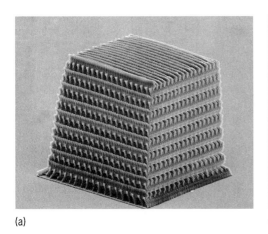

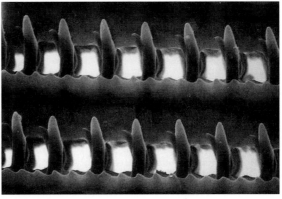

(a)   (b)

Fig. 3. Polymeric "stack-of-logs" photonic-bandgap-like structure made by 3D lithographic microfabrication. (a) Overall view; width of structure is comparable to that of a human hair. (b) Closeup side view showing 5-μm period.

lithographic microfabrication for a variety of polymeric microstructures. For many applications, materials with specific optical or mechanical properties, which may or may not be met with polymers, are needed. By using polymer structures fabricated by 3D lithographic microfabrication as a mold, it should be possible to produce 3D microstructures comprising a wide range of materials. Furthermore, extremely high density 3D optical memory (10^{12} bits/cm^3, compared to a conventional compact disk which holds approximately 10^{10} bits) has been demonstrated using two-photon initiated chemistry and other two-photon absorption induced reactions.

For background information *see* LASER; LASER SPECTROSCOPY; PHOTOEMISSION; PHOTON; QUASIELASTIC LIGHT SCATTERING; SELECTION RULES (PHYSICS) in the McGraw-Hill Encyclopedia of Science & Technology.

Seth R. Marder; Joseph W. Perry

Bibliography. M. Albota et al., *Science*, 281:1653, 1998; W. Denk, J. H. Strickler, and W. W. Webb, *Science*, 248:73, 1990; J. E. Ehrlich et al., *Opt. Lett.*, 22:1843, 1997; D. A. Parthenopoulos and P. M. Rentzepis, *Science*, 245:843, 1989; J. H. Strickler and W. W. Webb, *Opt. Lett.*, 16:1780, 1991.

Underground mining

To remain competitive, underground mine operators have adopted and updated technologies to decrease production costs and increase worker safety. The first part of this article discusses underground communications, a technology that has undergone rapid growth. Communication networks evolved from the simple mine telephone to fiber-optic- and radio-frequency-based systems allowing voice, data, and video communications throughout the mine workings. The second part of this article discusses seismic tomography and its use in mining. For example, unknown geologic or stress-related anomalies encountered during mining present a primary threat to uninterrupted operations, as well as creating hazardous conditions. Seismic tomography enables the mapping of these anomalous conditions ahead of mining.

Communication Networks

Mining equipment suppliers and operators typically implement communication technologies that are well proven in surface applications. The topology of the mine, the harshness of the environment, and the economics of mining all limit the speed with which communication networks are transferred underground; the time lag may be 10–20 years.

Topology. The mine's topology plays a large role in determining the communication network architecture. The challenge is to provide full coverage of the mine's workings while transferring information to the surface plant. The link between workings and plant is typically a vertical shaft blasted out of the rock to a depth of as much as 10,000 ft (3000 m).

The communication network must service the mine workings, which may spread out over a few square miles and be distributed on a number of stacked levels. The total length of tunnels excavated in a mine can total more than 100 mi (160 km).

Environment. Heat, humidity, and a highly corrosive environment require that communication electronics be properly adapted and packaged for use in mines. In coal mines the equipment must also meet stringent safety requirements to limit the risk of methane explosions. Underground electronics must be able to withstand concussion caused by production and development blasting.

Economics. An underground communication network does not generate any direct revenue for the mine. All economic benefits are derived from improved working conditions and higher productivity. It is often difficult to quantify these savings and justify the additional capital spending required. By implementing the latest technology, the international underground mining community has successfully updated its communication architectures to meet its voice, data, video, and real-time control needs.

Implementation. An underground mine's communication network involves three separate functions: transport network, infrastructure, and voice and data communications.

Transport network. The transport network is responsible for the transfer of information between the mine and the surface operations. It moves the information in the shaft and provides distribution to each level in the mine.

Initially, the transport network was composed of twisted pairs of cables suspended in the shaft. Each twisted pair provided a voice or low-speed data (1200–9600 baud) connection to the surface.

In the 1980s, mines began to replace their copper transport networks with fiber-optic cables. Fiber optics provides increased bandwidth and expandability, and greatly reduces the installation space required in the shafts.

Multiplexers installed at each level provide an interface to other media that can then be distributed throughout the mine. Despite the adoption of fiber optics in the shaft, copper continues to play a role for specialized transport needs. An example is the use of a copper cable to provide voice communications in the shaft.

In addition to simplifying the transfer of information in the shaft, fiber optics increases the communication bandwidth and allows surface computer networks to be extended underground. The bandwidth availability also provides cost-effective control of mobile production equipment from the surface. Such control is still very limited in practice.

Infrastructure. When developing a mine, an operator invests in significant underground infrastructure. This may include repair shops, electrical substations, pumping stations, ventilation fans, crushing plants, offices, and so on. Physically located near the shaft, this infrastructure often requires electrical connections to local area networks (LANs), telephones, programmable controllers, and data acquisition devices. Servicing these functions and areas of the mine is the responsibility of core distribution.

Because this infrastructure is located in a part of the mine that is regularly serviced and is usually designed for the life of the mine, the core distribution function uses technologies such as fiber, twisted pair, or coaxial cabling. Surface and industrial electronics are relatively easily installed. Thus, routers, bridges, programmable controllers, telephones, and personal computers equipped with Ethernet cards are all found in these areas.

Voice and data communications. A mine, unlike other industrial environments, is constantly changing. To meet production requirements, the miners must systematically increase the number and length of tunnels to follow the ore outline. As these tunnels extend, the mine workings' communication network must provide voice and data links to the workers and equipment in the active production and development headings.

Historically, a party-line telephone extending to key points throughout the mine workings provided the only means of communication to the outlying areas of the mine. Twisted-pair wiring was used for the very limited data communications in the active production areas.

In the late 1980s and early 1990s, communications to the working areas of the mine changed drastically. A radio-based communication system, using a cable that could propagate radio-frequency (RF) energy in the tunnels, was developed and was economically installed in the mine workings.

This radio-based network, using "leaky feeder" technology, is currently used in over 250 mines worldwide, providing multichannel voice, low-speed data (9600 baud), and video communications. Copper wire strung throughout the workings continues to provide data communications in an ever-increasing number of applications in the mine.

Challenges. To remain competitive, mines are implementing higher levels of automation and supervision in the development and production areas. This trend to widely distribute control and supervision will increase the amount of communication bandwidth that will be required in the outlying areas of the mine.

Although fiber and hybrid networks will meet the needs for stationary equipment, there remains a significant challenge to provide reliable and cost-effective communications to the mobile fleet of production equipment. The greatest challenge is to develop robust, cost-effective communication networks that provide real-time control of mobile equipment.

Andrew Dasys

Seismic Tomography

Seismic tomography is a method of generating images of structural features within a body of rock by propagating energy through the mass from multiple angles. The reconstructed image represents the interior of the target through the effect of that rock mass

on the passage of the original signal.

Velocity and attenuation tomography. The path followed by the seismic wave from the source to the receiver is represented as a ray. Velocity tomography is based on the time required for each ray to travel through the rock from source to receiver. Attenuation tomography is a representation of the reduction in measured power of the received signal. These two types of tomography provide different information on the physical makeup of the rocks tested. A known relationship among rock type, stress or fracturing, and seismic wave velocity and attenuation is fundamental to conducting studies. As stress increases, microfractures within the rock close, resulting in higher seismic wave velocities and lower attenuation. In areas where features such as fracture zones, faults, or clay-filled cavities exist, seismic waves generally travel at a lower velocity and have increased attenuation. This same behavior is noted in rocks of varying lithology, as harder, more consolidated materials propagate seismic waves at higher velocity and lower attenuation than softer, less consolidated rocks.

In velocity tomography, the source and receiver locations and corresponding travel times are recorded for each ray. The actual refracted ray path and velocity variations along the ray path within the medium are unknown. A forward velocity model is constructed to estimate travel time and the refraction path for each ray. This step is accomplished by approximating the velocity medium as a continuous grid mesh with assigned values. Travel times are estimated by propagating a finite difference wavefront across the mesh from a known source location. For low-velocity contrasts, straight rays are often assumed. With higher-velocity contrasts, rays bend or refract, resulting in longer travel paths. Differences between the estimated and measured travel times are used to iteratively update the velocity mesh in regions along the ray path. Refraction paths are also adjusted iteratively to account for changes in the velocity mesh. Iterations are repeated until the velocity mesh converges to a solution.

Attenuation tomography is performed in a similar manner to velocity tomography except that, instead of travel time, the reduction in transmitted power is computed through the use of a linearized ray-damping factor.

Applications. Attenuation tomography has been applied to map relative stress levels in longwall coal mining operations. Geophones are installed along the headgate and tailgate entries. As the longwall shearer moves along the face and cuts coal, it generates seismic energy that travels through the rock to geophone positions. The condition of the rock existing between the shearer and the receivers affects the amplitude of the received signal. Both rock type and stresses contribute to attenuative losses. The amplitude of the received signal can be processed to form a picture of the stress levels present in the coal panel ahead of the mining face (**Fig. 1**). The average amplitude of the received shearer noise is computed for each source location, and a distance-versus-amplitude plot is produced. If the material between source and receiver were homogeneous, these data would form a straight line. Under realistic conditions, deviations from this linear behavior exist, and these deviations provide information on material properties ahead of the mining operation. A tomographic inversion for attenuation is performed on data collected during the passage of the shearer from headgate to tailgate.

Several factors must be understood when interpreting tomograms. Geophones are normally attached to roof bolts, so tomograms represent energy transmitted through the immediate roof. Fea-

tures that remain stationary as mining progresses are likely geologic anomalies, whereas features that advance with the face are usually mining-induced conditions, presumably stress-related. As stress increases from virgin loading levels, microfractures within the roof close, allowing seismic energy to travel with less attenuation. As stress approaches the yield point of the rock, new fractures are formed, resulting in increased attenuation. Thus, darker shades on the tomograms indicate stresses higher than virgin loading but lower than yield. Areas that display lighter shades over the duration of monitoring are likely undergoing relatively minor increases in loading. Areas that were initially under high stress but later show lighter shades have likely yielded. Another component necessary for an accurate understanding of the tomogram is the concept of ray path coverage. Tomography works by traversing the study area with many different source-receiver ray paths. Areas which have a high number of ray paths and in which the ray paths are traveling at a variety of angles have the most accurate results. In the case of a longwall mine, areas with the best coverage, and thus the best results, include two triangular regions, each with the face as one edge and the headgate-tailgate geophones as the other edge. Rays tend to cluster in areas of low attenuation, where transmission is relatively easier than through highly attenuative material.

Commonly, a tomogram of a coal mine shows a region of relatively high stress at a distance, say 100 ft (30 m), ahead of the working face. The higher-stress area extends forward into the unmined coal and represents an abutment load caused by the extraction of coal and the formation of a caved or gob zone behind the face. The tomogram reveals that around the headgate and tailgate these regions are under relatively lower stress and so either have yielded or have not yet experienced elevated stresses. Several tomograms may be produced during each mining shift so that dynamic changes in loading conditions and the development of hazardous stress concentrations are monitored in a timely fashion.

In addition to monitoring stress conditions, tomographic imaging may be used to ascertain geologic conditions prior to excavation. As an example, data collected at a gold mine showed calcite-lined cavities varying in size and shape. Production above a void could cause a floor failure that might claim both personnel and equipment. In addition, gases within the voids could pose an immediate health risk if directly encountered and a long-term risk if ventilation air short-circuits through the voids and pushes gases into the fresh air supply. Velocity tomography was used to image the mine, providing a three-dimensional picture (**Fig. 2**). The volume of the rock was imaged, with mine workings and voids superimposed. Areas of very low acoustic velocity were interpretable as void spaces. The cavities imaged in this study agreed with both known and unknown features, and areas of highly stressed rock surrounding the cavities were evident.

Tomographic imaging principles have been developed within the earth sciences to the degree that

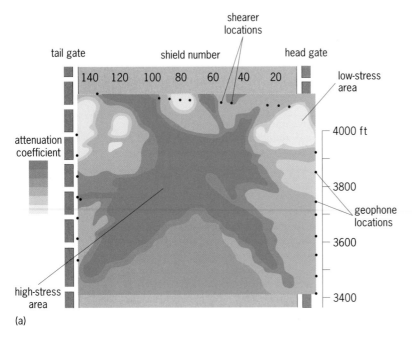

(a)

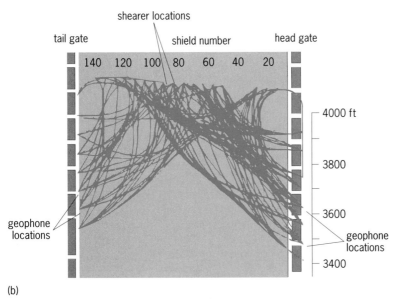

(b)

Fig. 1. Tomographic inversion of seismic attenuation ahead of a coal mine face. (*a***)** Noise from the coal shearer is recorded at discrete times. Lower attenuation in this application may be interpreted as regions of elevated relative stress. (***b***)** Ray path coverage used in the inversion. Attenuation along each ray is computed. Rays tend to concentrate in areas of lower attenuation, where transmission is more efficient.

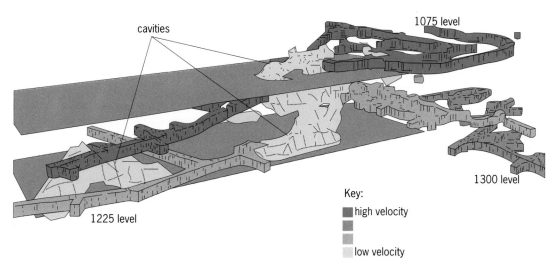

cavities

1075 level

1300 level

1225 level

Key:
■ high velocity
■
■
□ low velocity

Fig. 2. Three-dimensional, tomographic reconstruction of the velocity structure present in an underground gold mine. Areas of low seismic velocity correspond to large-scale voids.

routine application using personal computers has become a reality. When tomography is applied to mining situations, information on conditions that will be encountered in future excavation or conditions that are developing due to dynamic, mining-induced stress changes becomes readily available. This information provides a means for planning and avoiding conditions that pose a threat to all aspects of mining operations.

For background information *see* COAL MINING; COMPUTERIZED TOMOGRAPHY; DATA COMMUNICATIONS; LOCAL-AREA NETWORKS; MINING; ROCK BURST; ROCK MECHANICS; SEISMIC STRATIGRAPHY; SEISMOGRAPHIC INSTRUMENTATION; SEISMOLOGY; UNDERGROUND MINING in the McGraw-Hill Encyclopedia of Science & Technology.

David R. Hanson; David M. Neil

Bibliography. T. C. Bartee, *Digital Communications*, SAMS, 1986; T. N. Bishop et al., Tomographic determination of velocity and depth in laterally varying media, *Geophysics*, 50(6):903–923, 1985; A. J. Devaney, Geophysical diffraction tomography, *IEEE Trans. Geosci. Remote Sens.*, GE-22(1):3–13, 1984; K. A. Dines and J. R. Lytle, Computerized geophysical tomography, *Proc. IEEE*, 67(7):1065–1073, 1979; W. A. Hustrulid, *Underground Mining Methods Handbook*, Society of Mining Engineers of the American Institute of Mining, Metallurgical, and Petroleum Engineers, Inc., 1982; Jie Zhang et al., Nonlinear refraction and reflection travel time tomography, *J. Geophys. Res.*, 103(B12):29,743–29,757, 1998; *Proceedings of the 4th International Symposium on Mine Mechanisation and Automation*, Brisbane, July 1997.

Underwater vehicles

Autonomous underwater vehicles have been designed to make measurements, relay information, or perform tasks in a critical, inhospitable environment. Both large vehicles (transport-class platforms) and small vehicles (network-class platforms) have a role in exploring and working in the ocean. Historically, large vehicles have often been designed to address specific missions. Recently, small vehicles, enabled by advances in technology, are evolving more generically to address measurement of a variety of ocean state variables. Mapping ocean variability today and forecasting it tomorrow requires coupled modeling-sampling systems. In advanced nowcasting/forecasting systems, modeling and sampling are intimately linked by design. Sampling initializes a model which then produces three-dimensional fields used to guide further sampling. The feedback in this adaptive approach is tuned to minimize forecast error.

Ocean sampling has been based primarily on ships, satellites, floats, and moorings. The most comprehensive measurements from ships involve underway profilers such as thermistor chains, acoustic Doppler current profilers, and undulating towed bodies. These methods provide two-dimensional sections through temporally evolving fields. Satellites also provide two-dimensional images of the ocean surface. Arrays of moorings and floats provide simultaneous time series, but typically the spatial sampling is sparse due to cost. To properly resolve temporally evolving spatial gradients with a limited number of sensors, spatially adaptive sampling is necessary. Adaptive sampling and requirements for long duration and affordability dictate mobile, autonomous platforms with low unit cost that operate as an intelligent network. Such mandates form the genesis of the network-class platform.

Classification. Autonomous underwater vehicles (AUVs) span a wide range of sizes and capabilities, related to their intended missions. Each is a mobile instrumentation platform with propulsion, sensors, and on-board intelligence designed to complete sampling tasks autonomously. At the large end of the scale, transport-class platforms in the order of 10 m (33 ft) length and 10 metric ton (11 tons) weight in air have been designed for missions requiring long endurance, high speed, large payloads, or high-power sensors. At the small end of the scale, network-class platforms in the order of 1 m (3.3 ft) length and 100 kg (220 lb) weight in air address missions requiring portability, multiple platforms, adaptive spatial sampling, and sustained presence in a specific region. Overall, vehicles can also be categorized in terms of propulsion method: propeller-driven or buoyancy-driven. Historically, most vehicles have been propeller-driven, but several new, variable-buoyancy, underwater gliders are emerging in the network class. Vehicles can also be categorized in terms of their maximum operating depth: full ocean depth (4000–6000 m; 2.5–3.7 mi) and continental shelf depths (100–500 m; 330–1640 ft).

Over 50 autonomous underwater vehicles have been designed and built since 1963 by government laboratories, industry, and academia. Most vehicles in all classes are research prototypes. Commercial vehicles are becoming available, particularly in the network class. This trend will accelerate greatly in the next few years as capabilities improve and the need for cost-effective sampling grows. Reinforcing this trend are the Global Positioning System (GPS) and emerging global communication services enabling timely access to data and real-time control. Developments in microprocessors, memory, batteries, and fuel cells are leading to higher performance in smaller platforms. Fiber optics and micro-electro-mechanical machining are spawning a new generation of compact, high-precision, low-power sensors ideally suited for payloads on autonomous underwater vehicles.

This article reviews representative battery-operated, sampling vehicles. Ranges given are those specified by the developers for state-of-the-art batteries (typically silver-zinc, nickel-cadmium, lead-acid, or alkaline); no attempt has been made to normalize power sources. New lithium-based batteries and fuel cells will considerably extend these ranges.

Transport class. In the transport class, the Mobile Undersea Systems Test Laboratory is the world's largest autonomous underwater vehicle: 10 m (33 ft) long, 1.2 m (4 ft) in diameter, 9.2 metric tons (10 tons) in air, rated to 600 m (1970 ft) depth, with a 140 km (87 mi) range and 1 metric ton (1.1 ton) payload capacity. The Advanced Unmanned Search System is 5 m (16 ft) long, 0.8 m (2.6 ft) in diameter, rated to 6100 m (3.8 mi) depth, and has a 130 km (80 mi) range. Other transport-class vehicles rated to full ocean depth include the French Epaulard, the Russian MT-88, and the Chinese CR-01. Theseus, developed for the Canadian Department of National Defense, is rated to 1000 m (3280 ft) depth with an operating range of 750 km (466 mi) and an endurance of 100 h. In 1996, Theseus successfully completed two 350-km (218-mi), 50-h missions under the Arctic icecap.

Intermediate between transport class and network class, Autosub is 7 m (23 ft) long, 0.9 m (3 ft) in diameter, rated to 500 m (1640 ft) depth, and has an operating range of 260 km (162 mi) with a cubic meter (10 ft³) of payload space. The Marine Utility Vehicle System (MARIUS) is 4.5 m (14.8 ft) long, 0.6 m by 1.1 m (2 ft by 3.6 ft) in rectangular cross section, rated to 600 m (1970 ft) depth, and has an operating range of 50 km (31 mi). MARIUS was designed for seabed inspections and environmental surveys in coastal waters.

Network class. Typical high-performance, propeller-driven vehicles in the network class are the Autonomous Benthic Explorer (ABE), Odyssey, Ocean Explorer, and Remote Environmental Monitoring Units (REMUS). ABE is 450 kg (992 lb) in air, rated to 6000 m (3.7 mi) depth, and has an operating range of 100 km (62 mi). In 1997, ABE mapped magnetic anomalies associated with deep-ocean spreading zones in the North Pacific. Odyssey is 2.1 m (6.9 ft) long, has a maximum diameter of 0.6 m (2 ft), and is rated to 6700 m (4.2 mi) depth. Ocean Explorer is similar in form but modular, expandable to 3 m (9.8 ft) long, and is rated to 300 m (985 ft) depth. In both vehicles, a free-flooded plastic exoskeleton provides a low-drag, low-cost hull, resulting in an operating range of 60 km (37 mi). Inside the hull are selectable pressure cases for control, sensors, and propulsion. Odyssey has been deployed in the Arctic and Antarctic, Haro Strait, and the Labrador Sea, with plans for a 1000-km (620-mi) Arctic deployment in 2001 using a fuel cell. Ocean Explorer has performed numerous coastal mapping experiments, and will demonstrate acoustic communication capabilities in 1999 by rendezvousing with a nuclear submarine. A highly portable, shallow-water vehicle in this class, REMUS is 1.5 m (4.9 ft) long, 20 cm (8 in.) in diameter, 40 kg (88 lb) weight in air, with up to 80 km (50 mi) range (**Fig. 1**). REMUS has mapped water properties and performed mine reconnaissance to the edge of the surf zone.

Network-class, buoyancy-driven vehicles include Spray, Slocum, and the Virtual Mooring Glider (VIRMOG). These small, near-neutrally-buoyant platforms move vertically and horizontally through the water as a result of small changes in buoyancy. Slocum derives its propulsion power for a sawtooth ascent/descent through the water column from a thermodynamic

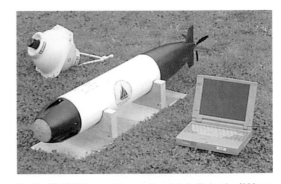

Fig. 1. REMUS is rated for continental shelf depths (200 m; 355 ft) and is compact and lightweight (about 154 kg; 70 lb). It is depicted with its laptop computer for downloading missions and uploading data, and one of the acoustic transponders used for navigation. (*Woods Hole Oceanographic Institution*)

Fig. 2. Spray. (*Scripps Institution of Oceanography*)

engine utilizing the vertical thermal gradient in temperate and tropical oceans. This vehicle is designed for deep-ocean sampling, and uses changes in buoyancy for propulsion (underwater glider). Two prototype versions exist: to change buoyancy, one uses batteries and one uses ambient ocean thermal gradients (thermoclines). By exploiting thermal energy from the environment, a 5-year, 40,000-km (24,855-mi) range is feasible. An interactive network of such vehicles is particularly suitable for sampling large-scale ocean processes. Spray and VIRMOG are fitted with battery-operated buoyancy controllers, and can operate with an estimated 40-day, 800-km (497-mi) range independent of temperature gradient. Spray and VIRMOG are designed for both deep- and shallow-ocean sampling, and use changes in buoyancy for propulsion (underwater glider). Batteries are used as the energy source for the variable-buoyancy engine (**Fig. 2**). One mission envisioned for VIRMOG is as a holding station at a fixed location, thus virtually emulating a mooring (**Fig. 3**).

Objectives. All state-of-the-art vehicles contain libraries of intelligent control algorithms to execute complex missions with abstract goals (for example, find and map an anomalous feature). Capabilities that are becoming part of a typical vehicle's intelligence include rendezvous and docking, survey, gradient following, obstacle avoidance, adaptive sampling, terrain following, and fault detection and recovery. Reconfiguration of vehicle software for a range of missions, sensor types, and performance will often be accomplished remotely through satellite or acoustic links.

Navigation. A universal, subsurface navigation system for autonomous underwater vehicles is not yet available. Presently used are radio and satellite systems; long, short, and ultrashort baseline acoustic systems; acoustic Doppler and correlation speed logs; inertial systems; and terrain-following techniques. Radio and satellite navigation requires the vehicle to be at or close to the surface, long-baseline acoustic navigation accuracy degrades at long ranges, and low-cost inertial systems accumulate significant errors in long-duration vehicles. Combined techniques yield improved performance, for example, Doppler-sonar-corrected inertial navigation. For long-baseline acoustic navigation, accuracy can be enhanced through timely feedback of measured sound speeds into on-board acoustic propagation models. Location precision of 1 m (3.3 ft) over a range of 10 km (6 mi) is typical, using multiple arrivals from acoustic navigation beacons.

Communication. Communication and energy are critical for all autonomous systems. For reliable vehicle control and monitoring, acoustic communication modems are being developed that adapt the signal phase to the environment or hop among frequencies to maintain a fixed bit error rate. State-of-the-art acoustic modems are capable of 10 Kbit/s data rates at 10-km (6-mi) range and 3 Kbit/s at 60-km (37-mi) range in favorable oceanic conditions. Energy efficiency varies from 1 Kbit/(joule)(km) with off-the-shelf commercial modems to 100 Kbit/(joule)(km) for research prototypes. Network communication protocols, such as those used on the Ocean Explorer, provide a scalable approach to remote monitoring and control of multiple vehicles. The acoustic modem becomes a node on the network, allowing for remote access to vehicle subsystems.

Energy. Energy, stored on-board for both propulsion and payload, is always limited. Additional subsea sources of energy may be distributed at storage stations, generating stations, or stations fixed to cables from shore or an overlying surface buoy. Current research efforts include (1) autonomous energy transfer from a mooring to a vehicle at high rate (200 W) and high efficiency; (2) in-situ recharge of batteries under ambient pressure; (3) intelligent power management; and (4) renewable power generation utilizing ocean waves and currents. An inductive coupler for power and data transfer has been installed in an Odyssey. Tests have demonstrated a continuous transfer of over 200 W into the vehicle with an 84% overall efficiency. Pressure-compensated lithium cells are being researched for use. Prototype, in-situ power sources using piezoelectric polymer materials flexed by ocean waves are being tested.

Future. Growth in network-class vehicles is fueled by size and cost reductions, modular scalable architectures, acoustic cellular navigation and communication, and efficient plug-and-play manufacturing techniques. Needs for long-term presence and endurance will be met through docking stations with available or renewable energy sources. Compact systems will facilitate surface, submarine, or air deployment. Coupling with environmental forecast models through the Internet will enable global, intelligent, and adaptive sampling.

Fig. 3. VIRMOG. (*University of Washington*)

For background information *see* ARTIFICIAL INTELLIGENCE; GEOGRAPHIC INFORMATION SYSTEMS; MARINE NAVIGATION; OCEANOGRAPHIC VESSELS; ROBOTICS; SATELLITE NAVIGATION SYSTEMS; UNDERSEA VEHICLES; UNDERWATER NAVIGATION in the McGraw-Hill Encyclopedia of Science & Technology.

Thomas B. Curtin

Bibliography. M. D. Ageev, The use of autonomous unmanned vehicles for deep-water search operations, *Subnotes*, September-October 1990; B. Grandvaux and J. Michel, Epaulard: An acoustically remote controlled vehicle for deep ocean survey, *Proc. Mar. Technol.*, 79:357–359, 1979; H. Li, New robotic vessel extends deep-ocean exploration, *Science*, 278: 1705, 1997; D. R. Yoerger, A. M. Bradley, and B. B. Walden, The Autonomous Benthic Explorer (ABE): An AUV optimized for deep seafloor studies, *Proceedings of the 7th International Symposium on Unmanned Untethered Submersible Technology*, University of New Hampshire, Doc. 91-9-01, 1991.

Vaccination

Traditional vaccines evolved into highly successful preventive medical strategies during the twentieth century, as exemplified by the eradication of smallpox through a global immunization program and the elimination of polio in the Americas. Vaccines have altered population dynamics and increased life expectancies in both industrialized and developing countries. Despite global benefits, developing countries still experience sustained mortality rates from the spread of infectious diseases. National vaccination programs rely on widespread delivery of traditional vaccines, which require preservation against heat inactivation. Manufacturing and purification costs for traditional vaccines are often prohibitive for nations of limited economies. In addition, more than a third of all current vaccination centers in developing countries report injection-related infections following syringe use. Therefore, novel strategies are evolving to provide alternative vaccination methods.

Characteristics of vaccines. Traditional vaccines consist of live or attenuated strains of a pathogenic virus or bacterium and are formulated with an adjuvant (inert material which stimulates the immune system). Common adjuvants consist of multiple components, including an oil emulsion for antigen encapsulation, and possibly additional killed cells from another pathogen (such as *Mycobacterium tuberculosis* in Freund's adjuvant), to enhance stimulation of the immune response at the site of administration.

In the last decade, subunit vaccines were introduced. These consist of one or more proteins derived from a pathogen, usually in the form of a recombinant protein produced in a yeast or bacterial system. The recombinant proteins are purified to remove contaminants and to concentrate the antigen.

Vaccines which comprise live or attenuated components or pathogen subunits require protection against heat inactivation, from manufacture until the time of administration. They are commonly stored in vials which carry heat-sensitive labels as an indicator of the individual exposure received. Vaccination programs in most developing countries commonly require extensive transport of vaccines produced in industrialized countries and further distribution across the breadth of the nation, often in tropical or arid climates. Ineffective immunization with heat-inactivated vaccine poses the threat of loss of faith in the entire vaccine program.

Continuing development focuses on production of vaccines which can be produced and administered through alternative methods and are heat-stable. Candidate techniques include the use of subunit vaccines, immunization via oral administration, and production in low-contamination systems such as transgenic plants.

Subunit vaccines. Subunit vaccines contain antigenic protein units which make up only part of the pathogenic organism or toxin. For instance, infectious viral particles or native bacterial toxins which cause human disease are generally composed of multiple protein units. Many viral particles are constructed from one or more shell layers (each layer constructed by a protein matrix), and many have additional spike proteins which are present on the external surface of the particle or spiral. Toxins secreted by bacterial infections also comprise one or more protein components which can be genetically reproduced independent of infection. Each of these viral or toxin protein subunits may contain specific determinants which, when presented to the immune system, can cause stimulation sufficient for protection against infection by the native pathogen.

Subunit vaccines can take several forms. They may be expressed in the simple form of linear bacterial or viral peptides or designed to form complex proteins which imitate the native form. When produced in an appropriate system, copies of the subunit protein may assemble to form empty particles which mimic the morphology, stability, and antigenic characteristics of the authentic virus. Full-length bacterial proteins can be produced in plant, animal, or bacterial systems capable of performing the three-dimensional processing necessary for antigenic authenticity. Subunit vaccines do not contain the genetic components (nucleic acids) inherent in live or attenuated vaccines, thereby removing any possibility of infection from vaccination. Modern genetic engineering facilitates the formation or fusion of different multiple subunit proteins (either from the same pathogen or from different pathogens) in the same production system. This allows the production of safe vaccines with multiple antigenic determinants against a specific pathogen, or for the construction of multicomponent proteins carrying determinants for vaccination against multiple diseases.

Oral immunization. The mucosal immune system is represented by regions of lymphoid tissue within the submucosal linings of the gastrointestinal system, the urogenital tracts, and the respiratory tract. These mucosa-associated lymphoid tissues (MALT) contain highly organized lymph systems which aggregate to form structures such as the tonsils, adenoids, Peyer's patches, or the lamina propria lining the small intestine. In these regions, specialized M cells capture foreign antigens and translocate them across the mucosal lining for presentation to B cells and T cells which, when activated, migrate to lymph nodes to mature. Stimulated B cells are capable of maturing into plasma cells responsible for the production of immunoglobulin A (IgA), which is secreted across the mucosal lining as the body's first defense against pathogenic organisms. The highly organized mucosal immune system secretes IgA across all the mucosal surfaces in response to an antigen detected at one of those surfaces.

Immunization by ingestion is designed to present vaccines to the MALT in a manner which stimulates IgA secretion sufficient for protection against native pathogen. The Peyer's patches are considered the most concentrated MALT. Oral immunization targeted to the Peyer's patches requires antigen protection against the extreme pH and the proteolytic enzymes found within the stomach. Antigen release within the intestinal tract can be achieved by use of controlled-release capsules or by natural encapsulation within the cellulose wall of plant cells. Alternatively, oral vaccines for general presentation to the MALT can be formulated from bacterial expression systems, where the purified antigen is coated onto ingested substances.

The mucosal linings which are stimulated by oral immunization also constitute the points of entry for pathogenic organisms. A mucosal immune response at these sites, therefore, provides appropriate protection. Oral vaccines must account for some inevitable loss of antigen through digestive processes. As a consequence, the antigen dose required for presentation to MALT for sufficient immunogenic stimulation is far more than that required by injection-based vaccines. As with traditional vaccines, oral vaccines generally require a multiple-dose regime and periodic boosting. Despite the ease of vaccination and the eradication of injection-related injury, oral vaccination necessitates controlled administration. If a foreign antigen is presented too frequently, the B and T cells may become subject to clonal silencing and the vaccine becomes ineffective. This silencing (or immune tolerance) is commonly induced by the vast array of commonly digested plant and animal proteins in foods.

The requirement for oral adjuvants which support subunit vaccines represents a paradigm shift in vaccine development. In contrast to traditional adjuvants, oral delivery requires protection against digestion rather than phagocytosis, and stimulation of specific cells and tissues rather than simply increasing the magnitude of leukocyte activity. Candidate oral adjuvants are typically inactivated forms of proteins derived from enteric pathogens, which

increase the sensitivity of M cells to foreign proteins.

Transgenic plants. Genetic engineering has facilitated the use of transgenic plants as candidate vaccine production and delivery systems. Plant expression vectors which consist of elements from plant nucleic acid sequences can be constructed to include the gene encoding for a subunit protein derived from a viral or bacterial pathogen. Plant transformation techniques allow the introduction of the vaccine expression cassette within individual plant cells. The cassette is stably incorporated into the plant's genome. As the plant cell divides and multiplies to form a mature plant, the introduced gene is replicated and the new protein is expressed along with other essential plant proteins in that plant and all of its progeny.

An alternative method utilizes recombinant plant viruses which are not harmful to humans as a source of genetic amplification within individual plant cells. Viral vectors can be constructed to contain the gene for a subunit vaccine, and gene expression is driven by the viral replicating mechanisms.

Both methods serve to accumulate the foreign protein within the plant cells. The expression of subunit proteins in edible tissues (such as fruits or vegetables) provides a clean production system with no requirement for purification and for removal of contaminants normally associated with yeast or bacterial production systems. Plant cells are composed of complex structures such as the endoplasmic reticulum and Golgi apparatus which are capable of the posttranslational processing required for antigen authenticity, and the rigid cell walls offer initial protection against digestive mechanisms.

Plants provide a cost-effective alternative for mass production, and they provide antigen in a prepackaged form ready for oral administration. While oral vaccines in transgenic plants do not require refrigeration, excessive heating (such as that required for cooking) would denature the vaccine, rendering it ineffective. Accordingly, the choice of the plant expression system is limited to those which, in addition to containing moderate protein levels, do not require cooking before consumption. Bananas are currently a focus for plant-based vaccine production since they can be consumed uncooked by both infants and adults.

For background information *see* BIOLOGICALS; POLIOMYELITIS; SMALLPOX; VACCINATION; VIRUS in the McGraw-Hill Encyclopedia of Science & Technology.

Charles J. Arntzen; Dwayne D. Kirk

Bibliography. T. Arakawa et al., Efficacy of a food plant-based oral cholera toxin B subunit vaccine, *Nature Biotechnol.*, 16:292–297, 1998; K. E. Palmer et al., Antigen delivery systems: Transgenic plants and recombinant plant viruses, *Mucosal Immunology*, pp. 793–807, 1998; C. O. Tacket et al., Immunogenicity in humans of a recombinant bacterial antigen delivered in a transgenic potato, *Nature Med.*, 4(5):607–609, 1998; S. Y. Wong et al., Edible vaccines, *Sci. Med.*, pp. 36–45, November–December 1998.

Vertebrate origins

An outstanding problem in vertebrate paleobiology is the origin of vertebrates, which has been an issue for well over a hundred years. How are vertebrates related to invertebrates; what characteristics are specific to vertebrates in terms of their body plan; and what genetic innovations were necessary in order to develop these vertebrate-specific features remain questions. Answers are not likely to be found in the fossil record of early vertebrates, but require tools that have come into use only recently.

Phylogenetic position. Higher animal phyla can be divided into two major groups: the protostomes and the deuterostomes. The names are derived from early features of the embryo—most often, the different fate of the blastopore (the opening of the digestive tract). In protosomes, this opening usually becomes the mouth; in deuterostomes, usually the anus. It is now becoming clear that each group is indeed a natural (monophyletic) assemblage of animal phyla. The deuterostomes are of concern here because this group includes the phylum Chordata which contains the subphylum Vertebrata (**Fig. 1**).

Deuterostomia comprises three phyla: the echinoderms (including sea urchins, brittle stars, sea stars, sea cucumbers, and sea lilies); the hemichordates (a small and little-known group which includes the acorn worms); and the chordates (Fig. 1a). It has become clear from molecular evidence, especially studies involving mitochondrial deoxyribonucleic acid (DNA), that echinoderms and hemichordates are sister taxa (each other's nearest relative). Collectively, these two taxa are then the sister group of the chordates.

There are three major groups of chordates: the urochordates (which include the sea squirt); the cephalochordates (the lancelet amphioxus); and the vertebrates. There is much morphological and molecular evidence to suggest that the nearest invertebrate relative to the vertebrates is amphioxus (Fig. 1b), and much of the studies discussed below concern genetic data from this animal. Within the vertebrates the phylogenetic relationships among the three major groups (the two jawless fish taxa, the hagfish and lamprey, and the jawed fish or gnathostomes) are equivocal. Morphological studies and mitochondrial DNA analyses suggest that lamprey and gnathostomes are sister taxa, whereas nuclear ribo-

somal DNA studies suggest that lamprey and hagfish are sister taxa. Because of the discord between these two lines of phylogenetic inquiry, these taxa are shown as a trichotomy (Fig. 1c).

The importance of these phylogenetic studies is twofold. First, they allow the position of the vertebrates to be firmly established within the deuterostomes as the sister taxon of amphioxus. Second, they allow the rejection of previous ideas proposed to explain the origin of vertebrates whose predicted phylogenetic relationships are not in accord with the phylogeny known today. For example, one proposal suggested that larvaceans, one of the three major groups of urochordates, are highly derived ascidians (these, along with salps, are the other two groups of urochordates). Hence, the life cycle seen in ascidians, with a swimming tadpole-like larva similar in morphology to a larvacean and a bottom-dwelling adult, was suggested to be primitive not only for urochordates but for chordates as a whole. However, a recent molecular analysis, coupled with older morphological data, strongly suggests that larvaceans are basal urochordates and not highly derived ascidians (Fig. 1d). This means that the ascidian life cycle is not a good substitute for understanding the origin of chordates. Since the basic phylogenetic framework of the scenario is in error, it is probable that the major evolutionary transitions proposed are flawed as well.

Origin of characters. One of the most distinctive features of the vertebrate body plan is the brain, which is tripartite in organization with forebrain, midbrain, and hindbrain regions. Although the size and complexity of the vertebrate brain are unique, the question here is whether the basic tripartite organization is a vertebrate novelty as well. This has been investigated by comparing the expression patterns of key developmental genes in both vertebrates and the other chordate groups (**Fig. 2**). The reasoning is as follows: if two structures are homologous (derived from a similar structure present in the latest common ancestor of the taxa under consideration), then each structure should utilize the same genes during its development (unless, of course, there are secondary modifications). It is known that a suite of genes is expressed in the vertebrate brain, many of which are expressed in only one region. For example, the gene *Orthodenticle* (*Otx*) is expressed in the vertebrate forebrain and midbrain, whereas the *Hox* genes

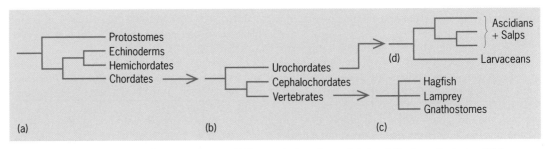

Fig. 1. Interrelationships among deuterostome taxa. (*a*) Among deuterostome phyla in relation to protostomes. (*b*) Among chordate subphyla. (*c*) Among vertebrate fish taxa. (*d*) Among urochordate classes.

(Fig. 2) are expressed in specific areas of the hindbrain and spinal cord. Examination of these same genes in ascidians and amphioxus shows the same relative pattern of expression, suggesting that at least two regions of the vertebrate brain (fore/midbrain and hindbrain) are more primitive than previously thought.

The second area of active research concerns the origin of the neural crest. The neural crest is responsible for the development of many vertebrate-specific features, including much of the head. However, its origin is obscure because nothing obviously like it exists in amphioxus or any other animal group. Given its tremendous evolutionary importance, several research groups have sought to identify the ancestral tissue of the neural crest in amphioxus. Their approach is similar to the one described above: if the neural crest expresses specific genes during its development, then identification of the same genes and their spatial localization in amphioxus may indicate the homolog of the neural crest. Two genes (among many) are expressed in the neural crest: *Distal-less* and *snail*. Examination of the expression patterns of these genes in amphioxus shows nonoverlapping expression in epidermal or neural cells, respectively. Hence, a simple gene expression study is not sufficient to identify a neural crest homolog in amphioxus, and the origin of the neural crest remains mysterious.

Gene duplication. It appears from the studies described above that some basic architectural similarities exist between amphioxus and vertebrates, especially with respect to the brain. However, there are clear novelties, both tissue type (such as the neural crest) and tissue complexity (cerebral cortex, cerebellum, and so on). The genetic source of these innovations has been investigated. Studies on several invertebrates, including ascidians, and vertebrates have shown a surprising phenomenon; there is a significant increase in the number of genes found in vertebrates in comparison to invertebrates (**Fig. 3**). The number of genes in vertebrates is estimated to be 60,000–100,000, whereas for invertebrates estimates range 10,000–20,000. In fact, the recently completed *Caenorhabditis elegans* (a nematode worm) genome project revealed 19,099 predicted protein-coding genes. This large increase in gene number is not just an increase in single genes but appears to represent two rounds of whole-genome duplication. For example, *Hox* genes are a group of about 10 genes physically linked on the chromosome [and the expression pattern of one of these genes, *Hox1* is indicated above (Fig. 2)]. Studies from insects, *C. elegans*, sea urchin, and amphioxus have shown that all of these organisms possess only a single *Hox* cluster (and many other studies from a variety of other invertebrate taxa are consistent with the presence of a single *Hox* cluster in these taxa as well). Vertebrates, however, have at least four clusters; hence the entire chromosome must have been duplicated, not just any specific *Hox* gene. These duplications happened early in vertebrate history, as the lamprey

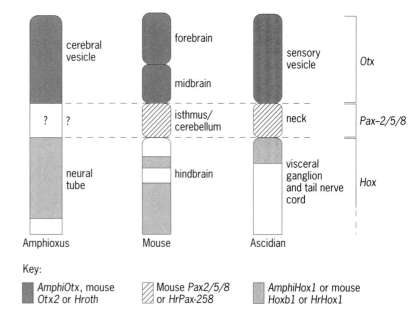

Fig. 2. Comparison of the three vertebrate brain regions and gene expression patterns with amphioxus and ascidian. Each gene expression pattern supports a hypothesis of homology for the fore/midbrain, isthmus/cerebellum, and hindbrain regions with the indicated regions of the amphioxus or ascidian brain. Region-specific genes are listed at the far right, and their expression domains are indicated according to the key. Note that the *Pax2/5/8* gene has not yet been reported from amphioxus. (*Modified from N. A. Williams and P. W. H. Holland, Molecular evolution of the brain of chordates, Brain Behav. Evol., 52:177–185, 1998*)

is estimated to have at least three *Hox* clusters. The fourfold amplification is not limited to just *Hox* genes. Many genes exist in three- to fourfold abundance over what is found in any invertebrate analyzed, and each gene duplicate remains in the same relative position on the chromosome with respect to other duplicated genes.

The importance of this fantastic increase in genomic information is exemplified when the spatial expression domains of these new genes are

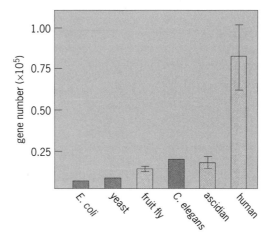

Fig. 3. Gene number estimates. Shaded bars indicate results from completed genome projects; open bars are estimated values. Note that the value for "human" is representative for other vertebrates examined, including mouse and pufferfish. (*Modified from M. W. Simmen et al., Gene number in an invertebrate chordate, Ciona intestinalis, Proc. Nat. Acad. Sci. USA, 95:4437–4440, 1998*)

examined. The expression pattern of *Otx* in the vertebrate was described above (Fig. 2). However, this was oversimplified. There are two *Otx* genes in the mouse: *Otx2*, which shows a similar expression profile to the single *Otx* gene in amphioxus and ascidian (and the one indicated in Fig. 2); and *Otx1*, which shows an overlapping but much more restricted expression profile. The importance of each gene to the development of the vertebrate body plan is best realized when these genes are knocked out; that is, their function during normal development is eliminated. Removal of *Otx2* results in embryonic lethality with the complete absence of head structures. Elimination of *Otx1*, however, results only in abnormalities in the cerebral cortex and cerebellum, and hence suggests that these are unique roles not shared with *Otx2*. Therefore, the duplication of *Otx* genes in the vertebrates left one gene performing the ancestral role (*Otx2*) and another gene (*Otx1*) recruited to new roles, specifically roles in the development of novel vertebrate features like the cerebral cortex and cerebellum within more primitive brain regions (that is, the forebrain and isthmus/cerebellum; Fig. 2).

Researchers are finally in a position to begin a mechanistic foray into the origin of the vertebrates. It is known how the vertebrates are related to the invertebrates, and scientists suspect that it was the recruitment of new gene duplicates to novel developmental pathways that was crucial for the origin of the vertebrate body plan. What remains to be determined is how these new genes are coopted into novel developmental pathways and how these new pathways translate into novel morphologies.

For background information *see* ANIMAL EVOLUTION; CHORDATA; GENE; NERVOUS SYSTEM; NEURAL CREST in the McGraw-Hill Encyclopedia of Science & Technology.   Kevin J. Peterson

Bibliography. H. Gee, *Before the Backbone: Views on the Origin of the Vertebrates*, Chapman and Hall, 1996; J. H. Postlethwait et al., Vertebrate genome evolution and the zebrafish gene map, *Nat. Genet.*, 18:345–349, 1998; N. A. Williams and P. W. H. Holland, Gene and domain duplication in the chordate *Otx* gene family: Insights from amphioxus *Otx*, *Mol. Biol. Evol.*, 15:600–607, 1998; N. A. Williams and P. W. H. Holland, Molecular evolution of the brain of chordates, *Brain Behav. Evol.*, 52:177–185, 1998.

Viral disease

Vesicular stomatitis is a viral disease, most often of horses, cattle, and pigs, characterized by fever and by vesicular and erosive lesions on the tongue, gums, lips, feet, and teats. Infection in dairy herds can result in substantial losses in milk production, and serious weight losses may occur in beef cattle and swine. Infection of horses restricts their use and movement, frequently resulting in disruption of equestrian events. A further concern with vesicular stomatitis in ruminants and pigs is the close clinical resemblance to foot-and-mouth disease, one of the world's most feared livestock diseases. Diagnosis of vesicular stomatitis often results in onerous and economically devastating quarantine measures. In humans, vesicular stomatitis virus may cause influenzalike symptoms, including fever, nausea, vomiting, headaches, and muscular pains. Essentially a disease of the Americas, vesicular stomatitis has in the past been transported to Europe and Africa but has failed to become established there.

Causes of vesicular stomatitis. The viruses causing vesicular stomatitis are members of the genus *Vesiculovirus* in the Rhabdoviridae, a very large family which includes viruses of mammals, birds, fish, insects, and plants. Several of the vertebrate and plant viruses are transmitted by insects. Vesicular stomatitis virus serotype New Jersey (VSV-NJ) has been by far the most frequent cause of vesicular stomatitis in the United States. Vesicular stomatitis virus serotype Indiana (VSV-I), a common cause of vesicular stomatitis in Central and South America, disappeared in the United States after 1965 but suddenly reappeared in 1997 in the Southwest.

Host range of VSV. The host range of vesicular stomatitis virus is extremely broad, and the widespread presence of antibodies to the virus in humans, livestock, pets, and many species of wildlife in endemic areas suggests that most vertebrates may be susceptible to the disease. Black flies, sand flies, and midges can support replication of the virus. Thus, vesicular stomatitis virus occurs in both vertebrates and invertebrates.

Transmission and ecology of vesicular stomatitis. The rhabdoviruses which cause VSV-NJ and VSV-I are among the most thoroughly researched of all animal viruses, and yet the ecology of the disease remains an enigma, with major portions of the maintenance and transmission cycles of the virus being unknown. Contact transmission of vesicular stomatitis can occur, but recent evidence suggests that the disease is only mildly contagious. In a recent experiment, susceptible animals stabled beside infected animals were fed contaminated feed and water but remained healthy. Epidemiological studies during the 1995 and 1997 outbreaks of vesicular stomatitis in the United States revealed no correlation between animal movement and spread of the disease.

The role of insects in the transmission of vesicular stomatitis has remained controversial for decades, but recent evidence has led to increasing recognition of the importance of insects not only as vectors but as possible reservoirs of the virus. Characteristics of vesicular stomatitis which support the contention that insects play a major role in its transmission include the seasonal nature of the disease, with sudden appearance in early summer and disappearance soon after the first killing frosts. The spread of the disease along river valleys (as opposed to centrifugal spread) suggests the involvement of insects, as does the fact that vesicular stomatitis is most explosive in areas with irrigation canals and other moving water. The infection rate in stabled animals is much lower than in animals on pasture. Spread of the disease along river valleys (usually northward) is often characterized by large jumps in the absence of evidence of animal movement. This too suggests arthropod involvement and raises the question of whether the disease is in fact moving northward or whether it is emerging from a hidden reservoir in insects or vertebrate carriers as the weather warms.

This strong circumstantial evidence supporting the role of insects in the ecology of vesicular stomatitis has been buttressed during the past decade by extensive experimental evidence that several species of biting flies may be involved in the ecology of the disease. During epidemics the virus has been isolated from naturally infected black flies, sand flies, and *Culicoides* midges as well as from nonbiting flies, including house flies. Extensive field studies on the ecology of vesicular stomatitis virus, combined with laboratory studies, have clearly demonstrated that sand flies (*Lutzomyia* spp.) are naturally infected with the virus, can transmit it to susceptible vertebrates, and can pass the virus transovarially from generation to generation. Isolation of virus from naturally infected male flies (plant feeders, not blood feeders) is proof either that transovarial transmission takes place in nature, or that the virus can be acquired from plants.

Since sand flies have very limited range and are not strong fliers, they are not good candidates for the role of epidemic vectors of vesicular stomatitis. The rapid movement of the disease, especially along watercourses, suggests that more aggressive and wide-ranging flies such as black flies (*Simulium* spp.), which breed in running water, might be the principal vectors during epidemics.

Recent laboratory findings in Arizona support the thesis that black flies can serve as epidemic vectors of vesicular stomatitis virus. Black flies were infected with vesicular stomatitis virus by feeding, the virus replicated in the black flies, and the flies transmitted the virus to laboratory and wild mice. These experimental findings, together with the evidence that the sand fly is the vector in the natural cycle of vesicular stomatitis in endemic areas where transovarial transmission of the virus in sand flies also occurs, suggest a central role for insects in the maintenance of this virus in nature.

Role of wildlife in vesicular stomatitis. Serological surveys for VSV-NJ antibodies in wild animals in Colorado, Arizona, Georgia, Mexico, and Central America have revealed silent vesicular stomatitis infection in many wildlife species before, during, and after viral epidemics. While none of these animal species has been identified as long-term reservoirs of the virus, they serve as useful indicators of viral activity. In spite of numerous attempts to identify a natural reservoir of the virus in vertebrates and invertebrates, none has been found.

For background information *see* ANIMAL VIRUS; VIRUS, DEFECTIVE; VIRUS CLASSIFICATION in the McGraw-Hill Encyclopedia of Science & Technology.

C. John Maré

Bibliography. J. A. Comer et al., Population dynamics of *Lutzomyia shannoni* (Diptera: Psychodidae) in relation to the epizootiology of vesicular stomatitis virus on Ossabaw Island, Georgia, *J. Med. Entomol.*, 31:850–854, 1994; R. P. Hanson, The natural history of vesicular stomatitis virus, *Bact. Rev.*, 16:179–204, 1952; D. G. Mead, C. J. Maré, and E. W. Cupp, Vector competence of select black fly species for vesicular stomatitis virus (New Jersey serotype), *Amer. J. Trop. Med. Hyg.*, 57:42–48, 1997; P. A. Webb and F. R. Holbrook, Vesicular stomatitis, in T. P. Monath (ed.), *The Arboviruses: Epidemiology and Ecology*, CRC Press, pp. 1–29, 1989.

Virtual reality

Virtual reality (the computer-generated simulation of an environment) is being used in the manufacturing world for the layout of facilities and the control of factory operations. This article explores such uses and discusses virtual reality technology as used in virtual systems.

Facilitation of Design and Control

The design, construction, and operation of a large automated manufacturing facility requires substantial capital investment. To increase the likelihood of a long-term gain from this investment, close attention must be given to the so-called life-cycle design of the factory. The four principal phases in life-cycle design of an automated factory are (1) system facility layout, (2) control of factory operation, (3) system maintenance and malfunction diagnostics, and (4) operator training.

Virtual reality involves a combination of computer hardware and software that provides a user (participant) with the perception of being immersed in a specific environment. The sense of immersion is obtained by giving a user environment-specific sensory cues and allowing the user to interact with the environment in a natural manner. For example, an overhead-crane virtual reality simulator provides this immersion by allowing a participant to "move" within a virtual factory and to raise, lower, and transport cargo by pushing buttons on a realistic control box.

All virtual reality simulators share several common functions. First, every simulator has an internal simulation program that models real-world behavior. Second, a virtual reality simulator must contain an extensive, organized database that describes the location, orientation, and dimensions of the three-dimensional (3D) virtual objects within the environment, the location and orientation of the participants immersed in the environment, and the commands from the user that alter the virtual environment. Third, a simulator must have a means to display the virtual objects with sufficient realism that a user accepts the virtual environment as a substitute for the real environment. Head-mounted displays, when coupled with head-tracking devices, enable a participant to have a 360° field of view.

Production simulation. Virtual reality simulators are used in optimizing manufacturing facility layout and material handling, an important component of

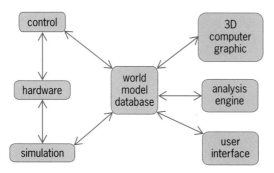

Fig. 1. Hierarchical real-time control system software architecture.

factory operation control. The main objective of a given layout, that is, machine types and their locations, is to provide an efficient means of processing materials. Given the uncertainty in production demands and associated machining operations over the life of a factory, it is generally impossible to identify a unique optimal layout. However, engineers designing manufacturing facilities can employ production simulation software to test the to-be-built factory operation under different types of production demands. The results of these simulations allow them to evaluate the efficiency of a given layout.

Production simulation software is limited in two respects. First, the software has limited interactivity: typically, a set of "initial conditions" is created, and the simulation is allowed to run with little or no user manipulation. Second, the output of the software is a collection of the data that must be postprocessed into graphs depicting machining queues or production delays.

The integration of the production simulation software and virtual reality interface can provide considerable improvement in the generation of a successful facility layout and a material handling system for an automated factory. With this integration, a facility designer can, in effect, walk through the virtual factory, examine equipment placement and clearance between adjacent machines, make inquiries regarding the status of production scheduling and manufacturing operations, and alter production schedules. A designer can interact with a virtual factory as if it were the actual building and search for trouble spots before the capital resources have been spent on construction.

Working simulators. The virtual testbed allows a user to navigate within a virtual factory and to reconfigure interactively the factory layout. The testbed employs a hierarchical real-time control system (HRCS), which coordinates the various software modules that make up the virtual testbed. The architecture of the hierarchical real-time control system specifies the information flow among the modules (**Fig. 1**). Each module in the testbed is a separate program, a feature that simplifies software development and maintenance. The "world model" database links the various modules.

The control and simulation modules are central to the operation of the virtual reality simulator. The control module coordinates virtual factory operation, including handling. For a prototypical factory, the control software is developed and tested with numerous device emulators and environmental simulators embedded in the simulation module. Once the software is thoroughly tested and debugged, it can be used to help operate the actual factory. The real-time simulation module replicates the behavior of the virtual factory. The device emulators must correspond closely to their real counterparts (that is, the actual machines). The emulators provide sensory feedback on the machine operation to the control module. The environmental simulators combine the operations of material handling equipment, such as conveyors and autonomous guided vehicles, with the physical characteristics of the materials, such as mass, size, gravitational effects, and frictional effects.

The virtual testbed has been used to simulate and analyze several United States post office facilities, which are a type of factory. The testbed has been configured to emulate several existing facilities and several planned facilities. A key feature in this application is the use of a touch screen and corresponding software for operator interactions and accelerated material (mail) handling; the same touch screen and software will then be used with an actual post office facility.

The virtual reality simulator employs a novel self-organization simulator (SOS) to guide the operation of a virtual automated manufacturing facility. The software architecture of this simulator specifies information flow (**Fig. 2**). The actuated positions result from the user (operator) issuing commands to objects (such as motorized pallets) within the virtual environment. The self-organization simulator responds to these commands and sends the response, such as a new position for a cart, to the virtual environment. Like the testbed described above, this simulator uses independent programs for the self-organization-simulator controller and the virtual reality model.

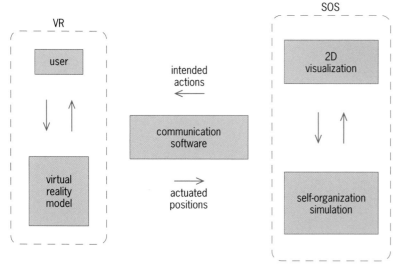

Fig. 2. Simulator and self-organization-simulator software architecture.

The simulator has been tested by building a model of a virtual factory for processing printed circuit boards. The processing consists of drilling holes in the boards and transporting them. The factory contains various drilling machines with different capabilities and carts to store and handle the boards.

The integration of factory simulation software and virtual reality technology holds the promise for improving automated manufacturing facility design and operation. Virtual reality technology has been successfully applied to areas such as situational awareness, crisis planning, medicine, maintenance, and engineering design. Integrated software design and virtual reality technology can be applied toward planning, evaluating, and operating automated manufacturing facilities.   B. H. Wilson; R. Mourant

Virtual Systems

Virtual systems is a multidisciplinary paradigm, bringing together computers, computer graphics, audio and visual perception, and mathematical sciences. One of its central themes is effective use of virtual reality technology in addressing problems at various stages of product or process design and manufacturing life cycles.

Virtual systems usage in industrial engineering covers a broad range of topics: designing products that can be evaluated and tested for manufacturability, including ergonomic functionality and reliability, without having to build actual scale models; designing products for esthetic value meeting individual customer preference; ensuring facility and equipment compliance with various standards; facilitating remote operation and control of equipment (telemanufacturing and telecollaboration); developing production plans and schedules and simulating their correctness; and educating employees about advanced manufacturing techniques.

In the virtual systems framework, concepts such as virtual reality telecollaboration for managing factories are increasingly relevant. Virtual reality telecollaboration is a form of collaborative virtual environment (CVE), a shared workspace wherein a group of people can collaborate with others in virtual environments. To accomplish this, methods must be found for coping with the increasing demands for graphically and geometrically complex environments while maintaining adequate performance and perceptual continuity. Users should be able to interact meaningfully with the objects and other participants within a virtual environment. This requires ways to specify the behavior and properties of those objects.

One attraction of virtual environments is that they allow people to indirectly experience a shared environment even when they are geographically separated. When compared to other media that have been used for collaboration in distant workplaces, such as phone, letter, e-mail, and fax, a virtual environment promises to be more flexible and interactive. It is superior to the traditional computer supported cooperative work (CSCW) environment, based on con-

Fig. 3. Virtual system telecollaboration experiment between the University of Illinois and the University of Tokyo.

ventional user interface devices such as a keyboard and mouse. For example, in a virtual reality telecollaboration experiment between the Electronic Visualization Laboratory at the University of Illinois and the University of Tokyo, a blue screen was used to isolate the person, data compression was used for transmission, and the results were much better than video conferencing because users were in their proper 3D scene positions (**Fig. 3**). The most common uses of collaborative virtual environment are currently design, education, and training.

Avatar representation. An avatar is a virtual representation for each participant in the collaborative virtual environment. It conveys a sense of someone's presence (known as telepresence) by providing the location (position and orientation) and identity. Some examples of avatars are the graphical human figure model, the talking head, and the real-time reproduction of a 3D human image. The avatar should be efficient to convey co-presence and awareness with minimum processing power, network bandwidth, and resources. Although full-body motions and video image schemes produce high-fidelity avatars, huge network bandwidth and fast network speed are required to transmit these data to each virtual environment. Therefore, a simplified avatar is usually adopted, which sends only the tracked hand or head position and orientation for conveying nonverbal information.

Latency, jitter, and bandwidth. Latency is the delay between performing an action and seeing the result of that action in virtual reality. Less than 100 milliseconds is usually required for a general virtual reality application. Network latency is the time required to get a unit of information from source to destination through a network. The issue is reducing the network latency to increase the interactivity and the dynamic nature of a collaborative virtual environment and to provide consistency (synchronized view) among virtual environments. A study of an acceptable network latency in a collaborative virtual environment demonstrated that humans tolerate a maximum of 200 ms network latency for real-time coordination. Networks also exhibit variable latencies, called jitter, which means that the packets do not arrive at a fixed delay.

Bandwidth is the available network capacity for data transfer. An evolving alternative is to take advantage of a high-speed fiber-optic broadband network such as asynchronous transfer mode (ATM), for which typical speeds of up to 155 megabits per second (Mbps) or 622 Mbps are common, depending upon the configuration. Asynchronous transfer mode supports low network latency and high bandwidth, but resources of such a high-performance network are still limited to a few research institutes and companies. Round-trip network latency is the delay between performing an action and getting the result of that action to loop back through the network. For example, to measure round-trip network latency in virtual systems a packet size of 55 bytes was used between two connected hosts. The 55 bytes of packet consisted of 3 bytes of header type, object, size; 12 bytes of object position X, Y, Z; 16 bytes of object orientation X, Y, Z, angle; 12 bytes of hand position X, Y, Z; and 12 bytes of hand orientation azimuth, elevation, roll.

Data structures. Virtual reality environments are used for detailed representation of objects that is next best to reality. However, such details make the environments too flexible from a representational standpoint and limit the potential use for design and planning. This difficulty is due to the very large search space from which solutions to design and planning problems are to be found. A formalism that is more restrictive from a representational standpoint but is more usable for design and planning is needed.

Instead of directly manipulating a virtual reality tree structure as is done in all current implementations, a tree with a lower number of nodes (abstract tree) representation is needed. Usually, the number of nodes in current implementation of virtual reality tree structure is large, making the search process inefficient for many problems. However, by using a lower number of nodes, the data model is mapped to an abstract tree, and the abstract tree is then mapped to a virtual reality tree (or virtual reality run-time database).

An example is a scene diagram in Virtual Reality Modeling Language (VRML 97). The VRML 97 format contains global nodes, prototype–instance of prototype, scene hierarchy–one root node or multiple root nodes (both allowed), scripts and interpolators, routes, and time sensors. These primitives are used to develop the run-time database, which results in an extremely large tree structure because of the amount of detail involved. Using these primitives, a scene diagram is developed. A scene diagram is a directed tree containing object geometry, topology, kinematic transformations, texture, lighting, and surface properties such as shininess and transparency.

A second example is a commonly used computer aided design/computer aided manufacturing (CAD/CAM) software program, ProEngineer, which builds an object tree with base object as root node and relationship among objects as child nodes. The data model can be progressively built by adding nodes to the bottom of the tree. ProEngineer can export the data model in .iv (Inventor) and .wrl (VRML), which are two common virtual reality data formats.

As in the above two examples, the emphasis of existing software is either to concentrate on the very detailed representational structure of the virtual reality tree or to provide the user with complete freedom to develop a virtual reality tree.

For background information *see* COMPUTER-AIDED DESIGN AND MANUFACTURING; COMPUTER GRAPHICS; COMPUTER VISION; CONCURRENT PROCESSING; HUMAN-COMPUTER INTERACTION; HUMAN-FACTORS ENGINEERING; MANUFACTURING ENGINEERING; MODEL THEORY; VIRTUAL REALITY in the McGraw-Hill Encyclopedia of Science & Technology.   Pat Banerjee

Bibliography. A. Banerjee et al., Assembly planning effectiveness using virtual reality, in *Presence: Teleoperators and Virtual Environments*, 8(2):204–217, 1999; C. J. J. Lu et al., A virtual testbed for the lifecycle design of automated manufactuing facilities, *Int. J. Adv. Manuf. Technol.*, 14(8):608–615, 1998; A. Johnson et al., CAVERN: The CAVE Research Network, in *Proceedings of the 1st International Symposium on Multimedia Virtual Laboratory*, Tokyo, March 25, 1998; L. J. Rosenblum, VR systems: Out from the laboratory, in *Proceedings of the 3d Annual IEEE International Conference on Virtual Systems and Multimedia*, Geneva, September 10–12, 1997; Special Issue on Virtual Manufacturing, *IIE Trans.*, 30(7):581–644, 1998; J. Vaario et al., Factory animation by self-organization principles, in *Proceedings of the 3d Annual IEEE International Conference on Virtual Systems and Multimedia*, Geneva, September 10–12, 1997.

Wireless emergency services

Fast response can be the key to saving lives in an emergency. In the 1960s, the U.S. government reserved 911 as the telephone number to be used for requesting emergency assistance, because this number can be easily memorized and dialed quickly. Since that time, public safety agencies in the United States have invested heavily in Enhanced 911 (E911) service, which provides for automatic display of the calling party's address and phone number on the public safety operator's console. However, in recent years wireless communication networks, such as cellular and personal communication services (PCS), which cannot locate emergency callers, have been deployed throughout the United States. This is a growing concern because nearly 100,000 wireless 911 calls are placed each day and that number continues to increase. Unfortunately, public safety officials report that many wireless emergency callers cannot accurately describe their location.

The Federal Communications Commission (FCC) in 1996 issued rules which require operators of wireless networks which interconnect with the public-switched telephone network to provide the capability to locate 911 callers. In the first phase, wireless network operators are required to identify the cell site from which a 911 call is placed and, if possible, the cell sector. Location accuracy based on cell site and sector information is highly variable, depending on the distance between cell sites; location estimates may have errors of 0.25–0.5 mi or more. In the second phase, wireless carriers are required to locate emergency callers within at least 125 m (410 ft) RMS (root-mean-square), that is, on at least two-thirds of all calls, and to pass the caller's location through to the responsible public safety answering point. Phase 1 was to be implemented in 1998 and phase 2 is to be implemented by October 2001, but in April 1999 only 5% of cellular and PCS networks in the United States had phase 1 in operation.

This article describes the principal technologies which are expected to be used to accurately locate wireless 911 callers. Wireless caller location systems proposed for E911 are generally divided into two categories: (1) network-based systems, which require the installation of some equipment at network cell sites to compute the caller's location based on transmissions from standard wireless phones; and (2) handset-based systems, which require the addition of the Global Positioning System (GPS) or other location technology to the wireless phone. Many of the systems for locating wireless callers are being developed in the United States, in response to the FCC's mandate. However, several foreign companies are also active in this area.

Network-based location systems. Most network-based location systems use either time-difference-of-arrival (TDOA) or angle-of-arrival (AOA) measurements, or a combination of both, to locate the caller. These technologies are often combined with proprietary location methods and techniques for reducing signal interference (noise) and echoes caused by reflected signals (multipath). One proprietary system employs a technology called location fingerprinting.

Time-difference-of-arrival. Most network-based location systems are based at least partially on time-difference-of-arrival technology. Such systems compute the caller's location by measuring the differences between the arrival times of wireless phone transmissions received at two or more individual base stations (cell sites). For example, if these transmissions arrive at base station A and base station B at exactly the same time, the caller is located somewhere on the straight line L_1, the perpendicular bisector of the line segment between the base stations (**Fig. 1**). However, if one base station receives the signal before the other, the caller is located somewhere along a hyperbola made up of points for which the difference in the arrival time of the signals at the two base stations would be the same. Receipt of the transmitted signal at the third base station allows another hyperbola to be formed from a second pair of time-difference-of-arrival measurements. The caller is located at the point of intersection of the two hyperbolas.

Time-difference-of-arrival systems require synchronization of the clocks at each base station and cal-

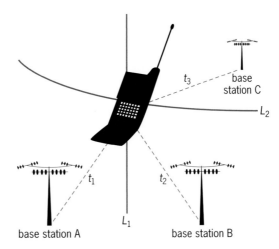

Fig. 1. Location of a wireless phone in a time-difference-of-arrival system. Time of arrival of the signal from the phone is the same at base stations A and B. Line L_1 represents all points for which t_1 (transmission time to base station A) equals t_2 (transmission time to base station B). Hyperbolic line L_2 represents all points for which t_1 minus t_3 (transmission time to base station C) is the same. The phone is located at the intersection of L_1 and L_2.

ibration of time delays within the receiver's electronics. To maintain time calibration to better than 100 nanoseconds (approximately equivalent to 30 m or 100 ft), some suppliers of time-difference-of-arrival location systems install a GPS receiver at each base station.

Developers of these systems claim to achieve position location accuracies ranging from 90 to 125 m (295 to 410 ft) RMS, depending on the location of the cell phones relative to the base stations, on the multipath environment, and on other factors such as interference and radio signal blockage that can degrade the accuracy of the timing measurements.

Angle-of-arrival. Angle-of-arrival systems compute the location of the caller by measuring the directions from which signals transmitted from the wireless phone arrive at the base stations. The known position of two or more base stations and the measured angle of arrival of the received signal are used to compute the caller's location (**Fig. 2**).

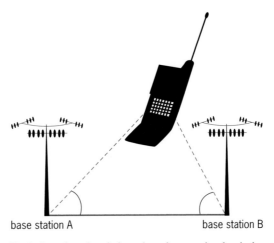

Fig. 2. Location of a wireless phone in an angle-of-arrival system. Angle-of-arrival measurements at two base stations of known position are used.

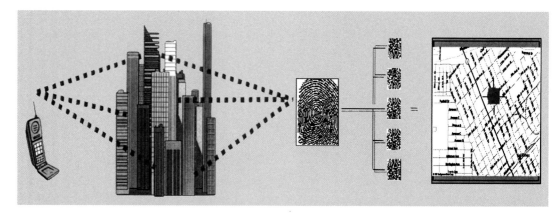

Fig. 3. Location of a wireless phone by location fingerprinting. The phone emits radio signals which bounce off buildings and other obstacles, reaching their destinations via multiple paths. Equipment at the base station analyzes the unique characteristics of the signal, including its multipath pattern, and compiles a fingerprint. The fingerprint is compared to a database of previously fingerprinted locations, with associated geographic coordinates, and a match is made.

The accuracy with which angle of arrival can be measured depends on factors such as the design of the receiving antenna, the number of cell sites used in the computation, the postprocessing software used to reduce random measurement errors, and the multipath environment (reflected signals). With angle-of-arrival, accuracy degrades as the phone's distance from the cell site increases. Another concern is that angle-of-arrival systems often require the installation of a multiple-array antenna at network cell sites. Some carriers object to this feature because it may require the service provider to obtain approvals from local zoning boards. While sensor measurement accuracy of 1–2° is often achievable, multipath and other error sources often degrade the accuracy to 5–10° in dense suburban environments, which equates to 460–920 ft at 1 mi.

Hybrid systems. Several companies have developed hybrid systems, which combine time-difference-of-arrival and angle-of-arrival technologies. They claim that a hybrid system can provide superior performance. The angle-of-arrival capability requires signal measurements at only two base stations, while systems based on time-difference-of-arrival alone require measurements at three or more base stations. Thus, angle-of-arrival can compute the location of callers a higher percentage of the time, particularly when only a single cell site covers a large rural area.

Developers also note that, under some conditions, time-difference-of-arrival offers advantages over angle-of-arrival, and that in a hybrid system the accuracy and the availability of a location solution can be improved 30% or more over systems based on time-difference-of-arrival or angle-of-arrival alone. This improvement is based on using time-difference-of-arrival measurements to offset the sensitivity of angle-of-arrival systems to distance from the base stations.

Location fingerprinting. One provider has developed a system which computes a caller's location using a technique referred to as location fingerprinting. When the system is initially installed, the provider's technicians map the cellular radio signal propagation characteristics throughout the network by driving the local roads and transmitting cellular signals. They record the received propagation and multipath characteristics and correlate them to accurately determined reference locations that are surveyed to within a few meters using differential GPS–based locations. When a cellular phone user places a 911 call, the system compares the received signal's propagation characteristics with the previously mapped signal propagation and multipath patterns to determine the caller's location (**Fig. 3**). This provider reports that the multipath signals prevalent in urban areas can aid the solution rather than impeding it. Location fingerprinting is reported to yield the best results in urban environments, where accuracy of approximately 50 m (160 ft) RMS has been reported in system trials.

Handset-based location systems. In handset-based location systems, the position location is determined within the handset, usually by adding a GPS satellite navigation capability. These systems require hardware or software modifications to the wireless phone, but do not require the investment in network infrastructure required by network-based systems. While it is doubtful that wireless carriers could convert all subscriber phones to location-capable configurations by the FCC's deadline of October 2001, the FCC has stated that it will consider granting waivers to carriers proposing handset-based solutions or may modify the rules to allow for such solutions.

For the incorporation of GPS in wireless phones to be practical for the mass market, the power consumption must be minimal, GPS location computation time must be rapid, and the GPS component must add very little to the cost of the phone. GPS will consume less power than the phone when both are operating at full power, and companies developing GPS technology for cellular phones report that GPS power consumption can be as little as 10–15 milliwatts in standby mode. By automatically updating GPS satellite precise orbital information (ephemeris) every hour, a GPS receiver in a cellular phone can compute the caller's location within 3 s or less when a 911 call is placed or other location-based services are requested. To minimize the incremental cost for including GPS, phone manufacturers will incorporate a 1.5-GHz receiver (GPS frequency band) into the phone but will use the phone's existing digital signal processor for computational chores.

Some suppliers of GPS technology for cellular phones propose a network-assisted architecture in which GPS ephemeris and other satellite data are passed to the phone through the cellular network. In another proposed mode, the GPS receiver in the phone will compute only the pseudoranges (distance measurements) to the satellites and will pass these data to a network-based server which will compute the caller's location. These GPS architectures are intended to minimize the cost of the GPS components in the phone and to provide for fast computation of the caller's location. They also make it possible to compute location with weaker satellite signals, for improved performance in dense urban areas and inside buildings. Finally, server-based architectures can improve GPS accuracy from approximately 50 m (160 ft) RMS to 10–20 m (30–60 ft). Further improvement is expected before 2006, when the U.S. Department of Defense is mandated to discontinue the intentional degradation of positioning by civil users of GPS that is known as selective availability. *See* GLOBAL POSITIONING SYSTEM (GPS).

For background information *see* DISTANCE-MEASURING EQUIPMENT; ELECTRONIC NAVIGATION SYSTEMS; HYPERBOLIC NAVIGATION SYSTEM; MOBILE RADIO; SATELLITE NAVIGATION SYSTEMS; TELEPHONE SERVICE in the McGraw-Hill Encyclopedia of Science & Technology.   Clement J. Driscoll

Bibliography. Federal Communications Commission, *Report and Order and Further Notice of Proposed Rulemaking*, Docket 94-102, July 26, 1996; TechnoCom Corporation (Encino, CA), *Report on Wireless E-911 Location Technologies*, 1999.

X-ray astronomy

In recent decades the scope of observational astronomy has extended over the entire electromagnetic spectrum, and now ranges from radio waves to high-energy photons with teraelectronvolt (1 TeV = 10^{12} eV) energies. The different bands reveal different aspects of the universe. In the x-ray domain, at photon energies between 0.1 and 500 keV, thermal radiation from high-temperature plasmas or nonthermal radiation produced by relativistic electrons interacting with magnetic fields (synchrotron radiation) or intense photon fields (the inverse Compton effect) is observed.

Since the first detection of solar x-rays in 1949 and the discovery of the first cosmic x-ray source in 1962, the discipline of astrophysics has evolved at an enormous pace, aided by numerous balloon-, rocket-, and satellite-borne instruments. Surprisingly, almost all kinds of astrophysical objects, from nearby comets to the most distant quasars, have turned out to be x-ray emitters.

This article discusses results obtained with the *RoentgenSatellite* (*ROSAT*), which was designed, built, and operated from 1990 to 1998 under the leadership of the Max-Planck-Institut für Extraterrestrische Physik (MPE) in Garching, Germany, and sponsored by Germany, the United States, and the United Kingdom.

ROSAT mission. *ROSAT* is a 2.4-metric-ton (2.6-short-ton) satellite carrying a large x-ray telescope of the Wolter type and a small extreme-ultraviolet (EUV) telescope. The x-ray telescope has two types of image detectors, which sit on a turret and operate in the energy range from 0.1 to 2.4 keV. The first one is a multiwire proportional counter (position-sensitive proportional counter or PSPC) providing colored (energy-resolved) images with 20 arcsecond resolution and a $2°$ field of view. The other detector is a multichannel plate detector (high-resolution imager or HRI) delivering images of high angular resolution (5 arcseconds) but low energy resolution ("black and white") over a field of view of about $0.5°$.

ROSAT was launched by a Delta rocket on June 1, 1990, and saw first light 2 weeks later. After 6 weeks of calibrations and instrument verifications, *ROSAT* performed a half-year all-sky survey, the first ever done with an imaging x-ray telescope, leading to the discovery of 80,000 x-ray sources. After that, the satellite was used in a pointed mode for more than 7 years. This program was completely open to guest observers. A total of 9000 fields in the sky were studied with an average observation time of 3 h. The longest pointing—the *ROSAT* Deep Survey—lasted 2 weeks.

The following summary of *ROSAT* highlights is organized in a geocentric fashion, ordering the different classes of objects according to their distance from the Earth.

Moon. *ROSAT* took the first x-ray picture of the Moon (**Fig. 1**). The sunlit side of the Moon shows a uniform brightness distribution as in optical light. It is due to solar coronal x-rays undergoing Thomson scattering in a very thin layer of the lunar surface. The PSPC spectrum shows a broad spectral bump at 0.6 keV which is due to fluorescent resonance scattering by oxygen. The effective reflectivity of the lunar surface in the *ROSAT* band is only 0.01%. The Moon casts a shadow on the x-ray sky. The small flux of x-rays apparently coming from the dark side of the Moon is probably produced in the Earth's upper atmosphere by charge-exchange processes of ions in the solar wind.

Comets. Comets are cold objects that have been described as dirty snowballs. Therefore the discovery with *ROSAT* of x-rays from Comet Hyakutake on March 27, 1996, was surprising. Later, another four comets were found in the *ROSAT* All Sky Survey archive, among them Comet Levy. Various physical processes have been proposed to explain the observed extended x-ray emission, including scattering of solar x-rays by cometary dust, x-rays produced by dust-dust scattering in the cometary coma, and bremsstrahlung x-rays from electrons accelerated at

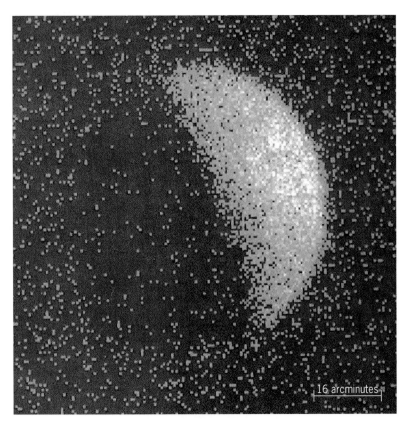

Fig. 1. X-ray image of the Moon, showing scattered solar x-rays from the sunlit side and shadowing of the x-ray sky background by the lunar disk.

the shock between the solar wind and the coma. All these models cannot explain the observations, but the ongoing research activities suggest that charge exchange between highly charged ions (such as C^{5+}, C^{6+}, O^{6+}, and O^{7+}) in the solar wind and neutral particles (such as water) in the cometary cloud is the dominant source of the observed x-ray emission.

Stars. *ROSAT* has tremendously fostered stellar astronomy. The study of a complete sample of stars of solar type has revealed the existence of a sharp lower bound to the x-ray flux, measured at the surface of the star. This minimum stellar x-ray flux is identical to the flux observed in the coronal holes of the Sun. This result suggests that the stars of minimal x-ray flux are completely surrounded by stellar analogs of solar coronal holes.

With ground-based optical follow-up observations of unidentified *ROSAT* All-Sky Survey sources, several hundred new T Tauri stars have been identified based on the H_α emission of hydrogen and on lithium absorption. Surprisingly, new T Tauri stars have been found even outside clouds, that is, outside regions of ongoing star formation. Such off-cloud T Tauri stars have been either ejected from their birthplaces in the clouds or formed in small cloudlets which have dispersed since then.

Deep *ROSAT* observations have led to the discovery of brown dwarfs showing x-ray emission. Brown dwarfs are stars having too small a mass to ignite hydrogen burning. The x-ray properties of these brown dwarfs are similar to those of nuclear burning stars of

low mass. This result suggests that the x-ray emission is produced by coronal processes.

Nuclear-burning white dwarfs. Early in the *ROSAT* mission a number of objects were discovered emitting extremely soft x-rays. They are very bright and show temperatures of a few hundred thousand degrees. It turned out that these sources are species which had been predicted to exist but which had not been found before *ROSAT*. They are white dwarfs in binary systems accreting matter from their companions at a rate just sufficient to sustain steady nuclear burning on the surface of the white dwarf. Thus, they represent a unique situation in which a thermonuclear reaction can be observed which is usually deeply buried in the interior of a star.

Supernova remnants. Stars much more massive than the Sun explode when the nuclear fuel in their core is exhausted. Their core collapses to a neutron star or a black hole while the shell of the star is expelled in a giant explosion. Supernova 1987A was the *ROSAT* first light target on June 16, 1990, but it was too faint to be seen at that time. It was discovered in soft x-rays with *ROSAT* in 1992 and has steadily become brighter since then. Some 200 supernova remnants have been found with *ROSAT*. Three of them are located in the Vela region (**Fig. 2**). One of them is the Vela Supernova Remnant, which, at a distance of 1500 light-years, is one of the closest supernova remnants. Its diameter is about 200 light-years, its age about 20,000 years. Protrusions discovered with *ROSAT* at the periphery of the shell are probably produced by fragments of the exploding star; they show different chemical compositions. The Puppis A remnant at the rim of the Vela Supernova Remnant is located at a much larger distance. It is younger and has a much higher temperature than the Vela Supernova Remnant. Recently, a third supernova remnant has been discovered in the Vela complex with ROSAT; it must be very young (680 years), since radioactive titanium-44 has been detected from it with the *Compton Gamma-Ray Observatory*. This radioisotope is produced in supernova explosions and has a mean lifetime of only 90 years. This supernova must have been very bright, and it is unclear why it has not been recorded in the Chinese and Japanese annals.

Neutron stars. Neutron stars shine in different ways. If they are highly magnetized and rapidly rotating, they appear as radio pulsars. These objects, discovered in 1967, emit beamed radiation produced by high-energy electrons (and positrons) that are accelerated in their magnetospheres. In about a half dozen young pulsars, optical and gamma-ray pulses have also been seen. With *ROSAT*, 34 radio pulsars have been detected through their magnetospheric x-ray emission, including the 89-ms Vela pulsar, which is also found pulsing in the optical and gamma-ray bands. Four of the radio pulsars seen with *ROSAT*, including the Vela pulsar, exhibit an additional thermal spectrum corresponding to a temperature of the order of 10^6 K, which is interpreted as the thermal radiation from the surface (photosphere) of the

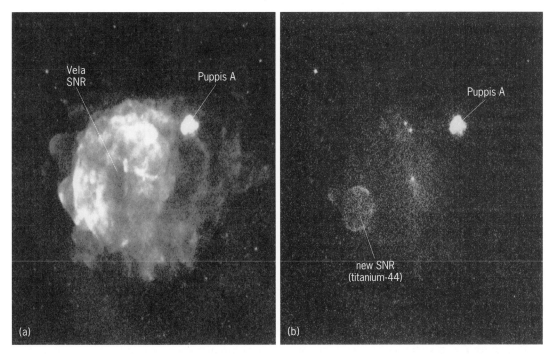

Fig. 2. *ROSAT* images of Vela region. (*a*) Image at 0.1–2.4 keV, showing Vela Supernova Remnant (SNR) and Puppis A. (*b*) Same image at high photon energies (greater than 1.3 keV), showing another supernova remnant, which has a circular shape with 2° diameter. Gamma-ray line emission from radioactive titanium −44 indicates that the explosion took place only 680 years ago.

neutron star. A few point sources discovered near the centers of young supernova remnants also show very soft x-ray emission, which must be attributed to photospheric emission from neutron stars. One of these sources, located in Puppis A (Fig. 2*a*), exhibits a 75-ms modulation of the x-ray flux, which confirms the identification with a neutron star. The importance of these observations lies in the fact that the surfaces of these tiny stars (with radii of 10 km or 6 mi) have become visible. Future x-ray spectroscopy should make it possible to measure their enormous gravity as well as their radii, which depend on the physical properties of matter at supranuclear densities.

Galaxies and active galactic nuclei. The pioneering *Uhuru* sky survey in 1971–1973 discovered a total of 334 x-ray sources, mostly members of the Milky Way Galaxy, in which the solar system resides. By contrast, the *ROSAT* survey of the Andromeda Galaxy (M31), a spiral galaxy 2 million light-years away, yielded 500 x-ray sources in this galaxy alone. As in the Milky Way Galaxy, the brightest of them are young supernova remnants or binary systems with neutron stars or black holes accreting matter from a companion. These bright source populations have been studied with *ROSAT* in many galaxies. In addition, the hot interstellar medium, which is heated by supernova explosions, has been investigated.

A small fraction of all galaxies have an active galactic nucleus emitting huge amounts of energy in all spectral bands. The radiation is variable, indicating that it is emitted from a small region, generally not more than a few light-weeks or light-months across.

Active galactic nuclei often display jets, originating in their core. It is generally believed that the central engine is a supermassive black hole swallowing matter at a high rate. Quasars are the most extreme

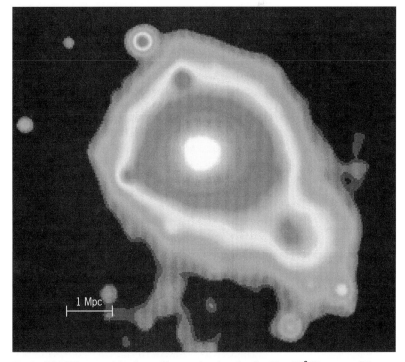

Fig. 3. *ROSAT* image of the Coma Cluster, showing clouds of hot (10^8 K) x-ray-emitting gas of the main cluster and of a group of galaxies (bottom right) falling onto it. X-ray observations allow a determination of the mass of the hot gas and of the dark matter needed to hold the cluster gas together. 1 megaparsec (Mpc) = 3.26×10^6 light-years.

representatives of the class of active galactic nuclear sources. The quasar 3C 273 contains a black hole of about 10^9 solar masses, accreting one earth mass per second.

These active galactic nuclei are very bright x-ray sources, and about 50% of all the 150,000 *ROSAT* sources belong to this class. Because of their enormous brightness, they can be detected at large distancies or redshifts. In the *ROSAT* Deep Survey with a total accumulated observation time of 1.4×10^6 seconds, about 1000 sources are detected per square degree in the sky. Most of them are quasars at cosmological distances. A current question is whether the galaxies or the supermassive black holes found in their centers evolved first.

Clusters of galaxies. Galaxies are not distributed randomly in the universe but form clusters or groups consisting of a dozen to thousands of galaxies which are gravitationally bound. With diameters of millions of light-years, they are the largest physical objects in the universe. As early as 1932, Fritz Zwicky observed that the velocities of galaxies in clusters were much larger than expected from the observed masses. The discrepancy is a factor of 30, and the invisible component is called dark matter. One early surprise of x-ray astronomy was the detection by *Uhuru* of large quantities of hot plasma in clusters shining in x-rays. *ROSAT* observations show that the mass of the hot plasma is typically a factor of 4 or 5 higher than that of the galaxies and represents some 20% of the cluster mass. Several thousand clusters have been found with *ROSAT*. Many of them are optically identified and have known redshifts, and thus known distances. Many clusters show double structures, indicating the merging of two clusters, or complicated inner structures which must be due to earlier merging or interactive processes (**Fig. 3**). Thus, it is possible to see how these large structures evolve in the universe. However, the evolution of the cluster population with redshift is smaller than expected in a high-density universe. In fact, the matter density inferred from cluster evolution is only about 30% of the "critical density" which is necessary to close the universe.

Prospects. The *ROSAT* mission concluded after $8^1/_2$ years of very successful scientific operations. The last astronomical observation was made on December 18, 1998, and on February 12, 1999, the satellite was switched off after a number of recovery actions had failed. At that time there were three other active x-ray satellites in orbit, and four large x-ray satellite missions were scheduled for launch in 1999 and 2000.

For background information *see* BROWN DWARF; COMET; GALAXY, EXTERNAL; GAMMA-RAY ASTRONOMY; INTERSTELLAR MATTER; MOON; NEUTRON STAR; PULSAR; SOLAR WIND; SUN; SUPERNOVA; T TAURI STAR; UNIVERSE; WHITE DWARF STAR; X-RAY ASTRONOMY; X-RAY TELESCOPE in the McGraw-Hill Encyclopedia of Science & Technology.   Joachim Trümper

Bibliography. B. Aschenbach, H. M. Hahn, and J. E. Trümper, *The Invisible Sky, ROSAT and the Age of X-ray Astronomy*, Springer-Verlag, 1998; P. A. Charles and F. D. Seward, *Exploring the X-ray Universe*, Cambridge University Press, 1995.

Zoonosis

Anthrax, a disease of antiquity, infects domestic and wild herbivorous animals, including sheep, goats, cattle, and swine. Humans become infected under natural conditions by contact with infected animals or contaminated animal products, such as wool, hair, hides, and bone meal. *Bacillus anthracis*, the causative agent of anthrax, has recently become of great concern as a biological weapon.

Microbiology and pathogenesis. Anthrax is caused by *Bacillus anthracis*, a large spore-forming bacillus (1×8 micrometers). The factors contributing to pathogenesis are a proteinaceous capsule surrounding the bacterium and two toxins, edema toxin and lethal toxin. The capsule and toxins are encoded by separate extrachromosomal elements (plasmids). After inoculation through the skin or the gastrointestinal mucosa, the spore is phagocytized and thought to germinate into the bacillus form within large phagocytic cells called macrophages. The bacillus subsequently grows extracellularly at the local site with production of its capsule and toxins, resulting in tissue death or necrosis with extensive tissue fluid accumulation (edema). Inhalational anthrax, the form that would occur after a biological warfare attack, develops after the inhaled spores are deposited in the alveoli, the deepest air sacs of the lung. There, the spores are phagocytized and transported by macrophages to lymph nodes situated in the mediastinum, the central portion of the chest cavity. In this location, the spore germinates into the bacillus, which replicates with production of the capsule and two toxins, resulting in a hemorrhagic, necrotic, and swollen lymph gland and subsequent spread to the entire mediastinum. The bacilli then enter the bloodstream and spread throughout the body. Death results from respiratory failure with pulmonary edema and is often associated with infection and hemorrhage in the brain.

On a cellular and molecular level, the capsule prevents the bacillus from being ingested and destroyed by phagocytic cells. The toxins are delivered into the cytoplasm of the host cells, where they exert toxic effects on specific target molecules. The two anthrax toxins are bipartite protein toxins, containing a common cell-binding protein, called protective antigen, and two distinct proteins that are enzymes. The protective antigen binds to specific host-cell receptors and serves to translocate the enzymatic protein of both edema and lethal toxin into the cell cytoplasm, where it exerts its toxic effect. The edema toxin consists of protective antigen combined with edema factor, an enzyme called adenylate cyclase, that raises the intracellular levels of cyclic adenosine monophosphate, resulting in the characteristic edema that often occurs during the infection. The increased cyclic adenosine monophosphate also interferes with the function of the host's phagocytes. The lethal toxin consists of the same protective antigen together with lethal factor, a metalloprotease that degrades a cytoplasmic target molecule essential for vital cell functions, possibly by inactivating protein kinases. The lethal toxin specifically kills phagocytic macrophages. In these cells, it increases reactive oxygen intermediates and releases bioactive molecules such as tumor necrosis factor and interleukin-1 that may result in severe toxicity and contribute to the death of the host. The exact molecular interactions resulting in death are under intensive investigation.

Clinical manifestations. Cutaneous anthrax occurs after an incubation period of 1–5 days, beginning with a small raised lesion that becomes fluid-filled over a day or two. The lesion ruptures, leaving a depressed ulcer, which evolves over a period of a week into a black-scab-covered lesion 1–3 cm in diameter. The lesion is usually painless and often surrounded by varying degrees of edema. Patients have fever and malaise. With treatment, mortality is less than 1%.

Oropharyngeal (involving the mouth and pharynx) anthrax presents as severe sore throat, pain on swallowing, and neck swelling. Gastrointestinal anthrax presents with severe abdominal pain, rapid onset of abdominal swelling due to fluid accumulation, and in some cases massive diarrhea. Mortality of 10–50% occurs in oropharyngeal and gastrointestinal cases.

Inhalational anthrax presents after an incubation period of 1–6 days with ill-defined symptoms of malaise, fatigue, and fever, often with a cough and some chest pain. These symptoms persist for 2–3 days, followed by the sudden onset of shortness of breath. This is usually followed by rapid progression to shock and death within 24–36 h. Essentially all cases result in death.

Meningitis (inflammation of the membranes lining the brain) may occur as a complication of any of the forms of anthrax.

Diagnosis. In all forms of anthrax, a high index of suspicion is necessary to elicit a history of exposure and establish the diagnosis. Inhalational anthrax will be suspected before death only if an epidemiological history of aerosol exposure is obtained. In a bioterrorist outbreak, the disease would manifest itself as a large outbreak of a flulike illness followed by severe shortness of breath and death within 2–3 days in previously healthy adults. The diagnosis can be made with rapid tests on blood to identify the protective antigen by an enzyme-linked immunosorbent assay and to detect bacterial deoxyribonucleic acid amplified by the polymerase chain reaction. Definitive diagnosis is established by growing the organism from blood or other tissues.

Epidemiology. *Bacillus anthracis* is a soil organism with a worldwide distribution. The spore is extremely resistant to the environment and can survive for long periods of time, resulting in recurrent outbreaks of disease, separated by decades, in domestic animals grazing on the same contaminated pasture. The organism persists in the soil after anthrax-infected animals die. Infection in humans is not part of its natural ecological cycle. Large outbreaks have occurred in animals, although the disease can be readily controlled by the use of a live attenuated veterinary vaccine.

In the developing world, anthrax remains a serious disease of livestock and potentially of humans. It is estimated that there are 2000 human cases per year worldwide. The largest reported outbreak occurred in Zimbabwe in 1979–1985, with 10,000 human cases and approximately 100 deaths. Essentially all cases involved the skin. The cutaneous form of human disease, responsible for more than 95% of cases, results from the introduction of spores through abrasions or cuts after direct contact with infected animals or contaminated animal products. Oropharyngeal anthrax and gastrointestinal anthrax develop after the ingestion of improperly cooked contaminated meat, and are very rare. Although outbreaks occur in developing countries, no cases have been reported in the United States.

Inhalational anthrax results from the inhalation of airborne spores. Under natural conditions, this occurred during the industrial processing of contaminated animal products, including goat hair, wool, and hides in an enclosed space such as a factory. This form of anthrax was first identified in the nineteenth century among mill workers in England, resulting in woolsorters' disease, and in Austria and Germany among workers processing animal hides. Its rarity is indicated by the fact that only 18 cases were reported in the United States in the twentieth century. The disease is not transmissible from person to person.

In 1979, the accidental release of anthrax spores into the air from a military microbiology laboratory facility in Sverdlovsk, in the Soviet Union, resulted in the largest known epidemic of inhalational anthrax, with approximately 70 cases reported. This event demonstrated the possible consequences of using anthrax as a biological weapon. While naturally occurring inhalational anthrax had been known since the nineteenth century, this episode established that large numbers of cases could develop from the release of anthrax spores into the atmosphere. Indeed, Iraq admitted that it weaponized anthrax spores for use during the 1991 Gulf War. In 1995, the Aum Shinrikyo cult in Japan attempted to use anthrax as a terrorist weapon, and in the last few years there have been a number of terrorist hoaxes involving anthrax in the United States.

Prevention and control. A protective antigen-based, nonliving vaccine has been licensed for human use in the United States since 1970. The vaccine should be given to workers exposed to contaminated animal products in high-risk industries. It is being given to the Armed Forces to protect against the use of anthrax as a biological weapon. Anthrax in animals can be controlled by the use of a live attenuated vaccine and appropriate disposal of infected carcasses.

Research. Currently, research is focused upon the early events of pathogenesis, and the mechanism of action of the toxins to develop therapeutic agents to block their harmful effects. Other areas concern early detection of aerosols of anthrax as an early warning system of biological attack and development of more rapid tests to diagnose the disease in humans.

A more easily manufactured, highly purified, and characterized protective antigen vaccine is being evaluated in animal models. Efforts are focused on identifying other protective compounds to add to the protective antigen. Vaccines based on nucleic acids, plant expression systems, and live attenuated bacteria are being investigated.

For background information *see* ANTHRAX; INFECTIOUS DISEASE; MEDICAL BACTERIOLOGY; TOXIN; ZOONOSIS in the McGraw-Hill Encyclopedia of Science & Technology.   Arthur M. Friedlander

Bibliography. P. S. Brachman and A. M. Friedlander, Anthrax, in S. A. Plotkin and W. A. Orenstein (eds.), *Vaccines*, 3d ed., W. B. Saunders, 1999; P. Hanna, Anthrax pathogenesis and host response, *Curr. Top. Microbiol. Immunol.*, 225:13–35, 1998; S. H. Leppla, Anthrax toxins, in J. Moss et al. (eds.), *Bacterial Toxins and Virulence Factors in Disease*, Marcel Dekker, 1995; S. F. Little and B. E. Ivins, Molecular pathogenesis of *Bacillus anthracis* infection, *Microbes Infect.*, 1:119–127, 1999.

Contributors

Contributors

The affiliation of each Yearbook contributor is given, followed by the title of his or her article. An article title with the notation "in part" indicates that the author independently prepared a section of an article; "coauthored" indicates that two or more authors jointly prepared an article or section.

A

Adam, Prof. Nabil R. *Center for Information Management, Integration and Connectivity, Rutgers University, Newark, New Jersey.* ELECTRONIC COMMERCE—coauthored.

Allwinn, Dr. Regina. *Institut für Medizinische Virologie, J.W. Goethe-Universität Frankfurt/Main, Frankfurt, Germany.* DENGUE FEVER—coauthored.

Anglin, Dr. Donald L. *Consultant, Automotive and Technical Writing, Charlottesville, Virginia.* ELECTRIC VEHICLE.

Arntzen, Dr. Charles J. *Boyce Thompson Institute for Plant Research, Cornell University, Ithaca, New York.* VACCINATION—coauthored.

Ashby, Dr. F. Gregory. *Department of Psychology, University of California, Santa Barbara.* MULTIDIMENSIONAL PSYCHOLOGY.

Atluri, Prof. Vijayalakshmi. *Management Science/Information System Society Department, and Center for Information Management, Integration and Connectivity, Rutgers University, Newark, New Jersey.* ELECTRONIC COMMERCE—coauthored.

B

Balzani, Prof. Vincenzo. *Dipartimento de Chimica, Università di Bologna, Bologna, Italy.* MOLECULAR MACHINES.

Banerjee, Dr. Pat. *Associate Professor, Mechanical Engineering, University of Illinois, Chicago.* VIRTUAL REALITY—in part.

Banks, Dr. Danny. *Biomedical Engineering, University of Surrey, United Kingdom.* NEUROTECHNOLOGY.

Barker, Dr. Kenneth R. *Professor Emeritus, Plant Pathology Department, North Carolina State University, Raleigh.* PEST MANAGEMENT—coauthored.

Barton, Dr. Jennifer Kehlet. *Biomedical Engineering Program, University of Arizona, Tucson.* OPTICAL COHERENCE TOMOGRAPHY.

Bascompte, Dr. Jordi. *National Center for Ecological Analysis and Synthesis, University of California, Santa Barbara.* POPULATION DYNAMICS.

Baxendale, Steve. *Director, Star Schools Program, Pacific Resources for Education and Learning, Association for Supervision and Curriculum Development, International Teleconferencing Association, National Staff Development Council, Honolulu, Hawaii.* DISTANCE EDUCATION.

Beach, Dr. Reginald. *Office of Naval Research, Arlington, Virginia.* OCEAN CURRENTS—in part.

Bennett, Dr. Keith D. *Kvartärgeologi, Institutionen för geovetenskaper, Uppsala Universitet, Uppsala, Sweden.* SPECIES STABILITY.

Bentley, Prof. Ronald. *Professor Emeritus, Department of Biological Sciences, University of Pittsburgh, Pennsylvania.* CHIRAL DRUGS.

Berger, Dr. Edward A. *The Laboratory of Viral Diseases, National Institute of Allergy and Infectious Diseases, National Institutes of Health, Bethesda, Maryland.* CORECEPTOR.

Berner, Prof. Elizabeth K. *Department of Geology and Geophysics, Yale University, New Haven, Connecticut.* GLOBAL WARMING.

Bondos, Dr. Sarah. *Department of Biochemistry and Cell Biology, Rice University, Houston, Texas.* PROTEIN—coauthored.

Bowring, Samuel A. *Associate Professor of Geology, Department of Earth, Atmospheric, and Planetary Sciences, Massachusetts Institute of Technology, Cambridge.* PERMOTRIASSIC MASS EXTINCTION.

Brittain, Scott T. *Department of Chemistry and Chemical Biology, Harvard University, Boston, Massachusetts.* MICROSTRUCTURES—coauthored.

Brooks, Dr. Paul D. *U.S. Geological Survey, Division of Water Resources, Boulder, Colorado.* SOIL CHEMISTRY.

Brown, Dr. Cecelia. *Assistant Professor and Chemistry-Mathematics Librarian, University Libraries, University of Oklahoma, Norman.* DEEP CRUSTAL MICROBES—coauthored.

Budd, Dr. Ann F. *Department of Geology, University of Iowa, Iowa City.* CORAL REEF COMMUNITIES.

Burns, Dr. Edith A. *Department of Medicine, Section Geriatrics, Medical College of Wisconsin, Milwaukee.* IMMUNE SYSTEM.

Byrom, Ted G. *Consulting Engineer, T. G. Byrom and Associates, Roanoke, Texas.* OIL PRODUCTION—in part.

C

Cabarle, Bruce J. *Director, Global Forest Program, World Wildlife Fund, Washington, D.C.* FOREST CERTIFICATION.

Campbell, Dr. Colin J. *Petroleum Consultant, Le Mas de l'Etanjou, France.* OIL RESERVES.

Campbell, Dr. Janet W. *Program Manager, Ocean Biology/Biogeochemistry, Office of Earth Sciences, Research Division, National Aeronautics and Space Administration, Washington, D.C.* OCEAN OPTICS—in part.

Camper, Dr. Sally. *Associate Professor, Department of Human Genetics, University of Michigan Medical School, Ann Arbor.* HEARING IMPAIRMENT—coauthored.

Carney, Prof. Bruce W. *Baron Professor of Astronomy, Department of Physics and Astronomy, University of North Carolina, Chapel Hill.* GLOBULAR CLUSTERS.

Caro, Dr. Jose F. *Lilly Research Laboratories, Eli Lilly and Company, Indianapolis, Indiana.* OBESITY.

Carpenter, Prof. Michael P. *Staff Physicist, Physics Division, Argonne National Laboratory, Argonne, Illinois.* SUPERDEFORMED NUCLEI.

Cesarsky, Dr. Catherine. *CEA/Saclay/DSM, Orme des Merisiers, France.* INFRARED ASTRONOMY—coauthored.

Cesarsky, Diego A. *IAS/CNRS, Université de Paris Sud, France.* INFRARED ASTRONOMY—coauthored.

Chalamala, Dr. Babu R. *Senior Scientist, Flat Panel Display Division, Motorola Inc., Tempe, Arizona.* FIELD-EMISSION DISPLAYS—coauthored.

Chambers, Dr. Jeff. *National Center for Ecological Analysis and Synthesis, Santa Barbara, California.* BIODIVERSITY.

Chapman, Dr. Clark R. *Institute Scientist, Department of Space Studies, Southwest Research Institute, Boulder, Colorado.* JUPITER.

Chapman, Dr. Piers. *World Ocean Circulation Experiment, Department of Oceanography, Texas A&M University, College Station.* OCEAN CURRENTS—in part.

Chong, Dr. Ken P. *Director, Mechanics and Materials Program, National Science Foundation, Arlington, Virginia.* SMART STURUCTURES AND MATERIALS.

Clarke, Prof. Keith C. *Department of Geography, University of California, Santa Barbara.* GEOGRAPHIC INFORMATION SYSTEMS.

Coles, Dr. Bryan W. *Raytheon Systems Company, Tewksbury, Massachusetts.* OCEAN OPTICS—in part.

Cook, Dr. P. Dan. *Vice President, Chemistry, Isis Pharmaceuticals, Carlsbad Research Center, Carlsbad, California.* ANTISENSE DRUGS.

Cott, Dr. Jerry M. *Adult Psychopharmacology Research Program, National Institute of Mental Health, Rockville, Maryland.* ANTIDEPRESSANT.

Crone, Dr. Elizabeth E. *Ecology Division, Biological Sciences Department, University of Calgary, Calgary, Canada.* RESTORATION ECOLOGY.

Curtin, Dr. Thomas B. *Office of Naval Research, Arlington, Virginia.* UNDERWATER VEHICLES.

D

Dasys, Dr. Andrew. *Manager Engineering, El-Equip Inc., Sudbury, Ontario, Canada.* UNDERGROUND MINING—in part.

De Micheli, Prof. Giovanni. *Professor of Electrical Engineering, Computer Systems Laboratory, Stanford University, California.* COMPUTER-AIDED CIRCUIT DESIGN.

Delcomyn, Dr. Fred. *Professor, Department of Entomology, University of Illinois, Urbana.* NEUROETHOLOGY.

Deschamps, Dr. Jacques. *General Manager, Thomson Plasma, France.* PLASMA DISPLAYS—coauthored.

Deslattes, Dr. Richard D. *Atomic Physics Division, National Institute of Standards and Technology, Gaithersburg, Maryland.* KILOGRAM.

Diaz, Prof. Arthur F. *Department of Materials Engineering, San Jose University, San Jose, California.* ORGANIC ELECTRICAL CONDUCTORS.

DiMichele, Dr. William A. *Research Paleontologist and Curator, Department of Paleobiology, Smithsonian Institution, National Museum of Natural History, Washington, D.C.* ECOSYSTEM DYNAMICS.

Doerr, Dr. H. W. *Institut für Medizinische Virologie, J.W. Goethe-Universität Frankfurt/Main, Frankfurt, Germany.* DENGUE FEVER—coauthored.

Doshi, Dr. Bharat T. *Bell Laboratories Fellow, Advanced Communications Technologies, Bell Laboratories, Lucent Technologies, Holmdel, New Jersey.* MULTISERVICE BROADBAND NETWORK TECHNOLOGY.

Doyeux, Henri. *Technical Research & Development Manager, Thomson Plasma, Morians, France.* PLASMA DISPLAYS—coauthored.

Doyle, Dr. Michael P. *Regents Professor and Director, College of Agricultural and Environmental Sciences, Center for Food Safety and Quality Enhancement, University of Georgia, Griffin.* FOOD-BORNE DISEASE.

Driscoll, Clement J. *Palos Verdes Peninsula, California.* WIRELESS EMERGENCY SERVICES.

Dugan, Dr. Frank M. *Research Plant Pathologist, USDA-ARS Western Regional Plant Introduction Station, Washington State University, Pullman.* AGRIBIOTECHNOLOGY.

E

Eberl, Dr. Karl. *Max-Planck-Institut für Festkörperforschung, Stuttgart, Germany.* QUANTUM-DOT LASERS—coauthored.

Edwards, Dr. Peter A. *Department of Biological Chemistry, University of California, Los Angeles.* CHOLESTEROL.

Eghball, Dr. Bahman. *U. S. Department of Agriculture, Soil and Water Conservation Research Unit, University of Nebraska, Lincoln.* ANIMAL WASTE MANAGEMENT.

Eliseev, Dr. Alexey V. *Assistant Professor, Department of Medicinal Chemistry, State University of New York at Buffalo.* COMBINATORIAL CHEMISTRY.

Erickson, Prof. Larry E. *Department of Chemical Engineering, Kansas State University, Manhattan.* SOIL REMEDIATION—in part.

Exton, Dr. John H. *Howard Hughes Medical Institute, Department of Molecular Physiology and Biophysics, Vanderbilt University School of Medicine, Nashville, Tennessee.* PHOSPHOLIPASES.

F

Family, Prof. Fereydoon. *Samuel Candler Dobbs Professor of Condensed Matter Physics, Department of Physics, Emory University, Atlanta, Georgia.* FRICTION—coauthored.

Filippenko, Prof. Alexei V. *Astronomy Department/Radio Astronomy Laboratory, Theoretical Astrophysics Center, University of California, Berkeley.* COSMOLOGICAL CONSTANT.

Fishwick, Dr. Paul A. *Department of Computer and Information Science and Engineering, University of Florida, Gainesville.* SIMULATION.

Foust, Dr. Jeffrey A. *Department of Earth, Atmospheric, and Planetary Sciences, Massachusetts Institute of Technology, Cambridge.* MOON.

Friedlander, Dr. Arthur M. *Science Advisor, U.S. Army Medical Research Institute of Infectious Diseases, Fort Detrick, Maryland.* ZOONOSIS.

Fuhrmann, Dr. Mark. *Environmental and Waste Technology Center, Brookhaven National Laboratory, Upton, New York.* SOIL REMEDIATION—in part.

G

Garzke, William H., Jr. *Chairman, Marine Forensic Panel, The Society of Naval Architects and Marine Engineers, Montclair, Virginia.* MARINE FORENSICS—coauthored.

Gaylor, Diane C. *Paul, Weiss, Rifkind, Wharton, & Garrison, Washington, D.C.* COMMUNICATIONS SATELLITE—coauthored.

Geissel, Dr. Hans. *GSI, Gesellschaft für Schwerionenforschung, Darmstadt, Germany.* EXOTIC NUCLEI—coauthored.

Gerber, Dr. Leah R. *Washington Cooperative Fish and Wildlife Research Unit, University of Washington, Seattle.* MARINE CONSERVATION.

Gimzewski, Dr. J. K. *IBM Research Division, Zurich Research Laboratory, Ruschlikon, Switzerland.* NANOELECTRONICS.

Glazer, Dr. Peter M. *Departments of Therapeutic Radiology and Genetics, Yale University School of Medicine,* New Haven, Connecticut. DEOXYRIBONUCLEIC ACID (DNA)—coauthored.

Gosnold, Dr. William, Jr. *Department of Geology and Geological Engineering, University of North Dakota, Grand Forks.* CLIMATE CHANGE.

Gourley, Dr. Mark F. *Attending Rheumatologist, Washington Hospital Center, Washington, D.C.* BIOCAVITY LASER—coauthored.

Gourley, Dr. Paul L. *Distinguished Member of the Technical Staff, Sandia National Laboratories, Albuquerque, New Mexico.* BIOCAVITY LASER—coauthored.

Gozani, Dr. Tsahi. *Ancore Corporation, Santa Clara, California.* EXPLOSIVES DETECTION.

Grätzel, Dr. Michael. *Professor of Chemistry, Département de Chimie, Institut de Photonique et Interfaces, École Polytechnique Fédérale de Lausanne, Lausanne, Switzerland.* ELECTRON TRANSFER.

Grisham, Dr. Larry R. *Plasma Physics Laboratory, Princeton University, New Jersey.* NUCLEAR FUSION.

H

Hall, Dr. Richard B. *Professor of Forestry and Genetics, Department of Forestry, Iowa State University of Science and Technology, Ames, Iowa.* FOREST GENETICS.

Ham, Dr. William E. *Compaq Computer Corporation, Shrewsbury, Massachusetts.* SMALL COMPUTER SYSTEM INTERFACE (SCSI).

Hanson, Dr. David R. *Manager Core Technology, NSA Engineering Inc., Golden, Colorado.* UNDERGROUND MINING—coauthored.

Harding, Dr. Elaine K. *Department of Environmental Studies, University of California, Santa Cruz.* CONSERVATION.

Hartmann, Dr. Joachim. *Physik-Department, Technische Universität München, Garching, Germany.* EXOTIC ATOMS.

Henderson, Dr. Dale. *Centre for Metrology, National Physical Laboratory, Teddington, Middlesex, United Kingdom.* FREQUENCY AND TIME STANDARDS.

Hinderliter, Hal. *Director, Center for Imaging Excellence, Graphic Arts Technical Foundation, Sewickley, Pennsylvania.* COMPUTER-TO-PLATE PRINTING.

Ho, Dr. Chih-Ming. *Benrich-Lockheed Martin Professor and Director of Center for Micro Systems, Mechanical and Aerospace Engineering Department, University of California, Los Angeles.* MICRO-ELECTRO-MECHANICAL SYSTEMS (MEMS).

Hoegh-Guldberg, Dr. Ove. *Associate Professor, School of Biological Sciences, University of Sydney, Australia.* CORAL DISEASES.

Holland, Dr. Steven M. *Department of Geology, University of Georgia, Athens.* FOSSIL.

Hoy, Dr. Ron. *Professor, Section of Neurobiology and Behavior, Cornell University, Ithaca, New York.* HEARING (INSECT).

Hu, Prof. Zhibing. *Department of Physics, University of North Texas, Denton.* SENSORS.

Hussey, Dr. Richard S. *Professor, Department of Plant Pathology, University of Georgia, Athens.* PEST MANAGEMENT—coauthored.

J

Jackman, Rebecca J. *Department of Chemical Engineering, Massachusetts Institute of Technology.* MICROSTRUCTURES—coauthored.

Jons, Otto P. *Executive Vice President, Nichols Advanced Marine, Arlington, Virginia.* COMPUTER-AIDED ENGINEERING.

Josephson, Prof. Edward S. *Adjunct Professor, Food Science and Nutrition Research Center, University of Rhode Island, West Kingston, Rhode Island.* FOOD IRRADIATION—coauthored.

Joslin, Dr. Ronald D. *NASA Langley Research Center, Hampton, Virgin Islands.* AIRCRAFT LAMINAR FLOW CONTROL.

K

Kanonik, Frank. *Associate Director of Training, Graphic Arts Technical Foundation, Sewickley, Pennsylvania.* DIGITAL PRESSES.

Kelley, Dr. Matthew W. *Assistant Professor, Department of Cell Biology, Georgetown University School of Medicine, Washington, D.C.* REGENERATION (EAR).

Kenrick, Dr. Paul. *Department of Palaeontology, The Natural History Museum, Cromwell Road, United Kingdom.* PLANT EVOLUTION.

Kinzie, Robert W. *Chairman, Iridium LLC, Washington, D.C.* COMMUNICATIONS SATELLITE—in part.

Kirk, Dwayne D. *Boyce Thompson Institute for Plant Research, Cornell University, Ithaca, New York.* VACCINATION—coauthored.

Klein, Dr. Spencer. *Staff Physicist, Nuclear Science Division, Lawrence Berkeley National Laboratory, Berkeley, California.* QUARK-GLUON PLASMA.

Kondo, Keiji. *Senior Research Scientist, Central Laboratorie for Key Technology, Kirin Brewery Co. Ltd., Yokohama-shi, Kanagawa, Japan.* CANDIDA (INDUSTRY).

Kowalewski, Dr. Michal. *Department of Geological Sciences, Virginia Polytechnic Institute and State University, Blacksburg.* PALEOECOLOGY.

Krim, Prof. Jacqueline. *Professor of Physics, Department of Physics, North Carolina State University, Raleigh.* FRICTION—coauthored.

Krumholz, Dr. Lee R. *Assistant Professor, Department of Botany and Microbiology, University of Oklahoma, Norman.* DEEP CRUSTAL MICROBES—coauthored.

Kück, Dr. Ulrich. *Lehrstuhl für Allgemeine Botanik, Ruhr-Universität Bochum, Bochum, Germany.* FILAMENTOUS FUNGI.

Kvenvolden, Dr. Keith A. *U.S. Geological Survey, U.S. Department of the Interior, Menlo Park, California.* GAS HYDRATE.

L

Laca, Prof. Emilio A. *Department of Agronomy and Range Science, University of California, Davis.* PRECISION AGRICULTURE—coauthored.

Larsson, Prof. Lars. *Department of Naval Architecture and Ocean Engineering, Chalmers University of Technology, Gothenburg, Sweden.* SHIP DESIGN—in part.

Li, Zongzhi. *Purdue University, School of Civil Engineering, West Lafayette, Indiana.* INTELLIGENT TRANSPORTATION SYSTEMS—coauthored.

M

McCarthy, Dr. Peter J. *Harbor Branch Oceanographic Institution, Inc., Division of Biomedical Marine Research, Fort Pierce, Florida.* MARINE NATURAL PRODUCTS—coauthored.

McDonald, Keith D. *President, Sat Tech Systems, and Technical Director, Navtech Seminars and GPS Supply, Alexandria, Virginia.* GLOBAL POSITIONING SYSTEM (GPS).

Macer, Dr. Darryl. *Institute of Biological Sciences, University of Tsukuba, Tsukuba Science City, Japan.* BIOETHICS.

McPhaden, Dr. Michael. *NOAA/Pacific Marine Environmental Laboratory, Seattle, Washington.* EL NIÑO.

Mapes, Dr. Gene. *Department of Environmental and Plant Biology, Ohio University, Athens.* CONIFER—coauthored.

Marder, Dr. Seth R. *Department of Chemistry, University of Arizona, Tucson.* TWO-PHOTON ABSORPTION—coauthored.

Maré, Dr. C. John. *Department of Veterinary Science and Microbiology, University of Arizona, Tucson.* VIRAL DISEASE.

Markle, Stephen P. *Lieutenant Commander, U.S. Navy, Deputy Director, Environmental Protection Systems Division, Naval Sea Systems Command, Arlington, Virginia.* ENVIRONMENTAL ENGINEERING.

Matthews, Dr. Kathleen. *Department of Biochemistry and Cell Biology, Rice University, Houston, Texas.* PROTEIN—coauthored.

Meehan, Dr. Nathan. *General Manager—Engineering, Union Pacific Resources Co., Fort Worth, Texas.* OIL PRODUCTION—in part.

Mendillo, Prof. Michael. *Center for Space Physics, Boston University, Boston, Massachusetts.* PLANETARY ATMOSPHERES.

Meyer-Berthaud, Dr. Brigitte. *Laboratorie de Paléobotanique, Institut des Sciences de l'Evolution de Montpellier, Place Bataillon, France.* FOREST ECOSYSTEM.

Miall, Dr. Andrew D. *Geology Department, University of Toronto, Toronto, Ontario, Canada.* STRATIGRAPHY.

Michelson, Prof. Robert. *Principal Research Engineer, Georgia Tech Research Institute, Adjunct Associate Professor, School of Aerospace Engineering, Georgia Institute of Technology, Smyrna.* AUTONOMOUS NAVIGATION.

Milgram, Prof. Jerome H. *Department of Ocean Engineering, Massachusetts Institute of Technology, Cambridge.* SHIP DESIGN—in part.

Miller, Dr. Clyde A. *Program Director for Risk Management, Free Flight Phase 1 Program, Washington, D.C.* AIR-TRAFFIC CONTROL.

Moldowan, Prof. J. Michael. *Professor (Research), Department of Geological and Environmental Sciences, School of Earth Sciences, Stanford University, California.* BIOMARKERS.

Morré, Dorothy M. *Department of Foods and Nutrition, Purdue University, West Lafayette, Indiana.* ANTIOXIDANT—coauthored.

Morré, Dr. D. James. *Medicinal Chemistry and Molecular Pharmacology Department, Purdue University, West Lafayette, Indiana.* ANTIOXIDANT—coauthored.

Moss, Dr. Cynthia F. *Associate Professor, Department of Psychology, University of Maryland, College Park.* ECHOLOCATION.

Moss, Dr. Gerard P. *Department of Chemistry, Queen Mary and Westfield College, University of London, United Kingdom.* CHEMICAL NOMENCLATURE.

Mourant, Prof. Ronald. *Department of Mechanical, Industrial, and Manufacturing Engineering, Northeastern University, Boston, Massachusetts.* VIRTUAL REALITY—coauthored.

Müenzenberg, Gottfried. *GSI, Gesellschaft für Schwerionenforschung, Darmstadt, Germany.* EXOTIC NUCLEI—coauthored.

Murphy, Dr. Barbara. *Director of Transplant Nephrology, Renal Division, Mount Sinai School of Medicine, New York, New York.* IMMUNOSUPPRESSION.

N

Nachenberg, Dr. Carey. *Chief Researcher, Symantec Anti-Virus Research Center, Santa Monica, California.* COMPUTER VIRUS.

Nakamura, Dr. Shuji. *Research & Development Department, Nichia Chemical Industries, Ltd., Japan.* BLUE LASERS.

Neil, David M. *President and CEO, NSA Engineering Inc., Golden, Colorado.* UNDERGROUND MINING—coauthored.

Nelson, Dr. Robert A. *President, Satellite Engineering Research Corporation, Bethesda, Maryland.* COMMUNICATIONS SATELLITE—in part.

Nielsen, Dr. Ruby I. *Novo Nordisk, Novo Allé, Denmark.* FUNGAL ENZYME.

Nof, Dr. Shimon Y. *Professor, School of Industrial Engineering, Purdue University, West Lafayette, Indiana.* INTELLIGENT COLLABORATIVE AGENTS.

O

Odin, Dr. G. S. *Laboratoire de Géochronologie et Sédimentologie Océanique (Géosédocé), Université P. et M. Curie, Paris.* GEOLOGICAL TIME SCALE.

Olah, Prof. George A. *Professor and Director, Loker Hydrocarbon Research Institute, University of Southern California, Los Angeles.* HYPERCARBON CHEMISTRY—coauthored.

Ostfeld, Dr. Richard S. *Institute of Ecosystem Studies, Millbrook, New York.* DISEASE.

P

Padian, Dr. Kevin. *Museum of Paleontology, University of California, Berkeley.* PTEROSAUR.

Park, Prof. Jeffrey. *Department of Geology and Geophysics, Yale University, New Haven, Connecticut.* NUCLEAR TESTING.

Paul, Dr. Willi. *ABB Corporate Research Ltd., Segelhof, Switzerland.* SUPERCONDUCTING DEVICES.

Perry, Dr. Joseph W. *Department of Chemistry, University of Arizona, Tucson.* TWO-PHOTON ABSORPTION—coauthored.

Perryman, Dr. Michael A. C. *European Space Agency, ESTEC (SCI-SA), The Netherlands.* ASTROMETRY.

Peterson, Dr. Kevin J. *Division of Biology, California Institute of Technology, Pasadena.* VERTEBRATE ORIGINS.

Pomponi, Dr. Shirley A. *Director, Harbor Branch Oceanographic Institution, Inc., Division of Biomedical Marine Research, Fort Pierce, Florida.* MARINE NATURAL PRODUCTS—coauthored.

Pope, Brian. *Senior Vice President, Alstom USA Inc., Aston, Pennsylvania.* PODDED PROPULSION.

Post, Dr. Richard F. *Lawrence Livermore National Laboratory, University of California, Livermore.* MAGNETIC LEVITATION.

Potter, Dr. Huntington. *Professor and Eric Pfeiffer Chair for Research on Alzheimer's Disease, Suncoast Gerontology Center, University of South Florida, Department of Biochemistry and Molecular Biology, Tampa.* ALZHEIMER'S DISEASE.

Prakash, Prof. G. K. Surya. *Loker Hydrocarbon Research Institute, University of Southern California, Los Angeles.* HYPERCARBON CHEMISTRY—coauthored.

Preiser, Wolfgang. *Institut für Medizinische Virologie, J.W. Goethe-Universität Frankfurt/Main, Frankfurt, Germany.* DENGUE FEVER—coauthored.

Probst, Frank J., III. *Department of Human Genetics, University of Michigan Medical School, Ann Arbor.* HEARING IMPAIRMENT—coauthored.

R

Ravishankara, Dr. A. R. *Aeronomy Laboratory, National Oceanic and Atmospheric Administration, Boulder, Colorado.* ATMOSPHERIC CHEMISTRY.

Reed, John K. *Harbor Branch Oceanographic Institution, Inc., Division of Biomedical Marine Research, Fort Pierce, Florida.* MARINE NATURAL PRODUCTS—coauthored.

Reuss, Robert H. *Flat Panel Display Division, Motorola Inc., Tempe, Arizona.* FIELD-EMISSION DISPLAYS—coauthored.

Rice, Dr. Louis B. *Infectious Disease, VA Medical Center, Case Western Reserve University, Cleveland, Ohio.* ANTIBIOTIC RESISTANCE.

Righter, Dr. Kevin. *Department of Planetary Sciences, Lunar and Planetary Laboratory, University of Arizona, Tucson.* RADIOISOTOPE DATING.

Robitaille, Dr. Pierre-Marie. *Advanced Biomedical Imaging, MRI Facility, Ohio State University Medical Center, Columbus.* MAGNETIC RESONANCE IMAGING.

Rothwell, Prof. Gar W. *Department of Environmental and Plant Biology, Ohio University, Athens.* CONIFER—coauthored.

Rowe, Dr. Nick. *Laboratoire de Paléobotanique, Université de Montpellier, Place Eugène Bataillon, France.* TREE—coauthored.

Ruesink, Dr. Jennifer. *Centre for Biodiversity Research, Department of Zoology, University of British Columbia, Vancouver.* INVASION ECOLOGY.

S

Sabota, Dr. Cathy. *Professor/Extension Horticulturist, Alabama A&M University, Alabama Cooperative Extension System, Normal.* MUSHROOM (MEDICINE).

Sato, Dr. Thomas N. *Internal Medicine Cardiology, University of Texas Southwestern Medical Center, Dallas.* ANGIOGENESIS.

Schiffer, Dr. P. *Physics Department, University of Notre Dame, Indiana.* MAGNETORESISTANCE—coauthored.

Shay, Dr. Jerry W. *Professor of Cell Biology, University of Texas Southwestern Medical Center, Dallas.* CELL SENESCENCE.

Sherman, Dr. Robert. *Applied Surface Technologies, New Providence, New Jersey.* CARBON DIOXIDE.

Silloway, Richard. *Montclair, Virginia.* MARINE FORENSICS—coauthored.

Sinha, Prof. Kumares C. *Olson Distinguished Professor of Civil Engineering and Head, Transportation and Infrastructure Systems Engineering, Purdue University, School of Civil Engineering, West Lafayette, Indiana.* INTELLIGENT TRANSPORTATION SYSTEMS—coauthored.

Slatt, Dr. Roger M. *Department of Geology and Geological Engineering, Colorado School of Mines, Golden.* TURBIDITE—coauthored.

Smith, Prof. Ian W. M. *School of Chemistry, University of Birmingham, Edgbaston, Birmingham, United Kingdom.* CHEMICAL REACTIONS.

Spalart, Dr. Philippe R. *Boeing Commercial Airplane Group, Seattle, Washington.* AIRPLANE TRAILING VORTICES.

Speck, Dr. T. *Botanischer Garten der Universität Freiburg, Germany.* TREE—coauthored.

Spector, Phillip L. *Paul, Weiss, Rifkind, Wharton, & Garrison, Washington, D.C.* COMMUNICATIONS SATELLITE—coauthored.

Stanton, Dr. Anthony P. *Director, Graphic Communications Management, Carnegie Mellon University, School of Industrial Administration, Pittsburgh, Pennsylvania.* DIGITAL PROOFING.

Stringer, Dr. Chris. *Department of Paleontology, The Natural History Museum, United Kingdom.* PHYSICAL ANTHROPOLOGY.

Suzuki, Dr. Yoichiro. *Kamioka Observatory, Institute for Cosmic Ray Research, University of Tokyo, Japan.* NEUTRINO.

Swint-Kruse, Dr. Liskin. *Department of Biochemistry and Cell Biology, Rice University, Houston, Texas.* PROTEIN—coauthored.

T

Tans, Dr. Sander J. *Department of Applied Physics and DIMES, Delft University of Technology, The Netherlands.* NANOCHEMISTRY.

Taub, Dr. Irwin A. *Senior Research Scientist, U.S. Army Soldier and Biological Chemical Command, Soldier System Center, Natick, Massachusetts.* FOOD IRRADIATION—coauthored.

Taylor, Dr. Thomas N. *Department of Ecology and Evolutionary Biology, University of Kansas, Lawrence.* PLANT-FUNGI INTERACTIONS (PALEOBOTANY).

Tonkay, Dr. Greg. *Department of Industrial and Manufacturing Systems, Lehigh University, Harold S. Mohler Laboratory, Bethlehem, Pennsylvania.* DATA MINING.

Trümper, Dr. Joachim. *Max-Planck-Institut für Extraterrestrische Physik, Garching, Germany.* X-RAY ASTRONOMY.

Tumanski, Prof. Slawomir. *Warsaw University of Technology, Institute of Electrical Theory and Measurements, Warsaw, Poland.* MAGNETOVISION.

Turner, Prof. Anthony P. F. *Institute of BioScience and Technology, Cranfield University, Bedfordshire, England.* BIOSENSOR.

Tytler, Dr. David R. *Center for Astrophysics and Space Science, University of California, San Diego.* DEUTERIUM.

U–V

Unsworth, Prof. Martyn. *Geophysics Program, University of Washington, Seattle.* MAGNETOTELLURICS.

van Dover, R. B. *Bell Laboratories, Lucent Technologies, Murray Hill, New Jersey.* MAGNETORESISTANCE—coauthored.

van Kessel, Prof. Chris. *Department of Agronomy and Range Science, University of California, Davis.* PRECISION AGRICULTURE—coauthored.

Vasquez, Dr. Karen M. *Departments of Therapeutic Radiology and Genetics, Yale University School of Medicine, New Haven, Connecticut.* DEOXYRIBONUCLEIC ACID (DNA)—coauthored.

Veit, Dr. Richard R. *Department of Biology, The College of Staten Island, New York.* CLIMATE.

Veitch, Dr. James. *Vice President, Business Development, Data Logic Division of Gene Logic, Berkeley, California.* DISTRIBUTED DYNAMIC SYSTEM.

von Puttkamer, Dr. Jesco. *Office of Space Flight, NASA Headquarters, Washington, D.C.* SPACE FLIGHT.

W

Walecka, Prof. John Dirk. *Chair, Department of Physics, College of William & Mary, Williamsburg, Virginia.* NUCLEAR STRUCTURE.

Walker, Dr. Graeme M. *Reader in Biotechnology, University of Abertay Dundee, School of Molecular and Life Sciences, Dundee, Scotland.* BIOTECHNOLOGY (YEAST).

Watkins, Paul A. *Associate Professor of Neurology, Johns Hopkins University School of Medicine, Kennedy Krieger Institute, Baltimore, Maryland.* PEROXISOME.

Wehmeyer, Dr. Lillian Biermann. *Chair, Department of Educational Administration, Director, Ed.D. in Educational Leadership and Administration, Azusa Pacific University, Azusa, California.* COMPUTERIZED SEARCHES.

Westrop, Dr. Stephen R. *School of Geology and Geophysics, University of Oklahoma, Norman.* TRILOBITA.

Whitesides, Prof. George M. *Mallinckrodt Professor of Chemistry, Harvard University Department of Chemistry, Boston, Massachusetts.* MICROSTRUCTURES—coauthored.

Whitfield, Robert J. *Manager, Electrical Engineering, John J. McMullen Associates, Inc., Alexandria, Virginia.* INTEGRATED ELECTRIC PROPULSION.

Wicks, Dr. Elin M. *Department of Industrial Engineering, University of Missouri, Columbia.* PRODUCTION LEAD TIME.

Wiemer, Paul. *Department of Geology and Energy Minerals and Applied Research Center, University of Colorado, Boulder.* TURBIDITE—coauthored.

Wiggins, Dr. Gary. *Head, Chemistry Library, Indiana University, Bloomington.* CHEMICAL INFORMATION SOURCES.

Wilson, Dr. Bruce. *Assistant Professor, Department of Mechanical, Industrial, and Manufacturing Engineering, Northeastern University, Boston, Massachusetts.* VIRTUAL REALITY—coauthored.

Winterer, Dr. Juliette. *Assistant Professor, Department of Biology, Franklin and Marshall College, Lancaster, Pennsylvania.* GENETIC ENGINEERING.

Wolfe, Dr. Cecily J. *Department of Terrestrial Magnetism, Carnegie Institution of Washington, Washington, D.C.* MID-OCEAN RIDGE STRUCTURES.

Wright, Dr. Amy E. *Harbor Branch Oceanographic Institution, Inc., Division of Biomedical Marine Research, Fort Pierce, Florida.* MARINE NATURAL PRODUCTS—coauthored.

Y–Z

Yesha, Prof. Yelena. *Computer Science Department, University of Maryland, Baltimore.* ELECTRONIC COMMERCE—coauthored.

Young, Dr. Michael W. *Professor and Head, Laboratory of Genetics, The Rockefeller University, New York.* CIRCADIAN RHYTHM.

Zettl, Prof. Alex. *Department of Physics, University of California, Berkeley.* NEW CARBON MATERIALS.

Zoschke, Les T. *Vice President, Modular Mining Systems, Tucson, Arizona.* OPEN PIT MINING.

Zundel, Markus K. *Max-Planck-Institut für Festkörperforschung, Stuttgart, Germany.* QUANTUM-DOT LASERS—coauthored.

Index

Index

X

Y

Z